Alexey V. Borisov, Ivan S. Mamaev

Rigid Body Dynamics

De Gruyter Studies in Mathematical Physics

Volume 52

Alexey V. Borisov, Ivan S. Mamaev

Rigid Body Dynamics

DE GRUYTER

Higher Education Press

Authors
Alexey V. Borisov
Moscow Institute of Physics and Technology (State University)
9 Institutskiy per.
Dolgoprudny, Moscow Region
141701 Russia
E-mail: borisov@rcd.ru

Ivan S. Mamaev
Kalashnikov Izhevsk State Technical University
7 Studencheskaya str
Izhevsk, Udmurt Republic
426069 Russia
E-mail: mamaev@rcd.ru

ISBN 978-3-11-054279-0
e-ISBN (E-BOOK) 978-3-11-054444-2
e-ISBN (EPUB) 978-3-11-054297-4

Library of Congress Control Number: 2018958374

Bibliografische Information der Deutschen Nationalbibliothek
The Deutsche Nationalbibliothek lists this publication in the Deutsche Nationalbibliografie; detailed bibliographic data are available on the Internet at http://dnb.dnb.de.

Printing and binding: CPI books GmbH, Leck

www.degruyter.com

Contents

Introduction

1 Euler – Poisson equations and integrable cases. In this introduction, we provide a brief outline of the main milestones in the development of rigid body dynamics. The equations of rigid body motion about a fixed point in various force fields were inspired by the work of Euler, d'Alembert, Clairaut, Poisson and Lagrange. The latter presented them in a fairly general and modern form in his *Analytical Mechanics*. Of particular interest was and still is the problem of the motion of a rigid body about a fixed point in a homogeneous gravitational field (i.e., of a heavy rigid body). A basis for research along these lines is the system of equations obtained by Euler and Poisson, which combine the elegance and simplicity of representation with the difficulty and even impossibility of obtaining a general solution in explicit form. Historically, integrable cases, i.e., cases in which a general solution can be obtained (in terms of quadratures, as the authors of classic works put it) were the first to be studied. Obviously, this requires that some system parameters or initial conditions be given. The most popular of them were found already by Euler (1758) and Lagrange (1788) when the general principles of dynamics were formed and developed.

A much more complicated case of integrability of the Euler – Poisson equations, which gave impetus to new research in the area of integrable systems, was found by S. V. Kovalevskaya in 1888. This result was highly regarded by the Paris Academy of Sciences, which awarded the Bordin prize to S. V. Kovalevskaya in 1888 for a memoir on the rotation of a rigid body about a fixed point. We note that the Academy of Sciences had announced a competition for research on the same topic twice before, but no one had been awarded the prize. In the spring of 1889, Kovalevskaya was awarded a prize by the Royal Swedish Academy of Sciences for her second memoir on the problem of rigid body rotation.

The integrability of the Euler and Lagrange cases is due to natural dynamical symmetries and preservation of the corresponding first integrals. S. V. Kovalevskaya found her own integrable case from nonobvious analytical considerations, using the well-developed theory of algebraic functions (elliptic functions are a particular case of these functions). This required uniqueness of the general solution on the complex plane of time, which laid the foundations for the Painlevé – Kovalevskaya test, one of the most advanced methods of integrability analysis of dynamical systems. The Kovalevskaya integral has no natural symmetry origin; its symmetries are said to be hidden, and the problem of motion description and explicit integration in this case is much more complicated.

After Kovalevskaya, no general integrable cases have been found for the Euler – Poisson equations, and it has even been proved that they do not exist. Nevertheless, the problem of obtaining various partial integrals and partial solutions, of which about two dozens have been obtained, has long been a topic of special interest. But this number does not cover even a small part of the full range of possible motions of a heavy rigid body.

2 Poincaré and Helmholtz equations. In the second half of the 19th century and in the early 20th century, integrable cases were found in rigid body dynamics for various versions of the problem of rigid body motion: motion of a body in a fluid, motion of a body containing cavities filled with fluid, gyrostats, and nonholonomic problems. The study of these problems was made possible by the development of the general formalism of rigid body dynamics, which culminated in the Poincaré equations allowing the equations of rigid body motion to be represented in terms of group variables.

Here, mention should be made of progress in the hydrodynamics of an ideal fluid and vortex theory, whose foundations were laid by H. Helmholtz.

In this way, equations for the vorticity vector, similar to dynamical equations for angular momentum, were obtained, and Poincaré studied for the first time the precession of the Earth axis, using as the Earth model a rigid body (mantle) containing cavities filled with an incompressible vortex fluid (core). Using simplified and integrable Poincaré equations, Volterra posed the model problem of gyrostats, which he applied to describe the motion of the Earth poles.

3 Integrable cases. The classical period. As mentioned above, for various forms of equations, finding cases that are fixed by restrictions on parameters and initial conditions and finding cases of explicit solvability of the problem in quadratures (i.e., integrable cases, as they are called nowadays) was considered to be of primary importance in the classical period.

Integrable cases are usually associated with the names of those who have discovered them. We have already mentioned the cases bearing the names of Euler, Lagrange and Kovalevskaya. Other names include those of the well-known Western mathematicians and mechanics researchers: G. Kirchhoff, A. Clebsch, P. Appell, F. Brun, V. Volterra; major advances are due to Russian scientists such as A. M. Lyapunov, V. A. Steklov, N. E. Zhukovsky, S. A. Chaplygin, and D. N. Goryachev. In this sense, rigid body dynamics can be regarded as an area that is the most rich in interesting, nontrivial integrable problems, which comprise a precious heritage of treasures for the study of modern dynamics.

In the classical period, great emphasis was placed not only on finding first integrals, but also on obtaining explicit solutions in various classes of functions (mainly elliptic ones). Particular progress along these lines was made by S. V. Kovalevskaya, V. Volterra, G. Halphen, F. Kötter, and their techniques are still being adopted.

4 Nonintegrability results by Poincaré and Kozlov. Qualitative analysis. In the first half of the 20th century, interest in searching for integrable cases waned somewhat. This is mainly due to the fact that many mathematicians gained insight into the results of H. Poincaré on the nonintegrability of a typical Hamiltonian dynamical system [474] (in particular, the well-known three-body problem). As a consequence, many results of the classical works lost much of their value in the eyes of mathematicians and this prompted them to develop new methods of perturbation theory: the principle of averaging, KAM-theory etc.

The basic equations of rigid body dynamics are generally nonintegrable as well, and hence they exhibit complicated, unpredictable behavior [474], which are the subject of a new area of research called *determined chaos*. A systematical treatment of the effects of nonintegrability in rigid body dynamics is given in the book by V. V. Kozlov [333]. We have gathered the main results on the nonintegrability of equations of rigid body dynamics in one of the appendices.

There is another fact which makes the book by V. V. Kozlov [333] so important. Unlike the authors of classical works who had an unnatural propensity to obtain explicit solutions telling us little about the real motion of the system, he raises the problem of qualitative analysis of integrable dynamical systems, and, using the Kovalevskaya and Goryachev – Chaplygin tops as examples, draws general conclusions on the behavior of the line of nodes and angles of proper rotation. The latest results were obtained by applying the Liouville – Arnold theorem and the Weyl theorem on uniform distribution.

5 Poisson structures and Lie algebras. In addition to the idea of a wide application of computer methods, we tried to present here in more detail the modern methods of Poisson dynamics and geometry, the theory of Lie groups and Lie algebras, which we had outlined only roughly in our previous book, *Poisson structures and Lie algebras in Hamiltonian mechanics* [95]. Rigid body dynamics plays a special role in developing these methods. As mentioned above, it is a good area for testing new mathematical theories, and its significance, particularly to the development of many branches of topology and nonlinear Poisson structures, nonholonomic geometry, the theory of symmetries and tensor invariants, can hardly be overestimated.

It may even be asserted that, just as the analysis of the three-body problem has enabled the understanding of the profound ideas of H. Poincaré about the nonintegrability of dynamical systems, the results and methods of Sophus Lie have been integrated into the general mathematical culture, as a consequence of their application to the dynamics of tops, which provide examples of mechanical realization of the most natural Lie groups and Lie algebras. Moreover, as opposed to celestial mechanics, vibration theory and vortex dynamics, rigid body dynamics contains, on the one hand, a number of nontrivial integrable cases and, on the other hand, in view of the compactness of configuration space, is more preferable for analysis of chaotic motions.

6 Topological analysis. The application of methods of topological analysis to the integration of problems of rigid body dynamics, namely, the study of rearrangements of Liouville tori when a parameter passes through critical values, was first proposed by M. P. Kharlamov [286] and developed further in the theory of topological invariants, which was created (mainly by the school of A. T. Fomenko) for the classification of integrable Hamiltonian systems with two degrees of freedom. Most of the results obtained by these techniques are presented in the book [66]. Complex methods which lead basically to the same results are presented in the book by M. Audin [25].

7 The L-A pair method. The development of the method of isospectral deformation (Lax representation, L-A pair) revived interest in integrable problems of rigid body dynamics in 1970–1990. As a result, quite a number of new integrable cases were found. As a rule, most publications of that period were concerned with multidimensional generalizations of well-known natural physical analogs. The development of this avenue of research is also due to the fact that the ideas of the theory of Lie groups and Lie algebras and ideas of analysis of general (nonlinear and degenerate) Poisson structures have found their way into rigid body dynamics. An account of the current status of research on these topics can be found in our book [87].

It turned out that many constructions of the Lie-algebraic approach and methods of qualitative analysis can be extended to nonholonomic problems of rigid body dynamics, where some new integrable systems [90] have been added for the last few decades.

8 Systems of analytical computations and first integrals. We also note that systems of analytical computations play a large role in the problem of finding high-degree integrals. One of the first integrable cases obtained by this method is the Sokolov case in the Kirchhoff equations (2001). Since then, rigid body dynamics has been enriched with a number of more complex first integrals, even those of degree 6, and the possibilities of finding them have obviously not been exhausted.

It has become standard practice to apply systems of analytical computations to stability analysis of fixed points and periodic solutions by normalization methods. This analysis is widely used in studies on rigid body dynamics, where there exist many interesting partial solutions which provide fertile ground for such activities (A. P. Markeev, [413, 414, 415, 416]).

9 The motion of a rigid body in an ideal fluid. One of the new problems addressed in this book and worthy of special note is that of a heavy rigid body moving in an ideal fluid. The first problems of the motion of a rigid body interacting with a liquid medium go back to Maxwell, Kirchhoff, Lamb, Zhukovsky and Chaplygin. However, in the general case such problems are very complicated and still remain unresolved. During a real motion one should take account of the lifting force and various drag forces due to vortex shedding and friction of the fluid with the surface of the body. Generally speaking, most interesting effects arising from interaction between the medium and the body are due to viscosity. One usually uses phenomenological models of interaction which incorporate viscosity as one of the force (as a rule, nonpotential) factors. However, even if viscosity is excluded altogether, the problem can be of great interest, since there remain factors of fluid resistance due to added masses and the inertia of fluid. In this setting, problems of a heavy rigid body falling in a fluid were considered as far back as the end of the 19th century by V. A. Steklov, S. A. Chaplygin, and D. N. Goryachev. The rapid development of aerohydrodynamics in the early 20th century and the explanation of the nature of the lifting force by Kutta, Zhukovsky and Chaplygin made it possible to introduce into these equations additional terms related to circulation. It turned out that all these formulations of the problem *keep us within the scope of Hamiltonian formalism* and can be used to explain the results of various experiments. We note that full-scale experiments with falling plates or disks are still being conducted in various hydrodynamical laboratories both in Russia and abroad, but none of the existing theories is capable of completely explaining the results of such experiments.

A more accurate description of the fall is possible beyond the scope of Hamiltonian formalism – dissipative factors need to be taken into account. However, serious research on this would have to be fairly extensive and, although there have been a number of outstanding publications on these topics, we do not discuss them in this book.

10 Physical applications. In recent decades, a few other lines of research have emerged that are concerned with the dynamics of tops. One of them arose in quantum mechanics from the analysis of systems of interacting spins with anisotropy (the Heisenberg chain or XYZ-model).

The classical model is here a basis for understanding the dynamics at the quantum level, which, in a sense, can also be integrable and chaotic. Research on quantum chaos is only beginning to emerge [560, 481], but evidently will soon develop into a separate branch of science, where a great deal of attention will be given to the quantum description of tops. This is primarily due to the

fact that the model of a top is the main one in the quantum theory of angular momentum,which is used in quantum chemistry and molecular spectroscopy.

It is also interesting to note that the integrability conditions and the integrals for the spin model which are presented in the modern literature on quantum mechanics (see, e.g., [481], 2004) are simplified results obtained in classical works (W. Frahm, F. Schottka) over a century ago. This is due to the fact that many of modern physicists who might have gone a bit too far in their abstract and intricate theories (like quantum field theory or the theory of gravitation) do not have a very good understanding of problems that have a natural origin and are related to the dynamics of a usual spinning top used as a toy.

11 Computer methods of analysis. In a sense, even in the analysis of an integrable situation for which in principle a complete classification of all solutions can be given, computers opened up a new era. Previously, integrable systems had been studied predominantly by analytical methods allowing one to obtain explicit quadratures and geometric interpretations, which in many cases looked rather artificial. A combination of the ideas of topological analysis (bifurcation diagrams), stability theory, the method of phase sections and immediate computer visualization of "particularly special solutions" can provide a better illustration of the particularity of an integrable situation and of the most distinctive features of motion, and can be useful in obtaining a number of new results even for seemingly well-understood problems (such as the Kovalevskaya and Goryachev – Chaplygin tops, and the Bobylev – Steklov solution). The thing is that these results are very difficult to spot in cumbersome analytical expressions. After the computer has identified the results, the proof of these facts can also be obtained analytically. Of special note is the fact that to date no analysis has been made of rigid body motion in absolute space.

Some interesting motions of integrable tops may be capable of giving rise to concrete ideas concerning their practical application. We recall, for example, that so far no application has been found for the Kovalevskaya top discovered over a century ago, since almost nothing has been known of its motion, in spite of the complete solution in terms of elliptic functions.

In this book we present analytical expressions for families of doubly asymptotic solutions generated by unstable periodic solutions. For integrable cases the behavior of the system is the most complicated and looks irregular even despite the existence of an additional integral (recall that in the general case the motion of an integrable system is regular). Under perturbations such solutions are the first to break down, and regions appear which are filled with "real" chaotic trajectories and are located in phase space near these solutions.

Computer analyses give ample cause for "revision" and make it possible to grasp the real significance of analytical investigations. While some analytical results (concerning, for example, the separation of variables) turn out to be useful for the study of bifurcations and various topological problems, it is as good as useless to "develop" them further to obtain explicit quadratures (in terms of θ-functions). These results are contained, for example, in [228, 162], but they can be used only as exercises in differential equations rather than as methods of dynamical analysis.

12 Transition to chaos. The value of the results of the classical works in rigid body dynamics was questioned as far back as the 1970s by K. Magnus [401]. The epoch of faith in boundless opportunities offered by computers led to the conviction that all these results are useless and that a sufficiently powerful computer is capable of forecasting the motion on any time interval with fair accuracy. However, in view of the exponentially fast divergence of trajectories (due to instability in some regions of phase space) in typical dynamical systems, which are nonintegrable, such a computation on sufficiently large time intervals has no physical meaning, since the initial conditions for specific (applied) systems are always given with some error.

It seems that numerical methods can be relied on only in an integrable situation in which no such divergence occurs. However, it turns out that conservative systems retain many elements of integrable dynamics even in a stochastic situation. Under small perturbations of the integrable problem, nondegenerate periodic orbits persist and most of quasi-periodic motions (KAM-theory) do not decay.

As perturbations grow further, both periodic orbits and invariant tori undergo various bifurcations, which follow some general patterns. They determine changes in the whole structure of the phase flow, which combines regions with regular and chaotic behavior, and define scenarios of transition to chaos. The best-known scenarios describing the evolution of the dynamical behavior of the system from regular to chaotic behavior include the homoclinic structure (discovered by H. Poincaré) and cascades of period doubling bifurcations (discovered by M. Feigenbaum) possessing the universality property. In Hamiltonian systems, as parameters change, these two effects coexist with each other, determining the evolution of the phase portrait. Evidently, in various classes of problems we have different combinations of the two scenarios, although general patterns can be observed as well. In rigid body dynamics (as opposed, for example, to celestial mechanics) there has been almost no research along these lines (which, by the way, is impossible without high-precision computer simulations). In this book we have gathered perhaps nearly all research results in this area, which, unfortunately, are very scarce.

13 Probabilistic effects. We also note that in rigid body dynamics, even in a Hamiltonian setting, there are a number of systems which cannot be described using the standard methods of Hamiltonian chaos. These systems arise in the description of a heavy rigid body falling in a fluid. The behavior of such systems, despite their Hamiltonian nature, is similar to that of dissipative systems and is due to the fact that asymptotic behavior is different under changing initial conditions. It turns out that in the general case the dependence of asymptotic behavior on the initial conditions is very complicated (fractal), so that only a probabilistic description is possible. It has already long been standard practice to use such a description to investigate the motion of a rigid body under the action of constant and dissipative moments [442]. However, there is no relevant theory to describe the motion of the body in a fluid, and so computer methods have to be applied.

Another interesting and nonclassical problem is that of rigid body motion (in vacuum or in a fluid) with the geometry and dynamical characteristics changing with time. This problem goes back to J. Liouville and is closely related to his equations of a variable rigid body. In this book we present a number of results pertaining to the hydrodynamical problem of self-propulsion (i.e., to the possibility of reaching a given point in space by controlling the geometry and the dynamical characteristics of the body) and to the problem of adiabatic (slow and periodic) behavior of such systems. By the way, the problem of self-propulsion and controllability of motion is closely related to problems concerned with the mechanism of locomotion of fish and to the popular problem of a falling cat, which rights itself as it falls to land to its feet,irrespective of its initial orientation.

14 It is beyond the scope of this book to discuss problems of the stability of special motions and most applied and engineering problems, since a complete and adequate treatment of these would require a separate book. However, even physicists and engineers can gain from this book an understanding of the general formalism of writing the basic dynamical equations, as well as the main aspects of regular and chaotic behavior in rigid body dynamics. This book may be regarded as a reference book on these topics, but it also elucidates, as far as possible, the derivation of the main results and sometimes presents complete proofs.

We have not included sections on nonholonomic systems and multidimensional generalizations of rigid body dynamics. Many of the topics mentioned here have been treated in the books [90, 95, 87].

Of special note are nonholonomic systems describing the rolling motion of rigid bodies on each other without slipping. Traditionally, discussions of rolling

problems are included in courses on rigid body dynamics (Routh [498], MacMillan [400], Magnus [401] and others). However, this area has flourished very recently; besides, many dynamical effects in nonholonomic systems have been found to differ so much from Hamiltonian ones that they require separate research (see, e.g., [90]).

15 A unique feature of this book is a wide use of numerical experiments and methods of computer visualization coupled with analytical methods. We also present the basic historical facts and a fairly detailed list of references.

Almost all modern and classical integrable cases were verified by using the computer software package MAPLE. It turned out that some well-known results are not quite correct, and the other results were found to be considerably simplified. The computer visualization of motion and numerical integration were performed by using the software package CHAOS, developed at the Institute of Computer Science.

Acknowledgents The authors would like to express their sincere gratitude to V. V. Kozlov, A. V. Bolsinov, I. N. Gashenenko, I. V. Komarov, S. M. Ramod-anov, V. V. Sokolov, Yu. N. Fedorov, A. V. Tsiganov, H. M. Yehia, A. A. Kilin, D. V. Treschev and I. A. Bizyaev. We also owe a special debt of gratitude to all who helped to improve the language of the manuscript: Radi Valiulin, Ian Marshall, as well as Denis Blackmore and Oliver O'Reilly whose advice has been invaluable in putting the book into final form.

The Creators of Rigid Body Dynamics

This section presents, in chronological order, brief biographies of some of the scientists whose names are mentioned throughout the book. In most cases we describe only their achievements in the development of rigid body dynamics, although the results obtained by them in other fields of mathematics and mechanics are often no less notable. These short essays can be useful for understanding the evolution of the main ideas and methods of rigid body dynamics, and we shall return to the historical aspects of its development in the course of the book.

Euler, Leonard (15.4.1707 – 18.9.1783) was a great mathematician and mechanician. He was born in Switzerland and spent a large part of his life in Russia (1727–1741, 1766–1783). Euler made significant contributions to almost all areas of mathematics, and it is difficult to give an overview of his output, which extends to more than 860 works. Euler's contribution to other sciences including shipbuilding, artillery, turbine theory, and material resistance is also considerable.

L. Euler

For the theory of rigid body dynamics, Euler elaborated the moment-of-inertia theory and obtained a formula for the velocity distribution in a rigid body. In 1750 he obtained the equations of motion in a fixed coordinate system, which turned out to be less useful for applications. In a number of works written during the period 1758–1765, Euler used the notion, invented by him, of a coordinate system fixed in the body to obtain the equations — later to be called the Euler–Poisson[1] equations — in the form in which they are still commonly studied today.

These equations are formulated by means of the so-called Euler angles. He obtained the kinematic relations for the motion of a heavy rigid body, now called the Euler relations, and found the integrability case for which the force of gravity is absent. Euler reduced this integrable case to quadratures and examined various particular solutions.

1 Probably, the contribution of Poisson reflected in their name is the systematical presentation given by him in his well-known course of mechanics.

J. L. Lagrange

Lagrange, Joseph Louis (25.1.1736 – 10.4.1813) was a great French mathematician, mechanician, and astronomer. In his famous two-volume treatise *Analytical Mechanics*, he developed a general formalism for dynamics. He presented the equations of rigid body motion in an arbitrary potential force field using a coordinate system attached to the body, the projections of angular momentum, and the direction cosines (Volume II). He also presented the integrability case, which is characterized by having axial symmetry and which was reduced by him to quadratures. Following his principle of avoidance of graphical representation, Lagrange did not provide a geometric study of the motion, and the pictures of the apex behavior, which entered almost all mechanics textbooks after him, made their first appearance in the work of Poisson (1815), who treated this problem as if it were a completely new one[2]. Lagrange simplified the solution for the Euler case and gave a direct proof of the existence of real roots for the third-order equation which determines the position of principal axes. He also developed perturbation theory, which Jacobi was to make use of later to obtain the system of "osculating" variables for studying perturbations of the Euler top.

L. Poinsot

Poinsot, Louis (3.1.1777 – 5.12.1859) was a French engineer, mechanician, and mathematician. He introduced the concepts of ellipsoid of inertia and instantaneous axis of rotation and gave a geometric interpretation of the Euler case in which there arise the curves of polhodes and herpolhodes introduced by him (1851). He presented a geometric stability analysis of the rotation of a rigid body about the principal axes of the ellipsoid of inertia. In contrast to Lagrange, Poinsot insisted on the advantage of geometric methods in mechanics as opposed to analytical methods: "In all these solutions, we see only calculations without any clear picture of the body motion" [475][3]. The ideas of Poinsot were later supported and developed by N. E. Zhukovskii and S. A. Chaplygin. Also, Poinsot used a geometric approach in studying statics (*Elements of statics*, 1803).

2 Poisson did improve the notation which previously had made it difficult to understand the treatises of D'Alembert, Euler, and Lagrange, and presented various particular cases of motion: in some textbooks, the Lagrange case is called the Lagrange–Poisson case.

3 Translated from French into English.

Kirchhoff, Gustav Robert (12.3.1824 – 17.10. 1887) was a German physicist and mathematician. In his *Lectures on Mathematical Physics* (1874–1894, Vols. 1–4) he laid the foundations of modern elasticity theory, hydrodynamics, optics, and electrodynamics. He showed the analogy between the Euler–Poisson equations and the equations of the elastic line bend. Developing ideas of Thomson and Tait, he was able to reduce the problem of rigid body motion in an ideal fluid to a system of ordinary differential equations. He found the integrable case, which is characterized by axial symmetry, presented its solution in elliptic functions and studied various particular motions.

G. R. Kirchhoff

Clebsch, Rudolph Friedrich Alfred (19.1. 1833 – 7.11.1872) was a German mathematician and mechanician. He founded the journal *Mathematische Annalen*, which is still one of the leading mathematical journals. He was a specialist in projective geometry and the theory of invariants of algebraic forms. He proposed a change of representation for the Kirchhoff equations, equivalent to passing from the Lagrangian description to the Hamiltonian one. Using the new equations, he was able to find a new case having an additional quadratic integral which, as became clear later, is identical to the Brun and Tisserand integrals.

R. F. A. Clebsch

Zhukovskii, Nikolai Egorovich (17.1.1847 – 17.3.1921) was a Russian mechanician, mathematician, and engineer. He was hailed by V. I. Lenin as the “father of Russian aviation”. In his master’s thesis (1876), he laid the foundations for the theory of the motion of a rigid body with cavities filled with an ideal incompressible fluid. For multiply connected cavities, he showed that the equations of motion of the system are equivalent to those modelling the motion of a gyrostat – a rigid body with a fly wheel. He introduced the appropriate dynamical characteristics and calculated them for cavities of various shapes. He showed the integrability of a free gyrostat, whose explicit solution was subsequently obtained by V. Volterra using elliptic functions (1899). He studied “plane” motions of a rigid body in Lobachevskii space. He put forward a geometrical interpretation of the Kovalevskaya case and his own method, making use of a certain auxiliary curvilinear coordinate system, for reducing it to quadratures. He observed that the center-of-mass motion in the Hess case is pendulum-like, and suggested an interesting geometric study for it. In connection with his studies in hydro- and aero-mechanics, he considered a number of

N. E. Zhukovskii

model statements of problems of the plane motions of plates under the action of the lift force due to circulation. In mechanics, following Poinsot, N. E. Zhukovskii held the geometrical picture of a motion to be the ideal form of solution, although his geometric interpretations of the motions of the free gyrostat and the heavy rigid body case of Kovalevskaya are neither very simple nor very natural.

S. V. Kovalevskaya

Kovalevskaya, Sof'ya Vasil'evna (15.1.1850 – 10.2.1891) was a famous mathematician. In 1874 she defended her dissertation in Göttingen and was granted a Ph.D. In 1884 she was appointed to a five year position as "Professor Extraordinarius" (Professor without Chair) at the University of Stockholm and became a member of the editorial board of the journal *Acta Mathematica*. In 1889 she was appointed Professor Ordinarius (Professorial Chair Holder) at the same university and elected a corresponding member of the Petersburg Academy of Sciences. She was the first woman in the world to become a professor of mathematics.

For the discovery of the third integrability case of Euler–Poisson equations (the first two being those of Euler and Lagrange), she was awarded the Bordin prize (1888), and for another work on rigid body rotation, she was awarded a prize by the Swedish Royal Academy of Sciences. In these works she proposed a new method, now known as the Kovalevskaya method and still widely used, which furnishes a test for integrability related to the behavior of the general solution on the complex plane of time. She then also obtained the explicit solution by quadratures using theta-functions of two variables. Up to the present day, the transformations carried out by Kovalevskaya are considered to be far from trivial and have never been improved upon.

Kovalevskaya also dealt with general problems of integrating partial differential equations (producing the so-called Cauchy–Kovalevskaya theorem), analysis of the stability of the annuli of Saturn, and of light propagation in crystals.

Having a literary talent and having lived through a fascinating historical period in Europe, she left us the additional inheritance of several novels and memoirs, which remain popular today.

H. Poincaré

Poincaré, Henry Jules (29.4.1854 – 17.7.1912) was a famous French mathematician, physicist, astronomer, and philosopher. In his three-volume treatise *New Methods of Celestial Mechanics*, examining the three-body problem, he initiated the qualitative study of dynamical systems and discussed obstructions to the existence of analytical first integrals for a wide class of dynamical systems. He proposed (without proving them) arguments for such obstructions appropriate to the Euler–Poisson equations. He established a new form of dynamical equations on a group, which systematized the particular results of Euler and Lagrange and turned out to be the most suitable point of view for various problems of rigid body dynamics. The Hamiltonian variant of these equations was suggested by N. G. Chetaev.

Poincaré applied the formalism developed by him to derive the equations of motion for a rigid body having cavities filled with a vortex ideal incompressible fluid. For these equations, he found the integrability case characterized by having a dynamical symmetry. Moreover, he obtained the elliptic quadrature and used it to explain various effects of the Earth's precession; he pictured the Earth as a rigid shell filled with a fluid nucleus. Also, he found explicit formulae for the frequencies of small oscillations and obtained the necessary conditions for stability.

A. M. Lyapunov

Lyapunov, Aleksandr Mikhailovich (6.6. 1857 – 3.11.1918) was a famous Russian mathematician and mechanician, the creator of the stability theory of a dynamical system. He discovered one of the integrability cases of the Kirchhoff equations for a rigid body moving in a fluid. In a vast memoir of 1888, he showed and studied the stability of the rigid body screw motions in a fluid. He clarified the Kovalevskaya method mentioned above, demonstrating that her arguments concerning the uniqueness of solutions in the integrable cases had been correct. He then proposed a method of his own (the so-called Kovalevskaya–Lyapunov method) consisting of the introduction of a small parameter and analysis of the variational equation.

V. A. Steklov

Steklov, Vladimir Andreevich (9.1.1864 – 30.5.1926) was a Russian mathematician and mechanician, a student of A. M. Lyapunov. In 1894 he defended his master's thesis *On Rigid Body Motion in a Fluid* (published in 1893), in which he gave a new integrability case of the Kirchhoff equations and proved the theorem on the impossibility of other cases for which the additional integral is quadratic.

He found the equivalence between the Clebsch integrable case of Kirchhoff's equations and the Brun problem. In 1909, he found a new integrable family for the problem of the motion of a rigid body with cavities filled with fluid (Poincaré–Zhukovskii equations). He presented two particular solutions of the Euler–Poisson equations (one of which was found simultaneously by D. K. Bobylev).

S. A. Chaplygin

Chaplygin, Sergei Alekseevich (5.4.1869 – 8.10.1942) was a Russian mathematician and mechanician, one of the founders of modern hydro-aerodynamics. He found the particular integrability case of the Euler–Poisson equations for zero area constant, generalizing the more special solution of D.G. Goryachev and also more particular solutions characterized by a system of linear invariant relations. For the Kirchhoff equations, he also found an analogous particular case of partial integrability and its generalizations, studied the screw motions, and gave a geometric interpretation of various other motions (in particular, for the Clebsch case). He derived the equations of motion of a heavy rigid body in a fluid and studied the cases of plane and axially-symmetric motions in detail.

His reputation is especially due to his work on nonholonomic mechanics. In this area he found a number of integrable problems of rigid body dynamics: the rolling of an axially-symmetric body on a plane, the Chaplygin ball, the Chaplygin sleigh, etc. Like N. E. Zhukovskii, he strove for geometric clarity in his masterly analytical calculations.

D. N. Goryachev

Goryachev, Dmitrii Nikanorovich (20.10. 1867–10.7.1949) was a Russian mechanician, a student of N. E. Zhukovskii. He found an integrability case and existence of particular solutions in the Euler–Poisson equations for zero area constant. He studied the equations of rigid body motion in several force fields. He found a certain general form of potential for the rigid body system, admitting integrals of third and fourth degrees. He worked on the problem of heavy rigid body motion in a fluid. His master's thesis *On Some Cases of Motions of Rectilinear Parallel Vortices* defended in 1899 became a classical work in vortex structure dynamics. He also worked on the well-known three-body problem from celestial mechanics.

V. V. Kozlov

Kozlov, Valerii Vasil'evich (was born 1.1. 1950) is a Russian mathematician and mechanician, a member of the Russian Academy of Sciences (since 2000). In a number of works collected in the book *Qualitative Analysis Methods in Rigid Body Dynamics* (MGU, 1980), he proved the nonexistence of analytical integrals of the Euler–Poisson equations and also showed the dynamical effects preventing the integrability of these equations: the separatrix splitting and the birth of a large number of nondegenerate periodic solutions. These studies "closed" the problem of Poincaré posed in his *New Methods of Celestial Mechanics* (Vol. 1), thus opening a new epoch in the theory of rigid body dynamics: now, rather than searching for particular solutions with a given algebraic structure, it is qualitative methods for the study of general solutions which are of prime interest.

Also, V. V. Kozlov suggested new methods for analyzing integrable systems based on the use of the geometric Liouville–Arnol'd theorem and Weyl's uniform distribution theorem. Justifying the Kovalevskaya method, V. V. Kozlov proved a number of assertions relating the branching of the general solution on the complex time plane to the nonexistence of single-valued first integrals (Painlevé–Golubev conjecture). To find periodic solutions for rigid body dynamics, he introduced the use of variational methods. V. V. Kozlov has written a large number of works related to the motion of a body in a fluid in which many classical problems were solved and also new models and methods based on qualitative analysis are presented. In recent years he has turned his attention to problems of statistical mechanics.

1 Rigid Body Equations of Motion and Their Integration

1.1 Poisson Brackets and Hamiltonian Formalism

1 Poisson manifolds

The majority of problems to be considered in this book can be written in canonical Hamiltonian form and also possess a first integral—the energy integral. However, in many cases, the equations of motion admit a more convenient algebraic form based on some different system of variables and simplifying the search for integrals and particular solutions, the analysis of stability, etc. In this algebraic form, many properties of canonical Hamiltonian systems are preserved, but also some characteristic distinctions appear; these are studied in the general theory of Poisson structures (for details, see [95, 577, 418]).

We briefly present some basic definitions and results necessary for studying problems of rigid body dynamics. Note that the development of the theory of Poisson structures was in many respects motivated by the dynamics of tops, which allows one to make abstract formulations of many theorems more evident and natural.

We refer the reader to the textbooks [18, 21, 170, 418] on differential and symplectic geometry. All results of this section can be represented in coordinate form; the reader can ignore formal mathematical terminology, which is based on simple dynamical facts, but may seem to be alien to them on first acquaintance.

Poisson brackets and their properties. The equations of dynamics can be written in standard *Hamiltonian form*

$$\dot{\boldsymbol{q}} = \frac{\partial H}{\partial \boldsymbol{p}}, \qquad \dot{\boldsymbol{p}} = -\frac{\partial H}{\partial \boldsymbol{q}}, \qquad H = H(\boldsymbol{q}, \boldsymbol{p}), \tag{1.1}$$

where the canonical coordinates $(\boldsymbol{q}, \boldsymbol{p})$ are defined on an even-dimensional manifold $(\boldsymbol{q}, \boldsymbol{p}) \in M^{2n}$, called the *phase space*. The function H is called the *Hamiltonian*. The number $n = \frac{\dim M}{2}$ is called the number of degrees of freedom of the Hamiltonian system (1.1).

The divergence of the vector field (1.1) vanishes, i.e., the phase flow is incompressible(*the Liouville theorem*).

If we introduce the *Poisson bracket* of two functions F and G by the formula

$$\{F, G\} = \sum_i \left(\frac{\partial F}{\partial q_i} \frac{\partial G}{\partial p_i} - \frac{\partial F}{\partial p_i} \frac{\partial G}{\partial q_i} \right), \tag{1.2}$$

then (1.1) can be rewritten as

$$\dot{q}_i = \{q_i, H\}, \qquad \dot{p}_i = \{p_i, H\}. \tag{1.3}$$

DOI 10.1515/9783110544442-001

The evolution of any differentiable function $F = F(\boldsymbol{q}, \boldsymbol{p})$ is governed by the Hamiltonian law:

$$\dot{F} = \{F, H\}. \tag{1.4}$$

Equations (1.1) are not invariant under arbitrary transformations of coordinates. Moreover, the form (1.1) of the main equations of rigid body dynamics does not possess good algebraic properties but has singularities that are not related to the nature of the problem considered (see Sec. 1.3). Before discussing more convenient forms of the equations of motion, which preserve the main properties of the canonical form, we present an invariant exposition of Hamiltonian dynamics.

In the invariant construction of the Hamiltonian formalism, following P. Dirac, we start from Eqs. (1.3) and postulate the following properties of *Poisson brackets* defined for functions given on a manifold M, of arbitrary dimension with coordinates $\boldsymbol{x} = (x^1, \dots, x^n)$:

1° $\{\lambda F_1 + \mu F_2, G\} = \lambda\{F_1, G\} + \mu\{F_2, G\}, \quad \lambda, \mu \in \mathbb{R}$ (*bilinearity*),
2° $\{F, G\} = -\{G, F\}$ (*skew symmetry*),
3° $\{F_1 F_2, G\} = F_1\{F_2, G\} + F_2\{F_1, G\}$ (*the Leibniz rule*),
4° $\{\{H, F\}, G\} + \{\{G, H\}, F\} + \{\{F, G\}, H\} = 0$ (*the Jacobi identity*).

The Poisson bracket $\{\cdot, \cdot\}$ is also called a *Poisson structure* and the manifold M on which it is defined is called a *Poisson manifold.*

In this definition, we omit the *nondegeneracy* property (i.e., for any function $F(\boldsymbol{x}) \not\equiv$ const, there exists $G \not\equiv$ const, $\{F, G\} \not\equiv 0$), which obviously holds for the canonical structure (1.2). This allows one, for example, to introduce a Poisson bracket for odd-dimensional systems. In our considerations, a Poisson structure may be *degenerate* and possess *Casimir functions* $F_k(\boldsymbol{x})$ that commute with all variables x_i and, therefore, with all functions $G(\boldsymbol{x})$ on M: $\{F_k, G\} = 0$. Casimir functions are also called *central functions, Casimirs, or annihilators.*

Properties 1° – 4° allow us to express the Poisson bracket of functions F and G in explicit form in coordinates

$$\{F, G\}(\boldsymbol{x}) = \sum_{i,j} \{x^i, x^j\} \frac{\partial F}{\partial x^i} \frac{\partial G}{\partial x^j}. \tag{1.5}$$

The basic brackets $J^{ij} = \{x^i, x^j\}$ are called *structure functions* of a Poisson manifold M with respect to a given (generally speaking, local) coordinate system $\boldsymbol{x} = (x_1, \dots, x_n)$ (see [21, 450]). They form the *structure matrix (tensor)* $\mathbf{J} = \|J^{ij}\|$ of size $n \times n$.

If

$$\mathbf{J} = \begin{pmatrix} 0 & \mathbf{E} \\ -\mathbf{E} & 0 \end{pmatrix}, \quad \mathbf{E} = \|\delta_i^j\|, \tag{1.6}$$

then we obtain the canonical Poisson bracket defined by formula (1.2).

The following conditions satisfied by the structure matrix $\mathbf{J}(\boldsymbol{x})$ can be deduced from 1° – 4°:

I. skew symmetry:

$$J^{ij}(\boldsymbol{x}) = -J^{ji}(\boldsymbol{x}), \tag{1.7}$$

II. the Jacobi identity:

$$\sum_{l=1}^{n} \left(J^{il}\frac{\partial J^{jk}}{\partial x^l} + J^{kl}\frac{\partial J^{ij}}{\partial x^l} + J^{jl}\frac{\partial J^{ki}}{\partial x^l} \right) = 0. \tag{1.8}$$

Therefore, for example, an arbitrary constant, skew-symmetric matrix $\|J^{ij}\|$ determines a Poisson structure.

The invariant object determined by the tensor $\mathbf{J}$ is a bivector (bivector field):

$$\mathbf{J}(dF, dG) = \sum J^{ij}(\boldsymbol{x})\frac{\partial F}{\partial x^i} \wedge \frac{\partial G}{\partial x^j},$$

where dF is a covector with components $\frac{\partial F}{\partial x^i}$.

A smooth function H on the manifold defines the vector field $\boldsymbol{X}_H = \{\boldsymbol{x}, H\}$, called a Hamiltonian system. It has the following coordinate representation:

$$\dot{x}^i = X_H^i = \{x^i, H\} = \sum_j J^{ij}(x)\frac{\partial H}{\partial x^j}. \tag{1.9}$$

The function $H = H(\boldsymbol{x})$ is called the *Hamiltonian* or *Hamiltonian function.*

The commutator of vector fields and the Poisson brackets are related by

$$[\boldsymbol{X}_H, \boldsymbol{X}_F] = -\boldsymbol{X}_{\{H,F\}}.$$

It is easy to verify that any Hamiltonian vector field generates a transformation (phase flow) preserving the Poisson brackets.

A function $F(\boldsymbol{x})$ is called *first integral* if its derivative along the vector field vanishes, $\dot{F} = \boldsymbol{X}_H(F) = 0$; this condition is equivalent to the relation $\{F, H\} = 0$.

The system

$$F_1(\boldsymbol{x}) = 0, \; \dots, \; F_k(\boldsymbol{x}) = 0 \tag{1.10}$$

determines a *system of invariant relations* (defining the *invariant manifold*) if $\{F_i, H\} = 0$ on the submanifold defined by conditions (1.10).

Nondegenerate brackets. Symplectic structures. If a Poisson bracket is nondegenerate, then we can uniquely assign to it a closed, nondegenerate 2-form. Indeed, for any smooth function F, the operation $\boldsymbol{X}_F = \{F, \cdot\}$ is a differentiation and defines a tangent vector at any point of M. Using $1° - 4°$, we can prove that any tangent vector at a point of the manifold can be represented in such a form.

We define the nondegenerate 2-form ω^2 by the rule

$$\omega^2(\boldsymbol{X}_G, \boldsymbol{X}_F) = \{F, G\}.$$

Axioms $1^\circ - 4^\circ$ imply that this 2-form is bilinear, skew-symmetric and closed. It is also nondegenerate. This 2-form is called a *symplectic structure* and the manifold M is a *symplectic manifold*.

In coordinates, ω^2 has the form $\sum\limits_{i,j} \omega_{ij} dx_i \wedge dx_j$, where $\|\omega_{ij}\| = \|J^{ij}\|^{-1}$; in the canonical case (1.6), $\omega^2 = \sum\limits_i dp_i \wedge dq_i$. Any symplectic structure can be locally reduced to this form; this is the *Darboux theorem* (see [450, 18, 21]). In the next section, we state this theorem in a more general form.

Symplectic foliations. Generalized Darboux theorem. If a Poisson bracket is degenerate, then the Poisson manifold (phase space) is foliated by *symplectic leaves* (*fibers*). The restriction of the Poisson structure to a symplectic leaf is nondegenerate. Usually these leaves are common level surfaces of all Casimir functions. On any leaf, the Darboux theorem holds and the equations of motion may be written in canonical form. However, the reduction to such a system is not always necessary in applied problems since this leads to the loss of algebraicity of differential equations and restrictions in the use of geometrical and topological methods.

REMARK. In rigid body dynamics, the algebraic form of the equations of motion is usually used to search for integrals, particular solutions and to perform stability analysis. This form is also preferable for numerical integration since the canonical form often gives rise to singularities related to degeneration of local variables at some points (for example, Euler angles at the poles of the Poisson sphere; see Sec. 1.3).

For problems of qualitative analysis and perturbation theory, however, the canonical form is usually preferred since the corresponding methods are most developed and algorithmized for this form.

The *rank of a Poisson structure at a point* $\boldsymbol{x} \in M$ is the rank of the structural tensor at this point (obviously, it is even). The *rank of a Poisson structure on the whole manifold* M is the maximal rank, which it has at some point $\boldsymbol{x} \in M$. For a symplectic manifold the rank of the Poisson structure is constant and maximal at each point.

We state the general *Darboux theorem* for arbitrary Poisson manifolds (see [95, 450]).

Theorem. *Let $(M, \{\cdot,\cdot\})$ be a Poisson manifold of dimension n, and at a point $\boldsymbol{x} \in M$, let the rank of the bracket $\{\cdot,\cdot\}$ be locally constant, maximal and equal to $2r$ = the rank of the Poisson structure. Then there exists a local system of canonical coordinates $x_1, \dots, x_r, y_1, \dots, y_r, z_1, \dots, z_{n-2r}$, in which the Poisson brackets have the form*

$$\{x_i, x_j\} = \{y_i, y_j\} = \{x_i, z_k\} = \{y_i, z_k\} = \{z_k, z_l\} = 0,$$

$$\{x_i, y_j\} = \delta_{ij},$$

where $1 \leqslant i, j \leqslant r, \quad 1 \leqslant k, l \leqslant n - 2r.$

In these coordinates, any symplectic leaf in some neighborhood of $\boldsymbol{x}$ is defined by the equations $z_i = c_i$ (c_i = const) and the symplectic structure on the leaf is defined by the form $\omega = \sum_i dx_i \wedge dy_i$. Note that by this theorem, a symplectic leaf $z_i = k_i$ is defined only locally, while the global Casimir functions may not exist.

Through points at which the rank of the Poisson bracket is not maximal (i.e., less than $2r$) there pass *singular symplectic leaves* (see, e.g., [95]). Systems on singular symplectic leaves often arise in mechanics (see [95, 462]).

2 The Lie – Poisson bracket

One of the most important examples of a Poisson structure is that associated with a *Lie algebra*. Let c_{ij}^k be the structure constants of a Lie algebra $\mathfrak{g}$ in a basis $\boldsymbol{v}_1, \dots, \boldsymbol{v}_n$. The *Lie – Poisson bracket* of functions F and H defined on another (generally speaking) vector space V with coordinates $\boldsymbol{x} = (x_1, \dots, x_n)$ and a basis $\omega^1, \dots, \omega^n$ is defined by the formula

$$\{F, H\} = \sum_{i,j=1}^{n} J_{ij}(\boldsymbol{x}) \frac{\partial F}{\partial x_i} \frac{\partial H}{\partial x_j}, \tag{1.11}$$

where the structure tensor $J_{ij}(\boldsymbol{x}) = \sum_k c_{ij}^k x_k$ is linear in x_k. All necessary identities 1° – 4° (see Sec. 1) for the structure tensor can be obtained from the properties of the structure constants of Lie algebras:

1. $c_{ij}^k = -c_{ji}^k$,
2. $\sum_m (c_{im}^l c_{jk}^m + c_{km}^l c_{ij}^m + c_{jm}^l c_{ki}^m) = 0.$

As is known from the theory of Lie algebras, symplectic leaves of a Lie – Poisson structure are exactly orbits of the coadjoint representation of the corresponding Lie group (see [18, 21, 450]). A formal presentation and the corresponding proof can be found, e.g., in [18]. Hamilton's equations for the Lie – Poisson structure in the coordinates $\boldsymbol{x}$ are

$$\dot{x}_i = \{x_i, H\} = \sum_{k,j} c_{ij}^k x_k \frac{\partial H}{\partial x^j}. \tag{1.12}$$

Remark. Equations (1.12) can be written in a more invariant, coordinate-free form as follows:

$$\dot{\boldsymbol{x}} = \mathrm{ad}_{dH}^*(\boldsymbol{x}), \quad \boldsymbol{x} \in \mathfrak{g}^*, \tag{1.13}$$

where ad_ξ^*, ($\boldsymbol{\xi} \in \mathfrak{g}$) is the operator of the coadjoint representation of the Lie algebra $\mathfrak{g}$: $\mathrm{ad}_\xi^* : \mathfrak{g}^* \to \mathfrak{g}^*$.

In rigid body dynamics, the Lie – Poisson bracket occurs very often. This is due to the fact that the configuration space of a system is, as a rule, some combination of natural Lie groups ($SO(3)$, $E(3), \dots$). However, reduction with respect to cyclic variables can give rise to nonlinear Poisson brackets (see Ch. 4, Sec. 4.1).

Next we derive the equations of rigid body motion from general dynamical principles.

1.2 Poincaré and Poincaré – Chetaev Equations

1 Poincaré Equations

The most natural and convenient form of the equations of motion of a rigid body is obtained from the general dynamical equations in *quasi-coordinates*. The Lagrangian form of these equations was obtained by *H. Poincaré* [473], and the Hamiltonian form by *N. G. Chetaev* [134]. Possible generalizations to the nonholonomic case (i.e., for nonintegrable constraints) were considered in [332, 510]. For rigid body dynamics, the Poincaré – Chetaev equations lead to Hamiltonian equations with linear structure tensor, i.e., to the Lie – Poisson structure described above (see Sec. 1.1). The Poincaré and Poincaré – Chetaev equations are derived in this section, since an account of this cannot be found in the standard literature.

Let us consider the equations of motion of a Lagrange dynamical system determined by generalized coordinates $\boldsymbol{q} = (q_1, \dots, q_n)$, with redundancy — in general, these are not independent, i.e., they satisfy $m < n$ *holonomic* constraints of the form $f_j(\boldsymbol{q}) = 0$, $j = 1, \dots, m$ and *quasi-velocities* $\boldsymbol{\omega} = (\omega_1, \dots, \omega_k)$, which are expressed in terms of the generalized velocities $\dot{q}_i$ by the formulas

$$\dot{q}_i = \sum_{s=1}^{k} v_i^s(\boldsymbol{q})\omega_s, \quad i = 1, \dots, n. \tag{1.14}$$

We also assume that all holonomic constraints are taken into account, thus

$$(\nabla f_j, \dot{\boldsymbol{q}}) = \sum_{i,s} v_i^s(\boldsymbol{q})\omega_s \frac{\partial f_j}{\partial q_i} \equiv 0, \quad j = 1, \dots, m.$$

In the case where $k > n - m$ this condition requires the quasi-velocities to satisfy additional relations linear in ω_i.

The quantities ω_s are called the *Poincaré parameters* and are the components of the velocity of the system in the (generally speaking) nonholonomic basis of vector fields

$$\boldsymbol{v}^s = \sum_i v_i^s(\boldsymbol{q}) \frac{\partial}{\partial q_i}. \tag{1.15}$$

From now on it will be assumed that the vector fields form a closed system

$$[\boldsymbol{v}^i, \boldsymbol{v}^j] = \sum_{s=1}^{k} c_{ij}^s(\boldsymbol{q})\boldsymbol{v}^s, \quad i, j, s = 1, \dots, k. \tag{1.16}$$

In the case where $k \leqslant n$ this conditions is implied by integrability of the constraints [450]. If all c_{ij}^s are constant, then the fields $\boldsymbol{v}^s$ define a finite-dimensional Lie algebra. The equations of motion in the variables $(q_1, \dots, q_n, \omega_1, \dots, \omega_k)$ have the Lagrangian form

$$\frac{d}{dt}\left(\frac{\partial L}{\partial \omega_i}\right) = \sum_{r,s} c_{ri}^s \omega_r \frac{\partial L}{\partial \omega_s} + \boldsymbol{v}^i(L), \quad i = 1, \dots, k, \tag{1.17}$$

and are called the *Poincaré equations*; together with (1.14) they form a complete system of equations of motion. In (1.17), the differentiation along the vector field $\mathbf{v}^i$ is defined by formula (1.15).

If the Lagrange function is a homogeneous quadratic form of angular velocities (for example, the kinetic energy), then $\mathbf{v}_i(L) = 0$ and the system (1.17) for $\boldsymbol{\omega}$ decouples and can be integrated independently. In this case, Eqs. (1.17) are called the *Euler – Poincaré equations*.

Poincaré obtained his equations using the Hamiltonian variational principle [473]. Here we restrict our attention to the case where the number of components of the quasi-velocity $\boldsymbol{\omega} = (\omega_1, \dots, \omega_k)$ coincides with the dimension of the configuration M^k-space defined by the constraints $f_j(\boldsymbol{q}) = 0$, $j = 1, \dots, m$, i.e., $k = n - m$. In this case Eqs. (1.17) can be obtained directly from the Euler – Lagrange equations.

Indeed, introducing on M^k the local coordinates x_i, the Euler – Lagrange equations can be written as

$$\frac{d}{dt}\left(\frac{\partial L}{\partial \dot{x}_i}\right) - \left(\frac{\partial L}{\partial x_i}\right) = 0, \quad i = 1, \dots, k. \tag{1.18}$$

By (1.14) and (1.15), the following relations hold:

$$\begin{gathered}\omega_s = \sum_{i=1}^{k} a_s^i \dot{x}_i, \quad \dot{x}_i = \sum_{s=1}^{k} b_i^s \omega_s, \\ \mathbf{v}^s = \sum_{i=1}^{k} b_i^s \frac{\partial}{\partial x_i}, \quad i,\, s = 1, \dots, k,\end{gathered} \tag{1.19}$$

where the $k \times k$ matrices $\mathbf{A} = \|a_s^i\|$, $\mathbf{B} = \|b_i^s\|$ are inverse to one another ($\mathbf{AB} = \mathbf{E}$).

Denote the Lagrange function expressed in terms of the quasi-velocities by

$$\tilde{L}(\boldsymbol{x}, \boldsymbol{\omega}) = L(\boldsymbol{x}, \dot{\boldsymbol{x}}). \tag{1.20}$$

Using (1.19) we obtain

$$\begin{gathered}\frac{\partial L}{\partial x_i} = \frac{\partial \tilde{L}}{\partial x_i} + \sum_{k,\, s} \dot{x}_k \frac{\partial \tilde{L}}{\partial \omega_s} \frac{\partial b_s^k}{\partial x_i}, \\ \frac{\partial L}{\partial \dot{x}_i} = \sum_{s} \frac{\partial \tilde{L}}{\partial \omega_s} b_s^i, \quad i = 1, \dots, k.\end{gathered} \tag{1.21}$$

Substitute Eqs. (1.21) into Eqs. (1.18) and multiply both sides by the matrix $\mathbf{A}$; make use of Eqs. (1.19) in the resulting system and apply the following representation of the structure coefficients in (1.16):

$$c_{sp}^r(x) = \sum_{k,\, i} a_r^k \left(b_i^s \frac{\partial b_k^p}{\partial x_i} - b_i^p \frac{\partial b_k^s}{\partial x_i} \right).$$

Collecting similar terms, we arrive at Eqs. (1.17). □

In the case where the number of quasi-velocities is greater than the dimension of the configuration space, the reasoning becomes more difficult since the matrices **A** and **B** are not square and have no inverses.

2 Poincaré – Chetaev equations

N. G. Chetaev modified the Poincaré equations (1.17) and (1.14) by applying the *Legendre transform*:

$$M_i = \frac{\partial L}{\partial \omega_i}, \qquad \sum_i \omega_i M_i - L \,|_{\omega \to M} = H(\boldsymbol{M}, \boldsymbol{q}). \tag{1.22}$$

The variables M_i should be seen as "*quasi-momenta*". One has $\omega_i = \partial H / \partial M_i$ and Eqs. (1.17) can be written as

$$\dot{M}_i = \sum_{rs} c_{ri}^s \frac{\partial H}{\partial M_r} M_s - \boldsymbol{v}^i(H), \quad i = 1, \ldots, k. \tag{1.23}$$

To obtain a closed system, we should combine (1.23) with Eqs. (1.14) written in the form

$$\dot{q}_i = \sum_s v_i^s(\boldsymbol{q}) \frac{\partial H}{\partial M_s}, \quad i = 1, \ldots, n. \tag{1.24}$$

The combination of Eqs. (1.23) and (1.24) forms a Hamiltonian system with (in general) a degenerate Poisson bracket defined for arbitrary functions $f(\boldsymbol{M}, \boldsymbol{q})$ and $g(\boldsymbol{M}, \boldsymbol{q})$ by the formula [134]

$$\{f, g\} = \sum_i \left(\frac{\partial g}{\partial M_i} v^i(f) - \frac{\partial f}{\partial M_i} v^i(g) \right) + \sum_{sij} c_{ij}^s \frac{\partial f}{\partial M_j} \frac{\partial g}{\partial M_i} M_s. \tag{1.25}$$

It is straightforward to verify that this bracket satisfies all the conditions $1^\circ - 4^\circ$ (see Sec. 1.1) required for it to be a Poisson structure. From relation (1.25) one can easily obtain the structure matrix J^{ij}:

$$\{M_i, M_j\} = \sum_s c_{ij}^s(\boldsymbol{q}) M_s, \qquad \{q_i, q_j\} = 0, \quad \{q_i, M_j\} = v_i^j(\boldsymbol{q}). \tag{1.26}$$

Historical comments. For dynamical equations of the form (1.23), (1.24) N. G. Chetaev [134] developed a theory of integration similar to the Hamilton – Jacobi method. In the canonical case, success in the separation of variables stems from special coordinate systems on the configuration space (e.g., elliptic or sphero-conic coordinates). However, for the algebraic form (1.23), (1.24), only trivial symmetries can be studied (for example, the symmetries that arise in the Lagrange case; see Ch. 2).

For similar reasons, Chetaev's considerations on the generalizations of the Routh theorem in the presence of a cyclic integral and reduction of order have not received further development either. For the Poincaré – Chetaev equations possessing first integrals of cyclic type, a new reduction procedure is presented in Ch. 4, Sec. 4.1, 4.2. It provides a way to obtain the equations of the reduced system in the most simple algebraic form and sometimes leads to nonlinear Poisson brackets.

3 Equations on Lie groups

As a rule, the configuration space for rigid body dynamics is a Lie group. For example, to describe the rotation of a rigid body about a fixed point the configuration space is the group $SO(3)$ and in the case of free motion of a rigid body it is the group $E(3) = SO(3) \otimes_s \mathbb{R}^3$, the semi-direct product of the rotation group $SO(3)$ and the commutative group of translations $\mathbb{R}^3$.

On a group, it is convenient to choose as a basis of the vector fields $\boldsymbol{v}^s$ (1.15) left-invariant (or right-invariant) vector fields which form the corresponding Lie algebra. In this case, the tensor c_{ij}^k is independent of the choice of coordinates and is determined by the structure constants of the Lie algebra. The brackets (1.25) define the so-called *canonical structure on the cotangent bundle* over the group [95, 466].

If the Hamiltonian H is independent of q_i, i.e., $(\boldsymbol{v}_i(H) = 0)$, then the equations for quasi-momenta $M_1, \ldots, M_k$ are closed and the Euler equations for the inertial motion of a rigid body can be obtained; in this case, the constants c_{ij}^s are defined by the Lie algebra $so(3)$. For arbitrary Lie algebra with structure constants c_{ij}^s, similar equations with quadratic Hamiltonian are also (as in subsection 1) called the Euler – Poincaré equations.

If the Hamiltonian H depends on coordinates but the redundant coordinates can be chosen such that all components of the left-invariant fields $v_r^s(\boldsymbol{q})$ are linear in $\boldsymbol{q}$, then the bracket (1.26) becomes the usual Lie – Poisson bracket and all geometric dependencies for redundant variables are its Casimir functions or invariant relations. This can be achieved by using the matrix realization of the Lie group taking the components of its matrices as redundant coordinates. The Lie – Poisson structure obtained in this case corresponds to the semi-direct sum $\mathfrak{g} \oplus_s \mathbb{R}^{n^2}$, where $\mathbb{R}^{n^2}$ is the space of $(n \times n)$-matrices and $\mathfrak{g}$ is the Lie algebra of the given Lie group; this structure is called the *natural canonical structure of the cotangent bundle* over the Lie group. This is the way, for example, to obtain the equations of motion of a rigid body in terms of direction cosines and angular momenta (see Sec. 1.4). The matrix realization of Lie groups is also used in the multidimensional rigid body dynamics [67, 95, 466].

The Hamilton equations on a Lie group with the natural canonical structure for problems of rigid body dynamics always possess a standard invariant measure (since in this case all groups are unimodular). This is an analog of the Liouville theorem on the solenoidality of the canonical Hamiltonian flow.

A detailed derivation of the equations of motion of a rigid body in an arbitrary potential field is considered in Sec. 1.4. More complicated equations, whose derivation is based on hydrodynamical principles, describing the motion of rigid bodies in a fluid or the motion of rigid bodies with cavities filled with fluid are considered in Ch. 3.

4 Comments

1. Thus, the Poincaré and Poincaré – Chetaev equations only represent a convenient tool for rewriting the equations of motion of a system in the Lagrangian and Hamiltonian form in an arbitrary (including redundant) system of variables. The possibility of such representation is related with the presence of a *tensor invariant*, i.e., a Poisson structure whose coordinate form depends on the choice of coordinates; moreover, for redundant variables, the Poisson structure is degenerate.

Note that the connection between the Lagrangian and Hamiltonian forms is clear for most mechanicians only when written in their canonical form. Thus, in [61], the Hamiltonian form of the dynamical equations is considered as found from some considerations that are not completely natural; in particular, with reference to [447], where the author, unfamiliar with the general formalism of dynamical equations, really "rediscovers" the Euler angles and conjugate momenta. In [61], some strange theorems are proved that state that the Lagrangian form can be obtained from the Hamiltonian form; some confusion appears since the components of the angular momentum commute equally with the components of the linear momentum and with the direction cosines, and the Kirchhoff equations can be treated as a part of momentum equations on the group $E(3)$, i.e., the Euler – Poincaré equations for $\boldsymbol{M}$ and $\boldsymbol{p}$. In the case where the potential is absent, this part can be separated from the position equations (i.e., equations for direction cosines). On the other hand, the Kirchhoff equations can be considered as Hamiltonian equations on $SO(3)$, and in this case, the components of the momentum force $\boldsymbol{p}$ must be treated as direction cosines. Note that the Steklov analogy consists of this [549] (see also Ch. 3, Sec. 3.1).

A priori, the non-Hamiltonian coordinate form of the Newton equations of the dynamics of a satellite is used in [33], where the presence of the energy integral becomes more clear.

2. There is an interesting result concerning the question of the existence of a global invariant measure for the Euler – Poincaré equations on a Lie algebra (it is assumed that the density of this measure is differentiable; see [335]). It should be emphasized that the result which follows speaks of a global measure: in general, so-called singular measures often occur; for such measures, their densities vanish or become infinite at some points (or on submanifolds of lower dimension).

Theorem. [335]. *The Euler – Poincaré equations possess a global invariant measure if and only if the Lie algebra $\mathfrak{g}$ corresponds to a unimodular group G.*

(In terms of the structure constants, the unimodularity criterion is $\sum_k c_{ik}^k = 0$).

Moreover, in the unimodular case the invariant measure is standard, i.e., it is constant on the whole Lie algebra g. This is a consequence of the following more general result: *a* system of ordinary differential equations with homogeneous right-hand sides possesses an invariant measure if and only if its phase flow preserves the standard measure.

In [335], the following incorrect conclusion was deduced from the absence of an invariant measure: "it is impossible to present the Euler – Poincaré equations in Hamiltonian form for an arbitrary Lie algebra". Actually, as was noted above, the Euler – Poincaré equations are a priori Hamiltonian equations, and on any symplectic leaf of the Lie algebra (which is an invariant manifold of the system), they can be reduced to canonical Hamiltonian form by the Darboux theorem. However, for a generic Lie algebra, symplectic leaves can have different dimensions and hence a global measure does not exist.

Let us consider the simplest example of a solvable two-dimensional Lie algebra, having the corresponding Poisson bracket $\{M_1, M_2\} = M_1$. The phase space (the plane of the variables M_1 and M_2) consists of two two-dimensional symplectic leaves, the upper and lower half-planes without the straight line $M_1 = 0$. Each point of this straight line is a zero-dimensional symplectic leaf. For an arbitrary quadratic Hamiltonian $H = \frac{1}{2}(a_{11}M_1^2 + a_{22}M_2^2 + 2a_{12}M_1M_2)$, the equations of motion have the form

$$\dot{M}_1 = M_1(a_{22}M_2 + a_{12}M_1), \quad \dot{M}_2 = -M_1(a_{11}M_1 + a_{12}M_2). \tag{1.27}$$

Since symplectic leaves are invariant, all points of the line are equilibria positions. Trajectories of the system are arcs of ellipses $H = \text{const}$; moving along them, a point approaches the line $M_1 = 0$. Obviously, the system (1.27) cannot possess an absolutely continuous global invariant measure, however, this does not contradict its Hamiltonian property. In each of the two half-planes, the system can be represented in canonical form, for example, by using the variables $q = \ln M_1$ and $p = M_2$:

$$\dot{q} = \frac{\partial H}{\partial p}, \quad \dot{p} = -\frac{\partial H}{\partial q}.$$

In this case, the equations and the Hamiltonian become nonalgebraic, $H = \frac{1}{2}(a_{11}e^{2q} + a_{22}p^2 + a_{12}pe^q)$ – the Hamiltonian of the simplest Toda lattice.

1.3 Various Systems of Variables in Rigid Body Dynamics

Various coordinate systems are used to describe the motion of rigid bodies. Each reveals its own advantages and disadvantages in each specific problem. For example, in the search for first integrals, study of stability, and topological analysis, variables in which the equations are polynomial (or even homogeneous) are the most convenient. For numerical simulation, it is usually more convenient to have a system of equations of lowest possible order. For qualitative, perturbative, and nonlinear normalizations problems, systems of canonical variables that reflect the specific character of the perturbed problem are preferred. We describe the most significant systems of variables used in rigid body dynamics. In practical applications, especially to gyroscopes, various combinations and modifications of these systems are also used.

1 Euler angles

Consider a rigid body rotating about a fixed point O in a potential field. The configuration space, i.e., the set of all possible positions of the body, is the Lie group $SO(3)$, and we can take three angles θ, φ, and ψ as coordinates defining the position of the body; for the standard choice these angles are called the *Euler angles* (see [16, 224]).

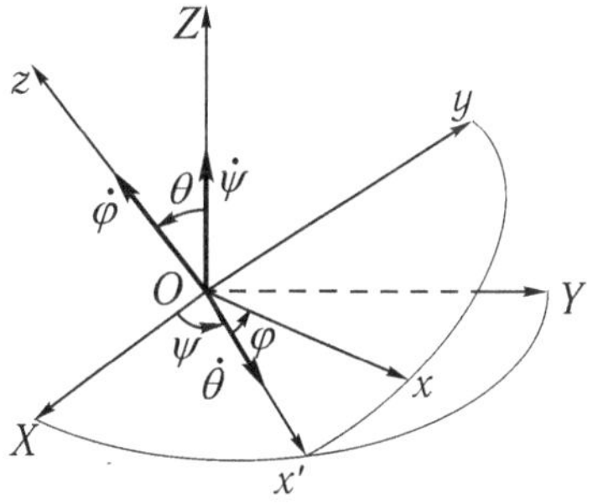

Fig. 1.1. Euler angles

At the point O, we place the vertices of two orthogonal frames: immovable $OXYZ$ and movable $Oxyz$, which is attached to the rotating body (see Fig. 1.1).

Rotation by the angle ψ (*precession angle*) about the axis OZ transfers the immovable frame into the position $Ox'y'z'$. Rotation by the angle θ (*nutation angle*) is made about the axis Ox' called the *line of nodes*. Rotation by the angle φ (*proper rotation angle*) about the axis Oz finally gives the moving frame $Oxyz$. The three rotations determined by the Euler angles θ, φ and ψ, completely define the position of the moving frame with respect to the immovable frame. The projections ω_1, ω_2, and ω_3 of the angular velocity $\boldsymbol{\omega}$ on the axes of the moving frame $Oxyz$ are expressed in terms of the Euler angles as follows:

$$\begin{aligned}\omega_1 &= \dot\psi \sin\theta \sin\varphi + \dot\theta \cos\varphi,\\ \omega_2 &= \dot\psi \sin\theta \cos\varphi - \dot\theta \sin\varphi,\\ \omega_3 &= \dot\psi \cos\theta + \dot\varphi.\end{aligned} \tag{1.28}$$

These relations are called the *kinematic Euler formulae*. Using (1.28), one can easily obtain the Lagrange function $L = L(\varphi, \psi, \theta, \dot{\varphi}, \dot{\psi}, \dot{\theta})$ of the system (see Sec. 1.6), leading to definition of the canonical momenta (by the Legendre transform)

$$p_\varphi = \frac{\partial L}{\partial \dot{\varphi}}, \quad p_\psi = \frac{\partial L}{\partial \dot{\psi}}, \quad p_\theta = \frac{\partial L}{\partial \dot{\theta}}. \tag{1.29}$$

2 Euler variables. Components of momentum and the direction cosines

For this system of variables $(\boldsymbol{M}, \boldsymbol{\alpha}, \boldsymbol{\beta}, \boldsymbol{\gamma})$, $\boldsymbol{M} = (M_1, M_2, M_3)$ are the components of the angular momentum in the moving frame $Oxyz$ attached to the body and $\boldsymbol{\alpha}, \boldsymbol{\beta}, \boldsymbol{\gamma}$ are orthogonal unit vectors of the immovable frame, written in the moving frame $Oxyz$. They are constructed as follows. The *matrix of direction cosines* (the rotation matrix) that determines the position of the body in the fixed frame

$$\mathbf{Q} = \begin{pmatrix} \alpha_1 & \beta_1 & \gamma_1 \\ \alpha_2 & \beta_2 & \gamma_2 \\ \alpha_3 & \beta_3 & \gamma_3 \end{pmatrix}, \tag{1.30}$$

is orthogonal and belongs to the group $SO(3)$.

That is,

$$\begin{aligned} (\boldsymbol{\alpha}, \boldsymbol{\alpha}) = (\boldsymbol{\beta}, \boldsymbol{\beta}) = (\boldsymbol{\gamma}, \boldsymbol{\gamma}) = 1, \\ (\boldsymbol{\alpha}, \boldsymbol{\beta}) = (\boldsymbol{\alpha}, \boldsymbol{\gamma}) = (\boldsymbol{\beta}, \boldsymbol{\gamma}) = 0, \end{aligned} \tag{1.31}$$

where here and in what follows, the parentheses (,) means the usual scalar product, and we find that the angular velocity $\boldsymbol{\omega} = (\omega_1, \omega_2, \omega_3)$ written in the moving frame $Oxyz$ can be represented by the skew-symmetric matrix $\tilde{\boldsymbol{\omega}} = \dot{\mathbf{Q}}\mathbf{Q}^T$, with entries $\tilde{\omega}_{ij} = \varepsilon_{ijk}\omega_k$.

Similarly, the angular velocity $\boldsymbol{\Omega} = (\Omega_1, \Omega_2, \Omega_3)$ written in the fixed frame $OXYZ$ is represented in the same fashion by the matrix $\tilde{\boldsymbol{\Omega}} = \mathbf{Q}^T\dot{\mathbf{Q}}$.

The directions of the vectors $\boldsymbol{\omega}$ and $\boldsymbol{\Omega}$ in the movable and immovable frames, respectively, define conic surfaces called the *movable and immovable Poinsot axoids*. The motion of the rigid body is represented as the rolling (without sliding) of the movable axoid on the immovable axoid; each axoid touches the other along the instantaneous rotation axis. Free motions of the body (without fixed points) can be interpreted as the rolling of the movable axoid on the immovable axoid with sliding along some axis that defines the instantaneous screw motion. If we mark instantaneous values of the angular velocities on the generatrices of the axoids, then we obtain the so-called *movable* and *immovable* hodographs; in the general case, they are spatial curves.

If the angular velocity as the function of time $\boldsymbol{\omega}(t)$ is known, the orientation of the rigid body can be found by solving the Poisson equation

$$\dot{\boldsymbol{\alpha}} = \boldsymbol{\alpha} \times \boldsymbol{\omega}, \qquad \dot{\boldsymbol{\beta}} = \boldsymbol{\beta} \times \boldsymbol{\omega}, \qquad \dot{\boldsymbol{\gamma}} = \boldsymbol{\gamma} \times \boldsymbol{\omega} \tag{1.32}$$

using the relations (1.31). These equations can be reduced to the Riccati equation as follows. For the unit sphere $\boldsymbol{y}^2 = 1$ we perform a stereographic projection onto the plane passing through the equator $y_3 = 0$, by the formulae

$$y_1 = \frac{2x}{x^2+y^2+1}, \qquad y_2 = \frac{2y}{x^2+y^2+1}, \qquad y_3 = \frac{x^2+y^2-1}{x^2+y^2+1}.$$

For the evolution of the coordinates x, y we obtain the equations

$$\dot{x} = \frac{1}{2}\omega_2 + \omega_3 y + \frac{1}{2}\omega_2(x^2-y^2) - \omega_1 xy, \qquad \dot{y} = -\frac{1}{2}\omega_1 - \omega_3 x + \frac{1}{2}\omega_1(x^2-y^2) - \omega_2 xy.$$

Writing in complex form, for the variable $\sigma = x + iy$ we obtain the Riccati equation (see [144])

$$\dot{\sigma} = -i\omega_3\sigma + \frac{1}{2}(\omega_2 - i\omega_1) + \frac{1}{2}(\omega_2 + i\omega_1)\sigma^2. \tag{1.33}$$

Remark. This equation is easy to obtain if one uses a complex representation for sphero-conic coordinates of the following form

$$x + iy = \frac{y_1 + iy_2}{1 - y_3}.$$

Angular momentum $\boldsymbol{M}$ is given by the formula

$$\boldsymbol{M} = \frac{\partial L}{\partial \boldsymbol{\omega}}, \tag{1.34}$$

where $L = L(\boldsymbol{\omega}, \boldsymbol{\alpha}, \boldsymbol{\beta}, \boldsymbol{\gamma})$ is the Lagrange function and ω is the angular velocity. $\boldsymbol{M}$ is related with the Euler variables φ, ψ, θ, p_φ, p_ψ, p_θ by the following relations, which are implied by the kinematic Euler equations (1.28) and (1.29):

$$\begin{aligned} M_1 &= \frac{\sin\varphi}{\sin\theta}(p_\psi - p_\varphi\cos\theta) + p_\theta\cos\varphi, \\ M_2 &= \frac{\cos\varphi}{\sin\theta}(p_\psi - p_\varphi\cos\theta) - p_\theta\sin\varphi, \\ M_3 &= p_\varphi. \end{aligned} \tag{1.35}$$

Remark 1. Our definition in (1.34) of angular momentum differs from the physical definition used in rigid body dynamics, $\boldsymbol{M} = \sum(\boldsymbol{r}_i \times m_i\boldsymbol{v}_i)$, but they coincide if $L = T$ is kinetic energy. Distinctions appear if gyroscopic forces occur in the system; such forces lead to terms linear in generalized velocities in the Lagrangian. In this case, definition (1.34) that follows from the Chetaev transform is more convenient.

Remark 2. The relation between the direction cosines (1.30) and the Euler angles is given by

$$\mathbf{Q} = \begin{pmatrix} \cos\varphi\cos\psi - \cos\theta\sin\psi\sin\varphi & \cos\varphi\sin\psi + \cos\theta\cos\psi\sin\varphi & \sin\varphi\sin\theta \\ -\sin\varphi\cos\psi - \cos\theta\sin\psi\cos\varphi & -\sin\varphi\sin\psi + \cos\theta\cos\psi\cos\varphi & \cos\varphi\sin\theta \\ \sin\theta\sin\psi & -\sin\theta\cos\psi & \cos\theta \end{pmatrix}.$$

3 Quaternion Rodrigues – Hamilton parameters

K. Gauss noted that the position of a rigid body can be uniquely defined by the set of quaternions $\lambda = \lambda_0 + \boldsymbol{i}\lambda_1 + \boldsymbol{j}\lambda_2 + \boldsymbol{k}\lambda_3$ with unit norm $\lambda_0^2 + \lambda_1^2 + \lambda_2^2 + \lambda_3^2 = 1$. They form the group $Sp(1)$, which is the universal covering of the group $SO(3)$ ($SO(3) \approx Sp(1)/\pm 1$) [170]. A detailed discussion of these redundant coordinates called the *Rodrigues – Hamilton parameters* can be found in Whittaker's monograph [611]. We clarify the geometrical meaning of the parameters λ_s [321, 611].

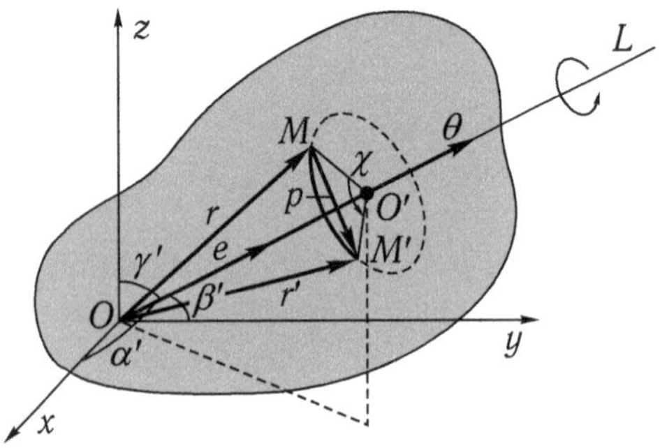

Fig. 1.2. Quaternion Rodrigues – Hamilton parameters.

It is known from kinematics that a rigid body with a fixed point O can be transferred from any position to another by a rotation through some angle χ about some axis OL attached to the body (see Fig. 1.2). The orientation of the axis OL is defined by a suitable unit vector $\boldsymbol{e}$. The position of a point M of the body is described by a vector $\overrightarrow{OM} = \boldsymbol{r}$, say. Suppose that the rotation moves the vector $\boldsymbol{r}$ to the position $\overrightarrow{OM}' = \boldsymbol{r}'$. The vector

$$\boldsymbol{p} = \overrightarrow{OM}' - \overrightarrow{OM} = \boldsymbol{r}' - \boldsymbol{r}$$

can be expressed in terms of $\boldsymbol{r}$, $\boldsymbol{e}$, and χ by the Rodrigues formula

$$\boldsymbol{p} = \frac{1}{1 + \frac{1}{4}\theta^2}\boldsymbol{\theta} \times \left(\boldsymbol{r} + \frac{1}{2}\boldsymbol{\theta} \times \boldsymbol{r}\right), \tag{1.36}$$

where the vector

$$\boldsymbol{\theta} = 2\tan\frac{\chi}{2}\boldsymbol{e} \tag{1.37}$$

is called the *vector of finite rotation*; its direction coincides with the direction of the unit vector $\boldsymbol{e}$ and its length is $2\tan(\chi/2)$.

Let

$$\boldsymbol{e} = \boldsymbol{i}\cos\alpha' + \boldsymbol{j}\cos\beta' + \boldsymbol{k}\cos\gamma', \tag{1.38}$$

where α', β', and γ' are the angles made by the vector $\boldsymbol{e}$ with the axes x, y, and z, respectively.

The quantities

$$\lambda_0 = \cos\frac{\chi}{2}, \qquad \lambda_1 = \cos\alpha' \sin\frac{\chi}{2},$$
$$\lambda_2 = \cos\beta' \sin\frac{\chi}{2}, \qquad \lambda_3 = \cos\gamma' \sin\frac{\chi}{2} \tag{1.39}$$

are the *Rodrigues – Hamilton parameters*. The parameter λ_0 is equal to the cosine of a half of the angle χ defining the finite rotation of the body. The parameters λ_1, λ_2, and λ_3 are proportional to the sine of $\chi/2$ multiplied by the direction cosines of the axis OL.

The relation of the Rodrigues – Hamilton parameters with the Euler angles θ, φ, ψ is

$$\lambda_0 = \cos\frac{\theta}{2}\cos\frac{\psi+\varphi}{2}, \qquad \lambda_1 = \sin\frac{\theta}{2}\cos\frac{\psi-\varphi}{2},$$
$$\lambda_2 = \sin\frac{\theta}{2}\sin\frac{\psi-\varphi}{2}, \qquad \lambda_3 = \cos\frac{\theta}{2}\sin\frac{\psi+\varphi}{2}. \tag{1.40}$$

The direction cosines $\boldsymbol{\alpha}$, $\boldsymbol{\beta}$, and $\boldsymbol{\gamma}$ are related with the quaternion parameters by the following quadratic relations that define the Cayley parametrization of the group $SO(3)$; we obtain a two-sheeted covering of $SO(3)$ by the three-dimensional sphere S^3: the same element of $SO(3)$ corresponds to two quaternions λ_i and $-\lambda_i$. In the quaternion representation the matrix of direction cosines (1.30) is

$$\mathbf{Q} = \begin{pmatrix} \lambda_0^2+\lambda_1^2-\lambda_2^2-\lambda_3^2 & 2(\lambda_0\lambda_3+\lambda_1\lambda_2) & 2(\lambda_1\lambda_3-\lambda_0\lambda_2) \\ 2(\lambda_1\lambda_2-\lambda_0\lambda_3) & \lambda_0^2-\lambda_1^2+\lambda_2^2-\lambda_3^2 & 2(\lambda_0\lambda_1+\lambda_2\lambda_3) \\ 2(\lambda_0\lambda_2+\lambda_1\lambda_3) & 2(\lambda_2\lambda_3-\lambda_0\lambda_1) & \lambda_0^2-\lambda_1^2-\lambda_2^2+\lambda_3^2 \end{pmatrix}. \tag{1.41}$$

In the index form, the entries of the matrix $\mathbf{Q} = \|Q_{ij}\|$ are expressed as

$$Q_{ij} = -2\left(\lambda_i\lambda_j + \left(\lambda_0^2 - \frac{1}{2}\right)\delta_{ij} - \lambda_0\lambda_k\varepsilon_{ijk}\right).$$

REMARK 3. The relation between the projections of the angular velocity $\boldsymbol{\omega}$ and the Rodrigues – Hamilton parameters has the form

$$\omega_1 = 2(\lambda_0\dot\lambda_1 + \lambda_3\dot\lambda_2 - \lambda_2\dot\lambda_3 - \lambda_1\dot\lambda_0),$$
$$\omega_2 = 2(-\lambda_3\dot\lambda_1 + \lambda_0\dot\lambda_2 + \lambda_1\dot\lambda_3 - \lambda_2\dot\lambda_0),$$
$$\omega_3 = 2(\lambda_2\dot\lambda_1 - \lambda_1\dot\lambda_2 + \lambda_0\dot\lambda_3 - \lambda_3\dot\lambda_0).$$

REMARK 4. As an alternative to the Rodrigues – Hamilton parameters, one can take complex quantities α, β, γ, and δ satisfying the condition

$$\alpha\delta - \beta\gamma = 1$$

and consider them to be the components of the complex rotation matrix

$$U = \begin{pmatrix} \alpha & \beta \\ \gamma & \delta \end{pmatrix}$$

with unit determinant. These parameters are called the *Cayley – Klein parameters*.

The relation between the Cayley – Klein and Rodrigues – Hamilton parameters is expressed by the formulae

$$\alpha = \lambda_0 + i\lambda_3, \quad \beta = -\lambda_2 + i\lambda_1, \quad \gamma = \lambda_2 + i\lambda_1, \quad \delta = \lambda_0 - i\lambda_3,$$

and the representation of the Cayley – Klein parameters in terms of the Euler angles has the form

$$\alpha = \cos\frac{\theta}{2}e^{i\frac{\psi+\varphi}{2}}, \qquad \beta = i\sin\frac{\theta}{2}e^{i\frac{\psi-\varphi}{2}},$$
$$\gamma = i\sin\frac{\theta}{2}e^{-i\frac{\psi-\varphi}{2}}, \qquad \delta = \cos\frac{\theta}{2}e^{-i\frac{\psi+\varphi}{2}}.$$

4 Andoyer variables

The Andoyer variables are the most convenient ones for perturbation theory applied to rigid body dynamics. Their geometric origin is shown in Fig. 1.3 (see also [150, 333, 95]).

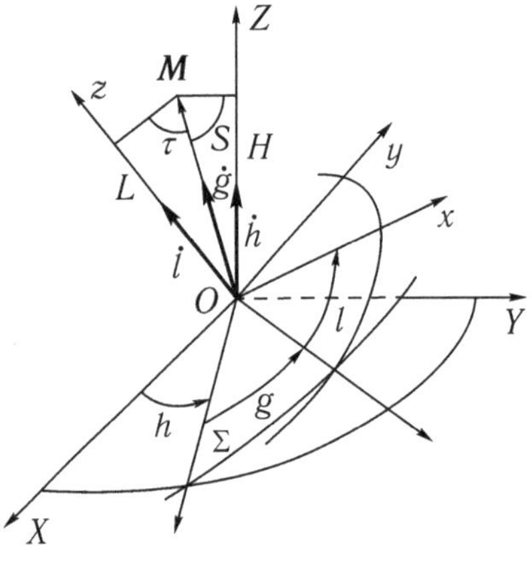

Fig. 1.3. Andoyer variables.

Here, $OXYZ$ is the immovable frame with origin at the fixed point, $Oxyz$ is the movable frame attached to the body with origin at the fixed point, and Σ is the plane passing through the fixed point and perpendicular to the angular momentum vector $\boldsymbol{M}$ (1.35) of the body. In this notation,

- L is the projection of $\boldsymbol{M}$ on the movable axis Oz;
- G is the value of $\boldsymbol{M}$;
- H is the projection of $\boldsymbol{M}$ on the fixed axis OZ;
- l is the angle between the axis Ox and the line of intersection of Σ with the plane Oxy;

g is the angle between the lines of intersection of Σ with the planes Oxy and OXY;
h is the angle between the axis OX and the line of intersection of Σ with the plane OXY.

The expressions of the components of the angular momentum in terms of the variables L, G, H, l, g, and h have the form

$$M_1 = \sqrt{G^2-L^2}\sin l, \quad M_2 = \sqrt{G^2-L^2}\cos l, \quad M_3 = L, \quad G^2 = \boldsymbol{M}^2, \tag{1.42}$$

i.e., L and l are cylindrical coordinates on the two-dimensional sphere in the space of momenta M_1, M_2, M_3.

The components of all direction cosines can be expressed as follows (as far as the authors know, not all of these relations are presented in the literature):

$$\begin{aligned}
\alpha_1 &= -\sin l \sin h \cos g \sin\tau \sin\zeta + \sin l \sin h \cos\tau \cos\zeta \\
&\quad - \sin l \sin g \cos h \sin\tau - \cos l \sin h \sin g \sin\zeta + \cos l \cos g \cos h, \\
\alpha_2 &= \cos l \cos g \sin h \sin\tau \sin\zeta - \cos l \sin h \cos\tau \cos\zeta \\
&\quad + \cos l \cos h \sin g \sin\tau - \sin l \sin g \sin\zeta \sin h + \sin l \cos h \cos g, \\
\alpha_3 &= \sin h \cos\tau \cos g \sin\zeta + \sin h \sin\tau \cos\zeta + \cos\tau \sin g \cos h, \\
\beta_1 &= -(\sin l \cos h \cos g \sin\tau \sin\zeta - \sin l \cos h \cos\zeta \cos\tau \\
&\quad - \sin l \sin g \sin h \sin\tau + \cos l \cos h \sin g \sin\zeta + \cos l \cos g \sin h), \\
\beta_2 &= \cos l \cos h \sin\tau \cos g \sin\zeta - \cos l \cos h \cos\zeta \cos\tau \\
&\quad - \cos l \sin g \sin h \sin\tau - \sin l \cos h \sin g \sin\zeta - \sin l \cos g \sin h, \\
\beta_3 &= -\sin h \cos\tau \sin g + \cos\tau \cos g \sin\zeta \cos h + \sin\tau \cos\zeta \cos h, \\
\gamma_1 &= (\sin\zeta \cos\tau + \sin\tau \cos\zeta \cos g) \sin l + \cos\zeta \sin g \cos l, \\
\gamma_2 &= (\sin\zeta \cos\tau + \sin\tau \cos\zeta \cos g) \cos l - \cos\zeta \sin g \sin l, \\
\gamma_3 &= \sin\zeta \sin\tau - \cos\tau \cos\zeta \cos g,
\end{aligned} \tag{1.43}$$

where $\sin\tau = \frac{L}{G}$, $\sin\zeta = \frac{H}{G}$.

Remark. The expressions of the direction cosines γ_i via the Andoyer variables can be found, for example, in [16, 333, 88]. The inverse formulae are $L = M_3$, $G = \sqrt{(\boldsymbol{M}, \boldsymbol{M})}$, $l = \arctan\frac{M_1}{M_2}$, $g = \arcsin\frac{M_2\gamma_1 - M_1\gamma_2}{\sqrt{M_1^2+M_2^2}}$. The formulae for α_3 and β_3 can be easily obtained from geometric considerations. The other direction cosines are implied by the commutation relations (1.59) (see the next section). The relations for λ_i^2 can be obtained via the formulae

$$\lambda_0^2 = \frac{1+\alpha_1+\beta_2+\gamma_3}{4}, \qquad \lambda_1^2 = \frac{1+\alpha_1-\beta_2-\gamma_3}{4},$$
$$\lambda_2^2 = \frac{1-\alpha_1+\beta_2-\gamma_3}{4}, \qquad \lambda_3^2 = \frac{1-\alpha_1-\beta_2+\gamma_3}{4},$$

and the λ_is themselves are defined up to a sign.

5 Comments

In contrast to the system of Euler angles with their adjoint canonical momenta, the system of Andoyer variables cannot be divided into natural position and momentum parts. They prove to be very convenient for perturbation analysis applied to rigid body systems: the variables G and L are integrals of motion for both of the best-known integrable problems of rigid body dynamics, the Euler and Lagrange cases, each of which is a natural candidate for an unperturbed system. Similar systems of "osculating elements" that are not necessarily canonical were used by Poisson, Charlier, Andoyer, and Tisserand in developing the theories of lunar librations and rotational motions of planets in celestial mechanics. The French geometer P. Serret was the first (1866) to use variables similar to the Andoyer variables [528]. Another similar system of variables was considered by Rado [476]. Andoyer introduced and systematically used these variables in his famous course in celestial mechanics [8]. They are sometimes called Serret – Andoyer variables. A. Deprit had independently introduced these variables in 1962 (see [150]) to clarify the phase geometry of the Euler case (see Sec. 2.2, Ch. 2). This allowed one to realize their universal character in rigid body dynamics; they were also used in qualitative analysis in [333] (they were called *special canonical variables* there) and for numerical simulations [88]. (Note that in the first edition of this book, we called them Andoyer – Deprit variables, but now it seems to us that this is not valid.)

A systematic study of the equations of motion for a heavy gyroscope in the Rodrigues – Hamilton quaternion variables (and also in the Cayley – Klein parameters) can be found in the remarkable book of F. Klein and A. Sommerfeld *Über die Theorie des Kreisels* [305] (indeed, the main results on this problem belong to F. Klein; see also [304]). At that time, the Hamiltonian structure of these equations (as equations on a Lie algebra) was not known; however, the Rodrigues – Hamilton variables proved to be very convenient for explicit integration in elliptic functions and for the analysis of numerous particular solutions. E. Study used a system of redundant variables (of the same type as Plücker coordinates) similar to quaternions in his book *Geometrie der Dynamen* and calculated the kinetic energy of a rigid body in these coordinates.

1.4 Different Forms of Equations of Motion

1 Equations of motion of a rigid body with a fixed point

In this section, we present the equations of motion of a rigid body in the various forms they take in different variables. The choice of variables to be used depends on the questions to be answered.

Euler – Poincaré equations on the group $SO(3)$. Consider the motion of a rigid body with one fixed point, i.e., this point is fixed in space (relative to some inertial reference frame). The configuration space is the group $SO(3)$. We represent elements of this group by orthogonal matrices of direction cosines: (1.30) (see Sec. 1.3, item 2)

$$\mathbf{Q} = \begin{pmatrix} \alpha_1 & \beta_1 & \gamma_1 \\ \alpha_2 & \beta_2 & \gamma_2 \\ \alpha_3 & \beta_3 & \gamma_3 \end{pmatrix} \in SO(3), \tag{1.44}$$

where, as above, $\boldsymbol{\alpha}$, $\boldsymbol{\beta}$, $\boldsymbol{\gamma}$ form an orthonormal basis in the immovable frame, but are represented in the moving frame attached to the body.

The angular velocity of the body $\boldsymbol{\omega} = (\omega_1, \omega_2, \omega_3)$ in the moving frame is found from the Poisson equations

$$\dot{\boldsymbol{\alpha}} = \boldsymbol{\alpha} \times \boldsymbol{\omega}, \qquad \dot{\boldsymbol{\beta}} = \boldsymbol{\beta} \times \boldsymbol{\omega}, \qquad \dot{\boldsymbol{\gamma}} = \boldsymbol{\gamma} \times \boldsymbol{\omega}, \tag{1.45}$$

which show that the vectors $\boldsymbol{\alpha}$, $\boldsymbol{\beta}$, and $\boldsymbol{\gamma}$ are constant in the absolute space. Rewriting (1.45) in matrix form, we obtain

$$\dot{\mathbf{Q}} = \widetilde{\boldsymbol{\omega}}\mathbf{Q}, \quad \widetilde{\boldsymbol{\omega}} = -\mathbf{Q}\dot{\mathbf{Q}}^{\mathrm{T}} = \dot{\mathbf{Q}}\mathbf{Q}^{\mathrm{T}}, \tag{1.46}$$

where

$$\widetilde{\boldsymbol{\omega}} = \begin{pmatrix} 0 & \omega_3 & -\omega_2 \\ -\omega_3 & 0 & \omega_1 \\ \omega_2 & -\omega_1 & 0 \end{pmatrix}.$$

From a group-theoretic viewpoint, the projections ω_i of the angular velocity in the body correspond to the components of the velocity of the point on the group $SO(3)$ in the basis of left-invariant vector fields. Similarly, the components Ω_i of the angular velocity in the fixed frame correspond to the representation of the angular velocity in the basis of right-invariant vector fields

$$\boldsymbol{\omega} = \sum_k \omega_k \boldsymbol{\xi}_k, \qquad \boldsymbol{\xi}_k = -\sum_{ij} \varepsilon_{kij} \Big(\alpha_i \frac{\partial}{\partial \alpha_j} + \beta_i \frac{\partial}{\partial \beta_j} + \gamma_i \frac{\partial}{\partial \gamma_j} \Big). \tag{1.47}$$

To find the fields $\boldsymbol{\xi}_k$, we write the time derivative taking (1.46) into account:

$$\frac{df}{dt} = \mathrm{Tr}\Big(\dot{\mathbf{Q}}^{\mathrm{T}} \frac{\partial f}{\partial \mathbf{Q}}\Big) = \mathrm{Tr}\Big((\widetilde{\boldsymbol{\omega}}\mathbf{Q})^{\mathrm{T}} \frac{\partial f}{\partial \mathbf{Q}}\Big), \quad \frac{\partial f}{\partial \mathbf{Q}} = \Big\| \frac{\partial f}{\partial Q_{ij}} \Big\|; \tag{1.48}$$

grouping terms containing ω_i, we obtain the vector fields $\boldsymbol{\xi}_i$ (1.47).

The commutation relations for the vector fields $\boldsymbol{\xi}_k$ have the form

$$[\boldsymbol{\xi}_i, \boldsymbol{\xi}_j] = \varepsilon_{ijk} \boldsymbol{\xi}_k, \tag{1.49}$$

where ε_{ijk} are the Levi-Civita symbols.

Substituting (1.47) and (1.49) in the Euler – Poincaré equations (1.17), we obtain the equations of motion in the form

$$\frac{d}{dt}\Big(\frac{\partial L}{\partial \boldsymbol{\omega}}\Big) = \frac{\partial L}{\partial \boldsymbol{\omega}} \times \boldsymbol{\omega} + \frac{\partial L}{\partial \boldsymbol{\alpha}} \times \boldsymbol{\alpha} + \frac{\partial L}{\partial \boldsymbol{\beta}} \times \boldsymbol{\beta} + \frac{\partial L}{\partial \boldsymbol{\gamma}} \times \boldsymbol{\gamma}, \tag{1.50}$$

which, together with (1.45), form the complete system of equations of motion for a body with a fixed point. System (1.45), (1.50) was obtained by J. Lagrange in the second part of his famous treatise *Méchanique analytique* [364].

REMARK. Eqs. (1.45), (1.50) may be represented in matrix form, which can be easily generalized to the higher-dimensional case:

$$\frac{d}{dt}\left(\frac{\partial L}{\partial \widetilde{\boldsymbol{\omega}}}\right) = \left[\widetilde{\boldsymbol{\omega}}, \frac{\partial L}{\partial \widetilde{\boldsymbol{\omega}}}\right] + \frac{\partial L}{\partial \mathbf{Q}}\mathbf{Q}^{\mathrm{T}} - \left(\frac{\partial L}{\partial \mathbf{Q}}\right)^{\mathrm{T}}\mathbf{Q}, \quad \dot{\mathbf{Q}} = \widetilde{\boldsymbol{\omega}}\mathbf{Q},$$

where $\frac{\partial L}{\partial \widetilde{\boldsymbol{\omega}}} = \left\|\frac{\partial L}{\partial \widetilde{\omega}_{ij}}\right\|$, $\frac{\partial L}{\partial \mathbf{Q}} = \left\|\frac{\partial L}{\partial Q_{ij}}\right\|$, and $[\cdot\,,\cdot]$ is the usual matrix commutator.

Equations of motion in angular velocities and quaternions. In Sec. 1.3 we presented the quaternion parametrization of the group $SO(3)$ for which the vector fields (1.47) are also linear functions of the coordinates. Indeed, it is easy to show that on the unit sphere $\lambda_0^2 + \boldsymbol{\lambda}^2 = 1$, $\boldsymbol{\lambda} = (\lambda_1, \lambda_2, \lambda_3)$, the components of the angular velocity (1.46) and the vector fields (1.47) have the form (see [335, 321])

$$\begin{aligned}
\omega_1 &= 2(\lambda_0\dot{\lambda}_1 - \lambda_1\dot{\lambda}_0 + \lambda_3\dot{\lambda}_2 - \lambda_2\dot{\lambda}_3),\\
\omega_2 &= 2(\lambda_0\dot{\lambda}_2 - \lambda_2\dot{\lambda}_0 + \lambda_1\dot{\lambda}_3 - \lambda_3\dot{\lambda}_1),\\
\omega_3 &= 2(\lambda_0\dot{\lambda}_3 - \lambda_3\dot{\lambda}_0 + \lambda_2\dot{\lambda}_1 - \lambda_1\dot{\lambda}_2),
\end{aligned}$$

$$\begin{aligned}
\boldsymbol{\xi}_1 &= \frac{1}{2}\left(\lambda_0\frac{\partial}{\partial\lambda_1} - \lambda_1\frac{\partial}{\partial\lambda_0} + \lambda_3\frac{\partial}{\partial\lambda_2} - \lambda_2\frac{\partial}{\partial\lambda_3}\right),\\
\boldsymbol{\xi}_2 &= \frac{1}{2}\left(\lambda_0\frac{\partial}{\partial\lambda_2} - \lambda_2\frac{\partial}{\partial\lambda_0} + \lambda_1\frac{\partial}{\partial\lambda_3} - \lambda_3\frac{\partial}{\partial\lambda_1}\right),\\
\boldsymbol{\xi}_3 &= \frac{1}{2}\left(\lambda_0\frac{\partial}{\partial\lambda_3} - \lambda_3\frac{\partial}{\partial\lambda_0} + \lambda_2\frac{\partial}{\partial\lambda_1} - \lambda_1\frac{\partial}{\partial\lambda_2}\right).
\end{aligned} \tag{1.51}$$

The commutation relations for the fields $\boldsymbol{\xi}_k$ also have the form (1.49).

The Poincaré equations (1.17), subject to (1.51), become

$$\begin{aligned}
\frac{d}{dt}\left(\frac{\partial L}{\partial \boldsymbol{\omega}}\right) &= \frac{\partial L}{\partial \boldsymbol{\omega}} \times \boldsymbol{\omega} + \frac{1}{2}\lambda_0\frac{\partial L}{\partial \boldsymbol{\lambda}} - \frac{1}{2}\boldsymbol{\lambda}\frac{\partial L}{\partial \lambda_0} + \frac{1}{2}\frac{\partial L}{\partial \boldsymbol{\lambda}} \times \boldsymbol{\lambda},\\
\dot{\lambda}_0 &= -\frac{1}{2}(\boldsymbol{\omega}, \boldsymbol{\lambda}), \qquad \dot{\boldsymbol{\lambda}} = \frac{1}{2}\lambda_0\boldsymbol{\omega} + \frac{1}{2}\boldsymbol{\lambda} \times \boldsymbol{\omega}.
\end{aligned} \tag{1.52}$$

Kinetic energy of a rigid body with a fixed point in vector and matrix forms can be represented as follows:

$$T = \frac{1}{2}(\boldsymbol{\omega}, \mathbf{I}\boldsymbol{\omega}) = -\frac{1}{2}\operatorname{Tr}(\widetilde{\boldsymbol{\omega}}\mathbf{J}\widetilde{\boldsymbol{\omega}}). \tag{1.53}$$

Here $\mathbf{I} = \|I_{ij}\|$ is the *tensor of inertia* of the rigid body with respect to the fixed point of the body; its components are defined by the formula

$$I_{ij} = \int_{\tau} (\boldsymbol{y}^2\delta_{ij} - y_i y_j)\rho(\boldsymbol{y})\,d^3\boldsymbol{y}, \tag{1.54}$$

where integration extends over all points $\boldsymbol{y}$ of the body τ and $\rho(\boldsymbol{y})$ is the density of the body at the point $\boldsymbol{y}$.

The tensor $\mathbf{J} = \|J_{ij}\|$ is also called the tensor of inertia, but it is defined by the formula

$$J_{ij} = \int\limits_{\tau} y_i y_j \rho(\boldsymbol{y})\, d^3\boldsymbol{y}; \tag{1.55}$$

this tensor is usually used for higher-dimensional generalizations.

The relation between **I** and **J** is

$$\mathbf{J} = \frac{1}{2}(\operatorname{Tr}\mathbf{I})\mathbf{E} - \mathbf{I}, \qquad \mathbf{I} = (\operatorname{Tr}\mathbf{J})\mathbf{E} - \mathbf{J}. \tag{1.56}$$

In the reference frame attached to the body, the tensors **I** and **J** are represented by constant symmetric matrices (in the fixed frame, they are represented by coordinate-dependent matrices); since they commute ($\mathbf{IJ} = \mathbf{JI}$), they can be simultaneously diagonalized. The corresponding coordinate system in the body is said to be *principal* and its axes are called the *principal axes* (*of inertia*). As is well known, the principal moments of inertia I_1, I_2, I_3 of any body obey the inequality of the triangle $I_i + I_j \geqslant I_k$. The equality takes place for the case of an infinitely thin plate.

2 Hamiltonian form of equations of motion for different systems of variables

Equations of motion in algebraic form. Equations (1.45), (1.50) can be represented in Hamiltonian form by using the Legendre transform

$$\boldsymbol{M} = \frac{\partial L}{\partial \boldsymbol{\omega}}, \quad H = \big[(\boldsymbol{M}, \boldsymbol{\omega}) - L\big]\big|_{\boldsymbol{\omega}\to\boldsymbol{M}}. \tag{1.57}$$

For a natural system with kinetic energy (1.53) and potential energy $U(\boldsymbol{\alpha}, \boldsymbol{\beta}, \boldsymbol{\gamma})$ we have

$$\boldsymbol{M} = \mathbf{I}\boldsymbol{\omega}, \quad H = \frac{1}{2}(\boldsymbol{M}, \mathbf{A}\boldsymbol{M}) + U(\boldsymbol{\alpha}, \boldsymbol{\beta}, \boldsymbol{\gamma}), \tag{1.58}$$

where $\mathbf{A} = \mathbf{I}^{-1}$, $\boldsymbol{M}$ are the components of the angular momentum referred to the moving axes, and $\boldsymbol{\alpha}$, $\boldsymbol{\beta}$, and $\boldsymbol{\gamma}$ are the components of the direction cosines.

The general formulae (1.26) and (1.49) imply that the Poisson bracket is defined by the algebra $so(3) \oplus_s (\mathbb{R}^3 \oplus \mathbb{R}^3 \oplus \mathbb{R}^3)$, which is the semi-direct sum of the algebra of rotations and three copies of the algebra of translations

$$\begin{aligned} &\{M_i, M_j\} = -\varepsilon_{ijk} M_k, \qquad \{M_i, \alpha_j\} = -\varepsilon_{ijk}\alpha_k, \\ &\{M_i, \beta_j\} = -\varepsilon_{ijk}\beta_k, \qquad \{M_i, \gamma_j\} = -\varepsilon_{ijk}\gamma_k, \\ &\{\alpha_i, \alpha_j\} = \{\beta_i, \beta_j\} = \{\gamma_i, \gamma_j\} = \{\alpha_i, \beta_j\} = \{\alpha_i, \gamma_j\} = \{\beta_i, \gamma_j\} = 0. \end{aligned} \tag{1.59}$$

The Hamiltonian equations are

$$\begin{gathered} \dot{\boldsymbol{M}} = \boldsymbol{M} \times \frac{\partial H}{\partial \boldsymbol{M}} + \boldsymbol{\alpha} \times \frac{\partial H}{\partial \boldsymbol{\alpha}} + \boldsymbol{\beta} \times \frac{\partial H}{\partial \boldsymbol{\beta}} + \boldsymbol{\gamma} \times \frac{\partial H}{\partial \boldsymbol{\gamma}}, \\ \dot{\boldsymbol{\alpha}} = \boldsymbol{\alpha} \times \frac{\partial H}{\partial \boldsymbol{M}}, \qquad \dot{\boldsymbol{\beta}} = \boldsymbol{\beta} \times \frac{\partial H}{\partial \boldsymbol{M}}, \qquad \dot{\boldsymbol{\gamma}} = \boldsymbol{\gamma} \times \frac{\partial H}{\partial \boldsymbol{M}}, \\ H = \frac{1}{2}(\boldsymbol{M}, \mathbf{A}\boldsymbol{M}) + U(\boldsymbol{\alpha}, \boldsymbol{\beta}, \boldsymbol{\gamma}). \end{gathered} \tag{1.60}$$

The equations of motion of a rigid body in a generalized potential field (e.g., a magnetic field) can also be represented in the form of (1.60). In this case, the Hamiltonian H contains terms linear in $\boldsymbol{M}$ (see below).

The Poisson bracket (1.59) is degenerate and possesses six Casimir functions:

$$\begin{aligned} f_1 &= (\boldsymbol{\alpha}, \boldsymbol{\alpha}), \quad & f_2 &= (\boldsymbol{\beta}, \boldsymbol{\beta}), \quad & f_3 &= (\boldsymbol{\gamma}, \boldsymbol{\gamma}), \\ f_4 &= (\boldsymbol{\alpha}, \boldsymbol{\beta}), \quad & f_5 &= (\boldsymbol{\alpha}, \boldsymbol{\gamma}), \quad & f_6 &= (\boldsymbol{\beta}, \boldsymbol{\gamma}). \end{aligned} \tag{1.61}$$

A generic symplectic leaf has dimension equal to six and is diffeomorphic to the cotangent bundle of the rotational group $SO(3)$. Hence there is no loss of generality in taking the representative example amongst these, for which $f_1 = f_2 = f_3 = 1$ and $f_4 = f_5 = f_6 = 0$, since the system (1.60) has three degrees of freedom.

In the immovable coordinate system, the position and velocity of a rigid body can be characterized by the projections on the fixed axes of the orthonormal basis vectors attached to the body, which can be expressed in terms of the rows of the matrix $\mathbf{Q}$, and the projections of the angular momentum vector on the same axes:

$$\begin{gathered} \boldsymbol{e}_1 = (\alpha_1, \beta_1, \gamma_1), \quad \boldsymbol{e}_2 = (\alpha_2, \beta_2, \gamma_2), \quad \boldsymbol{e}_3 = (\alpha_3, \beta_3, \gamma_3), \\ N_1 = (\boldsymbol{M}, \boldsymbol{\alpha}), \; N_2 = (\boldsymbol{M}, \boldsymbol{\beta}), \; N_3 = (\boldsymbol{M}, \boldsymbol{\gamma}). \end{gathered} \tag{1.62}$$

It is easy to show that the variables $\boldsymbol{N}$, $\boldsymbol{e}_1$, $\boldsymbol{e}_2$, and $\boldsymbol{e}_3$ also form a Lie – Poisson structure, which differs from (1.59) only by sign:

$$\begin{gathered} \{N_i, N_j\} = \varepsilon_{ijk} N_k, \\ \{N_i, e_{1j}\} = \varepsilon_{ijk} e_{1k}, \quad \{N_i, e_{2j}\} = \varepsilon_{ijk} e_{2k}, \quad \{N_i, e_{3k}\} = \varepsilon_{ijk} e_{3k}, \\ \{e_{ki}, e_{lj}\} = 0. \end{gathered} \tag{1.63}$$

For example, the spherical pendulum in a potential field is conveniently described in terms of the variables $\boldsymbol{N}$ and $\boldsymbol{e}_k$. The Hamiltonian has the form

$$H = \frac{1}{2ml^2} \boldsymbol{N}^2 + U(\boldsymbol{e}_3). \tag{1.64}$$

and the relation $(\boldsymbol{N}, \boldsymbol{e}_3) = 0$ holds (i.e., motion is on the zero orbit of $e(3)$). Here $\boldsymbol{e}_3$ is the unit vector connecting the fixed point and the center of gravity, $\boldsymbol{N} = ml^2 \boldsymbol{\omega}$, $\boldsymbol{\omega} = \boldsymbol{e}_3 \times \dot{\boldsymbol{e}}_3$ is the angular velocity, and l is the length of the pendulum. In other words, the spherical pendulum can be represented as the spherical top on the zero orbit of the algebra $e(3)$.

The generators $\boldsymbol{e}_i$ and $\boldsymbol{N}$ in (1.62) are also convenient for the description of reduction for systems admitting the Lagrange integral $F = M_3 = \text{const}$ as an integral of motion (see Sec. 4.1, 4.2, Ch. 4).

Quaternion representation of the equations of motion. In practice, the redundancy in Eqs. (1.60) is very inconvenient; for example, it may be expected, as often happens, that in numerical integration of these equations, the orthonormality relations (1.61) do not survive for a long time. This disadvantage is absent in the quaternion representation of the equations of motion presented by the authors in [92, 95]. The matrix of direction cosines in the quaternion representation has the form (1.41) and the corresponding commutation relations are

$$\begin{aligned} \{M_i, M_j\} &= -\varepsilon_{ijk} M_k, & \{M_i, \lambda_0\} &= \frac{1}{2}\lambda_i, \\ \{M_i, \lambda_j\} &= -\frac{1}{2}(\varepsilon_{ijk}\lambda_k + \delta_{ij}\lambda_0), & \{\lambda_\mu, \lambda_\nu\} &= 0. \end{aligned} \tag{1.65}$$

The corresponding Lie algebra is the semi-direct sum of the algebra of rotations $so(3)$ and the algebra of translations $\mathbb{R}^4$: $l(7) \approx so(3) \oplus_S \mathbb{R}^4$.

Bracket (1.65) is degenerate. It admits just one Casimir function

$$F(\lambda) = \lambda_0^2 + \lambda_1^2 + \lambda_2^2 + \lambda_3^2. \tag{1.66}$$

A generic symplectic leaf is diffeomorphic to the cotangent bundle T^*S^3 of the three-dimensional sphere and has dimension 6. The equations of motion have the form

$$\begin{aligned} \dot{\boldsymbol{M}} &= \boldsymbol{M} \times \frac{\partial H}{\partial \boldsymbol{M}} + \frac{1}{2}\boldsymbol{\lambda} \times \frac{\partial H}{\partial \boldsymbol{\lambda}} + \frac{1}{2}\frac{\partial H}{\partial \lambda_0}\boldsymbol{\lambda} - \frac{1}{2}\lambda_0 \frac{\partial H}{\partial \boldsymbol{\lambda}}, \\ \dot{\lambda}_0 &= -\frac{1}{2}\left(\boldsymbol{\lambda}, \frac{\partial H}{\partial \boldsymbol{M}}\right), \quad \dot{\boldsymbol{\lambda}} = \frac{1}{2}\boldsymbol{\lambda} \times \frac{\partial H}{\partial \boldsymbol{M}} + \frac{1}{2}\lambda_0 \frac{\partial H}{\partial \boldsymbol{M}}. \end{aligned} \tag{1.67}$$

For integrability, two additional involutive integrals are needed.

Remark. For a physical rigid body system, the Hamiltonian H is a single-valued function on $T^*SO(3)$. Since the covering of $SO(3)$ by quaternions (1.41) is twofold, the Hamiltonian function depends only on quadratic combinations $\lambda_i\lambda_j$. Nevertheless, systems having Hamiltonians with arbitrary dependence on quaternion components do occur in different branches of mechanics, for example, in celestial mechanics on a curved space, the Leggett system, quantum mechanics of spins (see Chs. 3, 4). Probably, the most value of the quaternionic representation (1.67) is in quantum mechanics, where some phenomena are essentially related to spin.

Canonical equations in Euler angles and Andoyer variables. In the Euler angles (θ, φ, ψ) and corresponding canonical momenta $p_\theta, p_\varphi, p_\psi$, the equations of motion have the usual Hamiltonian form

$$\dot{\boldsymbol{p}} = -\frac{\partial H}{\partial \boldsymbol{q}}, \quad \dot{\boldsymbol{q}} = \frac{\partial H}{\partial \boldsymbol{p}}, \quad \boldsymbol{q} = (\theta, \varphi, \psi), \quad \boldsymbol{p} = (p_\theta, p_\varphi, p_\psi), \tag{1.68}$$

obtained from the Lagrangian formalism in the variables $(\theta, \varphi, \psi, \dot{\theta}, \dot{\varphi}, \dot{\psi})$ by using the Legendre transform

$$\boldsymbol{p} = \frac{\partial L}{\partial \dot{\boldsymbol{q}}}, \qquad H(\boldsymbol{p}, \boldsymbol{q}) = \left[(\boldsymbol{p}, \dot{\boldsymbol{q}}) - L\right]\Big|_{\dot{\boldsymbol{q}}, \boldsymbol{q} \to \boldsymbol{p}, \boldsymbol{q}}.$$

Here L is the Lagrange function, which for natural systems has the form $L = T - U(\theta, \varphi, \psi)$. The kinetic energy of the body does not depend on the variable ψ and has the form

$$T = \frac{1}{2}(\mathbf{A}\boldsymbol{M}, \boldsymbol{M}) = \frac{1}{2}\Big[a_1\Big(\frac{\sin\varphi}{\sin\theta}(p_\psi - p_\varphi\cos\theta) + p_\theta\cos\varphi\Big)^2 + a_2\Big(\frac{\cos\varphi}{\sin\theta}(p_\psi - p_\varphi\cos\theta) - p_\theta\sin\varphi\Big)^2 + a_3p_\varphi^2\Big]. \tag{1.69}$$

The motion of a rigid body in a potential field is defined by a natural system, and the Hamiltonian has the form

$$H = T + U(\theta, \varphi, \psi). \tag{1.70}$$

If the potential energy is also independent of ψ $\left(\text{i.e. } \frac{\partial U}{\partial \psi} = 0\right)$, which corresponds to invariance of the force field with respect to rotations about the vertical axis in the immovable frame, then the variable ψ is cyclic and the generalized momentum $p_\psi = (\boldsymbol{M}, \boldsymbol{\gamma})$ is conserved. Upon Routhian reduction with respect to the precession angle ψ, we obtain a system describing the motion of a point on the sphere $\boldsymbol{\gamma}^2 = 1$ (where $\gamma_1 = \sin\theta\sin\varphi$, $\gamma_2 = \sin\theta\cos\varphi$, and $\gamma_3 = \cos\theta$), which is called the *Poisson sphere*. If $p_\psi \neq 0$, then the Hamiltonian contains terms linear in velocity (so-called gyroscopic terms) that cannot be eliminated by coordinate transforms and correspond to motion in a generalized potential field. The impossibility of eliminating gyroscopic terms is related to the global phenomenon of a "monopole", whose value is the integral of the gyroscopic 2-form over the Poisson sphere (see [447]). The "monopole" problem was raised for the first time by P. Dirac in connection with the quantization of the motion of a particle on a sphere. If $p_\psi = 0$, then the system discussed is natural, i.e., there are no gyroscopic terms.

If the dynamical symmetry condition $a_1 = a_2$ holds, the kinetic energy (1.69) becomes simpler, being also independent of the angle φ:

$$T = \frac{1}{2}\left(a_1\left(p_\theta^2 + \frac{(p_\psi - p_\varphi\cos\theta)^2}{\sin^2\theta}\right) + a_3p_\varphi^2\right). \tag{1.71}$$

If in addition the potential U is also independent of φ (i.e., $\frac{\partial U}{\partial \psi} = \frac{\partial U}{\partial \varphi} = 0$), in other words, $U = U(\theta) = U(\gamma_3)$, then there exists one more cyclic integral – the Lagrange integral $p_\varphi = M_3 = c_2 = \text{const}$, which corresponds to invariance of the system under rotations about the axis of dynamical symmetry. The (one-degree-of-freedom) system obtained after reduction is integrable (for details, see Sec. 2.3 Ch. 2). If $p_\psi = c_1 \neq 0$ but $p_\varphi = c_2 = 0$, then the equations describe the motion of the *spherical pendulum*.

In terms of Andoyer variables, the equations of motion also have the form (1.68), where $\boldsymbol{q} = (l, g, h)$ and $\boldsymbol{p} = (L, G, H)$. Since in the variables (L, G, H, l, g, h) purely position coordinates cannot be singled out, i.e., variables on the base and on the leaves of the cotangent bundle are mixed, the potential U depends in general on all variables, $U = U(L, G, H, l, g, h)$.

The kinetic energy T has the form

$$T = \frac{1}{2}\left[(G^2 - L^2)(a_1\sin^2 l + a_2\cos^2 l) + a_3L^2\right]. \tag{1.72}$$

It is easy to obtain the following facts:

1) if $\frac{\partial U}{\partial h} = 0$, then $H = p_\psi = (\boldsymbol{M}, \boldsymbol{\gamma})$ is preserved,
2) if $a_1 = a_2$ and $\frac{\partial U}{\partial l} = 0$, then $L = M_3$ is preserved.

$H = p_\psi = (\boldsymbol{M}, \boldsymbol{\gamma})$ is called the *area integral.*
$L = M_3$ is called the *Lagrange integral.*

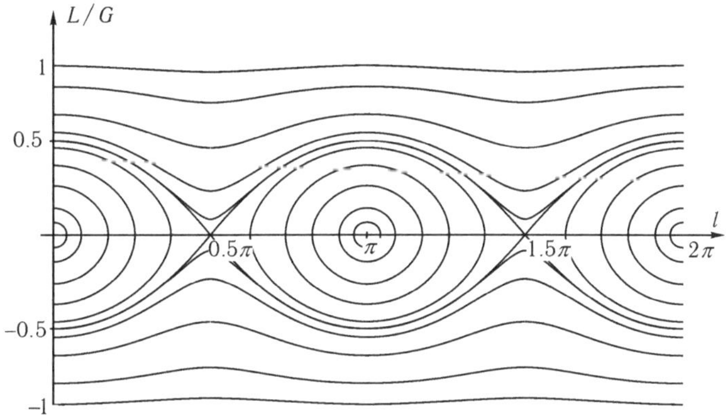

Fig. 1.4. The phase portrait of the Euler problem. Stable fixed points and the straight lines $|L| = G$ correspond to stable permanent rotations about the major and minor axes, unstable points correspond to rotations about the middle axis, separatrices are formed by double-asymptotic trajectories that connect unstable permanent rotations.

A feature of the representation of the kinetic energy in the form (1.72) is its independence of g. This allows one to integrate the Euler problem—the motion of a free top, for which $U \equiv 0$ (see Sec. 2.1, Ch. 2). The corresponding cyclic integral is $G = \text{const}$, which represents the angular momentum $G^2 = \boldsymbol{M}^2$. This makes the Andoyer variables useful for geometric interpretation and analysis of perturbation cases. The phase portrait for the Euler case on the cylindrical model of the sphere is shown in Fig. 1.4. A perturbation (such as a gravitational field) gives rise to chaotic motions on the phase portrait, in the neighborhood of separatrices joining unstable uniform rotations (Fig. 1.5). Next we describe methods to visualize of the phase flow in more detail.

3 Poincaré section and chaotic motions

The *Poincaré map* (Poincaré section, phase section) is useful for visualizing chaotic motions of a two-degree-of-freedom system. It turns the phase flow into a discrete, two-dimensional mapping.

We describe the method for construction of this mapping in the concrete case of rigid body dynamics. It is usually convenient to use the Andoyer variables , the section

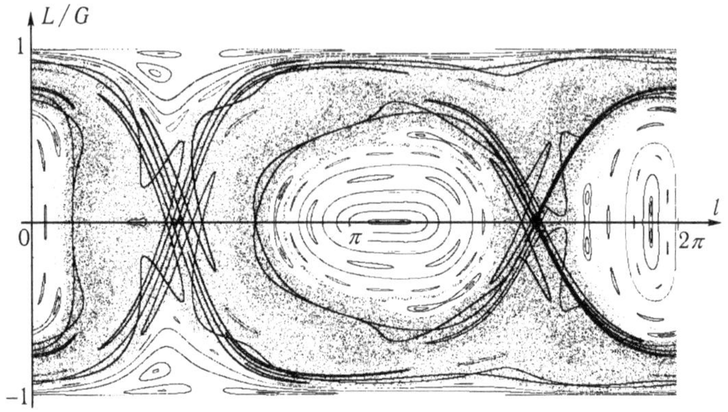

Fig. 1.5. The phase portrait (section formed by the intersection with the plane $g = \frac{\pi}{2}$) for the Euler – Poisson equation, fixing $h = 1.5$, $c = 1$, with parameters of the body: $\mathbf{I} = \mathrm{diag}(1.5; 1.2)$, $\boldsymbol{r} = (0.5, 0, 0)$. We see doubling of the period on the trajectory branching from permanent rotations near the points $(\pi, 0)$ and $(2\pi, 0)$ in Fig. 1.4 and splitting of the separatrices of unstable periodic orbits emanating from the points $(0.5\pi, 0)$ and $(1.5\pi, 0)$ in Fig. 1.4.

plane initially introduced in [205], and sometimes other section planes that clarify different aspects of the motion. In some cases, the Euler variables $(p_\theta, p_\varphi, \theta, \varphi)$ and other types of section may be useful [107, 175].

First, we fix an energy level $\mathscr{H}(L, G, H, l, g) = E = \mathrm{const}$. If we assume that the field is axially symmetric, then the variable h is cyclic and does not participate in the Hamiltonian, so the corresponding conjugate variable H, which is the constant of areas, can be considered as a parameter. Hence, on the energy level, we have a three-dimensional phase flow. Choose a section plane of the form $g = g_0 \bmod 2\pi$, $g_0 = \mathrm{const}$ (in the sequel, we sometimes use instead $l = l_0 \bmod 2\pi$, $l_0 = \mathrm{const}$) and consider the sequence of intersections of some trajectory with this plane, $\ldots, x_{n-1}, x_n, x_{n+1}, \ldots$, with the proviso that the crossings are all in the same direction, i.e., $\mathrm{sgn}\, \dot{g}(x_n) = \mathrm{sgn}\, \dot{g}(x_{n+1})$ (Fig. 1.6).

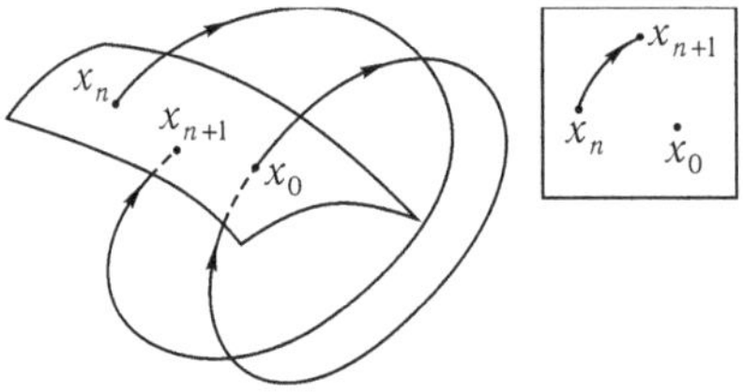

Fig. 1.6.

REMARK. The latter condition arises from the requirement that periodic orbits intersecting the plane $g = g_0$ at, generally speaking, two points, are to be fixed points of the map: $x_n = x_0$, $n = 1, \ldots$ (see Fig. 1.6).

For each point x_n, the Poincaré map puts into correspondence its iteration x_{n+1} belonging to the same trajectory. In general, this mapping is defined locally near some periodic solution since under the action of the phase flow, a point might leave the section plane and not return to it. However, this mapping is very helpful and illustrates different effects related with returning trajectories. It is also sometimes called the *first-return mapping.*

If we consider the Poincaré mapping globally, we have to identify domains on the section plane in which this mapping is well defined. Such domains are called the *domains of possible motions*. Usually they are defined by solutions of the energy equation $\mathscr{H}(\boldsymbol{p}, \boldsymbol{q}) = E$, $(\boldsymbol{p}, \boldsymbol{q}) \in \mathbb{R}^4$, $q = q_0 = \text{const}$ (in our case, $(\boldsymbol{p}, \boldsymbol{q}) = (L, G, l, g)$, $q_0 = g_0$). If the energy level surface is compact, then the Poincaré recurrence theorem is valid, and the point crosses the chosen plane infinitely many times. On the boundary of the domain of possible motions, the trajectory is tangent to the section plane, i.e., the transversality of the intersection is violated. Global Poincaré mappings are still little studied.

In the study of rigid body dynamics, we choose coordinates $(l \bmod 2\pi, L/G)$ on the section plane for reasons of compactness since $|L/G| \leqslant 1$ (see [205, 88]). Iterations of the mapping are found by numerical integration of the equations of motion in the variables $(\boldsymbol{M}, \boldsymbol{\gamma})$ and subsequent transformation to the section plane in the variables (L, G, l, g) by the formulae (1.42), (1.43):

$$M_1 = \sqrt{G^2 - L^2} \sin l, \quad M_2 = \sqrt{G^2 - L^2} \cos l, \quad M_3 = L,$$

$$\gamma_1 = \left(\frac{H}{G} \sqrt{1 - \left(\frac{L}{G}\right)^2} + \frac{L}{G} \sqrt{1 - \left(\frac{H}{G}\right)^2} \cos g \right) \sin l + \sqrt{1 - \left(\frac{H}{G}\right)^2} \sin g \cos l,$$

$$\gamma_2 = \left(\frac{H}{G} \sqrt{1 - \left(\frac{L}{G}\right)^2} + \frac{L}{G} \sqrt{1 - \left(\frac{H}{G}\right)^2} \cos g \right) \cos l - \sqrt{1 - \left(\frac{H}{G}\right)^2} \sin g \sin l,$$

$$\gamma_3 = \left(\frac{H}{G}\right)\left(\frac{L}{G}\right) - \sqrt{1 - \left(\frac{L}{G}\right)^2} \sqrt{1 - \left(\frac{H}{G}\right)^2} \cos g. \tag{1.73}$$

This is related to the problem of attaining sufficient accuracy of numeric integration and reduction of computation time. We also note that in the recent versions of the software we also use the quaternion equations in the variables $(\boldsymbol{M}, \lambda)$, which allow one to achieve high accuracy and also to find the absolute motion of the rigid body necessary for visualization of trajectories of different points of the body.

For integrable systems, consecutive iterations of the mapping lie on invariant curves consisting of periodic or quasi-periodic motions (see Sec. 1.7) and defined by the constant value of an additional integral (Fig. 1.4); in the general (nonintegrable) situation, a trajectory can fill, in a chaotic fashion, whole domains in the phase space (of the level surface $H = h$, Fig. 1.5).

The Poincaré mapping first appeared in the theory of nonintegrability and deterministic chaos where it continues to play an important role. It is also useful for the study of integrable cases because it allows one to represent visually the relative positions of different particular solutions in the phase space, among them some outstanding solutions, which are of great importance (see Ch. 2).

For the Euler case, the picture of the Poincaré mapping is well known (cf. Fig. 1.4). By the way, introducing the variables L, G, H, l, g, and h in [150], Deprit (1967) considered the possibility of visual interpretation of solutions to the Euler problem as their main merit; the visual interpretation can compete with the "geometric" Poinsot interpretation (Sec. 2.2, Ch. 2). A more detailed description of the Andoyer variables and their properties was presented by Deprit and Elipe in [152] (1993), where these variables are discussed in connection with the problem of full reduction and its applications to different problems of celestial mechanics. In the sequel, we use the Poincaré section as described above for the study of both integrable and nonintegrable cases.

1.5 Equations of Motion of a Rigid Body in Euclidean Space

1 Lagrangian formalism and Poincaré equations on the group $E(3)$

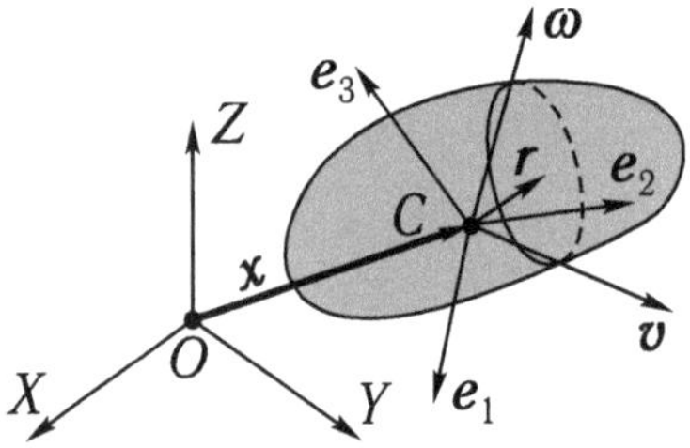

Fig. 1.7. A free rigid body.

Consider a rigid body moving in Euclidean space $\mathbb{R}^3$; its configuration space coincides with the group $E(3)$. In matrix form, elements of the group can be represented as

$$\mathbf{S} = \begin{pmatrix} \mathbf{Q}^{\mathrm{T}} & \begin{matrix} x_1 \\ x_2 \\ x_3 \end{matrix} \\ 0 & 1 \end{pmatrix} \in E(3),$$

where $\mathbf{Q} \in SO(3)$ is the matrix of direction cosines (1.30) and $\boldsymbol{x}$ is the position vector of some fixed point C of the body represented in the fixed frame (see Fig. 1.7).

We may write the equations of motion for the projections of the angular velocity $\boldsymbol{\omega} = (\omega_1, \omega_2, \omega_3)$ as well as the absolute velocity of the center of mass $\boldsymbol{v} = (v_1, v_2, v_3)$ on the axes attached to the body. Similarly to (1.46), we write the following obvious geometric relations:

$$\dot{\mathbf{Q}} = \widetilde{\boldsymbol{\omega}}\mathbf{Q}, \qquad \boldsymbol{v} = \mathbf{Q}\dot{\boldsymbol{x}}. \tag{1.74}$$

Now we find the corresponding basic left-invariant fields on the group $E(3)$. For this, consider the time derivative owing to Eqs. (1.74):

$$\frac{df}{dt} = \mathrm{Tr}\Big(\dot{\mathbf{Q}}^T \frac{\partial f}{\partial \mathbf{Q}}\Big) + \Big(\frac{\partial f}{\partial \boldsymbol{x}}, \dot{\boldsymbol{x}}\Big) = \mathrm{Tr}\Big((\widetilde{\boldsymbol{\omega}}\mathbf{Q})^T \frac{\partial f}{\partial \mathbf{Q}}\Big) + \Big(\mathbf{Q}\frac{\partial f}{\partial \boldsymbol{x}}, \boldsymbol{v}\Big),$$
$$\frac{\partial f}{\partial \mathbf{Q}} = \left\| \frac{\partial f}{\partial Q_{ij}} \right\|; \quad \frac{\partial f}{\partial \boldsymbol{x}} = \Big(\frac{\partial f}{\partial x_1}, \frac{\partial f}{\partial x_2}, \frac{\partial f}{\partial x_3}\Big).$$

Grouping terms with ω_i and v_i, we obtain

$$\begin{aligned} \boldsymbol{\omega} &= \omega_k \boldsymbol{\xi}_k, & \boldsymbol{\xi}_k &= -\sum_{ij} \varepsilon_{kij}\Big(\alpha_i \frac{\partial}{\partial \alpha_j} + \beta_i \frac{\partial}{\partial \beta_j} + \gamma_i \frac{\partial}{\partial \gamma_j}\Big), \\ \boldsymbol{v} &= v_i \boldsymbol{\zeta}_i, & \boldsymbol{\zeta}_i &= \alpha_i \frac{\partial}{\partial x_1} + \beta_i \frac{\partial}{\partial x_2} + \gamma_i \frac{\partial}{\partial x_3}. \end{aligned} \tag{1.75}$$

The commutators of the basic fields $\boldsymbol{\xi}_i$ and $\boldsymbol{\zeta}_j$ have the form

$$[\boldsymbol{\xi}_i, \boldsymbol{\xi}_j] = \varepsilon_{ijk}\boldsymbol{\xi}_k, \qquad [\boldsymbol{\xi}_i, \boldsymbol{\zeta}_j] = \varepsilon_{ijk}\boldsymbol{\zeta}_k, \qquad [\boldsymbol{\zeta}_i, \boldsymbol{\zeta}_j] = 0. \tag{1.76}$$

Using (1.75) and (1.76), we obtain the Poincaré equations of motion (1.17) for the dynamics of a free rigid body

$$\begin{aligned} \frac{d}{dt}\Big(\frac{\partial L}{\partial \boldsymbol{\omega}}\Big) &= \frac{\partial L}{\partial \boldsymbol{\omega}} \times \boldsymbol{\omega} + \frac{\partial L}{\partial \boldsymbol{v}} \times \boldsymbol{v} + \frac{\partial L}{\partial \boldsymbol{\alpha}} \times \boldsymbol{\alpha} + \frac{\partial L}{\partial \boldsymbol{\beta}} \times \boldsymbol{\beta} + \frac{\partial L}{\partial \boldsymbol{\gamma}} \times \boldsymbol{\gamma}, \\ \frac{d}{dt}\Big(\frac{\partial L}{\partial \boldsymbol{v}}\Big) &= \frac{\partial L}{\partial \boldsymbol{v}} \times \boldsymbol{\omega} + \frac{\partial L}{\partial x_1}\boldsymbol{\alpha} + \frac{\partial L}{\partial x_2}\boldsymbol{\beta} + \frac{\partial L}{\partial x_3}\boldsymbol{\gamma}, \\ \dot{\boldsymbol{\alpha}} = \boldsymbol{\alpha} \times \boldsymbol{\omega}, &\quad \dot{\boldsymbol{\beta}} = \boldsymbol{\beta} \times \boldsymbol{\omega}, \quad \dot{\boldsymbol{\gamma}} = \boldsymbol{\gamma} \times \boldsymbol{\omega}, \\ \dot{x}_1 = (\boldsymbol{\alpha}, \boldsymbol{v}), &\quad \dot{x}_2 = (\boldsymbol{\beta}, \boldsymbol{v}), \quad \dot{x}_3 = (\boldsymbol{\gamma}, \boldsymbol{v}). \end{aligned} \tag{1.77}$$

2 Kinetic energy of a rigid body in $\mathbb{R}^3$

The position vector of a point of a rigid body in the immovable coordinate system may be written in the form $\boldsymbol{q} = \mathbf{Q}^{\mathrm{T}}\boldsymbol{y} + \boldsymbol{x}$, where $\boldsymbol{y}$ is the time-independent position vector of the point in the body reference frame. Differentiating with respect to time, $\dot{\boldsymbol{q}} = \dot{\mathbf{Q}}^{\mathrm{T}}\boldsymbol{y} + \dot{\boldsymbol{x}}$, and integrating with respect to $\boldsymbol{y}$ over the body, we obtain the kinetic energy in vector and matrix form:

$$\begin{aligned} T &= \frac{1}{2}(\boldsymbol{\omega}, \mathbf{I}\boldsymbol{\omega}) + m(\boldsymbol{v}, \boldsymbol{r} \times \boldsymbol{\omega}) + \frac{1}{2}m\boldsymbol{v}^2 \\ &= -\frac{1}{2}\,\mathrm{Tr}(\widetilde{\boldsymbol{\omega}}\mathbf{J}\widetilde{\boldsymbol{\omega}}) + m(\boldsymbol{v}, \widetilde{\boldsymbol{\omega}}\boldsymbol{r}) + \frac{1}{2}m\boldsymbol{v}^2, \end{aligned} \tag{1.78}$$

where $m = \int_{\tau} \rho(\boldsymbol{y})\, d^3\boldsymbol{y}$ is the total mass of the body and $\boldsymbol{r} = \frac{1}{m}\int \boldsymbol{y}\rho(\boldsymbol{y})\, d^3 y$ is the position vector of the center of mass in the coordinate system attached to the body, $\rho(\boldsymbol{y})$ is the mass density of the body, and $\mathbf{I}$ and $\mathbf{J}$ are defined by relations (1.54) and (1.55).

If we choose the origin of the coordinate system attached to the body to be at *its center of mass*, then $\boldsymbol{r} = 0$ and the kinetic energy splits into the energy of the translational motion and the energy of rotation about the center of mass. This is the well-known *Bernoulli theorem.*

REMARK. For the motion of a rigid body in an ideal incompressible liquid (the Kirchhoff equations), in general, the kinetic energy cannot be divided into translational and rotational components.

3 Hamiltonian form of equations of motion of a rigid body in $\mathbb{R}^3$

For the transition to the Hamiltonian formalism (the Poincaré – Chetaev equations), we perform the Legendre transform

$$\boldsymbol{M} = \frac{\partial L}{\partial \boldsymbol{\omega}}, \quad \boldsymbol{p} = \frac{\partial L}{\partial \boldsymbol{v}}, \qquad H = (\boldsymbol{M}, \boldsymbol{\omega}) + (\boldsymbol{p}, \boldsymbol{v}) - L|_{\boldsymbol{\omega},\boldsymbol{v}\to\boldsymbol{M},\boldsymbol{p}}. \tag{1.79}$$

Here $\boldsymbol{M}$ is the angular momentum, $\boldsymbol{p}$ is the momentum of the body referred to the coordinate axes attached to the body.

The Poisson brackets of the variables $\boldsymbol{M}, \boldsymbol{p}, \boldsymbol{\alpha}, \boldsymbol{\beta}, \boldsymbol{\gamma}$, and $\boldsymbol{x}$ can be found by formula (1.25). It is completely defined by the form of fields (1.75) and their commutators (1.76) and is independent of the specific form of the Lagrange function. The only restriction is the requirement that the Lagrange function be nondegenerate with respect to velocities.

We find the following (nonzero) Poisson brackets:

$$\begin{gathered}\{M_i, M_j\} = -\varepsilon_{ijk}M_k, \quad \{M_i, p_j\} = -\varepsilon_{ijk}p_k,\\ \{M_i, \alpha_j\} = -\varepsilon_{ijk}\alpha_k, \quad \{M_i, \beta_j\} = -\varepsilon_{ijk}\beta_k, \quad \{M_i, \gamma_j\} = -\varepsilon_{ijk}\gamma_k,\\ \{p_i, x_1\} = -\alpha_i, \quad \{p_i, x_2\} = -\beta_i, \quad \{p_i, x_3\} = -\gamma_i.\end{gathered} \tag{1.80}$$

As was noted in Sec. 1.2, item 3, in this matrix realization, we obtain the Lie – Poisson bracket corresponding to the semi-direct sum $e(3) \oplus_s \mathbb{R}^{12}$.

REMARK. As follows from relations (1.80), in the quaternion parametrization of the rotation group, the Poisson structure in the variables $\boldsymbol{M}$, λ_0, $\boldsymbol{\lambda}$, $\boldsymbol{p}$, and $\boldsymbol{x}$ contains quadratic brackets since the direction cosines depend quadratically on quaternion variables.

In vector form, the Hamiltonian equations of motion can be written as

$$\begin{gathered}\dot{\boldsymbol{M}} = \boldsymbol{M} \times \frac{\partial H}{\partial \boldsymbol{M}} + \boldsymbol{p} \times \frac{\partial H}{\partial \boldsymbol{p}} + \boldsymbol{\alpha} \times \frac{\partial H}{\partial \boldsymbol{\alpha}} + \boldsymbol{\beta} \times \frac{\partial H}{\partial \boldsymbol{\beta}} + \boldsymbol{\gamma} \times \frac{\partial H}{\partial \boldsymbol{\gamma}},\\ \dot{\boldsymbol{p}} = \boldsymbol{p} \times \frac{\partial H}{\partial \boldsymbol{M}} - \frac{\partial H}{\partial x_1}\boldsymbol{\alpha} - \frac{\partial H}{\partial x_2}\boldsymbol{\beta} - \frac{\partial H}{\partial x_3}\boldsymbol{\gamma},\\ \dot{\boldsymbol{\alpha}} = \boldsymbol{\alpha} \times \frac{\partial H}{\partial \boldsymbol{M}}, \qquad \dot{\boldsymbol{\beta}} = \boldsymbol{\beta} \times \frac{\partial H}{\partial \boldsymbol{M}}, \qquad \dot{\boldsymbol{\gamma}} = \boldsymbol{\gamma} \times \frac{\partial H}{\partial \boldsymbol{M}},\\ \dot{x}_1 = \left(\boldsymbol{\alpha}, \frac{\partial H}{\partial \boldsymbol{p}}\right), \qquad \dot{x}_2 = \left(\boldsymbol{\beta}, \frac{\partial H}{\partial \boldsymbol{p}}\right), \qquad \dot{x}_3 = \left(\boldsymbol{\gamma}, \frac{\partial H}{\partial \boldsymbol{p}}\right).\end{gathered} \tag{1.81}$$

The motion of a free rigid body in a potential field in a coordinate system for which the center of mass is at the origin ($\boldsymbol{r} = 0$ in Eq. (1.78)) is described by a natural mechanical system with the Hamiltonian function

$$H = \frac{1}{2}(\boldsymbol{M}, \boldsymbol{A}\boldsymbol{M}) + \frac{1}{2m}\boldsymbol{p}^2 + U(\boldsymbol{\alpha}, \boldsymbol{\beta}, \boldsymbol{\gamma}, \boldsymbol{x}), \tag{1.82}$$

where $\mathbf{A} = \mathbf{I}^{-1}$ and the variables $\boldsymbol{M}$ and $\boldsymbol{p}$ are expressed in terms of the velocities of the body by the formulae

$$\boldsymbol{M} = \mathbf{I}\boldsymbol{\omega}, \qquad \boldsymbol{p} = m\boldsymbol{v}.$$

Remark. If the potential energy in (1.82) can be represented as

$$U = U_1(\boldsymbol{\alpha}, \boldsymbol{\beta}, \boldsymbol{\gamma}) + U_2(\boldsymbol{x}),$$

then the system of equations for the variables $\boldsymbol{M}$, $\boldsymbol{\alpha}$, $\boldsymbol{\beta}$, and $\boldsymbol{\gamma}$ in Eqs. (1.81) decouples; these variables describe the rotation of the body about the center of mass. In this case, if, instead of the projections of the momentum on the movable axes $\boldsymbol{p} = m\boldsymbol{v}$, one uses the momentum in the immovable space $\boldsymbol{P} = m\dot{\boldsymbol{x}}$, one can also decouple a system describing the motion of the center of mass in the canonical form

$$\dot{\boldsymbol{P}} = -\frac{\partial H_{\text{c.m.}}}{\partial \boldsymbol{x}}, \qquad \dot{\boldsymbol{x}} = \frac{\partial H_{\text{c.m.}}}{\partial \boldsymbol{P}}, \qquad H_{\text{c.m.}} = \frac{1}{2m}\boldsymbol{P}^2 + U_2(\boldsymbol{x}). \tag{1.83}$$

In other words, the Poisson structure in the variables $\boldsymbol{M}$, $\boldsymbol{\alpha}$, $\boldsymbol{\beta}$, $\boldsymbol{\gamma}$, $\boldsymbol{P}$, and $\boldsymbol{x}$ is not defined by the Lie – Poisson bracket since the bracket involving only the variables $\boldsymbol{P}$ and $\boldsymbol{x}$ is canonical.

1.6 Examples and Similar Problems

1 Motion of a rigid body with a fixed point in a superposition of constant uniform force fields

As was shown in [95], an arbitrary collection of fields can be reduced to the case of three mutually perpendicular fields. The Hamiltonian function has the form

$$H = \frac{1}{2}(\boldsymbol{M}, \mathbf{A}\boldsymbol{M}) - (\boldsymbol{r}_1, \boldsymbol{\alpha}) - (\boldsymbol{r}_2, \boldsymbol{\beta}) - (\boldsymbol{r}_3, \boldsymbol{\gamma}), \tag{1.84}$$

where $\boldsymbol{r}_1$, $\boldsymbol{r}_2$, and $\boldsymbol{r}_3$ are constant vectors in the body, defining three (generally speaking, different) centers of reduction—analogs of the center of mass. If $\boldsymbol{r}_1 = \boldsymbol{r}_2 = 0$, then the equations of motion for $\boldsymbol{M}$ and $\boldsymbol{\gamma}$ decouple; they are called the *Euler – Poisson equations.*

2 A free rigid body in a quadratic potential

Consider a rigid body moving in a single field with quadratic potential

$$\varphi(\boldsymbol{q}) = -\frac{1}{2}(\boldsymbol{q}, \mathbf{B}\boldsymbol{q}) - (\boldsymbol{g}, \boldsymbol{q}), \tag{1.85}$$

where $\mathbf{B}$ is a constant symmetric matrix and $\boldsymbol{g}$ is a constant vector. A potential of the form (1.85) appears, for example, in the expansion (up to order 2) of the gravitational potential near the surface of the Earth and the Coulomb potential of a charged body.

Representing the position vector of a point in the fixed space in the form $\boldsymbol{q} = \mathbf{Q}^{\mathrm{T}}\boldsymbol{y}+\boldsymbol{x}$, where $\boldsymbol{y}$ is the position vector of a point in the body, and integrating over the body, we obtain the potential energy (see also [61]):

$$\begin{aligned} U =& \frac{1}{2}\operatorname{Tr}(\mathbf{Q}^{\mathrm{T}}\mathbf{I}_1\mathbf{Q}\mathbf{B}) - \frac{1}{2}\mu_0(\boldsymbol{x}, \mathbf{B}\boldsymbol{x}) - \mu_0(\boldsymbol{g}, \boldsymbol{x}) \\ & - \frac{1}{2}\mu_0(\mathbf{Q}\boldsymbol{g}, \boldsymbol{r}_1) - \mu_0(\mathbf{Q}\mathbf{B}\boldsymbol{x}, \boldsymbol{r}_1). \end{aligned} \tag{1.86}$$

Here $\mu_0 = \int\limits_{\tau} \mu(\boldsymbol{y})\, d^3\boldsymbol{y}$ is the total “charge” of the body in the given field, $\mu(\boldsymbol{y})$ is its density, $\boldsymbol{r}_1 = \frac{1}{\mu_0}\int\limits_{\tau} \boldsymbol{y}\mu(\boldsymbol{y})\, d^3\boldsymbol{y}$ is the position vector of the center of reduction of the field, and $I_{1ij} = \int(\delta_{ij}\boldsymbol{y}^2 - y_iy_j)\mu(\boldsymbol{y})\, d^3\boldsymbol{y}$.

For the gravitational field $\mu(\boldsymbol{y})$ is the mass density, $\mu_0 = m$ is the mass of the body, $r_1 = \boldsymbol{r}$ is the position vector of the center of mass, and $\mathbf{I}_1 = \mathbf{I}$ is the tensor of inertia. If the fixed axes are chosen to be aligned with the eigenvectors of the matrix $\mathbf{B}$, then the Hamiltonian of the system, represented in coordinates having origin at the center of mass, can be written as

$$\begin{aligned} H =& \frac{1}{2}(\boldsymbol{M}, \boldsymbol{A}\boldsymbol{M}) + \frac{1}{2m}\boldsymbol{p}^2 + \frac{1}{2}(b_1(\boldsymbol{\alpha}, \mathbf{I}\boldsymbol{\alpha}) + b_2(\boldsymbol{\beta}, \mathbf{I}\boldsymbol{\beta}) + b_3(\boldsymbol{\gamma}, \mathbf{I}\boldsymbol{\gamma})) \\ & -\frac{1}{2}m(\boldsymbol{x}, \mathbf{B}\boldsymbol{x}) - m(\boldsymbol{g}, \boldsymbol{x}), \\ & \mathbf{B} = \operatorname{diag}(b_1, b_2, b_3). \end{aligned} \tag{1.87}$$

It follows that in this case the translational and rotational motions are separated and both systems can be integrated by quadratures [61] (Ch. 3, Sec. 3.4) (which we know to hold if the inertial and gravitational masses coincide, i.e., for the gravitational field). We also note that the translational and rotational motions are separated for an arbitrary field if the center of force application coincides with the center of mass.

3 Motion of a rigid body with a fixed point in a rotating coordinate system

Consider a rigid body moving in such a way that a certain point C traces out a circle of radius R, along which it moves uniformly with angular velocity Ω. We consider three coordinate systems:

(S1) an inertial coordinate system $OXYZ$ with origin at the center of the circle O, fixed in space,

(S2) a uniformly rotating system with center at the point C and the following basis vectors: the tangent vector $\boldsymbol{e}_\tau$ of the circle, the vector $\boldsymbol{e}_n$ of the normal to the plane of the circle, and the vector $\boldsymbol{e}_R$ from the point C to the center of the circle;

(S3) the system of axes $Cx_1x_2x_3$ attached to the body with origin at the point C (see Fig. 1.8).

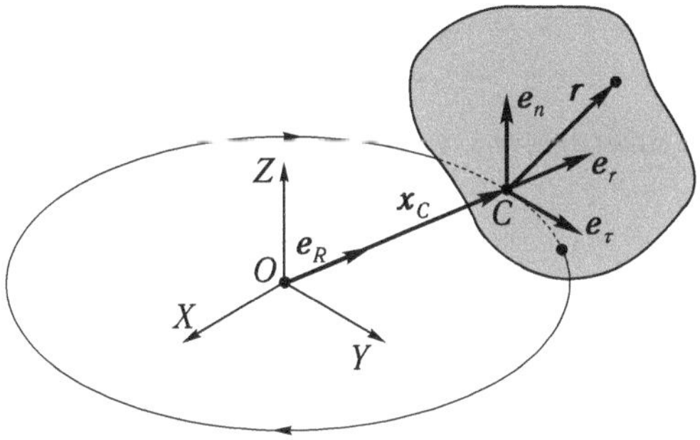

Fig. 1.8. Motion of a rigid body with a fixed point in a rotating coordinate system.

The configuration space of the system is the group $SO(3)$. An element of this group is a matrix $\mathbf{Q}$ defining transition from the system of axes (S3) attached to the body to the rotating coordinate system (S2). These matrices have the form (1.44), where $\boldsymbol{\alpha}$, $\boldsymbol{\beta}$, and $\boldsymbol{\gamma}$ are the projections of the vectors $\boldsymbol{e}_\tau$, $\boldsymbol{e}_n$, and $\boldsymbol{e}_R$ on the axes attached to the body. We also introduce the matrix $\mathbf{B}$ defining transition from the rotating coordinate system (S2) to the fixed coordinate system (S1) (the columns of the matrix $\mathbf{B}$ are the projections of the elements of an orthonormal basis of the fixed coordinate system on the vectors $\boldsymbol{e}_\tau$, $\boldsymbol{e}_n$, and $\boldsymbol{e}_R$).

The position of a point of the body with position vector $\boldsymbol{y}$ in the body is defined in the immovable space by the vector

$$\boldsymbol{q} = \mathbf{B}^{\mathrm{T}}(t)(\mathbf{Q}^{\mathrm{T}}(t)\boldsymbol{y} + \boldsymbol{x}_c), \tag{1.88}$$

where $\boldsymbol{x}_c$ is the position vector of the point C in the rotating coordinate system. Differentiating with respect to time, $\dot{\boldsymbol{q}} = \dot{\mathbf{B}}^{\mathrm{T}}(t)(\mathbf{Q}^{\mathrm{T}}(t)\boldsymbol{y} + \boldsymbol{x}_c) + \mathbf{B}^{\mathrm{T}}(t)\dot{\mathbf{Q}}^{\mathrm{T}}(t)\boldsymbol{y}$, and integrating $\frac{1}{2}m(\dot{\boldsymbol{q}}, \dot{\boldsymbol{q}})$ over the body, we obtain the kinetic energy in the form

$$T = \frac{1}{2}(\boldsymbol{\omega}, \mathbf{I}\boldsymbol{\omega}) + \Omega(\boldsymbol{\omega}, \mathbf{I}\boldsymbol{\beta}) - \mu(\boldsymbol{\omega}, \boldsymbol{r} \times \boldsymbol{\alpha}) + \frac{1}{2}\Omega^2(\boldsymbol{\beta}, \mathbf{I}\boldsymbol{\beta}) - \mu\Omega(\boldsymbol{r}, \boldsymbol{\gamma}), \tag{1.89}$$

where $\mu = 2mR\Omega$, $\mathbf{I}$ is the tensor of inertia of the body with respect to the point C, and $\boldsymbol{r}$ is the position vector of the center of mass referred to the coordinate axes of the body.

The motion of the body in a potential field is described by the Lagrange function

$$L = T(\boldsymbol{\omega}, \boldsymbol{\alpha}, \boldsymbol{\beta}, \boldsymbol{\gamma}) - U(\boldsymbol{\alpha}, \boldsymbol{\beta}, \boldsymbol{\gamma}), \tag{1.90}$$

where T is the kinetic energy (1.89) and U is the potential energy. The equations of motion (1.90) are defined by the Poincaré equations (1.45) and (1.50).

Performing the Legendre transform for system (1.90), we obtain

$$\begin{aligned} \boldsymbol{M} &= \frac{\partial L}{\partial \boldsymbol{\omega}} = \mathbf{I}(\boldsymbol{\omega} + \Omega\boldsymbol{\beta}) - \mu \boldsymbol{r} \times \boldsymbol{\alpha}, \\ H &= \frac{1}{2}(\boldsymbol{M}, \mathbf{A}\boldsymbol{M}) - \Omega(\boldsymbol{M}, \boldsymbol{\beta}) + \mu(\boldsymbol{M}, \mathbf{A}(\boldsymbol{r} \times \boldsymbol{\alpha})) \\ &+ \frac{1}{2}\mu^2 \left(\boldsymbol{r} \times \boldsymbol{\alpha}, \mathbf{A}(\boldsymbol{r} \times \boldsymbol{\alpha})\right) + U(\boldsymbol{\alpha}, \boldsymbol{\beta}, \boldsymbol{\gamma}), \end{aligned} \tag{1.91}$$

where $\mathbf{A} = \mathbf{I}^{-1}$. The equations of motion have the form (1.60).

The following two classical problems of rigid body dynamics reduce to the system (1.91).

Gyroscope and Foucault pendulum. In this case, $U = r_3\beta_3$, the body is axially symmetric,

$$a_1 = a_2 = 1, \qquad r_1 = r_2 = 0,$$

and the Hamiltonian is

$$H = \frac{1}{2}(M_1^2 + M_2^2 + a_3 M_3^2) - \Omega(\boldsymbol{M}, \boldsymbol{\beta}) - \mu r_3(M_1\alpha_2 - M_2\alpha_1) - \frac{1}{2}\mu^2 r_3^2 \alpha_3^2. \tag{1.92}$$

In this case, it is convenient to use the variables of the fixed space (1.62). Choosing appropriate units for length and mass and denoting $\boldsymbol{e}_3 = \boldsymbol{l}$, we can rewrite the Hamiltonian as

$$H = \frac{1}{2}\boldsymbol{N}^2 - \Omega N_2 + \mu(N_2 l_3 - N_3 l_2) - \frac{1}{2}\mu^2 l_1^2 - \mu l_3. \tag{1.93}$$

The system with the zero value of the integral $(\boldsymbol{N}, \boldsymbol{l}) = M_3 = 0$ corresponds to the gyroscope without proper rotation and is called the *Foucault pendulum*.

A satellite on a circular earth orbit. The center of mass coincides with the origin of the rotating system, i.e., $\boldsymbol{r} = 0$. The Newtonian potential in the quadratic approximation (expanded with respect to the ratio of the size of the satellite to the radius of the orbit) has the form

$$U = \frac{3}{2}\Omega^2(\boldsymbol{\gamma}, \mathbf{I}\boldsymbol{\gamma}), \quad \Omega^2 = \frac{GM}{R^3},$$

where G is the gravitational constant, M is the mass of the Earth, and R is the radius of the orbit. Thus,

$$H = \frac{1}{2}(\boldsymbol{M}, \mathbf{A}\boldsymbol{M}) - \Omega(\boldsymbol{M}, \boldsymbol{\beta}) + \frac{3}{2}\Omega^2(\boldsymbol{\gamma}, \mathbf{I}\boldsymbol{\gamma}). \tag{1.94}$$

Different dynamical effects in the motion of a satellite in a circular orbit are described in [33].

4 Relative motion of a rigid body with a fixed point

Consider a rigid body with a fixed point O moving in a reference frame (S0) with origin O, which, in its turn, moves and rotates in a prescribed manner.

Denoting by $\boldsymbol{\Omega}$ and $\boldsymbol{V}$ the angular and linear velocities of the moving reference frame (S0) referred to the coordinate axes attached to the body, we write the Lagrange function of the potential system in the form

$$L = \frac{1}{2}(\boldsymbol{\omega}, \mathbf{I}\boldsymbol{\omega}) + (\boldsymbol{\omega}, \mathbf{I}\boldsymbol{\Omega}) - m(\boldsymbol{W}, \boldsymbol{r}) - U(\boldsymbol{\alpha}, \boldsymbol{\beta}, \boldsymbol{\gamma}). \tag{1.95}$$

Here $\boldsymbol{\omega}$ is the angular velocity of the body, $\boldsymbol{W} = \frac{d}{dt}\boldsymbol{V}$ is the acceleration of the origin of the moving system, $\mathbf{I}$ is the tensor of inertia of the body with respect to the point O, m is the total mass, $\boldsymbol{r}$ is the position vector of the center of mass, and $\boldsymbol{\alpha}$, $\boldsymbol{\beta}$, and $\boldsymbol{\gamma}$ form an orthonormal basis in the moving frame (S0). All these vectors are referred to the coordinate axes attached to the body; moreover, $\boldsymbol{\Omega}$ and $\boldsymbol{V}$ can be considered as given functions of time.

The angular momentum and the Hamiltonian of system (1.95) are defined as follows:

$$\begin{gathered} \boldsymbol{M} = \frac{\partial L}{\partial \boldsymbol{\omega}} = \mathbf{I}(\boldsymbol{\omega} + \boldsymbol{\Omega}), \\ H = \frac{1}{2}(\boldsymbol{M}, \mathbf{A}\boldsymbol{M}) - (\boldsymbol{M}, \boldsymbol{\Omega}) + m(\boldsymbol{W}, \boldsymbol{r}) + U(\boldsymbol{\alpha}, \boldsymbol{\beta}, \boldsymbol{\gamma}), \end{gathered} \tag{1.96}$$

and the equations of motion have the form (1.81). Examples of such systems include gyroscopes and suspensions on aircraft and artificial satellites that perform various specified motions.

5 Motion of a rigid body sliding on a smooth plane

In addition to the Euler – Poisson equations, an interesting mechanical example in which the equations describing the evolution of the vectors $\boldsymbol{\omega}$ and $\boldsymbol{\gamma}$ (or $\boldsymbol{M}$ and $\boldsymbol{\gamma}$) decouple is the system governing the motion of a rigid body sliding without friction on a smooth plane in a potential field depending on the distance to this plane.

Generally speaking, in the absolute motion, the system has five degrees of freedom, but since for ideal sliding the reaction of the plane is perpendicular to the plane, two projections of the momentum of the system on this plane are conserved. Choosing the coordinate system attached to the body with origin at the center of mass (therefore, excluding the horizontal uniform rectilinear motion of the body), we obtain the following Lagrange function for the motion in the potential field $U(\boldsymbol{\gamma})$:

$$L = \frac{1}{2}(\boldsymbol{\omega}, \mathbf{I}\boldsymbol{\omega}) + \frac{1}{2}m(\boldsymbol{\omega}, \boldsymbol{r} \times \boldsymbol{\gamma})^2 - U(\boldsymbol{\gamma}), \tag{1.97}$$

where $\mathbf{I}$ is the tensor of inertia with respect to the center of mass, m is the mass of the body, $\boldsymbol{\omega}$ is the angular velocity referred to the coordinate axes attached to the body, $\boldsymbol{\gamma}$ is the normal vector to the plane in the same axes, and $\boldsymbol{r}$ is the vector from the contact point to the center of mass of the body (see Fig. 1.9).

If the body is convex everywhere, then it is always tangent to the plane at just one point and the vector $\boldsymbol{r}$ is uniquely expressed in terms of the vector $\boldsymbol{\gamma}$ by the Gauss projection

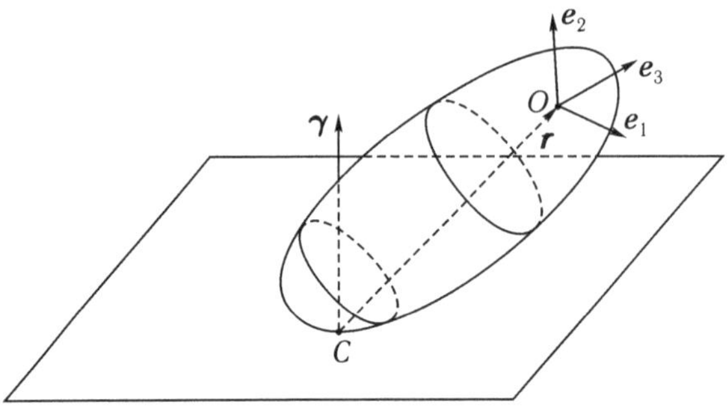

Fig. 1.9. Rigid body on a smooth plane.

of the surface of the body on the unit sphere

$$\gamma = -\frac{\operatorname{grad} F(\boldsymbol{r})}{|\operatorname{grad} F(\boldsymbol{r})|}, \tag{1.98}$$

where $F(\boldsymbol{r}) = 0$ is the equation of the surface of the body. For nonconvex bodies, Eq. (1.98) admits several solutions $\boldsymbol{r} = \boldsymbol{r}(\boldsymbol{\gamma})$ and, as a rule, one must consider additional equations for impacts.

For an ellipsoid with principal semi-axes b_1, b_2, and b_3, we easily obtain

$$\boldsymbol{r} = k(b_1^2\gamma_1,\ b_2^2\gamma_2,\ b_3^2\gamma_3), \quad k = (b_1^2\gamma_1^2 + b_2^2\gamma_2^2 + b_3^2\gamma_3^2)^{-1/2}. \tag{1.99}$$

After the Legendre transform, we obtain from (1.97)

$$\begin{gathered} \boldsymbol{M} = \frac{\partial L}{\partial \boldsymbol{\omega}} = \mathbf{J}\boldsymbol{\omega}, \quad \mathbf{J} = \mathbf{I} + m\boldsymbol{a} \otimes \boldsymbol{a}, \\ H = \frac{1}{2}(\mathbf{IA}\boldsymbol{M},\ \mathbf{A}\boldsymbol{M}) + \frac{1}{2}m(\boldsymbol{a},\ \mathbf{A}\boldsymbol{M})^2 + U(\boldsymbol{\gamma}), \end{gathered} \tag{1.100}$$

where $\boldsymbol{a} = \boldsymbol{r} \times \boldsymbol{\gamma}$ and $\mathbf{A} = \mathbf{J}^{-1}$. By (1.59), the Poisson bracket of $\boldsymbol{M}$ and $\boldsymbol{\gamma}$ is defined by the algebra $e(3)$.

For the gravitational field, the potential energy can be represented as

$$U(\boldsymbol{\gamma}) = mg(\boldsymbol{r},\ \boldsymbol{\gamma}),$$

where g is the free-fall acceleration. It is possible to generalize the system by adding to the body a rotor with gyrostatic momentum $\boldsymbol{K}$; in this case, terms linear in $\boldsymbol{M}$ appear in the Hamiltonian (1.100).

If the body is a sphere with an arbitrary ellipsoid of inertia but whose center of mass coincides with its geometric center, then we obtain either the Euler system (for $\boldsymbol{K} = 0$) (see Sec. 2.2, Ch. 2) or the Zhukovskii – Volterra system (for $\boldsymbol{K} \neq 0$) (see Sec. 2.7, Ch. 2).

Historical comments. The equations of motion of a heavy symmetric top spinning with its sharp end in contact with a smooth plane were first studied by S. D. Poisson in *Traité de Mécanique* (Vol. 2, Paris, 1833). In his studies, the plane can also move upward or downward with constant acceleration. In 1829 – 1831, A. A. Cournot considered this problem in three settings: (i) that not one, but several points of the body are in contact with the plane; (ii) that the body makes contact with the plane along a curve; (iii) that the body makes contact with the plane over a bounded area. In the mid-19th century, V. Puiseux made detailed investigations of the motion of a convex body on an absolutely smooth horizontal plane. He studied, in the linear case, the problem of the stability of uniform rotations of the body about the vertical axis and small oscillations near the equilibrium position on the plane. He also showed that if a heavy rigid body rotates sufficiently rapidly about its axis of symmetry, then the angle between this axis and the vertical always remains arbitrarily close to its initial value. Finally, in his famous treatise [305], F. Klein explicitly integrated in elliptic functions the problem of a heavy symmetric top that spins with its sharp end in contact with a smooth plane. A modern exposition with explicit formulae for the motion of the top can be found in [304]. In [541], integration of this problem in automorphic functions is discussed. It should be noted that these analytic investigations have almost nothing to do with the real (absolute) dynamics of the top. An analogy between the motion of a body on a smooth plane and the motion of a point on an immovable sphere is described in [434]. Qualitative problems concerning the motion of an integrable top are considered in [417].

REMARK. If a body slides on an arbitrary surface, the equations of motion cannot be represented in the form of a reduced Hamiltonian system with two degrees of freedom as discussed above. One of the simplest problems is the motion of a geometrically and dynamically axisymmetric rigid body on a sphere without sliding. This problem is analyzed in [297], where it is proved that by a nontrivial reduction, this problem can be turned into a system with two degrees of freedom and demonstrates a chaotic behavior.

Comment on nonholonimic systems. Another traditional model that describes the motion of a rigid body on a fixed surface is the model with total absence of slipping. In this model, the speed of the contact point vanishes and this constraint is nonholonomic. Although there is no dissipation of energy in this case (i.e., the system is conservative), the equations of motion cannot be written in Hamiltonian form and, in general, do not possess an invariant measure. This explains some unusual properties of such systems, which include, for example, the so-called rattlebacks. The majority of results on the integrability, stochasticity, stability, and other aspects of the dynamics of nonholonomic systems can be found in [90, 411, 280]. In [90], several dynamical systems of different types describing the rolling of one rigid body on another rigid body without slipping are presented. Depending on the existence of one or two integrals, symmetry fields, invariant measure etc., the presence of which is defined by the parameters of the system, a peculiar *hierarchy of dynamics* appears; at one end of this hierarchy, there are regular integrable systems; at the other end there are the most complicated systems such as rattlebacks. It was shown in [101] that the system modelling the behavior of a rattleback top under the most general conditions may give rise to the appearance of strange attractors, which are typical of dissipative systems.

Books on rigid body dynamics (e.g., Routh [498], MacMillan [400], Grammel [239]) usually also treat problems of nonholonomic rolling of rigid bodies. We depart here from this tradition since the study of the dynamics of nonholonomic systems has been greatly extended in modern times and justifies a separate treatment. In the present monograph, we confine our attention to problems that can be solved within the framework of the Hamiltonian formalism.

Nonholonomic systems became a subject of systematic independent study in the second half of the 19th century when Hertz [257] introduced the notion of nonintegrable, i.e., nonholonomic constraints in his treatise *Die Prinzipien der Mechanik in neuem Zusammenhange dargestellt.* Equations of motion for nonholonomic systems in different forms were obtained by Ferrers, Slesser, Appell, Routh, Boltzmann, Hamel, Voronets, Chaplygin, Suslov, etc. They also studied a series of integrable models that now constitute a kind of "gold reserve" of important problems for this branch of science. Modern methods in the theory of dynamical systems have led to a new development in this area.

6 Gyroscope in the Cardan suspension

The gyroscope in the Cardan suspension is a system of several bodies joined by cylindrical hinges (Fig. 1.10) (see, e.g., [401]).

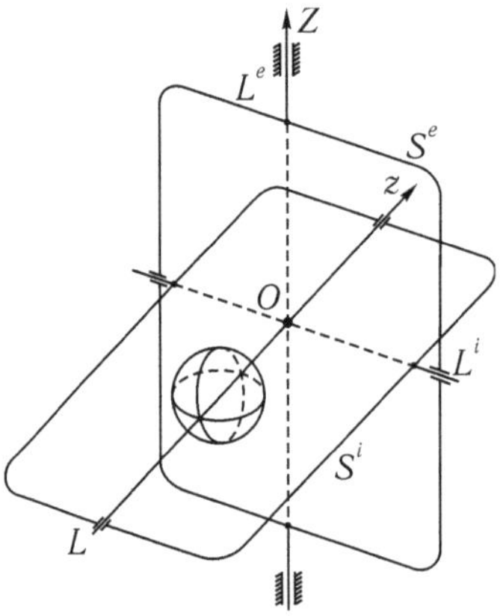

Fig. 1.10. Gyroscope in the Cardan suspension. The outer frame of the Cardan suspension S^e can rotate about the axis L^e, which is fixed in space. The axis L^i of the inner frame S^i is attached to the frame S^e. The axis L of the gyroscope is attached to the inner frame S^i.

We consider the case that is the most important in technical applications: the axes L_e and L_i and also L and L_i are mutually perpendicular and intersect at the same point O (see [401]). We choose a fixed coordinate system with origin at the point O with the axis OZ directed along the axis L^e. We then use the moving coordinate system with origin at the point O with the axis Oz directed along the axis L. Let $\boldsymbol{\alpha}$, $\boldsymbol{\beta}$, and $\boldsymbol{\gamma}$ be the projections of the orthonormal basis in the immovable space on the axes attached to the body, where $\boldsymbol{\gamma}$ is the vector corresponding to the axis OZ.

The Lagrange function for the gyroscope in a potential field can be written as

$$\begin{aligned} L = {} & \frac{1}{2}(\boldsymbol{\omega}, \mathbf{I}\boldsymbol{\omega}) + \frac{1}{2}I^e\left(\frac{\omega_1\gamma_1 + \omega_2\gamma_2}{\gamma_1^2 + \gamma_2^2}\right)^2 \\ & + \frac{1}{2(\gamma_1^2+\gamma_2^2)}\left[I_1^i(\omega_1\gamma_2-\omega_2\gamma_1)^2+(\omega_1\gamma_1+\omega_2\gamma_2)^2\left(I_2^i+I_3^i\frac{\gamma_3^2}{\gamma_1^2+\gamma_2^2}\right)\right] \\ & -U(\boldsymbol{\alpha},\boldsymbol{\beta},\boldsymbol{\gamma}), \end{aligned} \tag{1.101}$$

where $\boldsymbol{\omega} = (\omega_1, \omega_2, \omega_3)$ are the projections of the angular velocity on the axes attached to the body, $\mathbf{I}$ is the tensor of inertia of the body with respect to the point O, I^e is the moment of inertia of the frame S^e with respect to the axis L^e, and I_1^i, I_2^i, and I_3^i are the principal moments of inertia of the inner frame.

The Hamiltonian form of the system (1.101) is obtained by applying the Legendre transform (1.57). The Hamiltonian function of the system in the general case is very cumbersome; we present it here for the case where the body is dynamically symmetric about the axis L (i.e., $I_1 = I_2$):

$$\begin{aligned} H = {} & \frac{1}{2}a_3M_3^2 + \frac{1}{2}a_1k(M_1^2 + M_2^2) \\ & + \frac{1}{2}a_1^2k\left[I_1^i(M_1\gamma_1+M_2\gamma_2)^2 + \left(I^e+(I_3^i-I_2^i)\frac{\gamma_3^2}{\gamma_1^2+\gamma_2^2}\right)(M_1\gamma_2-M_2\gamma_1)^2\right] \\ & +U(\boldsymbol{\alpha}, \boldsymbol{\beta}, \boldsymbol{\gamma}), \\ k = {} & \left((1 + a_1I_1^i)\left(1 + a_1I^e + a_1(I_3^i - I_2^i)\frac{\gamma_3^2}{\gamma_1^2 + \gamma_2^2}\right)\right)^{-1}, \end{aligned} \tag{1.102}$$

where $\mathbf{A} = \mathbf{I}^{-1} = \operatorname{diag}(a_1, a_2, a_3)$.

Historical comment. The Foucault gyroscope and pendulum were both proposed by the famous French physicist L. Foucault (1819 – 1868) as devices to observe the rotation of the Earth in absolute space.

The pendulum has turned out to be the more practical of the two ideas and has become the standard lecture demonstration presented in physics courses. However, a complete analysis of the nonlinear model (only small oscillations are usually considered) is still outstanding; this model is nonintegrable. One of the first attempts to take into account the finiteness of amplitude is due to Kamerlingh-Onnes, the discoverer of superconductivity.

Experiments with the gyroscope performed by Foucault (1852) did not yield satisfactory results since, owing to friction, the gyroscope loses speed too quickly and chaotic precession of the axis of rotation sets in. The idea was that the axis of a symmetric gyroscope should remain constant in the immovable space, which would allow the measurement of the rotation of the Earth. However, Foucault did suggest a series of technical innovations for the construction of gyroscopes, which included the *Cardan suspension*, which, by the way, was known to the French architect Villard de Honnecourt in the 13th century before G. Cardano (1501 – 1576). Foucault also observed that if one deprives a gyroscope of one degree of freedom, then its rotation axis tends to coincide with the angular velocity of the bulk rotation of the suspension base related to the angular velocity of the Earth. This allows one to define the direction to the North pole and the latitude of the position of the device.

Foucault found that the two-degree-of-freedom gyroscope has two characteristic positions with respect to the surface of the rotating Earth and analysis of this fact led Foucault to invent two new devices, the *gyrocompass* and the *gyroscopic latitude measurer*, that found practical realizations only at the end of the 19th century (Aubrey, Sparry, Anschütz, etc.) in the designs of control devices for torpedoes and aircraft. Foucault coined the term *gyroscope*, which literally means *observation of rotation*. For a more detailed exposition of applications of gyroscopes see the books of Grammel [239] and Magnus [401].

In the next two sections we present, without proofs, the equations of motion for two remarkable problems concerning the motion of a rigid body in a fluid. A systematic treatment of these problems can be found in Ch. 3. Their full derivation is presented in Appendix A.

7 Motion of a rigid body in an ideal incompressible fluid and the Kirchhoff equations

In this case, the Hamiltonian has the form (see Appendix A)

$$H = \frac{1}{2}(\boldsymbol{M}, \mathbf{A}\boldsymbol{M}) + (\boldsymbol{M}, \mathbf{B}\boldsymbol{p}) + \frac{1}{2}(\boldsymbol{p}, \mathbf{C}\boldsymbol{p}) + U(\boldsymbol{\alpha}, \boldsymbol{\beta}, \boldsymbol{\gamma}, \boldsymbol{x}). \tag{1.103}$$

Here **A** and **C** are symmetric matrices (added moments of inertia and masses defined by the geometry of the body and its inertial properties) and **B** is an arbitrary matrix; for a body possessing three mutually perpendicular symmetry planes that intersect at the center of mass of the body, **B** may be chosen to be zero. The equations of motion have the form (1.81). Note that usually the term *Kirchhoff equation* is used for the particular case of (1.103) where $U(\boldsymbol{\alpha}, \boldsymbol{\beta}, \boldsymbol{\gamma}, \boldsymbol{x}) \equiv 0$, i.e., the case of inertial motion. In this case, the system of equations for $(\boldsymbol{M}, \boldsymbol{p})$ is closed (these are the Euler – Poincaré equations on $e(3)$) and their analysis is similar to the Euler – Poisson equations (for details, see Sec. 3.1, Ch. 3).

8 Falling rigid body in an ideal fluid

Consider the motion of a rigid body in a gravitational field in an infinite volume of vortex-free, ideal, incompressible fluid under the assumption that the fluid performs a potential (vortex-free) motion. This system generalizes the Kirchhoff equations that describe the inertial motion of a rigid body in a fluid. In Hamiltonian form, the equations of motion have the form (Ch. 1, Sec. 1.6):

$$\begin{gathered}\dot{\boldsymbol{M}} = \boldsymbol{M} \times \frac{\partial H}{\partial \boldsymbol{M}} + \boldsymbol{p} \times \frac{\partial H}{\partial \boldsymbol{p}} + \boldsymbol{\alpha} \times \frac{\partial H}{\partial \boldsymbol{\alpha}} + \boldsymbol{\beta} \times \frac{\partial H}{\partial \boldsymbol{\beta}} + \boldsymbol{\gamma} \times \frac{\partial H}{\partial \boldsymbol{\gamma}}, \\ \dot{\boldsymbol{p}} = \boldsymbol{p} \times \frac{\partial H}{\partial \boldsymbol{M}} - \frac{\partial H}{\partial x_1}\boldsymbol{\alpha} - \frac{\partial H}{\partial x_2}\boldsymbol{\beta} - \frac{\partial H}{\partial x_3}\boldsymbol{\gamma}, \\ \dot{\boldsymbol{\alpha}} = \boldsymbol{\alpha} \times \frac{\partial H}{\partial \boldsymbol{M}}, \quad \dot{\boldsymbol{\beta}} = \boldsymbol{\beta} \times \frac{\partial H}{\partial \boldsymbol{M}}, \quad \dot{\boldsymbol{\gamma}} = \boldsymbol{\gamma} \times \frac{\partial H}{\partial \boldsymbol{M}}, \\ \dot{x}_1 = \left(\boldsymbol{\alpha}, \frac{\partial H}{\partial \boldsymbol{p}}\right), \quad \dot{x}_2 = \left(\boldsymbol{\beta}, \frac{\partial H}{\partial \boldsymbol{p}}\right), \quad \dot{x}_3 = \left(\boldsymbol{\gamma}, \frac{\partial H}{\partial \boldsymbol{p}}\right),\end{gathered} \tag{1.104}$$

where the vectors $\boldsymbol{p}$, $\boldsymbol{M}$, $\boldsymbol{\alpha}$, $\boldsymbol{\beta}$, and $\boldsymbol{\gamma}$ are the projections of the impulsive force, impulsive momentum, and orthonormal basis in the fixed frame on the axes attached to the body and x_1, x_2, and x_3 are the projections of the position vector of the origin of the body-fixed reference frame onto the fixed axes.

The Hamiltonian of the system has the form

$$\begin{gathered}H = \frac{1}{2}(\mathbf{A}\boldsymbol{M}, \boldsymbol{M}) + (\mathbf{B}\boldsymbol{M}, \boldsymbol{p}) + \frac{1}{2}(\mathbf{C}\boldsymbol{p}, \boldsymbol{p}) + U, \\ U = \mu\big(x_3 + (\boldsymbol{r}, \boldsymbol{\gamma})\big), \quad \mu = \mu_b - \mu_l, \quad \boldsymbol{r} = \frac{\mu_b \boldsymbol{r}_b - \mu_l \boldsymbol{r}_l}{\mu_b - \mu_l},\end{gathered} \tag{1.105}$$

where **A**, **B**, and **C** are symmetric matrices defined by the geometry of the body and its inertial properties, μ_b and μ_l are the weights of the body and of the fluid displaced, and $\boldsymbol{r}_b$ and $\boldsymbol{r}_l$ are the position vectors of the center of mass and of the center of pressure in the moving frame.

A straightforward verification shows that there exist three integrals of motion (one of which depends explicitly on time):

$$(\boldsymbol{p}, \boldsymbol{\alpha}) = P_1, \quad (\boldsymbol{p}, \boldsymbol{\beta}) = P_2, \quad (\boldsymbol{p}, \boldsymbol{\gamma}) + \mu t = P_3.$$

This means that the momentum of the system can be represented as

$$\boldsymbol{p} = P_1\boldsymbol{\alpha} + P_2\boldsymbol{\beta} + (P_3 - \mu t)\boldsymbol{\gamma}, \tag{1.106}$$

i.e., the vector $\boldsymbol{P} = (P_1, P_2, P_3)$ represents the projections of the initial momentum on the fixed axes.

After an appropriate choice of time origin (for $\mu_b \neq \mu_l$) and a rotation of fixed axes, we get $P_2 = P_3 = 0$. From now on this assumption will be made.

Substituting (1.106) in the equations of motion (1.104), we obtain a system closed with respect to $\boldsymbol{M}$, $\boldsymbol{\alpha}$, $\boldsymbol{\beta}$, and $\boldsymbol{\gamma}$, which can be written in Hamiltonian form (Ch. 1, Sec. 1.5)

$$\begin{gathered}\dot{\boldsymbol{M}} = \boldsymbol{M} \times \frac{\partial \bar{H}}{\partial \boldsymbol{M}} + \boldsymbol{\alpha} \times \frac{\partial \bar{H}}{\partial \boldsymbol{\alpha}} + \boldsymbol{\beta} \times \frac{\partial \bar{H}}{\partial \boldsymbol{\beta}} + \boldsymbol{\gamma} \times \frac{\partial \bar{H}}{\partial \boldsymbol{\gamma}}, \\ \dot{\boldsymbol{\alpha}} = \boldsymbol{\alpha} \times \frac{\partial \bar{H}}{\partial \boldsymbol{M}}, \quad \dot{\boldsymbol{\beta}} = \boldsymbol{\beta} \times \frac{\partial \bar{H}}{\partial \boldsymbol{M}}, \quad \dot{\boldsymbol{\gamma}} = \boldsymbol{\gamma} \times \frac{\partial \bar{H}}{\partial \boldsymbol{M}}\end{gathered} \tag{1.107}$$

with the Hamiltonian explicitly depending on time,

$$\begin{aligned}\bar{H} &= \frac{1}{2}(\mathbf{A}\boldsymbol{M}, \boldsymbol{M}) + (\mathbf{B}\boldsymbol{M}, P_1\boldsymbol{\alpha} - \mu t\boldsymbol{\gamma}) \\ &\quad + \frac{1}{2}\big(\mathbf{C}(P_1\boldsymbol{\alpha} - \mu t\boldsymbol{\gamma}), P_1\boldsymbol{\alpha} - \mu t\boldsymbol{\gamma}\big) + \mu(\boldsymbol{r}, \boldsymbol{\gamma}).\end{aligned} \tag{1.108}$$

Historical comment. Equations (1.104) in different but equivalent forms can be found in papers of Steklov [557, 556], Goryachev [232], and Chaplygin [129]. Reduction of these equations to an elegant nonautonomous form using representation (1.106) was apparently first made by Kozlov [348] (albeit in the form of the Poincaré equations).

Now we point out some particular cases in which Eqs. (1.107) admit simplifications; they are given in [557].

I. Motion without an initial momentum. Suppose that the initial momentum is zero, $P_1 = 0$. The equations of motion for $\boldsymbol{M}$ and $\boldsymbol{\gamma}$ are closed; they form a (nonautonomous) Hamiltonian system on $e(3)$ with Hamiltonian (see Ch. 1, Sec. 1.7)

$$\bar{H} = \frac{1}{2}(\mathbf{A}\boldsymbol{M}, \boldsymbol{M}) - \mu t(\mathbf{B}\boldsymbol{M}, \boldsymbol{\gamma}) + \frac{1}{2}\mu^2 t^2(\mathbf{C}\boldsymbol{\gamma}, \boldsymbol{\gamma}) + \mu(\boldsymbol{r}, \boldsymbol{\gamma}).$$

If, in addition, the body has reflectional symmetry with respect to three planes intersecting at the center of mass, then this Hamiltonian admits further simplification: $\mathbf{B} = 0$, $\boldsymbol{r} = 0$.

II. Floating body. In [128], Chaplygin considered the case where the force of gravity is balanced by the buoyant force (the mean density of the body is equal to the density of the fluid and hence $\mu_b = \mu_l$), but the center of mass of the body does not coincide with the center of mass of displaced fluid. In this case, the body is acted on by a couple, and the total linear momentum of the body in absolute space is conserved, i.e.,

$$\boldsymbol{p} = P_1\boldsymbol{\alpha} + P_2\boldsymbol{\beta} + P_3\boldsymbol{\gamma},$$

where $\boldsymbol{P} = (P_1, P_2, P_3) = \text{const}$. As previously, after an appropriate choice of fixed axes we get $P_2 = 0$. Therefore, in this case the evolution of the vectors $\boldsymbol{M}$, $\boldsymbol{\alpha}$, $\boldsymbol{\beta}$, and $\boldsymbol{\gamma}$ is described by an autonomous Hamiltonian system with the Hamiltonian function

$$\begin{aligned}\bar{H} &= \frac{1}{2}(\mathbf{A}\boldsymbol{M}, \boldsymbol{M}) + (\mathbf{B}\boldsymbol{M}, P_1\boldsymbol{\alpha} + P_3\boldsymbol{\gamma}) \\ &\quad + \frac{1}{2}\big(\mathbf{C}(P_1\boldsymbol{\alpha} + P_3\boldsymbol{\gamma}), P_1\boldsymbol{\alpha} + P_3\boldsymbol{\gamma}\big) + \mu_b(\boldsymbol{r}, \boldsymbol{\gamma}),\end{aligned}$$

where $\boldsymbol{r}$ is the vector joining the center of mass of the body to the center of pressure.

If the initial momentum is directed along the vertical axis ($\boldsymbol{p} = P\boldsymbol{\gamma}$), then the evolution of the vectors $\boldsymbol{M}$ and $\boldsymbol{\gamma}$ ($\boldsymbol{\gamma}$ is directed along the gravitational field) is described by a system on $e(3)$ with the Hamiltonian function

$$\bar{H} = \frac{1}{2}(\boldsymbol{M}, \mathbf{A}\boldsymbol{M}) + P(\mathbf{B}\boldsymbol{M}, \boldsymbol{\gamma}) + \frac{1}{2}P^2(\mathbf{C}\boldsymbol{\gamma}, \boldsymbol{\gamma}) + \mu_b(\boldsymbol{r}, \boldsymbol{\gamma}). \tag{1.109}$$

9 The Levitron

We now consider a modern problem that is based on rigid body dynamics. A Levitron is a magnetic spinning top that hovers (levitates) in mid-air above a repelling magnetic fixed base. The Levitron was invented by Roy Harrigan (1983). A dynamical model of the Levitron based on adiabatic approximation was developed by Berry [40] in connection with the search for the range of stability of the relative equilibrium of a rotating top.

Assume that the top has an axis of dynamical symmetry, i.e., $I = \mathrm{diag}(I_1, I_1, I_3)$, defined by a unit vector $\boldsymbol{n}$, and the magnetic moment of the top has the form $\boldsymbol{m} = \mu\boldsymbol{n}$, where $\mu = \mathrm{const}$. In this case, it is convenient to write the equations of motion in the fixed frame (see item 2 in Sec. 1.4).

Let $\boldsymbol{N}$, $\boldsymbol{p}$, and $\boldsymbol{x}$ be the angular and linear momentum vectors and the position vector of the center of mass written in the fixed axes. Then the equations of motion of the top in the gravitational and magnetic fields have the form

$$\dot{\boldsymbol{N}} = \frac{\partial H}{\partial \boldsymbol{N}} \times \boldsymbol{N} + \frac{\partial H}{\partial \boldsymbol{n}} \times \boldsymbol{n}, \quad \dot{\boldsymbol{n}} = \frac{\partial H}{\partial \boldsymbol{N}} \times \boldsymbol{n}, \quad \dot{\boldsymbol{p}} = -\frac{\partial H}{\partial \boldsymbol{x}}, \quad \dot{\boldsymbol{x}} = \frac{\partial H}{\partial \boldsymbol{p}}.$$

The Hamiltonian becomes, up to constants,

$$H = \frac{1}{2m}\boldsymbol{p}^2 + \frac{1}{2I_1}\boldsymbol{N}^2 + mgx_3 - \big(\boldsymbol{B}(\boldsymbol{x}), \mu\boldsymbol{n}\big).$$

For domains outside the magnet, $\mathrm{rot}\,\boldsymbol{B} = 0$; therefore, one can represent the magnetic field in the form $\boldsymbol{B} = -\nabla\psi$, where $\psi(\boldsymbol{x})$ is the pseudoscalar potential satisfying the Laplace equation $\Delta\psi = 0$. For axisymmetric fields, it can be represented in series form as

$$\psi(\boldsymbol{x}) = \sum_{m=0}^{\infty} \frac{(-1)^m}{(2^m m!)^2}(x_1^2 + x_2^2)^m \frac{d^{2m}\phi(x_3)}{dx_3^{2m}},$$

where $\phi(x_3)$ is the potential on the symmetry axis Ox_3. For the simplest model of the magnet, a planar, uniformly charged disk of radius a, we have

$$\phi(x_3) = 2\pi\left(1 - \frac{x_3}{\sqrt{a^2 + x_3^2}}\right).$$

In this case, the Levitron has an obvious equilibrium position on the symmetry axis of the magnet which is defined by the relations

$$\boldsymbol{x} = (0, 0, z_s), \quad \boldsymbol{n} = (0, 0, 1), \quad \boldsymbol{p} = 0, \quad \boldsymbol{N} = (0, 0, N_z).$$

It is shown in [40] that for some values of the parameters, this (relative) equilibrium is linearly stable. However, since the Hessian of the Hamiltonian near this equilibrium is not positive (or negative) definite, this equilibrium is not Lyapunov stable. In [173], numerical estimates for the time spent by the top in a region near the equilibrium are given; it turns out that boundaries of regions of rapid escape have a very complex fractal pattern.

1.7 Theorems on Integrability and Methods of Integration

It is appropriate to divide differential equations, including Hamiltonian ones, into two classes, integrable and nonintegrable. At the same time, as Birkhoff [45] noted,

> When, however, one attempts to formulate a precise definition of integrability, many possibilities appear, each with a certain intrinsic theoretical interest.

This statement of Birkhoff, who considered that a dynamical problem is solved if some algorithm for the description of the behavior of all its trajectories is presented and there is a hint about a relation between the integrability and a specific, regular pattern of motion in the phase space. Such a regularity takes place if the system has a sufficient number of conservation laws — first integrals, symmetry fields, or other tensor invariants.

We state here several approaches allowing one to obtain a solution in terms of quadratures both for Hamiltonian and more general differential equations. To solve a system by quadratures means to present its solutions by a finite number of "algebraic" operations (including the inversion of functions) and "*quadratures*," i.e., calculation of integrals of known functions. Various aspects of integrability are discussed in reviews [449, 19, 335, 466, 171].

1 Hamiltonian systems. The Liouville – Arnold theorem

The following theorem links the integrability of a Hamiltonian system by quadratures with the existence of a sufficiently large number of its first integrals.

Theorem. *Assume that on a symplectic manifold* $M^{2n} = (\boldsymbol{p}, \boldsymbol{q}) = (p_1, \ldots, p_n, q_1, \ldots, q_n)$, *n functions in involution are given*:

$$F_1, \ldots, F_n\colon \{F_i, F_j\} \equiv 0, \ i, j = 1, \ldots, n.$$

We also assume that on the level manifold M_f *of the integrals* $\{\boldsymbol{x} \in M^{2n}\colon F_i = c_i, i = 1, \ldots, n\}$, *the n functions* F_i *are independent. Then*:

1. M_f *is a smooth manifold invariant under the phase flow with the Hamiltonian function* $H = F_1$.

2. *If the manifold M_f is connected and compact, then it is diffeomorphic to the n-dimensional torus* (Fig. 1.11)

$$T^n = \{(\varphi_1, \dots, \varphi_n) \bmod 2\pi\}.$$

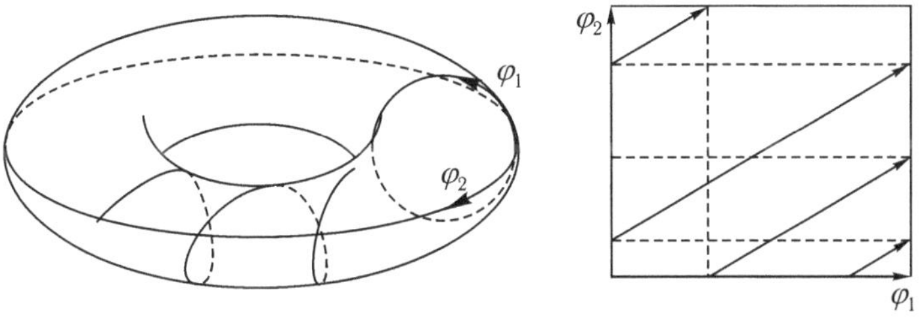

Fig. 1.11. Quasiperiodic motion on a torus and its development.

3. *The phase flow with the Hamiltonian function $H = F_1$ defines on M_f a conditionally periodic motion, i.e., in some angular coordinates $\boldsymbol{\varphi} = (\varphi_1, \dots, \varphi_n)$ we have the equations*

$$\frac{d\boldsymbol{\varphi}}{dt} = \boldsymbol{\omega}, \qquad \boldsymbol{\omega} = \boldsymbol{\omega}(c_1, \dots, c_n) = (\omega_1, \dots, \omega_n).$$

4. *The canonical equations with Hamiltonian function H are integrable by quadratures.*

Remark. The incomplete version of this theorem (with the assertion only on integrability by quadratures) was stated by Bour and generalized by J. Liouville. Its classical proof can be found, e.g., in Whittaker's treatise [611]. The statement above is due to Arnold [18].

In the case considered, the Hamiltonian system is called *Liouville-integrable* or *completely integrable*. It can be proved that for such a system, in a neighborhood of each torus, there exist variables (called *action-angle variables*) $(\boldsymbol{I}, \boldsymbol{\varphi} \bmod 2\pi) = (I_1, \dots, I_n, \varphi_1 \bmod 2\pi, \dots, \varphi_n \bmod 2\pi)$, in which the Hamiltonian $H(\boldsymbol{I})$ is independent of the angle variables $\boldsymbol{\varphi}$ and the equations of motion have the form

$$\dot{\boldsymbol{I}} = -\frac{\partial H}{\partial \boldsymbol{\varphi}} = 0, \qquad \dot{\boldsymbol{\varphi}} = \frac{\partial H}{\partial \boldsymbol{I}} = \boldsymbol{\omega}(\boldsymbol{I}).$$

Therefore, $\boldsymbol{I}(t) = \boldsymbol{I}_0$, $\boldsymbol{\omega}(\boldsymbol{I}) = \boldsymbol{\omega}(\boldsymbol{I}_0) = (\omega_1, \dots, \omega_n)$.

The action variables $\boldsymbol{I}$ parameterize invariant tori $\mathbb{T}^n = M_f$ in M^{2n} and the angle variables $\boldsymbol{\varphi}$ uniformly vary on them (generally speaking, with n different frequencies $\omega_1, \dots, \omega_n$. This motion is said to be *quasi-periodic*. The action-angle variables are very important in perturbation theory.

In some cases, the number of independent first integrals can be greater than $n = \frac{1}{2} \dim M^{2n}$; then not all of them are in involution. This leads to the *noncommutative integrability* of the system, and a general manifold M_f in the compact case is a torus of dimension less than n [443].

In rigid body dynamics, both commutative and noncommutative sets of integrals appear. If the set of integrals is noncommutative (for example, for dynamically symmetric tops), then such systems are called *degenerate* and possess redundant symmetries. In this case, such systems are also said to be *superintegrable*.

REMARK 1. By the Jacobi theorem that asserts that the Poisson bracket of two integrals is again an integral, their complete family forms a (generally speaking) infinite-dimensional Lie algebra. An example of this is considered in the appendix.

The study of the algebra of integrals is also necessary for various methods of system reduction, i.e., decreasing the number of degrees of freedom (Ch. 4). The relation between noncommutative integrability and Dirac reduction is discussed in [95] (see also [73]).

REMARK 2. The assumptions on the compactness and connectedness of M_f usually hold in rigid body dynamics owing to the compactness of the configuration space, which, for example, is the group $SO(3)$, and to restrictions on momenta imposed by the energy integral.

REMARK 3. Those invariant surfaces M_f where integrals become dependent are not usually smooth manifolds. In the space of the values of first integrals $(c_1, \dots, c_n)$, these points form *bifurcation surfaces* whose explicit form is well studied for most integrable systems [66, 286].

Theoretically, the integrability of a Hamiltonian system by quadratures is not necessarily associated with the existence of first integrals. It can be caused by symmetry vector fields, invariant p-forms, and other tensor conservation laws [95, 356, 335, 474]. However, for most physical examples only certain combinations of types of tensor invariants are studied. Now we consider a typical situation.

2 Theory of the last multiplier. Euler – Jacobi theorem

Many problems of rigid body dynamics can be integrated by another method that can be traced back to Euler and Jacobi. It is the theory of the last multiplier, in which for the integrability of a system by quadratures, in addition to a sufficient number of first integrals, the existence of some *invariant measure* is needed. An advantage of this method is that it can be applied not only to Hamiltonian systems, but, generally speaking, to arbitrary systems, for example, to nonholonomic ones. A series of nonholonomic systems possessing a nontrivial measure and integrable by the theory of the last multiplier was specified by Chaplygin [130]. We do not consider them in this book but note that in the 19th century, for the majority of problems in rigid body dynamics, "integrability" was understood to mean "Euler – Jacobi integrability", since the Hamiltonian structure, for example, even for the Euler – Poisson equations (see Sec. 2.1, Ch. 2), had not been clearly understood. We discuss this method now in more detail.

Consider an arbitrary autonomous system of differential equations in $\mathbb{R}^n$

$$\dot{\boldsymbol{x}} = \boldsymbol{v}(\boldsymbol{x}), \qquad \boldsymbol{x} \in \mathbb{R}^n; \tag{1.110}$$

let g^t be its phase flow. Generally, for integrability it is necessary to know at least $n-1$ first integrals. However, if Eq. (1.110) has an integral invariant with smooth density $\mu(\boldsymbol{x})$, i.e., for any measurable domain $D \subset \mathbb{R}^n$ and all t, the relation

$$\int\limits_{g^t(D)} \mu(\boldsymbol{x})\,d\boldsymbol{x} = \int\limits_{D} \mu(\boldsymbol{x})\,d\boldsymbol{x},$$

holds, then for the integrability of system (1.110) it suffices to know $n-2$ first integrals. Recall that by the Liouville theorem, a smooth function $\mu\colon \mathbb{R}^n \to \mathbb{R}$ is the density of an integral invariant $\int \mu(x)\,d\boldsymbol{x}$ if and only if

$$\operatorname{div}(\mu \boldsymbol{v}) = 0. \tag{1.111}$$

If $\mu(\boldsymbol{x}) > 0$ for all $\boldsymbol{x}$, then formula (1.111) defines a *measure* in $\mathbb{R}^n$ invariant under the action of g^t. The presence of a measure facilitates the integration of differential equations. Euler called μ the *integrating multiplier* (it is also called the *last multiplier*).

The following assertion is called the *Euler – Jacobi theorem* on the last multiplier [19, 332].

Theorem. *Assume that the system* (1.110) *with an integrating multiplier μ has $n-2$ first integrals $F_1, \ldots, F_{n-2}$. Suppose that the functions $F_1, \ldots, F_{n-2}$ are independent on the invariant manifold $M_c = \{\boldsymbol{x} \in \mathbb{R}^n\colon F_s(\boldsymbol{x}) = c_s,\ 1 \leqslant s \leqslant n-2\}$. Then the following assertions hold.*

1. *Solutions of Eq.* (1.110) *lying on M_c can be found by quadratures.*

If L_c is a connected compact component of the level set and $\boldsymbol{v}(x) \neq 0$ on L_c, then

2. *L_c is a smooth manifold diffeomorphic to a two-dimensional torus.*
3. *There exist angle coordinates $\varphi_1, \varphi_2 \bmod 2\pi$ on L_c such that Eq.* (1.110) *on L_c has the form*

$$\dot{\varphi}_1 = \frac{\lambda_1}{\Phi(\varphi_1, \varphi_2)}, \qquad \dot{\varphi}_2 = \frac{\lambda_2}{\Phi(\varphi_1, \varphi_2)},$$

where $\lambda_1, \lambda_2 = \mathrm{const}$ and Φ is a smooth, positive function that is 2π-periodic in φ_1 and φ_2.

Generally speaking, the function $\Phi(\varphi_1, \varphi_2)$ defining an invariant measure cannot be reduced to a constant and, although the motion on the torus occurs in rectilinear orbits (Fig. 1.11), it is not uniform. Note that in the case of a Hamiltonian system, such a reduction is always possible; this is a consequence of the Liouville-Arnold theorem.

As a rule, for general systems (1.110), for example, dissipative systems, a measure does not exist, and the proof of their integrability is a separate problem. Probably there are no general methods and, depending on the specific combination of conservation laws (tensor invariants), which are, in general, nonautonomous, the system can exhibit different behaviors.

3 Separation of variables. Hamilton – Jacobi method

In many cases, explicit solutions of a Hamiltonian equation in canonical form can be obtained by the *method of separation of variables* [274]. The problem of integration of an n-degree-of-freedom Hamiltonian system reduces to the search for a solution of the *Hamilton – Jacobi* partial differential equation

$$H\left(\frac{\partial S}{\partial \boldsymbol{q}}, \boldsymbol{q}\right) = \alpha_1, \tag{1.112}$$

which depends on n constants $(\alpha_1, \ldots, \alpha_n)$ and also satisfies the nondegeneracy condition

$$\det\left\|\frac{\partial^2 S}{\partial q_i \partial \alpha_j}\right\| \neq 0.$$

In this case, the function $S(\boldsymbol{q}, \alpha_1, \ldots, \alpha_n)$ is called a *complete integral* of Eq. (1.112). Consider the function $S(\boldsymbol{q}, \alpha_1, \ldots, \alpha_n)$ as the generating function of a canonical transformation $(\boldsymbol{q}, \boldsymbol{p}) \to (\boldsymbol{\beta}, \boldsymbol{\alpha})$:

$$\boldsymbol{p} = \frac{\partial S}{\partial \boldsymbol{q}}, \qquad \boldsymbol{\beta} = \frac{\partial S}{\partial \boldsymbol{\alpha}}. \tag{1.113}$$

For the new canonical variables $\boldsymbol{\alpha}$ and $\boldsymbol{\beta}$, by (1.112), we obtain the equations of motion in the form [274, 432]

$$\dot{\alpha}_i = -\frac{\partial H}{\partial \beta_i} = 0, \qquad \dot{\beta}_i = \frac{\partial H}{\partial \alpha_i} = \delta_{1i}, \qquad i = 1, \ldots, n,$$

where δ_{ij} is the Kronecker symbol. These equations can be easily integrated:

$$\alpha_i = \alpha_i^0, \quad \beta_i = \delta_{1i} t + \beta_i^0, \tag{1.114}$$

where α_i^0, β_i^0 = const. Thus, formulas (1.114) and (1.113) define a solution $\boldsymbol{q}(t), \boldsymbol{p}(t)$ of the canonical equations in the form of a system of algebraic equations.

The variables are separated if on the configuration space one can choose coordinates for which the complete integral is represented as

$$S(\boldsymbol{q}, \boldsymbol{\alpha}) = \sum_{k=1}^{n} S_k(q_k, \alpha_1, \ldots, \alpha_n). \tag{1.115}$$

By Jacobi, the method of separation of variables consists of a search for a (generally curvilinear) coordinate system in which (1.115) holds. Jacobi also found one remarkable change of variables which led him to the elliptic coordinates and allowed him to integrate the problem of finding geodesics on an ellipsoid, even in the multidimensional case. Jacobi also noted: "...We must thus follow the reverse path: having found some remarkable substitution, seek problems in which it can be successfully applied." [274].

REMARK. For degenerate systems (with a redundant set of integrals, several coordinate systems in which the variables are separated can exist, for example, the harmonic oscillator, the Kepler problem , etc.

As examples, we consider the classical Jacobi problem concerning geodesics on a three-axial ellipsoid and the Neumann problem concerning the motion of a point on a sphere in a quadratic potential. They are connected with two different, but reciprocal integrable Clebsch cases of the Kirchhoff equations (see Sec. 3.1, Ch. 3). Integration of this problem is based on applying elliptic and sphero-conical coordinates, which are universal in the theory of integrable systems. All well-known problems admitting separation of variables (on the configuration space) can be solved by using these coordinates or their degenerate cases.

Geodesic flow on the ellipsoid – Jacobi Problem [274]. Consider an ellipsoid in the three-dimensional space $\mathbb{R}^3$ with Cartesian coordinates x_1, x_2, x_3, which is defined by the equation

$$\frac{x_1^2}{a_1} + \frac{x_2^2}{a_2} + \frac{x_3^2}{a_3} = 1, \tag{1.116}$$

where $a_1 > a_2 > a_3 > 0$ are the squares of principal semi-axes.

The elliptic coordinates $\lambda_1, \lambda_2, \lambda_3$ in $\mathbb{R}^3$ are defined as the roots of the cubic equation

$$f(\lambda) = \frac{x_1^2}{a_1 - \lambda} + \frac{x_2^2}{a_2 - \lambda} + \frac{x_3^2}{a_3 - \lambda} = 1, \tag{1.117}$$

where $\lambda_3 < a_3 < \lambda_2 < a_2 < \lambda_1 < a_1$.

The Cartesian coordinates are expressed in terms of the elliptic coordinates by the residues of the function $f(\lambda)$ (1.117) at the points a_1, a_2, and a_3 by the formulas

$$x_1^2 = \frac{(a_1 - \lambda_1)(a_1 - \lambda_2)(a_1 - \lambda_3)}{(a_2 - a_1)(a_3 - a_1)}, \quad x_2^2 = \frac{(a_2 - \lambda_1)(a_2 - \lambda_2)(a_2 - \lambda_3)}{(a_1 - a_2)(a_3 - a_2)},$$
$$x_3^2 = \frac{(a_3 - \lambda_1)(a_3 - \lambda_2)(a_3 - \lambda_3)}{(a_1 - a_3)(a_2 - a_3)}.$$

In the new variables, the ellipsoid (1.116) is defined by the equation $\lambda_3 = 0$ and λ_1 and λ_2 define an orthogonal coordinate system on it. Rewriting the Hamiltonian of a free motion of a particle with unit mass on the ellipsoid (1.116) in these coordinates, we have

$$\begin{gathered} H = \frac{2}{\lambda_1 - \lambda_2}\left(\lambda_1 p_1^2 - \lambda_2 p_2^2\right), \\ g(\lambda) = (a_1 - \lambda)(a_2 - \lambda)(a_3 - \lambda), \end{gathered} \tag{1.118}$$

i.e., the variables are separated.

Using the expression for canonical momenta

$$p_1 = (\lambda_1 - \lambda_2)\frac{\lambda_1 \dot{\lambda}_1}{4(a_1 - \lambda_1)(a_2 - \lambda_1)(a_3 - \lambda_1)},$$
$$p_2 = (\lambda_2 - \lambda_1)\frac{\lambda_2 \dot{\lambda}_2}{4(a_1 - \lambda_2)(a_2 - \lambda_2)(a_3 - \lambda_2)},$$

we obtain the equations of motion in the form

$$\frac{d\lambda_1}{\sqrt{R(\lambda_1)}} = \frac{dt}{\lambda_1 - \lambda_2}, \quad \frac{d\lambda_2}{\sqrt{R(\lambda_2)}} = \frac{dt}{\lambda_2 - \lambda_1},$$
$$R(\lambda) = -\frac{(\lambda - \alpha_1)(\lambda - a_1)(\lambda - a_2)(\lambda - a_3)}{\lambda}, \tag{1.119}$$

where α_1 is a separation constant satisfying the inequalities $a_3<\alpha_1<a_1$. Equations (1.119) are usually associated with the names of Abel and Jacobi as well as Kovalevskaya who used them for integration in hyperelliptic functions.

Remark. The *Abel – Jacobi equations* (1.119) can also be written as

$$\frac{d\lambda_1}{\sqrt{R(\lambda_1)}} + \frac{d\lambda_2}{\sqrt{R(\lambda_2)}} = 0, \quad \frac{\lambda_1\, d\lambda_1}{\sqrt{R(\lambda_1)}} + \frac{\lambda_2\, d\lambda_2}{\sqrt{R(\lambda_2)}} = dt. \tag{1.120}$$

In this case, solutions of the Hamilton – Jacobi equation have the form

$$S(\lambda_1, \lambda_2, \alpha_1, \alpha_2) = \sqrt{\frac{\alpha_2}{2}}\Big(\int \frac{\lambda_1 - \alpha_1}{\sqrt{R(\lambda_1)}}\, d\lambda_1 + \int \frac{\lambda_2 - \alpha_1}{\sqrt{R(\lambda_2)}} d\lambda_2\Big), \tag{1.121}$$

where $\alpha_2 = h$ is the energy. The trajectory and the law of motion can be found from the algebraic equations

$$\frac{\partial S}{\partial \alpha_1} = \beta_1, \quad \frac{\partial S}{\partial \alpha_2} = t + \beta_2, \quad \beta_1, \beta_2 = \text{const}. \tag{1.122}$$

System with quadratic potential on a sphere – Neumann Problem. Consider a sphere in three-dimensional space given by the equation

$$x_1^2 + x_2^2 + x_3^2 = 1, \tag{1.123}$$

and suppose that the potential energy of a particle of unit mass in Cartesian coordinates has the form

$$U(\mathbf{x}) = \frac{1}{2}(a_1 x_1^2 + a_2 x_2^2 + a_3 x_3^2), \quad 0 < a_3 < a_2 < a_1. \tag{1.124}$$

The sphero-conic coordinates λ_1 and λ_2 on the sphere (1.123) are defined as the roots of the quadratic equation

$$f(\lambda) = \frac{x_1^2}{a_1 - \lambda} + \frac{x_2^2}{a_2 - \lambda} + \frac{x_3^2}{a_3 - \lambda} = 0, \tag{1.125}$$

satisfying the inequalities $a_3 < \lambda_2 < a_2 < \lambda_1 < a_1$. The expressions of the Cartesian coordinates in terms of the sphero-conic coordinates are

$$x_1^2 = \frac{(a_1-\lambda_1)(a_1-\lambda_2)}{(a_2-a_1)(a_3-a_1)}, \quad x_2^2 = \frac{(a_2-\lambda_1)(a_2-\lambda_2)}{(a_1-a_2)(a_3-a_2)}, \quad x_3^2 = \frac{(a_3-\lambda_1)(a_3-\lambda_2)}{(a_1-a_3)(a_2-a_3)}. \tag{1.126}$$

The Hamiltonian of the particle with potential (1.124) in the sphero-conic coordinates is

$$H = \left(\frac{2(\lambda_1 - a_1)(\lambda_1 - a_2)(\lambda_1 - a_3)}{\lambda_1 - \lambda_2} p_1^2 + \lambda_1\right) + \left(\frac{2(\lambda_2 - a_1)(\lambda_2 - a_2)(\lambda_2 - a_3)}{\lambda_2 - \lambda_1} p_2^2 + \lambda_2\right), \tag{1.127}$$

where the momenta p_1 and p_2 canonically conjugate to the variables λ_1 and λ_2 are expressed in terms of the velocities as

$$p_1 = \frac{(\lambda_1 - \lambda_2)\dot{\lambda}_1}{4(\lambda_1 - a_1)(\lambda_1 - a_2)(\lambda_1 - a_3)}, \quad p_2 = \frac{(\lambda_2 - \lambda_1)\dot{\lambda}_2}{4(\lambda_2 - a_1)(\lambda_2 - a_2)(\lambda_2 - a_3)}. \tag{1.128}$$

Separating the variables, we obtain the Abel – Jacobi equations governing the evolution of λ_1 and λ_2:

$$\frac{d\lambda_1}{\sqrt{R(\lambda_1)}} = \frac{dt}{\lambda_1 - \lambda_2}, \quad \frac{d\lambda_2}{\sqrt{R(\lambda_2)}} = \frac{dt}{\lambda_2 - \lambda_1},$$
$$R(\lambda) = -(\lambda^2 + 2\alpha_2\lambda + 2\alpha_1)(\lambda - a_1)(\lambda - a_2)(\lambda - a_3),$$

where α_1 and $\alpha_2 = h$ are separating constants. The solution of the Hamilton – Jacobi equation has the form

$$S(\lambda_1, \lambda_2, \alpha_1, \alpha_2) = \int \sqrt{-\frac{\lambda_1^2 + 2\alpha_2\lambda_1 + \alpha_1}{4(\lambda_1 - a_1)(\lambda_1 - a_2)(\lambda_1 - a_3)}}\, d\lambda_1 + \int \sqrt{-\frac{\lambda_2^2 + 2\alpha_2\lambda_2 + \alpha_1}{4(\lambda_2 - a_1)(\lambda_2 - a_2)(\lambda_2 - a_3)}}\, d\lambda_2.$$

Comments. **1.** As a general rule, but not always, for separating variables, the equations of motion can be represented in the Abel – Jacobi form (1.120). It is well known that any solution of such equations can be expressed as theta-functions (more formally, can be linearized by the Abel transformation on the Jacobian of an elliptic or hyperelliptic curve). Initially, the complete linearization was performed by Kovalevskaya for the case discovered by her. She used the theory of theta-functions of two variables that had very recently been developed by Rosenhain and Königsberger. A consequence of this linearization is the remarkable fact that the general solution of the system can be extended to single-valued holomorphic functions on the complex plane of time, i.e., it can have only poles as singularities.

The expression of the general solution for the majority of integrable problems of rigid body dynamics in terms of single-valued elliptic (in the complex sense) functions of time is caused by the fact that the common level set of first integrals representing the intersection of sufficiently simple algebraic surfaces (of quadric type) admits an extension on the complex plane up to Abelian varieties (Abelian tori) that have a parameterization by theta-functions. It is studied in projective and algebraic geometry and the systems themselves are said to be *algebraically integrable*. Thus, the general solution can be single-valued not on the complex plane of time but on its finite-sheeted covering (the Goryachev – Chaplygin case, Sec. 2.5, Ch. 2).

2. The separation of variables problem clearly stated by Jacobi in his *Vorlesungen über Dynamik* [274] is still a subject of serious research. J. Liouville and P. Stäckel found the most general forms of Hamiltonians that admit separation of variables. It turns out that if one uses only transformations of the configuration space (point transformations), then separation of variables is closely related to the presence of a complete set of integrals that are quadratic in momenta. Such results for natural systems with two degrees of freedom were initially obtained by Darboux, Whittaker, and Birkhoff (see [611, 45]). They are discussed in [451] from a modern, topological point of view.

Note that a similar result for systems with integrals linear in momenta is as follows: linear integrals are always connected with an easily identifiable symmetry group acting on the configuration space and with a cyclic variable. In this case, at least locally, the corresponding reduction of order is always possible.

If we consider not only coordinate but also momentum transformations (i.e., general transformations of the phase space), then the problem in some sense is always solvable: by the Liouville – Arnold theorem, near a nonsingular level set of first integrals, action – angle variables always exist, and they are separating variables. As a rule, these variables are distinct for different domains of the phase space which is partitioned by singular (critical) invariant tori, and their construction (given in the proof of the theorem) is not constructive. In practice, the situation is usually the opposite one: the action – angle variables can be constructed if some separating variables are found (see Sec. 5.2, Ch. 5).

The separated variables obtained by expanded phase transformations are known for the Kovalevskaya and Goryachev – Chaplygin cases (see Sec. 2.4, 2.5, Ch. 2, Sec. 5.2, Ch. 5). In these cases, the additional integrals have degree, respectively, three and four in momenta.

For natural Hamiltonian systems with two degrees of freedom possessing an additional quadratic integral, there exist general considerations (see, e.g., Whittaker [611], Birkhoff [45]) that allow one to construct separating variables explicitly. However, for nonnatural two-degree-of-freedom systems and also for systems possessing an additional integral with degree higher than two in momenta, separation of variables is a kind of art. For multidimensional systems, the problem of separation is even more complicated. There are some multidimensional generalizations of two-degree-of-freedom systems (Jacobi and Neumann type problems) for which analogs of elliptic and spheroconic coordinates exist. The problem of separation of variables on S^n is considered in more detail in [58, 616].

3. S. A. Chaplygin knew perfectly well special analytical methods that allowed him to find separation variables for a series of problems of rigid body dynamics including nonholonomic systems. The well-known works of S. V. Kowalewsky [329, 330] also remain brilliant examples of unsurpassed analytical skill. It is interesting that classical authors and their followers used these separated variables for the analysis of particular (as G. Appelrot had it, particularly remarkable) solutions and tried to give them dynamic applications. Separation of variables was regularly used for qualitative and topological analysis in [333, 286], respectively. Using separated variables, it is possible to explicitly construct the corresponding set of action-angle variables [310]. This is especially important for analysis of the perturbed case [474, 335, 333] and for quasi-classical quantization.

Not all of the integrable problems described in this book have been integrated by quadratures, which, in the generalized sense (on the whole phase space) is equivalent to the separation of variables. For some systems (systems of Clebsch, Shottky – Manakov, and the Kovalevskaya gyrostat), the separation of variables is very complicated and practically useless for dynamics owing to various reasons (e.g., implicit solution of algebraic equations or complex notation).

4. Another approach to integration is to search for the so-called **L**-**A**-pairs. According to this method, a problem is regarded as solved if there is a Lax representation [36, 61, 486] with spectral parameter λ, $\dot{L} = [L(\lambda), A(\lambda)]$, which in principle provides a general solution in terms of theta-functions. From the standpoint of algebraic geometry, the problem becomes that of a possible linearization of the flow on *Prym* (Jacobi) *varieties* based on the analysis of *polar decompositions* (*divisors*). This method consists in the construction of some functions on the Riemann surface Γ related with the spectral curve $\det(L(\lambda) - wE) = 0$ and analytic except for a finite number of points. These functions, called the *Baker – Akhiezer functions*, depend on the system parameters and, on the other hand, by the uniqueness theorem, are uniquely expressed by the Riemann θ-function of the surface Γ. This connection allows one to find a general solution in terms of the Riemann θ-functions.

5. There is another approach to explicit integration, which is also based on the commutation representation $\dot{L} = [L, A]$, where, however, L and A are (2×2)-matrices and contain the spectral parameter. This method of integration is more closely related to the separation of variables and is based on the method of r-matrix and Sklyanin quadratic algebras. From the standpoint of Poisson geometry [87, 587], the problem is to embed the initial integrable dynamical system in some universal quadratic algebra of Poisson brackets. Such a quadratic algebra possesses special symplectic coordinates that separate variables for some classes of dynamical systems (similarly to the use of elliptic and sphero-conic coordinates in the classical method of separation of variables). Recently (see [534, 539, 588]), by using this approach, some problems of rigid body dynamics that had long resisted integration by usual classical methods were solved.

6. The known separated variables for an integrable system allow one to study solutions with the simplest structure (degenerate cases or the Appelrot classes of remarkable motions of the Kovalevskaya top) and to perform the bifurcation (topological) and qualitative analyses [333, 286].

Explicit integration and the corresponding separation of variables for the majority of problems of rigid body dynamics, in a slightly modified form, is presented in Sec. 5.2 of Ch. 5. The problem of separation of variables for many newer systems (e.g., gyrostatic generalizations and multidimensional tops) still remains open. Probably, to solve this problem, one ought to modify the Jacobi method and make its scheme less “tough” (see [406]). For all we know, topological analysis and complex analytical methods may provide some additional and useful insight into this problem.

Indeed, for all known integrable systems, the critical level sets of integrals can be defined from the multiplicity condition for roots in the characteristic polynomial of the Abel – Jacobi equations or directly from the condition that the rank of the integral mapping be not maximal. This probably allows one, with some arbitrariness, to restore the separating transformation. Complex methods based on the study of full-parameter Laurent decompositions could probably be effective also [656, 376]. They, as well as Lax representations with a spectral parameter, are capable of giving a spectral curve in the hyperelliptic case; in this way, it is possible to restore the separating transformations and to obtain Abel – Jacobi equations (M. Adler, P. van Moerbeke [2, 4], P. Vanhaecke [591]). However, no new nontrivial explicit results in rigid body dynamics have been obtained by means of such an approach.

7. We note that there is a variety of problems involving geometrical, topological or qualitative analysis of integrable systems whose solution does not require separation of variables. For example, in his master’s thesis S. A. Chaplygin [131] established many remarkable geometrical facts concerning integrable problems in the Kirchhoff equations without separating variables.

Topological analysis, up to the construction of topological invariants, can also be performed by using only the conditions for singularity of the common level set of first integrals (this is used regularly in [66]). Moreover, for many systems, it is more convenient to calculate the action variables by numerical methods [74, 175] than to obtain them by intricate analytical formulae coming from separation of variables [406]. In any case, computer methods of dynamics play an increasing role in the study of classical problems of explicit integration.

2 The Euler – Poisson Equations and Their Generalizations

2.1 Euler – Poisson Equations and Integrable Cases

1 A rigid body with a fixed point

The Euler – Poisson equations governing the motion of a rigid body about a fixed point in a uniform gravitational field are

$$\begin{cases} \mathbf{I}\dot{\boldsymbol{\omega}} + \boldsymbol{\omega} \times \mathbf{I}\boldsymbol{\omega} = \mu \boldsymbol{r} \times \boldsymbol{\gamma}, \\ \dot{\boldsymbol{\gamma}} = \boldsymbol{\gamma} \times \boldsymbol{\omega}, \end{cases} \tag{2.1}$$

where $\boldsymbol{\omega} = (\omega_1, \omega_2, \omega_3)$ is the angular velocity vector, $\boldsymbol{r} = (r_1, r_2, r_3)$ is the position vector of the center of mass, and $\boldsymbol{\gamma} = (\gamma_1, \gamma_2, \gamma_3)$ is the unit vertical vector, all given in the coordinate system formed by the principal axes with origin at the fixed point; $\mathbf{I} = \operatorname{diag}(I_1, I_2, I_3)$ is the inertia tensor relative to the fixed point in these axes and $\mu = mg$ is the body's mass (Fig. 2.1).

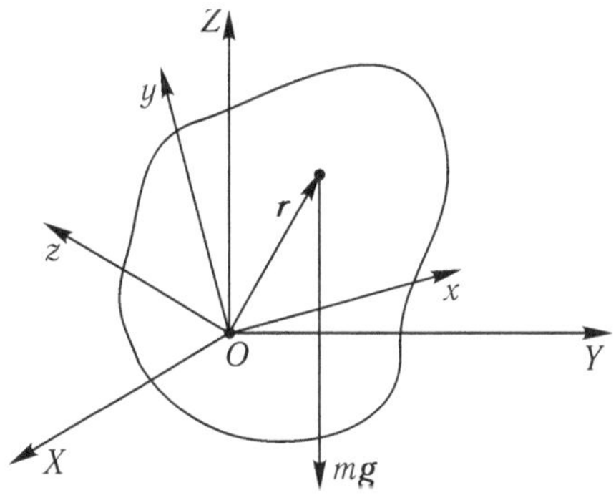

Fig. 2.1. A rigid body with a fixed point in a gravitational field.

Using the angular momentum vector $\boldsymbol{M} = \mathbf{I}\boldsymbol{\omega}$ specified in body coordinates, equations (2.1) can be represented in Hamiltonian form

$$\dot{M}_i = \{M_i, H\}, \quad \dot{\gamma}_i = \{\gamma_i, H\}, \quad i = 1, 2, 3. \tag{2.2}$$

Here the Lie – Poisson bracket is associated with the algebra $e(3)$

$$\{M_i, M_j\} = -\varepsilon_{ijk} M_k, \quad \{M_i, \gamma_j\} = -\varepsilon_{ijk}\gamma_k, \quad \{\gamma_i, \gamma_j\} = 0, \tag{2.3}$$

and the Hamiltonian is the total energy of the body

$$H = \frac{1}{2}(\mathbf{A}\boldsymbol{M}, \boldsymbol{M}) - \mu(\boldsymbol{r}, \boldsymbol{\gamma}), \tag{2.4}$$

where $\mathbf{A} = \mathbf{I}^{-1}$.

DOI 10.1515/9783110544442-002

REMARK. The equations of motion in the form (2.1) were already known to Euler (1758). For a rigid body moving freely ($\boldsymbol{r} = 0$) he also established the simplest case of integrability. The integrability of an axisymmetric top with the center of mass on the axis of symmetry was established by Lagrange and somewhat later by Poisson. Today equations (2.1) are called the *Euler – Poisson equations*.

The Lie – Poisson bracket (2.3) is degenerate, it possesses two Casimir functions which commute (with respect to the bracket (2.3)) with any function of $(\boldsymbol{M}, \boldsymbol{\gamma})$,

$$F_1 = (\boldsymbol{M}, \boldsymbol{\gamma}), \quad F_2 = \boldsymbol{\gamma}^2. \tag{2.5}$$

The vector form of (2.2) is

$$\begin{cases} \dot{\boldsymbol{M}} = \boldsymbol{M} \times \dfrac{\partial H}{\partial \boldsymbol{M}} + \boldsymbol{\gamma} \times \dfrac{\partial H}{\partial \boldsymbol{\gamma}}, \\ \dot{\boldsymbol{\gamma}} = \boldsymbol{\gamma} \times \dfrac{\partial H}{\partial \boldsymbol{M}}. \end{cases} \tag{2.6}$$

Equations (2.2) and (2.6) can be obtained by writing the Poincaré – Chetaev equations on $SO(3)$ (see Sec. 1.2, Ch. 1).

The functions F_1 and F_2 are integrals of equations (2.6) for an arbitrary Hamiltonian H. For the Euler – Poisson equations they admit a natural physical and geometrical interpretation. F_1 is the projection of angular momentum on a fixed vertical axis. In rigid body dynamics F_1 is called the *integral of areas* because of its symmetry with respect to rotations about the fixed vertical axis. The integral $F_2 = \text{const}$ has a simple geometric meaning: it is the square of the absolute value of the unit vertical vector. For real motions F_2 can always be normalized to 1.

When the bracket (2.3) is restricted to a common level set of both integrals F_1 and F_2, it becomes nondegenerate and by the Darboux Theorem (Sec. 1.1, Ch. 1) it can be represented in canonical form using appropriate symplectic coordinates. For various purposes the canonical Euler variables $(\theta, \varphi, \psi, p_\theta, p_\varphi, p_\psi)$ and the Andoyer variables (L, G, H, l, g, h) may be used. In both cases we get a canonical Hamiltonian system with two degrees of freedom on the symplectic leaf defined by $p_\psi = \text{const}$ (or $H = \text{const}$).

2 Hess equations

Hess [258] proposed another form for the Euler – Poisson equations (2.1), (2.4), and (2.6). This new form is obtained by expressing the components γ_1, γ_2, and γ_3 in terms of the constants of first integrals (energy, area and geometric integrals) and substituting the result into the equations for M_1, M_2, and M_3. To do this in the most symmetric manner, let us represent the vector $\boldsymbol{\gamma}$ as

$$\boldsymbol{\gamma} = \lambda_1 \boldsymbol{M} + \lambda_2 \boldsymbol{r} + \lambda_3 \boldsymbol{M} \times \boldsymbol{r}. \tag{2.7}$$

Using the first integrals, one gets

$$
\begin{aligned}
\lambda_1 &= \Delta^{-1}\left(c + \mu^{-1}(h - T)(\boldsymbol{M}, \boldsymbol{r})\right), \\
\lambda_2 &= \Delta^{-1}\left(-c(\boldsymbol{M}, \boldsymbol{r}) - \mu^{-1}(h - T)\boldsymbol{M}^2\right), \\
\lambda_3 &= \pm\Delta^{-1}\left(|\boldsymbol{M} \times \boldsymbol{r}|^2 - \left|\mu^{-1}(h - T)\boldsymbol{M} + c\boldsymbol{r}\right|^2\right)^{1/2} = \pm\sqrt{f},
\end{aligned}
\tag{2.8}
$$

where

$$
(\boldsymbol{M}, \boldsymbol{\gamma}) = c, \quad (\boldsymbol{\gamma}, \boldsymbol{\gamma}) = 1, \quad T = \frac{1}{2}(\mathbf{A}\boldsymbol{M}, \boldsymbol{M}), \quad T - \mu(\boldsymbol{r}, \boldsymbol{\gamma}) = h,
$$
$$
c, h = \text{const}, \quad \Delta = |\boldsymbol{M} \times \boldsymbol{r}|^2.
$$

The Hess equations [258] take the form

$$
\dot{\boldsymbol{M}} = \boldsymbol{M} \times \mathbf{A}\boldsymbol{M} + \mu\lambda_1 \boldsymbol{r} \times \boldsymbol{M} + \mu\lambda_3 \boldsymbol{r} \times (\boldsymbol{M} \times \boldsymbol{r}). \tag{2.9}
$$

They possess an invariant measure with density $\rho = \frac{1}{\sqrt{f}}$. Note that in the equation for λ_3 the sign changes as soon as $f = 0$. From equation (2.9) we obtain the evolution of the squared angular momentum $\boldsymbol{M}^2$ in the form

$$
\frac{d}{dt}\boldsymbol{M}^2 = \pm 2\mu\sqrt{f}.
$$

The sign changes when $\boldsymbol{M}^2$ attains an extremum. Some gyrostatic generalizations of the Hess equations (obtained by P. V. Kharlamov) are discussed in [213, 211].

3 The Kirchhoff analogy for an elastic rod

There is an equivalence between the Euler – Poisson equations and the equations describing the equilibrium configuration of an infinitely thin elastic rod with cylindrical cross-section discovered by G. Kirchhoff [302]. In some sense this similarity provides a “3D interpretation” for rigid body dynamics. This implies that the study of motions of the rigid body can be replaced by analysis of the shape of an elastic rod or, more exactly, by the study of the orientation of a rod-fixed frame in the absolute space.

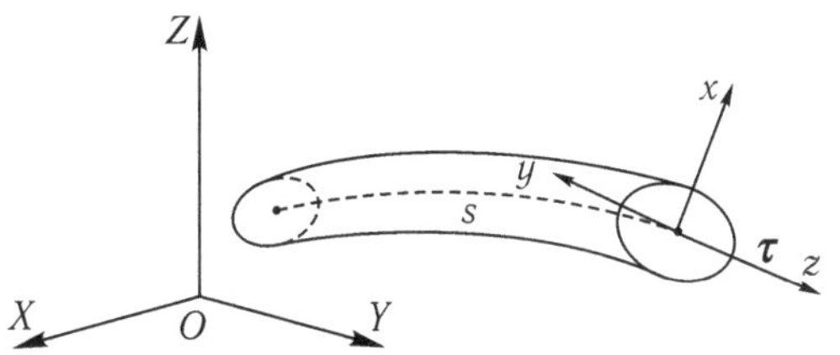

Fig. 2.2. Kirchhoff’s analogy for an elastic rod.

Consider an elastic rod to the ends of which a *constant force* and a *constant moment* are applied. Let s be the axial coordinate measured along the rod axis and ds a rod element. Choose a coordinate frame for each cross-section of the rod; let $\boldsymbol{P}$ and $\boldsymbol{M}$ be the vectors of force and moment in this frame (Fig. 2.2). Then, for each cross-section the end force and the moment satisfy the equilibrium equation

$$\frac{d\boldsymbol{M}}{ds} = \boldsymbol{M} \times \boldsymbol{\omega} + \boldsymbol{P} \times \boldsymbol{\tau}. \tag{2.10}$$

Here $\boldsymbol{\omega}$ is the vector of the "angular velocity" of the frame fixed in a cross-section (i.e., the rotational velocity of the coordinate frame fixed in the cross-section, depending on the axial coordinate s) and $\boldsymbol{\tau}$ is the unit vector tangent to the axis of the rod. Denote by $\boldsymbol{\gamma}$ the unit vector along the end force $\boldsymbol{P}$, and use the invariance of the force in the inertial frame to obtain the "kinematic equilibrium equation"

$$\frac{d\boldsymbol{\gamma}}{ds} = \boldsymbol{\gamma} \times \boldsymbol{\omega}. \tag{2.11}$$

For small deformations [302] the rotational velocity and the elastic moments in the cross-section obey Hooke's law

$$\boldsymbol{\omega} = \mathbf{A}\boldsymbol{M},$$

where $\mathbf{A}$ is a constant matrix.

Assigning to each cross-section a frame such that the axis Oz is directed along the tangent to the rod's axis and such that the two other axes are aligned with the axes of principal moments of inertia of the cross-section (Fig. 2.2), we have

$$\mathbf{A} = \operatorname{diag}(a_1, a_2, a_3), \qquad \boldsymbol{\tau} = (0, 0, 1).$$

Eqs. (2.10) and (2.11) can be represented in the Hamiltonian form (2.2) with the Hamiltonian

$$H = \frac{1}{2}(\boldsymbol{M}, \mathbf{A}\boldsymbol{M}) + \mu(\boldsymbol{\tau}, \boldsymbol{\gamma}), \tag{2.12}$$

where $\mu = |\boldsymbol{P}|$ is the absolute value of the end force. The system (2.6) with the Hamiltonian (2.12) is called the *Euler – Kirchhoff equations*. They are exactly the Euler – Poisson equations up to redefinition of the variables. This is the essence of Kirchhoff's analogy which allows application of the methods of rigid body dynamics to the analysis of elastic systems.

For the case of Kovalevskaya (see Table 2.1) the possible shapes of the corresponding elastic rod are studied in [225].

4 Integrable cases

The Hamiltonian (2.4) is a first integral of the system (2.1) and also of the system (2.6). For the Liouville integrability (see Sec. 1.7, Ch. 1) of these systems an additional integral

Table 2.1. The cases of integrability of the Euler – Poisson equations.

	Author	Additional conditions and the Hamiltonian / First integral
1	Euler (1758)	$\boldsymbol{r} = 0, \quad H = \frac{1}{2}(\mathbf{A}\boldsymbol{M}, \boldsymbol{M})$
		$F = \boldsymbol{M}^2 = \text{const}$
2	Lagrange (1788)	$a_1 = a_2 = 1,\ r_1 = r_2 = 0,$ $H = \frac{1}{2}(M_1^2 + M_2^2 + a_3 M_3^2) - r_3\gamma_3$
		$F = M_3 = \text{const}$
3	Kovalevskaya (1888)	$a_1 = a_2 = 1,\ a_3 = 2,\ r_2 = r_3 = 0,\ r_1 = x,$ $H = \frac{1}{2}(M_1^2 + M_2^2 + 2M_3^2) - x\gamma_1$
		$F = \left(\frac{M_1^2 - M_2^2}{2} + x\gamma_1\right)^2 + (M_1 M_2 + x\gamma_2)^2$
4	Goryachev, Chaplygin (1903)	$a_1 = a_2 = 1,\ \frac{a_3}{a_1} = 4,\ r_2 = r_3 = 0,\ r_1 = x,$ $H = \frac{1}{2}(M_1^2 + M_2^2 + 4M_3^2) - x\gamma_1,$ $(\boldsymbol{M}, \boldsymbol{\gamma}) = 0$
		$F = M_3(M_1^2 + M_2^2) + xM_1\gamma_3$
5	Hess (1890)	$r_2 = 0,\ r_1\sqrt{a_3 - a_2} \pm r_3\sqrt{a_2 - a_1} = 0,$ $H = \frac{1}{2}(\boldsymbol{M}, \mathbf{A}\boldsymbol{M}) + r_1\gamma_2 + r_3\gamma_3$
		$F = r_1 M_1 \mp r_3 M_3 = 0$

is required. Many leading mathematicians, especially in the nineteenth century, tried and failed to obtain this integral. In the general case it was not found.

In the general case, there are some fundamental dynamical effects preventing the Euler – Poisson equations from being integrable. We present all integrable cases known today. The Euler, Lagrange and Kovalevskaya cases (Table 2.1) are *absolutely integrable*, "absolutely" means that if the parameters (the matrix **A** and the vector $\boldsymbol{r}$) satisfy the specified relations an additional first integral exists for all initial conditions. The Goryachev – Chaplygin case is *partially integrable*, since an additional integral exists only if $F_1 = 0$. In the Hess case there exists an invariant relation $F = 0$, linear in $\boldsymbol{M}$, such that $\dot{F} = \lambda F = 0$. In this chapter we give a detailed analysis of all five cases, including qualitative and computer-assisted analyses of the system.

There exist other particular solutions to the Euler – Poisson equations representing as a rule some periodic or asymptotic motions. In what follows we consider the most interesting solutions, which are those having a clear mechanical meaning. In addition to these solutions, there has been a huge amount of excessively formal, intricate, questionable and even wrong research over the last two hundred years of search for an additional integral of the Euler – Poisson equation. For an overview of these investigations see the book [228] . Such results and methods (for example, hodographs) lead to very incomprehensible and voluminous solutions (some of them take up several pages) and have nothing in common with the modern techniques (qualitative and computer-assisted analyses of nonintegrable cases) used in the exploration of the Euler – Poisson equations.

At this point it is appropriate to cite K. Magnus [401]: "Around 1900 the search for new integrable cases of the so attractive nonlinear equations of motion of the heavy gyroscope became a kind of sport for mathematicians. Some investigators wandered away from the physical problem and studied some cases that were impossible physically, since the inequalities for I_1, I_2, and I_3 were violated, or geometrically, since the condition ($\boldsymbol{\gamma}^2 = 1$) failed. We have no possibility to consider these works here."

The main results concerning the nonexistence of analytical integrals and the nonintegrability of the Euler – Poisson equations are due to V. V. Kozlov, S. L. Ziglin, and S. V. Bolotin. Their results are discussed in the books [333, 335] and are associated with the splitting of asymptotic surfaces, branching of solutions on the complex plane of time and the birth of a large number of nondegenerate periodic solutions. A theorem demonstrating that the Euler, Lagrange, and Kovalevskaya cases are the only cases where an additional real-valued analytical integral of motion exists would be a culmination of research efforts along these lines. (If we also consider partial integrals, i.e., functions that are constant along not all but a family of solution curves, then the Goryachev – Chaplygin case may be added to this list) Unfortunately, such a conjecture has not been proved yet; some substantial advances, however, have been obtained in[335].

The algebraic integrability of the Euler – Poisson equations was studied by Husson (1906) [263] (see also [16]). He showed that an additional algebraic integral exists only for the cases of Euler, Lagrange, and Kovalevskaya.

The most complete computer analysis of stochasticity in the Euler – Poisson equations can be found in [205, 107, 88, 77, 208, 209]. The observation of transversal intersection of perturbed separatrices (see, for example, [88]) can serve as a "computer proof" of the nonexistence of an additional real-valued analytical integral. In [96] an infinite series of doublings of the period of permanent perturbed rotations in the Euler – Poinsot problem was obtained. This suggests that the Feigenbaum scenario is one of the possible scenarios of transition to chaos. Integrability and transition to chaos are discussed in detail in Ch. 6.

Remark. To integrate the system (2.1) by means of the last multiplier theory (Euler – Jacobi Theory, see Sec. 1.7, Ch. 1), an additional first integral is required. The system (2.1) has three first integrals and the standard invariant measure $\rho = \text{const}$. However, some natural generalizations of equations (2.1) (see Sec. 3.4, Ch. 3) cannot be integrated by this method. Such systems can, though, be integrated by means of the Hamiltonian formalism and applying the Liouville Theorem (Sec. 1.7, Ch. 1).

5 Absolute motion

Equations (2.6) govern the dynamics of the reduced system in which the precession of the rigid body about the vertical is ignored. To determine the absolute motion, one additional quadrature must be carried out

$$\dot{\psi} = \frac{\omega_1 \gamma_1 + \omega_2 \gamma_2}{\gamma_1^2 + \gamma_2^2} \tag{2.13}$$

or the corresponding Poisson equations for the direction cosines must be integrated

$$\dot{\boldsymbol{\alpha}} = \boldsymbol{\alpha} \times \boldsymbol{\omega}, \quad \dot{\boldsymbol{\beta}} = \boldsymbol{\beta} \times \boldsymbol{\omega}. \tag{2.14}$$

Neither method is optimal for numerical applications. In the first case the singularity near the poles $\gamma_3 = \pm 1$ gives rise to some problems, in the second case the vectors $\boldsymbol{\alpha}, \boldsymbol{\beta}, \boldsymbol{\gamma}$ fail to remain orthonormal due to the presence of dissipation in numerical schemes. Almost all computer illustrations in the present book are obtained by using the quaternion form of the equations of motion, see Sec. 1.4, Ch. 1. This system governs the absolute dynamics of a rigid body, has no singularities, and is not redundant, thus it is very convenient for application of numerical methods. In Sec. 3.4, Ch. 3 we use quaternions in the study of the dynamics when several potential fields are applied.

For a fixed value of the area integral the Euler – Poison equations (2.6) govern the dynamics of a particle on the Poisson sphere in a generalized potential field (see Sec. 4.1, Ch. 4).[1] For these equations several families of periodic and of asymptotic solutions are known. Almost all of these solutions, which mostly go back to classic treatments, are presented below. Let us consider in more detail some of the typical situations arising in analysis of the absolute motion.

Fixed points on the Poisson sphere (the Staude solutions and relative equilibria) represent uniform rotations of the body about a vertical axis.

Periodic solutions on the Poisson sphere do not usually correspond to periodic motions in the inertial frame. A necessary (and sufficient) condition for an absolute motion to be periodic is that the increment of the angle of precession $\Delta\psi$ is a rational multiple of π, that is (see (2.13)),

$$\Delta\psi = \int_t^{t+T} \frac{\omega_1 \gamma_1 + \omega_2 \gamma_2}{\gamma_1^2 + \gamma_2^2}\, dt = \frac{p}{q}\pi, \quad p, q \in \mathbb{N}. \tag{2.15}$$

Here the increment is calculated over a period T of the motion of the reduced system $\boldsymbol{\omega}(t)$, $\boldsymbol{\gamma}(t)$. In the generic case, $\Delta\psi$ is not a rational multiple of π, and the absolute motion has two frequencies and, thus, is quasi-periodic.

1 When this system is written in terms of the canonical coordinates on the cotangent bundle to the Poisson sphere, terms linear in momenta (magnetic monopole terms) appear in the Hamiltonian [447].

Special classes of motion. In rigid body dynamics, there are very special motions that cannot be classified as either periodic motions in the body-fixed coordinate frame or as periodic motions in the fixed space. In [227] periodic motions in the body-fixed coordinate frame are called α-periodic so as not to confuse with periodic motions in the fixed space. Examples of such special motions are precession and isoconic motions (for a body with a fixed point) and helical motions (for a freely moving body). A motion is called a *precessional motion* if there is an axis fixed in the body and an axis fixed in space between which the angle remains unchanged. If the angular velocity of such a motion is constant, then the precession is called *regular* (stability of regular precessions is studied in [110]). For such a general precession, absolute motion is not periodic because the the rate of proper rotation about the body-fixed axis and the rate of rotation of this axis in the fixed frame may not be rationally commensurate. G. G. Appelrot, who introduced the notion of regular precession for the Euler – Poisson equations [12], called them *nutation-free motions*. In 1947 G. Grioli [244] showed that a Lagrange gyroscope (Lagrange top) can have, in addition to the precession about the vertical axis, steady precessional motions about an axis other than the vertical. For such precessions he obtained a system of differential equations in the components of the angular velocity [245].

During the period of 1960 – 1970 precessional motions were systematically studied by the Italian scientific community (C. Totaro, R. Troilo, E. Bentsik, R. Balli, I. Bonzani, E. Pucci). In this connection, an interesting result due to G. Gorr, A. A. Iljukhin [226] (1974) and R. Troilo [578] (1976) should be mentioned. They proved that when the absolute value of the angular momentum is constant, any motion of a rigid body about a fixed point is either an irregular precession in the case of Hess (see below) a regular precession for the Lagrange top (or, of course, all motions in the Euler – Poinsot case).

Besides precessional motions, there is the class of *isoconic* motions of a rigid body about a fixed point. A motion is *isoconic* when the axoid of the angular velocity in the frame fixed in space and the axoid in the frame fixed in the body are symmetric to one another with respect to reflection in their common tangent plane (recall that that an axoid is the cone swept out by the angular velocity vector). For an isoconic motion one has

$$(\boldsymbol{\omega}, \boldsymbol{\gamma}) = (\boldsymbol{\omega}, \boldsymbol{a}),$$

where $\boldsymbol{a}$ is a unit vector fixed in the body frame. Isoconic motions are encountered in the families of solutions of Steklov, Grioli, Lagrange, and Zhukovskii – Volterra (see [227]).

For motion of a rigid body in the space (without a fixed point) it is customary to consider *uniform helical* motions, for which the translational velocity vector and the vector of angular velocity are collinear and have constant magnitude. For example, such motions were discovered and thoroughly explored for Kirchhoff's equations, which govern inertial motions of a rigid body in anideal fluid. If, in addition, gravity is taken into consideration, V. A. Steklov showed that there exists *a uniformly accelerated helical* motion characterized by growth linear in time of the translational and angular velocities. Stability of such motions is studied in detail by the authors in Sec. 6.7, Ch. 6.

Absolute motion: integrable and nonintegrable cases. Let us outline some general principles of experimental and computer studies of the motion of a rigid body around a fixed point. To understand the motion of a rigid body with a fixed point, we follow the motion of some typical points in the body, formally speaking, we choose some variables in the simplest and most natural manner and observe how these variables change with time. If some quantity x varies during the motion, that is, $x = x(t, x_0)$, then its representation in terms of

its spectrum of frequencies ω is constructed by means of the Fourier transform

$$\widehat{x}(\omega) = \lim_{T\to\infty} \int_{-T}^{T} x(t, x_0)\, e^{i\omega t} dt.$$

For integrable systems and for the motions of nonintegrable systems on invariant tori (not in a stochastic layer) x is a quasi-periodic function (according to the Liouville – Arnold theorem). In general it has n rationally independent frequencies (n is the number of degrees of freedom). For chaotic motions the spectrum is continuous, the observed motion alternates between regular behavior and strong chaotic pulsation.

REMARK. Frequencies $\omega_1, \ldots, \omega_n$ are called independent if $k_1\omega_1 + \cdots + k_n\omega_n = 0$, with $k_i \in \mathbb{Z}$, implies $k_1^2 + \cdots + k_n^2 = 0$.

For the motion of a rigid body $n = 3$, and therefore in the integrable cases the absolute motion is in general quasi-periodic with three frequencies. The motion of the reduced system is quasi-periodic with two frequencies if a linear integral (such as the area integral) exists. In this case, the third frequency for the absolute motion is obtained by integrating the angle of precession. Below the integrable cases of the Euler – Poisson equations will be considered.

There are some sets of coordinates for which the number of frequencies can be reduced. The reasons for this reduction are as follows.

a) The algebra of the integrals is redundant, for example, this is the case in the Euler – Poinsot problem.
Indeed, in the Euler case there exist three area integrals $N_1 = (\boldsymbol{M}, \boldsymbol{\alpha})$, $N_2 = (\boldsymbol{M}, \boldsymbol{\beta})$, and $N_3 = (\boldsymbol{M}, \boldsymbol{\gamma})$ generating the algebra $so(3)$: $\{N_i, N_j\} = \varepsilon_{ijk}N_k$. In this case the absolute phase space is foliated by two-dimensional but not three-dimensional tori. Note that the dynamics of the reduced system (on the Poisson sphere) has two frequencies also, that is, the dependence on t of Euler's angle of nutation ψ can be obtained without using quadratures.
In the Euler problem there are variables on phase space (the components M_i) that are periodic functions of time, i.e., we have a single-frequency motion. In practice these variables are almost impossible to measure in the laboratory.

b) The coordinates of certain special points of the body can vary in a simple way. For example, in the case of the Lagrange top the angle of nutation varies periodically. At the same time, the spectrum of the absolute motion consists of three frequencies, while that of the reduced system (in φ or in ψ) contains two frequencies. It is interesting to note that for $(\boldsymbol{M}, \boldsymbol{\gamma}) = 0$, the case of a spherical pendulum, the absolute motion is quasi-periodic with two frequencies.

c) Motions correspond to singular (critical) manifolds that determine bifurcation curves.
Stable and unstable one-dimensional invariant surfaces as well as asymptotic invariant surfaces of the reduced system in general correspond to two-frequency motions in the inertial frame. This is illustrated by the cases of Kovalevskaya and Goryachev – Chaplygin. In the latter case, for the singular Goryachev solution at a small value of energy a further degeneration arises, the motion in the inertial frame becomes periodic (see Sec. 2.5), and the body executes curious pendulum-like motions. For the Kovalevskaya top in the reduced phase space there is a set of three variables z_1, z_2, z_3 such that the motion with respect to these variables is a periodic motion along an ellipse (see Sec. 2.4). These variables depend in a highly nontrivial way on the components of both $\boldsymbol{M}$ and $\boldsymbol{\gamma}$.

Finally, there are the solutions for the Grioli and Hess systems. In the inertial frame the Grioli motion is highly degenerate, and the orbits of all points are periodic. For the motion of the Hess top a very simple behavior may typically be observed: the center of mass moves like a spherical pendulum (i.e., a two-frequency motion takes place). For small energies and for $(\boldsymbol{M}, \boldsymbol{\gamma}) = 0$ the body executes a periodic (one-frequency) motion, the center of mass oscillates within a plane (according to the law of the physical pendulum), and the apex of the mean axis moves periodically along a segment of a loxodrome. For $(\boldsymbol{M}, \boldsymbol{\gamma}) \neq 0$ the situation is again typical, i. e., the body is in motion with three frequencies. For large energies the motion has two frequencies, and for the reduced system (ψ is ignored) the orbit lies on a singular torus filled with doubly asymptotic motions (see Sec. 2.6, Ch. 2).

Note in conclusion that in the absolutely nonintegrable case the body executes complicated chaotic motions as well as various kinds of periodic and quasi-periodic motions. The chaotic motions should be studied not only with the help of frequency analysis, but also using more sophisticated techniques of statistical analysis, such as the correlation functions techniques, etc. The search for periodic and quasi-periodic motions in phase space is one of the most important problems of modern dynamics.

2.2 The Euler Case

1 Geometric – interpretation of Poinsot

In this section we look at some well-known analytic and geometric results in the Euler case. In this case there are no fields acting on the body ($\boldsymbol{r} = 0$). The Hamiltonian (the kinetic energy) and the additional integral are

$$H = \frac{1}{2}(\boldsymbol{M}, \mathbf{A}\boldsymbol{M}), \quad F_3 = \boldsymbol{M}^2 = f = \text{const.} \tag{2.16}$$

In the space of angular momentum $\boldsymbol{M}$, the intersections between the ellipsoid $H = h$ and the sphere $F_3 = f$ equations (2.16) are closed spatial curves called *polhodes*. Polhodes are shown in Fig. 2.3.

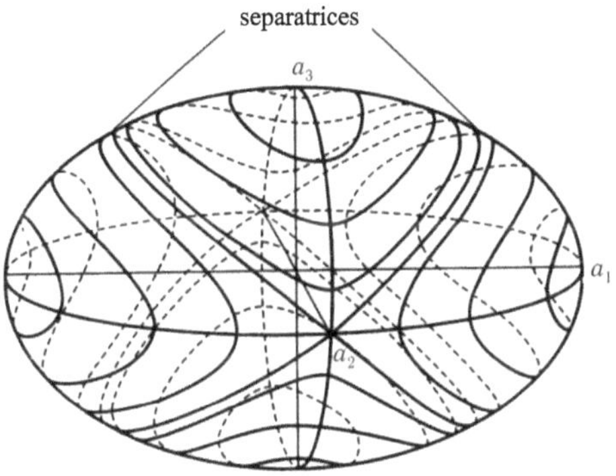

Fig. 2.3. Polhodes ($a_3 < a_2 < a_1$). In general, polhodes are spatial algebraic curves of the fourth order. There are two special cases where these curves are intersecting ellipses: these are projections onto the constant energy ellipsoid of the so called separatrices, which themselves are two-dimensional surfaces in the phase space. The separatrices are filled with doubly asymptotic motions which tend to unstable permanent rotations about the intermediate axis of the ellipsoid of inertia corresponding to the points of intersection of the ellipses. Polhodes degenerate into points that represent permanent rotations about the minor and major axes of the inertial ellipsoid. These rotations are known to be stable.

A geometric interpretation of the Euler case was given by L. Poinsot in 1851 [475]. The Poinsot construction is as follows: the inertial ellipsoid (the ellipsoid of energy) with the fixed center $\frac{1}{2}(I_1\omega_1^2 + I_2\omega_2^2 + I_3\omega_3^2) = h$ rolls without slipping on a plane fixed in the inertial frame and perpendicular to the angular momentum vector (Fig. 2.4). In this construction the instantaneous axis of rotation coincides with the vector joining the center of the ellip-

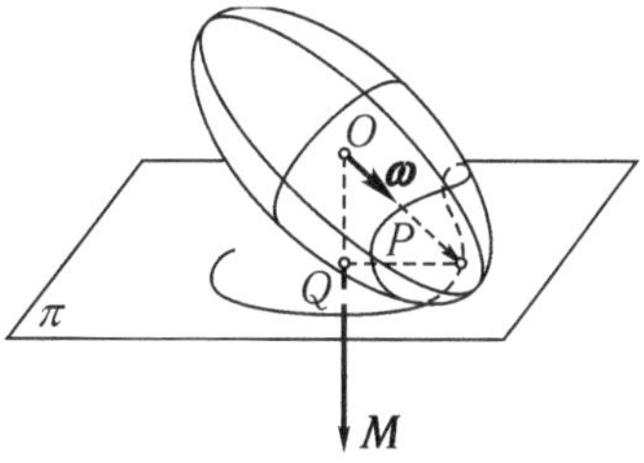

Fig. 2.4. Poinsot interpretation.

soid to the point of contact with the plane and the magnitude of the angular velocity is proportional to the magnitude of this vector The point of contact traces *polhodes* on the ellipsoid (Fig. 2.3) and *herpolhodes* on the plane (see Figs. 2.5 and 2.8).

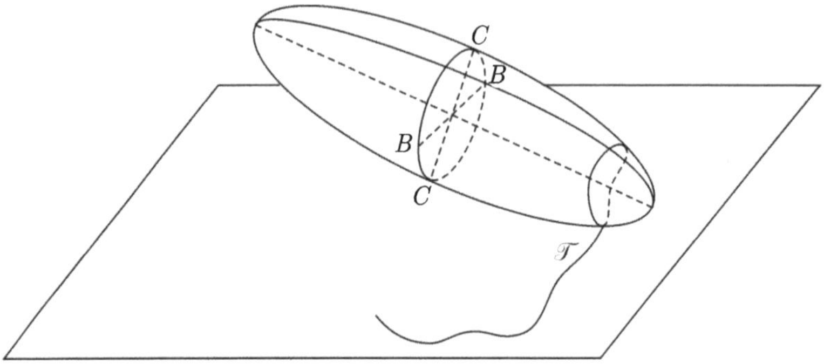

Fig. 2.5. Herpolhode. Although cited in many books, this picture, drawn by Poinsot himself, is *incorrect*: Hess[1] showed that a herpolhode has no inflection points or cusps. In general a herpolhode is a non-closed curve.

2 Explicit integration and bifurcation analysis

The equations of the Euler case are easily integrated by means of the Andoyer variables. Written in terms of these variables, the integral (2.16) becomes cyclic (for further details see Sec. 1.3, Ch. 1, where the phase portrait of the Euler case is also shown). The separatrices divide the inertial ellipsoid (the ellipsoid of constant energy) into four portions (see Fig. 2.3). For one of the portions, the components of the moment $\boldsymbol{M}$ are

$$M_1 = \sqrt{\frac{(f - 2hI_3)I_1}{I_1 - I_3}}\,\mathrm{cn}(\tau, k), \quad M_2 = \sqrt{\frac{(f - 2hI_3)I_2}{I_2 - I_3}}\,\mathrm{sn}(\tau, k),$$
$$M_3 = \sqrt{\frac{(2hI_1 - f)I_3}{I_1 - I_3}}\,\mathrm{dn}(\tau, k), \tag{2.17}$$

where

$$k^2 = \frac{(I_1 - I_2)(f - 2hI_3)}{(I_2 - I_3)(2hI_1 - f)}, \qquad \tau = \sqrt{\frac{(I_2 - I_3)(2hI_1 - f)}{I_1 I_2 I_3}}(t - t_0).$$

These formulae go back to Jacobi, Kirchhoff, and Greenhill (see, for example, [498]) To obtain the formulae for the other portions, some signs in (2.17) must be changed, dn should be replaced with cn and vice versa, and the expressions for the module k of the elliptic functions and for the uniformizing parameter τ must be modified.

Motion in the inertial frame. Expressions for the direction cosines as functions of time can be obtained in the following way. Choose a fixed frame with one unit vector directed along the vector of the angular momentum – which is constant in the inertial frame – and two other unit vectors orthogonal to it as well as to each other:

$$\boldsymbol{M} = \sqrt{f}\boldsymbol{\alpha}, \quad (\boldsymbol{M}, \boldsymbol{\beta}) = (\boldsymbol{M}, \boldsymbol{\gamma}) = 0, \tag{2.18}$$

where f is the constant from (2.16). Find relations for $\boldsymbol{\alpha}$ and $\boldsymbol{\gamma}$. Then $\boldsymbol{\beta}$ can be obtained through the orthogonality condition $\boldsymbol{\beta} = \boldsymbol{\gamma} \times \boldsymbol{\alpha}$.

Since f is constant, $\boldsymbol{\alpha}$ can be found from (2.17). To obtain relations for $\boldsymbol{\gamma}$, use sphero-conic coordinates on the Poisson sphere $\boldsymbol{\gamma}^2 = 1$. Then the area constant is equal to zero, and the Hamiltonian in the Euler case coincides with the additional integral in the Neumann problem with zero potential. In sphero-conic coordinates the variables separate, and the formulae (1.126) from Ch. 1 can be applied (see Sec. 1.7, Ch. 1 for more details).

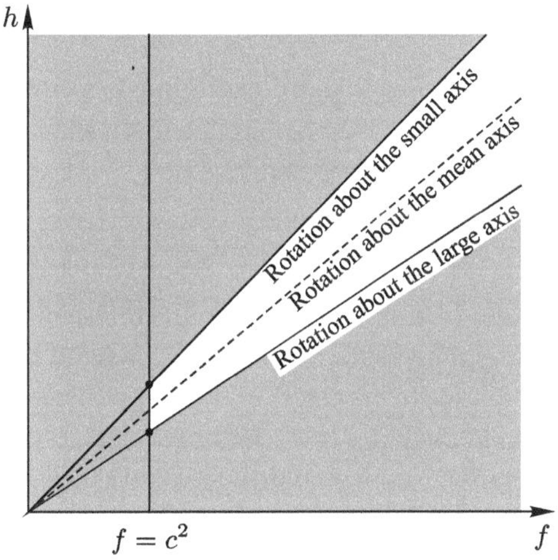

Fig. 2.6. Bifurcation diagram for the Euler – Poinsot case on the plane (h, f). Grey denotes domains where motions are impossible. (The domain of integrals $f < c^2$, $c = (\boldsymbol{M}, \boldsymbol{\gamma})$ is also inaccessible.)

Remark. If $(\boldsymbol{M}, \boldsymbol{\gamma}) = 0$, $\boldsymbol{M}$ can easily be expressed in terms of $\boldsymbol{\gamma}$ and $\dot{\boldsymbol{\gamma}}$ with the help of the Poisson equation $\dot{\boldsymbol{\gamma}} = \boldsymbol{\gamma} \times \mathbf{A}\boldsymbol{M}$:

$$\boldsymbol{M} = \mathbf{A}^{-1} \frac{\dot{\boldsymbol{\gamma}} \times \mathbf{A}^{-1}\boldsymbol{\gamma}}{(\boldsymbol{\gamma}, \mathbf{A}^{-1}\boldsymbol{\gamma})}.$$

The bifurcation diagram for the Euler – Poinsot case on the plane of the integrals h and f is shown in Fig. 2.6. It consists of three branches given by the equations $h = \frac{1}{2}a_i f$, $i = 1, 2, 3$, each branch representing rotations about a corresponding principal axis. Unstable rotations about the mean axis are shown as a dashed line. In this case, in the fixed space the mean axis of inertia of the body moves along a *loxodromic curve*

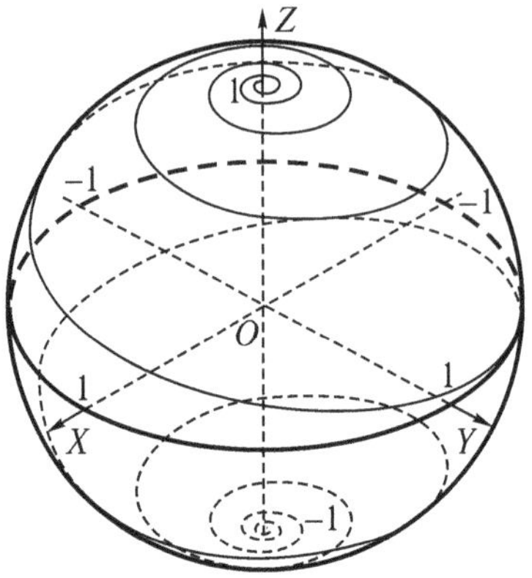

Fig. 2.7. Loxodromic trajectory of the apex of the intermediate inertial axis for motion of the body along a separatrix; $\mathbf{A} = \mathrm{diag}(1, 1.02, 2.0)$.

(loxodrome) on a sphere making a 180° turn (see Fig. 2.7). (A loxodrome is a curve which makes a constant angle with all meridians. For a more detailed discussion of doubly asymptotic motions of the body in the Euler case see Sec. **??**, Ch. 5.

For an elementary stability analysis (both linear and nonlinear) of the partial solutions considered below see [280, 502].

Herpolhode. Herpolhodes were first studied by Darboux. To obtain the differential equation for a herpolhode, we use a parametric representation of a polhode (with the square of the distance r^2 from a point of the polhode to the center of the ellipsoid taken as the parameter) and the equations of motion in the Euler case. Omitting calculations, we only give the final result (see [377] for details). In polar coordinates ρ and φ with the origin, Q, at the intersection of the angular momentum vector and the fixed plane (Fig. 2.4) the equation for the herpolhode is

$$\frac{d\zeta}{d\varphi} = \frac{2h}{f}\frac{\zeta + k}{\zeta\sqrt{P(\zeta)}}, \qquad \zeta = \rho^2, \tag{2.19}$$

where $k = \frac{1}{D}\prod_i (1 - a_i D)$ and $D = \frac{f}{2h}$. The function $P(\zeta)$ is a polynomial of the third degree

$$P = 8hl^2 \prod_{i,j,k} \Big[\frac{a_i^2}{(a_j - a_i)(a_k - a_i)}(\zeta - (a_j + a_k - Da_j a_k))\Big], \tag{2.20}$$

where $l = \sum_{i,j,k} \frac{(a_i - a_j)a_k^2}{a_1 a_2 a_3}$. Equation (2.19) can be solved in terms of elliptic quadratures. Qualitative investigations of $P(\zeta)$ show that the polhode lies between two concentric circles. The polhode touches them alternately at times when $\boldsymbol{\omega}$ passes through the principal planes of the inertial ellipsoid. A herpolhode has no points of inflection or cusps (Hess). For separatrix motions it forms a helix winding infinitely times about the center but of finite length equal to the length of the corresponding arc of the polhode. Typical herpolhode patterns are shown in Fig. 2.8. These patterns are studied in detail in a book of Grammel [239], where the relation is established between epicycloidal and pericycloidal Poinsot motions and points of the bifurcation diagram.

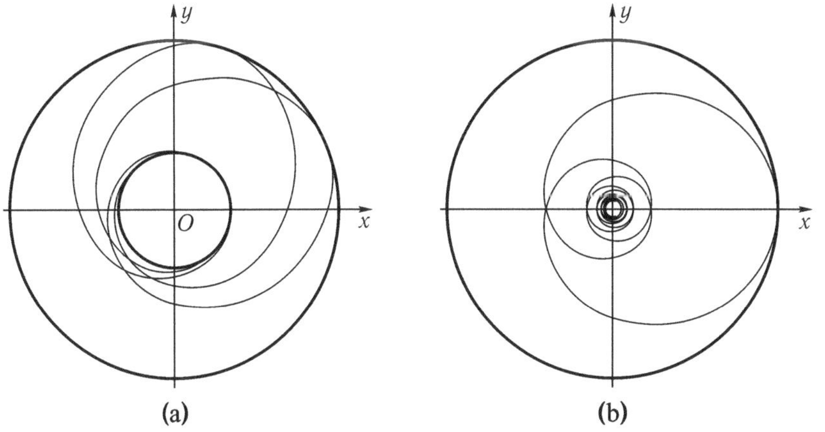

Fig. 2.8. Herpolhodes for the Euler case. (a) A herpolhode of general form is a quasi-periodic non-closed curve. (b) The herpolhode corresponding to a separatrix is a curve that tends to the origin and wraps on it infinitely many times.

Remark 1. Using KAM-theory, one can show that when a heavy body rotates rapidly, the absolute value $|\boldsymbol{M}|$ of its angular momentum (as well as the projection of the angular momentum vector on the vertical) change very little over an infinite span of time (A. N. Kolmogorov, V. I. Arnol'd [20]). At the same time, the vector $\boldsymbol{M}$ precesses slowly about the vertical, and the motion of the body relative to $\boldsymbol{M}$ is almost a Poinsot motion. In particular, if a body is launched into a fast spin about its major or minor axis, then in the body frame the angular velocity vector will always remain close to this axis, while for an inertial-frame observer the angular velocity will precess about the direction of gravity. These observations remain valid as well in the case of a weak gravitational field (which is equivalent to a fast spin).

Remark 2. V. I. Bogaevskii, and L. A. Ostrer [54] considered a rapidly spinning heavy body with $I_1 = I_2$ ("rapid" means that the body's potential energy is small compared to its kinetic energy). At the initial time, the resonance relations characteristic of the Euler – Poinsot interpretation are assumed to be satisfied (it is assumed that during time $t \sim 1$ the body makes $\frac{1}{\varepsilon}$ revolutions, where ε is small. They showed that in a finite time $t \sim 1$ a finite deviation from the initial inertial motion occurs. This is in contradiction with the just established "behavioral concept" of a fast spinning body: recall that the angular momentum $\boldsymbol{M}$ precesses slowly about the vertical, and the body executes a Poinsot-type motion about $\boldsymbol{M}$. This contradiction can be explained by the fact that the following two approximate integrals

$$2H = I_1(\omega_1^2 + \omega_2^2) + I_3\omega_3^2, \quad M^2 = I_1^2(\omega_1^2 + \omega_2^2) + I_3^2\omega_3^2,$$

are degenerate and for small ω_3 reduce to $\omega_1^2 + \omega_2^2 \approx \text{const}$. Explicit quadratures and a geometric interpretation for the motion discussed are given in [54].

Remark 3. In [187, 37, 38] various questions related to the motion of a rigid body in an inertial coordinate frame (a three-degree-of-freedom system) are discussed. According to N. Nekhoroshev [443], for the Euler – Poinsot case, which is known to be noncommutatively integrable, generalized action-angle variables can be introduced. Analytical details associated with these variables and some topological issues concerned with the foliation of the six-dimensional symplectic manifold by two-dimensional tori are considered in [187]. In [37, 38] a perturbed Euler – Poisson problem is studied; the perturbation, whose physical meaning is not clear, causes the body to spin fast. Both analytical and numerical estimates for the Arnold diffusion rate are obtained and are found not to be quite in agreement.

3 Comments

Various methods of explicit integration of the equations of motion for a free top were considered by Euler, Lagrange, Kirchhoff, Cayley, Greenhill, and Jacobi. The works of Jacobi [273, 272] are the most interesting ones. In these works the components of the angular velocity and all (nine) direction cosines are expressed in terms of elliptic functions. By applying these results to study a perturbed motion, Jacobi constructed a system of osculating variables close to action – angle variables. A modern presentation of these variables by Yu. A. Sadov (see [16, 333]) makes use of the phase space interpretation and the Andoyer variables. In his famous treatise [366] G. Lamb asserts that it was a certain Rueb (1834, [504]) who was the first to use elliptic functions in the integration of the equations of the Euler case.

In his *Analytical mechanics* Lagrange gave his own solution of the Euler problem, "I have tried to make this solution as clear and (if I may say so) as elegant as possible." Lagrange already considered this case too easy, ". . . I hope I will not be reproached for revisiting this problem." A remarkable point in his solution is that he showed for the first time the existence of the three principal axes of inertia for an arbitrary rigid body (a symmetric matrix can always be reduced to diagonal form). Note that this latter problem is not connected with the Euler case. Elliptic integrals make an appearance in the solution of Lagrange, but there is no mention there of their inversion. This idea was later developed in the works of Jacobi and was completed in the works of Weierstrass, Hermite, Halphen, and L.I.Somov.

The geometric construction of Poinsot is a good example of a geometric interpretation of motion in mechanics. Note that for the other integrable cases this construction is not so clear. Already Poinsot himself tried to improve his own interpretation. He considered another geometric interpretation with time taken into account in which a cone attached to the body rolls on a plane orthogonal to the angular momentum vector and rotating with a constant angular velocity. Using this interpretation, Darboux and Königs constructed an apparatus called a *herpolograph* which demonstrates inertial motion of a body. G.Darboux carried out a detailed analysis of the polhode and the herpolhode in [145]. The Poinsot interpretation was further developed by Jacobi, Silvester, and MacCullagh, see [377, 228, 562, 400], etc. MacCullagh introduced the reciprocal ellipsoid (the so-called ellipsoid of gyration) in the sense that the axes of the Poinsot ellipsoid are reciprocal to each other. His interpretation is less illustrative, and the arising curves of angular impulse are analogous to polhodes. All the above-mentioned results are more general but at the same time are more artificial and have only historical value. For more information on these interpretations, we refer the reader to [377, 228, 562, 400]. We also mention an extensive and interesting investigation by P. V. Voronets [606] on a free top.

To verify the laws of free motion, Maxwell invented a model of the top which now bears his name. His experiments are described in Webster's book on mathematical physics [610], where the theory of the top is given special attention: "this is a question of particular practical importance, especially for engineers, but it is often missed in the study of physics. Already Maxwell paid much attention to this question and constructed an apparatus for demonstration of these phenomena." An earlier apparatus for demonstration of the motion of a free top had been constructed by Bonnenberger (1817).

Using Maxwell's top one can study stability or instability of rotations about the different principal axes and also see that motions close to rotations about the mean axis (and hence close to a separatrix motion) are very complicated and seem to be non-ordered and chaotic. However, "proper chaos" appears as a result of addition of some perturbation, for example, a gravitational field.

2.3 The Lagrange Case

This case of integrability is of interest both in the study of classical mechanics and for technical applications. The basic laws of motion of the Lagrange top constitute the approximative (applied) theory of gyroscope.

In the Lagrange case a body possesses dynamical symmetry $a_1 = a_2$, and the center of mass lies on the axis of dynamical symmetry $r_1 = r_2 = 0$. The additional integral is $F_3 = M_3 = \text{const}$.

Reduction to a system with one degree of freedom. The simplest way to integrate the equations in the Lagrange case is to make use of the Euler angles θ, φ, and ψ and the conjugate canonical momenta p_θ, p_φ, and p_ψ. Looking at the Lagrange function (see (1.71), Ch. 1), we find that the variables φ and ψ are cyclic, and the corresponding linear momenta are integrals of motion:

$$p_\varphi = M_3 = \text{const}, \quad p_\psi = (\boldsymbol{M}, \boldsymbol{\gamma}) = \text{const}.$$

Eliminating the cyclic variables from the energy integral, we get

$$h = \frac{I_1}{2}\dot{\theta}^2 + \frac{p_\varphi^2}{2I_3} + \frac{(p_\psi - p_\varphi \cos\theta)^2}{2I_1 \sin^2\theta} + mgl\cos\theta, \tag{2.21}$$

where the values of the cyclic integrals $p_\psi = (\boldsymbol{M}, \boldsymbol{\gamma})$ and $p_\varphi = M_3$ may be considered as parameters.

Assuming without loss of generality that $I_1 = 1$ and $mgl = 1$ (this can be achieved by an appropriate choice of the units of length and time), (2.21) gives the quadrature for the angle of nutation

$$\dot{\theta}^2 = 2h - ap_\varphi^2 - \frac{(p_\psi - p_\varphi \cos\theta)^2}{\sin^2\theta} - 2\cos\theta, \quad a = I_3^{-1}. \tag{2.22}$$

For the variable $u = \gamma_3 = \cos\theta$ from (2.22) we obtain an elliptic quadrature (see also [401])

$$\begin{gathered} \dot{u} = \pm\sqrt{R(u)}, \\ R(u) = 2(h' - u)(1 - u^2) - (p_\psi - p_\varphi u)^2, \\ h' = h - \frac{ap_\varphi^2}{2} = \text{const}. \end{gathered} \tag{2.23}$$

Here $u(t)$ can be expressed in terms of elliptic functions. The *gyroscopic function* $f(u)$ is a cubic polynomial. A typical graph of f is shown in Fig. 2.9. A similar quadrature with a function $R(u)$ is possible for generalizations of the Lagrange case all admitting the integral $M_3 = \text{const}$ (as a rule, $R(u)$ is a polynomial of a higher degree).

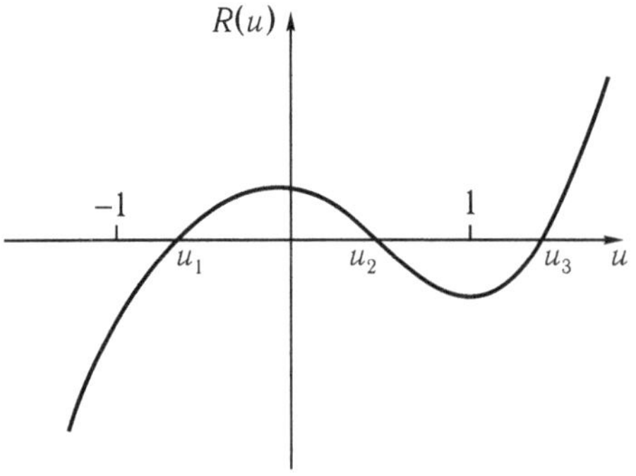

Fig. 2.9. The gyroscopic function for the Lagrange top. It can be easily shown that $u_3 \geqslant 1$.

Dynamics of the whole system. To determine the motion of the body, one should determine the angle of precession ψ and the angle of proper rotation φ using the known law of motion $u(t)$:

$$\dot{\psi} = \frac{p_\psi - p_\varphi u}{1 - u^2}, \quad \dot{\varphi} = (a-1)p_\varphi + \frac{p_\varphi - p_\psi u}{1 - u^2}. \tag{2.24}$$

The motion of the *apex*, a point on the axis of dynamical symmetry, is governed by the spherical angles θ and ψ. The trajectory of the apex always lies between two parallels with latitudes determined by the gyroscopic function (2.23) (see Fig. 2.9) and its shape is one of those shown in Figs. 2.10, 2.11, and 2.12.

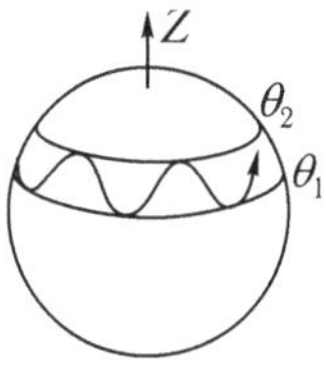

Fig. 2.10. The sign of $\dot{\psi}$ does not change during the motion and $\dot{\psi}$ never becomes zero.

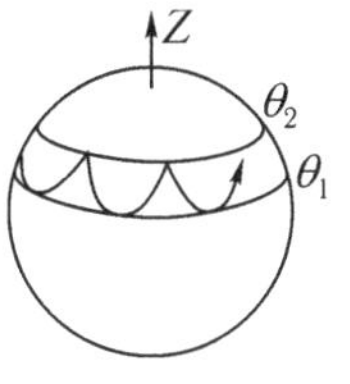

Fig. 2.11. The sign of $\dot{\psi}$ is preserved; $\dot{\psi}$ periodically takes the value of zero.

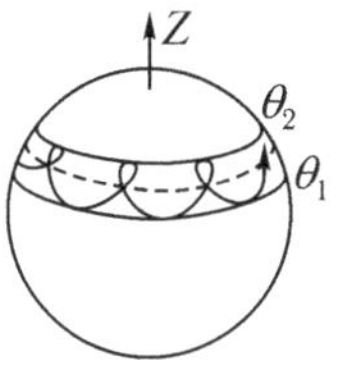

Fig. 2.12. The sign of $\dot{\psi}$ changes.

For some values of the integrals the graph of the gyroscopic function touches the horizontal axis at the point $u = 1$. Then we have an asymptotic (aperiodic) motion of the Lagrange top, the axis of symmetry tends to the vertical direction as $t \to \pm\infty$ (Fig. 2.13). Explicit formulae are given in Sec. **??**, Ch. 5.

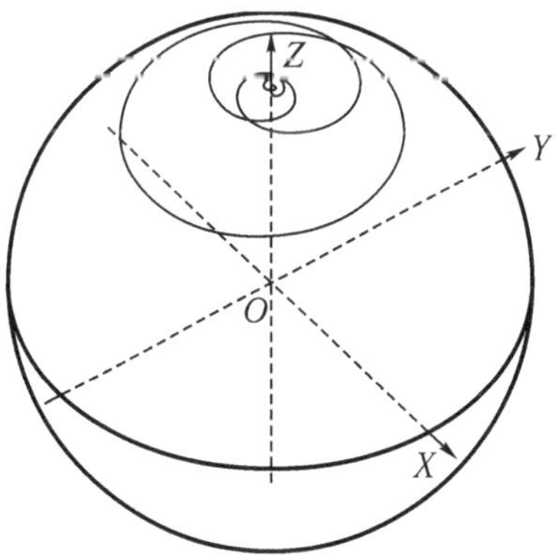

Fig. 2.13. The motion of the apex of the center of mass of the Lagrange top in the inertial frame for an asymptotic motion. This motion was pointed out by Klein and Sommerfeld [305].

1 Bifurcation diagram and geometric analysis of motion

The type of motion (the form of the trajectory) of the Lagrange top is completely specified by the values of the integrals of motion H, p_φ, and p_ψ. We will use the notation $H = h$, $p_\varphi = p$ and $p_\psi = c$ for the *values* (h, p, c) taken by the *functions* (H, p_φ, p_ψ).

Consider a three-dimensional space whose points represent the values of the first integrals (H, p_φ, p_ψ). Since the form of the trajectory of the Lagrange top is completely specified by (h, p, c), this space can be divided into domains, with each domain representing a particular type of motion. The region of "allowed values" of the integrals consists of the points for which the corresponding gyroscopic function (2.23) takes positive values for $u \in [-1, 1]$ (see Fig. 2.9). The boundary of this region is the so-called *surface of regular precessions*, for these values of the integrals the gyroscopic function (2.23) touches the axis Ou from below on the segment $[-1, 1]$. In this case the top executes a *regular precession*, the apex rotates uniformly around the vertical, the angle of inclination is constant, and the body rotates uniformly about its axis. The general form of the surface of regular precessions and its various cross-sections $p_\varphi = \text{const}$ and $p_\psi = \text{const}$ are shown in Figs. 2.13, 2.15, 2.16, and 2.14. On the planes $p_\varphi \pm p_\psi = 0$ two edges are clearly seen (one of the edges $p_\varphi - p_\psi = 0$ starts at the point $p_\varphi = p_\psi = 2$ and does not reach the origin). For small deviations from the condition for multiplicity of the roots a *pseudo-regular precession* appears, and nutation of the axis of dynamical symmetry also takes place.

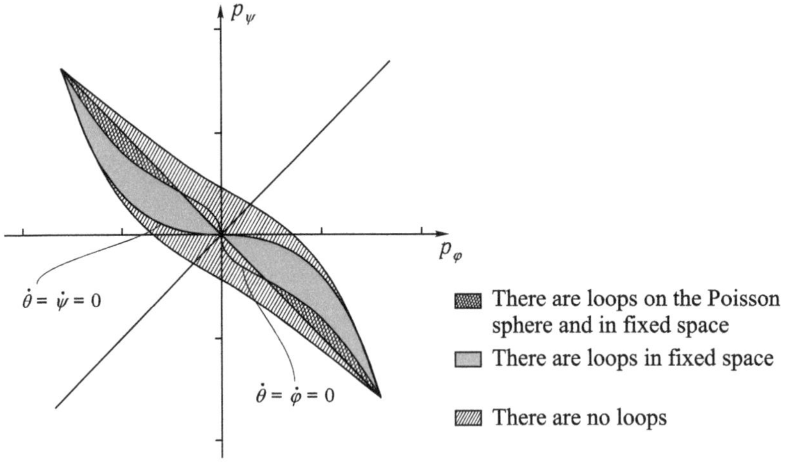

Fig. 2.14. The intersection of the surface of regular precessions (Fig. 2.13) and the plane $h = 0$ is presented. The lines of the return points (the points for which the apex reverses its motion) for the apex of the axis of dynamical symmetry ($\dot\theta = \dot\psi = 0$) and for the apex on the Poisson sphere ($\dot\theta = \dot\varphi = 0$) are shown. The narrow shaded domain represents solutions that have loops both on the Poisson sphere and in the absolute space. As for the other domains, loops may appear only in the absolute space or do not exist at all.

REMARK. The energy-momentum method is used in [378] for analysis of the motion of the Lagrange top. The paper also discusses stability of precessions about the vertical and tilted axes and outlines conditions for bifurcations. In this connection, we note the pioneering work of Routh[498], where for the first time the linear stability of precessions of a Lagrange top was studied, while nonlinear stability is discussed in a book of Rumyantsev [511].

2 Reduced systems (in ψ and φ)

Suppose that the canonical equations of motion for the top are written in terms of the Euler angles. Since the angle of proper rotation φ is cyclic, we can determine the motion of the apex of the Lagrange top in the inertial frame (Fig. 2.10) by applying the classical Routh procedure. Eliminating the angle of precession ψ from the equations of motion leads to the Euler – Poisson equations. They govern the evolution of the unit vector directed vertically upward in the frame fixed in the top and are interesting from the physical point of view. In other words, this system governs the motion of the "apexes" of the absolute space in the noninertial frame fixed in the moving top. The trajectories of this absolute apex on the corresponding sphere (the standard Poisson sphere) are similar to the trajectories of the moving apex (of the axis of dynamical symmetry) in the fixed space (Figs. 2.10, 2.11, and 2.12). However, in general, for a particular motion the apexes differ in behavior, this difference shows itself in the presence or absence of loops (see Figs. 2.10 and 2.12). More precisely, the cases shown in Fig. 2.14 are possible. The picture of motion in the body frame presented in this chapter helps to better understand the behavior of the angle of proper rotation ϕ (a question that is usually ignored in mechanics courses).

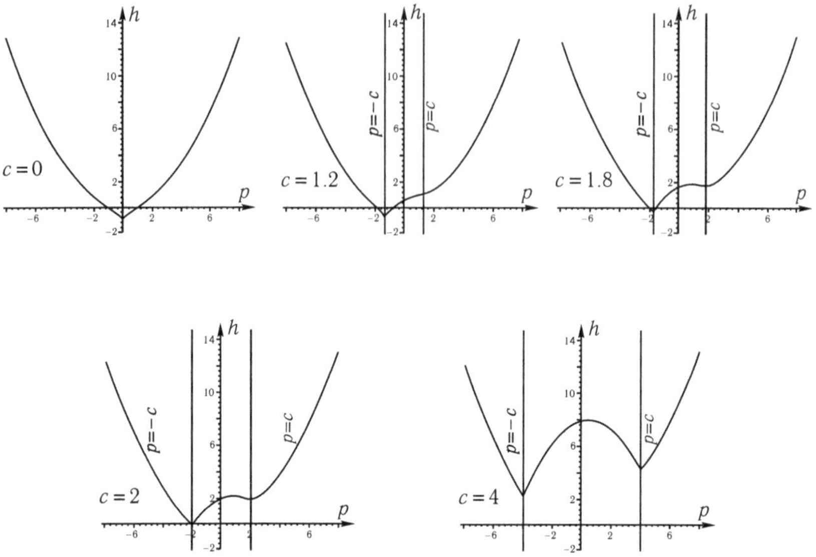

Fig. 2.15. Bifurcation diagram in the plane $(h,\ p)$ for various c $(I_1 = 1, I_3 = 5)$.

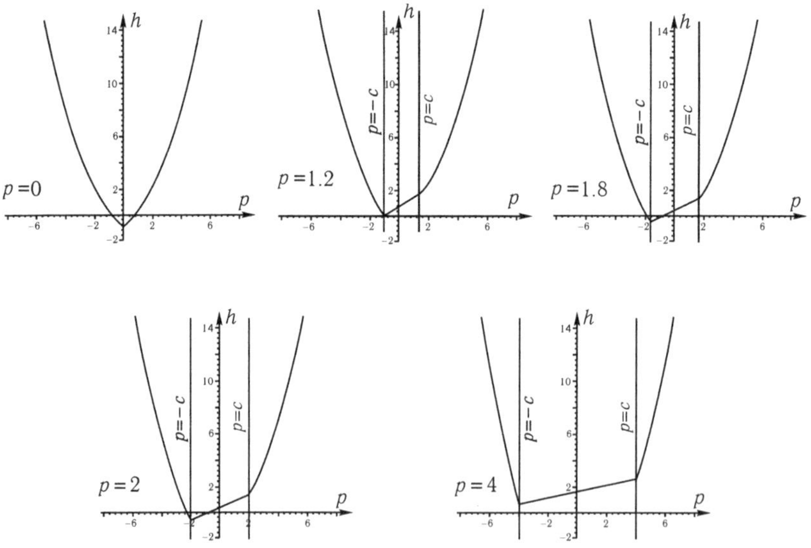

Fig. 2.16. Bifurcation diagram in the plane $(h,\ c)$ for various p $(I_1 = 1, I_3 = 5)$.

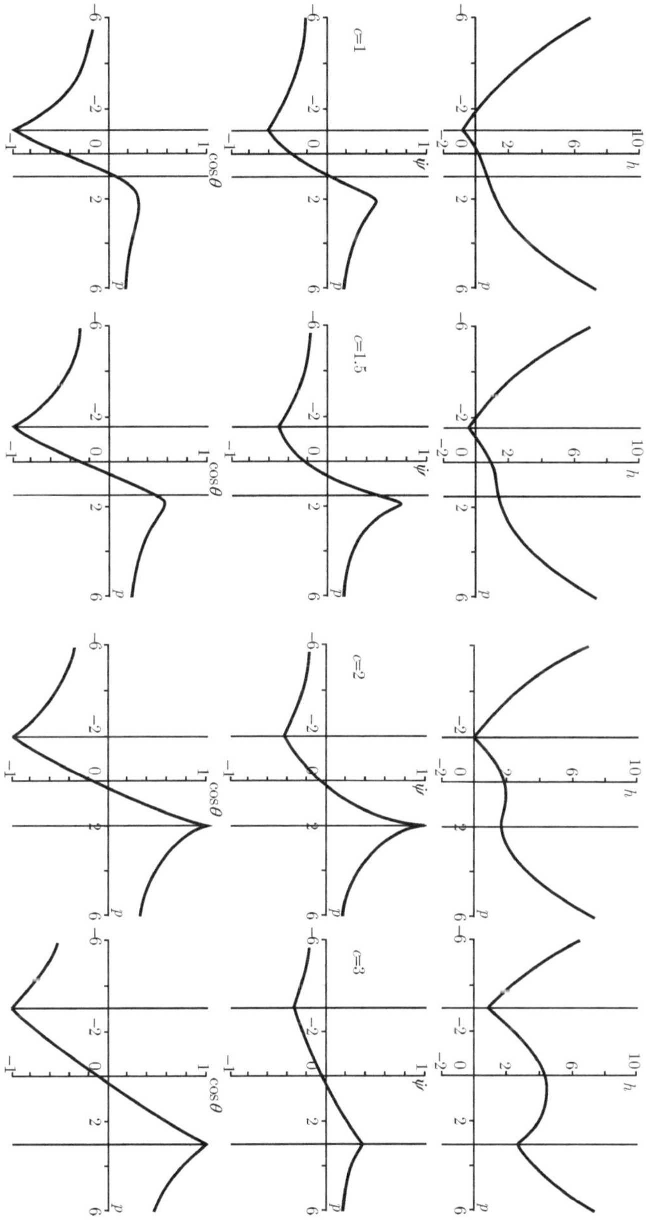

Fig. 2.17. The intersections of the surface of regular precessions (Fig. 1.7) and the planes $c = \text{const}$. The relationship between the angular velocity $\dot\psi(p)$, the latitude $\theta(p)$ of regular precessions and the value of the Lagrange integral p is represented. In the case of small friction, the constant p (the momentum of proper rotation of the top) diminishes. The diagram shows that during its initial motion the top approaches its vertical position $\cos\theta \to 1$, and if $c \geqslant 2$, the top becomes vertical for $p = c$ (the top "falls asleep"). As p decreases further, the direction of the precession (the sign of $\dot\psi$) is changed, this can be observed experimentally.

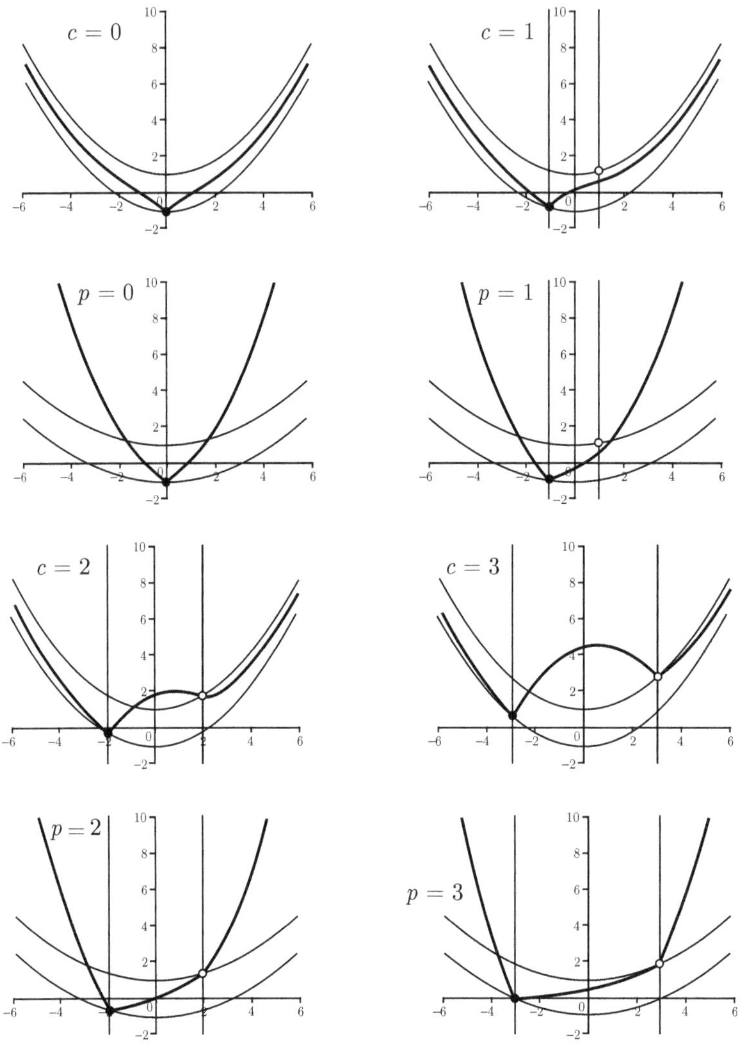

Fig. 2.18.

There are two very simple motions of the Lagrange top, namely, spin about the vertical with the center of mass above or below the fixed point. For brevity, these will be referred to as *upper motion* and *lower motion*. A top executing an upper motion is traditionally called a *sleeping top*. In the space of the first integrals these motions are governed by the equations: $p = c$ and $h = \frac{1}{2}ap^2 + 1$ for upper motion, $p = -c$ and $h = \frac{1}{2}ap^2 - 1$ for lower motion. Various sections of the domain of possible motion (DPM) and the corresponding partial solutions are shown in Fig. 2.18. The lower motion is seen to lie on the boundary of the DPM and therefore is always stable. The upper motion lies inside the DPM and for $p = c < 2$ it can be shown to be unstable [410], whilst for $p = c \geqslant 2$ the upper motion reaches the boundary and thus becomes stable. This constitutes the well-known *Maievsky criterion* for stability of a sleeping top. In Western Europe and the USA the name of Routh is attached to this criterion [498]. A rigorous analysis of the stability of a sleeping top was performed by Chetaev[135] (see also [136]).

To justify the noncompleteness of the standard analysis, the well-known *Hertz picture* [257] is useful. He proposed treating the angle of proper rotation as a nonobserved quantity and associating the corresponding cyclic motions with "hidden masses and parameters" leading to changes in the potential energy. For an observer in a noninertial frame of reference these hidden masses and parameters must be attributed to the inertial frame.

3 Bi-Hamiltonian property

The Lagrange case is characterized by an additional kind of symmetry: there exists another compatible Poisson structure (i. e., the system is bi-Hamiltonian). Indeed, the equations of motion can equally well be obtained if we define the Hamiltonian function as

$$H = (a-1)M_3\left(\frac{1}{2}(M_1^2 + M_2^2) + \gamma_3\right) + (M_1\gamma_1 + M_2\gamma_2 + M_3\gamma_3), \qquad (2.25)$$

$$a = \text{const}$$

and the Poisson structure as

$$\{\gamma_i, \gamma_j\} = -\varepsilon_{ijk}\gamma_k, \quad \{M_1, M_2\} = 1, \quad \{M_i, \gamma_j\} = 0, \qquad (2.26)$$

with two Casimir functions $F_1 = M_3$, $F_2 = (\boldsymbol{\gamma}, \boldsymbol{\gamma})$.

The Hamiltonian equations with the bracket (2.26) and the Hamiltonian (2.25) can be useful in the study of the Lagrange top with a generalized potential or dissipative perturbations. Another Poisson structure is found in [431], i.e., the problem of the Lagrange top is even tri-Hamiltonian.

4 Historical comments

Lagrange presented his case of integrability together with a general scheme for its integration in volume II of his *Analytical mechanics*. Somewhat later (1815) this problem was also solved by Poisson. He equipped Lagrange's analytical results with diagrams of the apex motion (similar to Figs. 2.10 – 2.12). Since then these diagrams have been reproduced in all textbooks on mechanics.

After the theory of Abelian functions had been developed and the case of Euler had been integrated in terms of these functions, Jacobi attempted to obtain similar quadratures for the Lagrange top. However, his work remained incomplete. Various forms of the general solution (i. e., expressions for the angular velocities and all the direction cosines or the Euler angles) in terms of theta functions can be found in the books of F. Klein and A. Sommerfeld [305], E. Whittaker [611], A. S. Domogarov [163], and W. D. Macmillan [400]. Most likely A. G. Greenhill [242] was the first to obtain the general solution. The quadratures he found are very complicated and have no practical use.

Jacobi also tried to obtain a complete geometric interpretation of the motion similar to the Poinsot construction for the Euler case. He formulated without proof a statement that the motion of the Lagrange top can be decomposed into two motions of Poinsot type: "forward motion and reverse motion". This statement was proved by E. Lottner in 1882, who published a posthumous edition of Jacobi's works. We will not discuss this result or improvements to it suggested by Darboux, Halphen, and Hess because of their excessive complexity and their artificial nature [400, 562]. These constructions can give neither a clear picture of motion nor analytical expressions.

The most complete and physically clear description of the motion of the Lagrange top can be found in the books of K. Magnus [401] and R. Grammel [239]. Here we represent their arguments in a more invariant form than they used and supply them with 3D illustrations. In some sense, these 3D images clearly demonstrate that the classification of the various motions of an axisymmetric top is a complicated problem.

It should also be noted that some of the bifurcation curves represented in [25, 567] do not coincide with those given by the authors of this book. As for [567], this "discordance" is due to the brevity of exposition (an exhaustive analysis of motion was not the author's goal). At the same time, some curves in the book by M. Audin [25] seem to be incorrect. The case of Lagrange is also examined from an algebraic point of view in [480], where the Lax equation is also obtained. An interesting geometrical and topological analysis of the Lagrange case is performed by I. N. Gashenenko [209] in the context of the theory of envelope surfaces. Several multi-dimensional generalizations of the Lagrange case with corresponding **L**-**A**-pairs are presently known and they can be found, for example, in our book [87].

Remark. The motion of a Lagrange top in the absence of gravity remains integrable when a particle moving along the axis of dynamical symmetry is added; the particle is subject to a potential force whose potential $V(z)$ is a function of the z-coordinate measured along this axis. In this case, the Lagrangian function reads

$$L = \frac{1}{2}\left[(I_1 + mz^2)(\dot\theta^2 + \sin^2\theta\dot\psi^2) + I_3(\dot\psi\cos\theta + \dot\varphi)^2 + \dot z^2\right] - V(z),$$

where m is the mass of the particle and θ, φ, ψ are the Euler angles. It is easy to see that the Lagrangian is separable [632].

2.4 The Kovalevskaya Case

The additional integrals in the cases of Euler and Lagrange have a natural physical origin, that is, the square of the absolute value of the angular momentum (Euler case) and its projection on the axis of dynamical symmetry (Lagrange case). For the integrability case found by S. V. Kovalevskaya (1888) the additional integral has no obvious symmetric interpretation. This case was found more than a century after either of the two previous cases and is much more complicated both for its explicit integration and for a qualitative analysis of the motion.

In the Kovalevskaya case the body possesses dynamical symmetry $a_1 = a_2$ and the center of mass lies in the equatorial plane of the ellipsoid of inertia $r_3 = 0$. In addition, the relations $\frac{a_3}{a_1} = \frac{I_1}{I_3} = 2$ hold. The Hamiltonian and the additional integrals found by Kovalevskaya are

$$
\begin{aligned}
H &= \frac{1}{2}(M_1^2 + M_2^2 + 2M_3^2) - x\gamma_1,\\
F_3 &= \left(\frac{M_1^2 - M_2^2}{2} + x\gamma_1\right)^2 + (M_1M_2 + x\gamma_2)^2 = k^2,
\end{aligned}
\tag{2.27}
$$

where the coordinates have been chosen without loss of generality in such a way that the position vector of the center of mass has components $\boldsymbol{r} = (x, 0, 0)$ and the weight of the body is assumed to be 1.

1 Explicit integration. The Kovalevskaya variables

Having found an additional integral, S. V. Kovalevskaya went on to find a remarkable change of variables transforming the equations of motion (2.1) to the canonical Abel–Jacobi form (see Sec. 1.7, Ch. 1). The transformed equations can be integrated in terms of theta functions (of two variables) according to a general scheme (see [329]). Here we present only the change of variables.

The Kovalevskaya variables s_1 and s_2 are defined by

$$
\begin{gathered}
s_1 = \frac{R - \sqrt{R_1R_2}}{2(z_1 - z_2)^2}, \qquad s_2 = \frac{R + \sqrt{R_1R_2}}{2(z_1 - z_2)^2},\\
z_1 = M_1 + iM_2, \qquad z_2 = M_1 - iM_2,\\
R = R(z_1, z_2) = \frac{1}{4}z_1^2z_2^2 - \frac{h}{2}(z_1^2 + z_2^2) + c(z_1 + z_2) + \frac{k^2}{4} - 1,\\
R_1 = R(z_1, z_1), \qquad R_2 = R(z_2, z_2),
\end{gathered}
\tag{2.28}
$$

where $F_1 = (M, \gamma) = c$ and $H = h$. To simplify calculations, we always assume that $x = 1$.

The equations of motion written in terms of the new variables are

$$
\frac{ds_1}{\sqrt{P(s_1)}} = \frac{dt}{s_1 - s_2}, \qquad \frac{ds_2}{\sqrt{P(s_2)}} = \frac{dt}{s_2 - s_1},
\tag{2.29}
$$

where

$$P(s) = \left(\left(2s + \frac{h}{2}\right)^2 - \frac{k^2}{16}\right)\left(4s^3 + 2hs^2 + \left(\frac{h^2}{4} - \frac{k^2}{16} + \frac{1}{4}\right)s + \frac{c^2}{16}\right).$$

Since the polynomial $P(s)$ is of degree five, the quadrature corresponding to (2.29) is called *ultraelliptic*, or alternatively *hyperelliptic*.

2 Bifurcation diagram and the Appelrot classes

The values of the integrals h, c, and k for which the polynomial $P(s)$ has multiple roots define in the space of these integrals a *bifurcation diagram*, i. e., a set of two-dimensional surfaces on which the motion changes type (Fig. 2.19). In this case ultraelliptic quadratures in (2.29) are reduced to elliptic quadratures, and trajectories of the corresponding motions (remarkable amongst all other motions) are called the *Appelrot classes* [14]. Different branches of the bifurcation diagram represent different Appelrot classes.

It can be shown that Appelrot classes determined by the multiplicities of the roots of the polynomial $P(s)$ coincide with the set of *singular integral manifolds* on which the integrals H, F_1, F_2, and F_3 are dependent, i. e., the rank of the Jacobi matrix $\left\|\frac{\partial(H, F_2, F_3, F_4)}{\partial(\boldsymbol{M}, \boldsymbol{\gamma})}\right\|$ is not maximal [286]. It is clear that these singular manifolds in the phase space of the reduced system (i. e., for the Euler – Poisson equations) determine stable and unstable periodic motions and also the asymptotic trajectories to these motions.

A bifurcation diagram with stability of the branches indicated is given in Fig. 2.19. Together with the Poincaré – section technique in the space of some canonical variables (for example, the Andoyer variables, Figs. 2.20 and 2.21), this diagram is very useful for the study of dynamics because it provides an illustrative representation of the qualitative behavior of the other solution curves of the system.

The topology of two-dimensional invariant manifolds separated by bifurcation curves is discussed by M. P. Kharlamov [286]. Bifurcations of these Liouville tori with the corresponding graphs and topological invariants are given in the book [66].

Explicit solutions for the Appelrot classes can be obtained straightforwardly, i. e., without resort to equations (2.29). The construction of these solutions, initiated by G. Appelrot [14] and found in their most complete form by A. I. Dokshevich [162], involves some highly nontrivial calculations. We will present some of these results mainly concerned with periodic and asymptotic motions which are the most important ones for dynamics. We will also try to clarify the mechanical meaning of these results.

There are four Appelrot classes.

I. The Delone solution [148]. For this case $k^2 = 0$ and $h > c^2$. Two invariant relations arise

$$\frac{M_1^2 - M_2^2}{2} + x\gamma_1 = 0, \quad M_1M_2 + x\gamma_2 = 0, \tag{2.30}$$

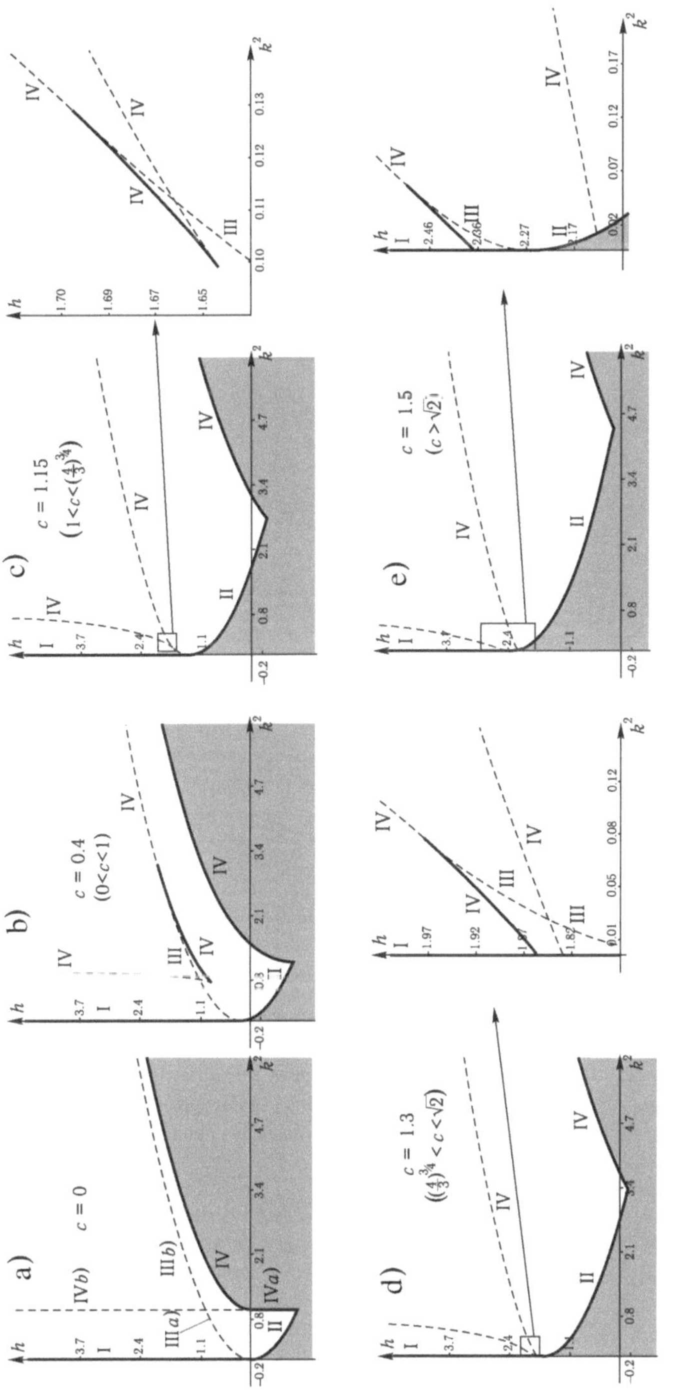

Fig. 2.19. Bifurcation diagrams in the Kovalevskaya case for various c. The Roman numbers indicate the Appelrot classes. The solid curves represent stable periodic solutions, the dashed lines correspond to unstable solutions and separatrices.

these relations determine a periodic solution of the Euler – Poisson equations.

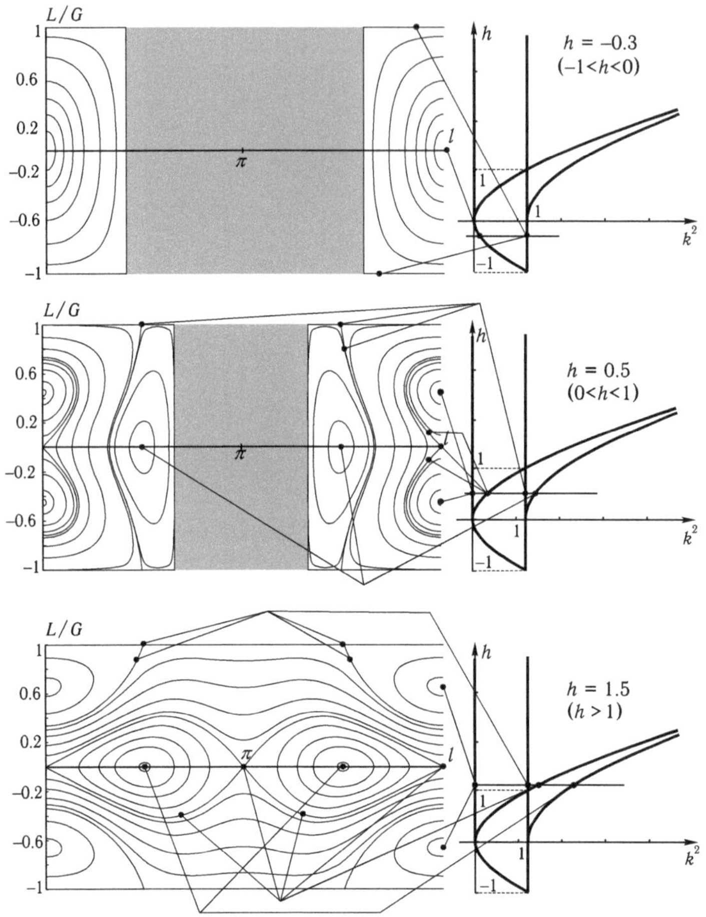

Fig. 2.20. The phase portrait (intersection with the plane $g = \pi/2$) for the Kovalevskaya case (the constant of the areas c is zero). Phase portraits of three different types are shown. It can be seen that the portrait structure changes and bifurcations of periodic solutions occur when crossing the critical levels of the energy $h = 0$ and $h = 1$. (The non-physical domain of the values l and L/G for fixed h and c is shown in gray.)

In this case for zero value of the area constant $c = 0$ the motion is periodic both for the reduced system (on the Poisson sphere) *and in the inertial frame* [227] (see Figs. 2.24 – 2.27).

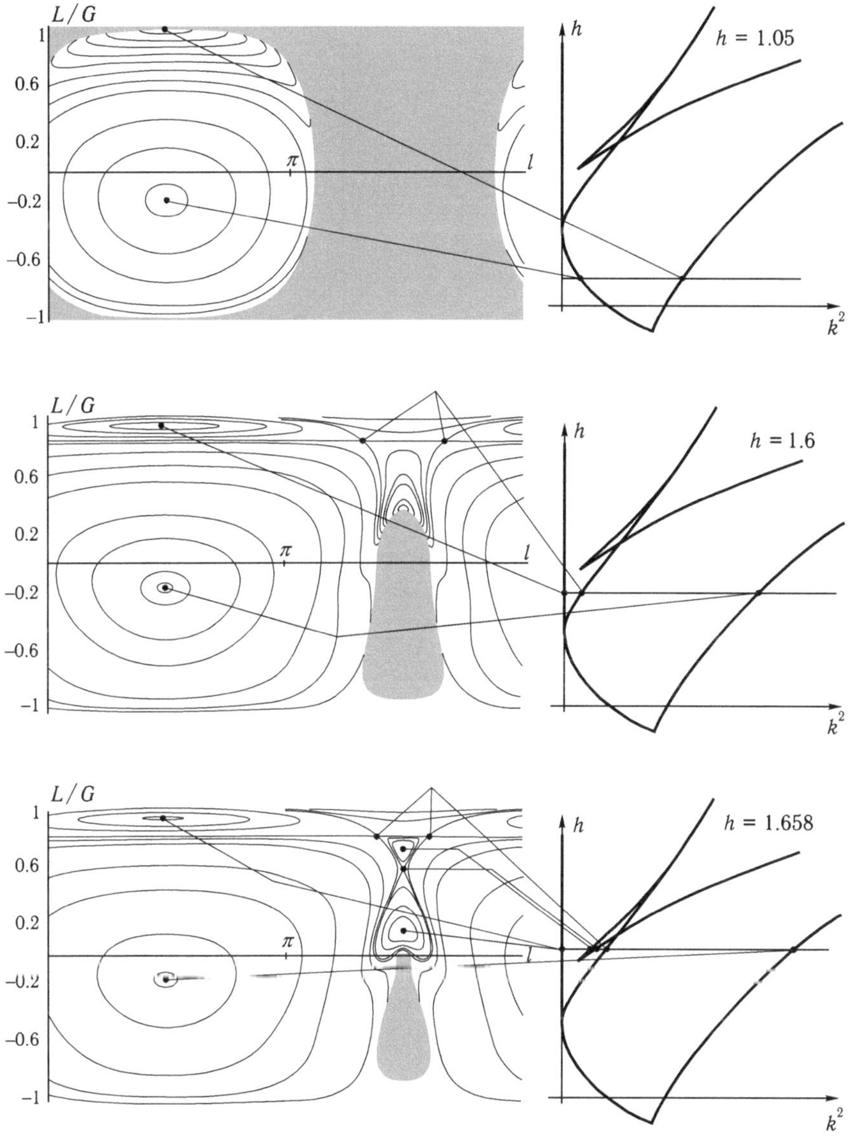

Fig. 2.21. Phase portrait (intersection with the plane $g = \pi$) in the Kovalevskaya case. Here $c = 1.15$ and the values of h are so chosen as to provide essentially different phase portraits. The coordinates l and L/G represent the sphere projected onto the surface of a cylinder. The phase portrait is symmetric relative to the meridian $l = \frac{\pi}{2}, \frac{3}{4}\pi$. (In the right figures, the bifurcation diagrams are shown only qualitatively but not quantitatively.)

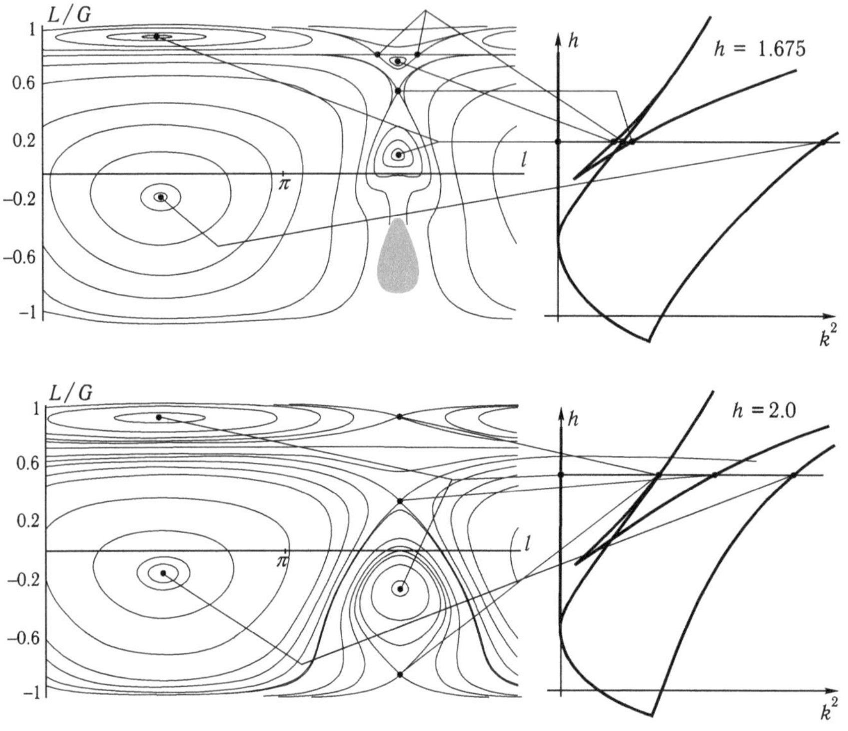

Fig. 2.21. Continued.

To obtain explicit formulae, we first express all the variables as functions of M_1 on a common level set of the integrals and the invariant relations (2.30)

$$\begin{gathered} M_2^2 = 2z - M_1^2, \quad M_3^2 = h - M_1^2, \\ xy_1 = -M_1^2 + z, \quad xy_2 = -M_1(2z - M_1)^{1/2}, \quad xy_3 = (x^2 - z^2)^{1/2}, \\ z = \frac{M_1^2 + M_2^2}{2} = (y_1^2 + y_2^2)^{1/2} = x\frac{-cM_1 \pm \sqrt{(h - c^2)(h - M_1^2)}}{h}. \end{gathered} \tag{2.31}$$

We get the equation for M_1:

$$\dot{M}_1 = M_2 M_3 = \left((h - M_1^2)(2z - M_1^2)\right)^{1/2}. \tag{2.32}$$

For $h = c^2$ this equation can be solved in elliptic functions. For $c = 0$ a simpler explicit solution can be obtained if M_3 is used instead of M_1.

Figure 2.19 shows that as soon as $c = \left(\frac{4}{3}\right)^{3/4}$, the branch of the Appelrot class IV "runs into" the Delone solution; if c increases further so that $c^2 < 2$, the branch divides the solution into three parts. For $c^2 = 2$ the branches of the four Appelrot classes merge at the point $h = 2$, $k^2 = 0$. To this point there corresponds an unstable fixed point on the Poisson sphere (the *Staude rotation*) (see Sec. 2.6, Ch. 2) and a one-dimensional asymptotic motion tending to this point. Using (2.32), for this motion we get

$$M_1 = \sqrt{2x}\,\frac{3+\cosh^2 u \pm 4\cosh u}{9-\cosh^2 u}, \qquad u = 2\sqrt{x}t. \tag{2.33}$$

For $c^2 > 2$ a branch of class IV also runs into the Delone solution, and the other branch intersects the segment of the parabola of class II.

II. The solutions of the second class lie on the lower branch of the parabola $(h-c^2)^2 = k^2$ with $\frac{1}{2}c^2 - 1 \leqslant h \leqslant c^2$. For $c = 0$ this class contains stable periodic trajectories, and the body executes planar oscillations in the meridional plane through the center of mass, moreover, $M_1 = M_3 = 0$ and $\gamma_2 = 0$.

For $c \neq 0$ there are additional invariant relations

$$M_3 = c\gamma_3, \quad M_1^2 + M_2^2 + \frac{M_1}{c} = k \tag{2.34}$$

and an explicit integration is done in [162]. For $c > \sqrt{2}$ the branches of classes II and IV intersect.

III. This class is represented by the branch of the parabola above the point at which it touches the axis $k^2 = 0$. The branch is defined by the equations

$$(h-c^2)^2 = k^2, \quad c^2 \leqslant h \leqslant c^2 + \frac{1}{2c^2}. \tag{2.35}$$

For $c = 0$ these relations define the entire upper branch of the parabola, for $c \neq 0$ this branch is bounded from above by one of the branches of class IV.

Physically, class III corresponds to unstable periodic solutions and solutions tending asymptotically to them. For $c = 0$ periodic motions of branch III for which $-1 < h < 1$ are oscillations of a physical pendulum in the meridional plane through the center of mass, and those for which $h > 1$ are rotations in this plane. These solutions meet at the point $h = 1$ which is the upper unstable equilibrium. The instability can be proved rigorously in various ways [502]. Our proof is based on an explicit determination of the asymptotic solution.

A common level surface of the integrals of motion corresponding to the third Appelrot class for zero value of the area constant $c = 0$ [162] can be parametrized as follows

$$\begin{aligned} M_1 &= \sqrt{M_1^2 + M_3^2}\sin\varphi, \quad & M_3 &= \sqrt{M_1^2 + M_3^2}\cos\varphi, \\ k_1 &= k\cos 2\theta, & k_2 &= k\sin 2\theta, \end{aligned} \tag{2.36}$$

where $k_1 = \gamma_1 + \frac{M_1^2 - M_2^2}{2}$ and $k_2 = \gamma_2 + M_1M_2$ (for $x = 1$). The Kovalevskaya integral is $k_1^2 + k_2^2 = k^2$.

Differentiating φ with respect to time gives

$$\dot{\varphi} = M_2 - \frac{M_1 k_2}{M_1^2 + M_3^2}. \tag{2.37}$$

Upon differentiation of (2.37) and elimination of M_2 with the help of (2.37) and the relation $h = k > 0$, we get

$$2\ddot{\varphi}\cos\varphi + \dot{\varphi}\sin\varphi = 2h\cos^2\varphi\sin\varphi. \tag{2.38}$$

Multiplying (2.38) by $\frac{\dot{\varphi}}{\cos^2\varphi}$ and integrating over time yields

$$\frac{\dot{\varphi}^2}{\cos\varphi} + 2h\cos\varphi = c_1 = \text{const.}$$

The constant of integration can be found from the condition $\varphi = 0$. This condition also implies that $M_1 = 0$ and $\dot{\varphi} = M_2$, hence $c_1^2 = 4x^2$. Thus,

$$\dot{\varphi}^2 = 2(x - k\cos\varphi)\cos\varphi, \quad k > 0. \tag{2.39}$$

REMARK. For $c \neq 0$ there exists a similar (but different) angular variable governed by the equation [162]

$$\dot{\varphi}^2 = 2(x - (k + c^2)\cos\varphi)\cos\varphi.$$

We obtain the equation for θ

$$\dot{\theta} = -M_3 = -\sqrt{M_1^2 + M_3^2}\cos\varphi.$$

Using the integral of energy $M_1^2 + M_3^2 - k_1 = h$ and the condition $h = k$ (which lead to the relation $\sqrt{M_1^2 + M_3^2} = \pm\sqrt{2k}\cos\theta$), we get

$$\dot{\theta} = \sqrt{2k}\cos\varphi\cos\theta.$$

Upon the change of variables $\cos\theta = (\cosh u)^{-1}$, we get

$$\dot{u} = \sqrt{2k}\cos\varphi.$$

The complete system of equations determining the asymptotic trajectories of the Appelrot class III for $c = 0$ and $h = k > 0$ is

$$\begin{aligned} 2\dot{\zeta} &= (1 - \zeta^2)(x - k + (x + k)\zeta^2), \quad \zeta = \tan\frac{\varphi}{2}, \\ \dot{u} &= \sqrt{2k}\cos\varphi, \quad \cosh u = (\cos\theta)^{-1}. \end{aligned} \tag{2.40}$$

The solutions are

$$\begin{aligned} &1.\quad k < x, \quad \zeta = \operatorname{cn}(\sqrt{x}t,\ k_0), \quad && k_0^2 = \frac{x + k}{2x}; \\ &2.\quad k > x, \quad \zeta = \operatorname{dn}\left(\sqrt{\frac{x + k}{2}}t,\ k_0\right), \quad && k_0^2 = \frac{2x}{x + k}; \\ &3.\quad k = x, \quad \zeta = (\cosh\sqrt{x}t)^{-1}, \end{aligned}$$

where k_0 is the modulus of the corresponding elliptic Jacobi functions.

Using I – III it can be shown that $\dot{u}$ does not change sign, i. e., these solutions correspond to asymptotic motions towards the periodic solution in cases I and II and to a fixed point in case III. (Analytical expressions for the case $c \neq 0$ are more cumbersome [162].)

REMARK. In [284] the invariant relations for the second and third Appelrot classes are presented in a slightly different algebraic form compared to (2.8)

$$M_3 = (\boldsymbol{M}, \boldsymbol{\gamma})\gamma_3, \quad 2\gamma_1\gamma_2 k_1 - (\gamma_1^2 - \gamma_2^2)k_2 = 0.$$

IV. This class consists of two branches (Fig. 2.19). One of them corresponds to stable periodic motions and the other to unstable periodic motions and to separatrices. For $c = 0$ both branches meet at the point $k^2 = x^2 = 1$, $h = 0$.

For $c \neq 0$ the parametric equations of the branches are as follows

$$k^2 = 1 + tc + \frac{t^4}{4}, \qquad h = \frac{t^2}{2} - \frac{c}{t}, \tag{2.41}$$

$$\begin{aligned} &t \in (-\infty, 0) \cup (c, +\infty), && \text{for} \quad c > 0, \\ &t \in (-\infty, +\infty) \setminus \{0\}, && \text{for} \quad c < 0, \end{aligned}$$

and for $c = 0$

1. $k^2 = x^2, \quad h < 0, \quad h^2 = k^2 + x^2$ (branch IVa);
2. $k^2 = x^2, \quad h > 0$ (branch IVb).

Stable and unstable periodic solutions for the Appelrot class IV in the Kovalevskaya case (and also in a more general case when the inertia tensor is $\mathbf{I} = \mathrm{diag}(1, a, 2)$, $a = \mathrm{const}$, and the solution itself does not depend on a) were obtained by D. K. Bobylev [53] and V. A. Steklov [559] (see also Sec. 2.6).

The Bobylev – Steklov solution. For this solution the relations

$$M_2 = 0, \quad M_1 = m = \mathrm{const}$$

always hold. Then we can express $\boldsymbol{\gamma}$ via M_3

$$\gamma_1 = \frac{c}{m} - M_3^2, \quad \gamma_2 = \left(k^2 - \left(\frac{1}{2}m^2 - \frac{c}{m} + M_3^2\right)^2\right)^{1/2}, \quad \gamma_3 = mM_3$$

and obtain an elliptic quadrature for M_3

$$\dot{M}_3 = -\left(k^2 - \left(\frac{1}{2}m^2 - \frac{c}{m} + M_3^2\right)^2\right)^{1/2}. \tag{2.42}$$

Here h and k^2 are determined by the parametric equations

$$h = \frac{1}{2}m^2 - \frac{c}{m}, \quad k^2 = 1 + \frac{1}{2}m^4 + cm$$

and therefore coincide with (2.41). For $c = 0$ in the fourth class there arise motions corresponding to oscillations and rotations obeying the law of a physical pendulum in the equatorial plane of the ellipsoid of inertia. For these solutions

$$M_1 = m = 0, \quad \gamma_3 = 0, \quad \dot{M}_3 = -(1 - (h - M_3^2)^2)^{1/2}.$$

The asymptotic solutions for an arbitrary $c \neq 0$ can be found in [162], but the formulae describing them are very bulky. We will calculate these solutions under additional assumptions

$$k^2 = \varkappa^2, \quad h > 0, \quad c = 0. \tag{2.43}$$

We use an interesting involutive transform $(\boldsymbol{M}, \boldsymbol{\gamma}) \mapsto (\boldsymbol{L}, \boldsymbol{s})$ (its square is the identity map) found by A. I. Dokshevich

$$\begin{aligned} L_1 &= -\frac{M_1}{M_1^2 + M_2^2}, & s_1 &= -\gamma_1 + 2\varkappa\gamma_3^2 \frac{M_1^2 - M_2^2}{(M_1^2 + M_2^2)^2}, \\ L_2 &= -\frac{M_2}{M_1^2 + M_2^2}, & s_2 &= -\gamma_2 + 4\varkappa\gamma_3^2 \frac{M_1 M_2}{(M_1^2 + M_2^2)^2}, \\ L_3 &= M_3 + 2\varkappa\gamma_3 \frac{M_1}{M_1^2 + M_2^2}, & s_3 &= \frac{\gamma_3}{M_1^2 + M_2^2}. \end{aligned} \tag{2.44}$$

Writing the equations of motion in terms of the new variables $(\boldsymbol{L}, \boldsymbol{s})$, we get

$$\begin{aligned} \dot{L}_1 &= L_2 L_3, & \dot{s}_1 &= 2L_3 s_2 - 4(k^2 - \varkappa^2) s_3 L_2, \\ \dot{L}_2 &= -L_1 L_3 - \varkappa s_3, & \dot{s}_2 &= -2L_3 s_1 + 4(k^2 - \varkappa^2) s_1 L_3, \\ \dot{L}_3 &= -2\varkappa c L_2 + \varkappa s_2, & \dot{s}_3 &= s_1 L_2 - s_2 L_1. \end{aligned} \tag{2.45}$$

Under the assumption (2.43) the equations for L_3, s_1, and s_2 decouple and can be solved straightforwardly

$$\begin{aligned} s_2 &= (1 - s_1^2)^{1/2}, \quad L_3 = (h + \varkappa s_1)^{1/2}, \\ \dot{s}_1 &= 2\sqrt{(h + \varkappa s_1)(1 - s_1^2)}. \end{aligned} \tag{2.46}$$

To solve the whole system (2.45), it is sufficient to solve the following linear equation of the second order with time-dependent coefficients

$$\begin{aligned} L_1 = s_1^{-1}\left(-L_3 s_3 \mp s_2 \sqrt{h s_3^2 - \frac{1}{4\varkappa} s_1} \right), \quad L_2 = \sqrt{h s_3^2 - \frac{1}{4\varkappa} s_1}, \\ \ddot{s}_3 = -\varkappa(1 + 2s_1) s_3. \end{aligned} \tag{2.47}$$

Equations (2.46) and (2.47) govern solutions tending asymptotically to the periodic motions under the assumptions (2.43) (see Fig. 2.33).

In the case $h = \varkappa$ (the energy of the upper unstable equilibrium) we obtain another solution (in addition to class III) expressed in terms of the elementary functions

$$s_1 = 1 - 2\tanh u, \quad s_2 = 2\frac{\tanh u}{\cosh u}, \quad L_3 = -\frac{\sqrt{2\varkappa}}{\cosh u}, \quad u = \sqrt{2\varkappa}t.$$

This solution tends asymptotically to the upper unstable equilibrium.

The Appelrot classes determine the simplest motions both in the reduced and in the absolute phase space. Other motions of the Kovalevskaya top have a quasi-periodic nature and depend on the corresponding domain of the bifurcation diagram.

For perturbations of the Kovalevskaya case, a stochastic layer appears near unstable solutions and their separatrices (Fig. 2.53). Unfortunately, the (asymptotic) solutions given above have still not been effective for the analytical study of the nonintegrability of the perturbed Kovalevskaya top (for $c = 0$ proof of the nonintegrability by means of variational methods is obtained in [64]).

3 Phase portrait and visualization of the most remarkable solutions

For any fixed value of the area constant $(\boldsymbol{M}, \boldsymbol{\gamma}) = c$ determining one of the various types of the bifurcation diagrams on the plane (k^2, h) there exists a set of different phase portraits. We recall that a phase portrait corresponds to a fixed level of the energy, h, and different types of portraits arise depending on a possible location of the straight line $h = \text{const}$ on the bifurcation diagram. Here we consider two series of phase portraits corresponding to the simplest ($c = 0$, Fig. 2.20) and the most complicated ($1 < c < \left(\frac{4}{3}\right)^{3/4}$, Fig. 2.21) bifurcation diagrams. We also illustrate some of the most remarkable solutions on the Poisson sphere and in the inertial frame.

Remark. In terms of the Euler variables $(\psi, \theta, \varphi, p_\psi, p_\theta, p_\varphi)$ the topology of invariant tori is studied in [107, 175] using the technique of Poincaré sections. In these works, questions concerned with how to introduce action-angle variables depending on the values of the first integrals are considered.

Phase portrait for $c = 0$. In this case the bifurcation diagram consists of two parabola pieces and two straight lines (see Fig. 2.19 a)). The physical meaning of the branches corresponding to the parabola $h^2 = k^2$ and to the straight line $k^2 = 1$ is obvious and was discussed above. The parabola contains solutions describing planar oscillations and rotations of the rigid body in the meridional plane. The rotations are executed around the axis Oy perpendicular to the axis Ox on which the center of mass lies. The straight line contains planar oscillations and rotations in the equatorial plane (around Oz). The branches $k^2 = 0$ and $h^2 = k^2 - 1$ contain the Delone and the Bobylev – Steklov solutions, respectively.

Phase portraits with indications of their locations in the bifurcation diagram were given above. Figure 2.19 a) shows that for the energy h there exist three intervals: $(-1, 0)$, $(0, 1)$, and $(1, \infty)$, each of them yielding a phase portrait of different type (Fig. 2.20).

Phase portrait for $c = 1.15$ $\left(1 < c < \left(\frac{4}{3}\right)^{3/4}\right)$. Using the bifurcation diagram (Fig. 2.19 c)) it can be shown that there are five intervals of the energy each of which corresponds to a different type of phase portrait (see Fig. 2.21). In this case the periodic solutions corresponding to the branches of the bifurcation diagram are not of such a simple form as those for $c = 0$ but are close to them when $h \gg c$.

Remark. To construct phase portraits, we use the Poincaré sections based on the Andoyer variables. These variables were introduced in Sec. 1.3, Ch. 1. For $c = 0$ the

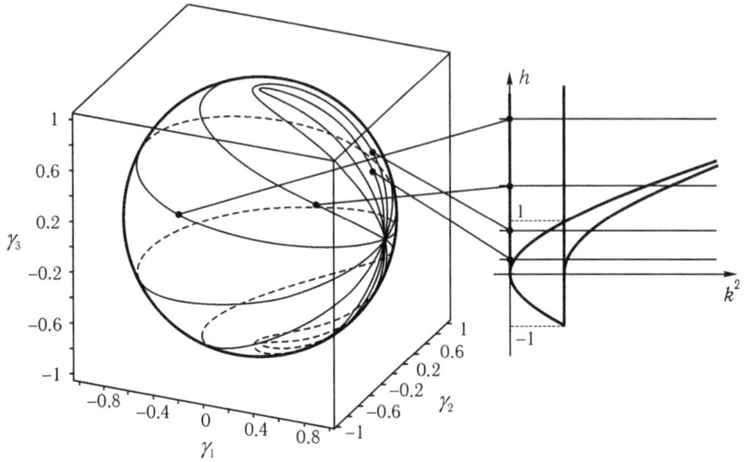

Fig. 2.22. The Delone solution. The motion of the unit vector $\boldsymbol{\gamma}$ at zero value of the area constant ($c = 0$) and for various values of the energy.

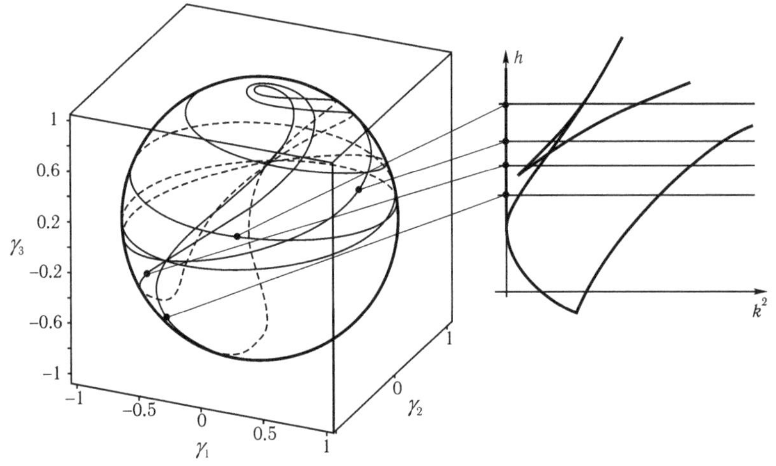

Fig. 2.23. The Delone solution. The motion of the unit vector $\boldsymbol{\gamma}$ at a nonzero value of the area constant ($c = 1.15$) and for various values of the energy h.

cross-section is $g = \frac{\pi}{2}$. For $c = 1.15$ we take $g = \pi$, since in this case not all periodic solutions intersect the plane $g = \frac{\pi}{2}$. The phase portraits have different types of symmetry on the sphere $(l, L/G)$. For example, for $g = \frac{\pi}{2}$ the portrait is symmetric with respect to the equator (the axis $L/G = 0$), and for $g = \pi$ it is symmetric with respect to the meridional plane ($l = \frac{\pi}{2}, \frac{3}{2}\pi$).

We proceed to the visualization of some of the most interesting motions of a rigid body in the reduced and in the absolute space.

The Delone solution ($k^2 = 0$). In this case the unit vertical vector $\boldsymbol{\gamma}$ traces out on the Poisson sphere a family of figure eight curves (see Figs. 2.22 and 2.23). For $c = 0$ (Fig. 2.22) the points of self-intersection of these curves coincide and this common point has coordinates $\boldsymbol{\gamma} = (1, 0, 0)$. This point determines the lowest position of the center of mass of the body. When c is increased, some irregular figure eight curves appear on the Poisson sphere, all of them are intersected at two points on the equator of the Poisson sphere (Fig. 2.23).

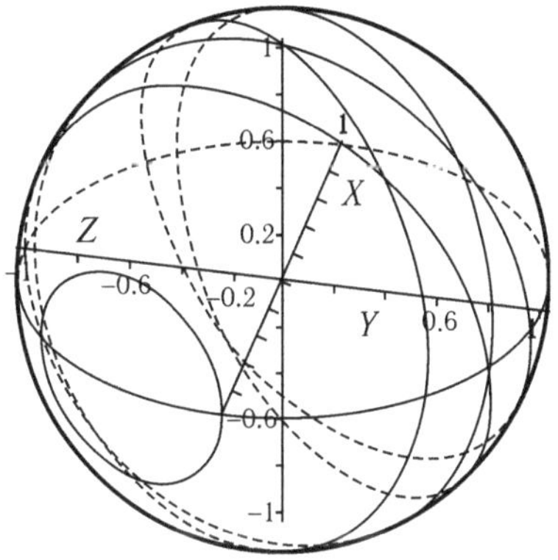

Fig. 2.24. The Delone solution. The motion of the apexes in the fixed frame for the zero value of the area constant ($c = 0$).

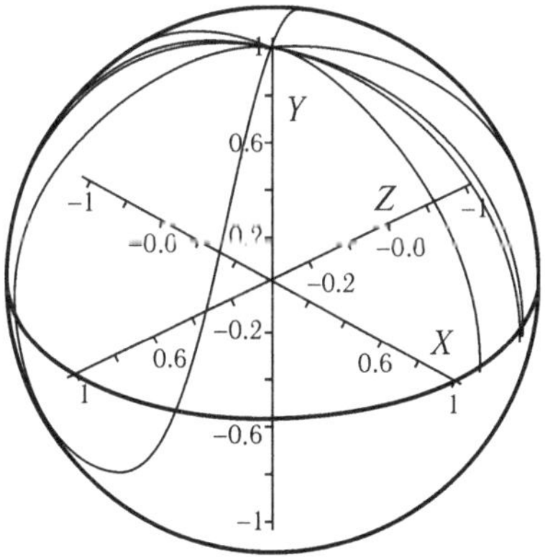

Fig. 2.25. The Delone solution. The motion of the apex of the center of mass for $c = 0$ and for various h.

It is known that for $c = 0$ the Delone solution determines periodic motions both in the reduced space and in the absolute space [228]. This is not the case for $c \neq 0$, the motion of the body in the absolute space is quasi-periodic. Figures 2.24 – 2.27 show the trajectories of the three apexes of the rigid body for $c = 0$ and for various values of the energy. The directions of the fixed axes $OXYZ$ have been chosen so as to make the picture as clear as possible.

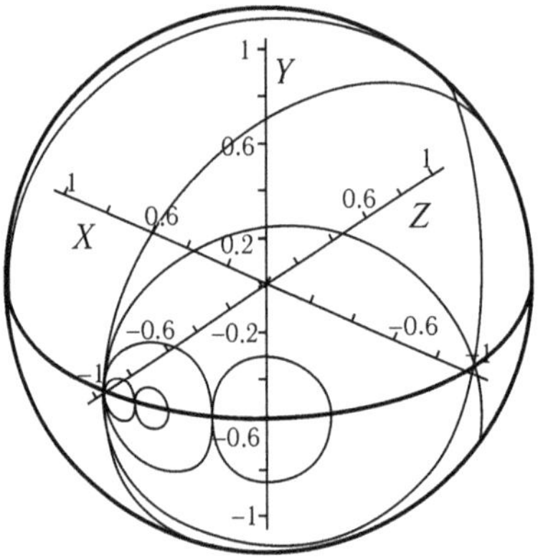

Fig. 2.26. The Delone solution. The motion of the apex lying in the equatorial plane perpendicular to the position vector of the center of mass for $c = 0$ and various h.

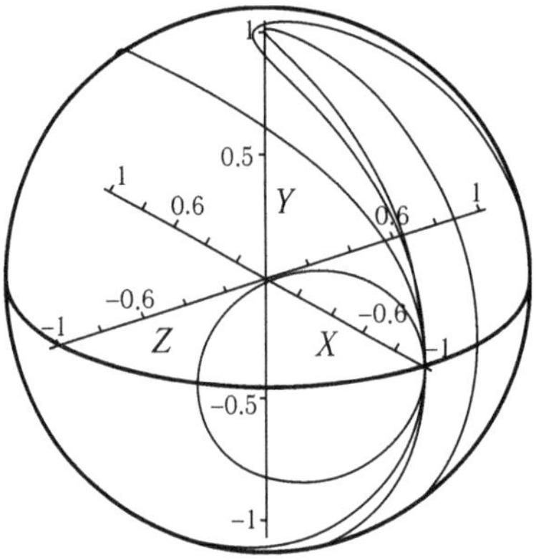

Fig. 2.27. The Delone solution. The motion of the apex of the axis of dynamical symmetry for $c = 0$ and for various h.

The Bobylev – Steklov solution. On the bifurcation diagram (see Fig. 2.19) the Bobylev – Steklov solution is situated on the lower right branch. This solution corresponds to a stable periodic solution on the Poisson sphere. (Figs. 2.28 and 2.29).

It is clearly seen from Fig. 2.28 that for $c = 0$ all trajectories on the Poisson the points of the equator $(0, 1, 0)$ and $(0, -1, 0)$ and do not cross the meridional plane $y_1 = 0$. In this case the center of mass executes a remarkable motion in the absolute space, *tracing out a curve with cusps lying on the equator for all values of the energy* (see Fig. 2.30).

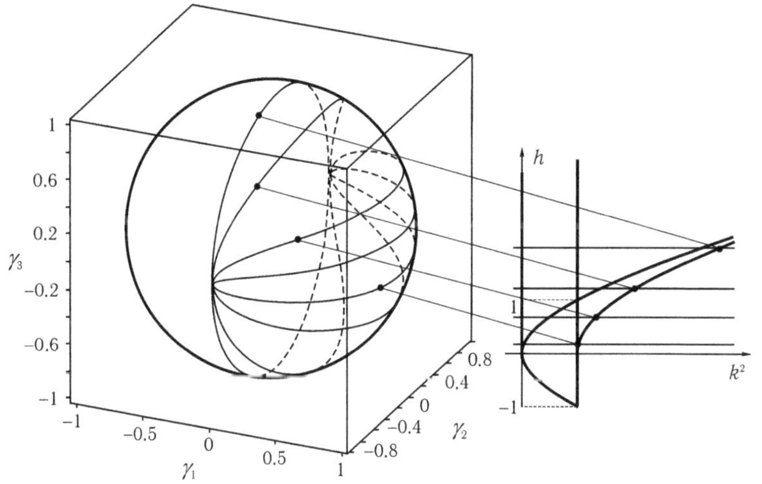

Fig. 2.28. The Bobylev – Steklov solution. The motion of the unit vertical vector on the Poisson sphere for $c = 0$ and various values of the energy.

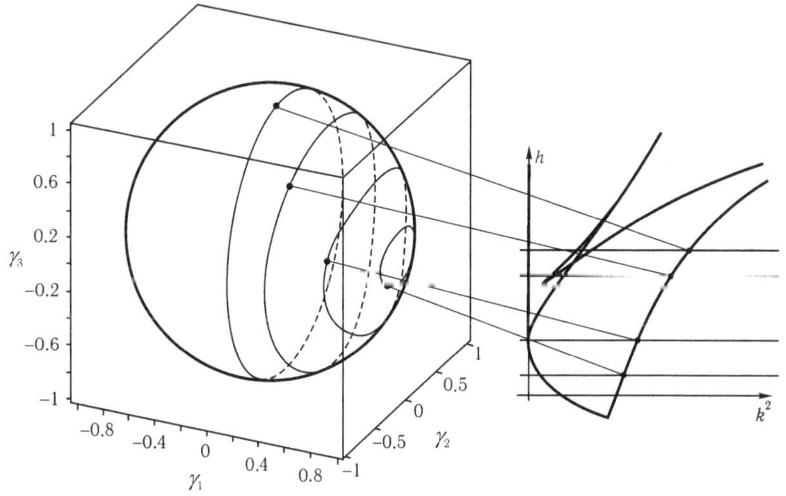

Fig. 2.29. The Bobylev – Steklov solution. The motion of the unit vertical vector on the Poisson sphere for $c \neq 0$ ($c = 1.15$) and for various values of the energy.

For $c \neq 0$ the trajectories on the Poisson sphere are shown in Fig. 2.29. In this case the apex of the center of mass moves in the fixed space along curves with cusps belonging to one latitude. This latitude depends on the constant value of the energy h (see Fig. 2.31). Physically, the Bobylev – Steklov solution can be obtained as follows:

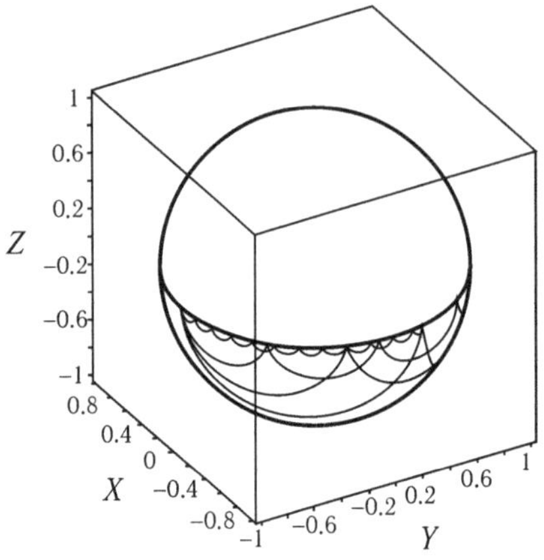

Fig. 2.30. The Bobylev – Steklov solution. The motion of the apex through the center of mass in the fixed space for $c = 0$ and various h.

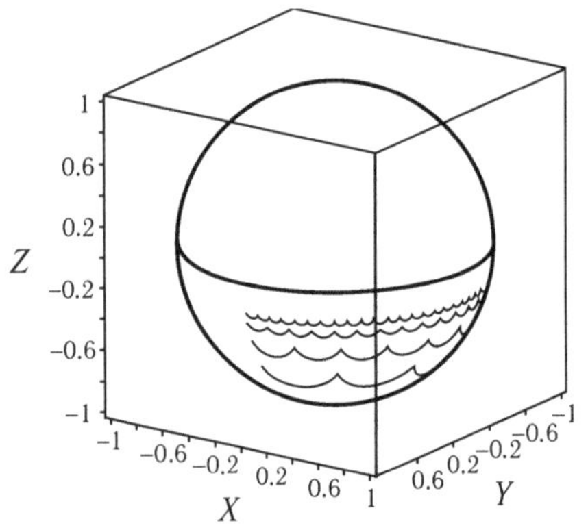

Fig. 2.31. The Bobylev – Steklov solution. The motion of the apex through the center of mass in the fixed space for $c \neq 0$ ($c = 1.15$) and for various h.

the body is set spinning about an arbitrary fixed axis through the center of mass and then is released without an initial push.

REMARK. The motions of the other apexes in the fixed space are rather too complicated for illustrative purposes and will not be covered here.

Unstable periodic solutions and separatrices for the Kovalevskaya case have an intricate shape both on the Poisson sphere and in the fixed space. Figure 2.33 shows the separatrix motions for $c \neq 0$ ($c = 1.15$) and for $h = 2$. It can be clearly seen that for most of the time the trajectory remains close to the periodic solution so that the curves are located more densely in this domain.

In some sense these trajectories show all the complexity of the integrable Kovalevskaya case — some motions seem to behave chaotically (the motion in the inertial frame seems even more chaotic).

REMARK 1. The Kovalevskaya integral may be represented in the form of a sum of squares as follows. Using the variables

$$S_1 = M_1\gamma_1 + M_2\gamma_2, \quad S_2 = M_1\gamma_2 - M_2\gamma_1,$$

it can be shown that the Kovalevskaya integral may be written as

$$F = \left(\frac{M_1^2 + M_2^2}{2}\right)^2 + x(M_1S_1 + M_2S_2) + x^2(\gamma_1^2 + \gamma_2^2).$$

Setting $\boldsymbol{S} = (S_1, S_2)$ and $\widetilde{\boldsymbol{M}} = (M_1, M_2)$, let λ be the angle between these vectors (see Fig. 2.32). Then, using $\gamma_1^2 + \gamma_2^2 = \sin^2\theta$ where θ is the angle between the vertical direction and the symmetry axis of the ellipsoid of inertia, we obtain

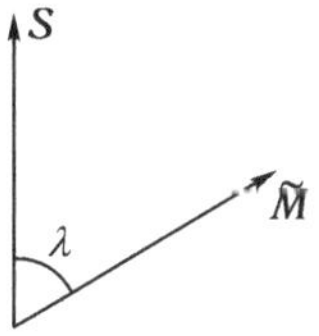

Fig. 2.32

$$F = \frac{1}{4}G^4\sin^2\lambda + \left(\frac{G^2\cos\lambda}{2} + x\sin\theta\right)^2 = k^2, \quad G^2 = \boldsymbol{M}^2.$$

REMARK 2. The following interesting nonlinear transformation preserves the Poisson bracket corresponding to the algebra $so(3)$

$$K_1 = \frac{M_1^2 - M_2^2}{2\sqrt{M_1^2 + M_2^2}}, \quad K_2 = \frac{M_1M_2}{\sqrt{M_1^2 + M_2^2}}, \quad K_3 = \frac{1}{2}M_3.$$

It can be shown that in terms of the canonical Andoyer variables this transformation corresponds to the canonical transformation $(L, l) \mapsto \left(\frac{L}{2}, 2l\right)$.

REMARK 3. In [250, 525] a family of integrable systems on the sphere S^2 is considered. These systems have an integral of degree four in momenta and cannot be reduced to the Kovalevskaya case (or to its generalization suggested by Goryachev). In [524] a similar construction is proposed for which the integral has degree three. In all of these works the additional integrals are not specified explicitly. The resulting families of integrable systems come from solving a certain differential equation, for which some existence theorems are proved. Various dynamical systems on S^2 with additional first integrals of degree 3 and 4 in momenta have been recently given by A. V. Tsiganov [579, 581].

REMARK 4. The analysis of the stability of oscillatory and rotational motions of the Kovalevskaya top about the axis of dynamical symmetry was performed in [415]; stability of the rotation about the vertical is discussed in [507].

REMARK 5. The motion of the line of nodes and the proper rotation is studied in [333] under the assumption that the first integrals are independent. In the present

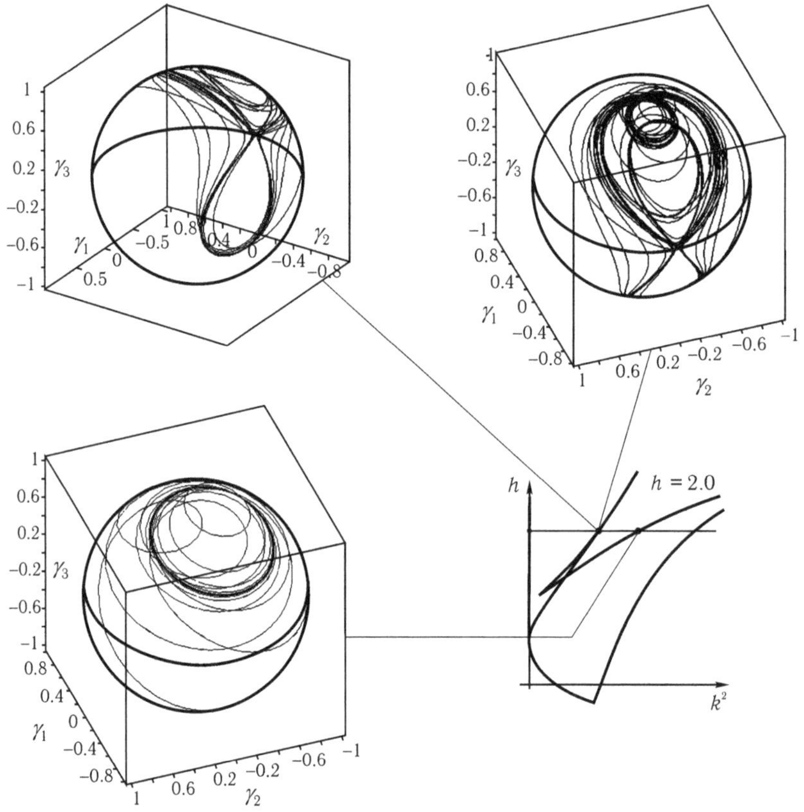

Fig. 2.33. Trajectories on the Poisson sphere of solutions tending asymptoticaly to unstable periodic solutions.

book the equations of motion for the Kovalevskaya case are reduced to the standard form on a torus. This reduction procedure has much in common with the introduction of canonical action-angle variables. Following [333], such methods are currently referred to as *qualitative analysis*.

4 Historical comments

The method of Kovalevskaya. S. V. Kovalevskaya was not guided by physical arguments when she discovered a new integrable case. She developed ideas of K. Weierstrass, P. Painlevé and H. Poincaré studying the analytical continuation of a solution of a system of ordinary differential equations to the complex plane of time. S. V. Kovalevskaya assumed that in the integrable cases the general solution on the complex plane has no singularities other than poles. This enabled her to find conditions for the existence of an additional integral. Along with the first integral, S. V. Kovalevskaya found a nontrivial system of variables which reduces the equations to the Abel – Jacobi form, and also obtained an explicit solution in terms of theta-functions.

A. M. Lyapunov in [394] refined the analysis made by Kovalevskaya in response to a criticism of Kovalevskaya by academician A. A. Markov. G. G. Appelrot [13] concerned himself with the same problem. Lyapunov assumed that the general solution as a complex function of time should not be branched (i.e., it should be meromorphic) and then used a perturbative approach to study solutions in a neighborhood of singularities (by analyzing the equations in variations). The Lyapunov method is different from the Kovalevskaya method. M. Adler and P. van Moerbeke returned to the Kovalevskaya method many years later. They connected the existence of a complete parametric family of one-valued Laurent (polar) expansions with the *algebraic* integrability of the system (in some narrow sense [2, 3]). The most extensive analysis of complete parametric expansions for the Euler – Poisson equations can be found in [376]. For a classical presentation of the results of Kovalevskaya and Lyapunov see [16, 224].

The ideas of Kovalevskaya laid the foundations for a new method for the analysis of integrability. They are also the first example of a search for obstructions to integrability; nowadays this is a full-fledged area of research [335]. Note that notwithstanding some rigorous results connecting branching of the general solution with nonexistence of first integrals [335], the method of Kovalevskaya remains only a test for integrability. This method is rather ambiguous and requires a certain amount of artistry and some additional reasoning in applications. In physical literature this test is usually called the *Painlevé* – Kovalevskaya test.

The Kovalevskaya case, its analysis and generalizations. A somewhat unnatural geometric interpretation of the Kovalevskaya case as well as a separate new method of reduction to quadratures was proposed by N. E. Zhukovskii [649]. He also used the Kovalevskaya variables to construct some interesting curvilinear coordinates on the plane (the plane M_1, M_2) corresponding to the separating variables of the Kovalevskaya top. His reasonings were simplified by V. Tannenberg and G. K. Suslov [562, 565].

F. Kötter also simplified the method of explicit integration of the Kovalevskaya case [323, 325]. He studied the motion in a frame uniformly rotating about the vertical axis. A modern approach to the Kovalevskaya variables and reduction to the Abel equations is discussed in [333]. A qualitative analysis of the motion of the axis of dynamical symmetry is given in [333], a topological and bifurcation analysis can be found in [286]. Action-angle variables for the Kovalevskaya top are constructed in [601] (see also [310, 176]). We introduce them in Sec. 5.2, Ch. 5. N. I. Mertsalov carried out some natural experiments which did not show any peculiarities in the motion of a top [570]. Under the Kovalevskaya conditions, some especially simple solutions expressed in terms of fractional rational functions were obtained by B. K. Mlodzieiowski [428].

G. V. Kolosov used some nonlinear transformation of the variables as well as time to integrate the Kovalevskaya case and reduce it to the problem of the motion of a point on the plane in a potential, and with variables that can be separated. This is the well-known Kolosov analogy, its classical version and some new generalizations are considered in Sec. 5.2, Ch. 5. In [316] G. V. Kolosov studied the trajectory of the angular momentum vector and found its regular singularities.

Methods of algebraic geometry were applied to the study of the structure of complex tori in [197, 25]. Bifurcation diagrams for the Kovalevskaya case in connection with the Kolosov analogy are considered in [220].

Quantization of the Kovalevskaya top has been discussed since the creation of quantum mechanics (Laporte, 1933). There is still no complete clarity in this issue [310, 477]. In [176] the Picard – Fuchs equation which appears in the integration of the Kovalevskaya case is derived. The first Lax representation for the n-dimensional Kovalevskaya case does not contain a spectral parameter and was constructed by A. M. Perelomov [461]. A Lax representation containing a spectral parameter for the general form (motion in two uniform fields) was proposed by A. G. Reyman and M. A. Semenov-Tian-Shansky [483]. This generalization of the Kovalevskaya case is still not enough studied (in particular, it has not been integrated by quadratures nor has been subjected to a topological and qualitative analysis).

2.5 The Goryachev – Chaplygin Case

Consider the partially integrable Goryachev – Chaplygin case for which the angular momentum vector is fixed to lie in the horizontal plane, i. e., $(\boldsymbol{M}, \boldsymbol{\gamma}) = 0$. This case is realized for almost the same restrictions on the dynamical parameters as in the Kovalevskaya case. However, the ratio of the moments of inertia now equals four $\frac{a_3}{a_1} = 4$, while in the Kovalevskaya case this ratio is two. The Hamiltonian and additional integral are

$$H = \frac{1}{2}(M_1^2 + M_2^2 + 4M_3^2) - x\gamma,$$
$$F = M_3(M_1^2 + M_2^2) + xM_1\gamma_3.$$

1 Explicit integration

S. A. Chaplygin [127] reduced the system to the Abel – Jacobi equations by introducing new variables of the Kovalevskaya type. These variables are defined by

$$M_1^2 + M_2^2 = 4uv, \qquad M_3 = u - v \tag{2.48}$$

and satisfy the equations

$$\begin{aligned} \frac{du}{\sqrt{P_1(u)}} - \frac{dv}{\sqrt{P_2(v)}} &= 0, \\ \frac{2u\,du}{\sqrt{P_1(u)}} + \frac{2v\,dv}{\sqrt{P_2(v)}} &= dt, \end{aligned} \tag{2.49}$$

with

$$P_1(u) = -\left(u^3 - \frac{1}{2}(h-x)u - \frac{1}{4}f\right)\left(u^3 - \frac{1}{2}(h+x)u - \frac{1}{4}f\right),$$
$$P_2(v) = -\left(v^3 - \frac{1}{2}(h-x)v + \frac{1}{4}f\right)\left(v^3 - \frac{1}{2}(h+x)v + \frac{1}{4}f\right),$$

where h is the energy and f is the constant value of the Chaplygin integral ($H = h$ and $F = f$).

REMARK 1. In fact the Chaplygin variables u and v are simply related to the Andoyer variables: $L = u - v$ and $G = u + v$ (see [333]). In Sec. 5.2, Ch. 5 we will construct a generalization of the Goryachev – Chaplygin case and find the corresponding separating variables via the analysis of the Andoyer variables for a bundle of the Lie – Poisson brackets including the algebras $so(4)$, $e(3)$, and $so(3, 1)$.

REMARK 2. An analogy between the Goryachev – Chaplygin top and a three-frequency closed Toda lattice was indicated in [32].

2 The bifurcation diagram and the phase portrait

Considering the multiple roots of the polynomials $P_1(u)$ and $P_2(v)$ it is easy to construct the bifurcation diagram [286]. On the plane (f, h) it consists of the three branches (Fig. 2.34):

$$\text{I.} \quad f = 0, \quad h > -1;$$
$$\text{II.} \quad h = \frac{3}{2}t^2 + 1, \qquad f = t^3, \quad t \in (-\infty, +\infty);$$
$$\text{III.} \quad h = \frac{3}{2}t^2 - 1, \qquad f = t^3, \quad t \in (-\infty, +\infty).$$

The first class (I) contains three periodic solutions

1) rotations and oscillations in the equatorial plane of the ellipsoid of inertia ($M_1 = M_3 = 0$, $\gamma_2 = 0$);
2) rotations and oscillations in the meridional plane of the ellipsoid of inertia ($M_1 = M_2 = 0$, $\gamma_3 = 0$);
3) particular Goryachev solutions with $f = 0$.

Unfortunately, the solutions lying on branches II and III are practically not studied. For some values of the energy, the phase portraits are shown in Figs. 2.35 and 2.36.

A description of possible topological types of Liouville foliation for the Goryachev – Chaplygin case together with those for the Kovalevskaya top is given in [286], while Fomenko graphs and topological invariants are discussed in [66]. The behaviors of the line of nodes and proper rotation in the generic case (the first integrals are functionally independent) are explored in [333]; the standard form of a system on a two-dimensional torus is also presented there. Using these results, a few theorems concerned with the hodograph of the angular velocity have been proved [217]. In [231] G. Gorr gave a

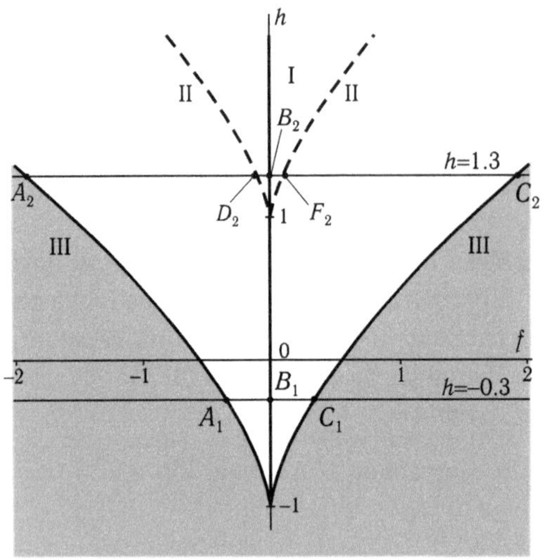

Fig. 2.34. Bifurcation diagram of the Goryachev – Chaplygin case. The non-physical domain of the integrals is shown in gray. Phase portraits for the two indicated levels of energy are constructed (see Figs. 2.35 and 2.36). Here and in the phase portraits, the letters A_i, B_i, C_i, ... denote the same periodic solutions and separatrices.

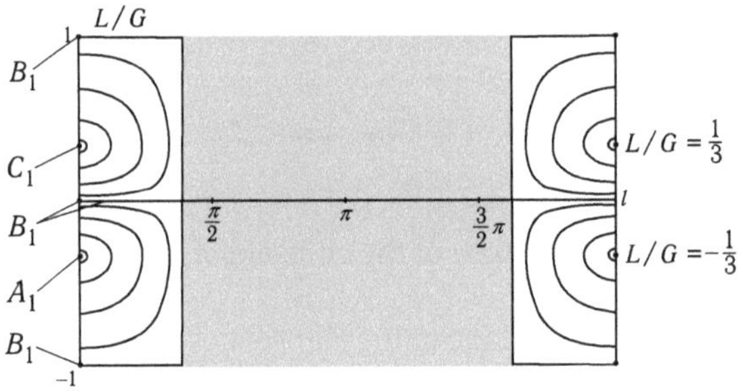

Fig. 2.35. The phase portrait of the Goryachev – Chaplygin case for $h = 0.3$ (section formed by the intersection with the plane $g = \pi/2$). The letters A_1, B_1, and C_1 denote periodic solutions located on the branches of the bifurcation diagram (Fig. 2.34). The point B_1 (with $f = 0$) on the bifurcation diagram corresponds to 1) two pendulum periodic solutions (on the phase portrait they are situated at the poles of the sphere $L/G = \pm 1$ and at the point $l = 0$, $L/G = 0$); 2) the whole straight line $L/G = 0$, $l \neq 0$ which is also filled with periodic solutions (the Goryachev solution) of pendulum type (see also Sec. 2.3).

qualitative picture of the body motion for some degenerate cases, when first integrals are not independent. L. N. Sretenskii [545] studied rapid rotations of the Goryachev – Chaplygin top about a horizontally oriented principal axis of the ellipsoid of inertia (the center of mass lies on this axis). Since the rotation was assumed to be rapid,

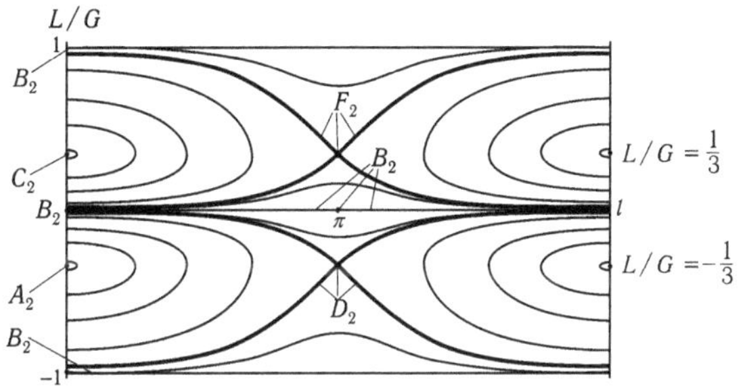

Fig. 2.36. The phase portrait of the Goryachev – Chaplygin case for $h = 1.3$ (section formed by the intersection with the plane $g = \pi/2$). The letters A_2, B_2, C_2, D_2, and F_2 denote the periodic solutions located on the branches of the bifurcation diagram (Fig. 2.34). In comparison with the previous portrait, the unstable solutions D_2 and F_2 (and their separatrices) are added. As above, the point B_2 on the bifurcation diagram represents four rotational periodic solutions, namely, rotations in the opposite directions in the equatorial and meridional planes. These rotations are the points $L/G = \pm 1$ and $l = 0, \pi$, $L/G = 0$, and also the straight line $L/G = 0$ whose points are periodic solutions (the Goryachev solutions) of the reduced system (see Sec. 2.3).

he introduced a small parameter via the standard procedure and studied the linear approximation equations.

Remark. The absence of explicit analytical expressions for the asymptotic solutions impedes the study of the perturbed system. N. I. Mertsalov in [422] asserted that the equations of the Goryachev – Chaplygin top for $c = (\boldsymbol{M}, \boldsymbol{\gamma}) \neq 0$ are integrable. The results of simulations presented in Fig. 2.37 show that this assertion is wrong, since for $c \neq 0$ a stochastic layer leading to nonintegrability occurs near the unstable manifolds.

3 Visualization of the most remarkable solutions

Among the periodic solutions of the Goryachev – Chaplygin problem, of special interest is *Goryachev's* solution. On the bifurcation diagram, it lies on the straight line $f = 0$. This line also contains the periodic solutions of the Euler – Poisson equations corresponding to the oscillations (for $h < 1$) and rotations ($h > 1$) of the rigid body in the planes Oxy and Oxz both obeying the law of the physical pendulum. Let us consider the Goryachev solution and the solutions on the branches II and III (see Fig. 2.34) in more detail.

The Goryachev solution [237]. For this solution there are two additional invariant relations [162]

$$M_1^2 + M_2^2 = bM_1^{2/3}, \quad f = M_3(M_1^2 + M_2^2) + M_1\gamma_3 = 0, \quad (b > 0). \tag{2.50}$$

These relations contain an arbitrary constant b and thereby define a one-parameter family of periodic solutions — a degenerate torus in the phase space filled with periodic

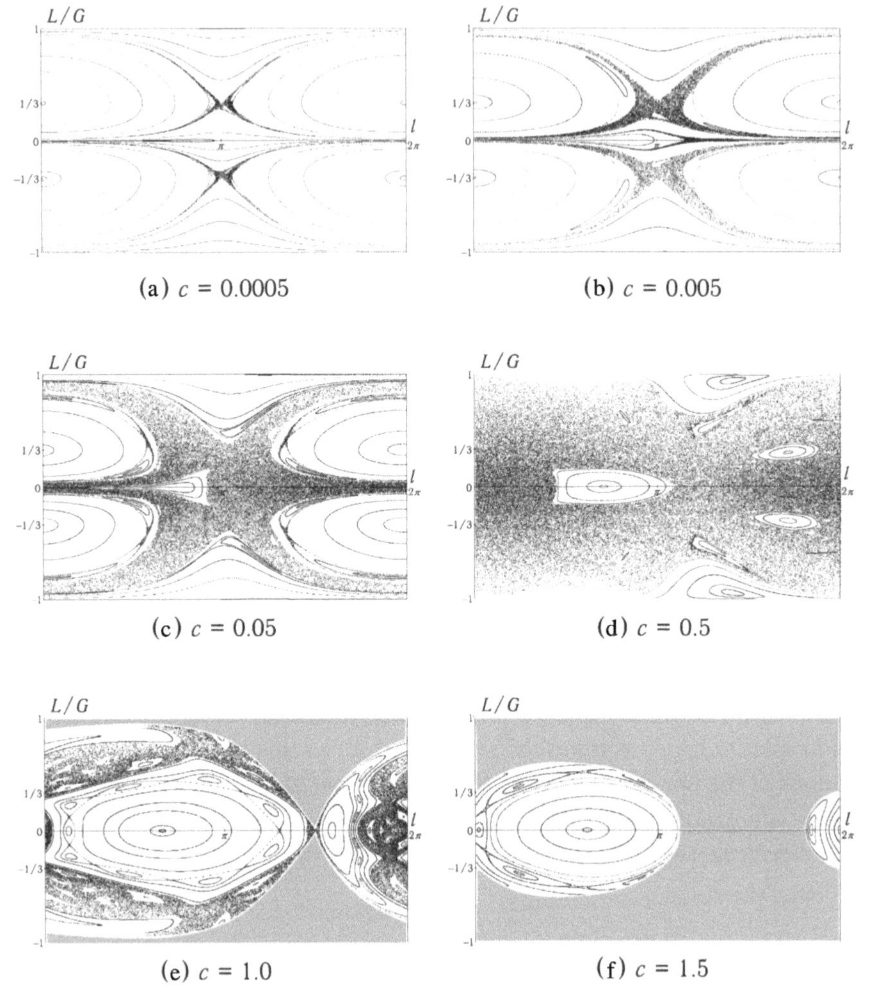

Fig. 2.37. Perturbation of the Goryachev – Chaplygin case for a fixed energy ($h = 1.5$) and increasing constant of the areas. Here the intersection plane is $g = \pi/2$. The nonphysical domains are shown in gray. The pictures show that near the separatrices a stochastic layer occurs. First, this layer widens but then diminishes together with the domain of possible motion. It is interesting that as c increases, the domain of possible motion and the stochastic layer gradually decrease and finally vanish completely.

solutions. The relations (2.50) were obtained by D. N. Goryachev. Then S. A. Chaplygin noticed that the condition $f = 0$ is excessively restrictive and obtained the solution (2.49) in the traditional form. If $h < 1$, then as b changes from 0 to b_{max}, the solution changes from oscillatory motion in the equatorial plane to oscillatory motion in the meridional plane (Fig. 2.38). In the phase portrait (see Fig. 2.35) this is the straight line $L/G = 0$ together with the meridian connecting this straight line with the poles. If $h > 1$, then as b changes from 0 to b_{max}, the solution gradually evolves from the rotation in the equatorial plane in one direction to the rotation in the same plane in the opposite direction (Fig. 2.36).

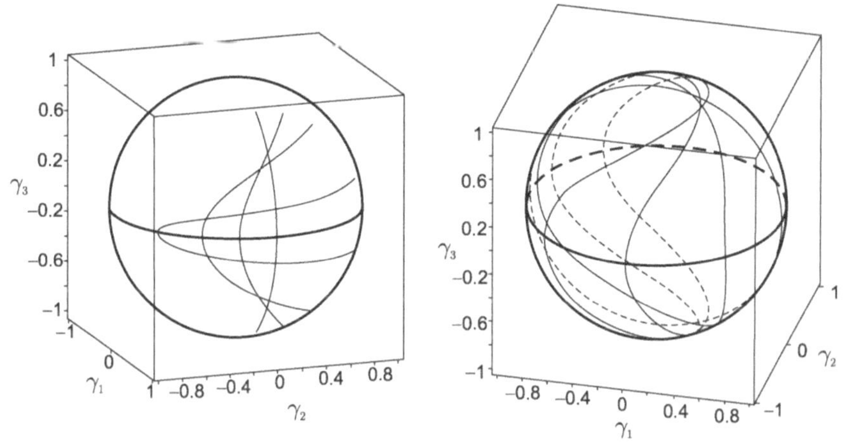

(a) The motion of γ for $f = 0$ and $h = -0.5$ (various types of pendulum motion).

(b) The motion of γ for $f = 0$ and $h = 1.5$ (various types of rotational motion).

Fig. 2.38. The "Goryachev solutions" form a torus filled with periodic solutions of the reduced system $\boldsymbol{M}$, $\boldsymbol{\gamma}$ (the so-called resonance 1 : 1). For $h < 1$ (Fig. a)) these are the pendulum type solutions and for $h > 1$ (Fig. b)) the rotational type solutions. This and the next figure show the trajectories on the Poisson sphere for various solutions from this torus.

The motion of the apex on the Poisson sphere is shown in Fig. 2.38. A remarkable phenomenon that has not been mentioned earlier is that in the inertial frame the Goryachev solutions correspond to periodic oscillatory motions for $h < 1$ (see Fig. 2.39) and to two-frequency quasi-periodic motions for $h > 1$ (Fig. 2.40).

All these facts cannot practically be derived directly from the very bulky analytical solution first obtained by Goryachev [237]. Despite some simplifications which can be found, for example, in [162], the explicit formulae provide only a vague impression of the motions which may be obtained by computer simulations. Under some additional restrictions on b and h Goryachev's solution was studied in [231]; in this work quasi-periodic and periodic solutions in the inertial frame (for all b except for a certain single, specific value $b \neq b^*$) are indicated.

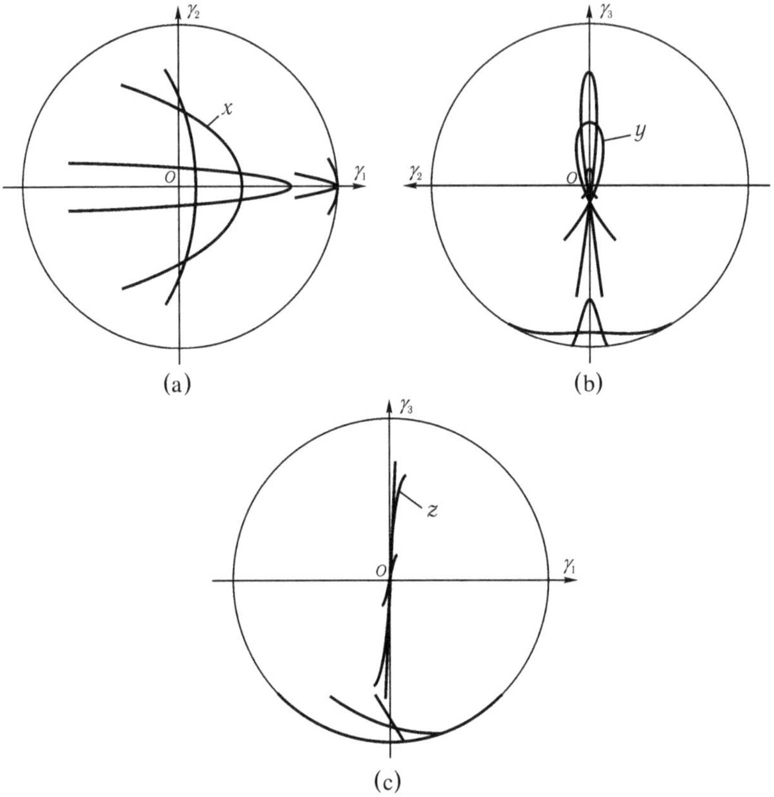

Fig. 2.39. For a fixed energy $h < 1$ ($h = -0.7$), this figure illustrates the behavior of the principal axes of a rigid body in a fixed frame of reference for the Goryachev solutions. It is well seen that these solutions are *periodic* solutions in the inertial frame which, as the parameter b changes, vary from oscillatory motions in the plane Oxy to oscillatory motions in the plane Oxz. (The axes fixed in the body are denoted by x, y, and z.)

Stable and unstable periodic solutions of the Euler – Poisson equations for the Goryachev – Chaplygin case lie, respectively, on branches III and II of the bifurcation diagram (Figs. 2.34 and 2.41 – 2.44). Numerical analysis shows that the motions of the whole system in the inertial frame corresponding to these solutions *are also periodic for arbitrary values of the energy* (see Figs. 2.43 and 2.44). This fact shows a peculiar feature of the rigid body dynamics for zero value of the area constant $(\boldsymbol{M}, \boldsymbol{\gamma}) = 0$. Instead of a formal proof we give a series of pictures which illustrate this statement (see [79]). These pictures show both the trajectories of the system on the Poisson sphere and the trajectories of the apexes in the inertial frame, most of them being extremely complicated.

Partial solution III was studied analytically in [229], where this solution was proved to be periodic in the inertial frame.

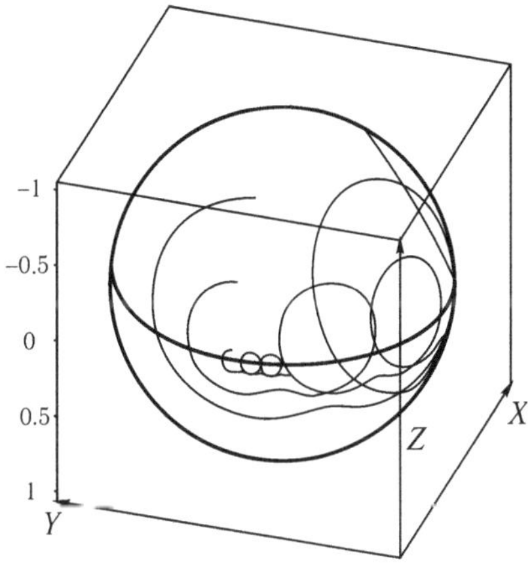

Fig. 2.40. The figure illustrates the quasi-periodic motion in the inertial frame (more exactly, the motion of the principal axis Oy) for the Goryachev solution for $h > 1$ ($h = 1.7$).

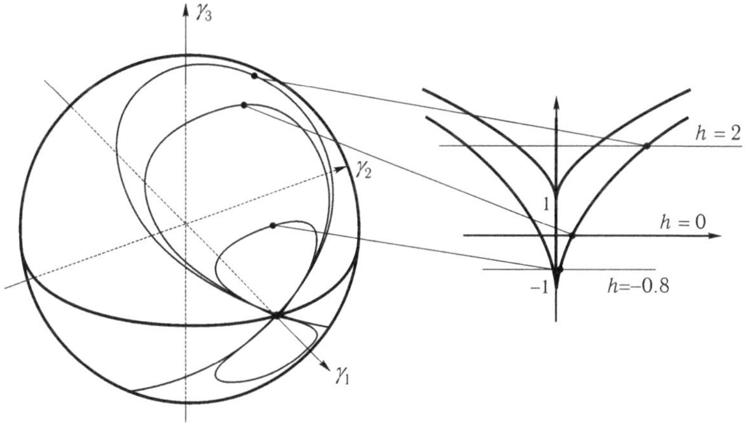

Fig. 2.41. Motion of the vertical unit vector $\boldsymbol{\gamma}$ on the Poisson sphere for a stable periodic motion in the Goryachev – Chaplygin case for various values of the energy.

Stability analysis of oscillations and rotations about the axis of dynamical symmetry (i.e., pendulum-type motions) in the Goryachev – Chaplygin case can be found in [414].

A general conclusion for the Goryachev – Chaplygin case is that in the course of its analysis we encounter interesting oscillatory and rotational motions *in the inertial frame*. It can be said that, in some sense, we deal with a "complicated pendulum". Nevertheless, the area of application of these oscillatory motions is still not clear. Note that the motion of the Goryachev – Chaplygin top is relatively simple compared to the Kovalevskaya top. The few analytical results obtained for the Goryachev – Chaplygin case fail to provide a vivid picture of the motion. On the contrary, computer-assisted

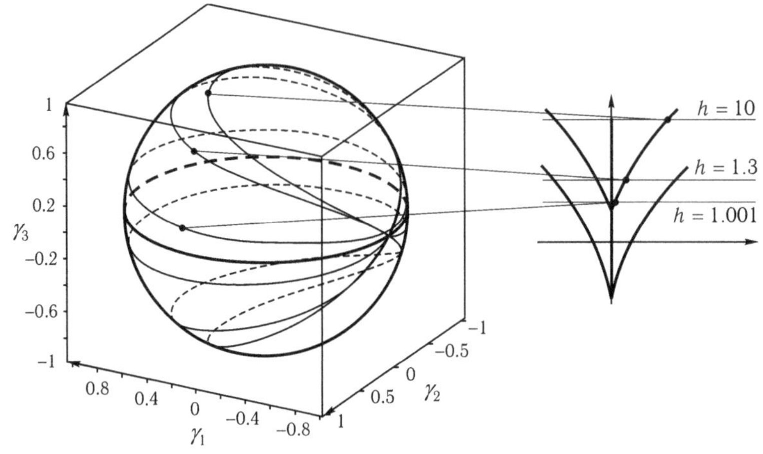

Fig. 2.42. Motion of the vertical unit vector **γ** on the Poisson sphere for an unstable periodic solution in the Goryachev – Chaplygin case for various energies.

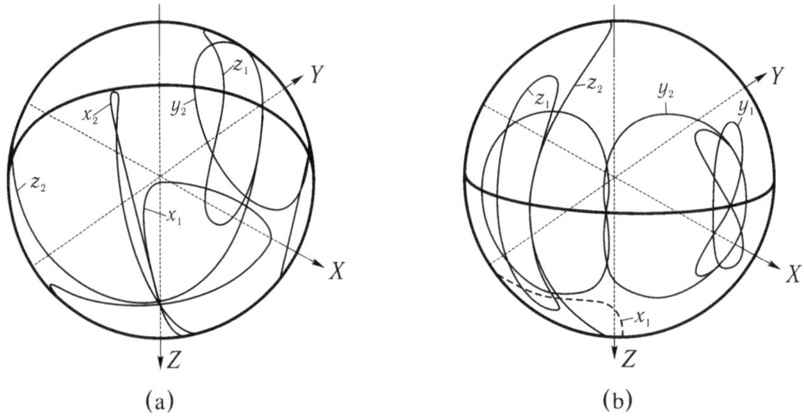

Fig. 2.43. Trajectories of the apexes of the principal axes x, y and z of the body in the fixed frame in the Goryachev – Chaplygin case for a stable periodic solution from branch III in Fig. 2.34 viewed from various positions. The trajectories are drawn for two different values of the energy h_1 and h_2. The index $i = 1,\ 2$ in the notation of the axes indicates the corresponding energy value.

analysis helps uncover remarkable properties of the motion which are also typical of some related integrable systems.

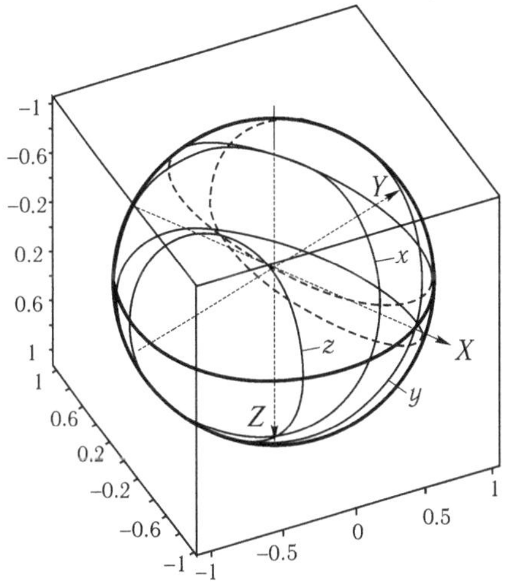

Fig. 2.44. Motion of the apexes of the principal axes of the body in the fixed space in the Goryachev – Chaplygin case for an unstable periodic solution on branch II in Fig. 2.34. The energy is fixed. The letters x, y, and z denote the trajectories of the corresponding axes. (The motions for other values of the energy are similar and therefore not shown.)

2.6 Partial Solutions of the Euler – Poisson Equations

Here we present several of the best known partial solutions of the Euler – Poisson equations (for a more complete list see [228, 162, 373]). A remarkable feature of these solutions is that they can be expressed without the use of excessively bulky quadratures and at the same time possess several interesting properties. However, these solutions are but a small part of the totality of asymptotic and periodic solutions that can be found numerically but can hardly be expressed analytically. Numerical simulation of all the solutions given below was performed in [212].

1 The Hess solution [258]

The Hess solution is a very special one in the sense that, in contrast to the previous cases (see Sec. 2.2 – Sec. 2.5), there is here only a one-parameter family of particular solutions satisfying the invariant relation (see Table 2.1)

$$r_1 M_1 + r_3 M_3 = 0. \qquad (2.51)$$

Thus, we have an isolated invariant manifold in the phase space (Fig. 2.46).

The conditions on the parameters in the Hess case

$$r_1\sqrt{a_3 - a_2} \pm r_3\sqrt{a_2 - a_1} = 0, \quad r_2 = 0 \qquad (2.52)$$

have the following physical meaning. Consider an *ellipsoid of gyration* , i. e., a level surface of the kinetic energy in the space of the angular momentum $\boldsymbol{M}$ (see Fig. 2.45)

$$\frac{1}{2}(\boldsymbol{M}, \mathbf{A}\boldsymbol{M}) = \text{const.} \qquad (2.53)$$

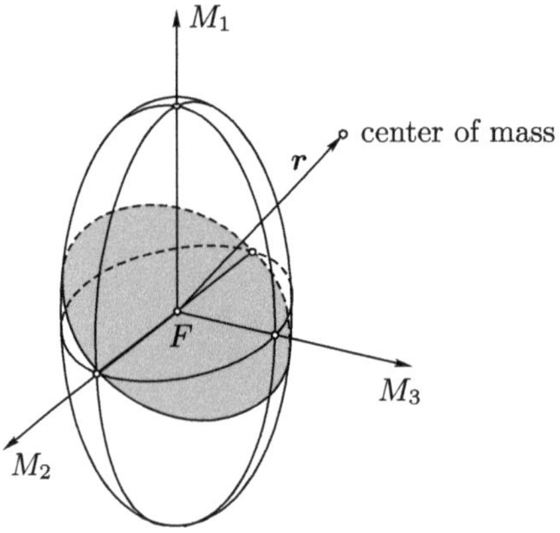

Fig. 2.45. The ellipsoid of gyration and location of the center of mass for the Hess case.

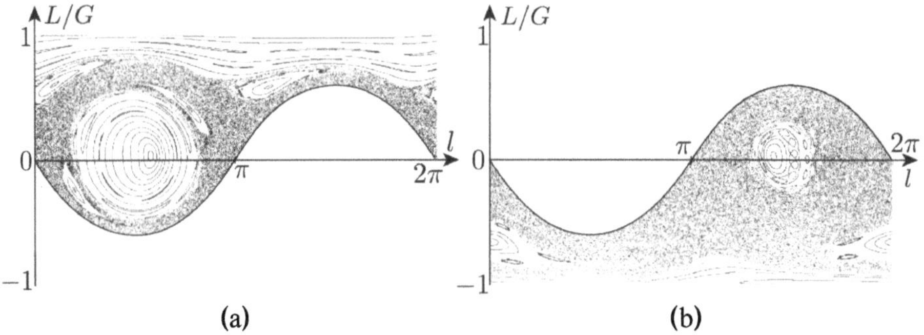

Fig. 2.46. Phase portrait (the intersection with the plane $g = \pi/2$) for the Hess case with $\mathbf{I} = \mathrm{diag}(1, 0.625, 0.375)$, $\boldsymbol{r} = (3, 0, 4)$, and $\mu = 1.995$. The constants of the integrals are $h = 50.0$ and $c = 5.0$. Figures (a) and (b) have been drawn for the same parameter values. Two stochastic layers divided by the double Hess separatrix may be clearly seen: the trajectories from one layer never penetrate into the other. Figure (b) also shows a *meander torus* that appears under these conditions (see Fig. 2.47).

Since all the eigenvalues of the matrix $\mathbf{A}$ are different, the ellipsoid of gyration possesses two circular sections through the mean axis. The conditions (2.52) mean

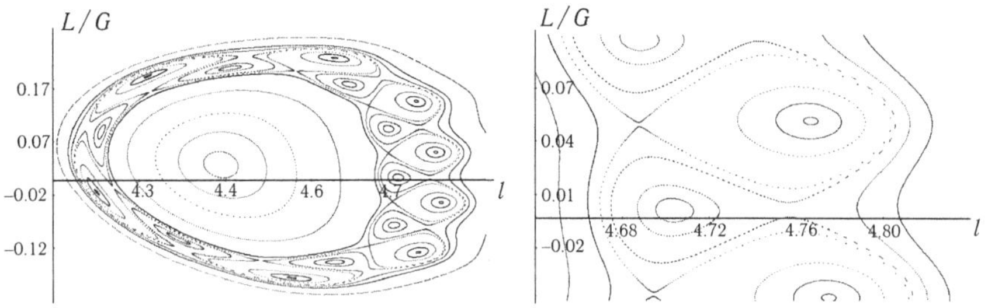

Fig. 2.47. Meander tori in the phase portrait of the Hess case (the values of parameters are as those for Fig. 2.46).

that the center of mass of the body lies on the axis perpendicular to one of the circular sections of the ellipsoid (2.53). The linear Hess integral (more precisely, the invariant relation) (2.51) means that the projection of the angular momentum onto this axis is zero. The behavior of solutions on this invariant manifold is, generally speaking, somewhat different from the classical quasi-periodic motions that occur when the conditions of the Liouville – Arnold theorem are fulfilled. In general terms, the equations for the Hess case cannot be integrated by quadratures on the manifold (2.51); however, a qualitative analysis of the motion can be carried out in full detail.

For certain values of the energy and area integrals, the Hess condition defines a pair of doubled (unsplit) separatrices (Fig. 2.46). The separatrices form a barrier between two chaotic zones (the presence of chaos indicates the lack of an additional integral for the Hess case). It is notable that *a meander torus* can arise in phase space for the Hess case (Fig. 2.47), though this it does not seem to be specific to this case (Sec. 6.5 Ch. 6).

The Hess case has much in common with the Lagrange case. To prove this, let us write the Hamiltonian (2.4) in a new coordinate system whose Ox_3-axis is perpendicular to the circular cross-section of the gyration ellipsoid (Fig. 2.45):

$$H = \frac{1}{2}\left(a_1'(M_1^2 + M_2^2) + a_3'M_3^2 + 2bM_3M_1\right) - \mu\gamma_3, \quad \mu = \text{const.} \tag{2.54}$$

This coordinate system is not a principal one. The components of the matrix of transformation of the principle coordinate system to this new one can be expressed in terms of the components of **A** as follows:

$$\mathbf{U} = \begin{pmatrix} \sqrt{\dfrac{a_3 - a_2}{a_3 - a_1}} & 0 & \mp\sqrt{\dfrac{a_2 - a_1}{a_3 - a_1}} \\ 0 & 1 & 0 \\ \pm\sqrt{\dfrac{a_2 - a_1}{a_3 - a_1}} & 0 & \sqrt{\dfrac{a_3 - a_2}{a_3 - a_1}} \end{pmatrix}. \tag{2.55}$$

The Hess invariant relation (2.51) takes the form

$$M_3 = 0. \tag{2.56}$$

Clearly, the Hamiltonian (2.54) on the level set $M_3 = 0$ coincides with the Hamiltonian for the Lagrange case.

A detailed analysis of the behavior of the solutions of the Euler – Poisson equations on the invariant submanifold (2.56) is given in Appendix D. Here we present only a number of phase portraits (see Figs. 2.48, 2.49), which illustrate the behavior of the system under the Hess condition (2.52) in the entire phase space. It is shown in [333] that for sufficiently large h the submanifold (2.56) is a pair of unsplit separatrices to unstable periodic solutions, which can be obtained by continuation from unstable permanent rotations of the Euler – Poinsot case.

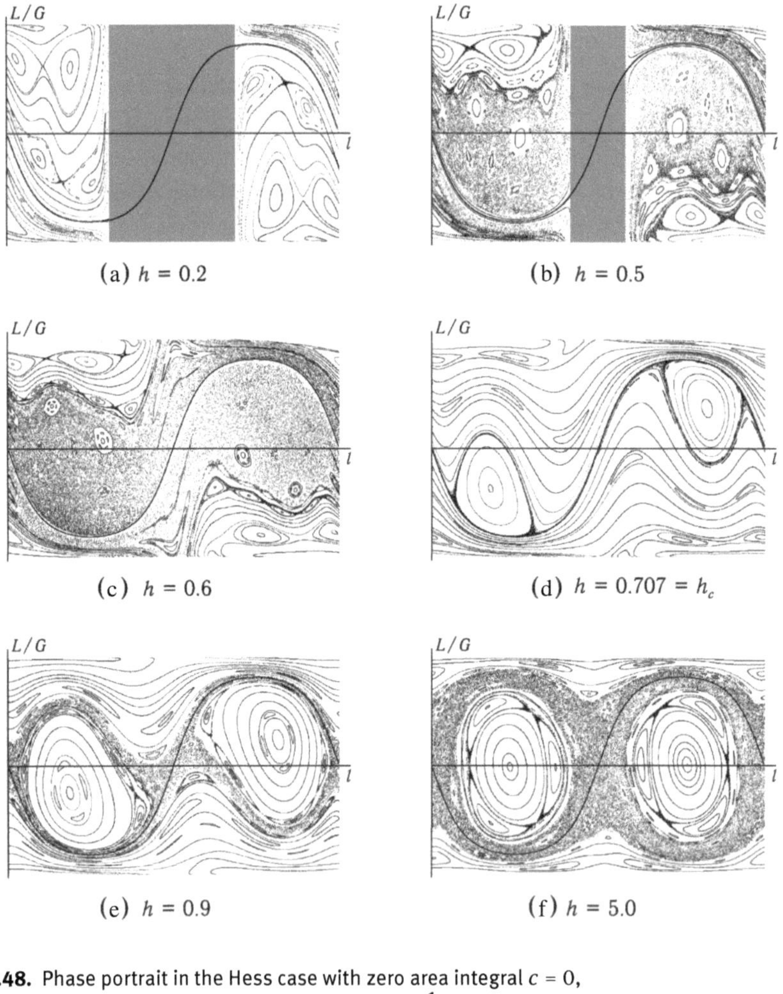

Fig. 2.48. Phase portrait in the Hess case with zero area integral $c = 0$, $H = \frac{1}{2}\left(M_1^2 + \frac{2}{3}M_2^2 + \frac{1}{2}M_3^2\right) + \frac{1}{\sqrt{3}}\gamma_1 + \frac{1}{\sqrt{6}}\gamma_3$, $h_c = \mu = \dfrac{1}{\sqrt{2}}$. For small values of the energy, the torus that corresponds to the Hess integral lies in a region regularly foliated by invariant tori. The nonphysical region is shown in gray.

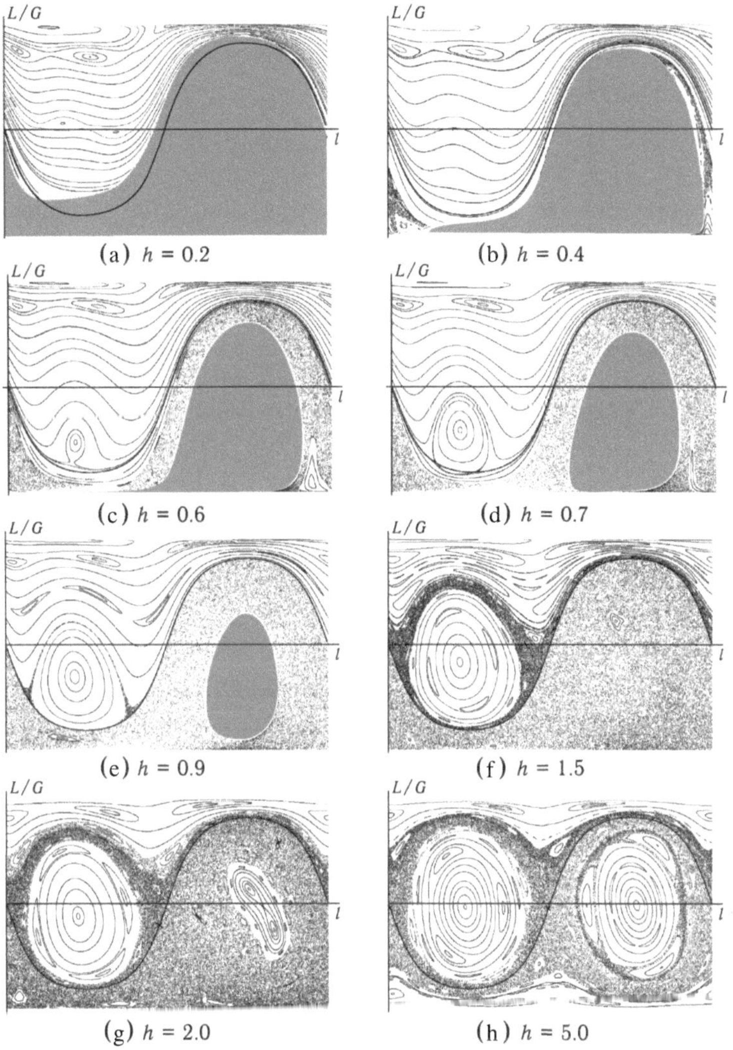

Fig. 2.49. Phase portrait in the Hess case with nonzero area integral $c = 1$, $H = \frac{1}{2}\left(M_1^2 + \frac{2}{3}M_2^2 + \frac{1}{2}M_3^2\right) + \frac{1}{\sqrt{3}}\gamma_1 + \frac{1}{\sqrt{6}}\gamma_3$. As before, for large h the Hess solution separates two stochastic layers and for small h the solution lies in a region regularly foliated by invariant tori.

In addition, for the Hamiltonian (2.54) we consider here in more detail the evolution of the variables γ_3 and ψ, which specify the position of the center of mass of the body. Without loss of generality it may be assumed that $a_1' = 1$, $a_3' = a_3$, $b = a_{13}$. Then

$$\dot{\gamma}_3 = \pm\sqrt{2(1-\gamma_3^2)(h - U_*)}, \quad \dot{\psi} = \frac{c}{1-\gamma_3^2}, \quad U_* = \frac{1}{2}\frac{c^2}{1-\gamma_3^2} - \mu\gamma_3, \tag{2.57}$$

where $(\boldsymbol{M}, \boldsymbol{\gamma}) = c$. As N. E. Zhukovskii noted in [646], these equations coincide with the equations of motion of a spherical pendulum.

Zhukovskii [646] also proved that in the course of motion the trajectory of the mean axis of the gyration ellipsoid makes a constant angle θ with the plane of the circular cross-section. This angle is given by

$$\sin\theta = \frac{a_2}{\sqrt{a_2(a_1 + a_3) - a_1 a_3}}. \tag{2.58}$$

Using this result, one can show that for vanishing value of the area integral $c = 0$ the mean axis of inertia follows a loxodrome. Zhukovskii termed this specific motion *a loxodromic pendulum* (which now can be justly referred to as the Hess loxodromic pendulum), found conditions under which such a motion occurs and suggested an experimental setup for its visualization [646].

Consider the loxodromic pendulum case ($c = 0$) in greater detail (Fig. 2.48). From (2.57) it follows that

$$\dot{\gamma}_3^2 = 2(h - \mu\gamma_3)(1 - \gamma_3^2), \quad \dot{\psi} = 0. \tag{2.59}$$

There are two essentially different cases.

***h* > *μ*.** The center of mass moves along a great circle (since $\psi = \text{const}$), and the mean axis of inertia follows a loxodrome (Fig. 2.7). The phase portrait contains chaotic trajectories and the Hess solution separates two immiscible stochastic layers (see also Fig. 2.46)). The Hess motion cannot be observed physically because due to instability the system inevitably enters one of the stochastic layers.

In the limit as $h \to \infty$ (or $\mu \to 0$), we have the well-known Euler case. The Hess solution transforms into one of the separatrices corresponding to unstable rotation about the mean axis of inertia [333].

***h* < *μ*.** The center of mass executes pendulum-type oscillations, while the mean axis follows a segment of a loxodrome according to (2.58). In the inertial frame the corresponding motion is periodic (like the Goryachev solution, this motion has a single frequency, see Sec. 2.5 Ch. 2). On the phase portrait (Figs. 2.48 a,b,c) the Hess condition defines an invariant curve consisting of fixed points; this curve is located in the domain filled with KAM tori.

REMARK. As mentioned above, a perturbation of the Euler – Poinsot problem which obeys the Hess conditions leads to the nonsplitting of one of the two pairs of the separatrices emanating from unstable permanent rotations, [333] (Figs. 2.48 f,2.49 h). The integral (2.51) defines a singular torus filled with doubly asymptotic orbits each of which tends to an unstable periodic solution as $t \to \infty$. As $\tilde{\mu} \to 0$, each of these periodic solutions turns into a permanent rotation about the mean axis of inertia. This description of the system's behavior does not contradict Zhukovskii's proposition on quasi-periodic motion of the center of mass [646] because the equations governing the center-of-mass dynamics are obtained by ignoring not the precession angle, but the angle of proper rotation about the axis perpendicular to a circular cross-section of the inertial ellipsoid.

Historical comment. Hess himself discovered the integral now bearing his name [258] (1890) when attempting to rewrite the equations of motion of a heavy rigid body in a more convenient form – compared to that of the traditional Euler – Poisson equations. The idea of employing nonprincipal axes is due to P. V. Kharlamov (see [16, 293]), who used it in the analysis of several mechanical problems. He is also the author of a special form of equations of motion in nonprincipal axes. Conditions for the Hess integral to exist were discovered by G. G. Appelrot [13] in his attempts to fill gaps in Kovalevskaya's analysis, where the Hess case was not mentioned. This does not imply that she had committed a mistake because solutions in the case of Hess exhibit branching in the plane of complex time [13]. The Hess case is sometimes referred to as the Hess – Appelrot case.

Nekrasov produced an extensive volume concerned with analytical solution of the equations of motion (reduction to the Ricatti equation) [445], whilst, as already mentioned, a geometric analysis and a physical, working model of the Hess top was proposed by Zhukovskii [646]. The Hess integral was independently rediscovered by Roger Liouville [2] [386] (1895). In [132] S. A. Chaplygin proved that any rigid body can execute a motion of Hess type provided the principal central moments of inertia are all different. V. V. Kozlov [333] found a connection between the Hess invariant relation and existence of a pair of nonsplit separatrices of a perturbed Euler – Poinsot problem.

A counterpart of the Hess case for the Kirchhoff equations was discussed by Chaplygin [131], who from the outset of his analysis used nonprincipal axes; using the separatrix-splitting condition, an equivalent result was obtained in [350].

2 Permanent Staude rotations

Let us consider an equilibrium point for the Hamilton equations on $e(3)$ with an arbitrary potential depending on $\boldsymbol{\gamma}$,

$$H = \frac{1}{2}(\mathbf{A}\boldsymbol{M},\ \boldsymbol{M}) + V(\boldsymbol{\gamma}). \tag{2.60}$$

Such an equilibrium on the Poisson sphere, $(\boldsymbol{\gamma}, \boldsymbol{\gamma}) = 1$, corresponds to a *relative equilibrium* of the unreduced system. The condition for an equilibrium, $\dot{\boldsymbol{M}} = \dot{\boldsymbol{\gamma}} = 0$, and invariance of the the area integral, $(\boldsymbol{M}, \boldsymbol{\gamma}) = c$, imply $\mathbf{A}\boldsymbol{M} = \lambda\boldsymbol{\gamma}$ and $\lambda = \frac{c}{(\mathbf{A}^{-1}\boldsymbol{\gamma}, \boldsymbol{\gamma})}$. The equations of motion yield

$$\begin{gathered} c^2(\mathbf{A}^{-1}\boldsymbol{\gamma} \times \boldsymbol{\gamma}) + \left(\boldsymbol{\gamma} \times \frac{\partial V}{\partial \boldsymbol{\gamma}}\right)(\mathbf{A}^{-1}\boldsymbol{\gamma},\ \boldsymbol{\gamma})^2 = 0, \\ \boldsymbol{\gamma}^2 = 1. \end{gathered} \tag{2.61}$$

This result was obtained by O. Staude in 1894 [548] and B. C. Mlodzieiowski [427] independently and simultaneously. In a moving frame of reference the first equation

2 By the way, his note in Comp. Rend. Acad. Sc. was communicated by H. Poincaré.

of (2.61) defines the so-called *Staude cone*. With respect to each generatrix of the cone the body rotates uniformly about the symmetry axis of the force field (about the vertical axis for the gravitational field) with angular velocity $|\boldsymbol{\omega}| = \frac{|c|}{(\mathbf{A}^{-1}\boldsymbol{\gamma}, \boldsymbol{\gamma})}$.

It is easy to verify [566] that uniform Staude rotations correspond to equilibria for the Hamiltonian (2.60), and these rotations, isoenergetic manifolds and the critical points of the reduced potential,

$$V_c(\boldsymbol{\gamma}) = V(\boldsymbol{\gamma}) + \frac{c^2}{2\,(A^{-1}\boldsymbol{\gamma}, \boldsymbol{\gamma})}, \tag{2.62}$$

are all equivalent to one another.

Indeed, the critical points of the reduced potential coincide with the critical points of the *momentum map*

$$F\colon\ (\boldsymbol{M}, \boldsymbol{\gamma}) \to (h, c), \tag{2.63}$$

where h and c are the values of the energy and area integrals, respectively. Each pair (h, c) defines a three-dimensional compact isoenergetic manifold in the phase space

$$\mathscr{E}^3_{h,c} = \{H = h,\ (\boldsymbol{M}, \boldsymbol{\gamma}) = c,\ (\boldsymbol{\gamma}, \boldsymbol{\gamma}) = 1\} \subset \mathbb{R}^3(\boldsymbol{M}) \times \mathbb{R}^3(\boldsymbol{\gamma}). \tag{2.64}$$

For regular values of the momentum map F the manifold (2.64) is smooth. The topological type of $\mathscr{E}^3_{h,c}$ changes at critical values (h, c), which form a subset $\Sigma_{crit} \subset \mathbb{R}^2(h, c)$ and define *the bifurcation diagram* in the plane of the first integrals. As a rule, Σ_{crit} is a union of regular curves in $\mathbb{R}^2$, called *bifurcation curves*. This diagram is also referred to as a Smale diagram. Smale suggested a general method for topological analysis of isoenergetic manifolds; he successfully applied this method to the planar three-body problem. At points on the bifurcation curves, not only does an isoenergetic manifold $\mathscr{E}^3_{h,c}$ undergo a change of topological type, but so also does its projection

$$\mathscr{E}^3_{h,c} \to U_{h,c} = \{V_c(\boldsymbol{\gamma}) \leqslant h\} \subset S^2 \tag{2.65}$$

onto the Poisson sphere. These projections are called *domains of possible motion* or DPM.

The fiber $\{(\boldsymbol{M}, \boldsymbol{\gamma})\} \subset \mathscr{E}^3_{h,c}$ above a point $\boldsymbol{\gamma} \in S^2$ is diffeomorphic to a circle (if $\boldsymbol{\gamma} \in U_{h,c} \setminus \partial U_{h,c}$), or it is a point (if $\boldsymbol{\gamma} \subset \partial U_{h,c}$), or it is empty (for $\boldsymbol{\gamma} \notin U_{h,c}$). Thus, topologically an isoenergetic manifold can be described in the following way [66]. If the projection $U_{h,c}$ is the entire Poisson sphere, then $\mathscr{E}^3_{h,c}$ is a S^1-bundle over the sphere, which is topologically equivalent to a bundle of unit tangent vectors. If $U_{h,c}$ is not the entire sphere, meaning there are holes, then $\mathscr{E}^3_{h,c}$ can be obtained from the Cartesian product $U_{h,c} \times S^1$ by shrinking to a point the S^1- fiber above each boundary point of the projection $U_{h,c}$.

The following statements hold:

- if h is less than the minimum value of the function $V_c(\gamma)$, then $\mathscr{E}^3_{h,c}$ is empty;
- if h exceeds the maximum value of the function $V_c(\gamma)$, then $\mathscr{E}^3_{h,c}$ is diffeomorphic to the projective space $\mathbb{R}P^3$;

– if h is not a critical value of the function $V_c(\gamma)$, then $\mathscr{E}^3_{h,c}$ is a smooth three-dimensional orientable manifold. Here the set $U_{h,c}$ is the union of pairwise disjoint two-dimensional manifolds with edges, that is, $D^2_{i_1} \cup \cdots \cup D^2_{i_m}$, where D^2_k is a 2-disk with k holes. The manifold $\mathscr{E}^3_{h,c}$ is diffeomorphic to the disjoint union $N^3_{i_1} \cup \cdots \cup N^3_{i_m}$ of three-dimensional manifolds, where $N^3_0 \equiv S^3$ is the three-dimensional sphere and N^3_k (for $k \geqslant 1$) is the connected sum of k copies of $S^1 \times S^2$.

For the Euler – Poisson equations the reduced potential reads

$$V_c(\gamma) = \frac{c^2}{2\left(A^{-1}\boldsymbol{\gamma}, \boldsymbol{\gamma}\right)} - (\boldsymbol{r}, \boldsymbol{\gamma}). \tag{2.66}$$

For the potential (2.66), the isoenergetic manifolds and their bifurcations under additional restrictions were studied by A. Iacob [264], S. B. Katok [281], Ya. V. Tatarinov [566] (see also a paper by R. P. Kuzmina [359]). The most comprehensive results on the subject have been recently obtained by I. N. Gashenenko. In the case of dynamical asymmetry, bifurcation curves for the Euler – Poisson equations are shown in Fig. 2.50 for various locations of the center of mass. The curves were drawn using numerical methods.

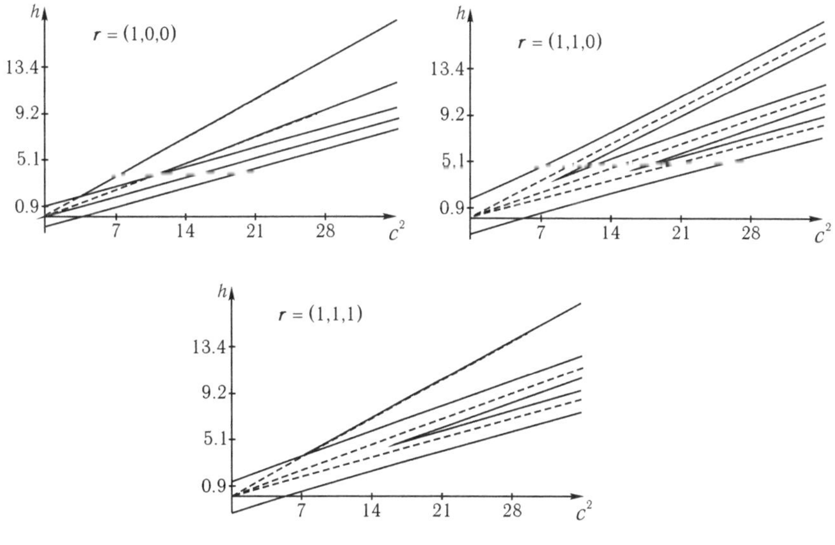

Fig. 2.50. Smale diagrams for various locations of the center of mass $\boldsymbol{r}$ ($\mathbf{I} = \mathrm{diag}(2,\ 1.5,\ 1)$).

3 Topological types of three-dimensional isoenergetic manifolds in the case when the center of mass lies in a principal plane

In [281] S. B. Katok performed a complete topological analysis of isoenergetic manifolds of the Euler – Poisson equations, assuming that the center of mass lies on a principal axis of inertia. She found that a connected component of the nonsingular isoenergetic manifold is one of the following four types: 1) S^3 (three-dimensional sphere), 2) $\mathbb{R}P^3$ (real projective space), 3) the product $S^1 \times S^2$ and 4) the connected sum $N_2^3 = (S^1 \times S^2)\#(S^1 \times S^2)$. The last one can be obtained by attaching two handles, each of the form $S^1 \times D^2$ (D^2 is a two-dimensional disk) to a three-dimensional sphere [18]. Ya. V. Tatarinov [566] made an attempt to generalize the results of S. B. Katok to the case where there are no restrictions on the location of the center of mass. Unfortunately, he missed one new type of isoenergetic integral manifold that may arise. A new connected component of an isoenergetic manifold was discovered by I. N. Gashenenko. [212] in the case where the center of mass lies in a principal plane of inertia. He obtained 46 types of nondegenerate bifurcation diagrams and found a new smooth integral manifold $\mathscr{E}_{h,c}^3 = N_3^3$ which is the connected sum of three copies of $S^1 \times S^2$. Let us consider the results of [212] in greater detail.

If the center of mass lies in a principal plane of inertia, then without loss of generality we may put

$$|\boldsymbol{r}| = 1, \quad r_1 > 0, \quad r_2 > 0, \quad r_3 = 0, \quad I_1 > I_2. \tag{2.67}$$

Under these conditions the Staude cone (2.61) degenerates into a pair of perpendicular planes

$$\gamma_3 = 0, \quad (I_2 - I_3)\gamma_2 r_1 + (I_3 - I_1)\gamma_1 r_2 = 0, \tag{2.68}$$

and the spherical curve, which is the intersection of the cone and the Poisson sphere, consists of two perpendicular great circles. In the case of uniform rotations, the components of the vectors $\boldsymbol{\omega}$ and $\boldsymbol{\gamma}$ can be represented in terms of two one-parameter families (the parameters are denoted by τ and σ)

$$\begin{aligned} 1.\quad & \gamma_3 = \omega_3 = 0, \quad \gamma_1 = -(1+\tau^2)^{-1/2}, \quad \tau \in \mathbb{R}, \\ & \omega_1 = \mu\gamma_1, \quad \omega_2 = \mu\gamma_2, \\ & \mu = \left| \frac{(r_1\tau + r_2)(1+\tau^2)^{1/2}}{(I_1 - I_2)\tau} \right|^{1/2}; \end{aligned} \tag{2.69}$$

$$\begin{aligned} 2.\quad & \gamma_1 = \frac{r_1\sigma^2}{I_3 - I_1}, \quad \gamma_2 = \frac{r_2\sigma^2}{I_3 - I_2}, \quad \sigma \in \mathbb{R}, \\ & \gamma_3 = \pm(1 - \sigma_0\sigma^4)^{1/2}, \\ & \omega_i = \sigma^{-1}\gamma_i, \quad \sigma_0 = \frac{r_1^2}{(I_1 - I_3)^2} + \frac{r_2^2}{(I_2 - I_3)^2}. \end{aligned} \tag{2.70}$$

The function $(1+\tau^2)^{1/2}$ and the expression $\tau/(r_1\tau+r_2)$ must have the same sign. If the ellipsoid of inertia has an axial symmetry ($I_1 = I_3$ or $I_2 = I_3$), then the family (2.70) does not exist.

Substituting (2.69) and (2.70) into the first integrals, we obtain a parametric representation for the bifurcation set $\Sigma \subset \mathbb{R}^2(h, c)$. The bifurcation curve B_1 corresponding to the set of critical points (2.69) can be written as

$$\begin{aligned} h &= \frac{I_2 r_1\tau^2 + (3I_2 - 2I_1)r_2\tau^2 + (3I_1 - 2I_2)r_1\tau + I_1 r_2}{2(I_1 - I_2)\tau(\tau^2+1)^{1/2}}, \\ c &= \frac{(I_1 + I_2\tau^2)|r_1\tau + r_2|^{1/2}}{|\tau|^{1/2}(I_1 - I_2)^{1/2}(1+\tau)^{3/4}}. \end{aligned} \tag{2.71}$$

Here $\tau \in (-\infty, -r_2 r_1^{-1}] \cup [-r_2 r_1^{-1}, 0) \cup (0, \infty)$. The bifurcation curve B_2 that corresponds to the set of critical points (2.70) is given by

$$\begin{aligned} h &= \frac{1}{2}\frac{I_3}{\sigma^2} + \frac{3}{2}\sigma^2, \\ c &= \frac{I_3}{\sigma} + \sigma^3\sigma_\star, \quad 0 < |\sigma| < \sigma_0^{-1/4}, \\ &\text{where } \sigma_\star = r_1^2(I_1 - I_3)^{-1} + r_2^2(I_2 - I_3)^{-1}. \end{aligned} \tag{2.72}$$

Equations (2.71) and (2.72) define a three-parameter family of bifurcation diagrams in the plane (h, c). The essential parameters are

$$\alpha = \frac{I_2}{I_1}, \quad \beta = \frac{I_3}{I_1}, \quad r_\star = \frac{r_2}{r_1}. \tag{2.73}$$

In the generic case, the bifurcation diagram Σ is comprised of smooth arcs, which may touch or intersect each other at critical points. A detailed analysis of these bifurcation diagrams was performed in [212].

If the center of mass lies on a principal axis of inertia, then generically, according to the results of S. B. Katok [281], there are seven types of generic bifurcation diagrams. The case where the center of mass is near a principal axis of inertia but $r_1r_2r_3 \neq 0$ is discussed in [566]. In this paper some of the types of the manifolds $\mathcal{E}^3_{h,c}$, $U_{h,c}$ are overlooked: the number of the types indicated is fewer than that in the limit case, which is considered in [281]. All the inaccuracies and inconsistencies were corrected in [212], where, in particular, the following was shown: if the center of mass is in a principal plane of inertia, then – as already mentioned – the bifurcation set of the momentum map (given by (2.71) and (2.72)) contains 46 essentially different nondegenerate bifurcation diagrams in the plane $\mathbb{R}^2(h, c)$.

In [212] it is also shown that under the condition (2.67) the momentum map F has at most four critical points for each singular level surface $\mathcal{E}^3_{h,c}$. Moreover, under the condition (2.67), for each fixed value c of the area integral $(\boldsymbol{M}, \boldsymbol{\gamma})$ the reduced potential $V_c(\boldsymbol{\gamma})$ has between two and ten critical points on the Poisson sphere.

Using these conclusions, it is easy to find the number of uniform Staude rotations when the integral of areas c is fixed: for an asymmetric body (under the condition (2.67))

genus of components l	0	1	2	3	4	11	12	111
$U_{h,c}$	S^2	D^2	$D^1 \times S^1$	D_2^2	D_3^2	$D^2 \cup D^2$	$D^2 \cup (D^1 \times S^1)$	$D^2 \cup D^2 \cup D^2$
$\mathscr{E}^3_{h,c}$	$\mathbb{R}P^3$	S^3	$S^1 \times S^2$	N_2^3	N_3^3	$S^3 \cup S^3$	$S^3 \cup (S^1 \times S^2)$	$S^3 \cup S^3 \cup S^3$

this number ranges from two to ten, while for an axisymmetric body whose center of mass does not lie on the axis of dynamical symmetry the number of rotations varies between two and six.

Following [566], we say that the connected manifolds $U_{h,c}$ and $\mathscr{E}^3_{h,c}$ have genus l if one has to remove l disks D^2 from a sphere to obtain $U_{h,c}$. If an integral manifold is not connected, then its genus is necessarily multivalued, namely, $l = l_1 l_2 \dots$, where l_i is the genus of its i-th connected component. Within a single domain from $\mathbb{R}^2(h, c) \setminus \Sigma$ the multivalued genus is constant; the genus is preserved under transformations of the bifurcation diagram until the domain shrinks to a curve or to a point.

I. N. Gashenenko [212] proved that when the condition (2.67) holds, the topological structure of the nonsingular manifolds $\mathscr{E}^3_{h,c}$, $U_{h,c}$ is as follows

As mentioned previously, the manifold N_3^3 is a new one here as compared with the case considered by S. B. Katok. N_3^3 arises, for example, when

$$\mathbf{I} = \mathrm{diag}(2.0, 1.18, 1.13), \quad \boldsymbol{r} = (0.99849, 0.054917, 0),$$
$$h = 2.034237, \quad c = 2.02485.$$

The domain of those points (h, c) for which $\mathscr{E}^3_{h,c} = N_3^3$ is always very small, so it is not surprising it was overlooked in [566]. It should be noted that I. N. Gashenenko [211, 209, 214, 208, 213] made the proposal to use not only the projection π of the isoenergetic manifold $\mathscr{E}^3_{h,c}$ onto the Poisson sphere, but also its projection onto the space of angular velocity components ω_1, ω_2, and ω_3. This projection has many remarkable features, among them the possibility of obtaining global Poincaré sections in the space of ω_1, ω_2, ω_3. Such sections are extremely useful for computer-aided analysis of the global dynamics; this topic will be discussed in greater detail in Ch. 6.

4 Stability of Staude rotations

Stability of the Staude rotations depends on the types of the critical points of the reduced potential. If $V_c(\boldsymbol{\gamma})$ has a minimum, then the corresponding uniform rotation is stable, the uniform rotation that corresponds to a saddle point of $V_c(\boldsymbol{\gamma})$ is unstable, and nothing can be said *a priori* about stability of the rotations that correspond to the maxima of $V_c(\boldsymbol{\gamma})$. It was shown in [281, 566] that the smooth curves corresponding to the minima, maxima and saddle points of the reduced potential are easy to locate in the bifurcation diagram. For a fixed value c of the area integral the number of minima and maxima differs from the number of saddle points by the Euler characteristic of the sphere, which is 2.

N. G. Chetaev and V. V. Rumiantsev [508] were the first to use Lyapunov functions constructed from first integrals for the purposes of stability analysis of Staude rotations. Stability regions on the Staude cone are shown in [508]: the boundaries of these regions correspond to infinitely remote points, intersection points and cusp points of the bifurcation curves. The stable uniform rotations found in [508] correspond to minima of the reduced potential. Further analysis of stability, partly based on ideas of KAM-theory, was performed in [511, 527, 328].

An elementary analysis of linear stability can be found in the monographs by R. Grammel [239] and K. Magnus [401]. However, the stability analysis can hardly be said to have been completed and the problem in full generality remains difficult.

The bifurcation curves for the Euler case (which correspond to permanent rotations about principal axes) are shown in Fig. 2.50 as dotted lines. When gravity is switched on, some of these curves split, namely, those for which the projection of the radius vector onto the corresponding principal axis is different from zero.

REMARK 1. In the case of a uniform gravitational field the Staude cone is an ordinary quadric cone. Under some assumptions about the parameters a_i and r_i this cone can degenerate into a pair of planes (distinct or coinciding) or it can become undefined. It is clear that the following five straight lines determine the cone completely:
1) the three principal axes of inertia with respect to the fixed point,
2) the straight line passing through the fixed point and the center of mass,
3) the straight line along the vector $\mathbf{A}\boldsymbol{r}$; this line is aligned with the vector $\boldsymbol{\omega}$, while $\boldsymbol{M} = \mathbf{I}\boldsymbol{\omega}$ is collinear with $\boldsymbol{r}$.

REMARK 2. As W. van der Woude noted [592], for a uniform gravitational field the Staude cone is a cone formed by the straight lines through the fixed point in the body with the following property: on each line there exists a point such that the line is aligned with one of the major axes of the ellipsoid of inertia with the center at this point. Exactly such a cone was considered by A. M. Ampère [7]. He studied the geometry of masses in a rigid body without considering the gravity force. For other potentials in (2.60) this result is obviously incorrect.

REMARK 3. Besides Staude rotations, it is customary within the framework of rigid body dynamics to study the stability of plane motions of pendulum type (see, for example, [618, 409, 637, 166]). There is an extensive literature on this issue, which we do not mention here because such plane motions are primarily studied within the realm of the theory of vibrations and advanced theory of stability.

5 Regular Grioli precessions

The Staude solutions are regular precessions around a vertical axis. These solutions exist for any mass distribution in the body.

A more general definition of regular precessions states that there should exist two axes, one of them fixed in space and the other fixed in the body, such that the angle between them is fixed. For example, for the Lagrange top, there can arise precessions of the apex of the axis of dynamical symmetry around the vertical (see Sec. 2.3). As the Italian scientist G. Grioli showed in 1947 [243], for the Euler – Poisson equations some "nonvertical" precessions are possible. However, for such precessions to exist the moments of inertia and the location of the center of mass must satisfy some additional equations.

For the nonvertical precessions of Grioli the center of mass is located on the perpendicular from the fixed point to the circular section of the *ellipsoid of inertia*. In this sense the Grioli case is dual to the Hess case, for which the center of mass is located on the perpendicular from the fixed point to a circular section of the *ellipsoid of gyration*. This connection with the ellipsoid of inertia shows that in the Grioli case it is convenient to use the angular velocity $\boldsymbol{\omega}$ instead of the angular momentum $\boldsymbol{M}$.

For the Grioli case the simplest explicit analytical expressions can be obtained by using a nonprincipal moving frame of reference with the axis Oz passing through the center of mass of the body. Various derivations of the expressions presented in [243, 228, 162] are not complete. Grioli himself gave a rather cumbersome derivation based on the use of the Euler angles; the expressions for the direction cosines were obtained by M.P.Gulyaev [249].

In the frame of reference mentioned above the Hamiltonian H is

$$H = \frac{1}{2}(\boldsymbol{M}, \mathbf{A}\boldsymbol{M}) - x\gamma_3 = \frac{1}{2}(\mathbf{I}\boldsymbol{\omega}, \boldsymbol{\omega}) - x\gamma_3, \qquad \boldsymbol{M} = \mathbf{I}\boldsymbol{\omega}, \qquad \mathbf{A} = \mathbf{I}^{-1}, \tag{2.74}$$

where $x = \text{const.}$ and the inertia tensor has the form

$$\mathbf{I} = \begin{pmatrix} I_1 & 0 & I_{13} \\ 0 & I_1 & 0 \\ I_{13} & 0 & I_3 \end{pmatrix}.$$

We want to find the conditions under which the projection of the angular velocity onto the radius vector of the center of mass is constant, i.e., $\omega_3 = \text{const.}$ Differentiating this relation along the Hamiltonian flow defined by (2.74), we obtain the four independent additional invariant relations

$$\begin{gathered} \omega_1^2 + \omega_2^2 = \omega_3^2, \\ x\gamma_1 + I_{13}(\omega_1^2 - \omega_3^2) + I_3\omega_1\omega_3 = 0, \\ x\gamma_2 + \omega_2(I_{13}\omega_1 + I_3\omega_3) = 0, \\ x\gamma_3 - I_{13}\omega_1\omega_3 = 0. \end{gathered} \tag{2.75}$$

These relations define the sought-for periodic solutions in the reduced phase space. It follows from (2.75) that $\boldsymbol{\omega}^2 = 2\omega_3^2 = \text{const.}$ and also $\omega_2 = \omega_3 \sin\tau$, $\omega_2 = \omega_3 \cos\tau$, and $\tau = \omega_3(t - t_0)$. The constants of the first integrals can be expressed in terms of ω_3 using (2.75)

$$H = \frac{1}{2}(I_1 + I_3)\omega_3^2 = h, \quad (\boldsymbol{M}, \boldsymbol{\gamma}) = \frac{I_{13}^2 - I_1 I_3}{x}\omega_3^2 = c, \tag{2.76}$$

and ω_3 itself is found from the condition

$$\boldsymbol{\gamma}^2 = \frac{I_3^2 + I_{13}^2}{x^2}\omega_3^4 = 1.$$

Thus, for a given value (I_1, I_3, I_{13}, x) of the set of parameters of the body there exists a unique value of $|\omega_3|$ and all the other constants of the integrals, for which the Grioli solution exists.

Having obtained explicit expressions for ω_1, ω_2, and ω_3 as functions of time, the direction cosines $\boldsymbol{\alpha}$, $\boldsymbol{\beta}$, and $\boldsymbol{\gamma}$ can easily be found. Thus, we determine the motion of the rigid body in the inertial frame. The kinematic Poisson equations for the center of mass with coordinates $(\alpha_3, \beta_3, \gamma_3)$ imply $\alpha_3'' = -\alpha_3$ and $\beta_3'' = -\beta_3$ (where the prime means differentiation with respect to τ). Upon integration, we obtain $\alpha_3 = \cos\tau$ and $\beta_3 = \frac{I_3}{\sqrt{I_3^2+I_{13}^2}}\sin\tau$. It also follows from (2.75) that $\gamma_2 = \frac{I_{13}}{\sqrt{I_3^2+I_{13}^2}}\sin\tau$.

With a knowledge of $\alpha_3, \beta_3, \gamma_1, \gamma_2$ and γ_3, the other direction cosines can be found

$$\alpha_1 = -\frac{\alpha_3\gamma_3\gamma_1 \pm \gamma_2\sqrt{\gamma_1^2 + \gamma_2^2 - \alpha_3^2(\boldsymbol{\gamma}, \boldsymbol{\gamma})}}{\gamma_1^2 + \gamma_2^2},$$

$$\alpha_2 = -\frac{\alpha_3\gamma_3\gamma_2 \mp \gamma_1\sqrt{\gamma_1^2 + \gamma_2^2 - \alpha_3^2(\boldsymbol{\gamma}, \boldsymbol{\gamma})}}{\gamma_1^2 + \gamma_2^2}.$$

Similar equations hold for β_1 and β_2.

We see that the general Grioli solution is periodic in the inertial frame; the Euler angles are periodic functions of time with the same period, and the apexes on the Poisson sphere move periodically with this period. For this reason, such a regular precession is strongly degenerate. The center of mass moves uniformly along the great circle perpendicular to the axis tilted at an angle θ_0 to the vertical, where $\tan\theta_0 = \frac{\gamma_3}{\beta_3} = \frac{I_{13}}{I_3}$ (Fig. 2.51). In this sense the Grioli solution is closer not to regular precessions but to rotational motions of pendulum type. Under the conditions (2.76) a typical phase portrait is shown in Fig. 2.52. Stochastic layers are clearly seen, and the Grioli solution corresponds to the fixed point of stable type. A numerical analysis of the stability of the Grioli solution has been carried out in [575, 413, 416]. A few typical (closed) trajectories of the apexes are shown in Fig. 2.51.

Remark. Some authors (see, for example, [239]) think of precession as the motion of a body when the line of nodes rotates uniformly about some axis fixed in the body.

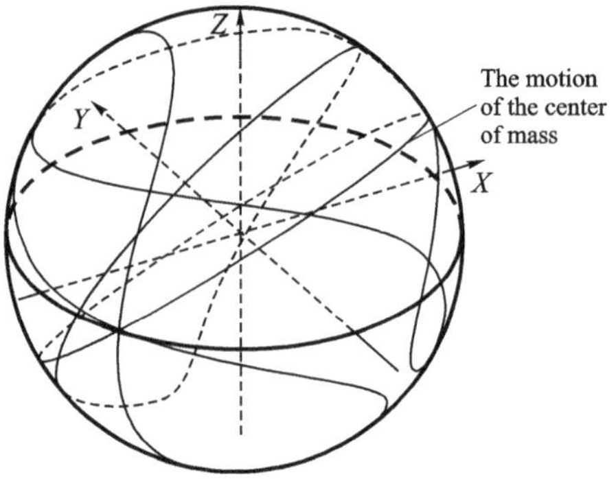

Fig. 2.51. Motion of the principal axes of the body and of the center of mass for the Grioli solution for $I_1 = 1, I_3 = \frac{1}{2}, I_{13} = 0{,}4$, and $\boldsymbol{r} = (0, 0, -1)$ (the center of mass is moving along a great circle).

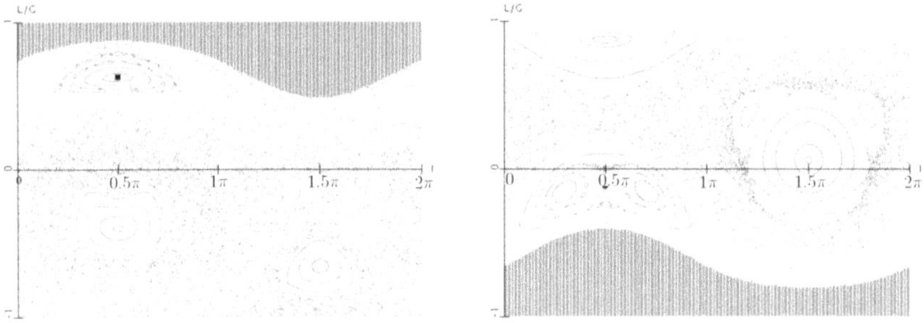

(a) $x = 1, I_1 = 1, I_{13} = 0.4, I_3 = 0.5$ (b) $x = 1, I_1 = 1, I_{13} = 0.25, I_3 = 0.75$

Fig. 2.52. Poincaré sections when a Grioli solution exists. In the case a the Grioli solution is stable and in the case b it is unstable. (The chosen section has been formed by the intersection with the plane $g = \pi$.)

6 The Bobylev – Steklov solution (1896) [53, 559]

Another particular solution which can be expressed in terms of elliptic quadratures is the Bobylev-Steklov solution. Under an additional assumption this solution coincides with the particular solution of the Kovalevskaya top defined by the fourth Appelrot class. For this solution the Hamiltonian H is

$$H = \frac{1}{2}(M_1^2 + aM_2^2 + 2M_3^2) + r\gamma_1, \quad a = \text{const.}, \quad r = \text{const.},$$

i. e., in contrast to the Kovalevskaya case, the rotational symmetry of the ellipsoid of inertia is not required ($a \neq 1$). In this case the invariant manifold is given by the equations

$$M_1 = a_0, \;\; M_2 = 0, \;\; \gamma_1 = \zeta_0 + \zeta_1 M_3^2, \;\; \gamma_2^2 = \eta_0 + \eta_1 M_3^2 + \eta_2 M_3^4, \;\; \gamma_3 = \xi_0 M_3,$$

where a_0, ζ_i, η_i, and ξ_0 depend on h and c. If the constants h and c satisfy the inequality $h \leqslant \frac{c^2}{2} - r_1$, then the Bobylev – Steklov solution does not exist.

Assuming $M_2 = 0$ and $M_1 = m = \text{const.}$, we easily get $M_3 = -\frac{r}{m}\gamma_3$. Using the area integral $(\boldsymbol{M}, \boldsymbol{\gamma}) = c$ and the geometric integral $(\boldsymbol{\gamma}, \boldsymbol{\gamma}) = 1$, we see that γ_3 can be expressed in terms of elliptic quadratures, that is,

$$\dot{\gamma}_3 = -m\sqrt{1 - \gamma_3^2 - \left(\frac{cm + r\gamma_3^2}{m^2}\right)^2}.$$

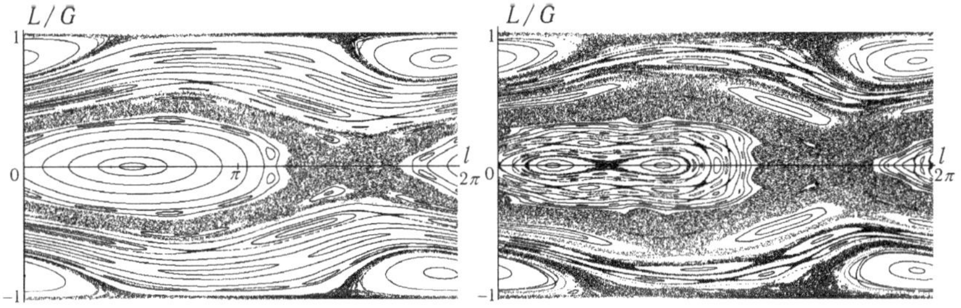

(a) $a = 1.03$ (b) $a = 1.07$

Fig. 2.53. Instability of the integrable Kovalevskaya case. The phase portrait (section formed by the intersection with the plane $g = \pi/2$) for the perturbed Kovalevskaya case with a small deviation from the dynamical symmetry $\mathbf{A} = \text{diag}(1, a, 2)$. The periodic Bobylev – Steklov solution is preserved for any a and in the phase portrait this solution corresponds to the fixed point $l = \pi/2$, $L/G = 0$. The values of the energy and area integrals are $h = 4$ and $c = 1$. (A period doubling bifurcation is seen.) It can be shown that for the Bobylev – Steklov solution the vector of angular velocity ω traces out a straight line segment in the body that is parallel to the third coordinate axis.

Figure 2.53 shows that as a increases, this solution loses stability and bifurcates, namely, one stable periodic solution gives birth to two stable solutions and one unstable one. For the unstable solution the center moves approximately as in Fig. 2.31, while in its vicinity a stochastic layer appears. As a increases, this stochastic layer extends, thereby indicating further general chaotization of the phase flow. A more complete computer analysis is beyond the scope of this book. It is interesting to note that even a very small deviation (about one percent) from the dynamical symmetry, i. e., from the Kovalevskaya case, leads to a perceptible chaotization of the portrait. This illustrates that the integrability of the Kovalevskaya case may be called “unstable”, since in practice a perfectly dynamically symmetric body is hard to make. Note that in his natural experiments N. I. Mertsalov used a top with comparatively poor dimensional accuracy; the initial conditions also could not be set with sufficient accuracy. That is why the photographs he made explain nothing [570]. For a more extensive discussion of the solution considered see [214].

7 Steklov's solution (1899) [555]

If the center of mass lies on a principal axis (e.g., $r_2 = r_3 = 0$) and $(I_1 - 2I_2)(I_1 - 2I_3) < 0$, then Steklov's partial solution is given by

$$\begin{gathered} M_2^2 = a_0 + a_1 M_1^2, \quad M_3^2 = b_0 + b_1 M_1^2, \\ \gamma_1 = \zeta_0 + \zeta_1 M_1^2, \quad \gamma_2 = \eta_0 M_1 M_2, \quad \gamma_3 = \xi_0 M_1 M_3, \end{gathered} \tag{2.77}$$

where a_i, b_i, ζ_i, η_0, and ξ_0 depend on the two dimensionless parameters I_2/I_1 and I_3/I_1; the dependence of M_1 on t can be expressed in terms of a Jacobi elliptic function. The Steklov solution defines a family of trajectories, one for each different choice of the parameters. Every isoenergetic surface contains two trajectories from this family. Moreover, each such surface is diffeomorphic to real projective space $\mathbb{R}P^3$ [212]. For Steklov's solution it is always true that $(\boldsymbol{M}, \boldsymbol{\gamma}) = 0$ and $h = \left| r_1 + \dfrac{I_1^2 r_1}{2(I_1 - I_2)(I_3 - I_1)} \right| \in (|r_1|, +\infty)$. Steklov's solution belongs to the class of isoconic motions and are periodic in the inertial frame [290].

V. A. Steklov himself indicated only one set of M_i, γ_i expressed in elliptic functions of t which satisfies equations (2.77). An alternative set was obtained by P. A. Kuzmin [358]. Orbital stability of these two solutions was studied in [357]. For a rigorous nonlinear stability analysis see [412].

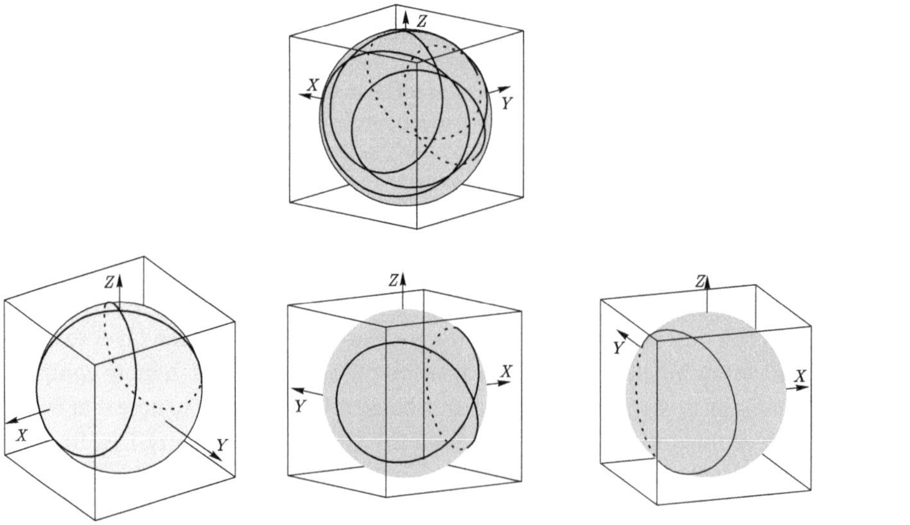

Fig. 2.54. Trajectories of apices in the inertial frame of reference for Steklov's solutions.

For Steklov's solutions, two of the three apexes and the radius vector, which joins the fixed point to the center of mass, all trace out figure eight curves in the fixed space

(Fig. 2.54). It is interesting to note that Steklov's solutions, which have an explicit analytic representation, are members of a wider class of solutions with the following properties: 1) they cannot generally be expressed analytically, 2) the motions are periodic in the fixed space, 3) two of the three apexes trace out figure eight curves. This class is discussed in detail in Sec. 6.4 Ch. 6.

8 Goryachev's solution (1899) [234]

Suppose that $I_3 = \lambda I_2$, $I_1 = \frac{16\lambda(1-\lambda)}{9-8\lambda} I_2$, $\lambda \in \left(\frac{3}{8}, \frac{1}{2}\right)$ (or $8(I_1 - 2I_3)(I_2 - I_3) + I_1 I_2 = 0$), and $r_2 = r_3 = 0$. Then there exist partial solutions for which

$$\begin{gathered} M_2^2 = a_0 + a_1 M_1^2, \quad M_3^2 = b_0 + b_1 M_1^2 + b_2 M_1^4, \\ \gamma_1 = \zeta_0 + \zeta_1 M_1^2 + \zeta_2 M_1^4, \quad \gamma_2 = (\eta_0 + \eta_1 M_1^2) M_1 M_2, \quad \gamma_3 = \xi_0 M_1 M_3, \end{gathered} \tag{2.78}$$

where a_i, b_i, ζ_i, η_i, ξ_0 depend on h, c. To these solutions there correspond two closed trajectories on a common level surface $\mathscr{E}^3_{h,c}$ of the first integrals, and the dependence of $z = M_1^2$ on time is given by a Jacobi elliptic function. Stability of Goryachev's solution is studied in [215]. Its stability depends on the value of the dimensionless parameter λ. For example, if $\lambda \in \left(\frac{3}{8}, 0.38492\right)$, then Goryachev's solution is linearly unstable.

9 Chaplygin's solution (1904) [127]

This solution occurs when

$$I_2 = \lambda I_1, \quad I_3 = \frac{9(2\lambda - 1)}{2(16\lambda - 9)} I_1, \quad \lambda \in \left(\frac{3}{2}, \frac{17 + \sqrt{73}}{16}\right),$$

and $r_2 = r_3 = 0$. In this case the following invariant relations hold

$$\begin{gathered} M_2^2 = a_0 M_1^{2/3} + a_1 M_1^2, \quad M_3^2 = b_0 M_1^{2/3} + b_1 M_1^2, \\ \gamma_1 = \zeta_0 M_1^{2/3} + \zeta_1 M_1^2, \quad \gamma_2 = (\eta_0 + \eta_1 M_1^{-4/3}) M_1 M_2, \\ \gamma_3 = (\xi_0 + \xi_1 M_1^{-4/3}) M_1 M_3. \end{gathered} \tag{2.79}$$

Here a_i, b_i, ζ_i, η_i, and ξ_i depend on λ. For this solution both the energy and the area integral vanish ($h = c = 0$), and the three-dimensional isoenergetic surface $\mathscr{E}^3_{h,c}$ is diffeomorphic to S^3. The dependence of $z = M_1^{2/3}$ on time is given by a quadrature of hyperelliptic type. Chaplygin's solution is unstable even linearly [357].

10 Kowalewski's solution (1908) [331]

Let $I_3 = \lambda I_2$, $I_1 = \dfrac{18\lambda(1-\lambda)}{10-9\lambda} I_2$, $\lambda \in \left(\dfrac{10}{27}, \lambda_\star\right)$, $\gamma_\star \approx 0.6219$, and $r_2 = r_3 = 0$. Then the partial solution is given by the equations

$$\begin{gathered} M_2^2 = a_0 + a_1 M_1 + a_2 M_1^2, \quad M_3^2 = b_0 + b_1 M_1 + b_2 M_1^2 + b_3 M_1^3, \\ \gamma_1 = \zeta_0 + \zeta_1 M_1 + \zeta_2 M_1^2 + \zeta_3 M_1^3, \quad \gamma_2 = (\eta_0 + \eta_1 M_1 + \eta_2 M_2^2) M_2, \\ \gamma_3 = (\xi_0 + \xi_1 M_1) M_3, \end{gathered} \tag{2.80}$$

where a_i, b_i, ζ_i, η_i, and ξ_i depend on λ. The values of the integrals h and c also depend on λ but in an essentially nontrivial way. The function $\omega_1(t)$ is hyperelliptic. As $\lambda \to \lambda_\star$, Kowalewski's solution tends to a stable uniform rotation about the vertical. The stability analysis of the solution (2.80) depending on λ can be found in [215].

In addition to the above-mentioned partial solutions, there are several others of similar form due to Dokshevich, Konosevich, Pozdniakovich etc. For a fuller treatment, see [228, 162, 212, 373].

2.7 Equations of Motion of a Heavy Gyrostat

1 A gyrostat

We can generalize the Euler – Poisson equations (2.6) by introducing a constant gyrostatic momentum. This momentum can be modeled, for example, by a balanced rotor rotating with constant angular velocity about an axis fixed in the body. Such a system is called a *balanced gyrostat*. A similar momentum appears during the motion of a rigid body with multiply connected cavities containing an ideal incompressible fluid with a nonzero circulation [647] (see Appendix A).

In such a generalization equations (2.6) remain unchanged, while a term linear in the momenta appears in the Hamiltonian (2.4)

$$H = \frac{1}{2}(\boldsymbol{M}, \mathbf{A}\boldsymbol{M}) - (\boldsymbol{r}, \boldsymbol{\gamma}) - (\boldsymbol{k}, \boldsymbol{M}). \tag{2.81}$$

Here $\boldsymbol{k}$ is some constant vector due to the rotor.

Remark. The equations of motion of a gyrostat are Hamiltonian equations on $e(3)$ with Hamiltonian (2.81). In physical terms, they may be obtained from the balance of angular momentum for the whole system $\mathfrak{M}$. The angular momentum of $\mathfrak{M}$ has two components:

$$\mathfrak{M} = \boldsymbol{M} + \boldsymbol{k},$$

where $\boldsymbol{M}$ is the angular momentum of the rigid body without rotor, $\boldsymbol{k}$ is the angular momentum of the rotor. The evolution of $\mathfrak{M}$ is given by

$$\frac{\tilde{d}\mathfrak{M}}{dt} = \frac{d\mathfrak{M}}{dt} + \boldsymbol{\omega} \times \mathfrak{M} = \boldsymbol{F}, \tag{2.82}$$

Table 2.2.

Generalization of the case	Author	The Hamiltonian and integral
Euler – Poinsot	Zhukovskii (1885), Volterra (1899)	$H = \frac{1}{2}\left(\boldsymbol{M} - \boldsymbol{k},\ \mathbf{A}(\boldsymbol{M} - \boldsymbol{k})\right)$ $F = \boldsymbol{M}^2$
Lagrange		$H = \frac{1}{2}(M_1^2 + M_2^2 + aM_3^2) + r_3\gamma_3 + k_3M_3$ $F = M_3$
Kovalevskaya	Yehia (1987), Komarov (1987)	$H = \frac{1}{2}\left(M_1^2 + M_2^2 + 2\left(M_3 - \frac{\lambda}{2}\right)^2\right) + r_1\gamma_1$ $F = (M_1^2 - M_2^2 - 2r_1\gamma_1)^2 + (2M_1M_2 - 2r_1\gamma_2)^2 + 4\lambda(M_3 - \lambda)(M_1^2 + M_2^2) - 8r_1\lambda M_1\gamma_3$
Goryachev – Chaplygin	Sretenskii (1963)	$H = \frac{1}{2}\left(M_1^2 + M_2^2 + 4\left(M_3 - \frac{k}{2}\right)^2\right) + r_1\gamma_1$ $F = (M_3 - k)(M_1^2 + M_2^2) - r_1M_1\gamma_3$
Hess	Sretenskii (1963)	$H = \frac{1}{2}\left(a_1(M_1+k_1)^2 + a_2M_2^2 + a_3(M_3+k_3)^2\right) + r_1\gamma_1 + r_3\gamma_3,$ $r_1\sqrt{a_3 - a_2} = r_3\sqrt{a_2 - a_1}$ $F = \sqrt{(a_2 - a_1)(a_3 - a_2)}(r_1M_1 + r_3M_3) + r_1a_3k_3 - r_3a_1k_1 = 0,$ $\dot{F}\big\vert_{F=0} = 0$

where $\boldsymbol{F}$ is the moment of external forces, and $\frac{\widetilde{d}}{dt}$, $\frac{d}{dt}$ denote derivatives of a vector in the fixed and in the moving frames of reference. Here $\boldsymbol{k} = \boldsymbol{k}(t)$ is some fixed function of time (this can be achieved, for example, with the help of some small electric motors), nevertheless, the theorem represented in (2.82) still applies. However, the law of conservation of energy does not hold, because forced rotation of the rotor requires an input of energy. In the gravitational field $\boldsymbol{F} = \boldsymbol{r} \times \boldsymbol{\gamma}$ and the rotor is revolving with a constant velocity $\boldsymbol{k} = \mathbf{I}\boldsymbol{\omega}_0 = \text{const}$. A more detailed discussion of gyrostats (systems with internal cyclic motion) can be found in the books [377, 615].

It turns out that under some additional constraints on the vector $\boldsymbol{k}$, which describes the position of the gyrostat in the body, all the cases presented in Table 2.1 Sec. 2.2 allow an "integrable" generalization (see Table 2.2).

2 The Zhukovskii – Volterra case

Let us consider in more detail the dynamics of a rigid body with an attached gyrostat when no field is applied. In this case the equations for $\boldsymbol{M}$ decouple and can be integrated independently. Let us write them in the form

$$\dot{\boldsymbol{M}} = \boldsymbol{M} \times \mathbf{A}(\boldsymbol{M} - \boldsymbol{k}).$$

The Hamiltonian and the additional integral do not depend on the configuration variables and can be represented as

$$H = \frac{1}{2}\big(\boldsymbol{M} - \boldsymbol{k}, \mathbf{A}(\boldsymbol{M} - \boldsymbol{k})\big) = h, \quad F = \boldsymbol{M}^2 = f \tag{2.83}$$

(the Hamiltonian (2.83) differs from (2.81) in the addition of the constant term and substitution $\boldsymbol{k} \to \mathbf{A}\boldsymbol{k}$). Therefore, in the space (M_1, M_2, M_3) each trajectory is the intersection between a sphere and an ellipsoid, whose centers do not necessarily coincide. These curves may be considered a direct generalization of the polhodes of the Euler problem (Sec. 2.2, Ch. 2) despite their much more complicated form (Fig. 2.55).

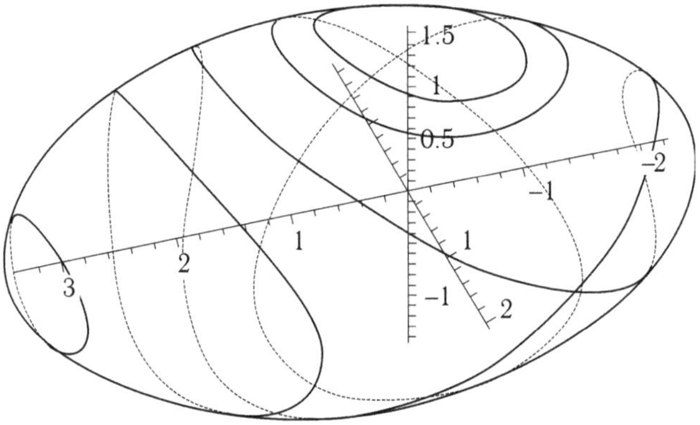

Fig. 2.55. Polhodes in the Zhukovskii – Volterra problem.

It is convenient to use a parametric representation of the branches of the bifurcation diagram on the plane of the integrals (2.83) (h, f). This representation can be easily obtained by using the condition that the integrals are functionally dependent (2.83) [286]

$$\begin{aligned} h &= \frac{t^2}{2}\left(\frac{a_1k_1^2}{(a_1-t)^2} + \frac{a_2k_2^2}{(a_2-t)^2} + \frac{a_3k_3^2}{(a_3-t)^2}\right), \\ f &= \frac{a_1^2k_1^2}{(a_1-t)^2} + \frac{a_2^2k_2^2}{(a_2-t)^2} + \frac{a_3^2k_3^2}{(a_3-t)^2}. \end{aligned} \tag{2.84}$$

Suppose that $a_1 > a_2 > a_3 > 0$; then for t varying from $-\infty$ to $+\infty$ the bifurcational curve is divided into four curves (see Fig. 2.56):

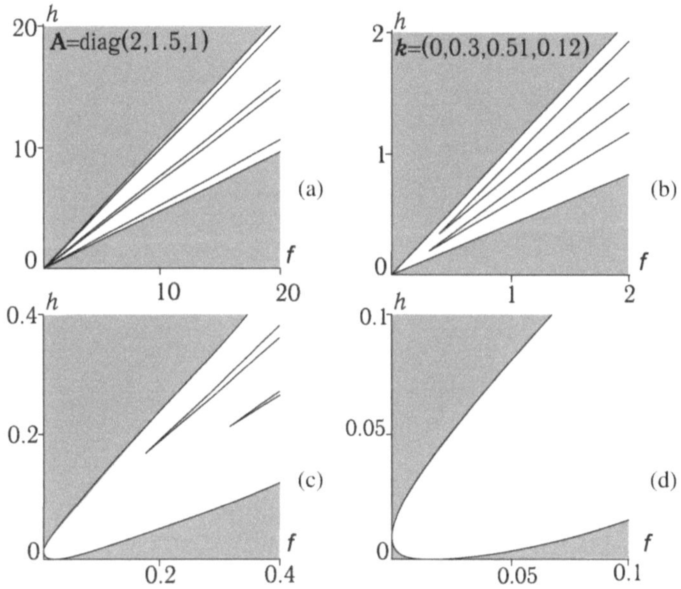

Fig. 2.56. Variously scaled images of the bifurcation diagram of the Zhukovskii – Volterra case on the plane of the integrals $h = H$ and $f = \boldsymbol{M}^2$. The domain of nonphysical values of the integrals is shown in gray. In addition to the curves marked in the images, the domain of possible motion is bounded from the left by the vertical straight line $f = c^2$, $c = (\boldsymbol{M}, \boldsymbol{\gamma})$, so that $f > c^2$. Stable curves are shown as solid lines and unstable curves as dashed lines.

I. $t \in (-\infty, a_3)$, the lower curve;
II. $t \in (a_3, a_2)$, the second curve from below with a cusp;
III. $t \in (a_2, a_1)$, the third curve from below also with a cusp;
IV. $t \in (a_1, \infty)$, the upper curve.

The upper and the lower curves meet continuously at the point $t = \infty$.

Figure 2.56 a) shows the transformation of the diagram into the diagram of the Euler – Poinsot case as $\boldsymbol{k} \to 0$ (see Fig. 2.6).

For the equations of motion in M_1, M_2, and M_3 (which are decoupled) without reference to the position variables $\boldsymbol{\gamma}$, the bifurcation diagram consists only of the branches mentioned above. For the whole system $\boldsymbol{M}, \boldsymbol{\gamma}$ the vertical line $f = c^2$, $c = (\boldsymbol{M}, \boldsymbol{\gamma})$, should be added to the diagram. The motion is possible only for $f > c^2$.

The line $f = c^2$ contains the motions of the rigid body for which the angular momentum of the body in the fixed space is vertical:

$$\boldsymbol{M} = c\boldsymbol{\gamma}. \tag{2.85}$$

It follows from this relation that the trajectories of the vector $\boldsymbol{\gamma}$ on the Poisson sphere also form polhodes congruent to those in Fig. 2.55 and obtained as the intersection of a

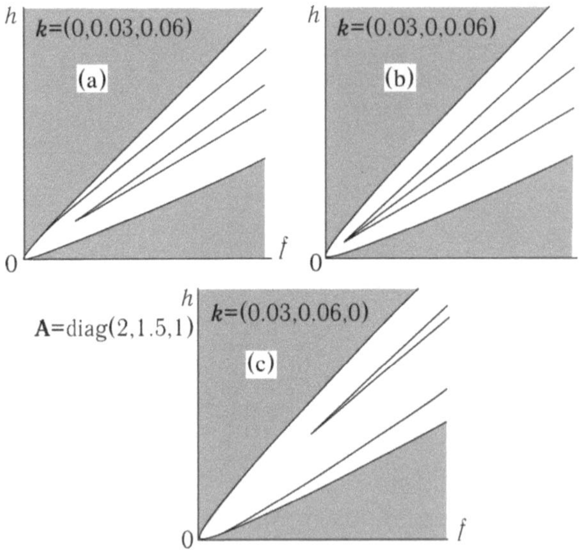

Fig. 2.57. Bifurcation diagram for the cases where the vector of the gyrostatic momentum lies in the principal plane.

(a) $k_1 = 0$, in this case the upper part of curve III merges with curve IV (i. e., curve III starts at the "middle" of curve IV);

(b) $k_2 = 0$, the two intermediate parts of the curves II and III merge;

(c) $k_3 = 0$, the two lower curves merge (similar to the case a).

sphere with an ellipsoid:

$$\frac{1}{2}\left(\boldsymbol{\gamma}-\frac{\boldsymbol{k}}{c}, \mathbf{A}\left(\boldsymbol{\gamma}-\frac{\boldsymbol{k}}{c}\right)\right)=\frac{h}{c^2}, \quad \boldsymbol{\gamma}^2 = 1. \tag{2.86}$$

If $\boldsymbol{k}$ belongs to one of the principal planes, the corresponding pair of branches on the bifurcation diagram merge (Fig. 2.57). If $\boldsymbol{k}$ is directed along a principal axis of the ellipsoid, then the two pairs of curves merge.

The stability of the curves is shown in Fig. 2.56. Linear stability was investigated already by V. Volterra and the most complete results are obtained in [500, 615]. A general conclusion on the stability is that due to the rotor the number of stationary motions (both stable and unstable) is doubled. The unstable motions disappear for small h and c (i.e., when the rotation of the rotor is fast).

The topology of the level surface of the first integrals in the Zhukovskii – Volterra problem is studied in [286, 66]. Various issues concerned with stability and bifurcations in the problem of the motion of a gyrostat are discussed in the series of works [180, 181, 182], where it is assumed that the gyrostatic momentum is aligned with a principal axis of inertia.

Separation of variables in the Zhukovskii – Volterra case. The Zhukovskii – Volterra case was integrated in terms of elliptic functions by V. Volterra in [604] (see also [615]). The contribution of N. E. Zhukovskii had been to find an additional integral and to make a study of various real mechanical formulations of the problem [647] (see also [429]). The easiest way to separate variables is to use the *Andoyer variables* [269], because then the Hamiltonian (2.81) for $r = 0$ is

$$\begin{aligned} H &= \frac{1}{2}(L^2 + \delta(G^2 - L^2)\cos^2 l) \\ &\quad -\kappa_1\sqrt{G^2 - L^2}\sin l - \kappa_2\sqrt{G^2 - L^2}\cos l - \kappa_3 L, \end{aligned} \tag{2.87}$$

where $\delta = \frac{a_2 - a_1}{a_3 - a_1}$ and $\kappa_i = \frac{a_i k_i}{a_3 - a_1}$, $i = 1, 2, 3$.

It follows from (2.87) that g is a cyclic variable. The explicit solution can be obtained in terms of quadratures containing a polynomial. These quadratures can be reduced to standard elliptic integrals. A more geometric procedure for obtaining an explicit solution is considered in Sec. 5.2, Ch. 5.

Remark. The equations of motion for a free gyrostat were considered in the early days of quantum mechanics in the context of the problem of describing the spectra of molecules. In the book of M. Born [103] it is stated that “the adequate molecular model is not simply a top, but a rigid body, in which is situated a fly-wheel with fixed bearings.” The rigid body and the flywheel play the roles of the system of nuclei and linear momentum of the electrons, respectively. Using this model, Kramers and Pauli made an attempt (not very successful) to construct a theory for the spectra of molecules for which the linear momentum of the electrons is arbitrarily distributed.

At the same time M. Born and V. Heisenberg showed that in a molecule consisting of a large number of nuclei and electrons the nuclei move, to a high approximation, like rigid systems. However, the total angular momentum of molecules differs from the rotational momentum of the nuclei because the electrons have their own angular momentum with respect to the nuclei, and this momentum is of the same order of magnitude.

These works on classical quantum theory gave birth to an entire branch of scientific activity focused on quantization of various dynamical systems concerned with motion of tops (the Gaudin magnet, the XYZ-model etc.).

The explicit Volterra solution. To obtain an explicit solution in terms of elliptic functions, V. Volterra used the projective coordinates

$$M_i = \frac{z_i}{z_4}, \quad i = 1, 2, 3 \tag{2.88}$$

and the linear nondegenerate transformation

$$z_r = \sum_{s=1}^{4} C_{rs}\xi_s, \quad r = 1, 2, 3, 4, \quad \text{with } C = \det\|C_{rs}\| \neq 0, \tag{2.89}$$

Fig. 2.58. V. Pauli and N. Born watch a spinning Thomson's top. The well-known effect of reversal of the axis of stable rotation cannot be explained unless friction is taken into account.

leading, for a suitable choice of the matrix $||C_{rs}||$, to the following equations of motion for the system

$$\begin{aligned} \xi_4\dot{\xi}_i - \xi_i\dot{\xi}_4 &= (\lambda_k - \lambda_j)\xi_j\xi_k/C, \\ \dot{\xi}_i\xi_j - \dot{\xi}_j\xi_i &= (\lambda_k - \lambda_4)\xi_k\xi_4/C. \end{aligned} \tag{2.90}$$

where (ijk) is an even permutation of (123). Here the coefficients C_{rs} are expressed in terms of the roots of some fourth-degree polynomial which contains system parameters and the values of the first integrals.

The system (2.90) has the same structure as the differential relations for the four Weierstrass sigma-functions $\sigma_1(u)$, $\sigma_2(u)$, $\sigma_3(u)$, and $\sigma_4(u)$ of the complex variable u. The λ_i are then suitably expressed in terms of the parameters of the differential equation for the Weierstrass $\wp$-function.

This is the key idea in the extensive work by V. Volterra [604], where the main goal was to explicitly obtain all direction cosines (i.e., the absolute motion). We do not give here all the calculations and discuss only some of the drawbacks of such "explicit" solutions. The equation of degree four for the coefficients C_{rs} determining the transform (2.89) cannot be solved explicitly. As a consequence, such a representation of the solution is quite formal and essentially close in spirit to the proofs of existence theorems. It is difficult to use such representations to make any useful dynamical conclusions. None of the results obtained after V. Volterra (stability, topological analysis, etc.) [615, 500] use his explicit quadratures. The tendency to reduce the problem to elliptic functions at any cost does not seem to be the most intelligent strategy, because elliptic functions are of little use for problems such as these. The Kötter "solutions" [324, 326] for the Clebsch and Steklov cases share the same disadvantages. Though these results must necessarily be referenced when writing a work concerned with rigid bodies, they are absolutely useless for understanding the dynamics and have no practical use. In general, the existing complex methods of analysis, when used blindly, can turn even a trivial mechanical problem into an overcomplicated and unsolvable problem of algebraic geometry [25].

3 Explicit solutions in the other cases

In generalizations of the Kovalevskaya and Goryachev – Chaplygin cases the gyrostatic momentum is directed along the axis of dynamical symmetry. Separation of variables for the Sretenskii case (a generalization of the Goryachev – Chaplygin case) is considered in [543, 542]. In Sec. 5.1 we obtain this result in a different way, making use of a pencil of Lie – Poisson brackets.

The case of Sretenskii is studied in [283]. The gyrostatic generalization of the Kovalevskaya case was obtained almost simultaneously by H. Yehia (1986) [629, 619], I. V. Komarov (1987) [310, 307], and L. N. Gavrilov (1987) [218]. The equations in this generalized case are integrated in terms of quadratures only when the area integral c is zero [588]. I. V. Komarov and A. V. Tsiganov proved that for $c \neq 0$ the trajectories of this system are isomorphic to those for the Clebsch case, which was integrated by F. Kötter in the 19th century using complex variables (see Ch. 3). For modern works on this subject see the comment in Sec. 3.1 Ch. 3).

In Sec. 5.1, we will extend this case to a pencil of Lie – Poisson brackets and present the corresponding additional integrals. For nonzero values of the area integral but under the condition that a certain characteristic polynomial $P(s)$ is degenerate, I. N. Gashenenko found a solution expressed in terms of elliptic functions [216]. This solution is a generalization of the fourth Appelrot class.

The second Sretenskii case, which generalizes the Hess integral, can be integrated according to the general scheme discussed in item 10, Sec. 3.2. A note should be made of the result of L. N. Gavrilov [219] stating that there are no absolutely integrable cases having an additional algebraic integral of motion other than those listed in Table (2.2).

Some issues related to the topological structure of three-dimensional isoenergetic manifolds for the general system (2.81) have been considered in recent papers by I. N. Gashenenko [213, 211].

The comprehensive classification of bifurcations of isoenergetic manifolds with respect to the values of the first integrals and parameters of the system is far from being complete.

2.8 Systems of Linked Rigid Bodies, a Rotator

We now give various formulations for the motion of a system of two (or even several) linked rigid bodies. A particular case of this problem was considered in the previous section.

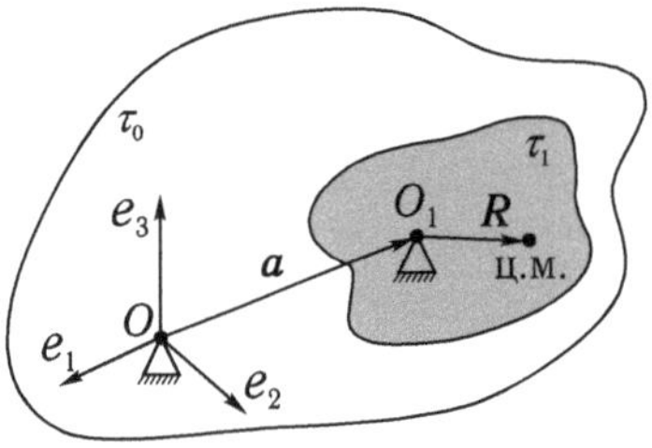

Fig. 2.59.

Two linked tops. Consider a system consisting of a "carrying" rigid body τ_0 with a fixed point O and a "carried" rigid body τ_1 fixed at a point O_1 in the carrying body (Fig. 2.59). In general, the mass distribution of the system changes as the carried body rotates.

Denote by $\boldsymbol{a} = (a_1, a_2, a_3)$ the vector from O to O_1 referred to the coordinate axes fixed in the carrying body and by $\boldsymbol{R} = (R_1, R_2, R_3)$ the vector from O_1 to the center of mass of the carried body referred to the same frame (Fig. 2.59).

The kinetic energy of the system is

$$T = \frac{1}{2}(\boldsymbol{\omega}, \mathbf{U}\boldsymbol{\omega}) + (\boldsymbol{\omega}, \mathbf{V}\boldsymbol{\omega}_1) + \frac{1}{2}(\boldsymbol{\omega}_1, \mathbf{W}\boldsymbol{\omega}_1),$$
$$\mathbf{U} = \mathbf{I}_0 + \mathbf{I}_a + \mathbf{I}_1 + \frac{1}{2}(\mathbf{I}_2 + \mathbf{I}_2^T), \quad \mathbf{V} = \mathbf{I}_1 + \mathbf{I}_2, \quad \mathbf{W} = \mathbf{I}_1, \tag{2.91}$$

where $\boldsymbol{\omega}$ is the angular velocity of the body τ_0, $\boldsymbol{\omega}_1$ is the angular velocity of the carried body with respect to the carrying body, $\mathbf{I}_0$ is the inertia tensor of the carrying body relative to the point O, $\mathbf{I}_1$ is the inertia tensor of the carried body relative to O_1, $\mathbf{I}_a = ||\delta_{ij}\boldsymbol{a}^2 - a_i a_j||$, and $\mathbf{I}_2 = ||\delta_{ij}(\boldsymbol{a}, \boldsymbol{R}) - a_i R_j||$.

The Legendre transform

$$\boldsymbol{M} = \frac{\partial T}{\partial \boldsymbol{\omega}} = \mathbf{U}\boldsymbol{\omega} + \mathbf{V}\boldsymbol{\omega}_1, \quad \boldsymbol{M}_1 = \frac{\partial T}{\partial \boldsymbol{\omega}_1} = \mathbf{V}\boldsymbol{\omega} + \mathbf{W}\boldsymbol{\omega}_1,$$
$$H = (\boldsymbol{M}, \boldsymbol{\omega}) + (\boldsymbol{M}_1, \boldsymbol{\omega}_1) - T\Big|_{\boldsymbol{\omega}, \boldsymbol{\omega}_1 \to \boldsymbol{M}, \boldsymbol{M}_1} \tag{2.92}$$

yields the Hamiltonian as a homogeneous quadratic function of the angular momenta $\boldsymbol{M}$ and $\boldsymbol{M}_1$

$$H = \frac{1}{2}(\boldsymbol{M}, \mathbf{A}\boldsymbol{M}) + (\boldsymbol{M}, \mathbf{B}\boldsymbol{M}) + \frac{1}{2}(\boldsymbol{M}_1, \mathbf{C}\boldsymbol{M}_1). \tag{2.93}$$

In contrast to the Poincaré – Zhukovskii equations governing the motion of a body with a cavity filled with vortical fluid (see Chapter 3, Sec. 3.2), the matrices **A**, **B**, and **C** depend on the position variables determining the location of the carried body with respect to the carrying body. This location is determined by an element of $SO(3)$. As position variables, we can choose the Euler angles, or the direction cosines, or any other set of local coordinates on $SO(3)$.

When no external field is applied, the position variables of the carrying body τ_0 do not appear in the Hamiltonian (2.93). Taking the direction cosines $\boldsymbol{\alpha}$, $\boldsymbol{\beta}$, and $\boldsymbol{\gamma}$ for the position variables of the carried body, we can write the equations of motion of the system (2.93) in Hamiltonian form with a bracket defined by the Lie algebra $so(3) \oplus (so(3) \oplus_s \mathbb{R}^9)$, the first term corresponds to angular momentum $\boldsymbol{M}$, the second to $\boldsymbol{M}_1$, and the third to the position variables of the body τ_1.

In coordinates the Lie – Poisson structure reads

$$\{M_i, M_j\} = -\varepsilon_{ijk}M_k, \quad \{M_{1i}, M_{1j}\} = -\varepsilon_{ijk}M_{1k},$$
$$\{M_i, \alpha_j\} = -\varepsilon_{ijk}\alpha_k, \quad \{M_{1i}, \beta_i\} = -\varepsilon_{ijk}\beta_k, \quad \{M_{1i}, \gamma_j\} = -\varepsilon_{ijk}\gamma_k.$$

The other brackets are zero. This system has four degrees of freedom (or six degrees of freedom if an external field is applied).

A body with a rotator is a system consisting of a carrying body τ_0 with a fixed point O and a carried body, the *rotator*, which rotates freely about an axis fixed in the carrying body. Denote by β the angle of rotation of the carried body about its axis of rotation.

Consider a particular case when the fixed point lies on the axis of the rotator (Fig. 2.60). A more general statement of the problem can be found in [270]. The kinetic energy is

$$T = \frac{1}{2}(\boldsymbol{\omega}, \mathbf{I}_0\boldsymbol{\omega}) + \frac{1}{2}(\boldsymbol{\omega} + \dot{\beta}\boldsymbol{n}, \mathbf{I}_1(\boldsymbol{\omega} + \dot{\beta}\boldsymbol{n})), \tag{2.94}$$

where $\boldsymbol{\omega}$ is the angular velocity of the carrying body, $\boldsymbol{n}$ is the unit vector along the axis of the rotator, $\mathbf{I}_0$ is the inertia tensor of the carrying body, and $\mathbf{I}_1$ is the inertia tensor of the rotator. If the rotator is not balanced, $\mathbf{I}_1$ depends on β.

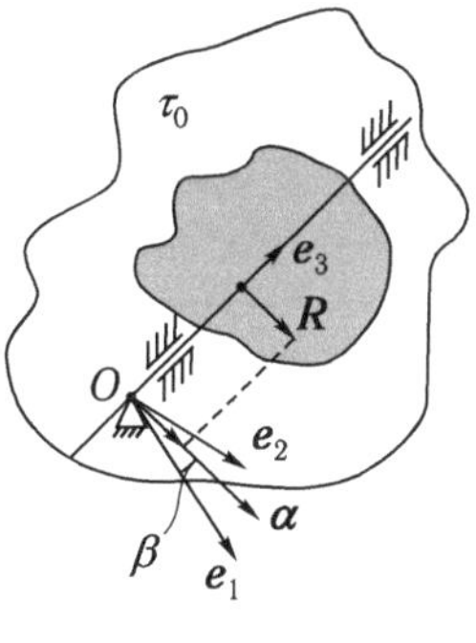

Fig. 2.60. A body with a rotator.

Choose a frame of reference fixed in the carrying body such that the axis Ox_3 is aligned with the axis of the rotator $\boldsymbol{n}$. Let $\widetilde{\boldsymbol{\alpha}}$ be the orthogonal projection of the position vector of the center of mass onto the plane x_1x_2 and let $\boldsymbol{\alpha}$ be the unit vector in the direction of $\widetilde{\boldsymbol{\alpha}}$.

$$\boldsymbol{M} = \frac{\partial T}{\partial \boldsymbol{\omega}} = (\mathbf{I}_0 + \mathbf{I}_1)\boldsymbol{\omega} + \mathbf{I}_1\dot{\beta}\boldsymbol{n}, \quad L = \frac{\partial T}{\partial \dot{\beta}} = (\boldsymbol{n}, \mathbf{I}_1(\boldsymbol{\omega} + \dot{\beta}\boldsymbol{n})),$$
$$H = \frac{1}{2}(\boldsymbol{M} - L\boldsymbol{m}, \mathbf{J}^{-1}(\boldsymbol{M} - L\boldsymbol{m})) + \frac{L^2}{2(\boldsymbol{n}, \mathbf{I}_1\boldsymbol{n})}, \tag{2.95}$$
$$\boldsymbol{m} = \frac{\mathbf{I}_1\boldsymbol{n}}{(\mathbf{I}_1\boldsymbol{n}, \boldsymbol{n})}, \quad \mathbf{J} = \mathbf{I}_0 + \mathbf{I}_1 - (\boldsymbol{n}, \mathbf{I}_1\boldsymbol{n})\boldsymbol{m} \otimes \boldsymbol{m}.$$

The brackets between the components of $\boldsymbol{M}$, L, and $\boldsymbol{\alpha}$ look like

$$\begin{gathered}\{M_i, M_j\} = -\varepsilon_{ijk}M_k, \quad \{L, \alpha_1\} = \alpha_2, \quad \{L, \alpha_2\} = -\alpha_1, \\ \{M_i, L\} = \{M_i, \alpha_1\} = \{M_i, \alpha_2\} = \{\alpha_1, \alpha_2\} = 0,\end{gathered} \tag{2.96}$$

as would be expected on intuitive grounds. The bracket (2.96) corresponds to the direct sum $so(2) \oplus e(2)$ and possesses two Casimir functions

$$F_1 = M_1^2 + M_2^2 + M_3^2, \quad F_2 = \alpha_1^2 + \alpha_2^2 = 1.$$

As a result, we have a system with two degrees of freedom. In general it is nonintegrable.

REMARK 1. This system is traditionally written in Hamiltonian form in terms of other variables. Instead of the position variables $\boldsymbol{\alpha}$, the position of the rotator in the carrying body is defined by the angle β. Denoting $p_\beta = L$, we can write the Poisson bracket for these generatrices as follows:

$$\{M_i, M_j\} = -\varepsilon_{ijk}M_k, \quad \{\beta, p_\beta\} = 1, \quad \{M_i, \beta\} = \{M_i, p_\beta\} = 0. \tag{2.97}$$

This corresponds to the direct sum $so(3) \oplus C_2$, where C_2 is the Poisson manifold with the constant structural tensor of the canonical variables β and p_β. The Poisson structure (2.97) has a single Casimir function $F = \boldsymbol{M}^2$.

Consider two known integrable cases of the system (2.95) with a cyclic integral. These cases, found by E. A. Ivin in [270], are mechanical realizations of the Zhukovskii – Volterra system considered above.

1. If $\mathbf{I}_1$ does not depend on $\boldsymbol{\alpha}$ (a *balanced rotator*), then β is a cyclic variable and L is the cyclic integral.

In this case, on the algebra $so(3) = \{(M_1, M_2, M_3)\}$ the Hamiltonian (2.95) defines a Zhukovskii – Volterra system with the vector of gyrostatic momentum $\boldsymbol{k} = L\boldsymbol{m}$.

2. *Assume that the carried body possesses dynamical symmetry with the axis of symmetry coinciding with the axis of the rotator,* $\mathbf{I}_0 = \operatorname{diag}(I_1, I_1, I_3)$. In this case the cyclic integral is

$$F = M_3 - L. \tag{2.98}$$

This integral corresponds to the cyclic variable $\beta - \varphi$, where φ is the angle of proper rotation (the angle of precession ψ is also cyclic).

Consider the system of variables commuting with the integral (2.98)

$$K_1 = M_1\alpha_1 + M_2\alpha_2, \quad K_2 = -M_1\alpha_2 + M_2\alpha_1, \quad K_3 = M_3. \tag{2.99}$$

They form the algebra $so(3)$:

$$\{K_i, K_j\} = -\varepsilon_{ijk}K_k. \tag{2.100}$$

The physical meaning of the variables (2.99) is transparent. They are the projections of the angular momentum of the carrying body onto axes fixed in the rotator. Writing the Hamiltonian (2.95) in terms of the new variables (2.99), we get

$$H = \frac{1}{2}\left(\boldsymbol{K} - F\boldsymbol{e}_3, (\widetilde{\mathbf{I}}_1^0)^{-1}(\boldsymbol{K} - F\boldsymbol{e}_3)\right) + \frac{1}{2I_3}F^2, \quad \boldsymbol{e}_3 = (0, 0, 1), \tag{2.101}$$

$$\widetilde{\mathbf{I}}_1^0 = \begin{pmatrix} x_1 + I_1 & 0 & z_1 \\ 0 & x_2 + I_1 & z_2 \\ z_1 & z_2 & x_3 \end{pmatrix},$$

where x_i and z_i are the components of the inertia tensor of the rotator relative to the point O in the frame fixed in the rotator. Thus, we again obtain the Zhukovskii – Volterra system with Hamiltonian (2.101).

Comment. The integrable cases considered above were presented in a paper by E. A. Ivin [270] as well as in his thesis, but in a cumbersome and somewhat unclear form. This was mostly due to the lack of a suitable algebraization of the equations of motion for the rotator. We have established such an algebraization using the general formalism of Poisson structures [95], thereby revealing the equivalence of this system with the Zhukovskii – Volterra system. This equivalence had not previously been explicitly pointed out. It must be noted that the dynamics of systems of linked rigid bodies is still poorly understood.

The Liouville equations govern the motion of a free rigid body whose dynamical parameters are given functions of time. These equations were obtained by J. Liouville in [383] and considered in greater detail in the treatise of F. Tisserand [574]. In this treatise, various applications of the equations to the problem of motion of celestial bodies with periodically changing parameters (due to thawing of glaciers, tidal effects, etc.) are considered. The equations of motion for such a system have the form (2.92), where $\boldsymbol{k}(t)$ is a known function of time. This shows that they are a particular case of the equations for a gyrostat with an imbalanced rotor moving with a prescribed motion in the body (i. e., there are no additional degrees of freedom due to the rotator).

We obtain the equations in Hamiltonian form on $so(3)$, i.e.,

$$\{M_i,\, M_j\} = -\varepsilon_{ijk} M_k \tag{2.102}$$

with Hamiltonian

$$H = \frac{1}{2}(\boldsymbol{M},\, \mathbf{A}(t)\boldsymbol{M}) - (\boldsymbol{k}(t),\, \boldsymbol{M}). \tag{2.103}$$

Using the separatrix splitting method, nonintegrability of the Liouville equations in the case where the moments of inertia and the gyrostatic momentum are periodic functions of time was first proved in [93] (1991). A similar result for adiabatic (i.e., slow and periodic) perturbations was obtained in [74] (1997). This paper also deals with diffusion of the adiabatic invariant at the crossing of separatrices. The property of nonintegrability was established independently in [179] (2002) for the case where only one moment of inertia evolves periodically in time in a rather specific manner. Numerical analysis of chaotic regimes of the Liouville equations can be found in [88]. For the case of a slow prescribed evolution of the parameters such an analysis will be presented in Sec. 6.6 Ch. 6.

Suppose that the orbital motion of a planet's center of mass is known; then the precession of the planet is governed by the Liouville equations. If the planet possesses axial symmetry, then one of the most well-known models for its rotational motion is the following one-and-half degree of freedom Hamiltonian system [370]

$$\frac{d\psi}{dt} = \frac{\partial H}{\partial X}, \quad \frac{dX}{dt} = -\frac{\partial H}{\partial \psi}$$

where

$$H(X, \psi, t) = \frac{\alpha}{2}\left(1 - e(t)^2\right)^{-3/2} X^2 + \sqrt{1 - X^2}\left(A(t)\sin\psi + B(t)\cos\psi\right).$$

Here X and ψ are the inclination and the angle of precession; the functions $A(t)$, $B(t)$ and the eccentricity $e(t)$ of the orbital motion are known functions of time. Depending on the accuracy required these functions may be chosen to be periodic, quasi-periodic or chaotic. If we set $e(t) = \text{const}$ and retain only periodic terms in the expansion of $A(t)$ and $B(t)$, then the resulting system is integrable. It is known as Colombo's top and its equilibria are called Cassini states. In a series of publications (see, for example, [370]), J. Laskar studied more general dependences of A and B on t which result from integration of multi-degree-of-freedom systems of planetary dynamics. (His objective was to investigate the dynamical restrictions which accompanied the evolution of the Solar System). The papers [370, 141] present general models for chaotization of the rotational motion of planets during the long-periodic evolution using the capture of Venus into 1:1 spin-orbital resonance as an example. These models also account for tidal effects, planetary oblateness, viscous friction and the interaction between the crust and the mantle. The above-mentioned papers also contain a detailed list of references on this subject.

3 The Kirchhoff Equations and Related Problems of Rigid Body Dynamics

3.1 Kirchhoff Equations

This section is entitled *Kirchhoff Equations* as historically the original prototype of such a system was the system invented by Kirchhoff to model the dynamics of a rigid body in an ideal incompressible fluid.

The label *Kirchhoff equations* is used here to encompass all Hamiltonian systems defined on the Lie-Poisson space $e(3)$ with Hamiltonian — a homogeneous quadratic function with respect to natural variables on $e(3)$. For most physical examples the requirement that the Hamiltonian be a positive definite form must be imposed.

At the end of the section there is a shorter discussion of examples for which the Hamiltonian is the sum of a quadratic function and a linear function with respect to natural variables on $e(3)$.

1 Equations of motion and physical interpretations

Dynamics of a rigid body in a fluid. The model of interest in this section describes a rigid body in an ideal fluid, meaning irrotational, incompressible, inviscid and at rest at infinity. The velocity potential of the fluid is single-valued. Remarkably the equations of motion of the rigid body decouple from the partial differential equations describing the fluid motion, and the resulting model is a system of six ordinary differential equations. A derivation of this model is given in Appendix A.

The equations of rigid body motion were obtained and studied by G. Kirchhoff. They can be written in Hamiltonian form on the algebra $e(3) = so(3)\oplus_s\mathbb{R}^3$ (see relations (2.3) in Ch. 2) and, making suitable identifications between the two sets of variables in the two different models, they take the same form as the Euler – Poisson equations (Sec. 2.1 of Ch. 2):

$$\begin{cases} \dot{\boldsymbol{M}} = \boldsymbol{M} \times \dfrac{\partial H}{\partial \boldsymbol{M}} + \boldsymbol{\gamma} \times \dfrac{\partial H}{\partial \boldsymbol{\gamma}}, \\ \dot{\boldsymbol{\gamma}} = \boldsymbol{\gamma} \times \dfrac{\partial H}{\partial \boldsymbol{M}}. \end{cases} \tag{3.1}$$

Here $\boldsymbol{M}$ amd $\boldsymbol{\gamma}$ are the three-dimensional vectors of “*impulsive momentum*” and “*impulsive force*”, respectively, in the reference frame fixed with respect to the body [302] (see also [95]).

The Hamiltonian H is a quadratic form in the variables $\boldsymbol{M}$ and $\boldsymbol{\gamma}$:

$$H = \frac{1}{2}(\mathbf{A}\boldsymbol{M}, \boldsymbol{M}) + (\mathbf{B}\boldsymbol{M}, \boldsymbol{\gamma}) + \frac{1}{2}(\mathbf{C}\boldsymbol{\gamma}, \boldsymbol{\gamma}), \tag{3.2}$$

where the matrices $\mathbf{A}$ and $\mathbf{C}$ are symmetric and $\mathbf{B}$ is arbitrary. From physical considerations H is the kinetic energy of the “body+fluid” system and $\mathbf{A}$, $\mathbf{B}$ and $\mathbf{C}$ must satisfy the

DOI 10.1515/9783110544442-003

requirement that H be positive definite. The presentation of the Kirchhoff equations in the forms (3.1) and (3.2) is due to A. Clebsch [139].

REMARK. G. Kirchhoff used an argument very similar to one previously used by W. Thomson, [571]. He obtained Eqs. (3.1) in Lagrangian form (see Appendix A):

$$\frac{d}{dt}\left(\frac{\partial L}{\partial \boldsymbol{\omega}}\right) = \frac{\partial L}{\partial \boldsymbol{\omega}} \times \boldsymbol{\omega} + \frac{\partial L}{\partial \boldsymbol{v}} \times \boldsymbol{v},$$
$$\frac{d}{dt}\left(\frac{\partial L}{\partial \boldsymbol{v}}\right) = \frac{\partial L}{\partial \boldsymbol{v}} \times \boldsymbol{\omega}.$$

The Lagrangian L, being the kinetic energy, is also a positive definite quadratic form in the velocity $\boldsymbol{v}$ and the angular velocity $\boldsymbol{\omega}$.

Equations (3.1) always possess the following integrals:

$$F_1 = (\boldsymbol{M}, \boldsymbol{\gamma}) = c_1, \quad F_2 = \boldsymbol{\gamma}^2 = c_2, \quad \text{and} \quad F_3 = H = h. \tag{3.3}$$

The functions F_1 and F_2, which are called, respectively, the *integrals of impulsive momentum and impulsive force*, are Casimir functions and fix a symplectic leaf. In what follows, by analogy with the Euler – Poisson equations, the integral F_1 will also be called the *area integral.* For the integrability of the Hamiltonian system with Hamiltonian (3.2) on the leaf, we need one more additional integral (this also follows from the theory of the last multiplier of Jacobi because of the existence of an invariant measure). In general the Kirchhoff equations are not integrable. The nonintegrability and stochastic properties of this system are discussed, e.g., in [95].

By contrast with the Euler – Poisson equations, the constant value c_2 of the integral F_2, which expresses the constancy of the impulsive force, is not necessarily equal to 1.

The physical meaning of the matrices **A**, **B**, and **C** is explained in Appendix A: they are related to added masses and moments of inertia of the body in the fluid considered. By a suitable choice of the coordinate system attached to the body, we can reduce the matrix **A** to diagonal form and **B** to symmetric form. In what follows, we always assume that such a reduction has been carried out, which allows the total number of parameters of system (3.2) to be reduced to 15.

Since any linear combination $\alpha F_1 + \beta F_2$ of the Casimir functions generates a zero Hamiltonian vector field, it follows that such a combination can be added to the Hamiltonian without changing the equations of motion. In this way the number of parameters in the Hamiltonian can be further reduced by 2. In particular, the conditions $\mathbf{B} = \lambda\mathbf{E}$ and $\mathbf{B} = 0$ (also $\mathbf{C} = \lambda\mathbf{E}$ and $\mathbf{C} = 0$) are equivalent. Yet another parameter may be removed by rescaling the time $t \to t/\alpha$, which is equivalent to multiplication of the Hamiltonian by an arbitrary constant: $H \to \alpha H$, α =const. Consequently, the number of parameters defining the family (3.2) is equal to 12.

Brun problem. Consider the motion of a rigid body about a fixed point in a linear force field, for example, where the force acting on each particle of the body is proportional to the distance from a fixed plane. This system may be represented as a Hamiltonian system of the form (3.1) with quadratic Hamiltonian (3.2). It is easy to show that the Hamiltonian H has the following form:

$$H = \frac{1}{2}(\mathbf{A}\boldsymbol{M}, \boldsymbol{M}) + \mu(\mathbf{I}\boldsymbol{\gamma}, \boldsymbol{\gamma}), \qquad \mathbf{A} = \mathbf{I}^{-1}, \tag{3.4}$$

where $\mathbf{I}$ is the tensor of inertia.

This problem was considered by Brun [108]. F. Tisserand arrived at the same problem when considering the motion of a rigid body under the action of a Newtonian gravitational source [574]. In this case, the quadratic potential in (3.4) appears as a quadrupole approximation in the expansion of the Newtonian potential as a series in powers of the ratio between the body size and the distance from the Newtonian center. It turns out that the Brun problem is equivalent to the Clebsch integrable case of the Kirchhoff equations (see Sec. 3.4 of Ch. 3). This analogy (3.4) was observed by V. A. Stekloff [549].

Grioli problem. This problem arises from modelling the motion of a charged rigid body with stationary charge distribution (dielectric) rotating about a fixed point in a constant magnetic field [244, 246, 39]. The Hamiltonian of the system involves cross terms in $\boldsymbol{M}$ and $\boldsymbol{\gamma}$ (generalized potential) and has the form

$$H = \frac{1}{2}(\mathbf{A}\boldsymbol{M}, \boldsymbol{M}) - \frac{1}{2}(\mathbf{J}\boldsymbol{\gamma}, \mathbf{A}\boldsymbol{M}), \qquad \mathbf{A} = \mathbf{I}^{-1}.$$

Here $\mathbf{I}$ is the tensor of inertia and $\mathbf{J}$ is the symmetric tensor of charge distribution.

We can also consider a more general force field composed of gyroscopic and potential forces, both quadratic in $\boldsymbol{M}$ and $\boldsymbol{\gamma}$. The equations of motion of such a system also reduce to the Kirchhoff equations. The equivalence between these problems was indicated in several sources [61, 454]. However, this equivalence is highly complicated as it is described at the level of equations of motion but not at the level of Hamiltonians and the corresponding Poisson brackets. The latter is more natural and the equivalence in such terms was established in [26].

In numerous works, which are listed in [621], H. Yehia suggested a family of generalizations of classical integrable cases of rigid body dynamics by using transformations of the type $\boldsymbol{\omega} = \boldsymbol{\omega}' + f(\boldsymbol{\gamma})\boldsymbol{\gamma}$ ($\boldsymbol{\omega}'$ is a new angular velocity and $f(\boldsymbol{\gamma})$ is an arbitrary function). In the case $f(\boldsymbol{\gamma})$ =const. such a transformation was already used by F. Tisserand. In this case, it is assumed that additional gyroscopic forces are present and we deal with a physically new formulation of the problem. Such a transformation is not necessary if we write both problems (the Kirchhoff equation and the Grioli problem) in universal Hamiltonian form on $e(3)$. However, in some works of H. Yehia the use of this transformation, which means the passage to a rotating coordinate system, helps to form a better picture of the dynamical laws of motion.

Neumann system [446]. The classical Neumann integrable problem describes the motion of a material point on the surface of a sphere in a force field with quadratic potential $U = \frac{1}{2}(\mathbf{B}\boldsymbol{q}, \boldsymbol{q})$, $\mathbf{B} = \operatorname{diag}(b_1, b_2, b_3)$. The equations of motion are

$$\begin{aligned} \ddot{q}_i &= b_i q_i + \lambda q_i, \quad i = 1, 2, 3, \\ \lambda &= -(\mathbf{B}\boldsymbol{q}, \boldsymbol{q}) - \dot{\boldsymbol{q}}^2, \end{aligned} \tag{3.5}$$

where q_i are extra Cartesian coordinates of the point on the sphere $\boldsymbol{q}^2 = 1$ and λ is a Lagrange multiplier. Passing to the variables $\boldsymbol{M} = \boldsymbol{q} \times \dot{\boldsymbol{q}}$ and $\boldsymbol{\gamma} = \boldsymbol{q}$, we can rewrite Eqs. (3.5) as

$$\dot{\boldsymbol{M}} = \boldsymbol{\gamma} \times \mathbf{B}\boldsymbol{\gamma}, \quad \dot{\boldsymbol{\gamma}} = \boldsymbol{\gamma} \times \boldsymbol{M}, \tag{3.6}$$

that is, we can represent them as a Hamiltonian system on $e(3)$ with the Hamiltonian function

$$H = \frac{1}{2}(\boldsymbol{M}, \boldsymbol{M}) + \frac{1}{2}(\mathbf{B}\boldsymbol{\gamma}, \boldsymbol{\gamma}).$$

The flow should be restricted to the level set $(\boldsymbol{M}, \boldsymbol{\gamma}) = 0$, which follows from the definition of $\boldsymbol{M}$ and $\boldsymbol{\gamma}$ for this problem.

This Hamiltonian system corresponds to the Clebsch case (see below) under the additional condition $(\boldsymbol{M}, \boldsymbol{\gamma}) = 0$, i.e., on the zero level set of the Casimir function F_2 (3.3). This analogy between the motion of a point on a sphere and the rigid body motion also holds in the n-dimensional situation (see [99]). The connection of the Neumann problem and the Clebsch case with self-similar solutions of the Landau–Lifshitz equations is considered in Appendix F.

Jacobi problem of geodesics on an ellipsoid [274]. Consider an ellipsoid in a three-dimensional space given by the equation

$$(\boldsymbol{q}, \mathbf{B}\boldsymbol{q}) = 1, \quad \mathbf{B} = \operatorname{diag}(b_1, b_2, b_3), \qquad b_i > 0.$$

On this ellipsoid, the dynamics of a free particle is governed by the equations

$$\ddot{\boldsymbol{q}} = \lambda \mathbf{B}\boldsymbol{q}, \quad \lambda = -\frac{(\dot{\boldsymbol{q}}, \mathbf{B}\boldsymbol{q})}{(\mathbf{B}\boldsymbol{q}, \mathbf{B}\boldsymbol{q})}\mathbf{B}\boldsymbol{q}. \tag{3.7}$$

Passing to the new variables

$$\boldsymbol{\gamma} = \mathbf{B}^{1/2}\boldsymbol{q}, \quad \boldsymbol{M} = (\mathbf{A}\dot{\boldsymbol{\gamma}}) \times \boldsymbol{\gamma}, \quad \text{with } \mathbf{A} = \mathbf{B}^{-1},$$

we can write the equations of motion (3.7) in Hamiltonian form (3.1) on the algebra $e(3)$ with the Hamiltonian function

$$H = \frac{1}{2} \det \mathbf{B} \frac{(\boldsymbol{M}, \mathbf{A}\boldsymbol{M})}{(\boldsymbol{\gamma}, \mathbf{B}\boldsymbol{\gamma})} \tag{3.8}$$

on the zero level set $(\boldsymbol{M}, \boldsymbol{\gamma}) = 0$ of the area constant.

After rescaling time $\frac{\det \mathbf{B}}{(\boldsymbol{\gamma}, \mathbf{B}\boldsymbol{\gamma})} dt = d\tau$, the system (3.8) on the energy level $H = c\frac{\det \mathbf{B}}{2}$ reduces to the Clebsch system (see below) with the Hamiltonian function

$$H' = \frac{1}{2}(\boldsymbol{M}, \mathbf{A}\boldsymbol{M}) - \frac{1}{2}c(\boldsymbol{\gamma}, \mathbf{B}\boldsymbol{\gamma}),$$

on the zero energy level $H' = 0$ (V. V. Kozlov [337]). Here, c is an arbitrary constant.

An analogous relation also holds in the multidimensional case [99]. In [95], the corresponding isomorphism was examined in detail for the case of quaternion equations of rigid body dynamics.

REMARK 1. K. Jacobi showed that the problem of the motion of a material point on the surface of an ellipsoid in a field with quadratic potential

$$U(\boldsymbol{q}) = \frac{1}{2}k\boldsymbol{q}^2 \tag{3.9}$$

is also integrable. This models, for example, the system on the ellipsoid for which the point considered is attached to the center of the ellipsoid by a spring obeying Hooke's law [274].

In this case we can represent the Hamiltonian in the form

$$H = \frac{1}{2}\det\mathbf{B}\frac{(\boldsymbol{M}, \mathbf{A}\boldsymbol{M})}{(\boldsymbol{\gamma}, \mathbf{B}\boldsymbol{\gamma})} + \frac{1}{2}k(\boldsymbol{\gamma}, \mathbf{A}\boldsymbol{\gamma}). \tag{3.10}$$

After rescaling time we obtain an integrable system of rigid body dynamics with fourth-order potential on the level set $(\boldsymbol{M}, \boldsymbol{\gamma}) = 0$:

$$H' = \frac{1}{2}(\boldsymbol{M}, \mathbf{A}\boldsymbol{M}) + \frac{1}{2}(\boldsymbol{\gamma}, \mathbf{B}\boldsymbol{\gamma})(k'(\boldsymbol{\gamma}, \mathbf{A}\boldsymbol{\gamma}) - c), \quad k' = \frac{k}{\det\mathbf{B}}. \tag{3.11}$$

On the levels $H' = 0$ and $H = \frac{1}{2}c\det\mathbf{B}$, the two systems given by (3.10) and (3.11) are isomorphic.

The integrable system (3.11) also appears in studies of separation of variables for polynomial potentials on the sphere [58, 616].

REMARK 2. The author of [599] found a connection between the n-dimensional Jacobi geodesic problem and the analysis of the stable equilibrium state of a Hill type linear equation with periodic coefficients. It turns out that the number of resonance zones is finite and does not exceed the ellipsoid dimension iff the periodic function $R(t)$ in the equation $\ddot{x} = -R(t)x$ is the Lagrange multiplier for a certain geodesic on the ellipsoid (more precisely, $R(t) = -\lambda(t)$).

Table 3.1. Integrable cases and invariant relations of the Kirchhoff equations

	Author	Conditions for the parameters and the first integral F
1	Kirchhoff (1870) [302]	$\mathbf{A} = \mathrm{diag}(a_1,\ a_2,\ a_3)$, $\mathbf{B} = \mathrm{diag}(b_1,\ b_2,\ b_3)$, $\mathbf{C} = \mathrm{diag}(c_1,\ c_2,\ c_3)$, $a_1 = a_2,\ b_1 = b_2,\ c_1 = c_2$, $F = M_3$ (an analog of the Lagrange case)
2	Clebsch (1871) [139]	I $\begin{cases} \mathbf{A} = \mathrm{diag}(a_1,\ a_2,\ a_3),\ \mathbf{B} = 0,\ \mathbf{C} = \mathrm{diag}(c_1,\ c_2,\ c_3), \\ \frac{c_2-c_3}{a_1} + \frac{c_3-c_1}{a_2} + \frac{c_1-c_2}{a_3} = 0,\ a_1 \neq a_2 \neq a_3 \neq a_1, \\ F = \boldsymbol{M}^2 - (\mathbf{A}\boldsymbol{\gamma},\ \boldsymbol{\gamma}), \end{cases}$ II $\begin{cases} \mathbf{B} = 0,\ a_1 = a_2 = a_3, \quad \mathbf{C} \text{ is arbitrary}, \\ F = (\boldsymbol{M}, \mathbf{C}\boldsymbol{M}) - \mu(\boldsymbol{\gamma}, \mathbf{C}^{-1}\boldsymbol{\gamma}), \quad \mu = \mathrm{const.} \end{cases}$
3	Lyapunov (1893) [392]	$\mathbf{A} = \mathbf{E} = \|\delta_{ij}\|$, $\mathbf{B} = \mathrm{diag}(b_1, b_2, b_3)$, $\mathbf{C} = \mathrm{diag}(c_1, c_2, c_3)$, $b_j = -2\mu(d_j + v), \quad c_k = (d_i - d_j)^2 + v', \ldots, \quad d_i = \mathrm{const.}$ $F = \sum \left(d_i M_i^2 + \left(2\mu \frac{d_1 d_2 d_3}{d_i} + v \right) M_i \gamma_i \right) + \mu^2 d_1 (d_2 - d_3)^2 \gamma_1^2 + \ldots$
4	Steklov (1893) [556]	$\mathbf{B} = \mathrm{diag}(b_1,\ b_2,\ b_3)$, $\mathbf{C} = \mathrm{diag}(c_1,\ c_2,\ c_3)$, $b_j = \mu \frac{a_1 a_2 a_3}{a_j} + v, \quad c_k = \mu^2 a_k (a_i - a_j)^2 + v', \mu,\ v,\ v' = \mathrm{const.}$ $F = \sum_j (M_j^2 - 2\mu(a_j + v) M_j \gamma_j) + \mu^2((a_2 - a_3)^2 + v')\gamma_1^2 + \cdots$
5	Chaplygin(1902) [131]	$\mathbf{B} = 0, \mathbf{C} = \mathrm{diag}(c, -c, 0), \mathbf{A} = \mathrm{diag}(a, a, 2a)$, $F = (M_1^2 - M_2^2 + \frac{c}{a}\gamma_3^2)^2 + 4M_1^2 M_2^2, \quad (\boldsymbol{M}, \boldsymbol{\gamma}) = 0$
6	Sokolov (2001) [537]	$\mathbf{A} = \mathrm{diag}(1, 1, 2)$, $b_{13} = \alpha$, $b_{11} = b_{22} = b_{33} = b_{12} = b_{23} = 0$, $c_{22} = 2\alpha^2$, $c_{33} = -2\alpha^2$, $c_{11} = c_{12} = c_{13} = c_{23} = 0$, $F = (M_3 - \alpha\gamma_1)\big[(M_3 - \alpha\gamma_1)(\boldsymbol{M}^2 + 4\alpha(M_3\gamma_1 - M_1\gamma_3)$ $+ 4\alpha^2(\gamma_1^2 + \gamma_3^2)) + 6\alpha(M_1 - 2\alpha\gamma_3)(\boldsymbol{M}, \boldsymbol{\gamma})\big]$
7	Chaplygin (1897) [131] Kozlov, Onoshchenko (1982) [350]	$\mathbf{A} = \mathrm{diag}(a_1,\ a_2,\ a_3)$, $a_1 > a_2 > a_3$, $b_{12} = b_{23} = 0, \quad c_{12} = c_{23} = 0$, $b_{11} = b_{22} \mp b_{13}\sqrt{\frac{a_1-a_2}{a_2-a_3}}, \quad b_{33} = b_{22} \mp b_{13}\sqrt{\frac{a_2-a_3}{a_1-a_2}}$, $c_{11} = c_{22} \mp c_{13}\sqrt{\frac{a_1-a_2}{a_2-a_3}}, \quad c_{33} = c_{22} \mp c_{13}\sqrt{\frac{a_2-a_3}{a_1-a_2}}$, $F_{\mp} = M_1\sqrt{a_1 - a_2} \mp M_3\sqrt{a_2 - a_3} = 0$ (i. e., $\dot{F}\vert_{F=0} = 0$)
8	Tsiganov, Sokolov (2002) [540]	$A = \mathrm{diag}(1, 1, 4), \quad b_{31} = 2a_1, \quad b_{13} = -a_1$, $b_{23} = 2a_2, \quad b_{32} = -a_2, \quad a_1, a_2 = \mathrm{const.} \quad (\boldsymbol{M}, \boldsymbol{\gamma}) = 0$, $c_{ij} = 0, \quad b_{11} = b_{22} = b_{33} = b_{12} = b_{21} = 0$, $F = (M_3 + a_1\gamma_1 + a_2\gamma_2)(M_1^2 + M_2^2)$

2 Integrable cases

A list of all known integrable cases of the Kirchhoff equations is presented in Table 3.1. Cases 1, 2, 3, and 4 are general integrable cases. Cases 5 and 6 are partially integrable cases: for these not only do the parameters take particular values, but additional restrictions are imposed on the values of integrals (in physical terms, on the initial conditions).

The necessary and also sufficient conditions for integrability of the Kirchhoff equations are discussed in [26, 91].

V. A. Steklov [556] showed that if the Hamiltonian (3.2) is a positive definite form, then the additional independent integral is made up of linear and quadratic terms in $\boldsymbol{M}, \boldsymbol{\gamma}$ only for the cases 1,2,3 and 4 of Table 3.1 (the Hamiltonian for a body moving in a fluid is kinetic energy and therefore is a positive definite form). In [501, 453], the result of Steklov was reverified and generalized to the case where the Hamiltonian includes additional linear terms.

Comments. **1.** The problem of algebraic integrability in the case where the matrices **B** and **C** in Hamiltonian (3.2) are not diagonal was studied by Roger Liouville [385] (one should not confuse him with the outstanding 19th century mathematician, Joseph Liouville). In this work R. Liouville claimed to have found the conditions for existence of an additional integral in the case where $b_{ij} \neq 0$ for $i \neq j$. However, this integral was not explicitly written. S. A. Chaplygin tried to obtain the integral, but was not successful. Numerical experiments carried out by the authors show chaotic behavior of the system under the general Liouville conditions, which shows the inaccuracy of the conclusions of [385].

Before publishing the first edition of this book (in the summer of 2001), the authors became familiar with the discovery by V. V. Sokolov of a new general integrable case of Eqs. (3.1) and (3.2) [537]. In this case, there is an additional fourth-order integral. The authors were able to extend this case to the Poincaré–Zhukovskii equations on $SO(4)$ (see Sec. 3.2) [75]. After some time, V. V. Sokolov and A. V. Tsiganov succeeded in finding a relation connecting the authors' case with the Kovalevskaya case: it turned out that their joint combination is integrable and they admit a reduction to the Abel–Jacobi equation in an analogous fashion to what happens in the classical Kovalevskaya case.

2. Along with the integrable cases shown in Table 3.1, there is one more general integrable case with an additional quadratic integral. It is realized for $\mathbf{A} \equiv 0$, which does not correspond to any real physical system. The additional integral $F = (\mathbf{B}\boldsymbol{\gamma}, \boldsymbol{\gamma})$ allows one to reduce the system to quadratures, which is performed particularly simply by using the so-called screw calculus (A. A. Burov, V. N. Rubanovskii [120]).

3. A further degenerate integrable case was found by V. V. Sokolov, see [333]. In this case, the Hamiltonian has the form $H = aM_1^2 - M_1\gamma_2 - M_2\gamma_1$ and again the additional integral is of degree four in momenta. It is a particular representative of a general family corresponding to a certain pencil of Poisson brackets. A mechanical meaning can be given to this family, and this is discussed in Ch. 5.

3 The case of axial symmetry

This case was discovered by G. Kirchhoff for a dynamically symmetric body of revolution moving in an ideal fluid. He also integrated the equations of motion in terms of elliptic functions.

This case of integrability is analogous to the Lagrange case of the Euler – Poisson equations (Sec. 2.3 of Ch. 2), and the additional integral $F = M_3$ is related to the existence of a cyclic coordinate (angle of proper rotation). The reduction to the system with one degree of freedom and explicit integration is described in Sec. 4.1 of Ch. 4.

The plane motions and partial solutions (of helical type) of a rigid body under the Kirchhoff integrability conditions are described in the book [365] of Lamb.

4 Clebsch case

A. Clebsch found two related integrable cases from the conditions for existence of an additional quadratic integral. Each of them is reciprocal to the other, i.e., the Hamiltonian of one of them can be taken as an integral of the other. In fact, they generate a uniform integrable family of quadratic Hamiltonians in which there are no cross terms, i.e., $\mathbf{B} = \mathbf{0}$ in (3.2).

In the table, the Clebsch integrals are given in their classical forms. This integrable family can also be represented in the following more symmetric form by using three involutive integrals:

$$\begin{aligned} \widetilde{G}_1 &= \mu\gamma_1^2 + \frac{M_3^2}{\lambda_1^2 - \lambda_2^2} + \frac{M_2^2}{\lambda_1^2 - \lambda_3^2}, \\ \widetilde{G}_2 &= \mu\gamma_2^2 + \frac{M_3^2}{\lambda_2^2 - \lambda_1^2} + \frac{M_1^2}{\lambda_2^2 - \lambda_3^2}, \\ \widetilde{G}_3 &= \mu\gamma_3^2 + \frac{M_2^2}{\lambda_3^2 - \lambda_1^2} + \frac{M_1^2}{\lambda_3^2 - \lambda_2^2}. \end{aligned} \tag{3.12}$$

Then one has

$$\sum_{i=1}^{3} \widetilde{G}_i = \mu\gamma^2, \quad \sum_{i=1}^{3} \lambda_i \widetilde{G}_i = H_I, \quad \sum_{i=1}^{3} \lambda_i^2 \widetilde{G}_i = H_{II},$$

where H_I and H_{II} are the Hamiltonians of two reciprocal Clebsch cases, respectively.

The form (3.12) of the integrals of motion was discovered by K. Uhlenbeck [589] in 1975, (and by R. L. Devaney [158]) shortly afterwards) in studying the Neumann problem, integrated by C. Neumann in 1859 by separating the variables (see Sec. 1.7 of Ch. 1). They admit a straightforward generalization to the multidimensional case [432].

In the two-dimensional case of S^2 the integrals (3.12) were already known to H. Weber (1878) [609].

A generalization of the Clebsch integrable cases to the situation where one has a pencil of Poisson brackets (in particular, the Schottky – Manakov system) is presented in Sec. 3.2 of Ch. 3, where we also present the contraction and the linear isomorphism between these cases. The Clebsch integrable family admits two different Lax representations with a spectral parameter; both are presented in the book [95].

Remark 1. F. Kötter [326] found a representation of an integrable Clebsch family in the symmetric form containing an arbitrary (spectral) parameter

$$Q(s) = \sum_{i=1}^{3} \left(\sqrt{s - D_i} M_i + \sqrt{(s - D_j)(s - D_k)} \gamma_i \right)^2,$$

where the D_i are arbitrary constants and s is the parameter. In the above formula (ijk) is an even permutation of (123). The connection of this representation with the existence of the **L**-**A**-pair on Lie pencils is considered in [95].

Remark 2. In [425], H. Minkowski found an analogy between the Clebsch case and the Jacobi problem of geodesics on an ellipsoid and thus was able to suggest his own method for its integration. A development of this analogy is presented above in Subsec. 1 of this section (see also [99]).

Remark 3. The first geometric interpretation of the motion in the Clebsch case for $(\boldsymbol{M}, \boldsymbol{\gamma}) = 0$ was given by S. A. Chaplygin [124], who represented the motion as the rolling without sliding of a hyperboloid along a helical surface. In [294] E. I. Kharlamova suggested a more natural interpretation for $(\boldsymbol{M}, \boldsymbol{\gamma}) = 0$, the corresponding motion can be viewed as a generalization of the Poinsot interpretation: the ellipsoid of inertia rolls without sliding along the surface of an elliptic cylinder, whose axis is directed along the vector $\boldsymbol{\gamma}$ and passes through the fixed point of the body.

Remark 4. In [636], it was shown that the Hamiltonian equations on $e(3)$ with Hamiltonian of the form $H = \frac{1}{2}(\boldsymbol{M}, \mathbf{A}\boldsymbol{M}) + U(\boldsymbol{\gamma})$ admit an additional quadratic integral for arbitrary values of the energy and area integrals only in the case of the Brun potential.

5 Steklov – Lyapunov family

The analysis by A. Clebsch of the conditions for the existence of quadratic integrals was incomplete. In his master's thesis, which was supervised by Lyapunov and appeared in 1893 in the form of a book [556], V. A. Steklov corrected Clebsch's analysis and found an integrable case with quadratic integral whose Hamiltonian contains cross terms in $\boldsymbol{M}$ and $\boldsymbol{\gamma}$. A. M. Lyapunov [392] subsequently found the reciprocal case, which was an extension of Steklov's result. In Table 3.1, the Hamiltonians and integrals are presented in the classical forms found by V. A. Steklov and A. M. Lyapunov.

The Steklov – Lyapunov family can be written in symmetric form by using three involutive integrals:

$$\widetilde{G}_i = \sum_{j\neq i}^{3} \frac{\left(M_{ij} + \frac{1}{2}(\lambda_i - \lambda_j)P_{ij}\right)^2}{\lambda_i - \lambda_j}, \qquad i = 1, 2, 3, \tag{3.13}$$

where the entries of the matrices $\mathbf{M} = \|M_{ij}\|$ and $\mathbf{P} = \|P_{ij}\|$ are related to the vectors $\boldsymbol{M}$ and $\boldsymbol{\gamma}$ by the formulas

$$M_{ij} = -\varepsilon_{ijk}M_k, \quad P_{ij} = -\varepsilon_{ijk}\gamma_k.$$

For example, $\widetilde{G}_1$ has the form

$$\widetilde{G}_1 = \frac{\left(M_3 + \frac{1}{2}(\lambda_1 - \lambda_2)\gamma_3\right)^2}{\lambda_1 - \lambda_2} + \frac{\left(M_2 + \frac{1}{2}(\lambda_1 - \lambda_3)\gamma_2\right)^2}{\lambda_1 - \lambda_3},$$

and the other $\widetilde{G}_i$ are obtained by cyclic permutations. The sum of these functions is a Casimir,

$$\sum_{i=1}^{3} \widetilde{G}_i = 2(\boldsymbol{M}, \boldsymbol{\gamma}).$$

REMARK. The Hamiltonian for the Lyapunov case (see Table 3.1) can be written as

$$H_{\mathrm{L}} = \frac{1}{2}\sum_{i=1}^{3} \lambda_i \widetilde{G}_i = \frac{1}{2}\boldsymbol{M}^2 + (\boldsymbol{M}, \mathbf{B}\boldsymbol{\gamma}) + \frac{1}{2}(\boldsymbol{\gamma}, \mathbf{C}\boldsymbol{\gamma}),$$

where $\mathbf{B} = \operatorname{diag}(b_1, b_2, b_3)$, $\mathbf{C} = \operatorname{diag}\left((b_2 - b_3)^2, (b_3 - b_1)^2, (b_1 - b_2)^2\right)$ and $b_k = \frac{1}{2}(\lambda_i + \lambda_j)$, $i, j, k = 1, 2, 3$.

The Hamiltonian for the Steklov case can be written as

$$\begin{aligned} H_{\mathrm{C}} &= \frac{1}{2}\sum_{i=1}^{3} \lambda_i^2 \widetilde{G}_i - \frac{1}{4}\left(\sum \lambda_i^2 + \left(\sum \lambda_i\right)^2\right)(\boldsymbol{M}, \boldsymbol{\gamma}) \\ &= \frac{1}{2}\sum_{\text{cycle}} (\lambda_i + \lambda_j)M_k^2 - \frac{1}{2}(\lambda_i + \lambda_k)(\lambda_j + \lambda_k)M_k\gamma_k + \frac{1}{4}(\lambda_i + \lambda_j)(\lambda_i - \lambda_j)^2\gamma_k^2, \end{aligned}$$

by using the change of variables $a_k = (\lambda_i + \lambda_j)$ it is reduced to the standard form (see Table 3.1) with $\mu = -1/2$.

The formula (3.13) produces the most symmetric representation of the *Steklov* and *Lyapunov* cases (see also Sec. 3.2 of Ch. 3) [2]. It can be obtained (as in the multidimensional case) by using an L-A pair with hyperelliptic spectral parameter of [65, 95]

.

A generalization of this family to the Poincaré – Zhukovskii family is presented in Sec. 3.2 of Ch. 3. Moreover, in contrast to the Clebsch case, this family also admits the addition of terms linear in $\boldsymbol{M}$ and $\boldsymbol{\gamma}$ (gyroscopic additions – see below) without losing the integrability property.

Remark. In [317] G. V. Kolosov found a fourth integral of motion of the system with Hamiltonian

$$H = \frac{1}{2}\Big(c_1M_1^2 + c_2M_2^2 + c_3M_3^2 + 2b_1M_1\gamma_1 + 2b_2M_2\gamma_2 + 2b_3M_3\gamma_3 + a_1\gamma_1^2 + a_2\gamma_2^2 + a_3\gamma_3^2\Big),$$

under the conditions

$$\begin{gathered}\frac{c_1(c_2-c_3)}{b_3-b_2} = \frac{c_2(c_3-c_1)}{b_1-b_3} = \frac{c_3(c_1-c_2)}{b_2-b_1},\\ a_1 - \frac{(b_2-b_3)^2}{c_1} = a_2 - \frac{(b_3-b_1)^2}{c_2} = a_3 - \frac{(b_1-b_2)^2}{c_3}\end{gathered} \tag{3.14}$$

on the constants a_i, b_i, c_i (i=1, 2, 3), in the form

$$F = \frac{b_3-b_1}{c_2}\left(M_1 - \frac{b_3-b_2}{c_1}\gamma_1\right)^2 + \frac{b_3-b_2}{c_1}\left(M_2 - \frac{b_3-b_1}{c_2}\gamma_2\right)^2$$

and showed that the Lyapunov and Steklov cases are particular cases of conditions (3.14). Thus, these two cases were included in a general integrable family whose particular representatives are integrals (3.13). Sometimes this family is called the *Lyapunov – Steklov – Kolosov case.*

Comments. **1.** To study the Clebsch and Steklov – Lyapunov cases, from the moment of their discovery, in keeping with the general mathematical ideology of the 19th century, various scientists tried to integrate them in terms of elliptic functions, notably G. Weber, G. G. Halphen, and F. Kötter. G. Weber integrated the second Clebsch case [609] for $(\boldsymbol{M}, \boldsymbol{\gamma}) = 0$, i.e., in essence, the Neumann problem. G. Halphen [254] considered in detail the case of dynamical symmetry whose integrability in elliptic functions is performed in a fashion analogous to that which had been applied to the Lagrange top. F. Kötter suggested his own method of integration for two Clebsch cases with nonzero area constant $(\boldsymbol{M}, \boldsymbol{\gamma}) \neq 0$ [326]. The work of F. Kötter was widely misunderstood by his contemporaries because of its complexity. V. A. Steklov and M. A. Tikhomandritskii pointed out some errors in Kötter's arguments [556, 228]. S. A. Chaplygin, who also considered that Kötter's analysis was incomplete [131], tried to integrate the Clebsch case independently [126].

In the Sitzungsberichte of the Prussian Royal Academy of Sciences, Kötter announced an explicit integration of the Steklov and Lyapunov cases. At first sight, the huge amount of calculations needed for their verification seems hopelessly impracticable. Recently, Yu. N. Fedorov suggested his own interpretation of Kötter's substitutions, for the Steklov-Lyapunov case, based on constructing an L-A pair of 2 × 2 matrices [70]. In our book [87], these arguments are presented in a modified form. Note that it is still unclear in which way F. Kötter arrived at his changes of variables. One of the possible explanations is that students of K. Weierstrass tended to have a mastery of the theory of theta-functions, which attained the highest development at the end of the 19th century, as well as the emerging theory of analytic manifolds and algebraic geometry. We also mention the book [192], in which geometric arguments partially explaining the idea of Kötter's changes are presented.

It is especially interesting to note that in the course of integrating the Steklov – Lyapunov case, Kötter in fact introduced what may be seen to be historically the first L-A pair with spectral parameter (see [95]) and the symmetric one-parameter representation

$$Q(s) = \sum_{i=1}^{3} (s - b_i)(z_i + s\gamma_i)^2$$

of integrals, where $2z_i = M_i - (d_j + d_k)\gamma_i$ and $d_i = \text{const}$.

In the works [324, 326] of Kötter, there are no explicit expressions for the characteristic polynomials in the Abel – Jacobi equations on a level surface of the first integrals. Such an "implicit" solution is of no practical use, since it does not allow one to construct bifurcation diagrams or to find especially remarkable solutions (see Ch. 2), etc. In other words such a solution does not help to perform qualitative and geometric analyses of the motion.

2. There have been several more modern results dealing with the explicit integration of the Clebsch and Steklov cases and related systems. The methods for reducing the Clebsch case to quadratures presented in [306, 294] (and in the book [16]), due to G. Kobb and E. I. Kharlamova, are based on the formalism of obtaining a complete integral of Hamilton – Jacobi equations (the first reduction of the Clebsch case to the Hamilton – Jacobi equations is due to H. Minkowski [425]). G. Kobb wrote down the Hamiltonian of the system in terms of the Euler angles [306], and E. I. Kharlamova did the same in sphero-conic coordinates [299]. It should be noted that in the unpublished manuscripts [126], S. A. Chaplygin used the Hamilton – Jacobi method for integrating two Clebsch cases in sphero-conic coordinates. However, he was successful only when the area constant is zero. Moreover, he suggested an analogous procedure for integrating the complete (i.e., for $\boldsymbol{M}$, $\boldsymbol{\gamma}$) system of equations for the Euler – Poinsot case.

In recent years, a number of papers have been written devoted to integrating the Clebsch case (and also the Schottky – Manakov case isomorphic to it; see below). The work [406] of V. G. Marikhin and V. V. Sokolov, which is the most general among them, modifies and strengthens the results of Kobb and Kharlamova. In this work, the complete integral of the corresponding Hamilton – Jacobi equation is presented and the action variables are constructed. However, the authors of [406] also failed to obtain the Abel – Jacobi equations and the expressions for characteristic polynomials, and this narrows the possibility of using these results (for qualitative analysis, say). Also note that the separation in [406] differs in an essential way from the Kötter separation, as can be seen from the fact that the genus of the corresponding hyperelliptic curves is different in each case.

Recently, A. M. Perelomov [464] has suggested an "elementary" approach to integrating the Clebsch case (more precisely, the top on $so(4)$ which is isomorphic to it). This solution is based on the ideas of Kötter and Schottky. It makes use of theta-functions of two variables and is essentially complex. This renders its role for analysis of dynamics also of little use. Mention should be made of another very general separation method [534] of E. K. Sklyanin and T. Takebe. It is assumed that this method makes it easy to resolve the Clebsch case. However, it has not yet been applied explicitly.

It seems that all possible integration methods for the Clebsch case are very complicated and, from a practical point of view, useless for applications. This accords with S. A. Chaplygin's opinion [131] that "the formulae which analytically resolve the problems (for the Clebsch and Steklov cases — *auth.*) are highly complicated, and hence they are of no use for studying the sequential course of motion." We also agree completely with K. Magnus [401] who remarked that in many respects problems of such a kind belong to the mathematical sports field and they add nothing to the description of motion.

In the book [36], general formulae using theta-functions are given as solutions of the Steklov problem. They allow us only partially to verify the Kötter results and give only a general idea of the general solution structure. Such general formulae are useless for dynamics.

An essential simplification of the explicit Kötter integration for the Steklov case was recently obtained by A. V. Tsiganov [584], who completed the Kötter transformations up to a certain transformation of the Poisson bracket of the algebra $e(3)$ and pointed out an explicit isomorphism between the Steklov system and a certain system on the sphere separating in sphero-conic coordinates [584]. These results are presented in Sec. 5.3 of Ch. 5.

Problems concerning the topological analysis of the Clebsch case were discussed by M. P. Kharlamov, T. I. Pogosyan and A. A. Oshemkov (see the survey [455]), and those of the Steklov case were discussed by O. I. Bogoyavlenskii and G. F. Ivakh (by the way, these studies were carried out without using separation of variables). The calculation of topological invariants for these cases is contained in the book [66] in which all necessary details can be found.

6 Chaplygin case (I)

S. A. Chaplygin found a particular integrable case on the zero level set of the area constant $(\boldsymbol{M}, \boldsymbol{\gamma}) = 0$ having an integral of order four. The Hamiltonian H and the integral F are

$$\begin{aligned} H &= \frac{1}{2}(M_1^2 + M_2^2 + 2M_3^2) + \frac{1}{2}c(\gamma_1^2 - \gamma_2^2), \\ F &= \left(M_1^2 - M_2^2 + c\gamma_3^2\right)^2 + 4M_1^2M_2^2. \end{aligned} \tag{3.15}$$

This system is related to the Kovalevskaya case for the Euler – Poisson equations. Its explicit integration was also carried out by S. A. Chaplygin [128].

A generalization of the Chaplygin case to the Poincaré – Zhukovskii equations was carried out by O. I. Bogoyavlenskii (see Sec. 3.2 of Ch. 3); in this case the dynamical symmetry in the Hamiltonian (3.15) is broken. In Sec. 5.2 of Ch. 5, we present a generalization of these cases to a pencil of brackets and their explicit integration.

The connection of this system with the dynamics of a rigid body in a superposition of homogeneous fields is shown in Sec. 4.1 of Ch. 4. There it is also shown in which way this case can be extended to the general quaternionic integrable case, which is a direct generalization of the Kovalevskaya case. The Chaplygin case also admits the addition of a gyrostatic term along the axis of dynamical symmetry (see Subsec. 10). Moreover, on the zero level set of the area constant, it is possible to integrate the system whose potential energy is a combination of those of the Chaplygin and Kovalevskaya cases (Ch. 5).

7 Chaplygin case (II)

This case is analogous to the Hess case in the Euler – Poisson equations and is characterized by the existence of an invariant relation of the form

$$M_1\sqrt{a_2 - a_1} \mp M_3\sqrt{a_3 - a_2} = 0, \qquad a_1 < a_2 < a_3. \tag{3.16}$$

The somewhat bulky collection of conditions presented in Table 3.1 has a simple geometric meaning. As for the Euler – Poisson equations, the gyration ellipsoid defined by the equation $(\boldsymbol{x}, \mathbf{A}\boldsymbol{x}) = 1$ has a pair of circular sections whose normal vectors are $\boldsymbol{n}_\mp = (\sqrt{a_1 - a_2}, 0, \mp\sqrt{a_2 - a_3})$; it is easy to show that under the conditions presented in Table 3.1, the function $H' = (\mathbf{B}\boldsymbol{M}, \boldsymbol{\gamma}) + \frac{1}{2}(\boldsymbol{\gamma}, \mathbf{C}\boldsymbol{\gamma})$ is invariant under rotations of the vectors $\boldsymbol{M}, \boldsymbol{\gamma}$ around $\boldsymbol{n}_-$ (or $\boldsymbol{n}_+$) (compare with Sec. 2.6 of Ch. 2).

S. A. Chaplygin presented the conditions and the method for explicit integration for this case in his master's thesis (1897) [131]. However, he did not notice its connections with the Hess case. In 1982, independently and in a more general form, this case was discovered by V. V. Kozlov and D. A. Onischenko [350], who obtained it from the conditions for separatrix splittings. It turned out that in this case, as in the Hess case, one pair of separatrices of the system (given by the condition (3.16)) is doubled and defines a one-parametric family of doubly-asymptotic motions.

The connection of this case with the existence of a cyclic variable on the level set given by (3.16) – the angle of rotation around the perpendicular to the circular section of the gyration ellipsoid – and a possible elimination of this cyclic variable by reduction are discussed in detail in Ch. 4.

8 Sokolov case [537]

In this case, the Hamiltonian H and the additional integral F can be represented as

$$H=\frac{1}{2}(M_1^2+M_2^2+2M_3^2)+M_3\alpha\gamma_1-\frac{1}{2}\alpha^2\gamma_3^2,\quad \alpha=\mathrm{const},$$

$$F=k_1k_2,\quad k_1=M_3,$$

$$k_2=M_3\big(\boldsymbol{M}^2+2\alpha(M_3\gamma_1-M_1\gamma_3)+\alpha^2(\gamma_1^2+\gamma_3^2)\big)+2\alpha(\boldsymbol{M},\boldsymbol{\gamma})(M_1-\alpha\gamma_3).$$

For the functions k_1 and k_2, the relations

$$\dot{k}_1=\alpha\gamma_2k_1,\quad \dot{k}_2=-\alpha\gamma_2k_2,$$

hold. It follows that the conditions $k_1=0$ and $k_2=0$ are *invariant conditions*. Also, the integral F can be represented as the sum of two squares: $F=\widetilde{k}_1^2+\widetilde{k}_2^2$, where $\widetilde{k}_1=0$ and $\widetilde{k}_2=0$ are a system of invariant relations. For this purpose, it is convenient to represent the Hamiltonian and the integral in the form

$$H=\frac{1}{2}(M_1^2+M_2^2+2M_3^2)+c(M_3\gamma_1-\gamma_3M_1),\quad c=\mathrm{const.},$$

$$F=(M_1^2-M_2^2+2c(M_3\gamma_1-\gamma_3M_1)+c^2\boldsymbol{\gamma}^2)^2+4(M_1M_2-c(M_3\gamma_2-M_2\gamma_3))^2.$$

This representation demonstrates a similarity between this case and the Kovalevskaya top. As was observed by V. V. Sokolov, for such a form, it is possible to directly apply the procedure of explicit integration for the Kovalevskaya case to the case considered and thus to obtain its solutions in terms of quadratures.

9 Sokolov – Tsiganov case [540, 539]

This is an analog of the Goryachev – Chaplygin case for the Euler – Poisson system, and it also admits a cubic integral of motion on the zero level set of the area integral $(M,\gamma)=0$:

$$H=\frac{1}{2}(M_1^2+M_2^2+4M_3^2)+a_1(2M_3\gamma_1-M_1\gamma_3)+a_2(2M_3\gamma_2-M_2\gamma_3),$$

$$F=(M_3+a_1\gamma_1+a_2\gamma_2)(M_1^2+M_2^2).$$

The separation of variables can be achieved by using the Andoyer variables. A system consisting of a combination of the Sokolov – Tsiganov case and the Goryachev – Chaplygin case is also discussed in [540, 539].

10 Integrable generalizations with linear terms in the Hamiltonian

Along with the previously considered integrable cases for which the Hamiltonian (3.2) is a homogeneous quadratic form in the variables $\boldsymbol{M},\boldsymbol{\gamma}$, the cases where terms linear in $\boldsymbol{M}$ and $\boldsymbol{\gamma}$ are added to the Hamiltonian are of considerable physical and mechanical interest. For the different models presented in Subsec. 1, the physical interpretations of these additions are also different. For example, for the system modelling the dynamics of a rigid body in a fluid, linear terms arise if we consider a nonsimply connected rigid body (see Appendix A). For the Brun system, adding linear terms corresponds to the presence of a rotor and a homogeneous constant force field. For the dynamics of a point on the sphere, adding linear terms corresponds to the presence of a constant electric (magnetic) field.

Equations of motion of a multiply connected body. If a body moving under the Kirchhoff conditions has holes, i.e., when it is multiply connected, then its equations of motion also have the form (3.1), but now the Hamiltonian has additional terms linear in $\boldsymbol{M}$ and $\boldsymbol{\gamma}$ [365, 291]:

$$H = \frac{1}{2}(\mathbf{A}\boldsymbol{M}, \boldsymbol{M}) + (\mathbf{B}\boldsymbol{M}, \boldsymbol{\gamma}) + \frac{1}{2}(\mathbf{C}\boldsymbol{\gamma}, \boldsymbol{\gamma}) + (\boldsymbol{a}, \boldsymbol{M}) + (\boldsymbol{b}, \boldsymbol{\gamma}), \tag{3.17}$$

where $\boldsymbol{a}$ and $\boldsymbol{b}$ are constant vectors that are linearly expressed in terms of the circulations of the fluid velocity around the contours surrounding the holes in the body (see Appendix A). Integrability conditions for such systems were studied in [499, 503].

The cases of Kirchhoff and Chaplygin (II) admit trivial generalizations (Table 3.1; see also Sec. 2.7, Ch. 2, Sec. 4.1, and Sec. 4.2, Ch. 4). Here, a constant gyrostatic momentum is added in the direction of a particular axis: for the Kirchhoff case, it is the dynamical symmetry axis; for the Chaplygin case (II), it is the axis perpendicular to the circular section of the gyration ellipsoid).

An integrable generalization of the Clebsch case by addition of linear terms is not known. A generalization of the Stekloff – Lyapunov family was obtained by V. N. Rubanovskii in [499], and the corresponding Lax representation was given in [191]. The gyrostatic generalization of the Chaplygin case (I) was obtained by H. Yehia [630] (it is also considered in Sec. 5.1 of Ch. 5).

Rubanovskii generalization of the Steklov – Lyapunov integrable family. In [499] it is shown that the Steklov – Lyapunov integrable family admits an integrable generalization in which linear terms are added to the Hamiltonian. Let us present this case of integrability by specifying the family of three involutive integrals

$$\begin{aligned}\widetilde{J}_s = \widetilde{G}_s + \frac{1}{(\lambda_s - \lambda_m)(\lambda_s - \lambda_n)} \bigg[r_s \left(M_s - \frac{\lambda_m + \lambda_n - 2\lambda_s}{2} \gamma_s \right) \\ + r_m \left(M_m - \frac{\lambda_n - \lambda_s}{2} \gamma_m \right) + r_n \left(M_n - \frac{\lambda_m - \lambda_s}{2} \gamma_n \right) \bigg]\end{aligned}$$

where (smn) is an even permutation of (123). For example, $\widetilde{J}_1$ has the following form (the others are obtained by a cyclic permutation of subscripts):

$$\begin{aligned}\widetilde{J}_1 = \widetilde{G}_1 + \frac{1}{(\lambda_1 - \lambda_2)(\lambda_1 - \lambda_3)} \bigg(r_1 \left(M_1 - \frac{\lambda_2 + \lambda_3 - 2\lambda_1}{2} \gamma_1 \right) \\ + r_2 \left(M_2 - \frac{\lambda_3 - \lambda_1}{2} \gamma_2 \right) + r_3 \left(M_3 - \frac{\lambda_2 - \lambda_1}{2} \gamma_3 \right) \bigg),\end{aligned} \tag{3.18}$$

where $\widetilde{G}_i$, $i = 1, 2, 3$ are the Steklov – Lyapunov integrals (3.13) and, as before, their sum is a Casimir:

$$\sum_{i=1}^{3} \widetilde{J}_i = 2(\boldsymbol{M}, \boldsymbol{\gamma}).$$

The Hamiltonian and the integral found by V. N. Rubanovskii [499] can be written in terms of the $\widetilde{J}_i$ as follows:

$$\begin{aligned} H &= \frac{1}{2}\sum_{i=1}^{3} \lambda_i^2 \widetilde{J}_i - \frac{1}{4}\left(\sum_{i=1}^{3} \lambda_i^2 + \left(\sum_{i=1}^{3} \lambda_i\right)^2\right)(\boldsymbol{M}, \boldsymbol{\gamma}) \\ &= \frac{1}{2}(\boldsymbol{M}, \mathbf{A}\boldsymbol{M}) + (\boldsymbol{M}, \mathbf{B}\boldsymbol{\gamma}) + \frac{1}{2}(\boldsymbol{\gamma}, \mathbf{C}\boldsymbol{\gamma}) + \frac{1}{2}(\boldsymbol{r}, \boldsymbol{M}) \\ &\quad - \frac{1}{4}(\boldsymbol{r}, (\mathbf{A} - (\operatorname{Tr}\mathbf{A})\mathbf{E})\boldsymbol{\gamma}), \\ F &= \sum_{i=1}^{3} \lambda_i \widetilde{J}_i = \boldsymbol{M}^2 + (\boldsymbol{M}, \mathbf{A}\boldsymbol{\gamma}) + \frac{1}{4}(\boldsymbol{\gamma}, \mathbf{C}\boldsymbol{\gamma}) + (\boldsymbol{r}, \boldsymbol{\gamma}), \end{aligned} \tag{3.19}$$

$$\mathbf{A} = \operatorname{diag}(a_1, a_2, a_3), \quad a_k = \lambda_i + \lambda_j, \quad \mathbf{B} = -\frac{\det \mathbf{A}}{2}\mathbf{A}^{-1},$$
$$\mathbf{C} = \operatorname{diag}\left(\frac{a_1(a_2 - a_3)^2}{4}, \frac{a_2(a_3 - a_1)^2}{4}, \frac{a_3(a_1 - a_2)^2}{4}\right),$$
$$\widetilde{\mathbf{C}} = \operatorname{diag}\left((a_2 - a_3)^2, (a_3 - a_1)^2, (a_1 - a_2)^2\right).$$

The Lax representation for this integrable case is presented in [191].

Remark. Integrals (3.18) can be also obtained by a contraction from an analogous generalization of (3.49) to $so(4)$ (Sec. 3.2 of Ch. 3).

Generalization of the Chaplygin case (I). This case of integrability can be generalized by adding a constant gyrostatic momentum along the dynamical symmetry axis (H. Yehia [630]) (1987). The Hamiltonian and the integral can be represented as

$$\begin{aligned} H &= \frac{1}{2}\left(M_1^2 + M_2^2 + 2\left(M_3 - \frac{\lambda}{2}\right)^2\right) + \frac{c}{2}(\gamma_1^2 - \gamma_2^2), \\ F &= \left(M_1^2 - M_2^2 + c\gamma_3^2\right) + 4M_1^2M_2^2 \\ &\quad + 4\lambda\left(M_3(M_1^2 + M_2^2) - c\gamma_3(M_1\gamma_1 - M_2\gamma_2)\right) \\ &\quad - 4\lambda^2(M_1^2 + M_2^2). \end{aligned} \tag{3.20}$$

In Ch. 5, we consider a generalization of this case to a pencil of Poisson brackets and to the case of addition of terms linear in γ_i to the Hamiltonian H (3.20).

Generalization of the Sokolov case. It turns out that it is possible to represent the Kovalevskaya and Sokolov cases as one family and to add the gyrostatic term [535]. In this case the Hamiltonian and the additional integral can be written as

$$H = \frac{1}{2}(M_1^2 + M_2^2 + 2M_3^2) - aM_3\gamma_1 - \frac{a^2}{2}\gamma_3^2 - kM_3 + b\gamma_2 + ka\gamma_1,$$

$$a,\ b,\ k = \text{const.},$$

$$\begin{aligned} F = {} & \left(M_1^2 - M_2^2 + 2a\gamma_1\left(M_2 - \frac{a}{2}\gamma_1 - k\right) + a^2\gamma_2^2 + 2b\gamma_2\right)^2 \\ & + 4\left(M_1M_2 + a\gamma_2\left(M_3 - \frac{a}{2}\gamma_1 - k\right) - \frac{a^2}{2}\gamma_1\gamma_2 - b\gamma_1^2\right)^2 \\ & + 4k(M_3 - k)\big(M_1^2 + M_2^2 + 2aM_1\gamma_3 + a^2(\gamma_1^2 + \gamma_2^2 + 2\gamma_3^2)\big) \\ & - 8bkM_2\gamma_3 + 4a^2k\gamma_3\big(b\gamma_3 - 2(\boldsymbol{M}, \boldsymbol{\gamma})\big). \end{aligned}$$

For $k = 0$ the integral is a sum of squares, and the separation of the Kovalevskaya variables is also generalized.

Generalization of the Sokolov – Tsiganov case. Here, we can also add the integrable linear potential of the Goryachev – Chaplygin case (on the level set $(\boldsymbol{M}, \boldsymbol{\gamma}) = 0$) to the quadratic terms:

$$\begin{aligned} H = {} & \frac{1}{2}(M_1^2 + M_2^2 + 4M_3^2) + \lambda M_3 + \mu_1\gamma_1 + \mu_2\gamma_2 \\ & + a_1(2M_3\gamma_1 - M_1\gamma_3) + a_2(2M_3\gamma_2 - M_2\gamma_3), \end{aligned}$$

$$F = \left(M_3 + a_1\gamma_1 + a_2\gamma_2 + \frac{\lambda}{2}\right)(M_1^2 + M_2^2) - (\mu_1M_1 + \mu_2M_2)\gamma_3,$$

$$\lambda, \mu_1, \mu_2, a_1, a_2 = \text{const.}.$$

This system is separated in the Andoyer variables.

Steklov particular solutions. In contrast to the Euler – Poisson equations, not so many particular solutions have been found for the Kirchhoff equations. The simplest of them are due to Lamb [365], whereas in [556] Steklov presented particular motions for a body having one symmetry plane.

A large amount of work on geometric interpretations of various particular cases is due to S. A. Chaplygin [131] (see also [124]). Results on the stability of uniform screw motions are due to A. M. Lyapunov [393]. In [131] S. A. Chaplygin also presented various conditions for the existence of integrals and particular solutions to the Kirchhoff equations.

Interesting particular solutions of the Kirchhoff equations were found by V. A. Steklov in a later work [558]. In this work, he considered the Hamiltonian (3.2) with

I $\mathbf{A} = \operatorname{diag}(a_1, a_2, a_3),\ \mathbf{B} = \operatorname{diag}(b_1, b_2, b_3),\ \mathbf{C} = \operatorname{diag}(c_1, c_2, c_3);$

II $\mathbf{A} = \operatorname{diag}(a_1, a_2, a_3),\ \mathbf{C} = \operatorname{diag}(c_1, c_2, c_3),\ \mathbf{B} \equiv 0$

and assumed also that one of the following systems of invariant relations holds:

1. $M_i = \lambda_i \gamma_i$, $\lambda_i = \text{const.}$, $i = 1, 2, 3$;
2. $M_1 = \lambda_1 \gamma_1$, $M_2 = \lambda_2 \gamma_2$, $M_3 = 0$, $\lambda_i = \text{const.}$, $i = 1, 2$;
3. $M_1 = \lambda_1 \gamma_2$, $M_2 = \lambda_2 \gamma_1$, $M_3 = 0$, $\lambda_i = \text{const.}$, $i = 1, 2$.

V. A. Steklov [558] obtained conditions for the existence of all such motions and also reduced the problem to quadratures under each of these conditions. For some of the cases the solutions turned out to be expressed in terms of elliptic functions of time; others required more complicated quadratures (which reduce to elliptic and ultraelliptic ones only in some particular cases). In this work, he also presented some geometric arguments which help to elucidate the behavior of the rigid body. Unfortunately, these motions are still poorly understood. Recently another class of particular solutions of the Kirchhoff equations was obtained in [590].

Remark. In a recent work [267], some partial solutions of screw motion type were found for the Kirchhoff equations in the Sokolov case. By using Mathematica necessary and sufficient conditions for stability of these particular solutions were found.

11 On the Kharlamov equations

In [285] M. P. Kharlamov found a generalization of the classical equations for the dynamics of a rigid body with a fixed point and in the presence of gyroscopic forces (for example, in a magnetic field). These equations have the form

$$\mathbf{I}\dot{\boldsymbol{\omega}} = (\mathbf{I}\boldsymbol{\omega} + \boldsymbol{\varkappa}) \times \boldsymbol{\omega} + \boldsymbol{\gamma} \times \frac{\partial \widetilde{U}}{\partial \boldsymbol{\gamma}}, \quad \dot{\boldsymbol{\gamma}} = \boldsymbol{\gamma} \times \boldsymbol{\omega}, \tag{3.21}$$

where $\mathbf{I} = \operatorname{diag}(I_1, I_2, I_3)$ and $\boldsymbol{\varkappa} = \boldsymbol{\gamma} F + \frac{\partial f}{\partial \boldsymbol{\gamma}}$ with F and f being arbitrary functions of $\boldsymbol{\gamma}$. Equations (3.21) admit the classical integrals

$$H = \frac{1}{2}(\mathbf{I}\boldsymbol{\omega}, \boldsymbol{\omega}) + \widetilde{U}(\boldsymbol{\gamma}), \quad F_1 = (\mathbf{I}\boldsymbol{\omega}, \boldsymbol{\gamma}) + f(\boldsymbol{\gamma}), \quad F_2 = \boldsymbol{\gamma}^2 = 1. \tag{3.22}$$

Equations (3.21) are Hamiltonian. It is convenient to represent the Poisson bracket in the variables $\boldsymbol{L} = \mathbf{I}\boldsymbol{\omega}$ and $\boldsymbol{\gamma}$:

$$\{L_i, L_j\} = -\varepsilon_{ijk}(L_k + \varkappa_k), \quad \{L_i, \gamma_j\} = -\varepsilon_{ijk}\gamma_k, \quad \{\gamma_i, \gamma_j\} = 0, \tag{3.23}$$

whilst the Hamiltonian coincides with the function H from (3.22). The integrals F_1 and F_2 are Casimir functions of the bracket (3.23).

Remark. If we assign *a priori* the bracket in the form (3.23) with some vector $\boldsymbol{\varkappa}(\boldsymbol{\gamma})$, then the Jacobi identity is equivalent to the equation

$$(\boldsymbol{\gamma} \times \nabla, \boldsymbol{\varkappa}) = 0,$$

whose general solution can be represented in the form $\boldsymbol{\varkappa} = \boldsymbol{\gamma} F + \nabla f$ mentioned above.

The rank of the bracket (3.23) is equal to 2, and its (nonsingular) symplectic leaf is diffeomorphic to the (co)tangent bundle of S^2 (if f has no singularities on the sphere). As a rule, in this case, we can represent the system in canonical[1]Hamiltonian form on T^*S^2 but with the Hamiltonian having additional singularities [95].

As is well known, there is a symplectomorphism of T^*S^2 onto an orbit with zero area constant of the (co)algebra $e(3)$. Therefore, let us consider in more detail the problem of the possibility of reducing the structure (3.23) to the Lie – Poisson bracket defined by the algebra $e(3)$ by using transformations of the form $\boldsymbol{M} = \boldsymbol{L} + \boldsymbol{\xi}(\boldsymbol{\gamma})$. We obtain the following system of partial differential equations:

$$\operatorname{grad}(\boldsymbol{\gamma}, \boldsymbol{\xi}) - \boldsymbol{\gamma} \operatorname{div} \boldsymbol{\xi} = \varkappa = \nabla f + \boldsymbol{\gamma} F.$$

This system is solvable on the sphere $\boldsymbol{\gamma}^2 = 1$. Setting $\boldsymbol{\xi} = \nabla\varphi$, we obtain the following Poisson problem with Neumann boundary conditions for the potential φ:

$$\Delta\varphi = F, \quad \left.\frac{\partial\varphi}{\partial\boldsymbol{n}}\right|_{S^2} = f(\boldsymbol{\gamma}).$$

Solving it in a neighborhood of S^2 and then restricting $\xi = \nabla\varphi$ to the sphere, we obtain the desired solution. However, this solution is not always represented in algebraic form for arbitrary F and f.

Thus, the equations in the Kharlamov form are equivalent to the standard equations of rigid body dynamics but with the Hamiltonian containing terms linear in the momenta M_i.

Remark. In certain cases, such nonstandard representation of the equations of rigid body motion with gyroscopic (or magnetic) forces leads to the discovery of "new" integrable cases of rigid body dynamics, which turn out to be equivalent to cases already known after reducing the equations to standard form on $e(3)$. The "nontrivial and complicated" integrability cases (found in the work of L. N. Oreshkina [452] and L. N. Kharlamova [296]), where equations (3.21) have an additional quadratic integral can serve as an example. We can show that after reducing all the integrals and functions f and F to homogeneous form by a suitable shift $\boldsymbol{\xi}(\boldsymbol{\gamma})$ (satisfying the equations $(\boldsymbol{\xi}, \boldsymbol{\gamma}) = f$ and $\operatorname{div}\boldsymbol{\xi} = -F$), these cases *reduce to the Clebsch case of the Kirchhoff equations.*

It is interesting to note that an insufficient use of Poisson transformation theory in the Donetsk school led to the same mistakes in finding new integrable cases and this was repeatedly manifested in connection with "new" integrable cases discovered by numerous authors (V. V. Lunev, H. Yehia and others.) [295]. Poisson structure theory allows us to write the equations of motion in a universal form in order to exclude the possibility of finding the same solution for equivalent (but different) forms of the equations.

1 Meaning in terms of natural canonical "position and momentum" variables on T^*S^2.

12 Generalization of Liouville equations. The problem of self-propulsion and the "falling cat"

In [206, 207, 424, 351, 352], the equations governing the spatial motions of a rigid body with variable mass geometry were obtained. These can also be represented in Kirchhoff form, i.e., in Hamiltonian form on the Poisson space corresponding to the algebra $e(3)$. In [206, 207, 352] these equations are applied to studying a deformable rigid body in an infinite volume of ideal fluid undergoing vortex motion, which is in a steady state at infinity.

It is assumed that variations in the mass geometry and shape of the body are caused by the action of internal forces, and in a certain movable coordinate system the movements of particles of the body are known functions of time. In [351, 355] it is assumed that the shell of the body is rigid and that the movement of the body is executed only due to a given, but variable, mass geometry.

Generalizing the arguments used to derive the Kirchhoff equations, we can easily show that, under the above conditions, the equations of motion have the form (3.1), (3.17), where the entries of the matrices **A**, **B**, and **C** and the components of the vectors $\boldsymbol{a}$, $\boldsymbol{b}$, and $\boldsymbol{c}$ are given functions of time.

Self-propulsion. The study of such a system is a particular case of the general problem of a deformable body moving in a fluid; it arises in connection with the study of the swimming motion of fish and also with the cavitation phenomenon. The first results in this direction were obtained by Taylor and Lighthill, who considered the motion in a viscous fluid and assumed that the energy exchange between the body and the fluid is due to vortex shedding from sharp edges of the body and also to the inertial action of the fluid on the body.

The problem of the motion of a rigid body due to deformations of its outer boundary in an ideal fluid executing vortex-free motion was considered by P. O. Saffman [516] in connection with the self-propulsion problem, i.e., the possibility of producing a directed propelling force. It turned out that in some cases this problem can be solved. The general approach to the self-propulsion problem, including suitable control of the mass geometry (without deformation of the outer boundary), under which the body can be transferred from one position to another was suggested in [351, 352]. It turned out that if the body has three mutually orthogonal symmetry planes, it has the complete controllability property if not all added masses (depending only on the form of the boundary) are equal to each other. It fails to have the complete controllability property if all the added masses coincide. In [351, 352], the Rashevskii – Chow theorem from nonholonomic geometry is used to prove the results; however, this does not allow one to find an explicit control method in the general case. Nevertheless, in the simplest cases (for example, in the plane), we can control the body using simple piecewise-linear motions of a point mass inside the body.

We also note that in [352], explicit integration by quadratures was considered for the planar case. In this case, the body periodically changes its mass geometry and shape. A resonance condition was given under which a propelling force can be created and the body can be made to leave a bounded domain. This result, which is of particular interest, was obtained in a slightly different way in [363] for the model example for which the motion of the body has periodic form. The planar motion of a rigid body with a rigid shell and periodically changing mass geometry was studied in detail in [355]; in this work, the trajectories of the rigid body are obtained by numerical integration. In [355], the authors considered the mechanism of acceleration of a rigid body, as well as its motion in a uniform field and in the absence of external forces but with random movement of a point inside the body. In the latter case, the behavior of the rigid body exhibits all the properties of a random walk on the plane.

We also mention a series of works [424, 207, 206] in which the authors examine various hydrodynamic statements of problems related to self-propulsion in an ideal fluid due to deformation of the outer shell of the body. In [207], the authors address issues regarding Hamiltonian structure, integrability, and stochastization of the corresponding equations of motion and also analyze the stability of some simple partial solutions.

If we study the mechanism of fish movement, then, along with the above results obtained by using Eqs. (3.1) and (3.17) with time-varying coefficients, it is possible to use other mechanical models (as is easily seen, they are closely related to each other). One of them is a system of linked rigid bodies moving in an ideal vortex-free fluid. For the planar case, this model was considered in [278]. However, one should keep in mind that all models that do not take into account the vorticity of the fluid bear little relation to the actual swimming motion of fish. In biomechanics there exist many experimental and theoretical data showing that both the vortices created by the fish itself and those created by fixed boundaries play an essential role in fish movement. It turns out that by making use of vortex shedding, a fish economizes its expenditure of muscle energy, creating a vortex street similar to the Kármán vortex street [380].

One of the model problems describing this interaction is the study of the joint motion of a rigid body and a vortex system in an ideal fluid. In the planar case, such an interaction was studied in [530, 81, 78, 529]. In particular, in [81, 245], it was shown that the problem of the planar motion of a circular cylinder and a single point vortex is *integrable*.

Falling cat problem. One of the possible models to describe the well-known effect that a cat falling from an arbitrary state always lands on its feet (and never falls on its back) (Fig. 3.1) consists in using generalizations of Eqs. (3.1) and (3.17) containing terms linear in $\boldsymbol{M}$ and $\boldsymbol{\gamma}$. Indeed, we can assume that in the falling process, the cat changes its geometry and redistributes its mass. In the first publication devoted to the study of this effect [405] (1894), photos of the process are presented (similar to Fig. 3.1); they show the falling cat changing its orientation in midair. In [339], the control is performed by flywheels clamped in the body considered. In particular, it is shown that by controlling only two flywheels (having different axes), it is possible to turn a body in such a way that it changes its initial orientation to any other given orientation.

Fig. 3.1.

The cat sharply turns the front part of its body to face downward, and this causes the back part to turn in the opposite direction. Tucking in its front paws and stretching out its back paws, it decreases the moment of inertia of the front part of its body and increases the moment of inertia of the back part. The angular momentum is equal to the product of the moment of inertia and the angular velocity, and, therefore, despite the fact that the angular momentum of the front part of the body is equal to that of the back part, the angular velocity of the front part is greater than that of the back part. As a result, the turning of the front part forward is greater than that of the back part backward. Next, the cat repeats the same trick in the opposite direction, now tucking in its back paws and stretching out its front ones; because of this, the angle turned through by the back part of the body is greater than that of the front part. When both stages are finished, the cat restores its shape and has turned through the desired angle. The cat executes these manipulations by changing its shape.

In [184, 277], the authors discuss the most conventional model of a falling cat (Shane and Cayne model). To explain the peculiarities of the fall, the cat is modelled by two Lagrange gyroscopes linked freely, i.e., the configuration space is the direct product $SO(3) \times SO(3)$ (Fig. 3.1). In [8, 277], the time-optimal control problem is also considered. In [430], the optimal control problem is related to the problem of choosing a good metric and to the general minimization problem of geodesics in sub-Riemannian geometry.

3.2 Poincaré – Zhukovskii Equations

1 Equations of motions and physical interpretations

Poisson structure and equations of motion. We start by considering a formal Hamiltonian system on the algebra $so(4)$, and before looking at any particular dynamical descriptions, present a number of consequences from the theory of Lie algebras. Depending on the dynamical origin of the equations considered, it is more convenient to use one of two different systems of variables, denoted by $(\boldsymbol{M}, \boldsymbol{p})$ or $(\boldsymbol{K}, \boldsymbol{S})$ and connected with each other by the following simple relations:

$$\boldsymbol{M} = \frac{\boldsymbol{K}+\boldsymbol{S}}{2}, \quad \boldsymbol{p} = \frac{\boldsymbol{K}-\boldsymbol{S}}{2}.$$

The variables $(\boldsymbol{M}, \boldsymbol{p})$ correspond to the "standard" (matrix) representation of $so(4)$, where the commutator relations have the form

$$\{M_i, M_j\} = -\varepsilon_{ijk}M_k, \quad \{M_i, p_j\} = -\varepsilon_{ijk}p_k, \quad \{p_i, p_j\} = -\varepsilon_{ijk}M_k. \tag{3.24}$$

The Casimir functions of the Poisson structure (3.24) are

$$F_1 = \boldsymbol{M}^2 + \boldsymbol{p}^2, \quad F_2 = (\boldsymbol{M}, \boldsymbol{p}). \tag{3.25}$$

The level surfaces of these integrals are diffeomorphic to $S^2 \times S^2$. This becomes obvious if we write the integrals in the variables $(\boldsymbol{K}, \boldsymbol{S})$ (see (3.29)).

The equations of motion can be represented in the vector form

$$\dot{\boldsymbol{M}} = \boldsymbol{M} \times \frac{\partial H}{\partial \boldsymbol{M}} + \boldsymbol{p} \times \frac{\partial H}{\partial \boldsymbol{p}}, \qquad \dot{\boldsymbol{p}} = \boldsymbol{p} \times \frac{\partial H}{\partial \boldsymbol{M}} + \boldsymbol{M} \times \frac{\partial H}{\partial \boldsymbol{p}}. \tag{3.26}$$

This system of variables is also convenient for describing the *linear pencil of Poisson structures* $\mathscr{L}_x$ (in what follows we repeatedly use it) which includes the algebras $so(4)$, $e(3)$, and $so(3,1)$. The commutator relations for $\mathscr{L}_x$ are given in terms of the variables $(\boldsymbol{M}, \boldsymbol{\gamma})$ by

$$\{M_i, M_j\} = -\varepsilon_{ijk}M_k, \quad \{M_i, p_j\} = -\varepsilon_{ijk}p_k, \quad \{p_i, p_j\} = -x\varepsilon_{ijk}M_k, \tag{3.27}$$

where x is an arbitrary constant. For $x > 0$ the commutator relations (3.27) define an algebra isomorphic to the algebra $so(4)$, for $x = 0$ they define the algebra $e(3)$, and for $x < 0$ an algebra isomorphic the algebra $so(3,1)$. Indeed, for $x \neq 0$ the transformation $\boldsymbol{M} \to \boldsymbol{M}, \boldsymbol{p} \to |x|^{1/2}\boldsymbol{p}$ maps the relations in (3.27) either to those in (3.24) or to the corresponding commutation relations for $so(3,1)$.

Under passage to the limit $x \to 0$ in the commutator relations (3.27), we obtain the algebra $e(3)$. This procedure is called *contraction* of Lie algebras, and in some cases, it allows us to find a relation between the integrable cases existing for the equations on different representatives of the pencil $\mathscr{L}_x$ (see also [447]).

The variables $(\boldsymbol{K}, \boldsymbol{S})$ correspond to the "canonical" decomposition of the Lie algebra $so(4)$ into the direct sum $so(3) \oplus so(3)$:

$$\{K_i, K_j\} = -\varepsilon_{ijk}K_k, \quad \{K_i, S_j\} = 0, \quad \{S_i, S_j\} = -\varepsilon_{ijk}S_k. \tag{3.28}$$

In the $(\boldsymbol{K}, \boldsymbol{S})$ variables, the Casimir functions take the form

$$F_1 = (\boldsymbol{K}, \boldsymbol{K}), \quad F_2 = (\boldsymbol{S}, \boldsymbol{S}). \tag{3.29}$$

The equations of motion are

$$\dot{\boldsymbol{K}} = \boldsymbol{K} \times \frac{\partial H}{\partial \boldsymbol{K}}, \qquad \dot{\boldsymbol{S}} = \boldsymbol{S} \times \frac{\partial H}{\partial \boldsymbol{S}}, \tag{3.30}$$

and in the case where the Hamiltonian H is quadratic, they can be interpreted as the equations of two coupled three-dimensional Euler tops – each one on $so(3)$.

Poincaré – Zhukovskii equations. By these equations we mean the Hamiltonian system on $so(4)$ with quadratic Hamiltonian (Euler – Poincaré equations on $so(4)$; see Sec. 1.2 of Ch. 1). In the vector representation, the Hamiltonian function can be represented in the following two equivalent forms:

$$H = \frac{1}{2}(\boldsymbol{M}, \mathbf{A}\boldsymbol{M}) + (\boldsymbol{M}, \mathbf{B}\boldsymbol{p}) + \frac{1}{2}(\boldsymbol{p}, \mathbf{C}\boldsymbol{p}) \tag{3.31}$$

or

$$H = \frac{1}{2}\left(\boldsymbol{K}, \mathbf{A}'\boldsymbol{K}\right) + \left(\boldsymbol{K}, \mathbf{B}'\boldsymbol{S}\right) + \frac{1}{2}\left(\boldsymbol{S}, \mathbf{C}'\boldsymbol{S}\right), \tag{3.32}$$

where $\mathbf{A}$, $\mathbf{A}'$, $\mathbf{C}$, and $\mathbf{C}'$ are symmetric matrices and $\mathbf{B}$ and $\mathbf{B}'$ are arbitrary matrices. There is the following obvious dependence between these matrices:

$$\mathbf{A} = \mathbf{A}' + \mathbf{B}' + \mathbf{B}'^T + \mathbf{C}', \quad \mathbf{B} = \mathbf{A}' - \mathbf{B}' + \mathbf{B}'^T - \mathbf{C}',$$
$$\mathbf{C} = \mathbf{A}' - \mathbf{B}' - \mathbf{B}'^T + \mathbf{C}'.$$

For definiteness, this notation will be used consistently in what follows.

The Hamiltonian (3.31), (3.32) depends on 21 parameters. There are three types of simple transformations that change (and in particular, exclude) some parameters in the Hamiltonian without changing the equations of motion. The first type consists of the group of transformations $SO(3) \times SO(3)$. By using this group, the matrices $\mathbf{A}'$ and $\mathbf{C}'$ in the representation (3.32) can be simultaneously brought to diagonal form. The addition of an arbitrary linear combination of the Casimir functions F_1 and F_2, which are homogeneous quadratic functions, to the Hamiltonian allows us to exclude two more parameters. Multiplying the Hamiltonian by an arbitrary constant, $H \to \alpha H$, and rescaling time by $t \to \frac{1}{\alpha}t$ allows us to further decrease the number of parameters by one more. *Hence, the family of quadratic Hamiltonians* (3.31) (*or* (3.32)) *is determined by* 12 *parameters.*

In what follows, in calculating the number of parameters in integrable families, time rescaling is excluded, and so the number of free parameters of the corresponding families will increase by one.

Historical comments. **1.** The system (3.31) is associated with the names of H. Poincaré and N. E. Zhukovskii. Both of them obtained it by considering the problem of motion of a body with cavities filled with a vortex fluid. This model is considered in the next subsection, and a detailed derivation of the equations of motion using fundamental hydrodynamical principles is contained in Appendix A (In his well-known treatise [365], H. Lamb presented the derivation of the equations and some stability results of H. Poincaré). Later on, it turned out that the same equations also describe other mechanical and physical systems.

2. H. Poincaré's derivation of Eqs. (3.26) and (3.31) in his work [472] was a completely modern one, using the formalism (developed by him) of general equations of motion on a Lie group. Also, he explicitly pointed out the reduction to elliptic quadratures for the axisymmetric case and considered the stability of regular precessions. It is interesting to view his disagreement with W. Thomson (Lord Kelvin) concerning the frequency and stability of the precession of a body with a fluid-filled cavity in this light: H. Poincaré used the system (3.30) to describe the motion of the Earth, which consists of a rigid shell (crust and mantle) and a liquid core. Later on, this model was also studied by V. A. Steklov, who presented the integrable cases discovered by him [550].

3. N. E. Zhukovskii obtained Eqs. (3.30) in his master's thesis [647], using mechanical and hydrodynamical reasoning. Later on, he focused on calculating the physical system parameters for cavities having various geometries. In considering multiply connected cavities admitting circulation flows, N. E. Zhukovskii found an integrable case, which shortly afterwards was integrated in terms of elliptic functions by V. Volterra [604] (see Sec. 2.7 of Ch. 2 and Sec. 5.2 of Ch. 5). The circulation flows in cavities lead to the appearance of linear terms in the Hamiltonian (3.31).

Dynamics of a rigid body with a cavity containing a fluid. *The Poincaré-Zhukovskii* equations (3.30), (3.32) describe the motion around a fixed point of a body having an ellipsoidal cavity completely filled with a homogeneous ideal fluid undergoing vortex motion [365, 421, 291, 429]; a detailed derivation of these equations is presented in Appendix A.

Let us choose a coordinate system which is rigidly attached to the shell and whose axes are aligned with the principal axes of the ellipsoidal cavity. In the representation $(\boldsymbol{K}, \boldsymbol{S})$, the vector $\boldsymbol{S}$ is proportional to the *fluid vorticity* $\boldsymbol{\Omega} = \frac{1}{2}\operatorname{rot}\boldsymbol{v}$, and its components in the frame fixed to the shell have the form

$$S_1 = \frac{2}{5}m_0 d_2 d_3 \Omega_1, \quad S_2 = \frac{2}{5}m_0 d_1 d_3 \Omega_2, \quad S_1 = \frac{2}{5}m_0 d_1 d_2 \Omega_3,$$

where d_1, d_2, and d_3 are the semiaxes of the cavity and m_0 is the fluid mass. Its evolution is governed by the *Helmholtz hydrodynamic equations* [365].

The vector $\boldsymbol{K}$is the angular momentum of the "body+fluid" system and is equal to

$$\boldsymbol{K} = \mathbf{I}\boldsymbol{\omega} + \mathbf{J}\boldsymbol{\Omega},$$

where $\mathbf{I}$ is the tensor of inertia of the "body+fluid" system, the entries of the matrix $\mathbf{J} = \operatorname{diag}(J_1, J_2, J_3)$ have the form

$$J_i = \frac{4}{5}m_0\varepsilon_{ijk}\frac{d_j^2 d_k^2}{d_j^2 + d_k^2}$$

and $\boldsymbol{\omega}$ is the angular velocity of the rigid shell.

The Hamiltonian is the kinetic energy expressed in terms of the variables $(\boldsymbol{K}, \boldsymbol{S})$ [429]:

$$H = \frac{1}{2}\left(\boldsymbol{K}, \mathbf{A}'\boldsymbol{K}\right) + \left(\boldsymbol{K}, \mathbf{B}'\boldsymbol{S}\right) + \frac{1}{2}\left(\boldsymbol{S}, \mathbf{C}'\boldsymbol{S}\right), \tag{3.33}$$

where $\mathbf{A}' = \mathbf{I}^{-1}$, $\mathbf{B}' = -\mathbf{D}\mathbf{I}^{-1}$, $\mathbf{C}' = \mathbf{D}(\mathbf{I}^{-1} + \mathbf{J}^{-1})\mathbf{D}$, and

$$\mathbf{D} = \operatorname{diag}\left(\frac{2d_2d_3}{d_2^2 + d_3^2}, \frac{2d_1d_3}{d_1^2 + d_3^2}, \frac{2d_1d_2}{d_1^2 + d_2^2}\right).$$

The function (3.33) *depends on* 9 *parameters*: 6 moments of inertia of the shell and ratios between the principal semiaxes of the cavity and the mass of the fluid.

REMARK. A generalization of the Poincaré – Zhukovskii equations to the case of existence of a force field was considered in [603]. In this case, a Hamiltonian system on the direct sum $e(3) \oplus so(3)$ is obtained. In [603], Veselova presents without proof some necessary conditions for existence of additional analytic and polynomial integrals and presents a trivial analog of the Lagrange case, which necessarily exists for such systems.

Rigid body dynamics in $\mathbb{R}^4$. Four-dimensional Euler top. The equations governing the motion about a fixed point of a free four-dimensional rigid body in a coordinate system fixed in the body have a similar but less general form. From this viewpoint, the problem was considered in the 19th century by W. Frahm (1875) and F. Schottky (1891) [196, 522, 61] (see Appendix A). The statement of the problem of the motion of a four-dimensional rigid body goes back to A. Kelley. For the n-dimensional case, the equations of motion were obtained by W. Frahm [196], who also found a number of first integrals. He noted that for $n = 4$, i.e., for the system (3.26) with (3.31) it suffices to find one more integral in order to integrate the system by quadratures.

Let us choose a system of principal axes of the body; in this system, the moment of inertia tensor $\mathbf{J} = \|J_{\mu\nu}\| = \|\int x_\mu x_\nu dm\|$ is of diagonal form $\mathbf{J} = \operatorname{diag}(\lambda_0, \lambda_1, \lambda_2, \lambda_3)$. In this case, the Hamiltonian can be represented as

$$H = \frac{1}{2}(\boldsymbol{M}, \mathbf{A}\boldsymbol{M}) + \frac{1}{2}(\boldsymbol{p}, \mathbf{C}\boldsymbol{p}), \tag{3.34}$$

where

$$\begin{aligned} \mathbf{A} &= \operatorname{diag}\left(\frac{1}{\lambda_2+\lambda_3}, \frac{1}{\lambda_1+\lambda_3}, \frac{1}{\lambda_1+\lambda_2}\right), \\ \mathbf{C} &= \operatorname{diag}\left(\frac{1}{\lambda_0+\lambda_1}, \frac{1}{\lambda_0+\lambda_2}, \frac{1}{\lambda_0+\lambda_3}\right). \end{aligned} \tag{3.35}$$

The matrix $\mathbf{X} \in so^*(4)$ of the angular momentum of the rigid body is related to its angular velocity $\boldsymbol{\Omega} \in so(4)$ by the formula

$$\mathbf{X} = \frac{1}{2}(\mathbf{J}\boldsymbol{\Omega} + \boldsymbol{\Omega}\mathbf{J}),$$

where

$$\begin{aligned} \mathbf{X} &= \begin{pmatrix} 0 & p_1 & p_2 & p_3 \\ -p_1 & 0 & -M_3 & M_2 \\ -p_2 & M_3 & 0 & -M_1 \\ -p_3 & -M_2 & M_1 & 0 \end{pmatrix}, \\ \boldsymbol{\Omega} &= \begin{pmatrix} 0 & \frac{p_1}{\lambda_0+\lambda_1} & \frac{p_2}{\lambda_0+\lambda_2} & \frac{p_3}{\lambda_0+\lambda_3} \\ -\frac{p_1}{\lambda_0+\lambda_1} & 0 & -\frac{M_3}{\lambda_1+\lambda_2} & \frac{M_2}{\lambda_1+\lambda_3} \\ -\frac{p_2}{\lambda_0+\lambda_2} & \frac{M_3}{\lambda_1+\lambda_2} & 0 & -\frac{M_1}{\lambda_2+\lambda_3} \\ -\frac{p_3}{\lambda_0+\lambda_3} & -\frac{M_2}{\lambda_1+\lambda_3} & \frac{M_1}{\lambda_2+\lambda_3} & 0 \end{pmatrix}. \end{aligned} \tag{3.36}$$

As shown below, this system is integrable (Schottky – Manakov case). The system (3.34) also describes the integrable geodesic flow of a certain metric on the group $SO(4)$ [23].

Rigid body in curved space. We can also write in the forms (3.26) and (3.31) the equations of free motion of a three-dimensional rigid body in the space S^3 of constant positive curvature [95]. This is a consequence of the direct relation of this problem to that of the motion of a four-dimensional rigid body, which can be more easily imagined in the case of the motion of a "flat" rigid body (plate) in S^2. Indeed, we can assume that the plate on the sphere is equivalent to a rigid body in $\mathbb{R}^3$ with a fixed point at the center of the sphere, which is attached to the plate by "weightless" generatrices.

REMARK. The study of the kinematics, statics and dynamics of a rigid body (and also a system of material points) in curved space goes back to W. K. Clifford, R. S. Ball, and R. S. Heath (see [322]), who developed the theory of screws, motors, and bi-quaternions. As a whole, these studies have not yielded any real results and are only of historical interest. In their modern form, the equations of motion of a rigid body for n-dimensional spaces of constant curvature (S^n and L^n, the n-dimensional sphere and pseudo-sphere) were obtained by W. Killing in a major work [299]. As mentioned above, these equations coincide with those of a "free" $(n+1)$-dimensional body in Euclidean space. W. Killing could not integrate his equations even in the case $n = 4$ and restricted himself to studying particular solutions.

Rigid body in S^3 and in a fluid. If we consider the free motion of a rigid body in the curved space S^3 (three-dimensional sphere) which is filled with a homogeneous incompressible ideal fluid (analog of the Kirchhoff equations (3.1) on $e(3)$), then the Hamiltonian has a more general form in comparison with (3.34):

$$H = \frac{1}{2}(\boldsymbol{M}, \mathbf{A}\boldsymbol{M}) + (\boldsymbol{M}, \mathbf{B}\boldsymbol{p}) + \frac{1}{2}(\boldsymbol{p}, \mathbf{C}\boldsymbol{p}). \tag{3.37}$$

Here $\mathbf{A}$ and $\mathbf{C}$ are symmetric matrices and $\mathbf{B}$ is arbitrary. The quadratic form (3.37) is the kinetic energy and so must be positive-definite. The entries of the matrices $\mathbf{A}$, $\mathbf{B}$, $\mathbf{C}$ depend on the added masses and moments of inertia of the body.

In a similar manner, we can write the equations of motion of a rigid body in a fluid (or vacuum) in the Lobachevskii space L^3. This problem was considered by G. Birkhoff in his book [46] and also N. E. Zhukovskii in the two-dimensional case [644]. In this case, a Hamiltonian system on the algebra $so(3,1)$ with Hamiltonian (3.37) is obtained. In our book [95], we present the derivation of the equations of motion for a rigid body with a gyrostat in the spaces of constant curvature S^3 and L^3. In this case, terms linear in $\boldsymbol{M}$ and $\boldsymbol{p}$ are added to the Hamiltonian (3.37).

System of interacting spins. The classical dynamics of two interacting *spins* (*spherical rotators*) (corresponding to a vector representation of the rotation group) are also described by a Hamiltonian system on $so(4)$ [402, 194, 546, 547]. Passing from the spin operators $\widehat{\boldsymbol{S}}_1$ and $\widehat{\boldsymbol{S}}_2$ to the classical vectors $\boldsymbol{K} = \boldsymbol{S}_1$ and $\boldsymbol{S} = \boldsymbol{S}_2$, we obtain the dynamical system (3.30). Hamiltonians of the form (3.32) correspond to different spin systems. In this case the cross terms describe the so-called *spin exchange* *interaction*. For spins in an external magnetic field it is necessary to add linear terms to the Hamiltonian.

The most general two-spin system of this type is given by a Hamiltonian [402] of the following form:

$$H = -(\mathbf{B}'\boldsymbol{K},\ \boldsymbol{S}) + \frac{1}{2}(\mathbf{A}'\boldsymbol{K},\ \boldsymbol{K}) + \frac{1}{2}(\mathbf{A}'\boldsymbol{S},\ \boldsymbol{S}), \tag{3.38}$$

where $\mathbf{A}'$ and $\mathbf{B}'$ are diagonal matrices.

The case $\mathbf{A}' = 0$ is the so-called *two-spin XYZ-model*. The case $a'_{33} = b'_{33} = 0$ is known as the *generalized two-spin XY-model* (Heisenberg model; see Appendix F).

The system (3.38) also describes the dynamics of two coupled classical tops (linked bodies; see Sec. 2.8 of Ch. 2)) whose interaction energy depends only on the components of angular momenta and is independent of positional variables.

2 Integrability cases

Since for the Poisson structure (3.24), (3.28) there are two Casimir functions, it follows that for the integrability of the corresponding equations of motion one more first integral is needed. In the general case, no additional integral exists and there will be domains in the phase space where the system exhibits chaotic behavior.

The known integrable cases of the systems (3.31), (3.32) are presented in Table **??**.

The algebra $so(4)$ admits the standard and canonical representations. The conditions for the parameters in Table **??** are presented in each case in the representation in which they are simplest.

REMARK 1. The entries of the matrices **A**, **B** and **C** for the Poincaré – Zhukovskii system cannot be arbitrary and consequently not all integrable cases presented in Table **??** have a physical meaning.

REMARK 2. The integrable cases of equations on the algebra $e(3)$ whose additional integral depends only on the variables $\boldsymbol{M}$ (of the type of the Lagrange and Hess cases for the Euler – Poisson equations, or of the type of the Kirchhoff and Chaplygin (II) cases for the Kirchhoff equations) are obviously transferred to systems on the bracket pencil (3.27), including the algebra $so(4)$ for $x = 1$. This is due to the fact that the equations for $\dot{\boldsymbol{M}}$ coincide for all brackets of the pencil (see below).

REMARK 3. The relative equilibria of systems (3.26) for which $\dot{\boldsymbol{K}} = \dot{\boldsymbol{S}} = \dot{\boldsymbol{M}} = \dot{\boldsymbol{\gamma}} = 0$ are interpreted in different ways depending on the physical statement of the problem. For the motion of a body with vortex holes, they define particular solutions for which the motion of the body is a uniform rotation about a certain axis, and the vorticity vector "is frozen" in the body. The study of stationary configurations for the model of coupled tops defining the dynamics of a spin chain is of especial interest: such configurations define a certain coherent state and are of considerable significance in quantum physics; they are considered in Ch. 6 for the finite-dimensional and infinite-dimensional cases.

3 The case of axial symmetry (H. Poincaré)

This is the simplest case of integrability, where, for each of the diagonal matrices **A**, **B**, **C** (or $\mathbf{A}'$, $\mathbf{B}'$, $\mathbf{C}'$) the same pair of eigenvalues coincides: that is, for example, $a_{11} = a_{22}$, $b_{11} = b_{22}$ and $c_{11} = c_{22}$. After dropping the functions (3.25), the Hamiltonian can be written as

$$H = \frac{1}{2}\left(M_1^2 + M_2^2 + aM_3^2\right) + b\left(M_1p_1 + M_2p_2\right) + \frac{1}{2}c(p_1^2 + p_2^2).$$

The additional integral has the form $M_3 = \text{const.}$ or $K_3 + S_3 = \text{const.}$ This integral can be referred to as being of Lagrange type. The reduction to quadratures, which leads to the solution in terms of elliptic functions, was performed by H. Poincaré in [472] (see also Sec. 4.2 of Ch. 4).

4 Schottky – Manakov case

In 1891, F. Schottky found the first integrable case of the system (3.31) [522] and noted its connection with the Clebsch case of the Kirchhoff equations. For this case, $\mathbf{B} = 0$ and the Hamiltonian is given by the formula (3.34), with the entries of the matrices **A**, **C** given by relations (3.35) with arbitrary parameters λ_μ, $\mu = 0, \dots, 3$. One also usually associates this case with the name of S. V. Manakov, who showed the integrability of its n-dimensional analog (1976, [404]).

In the representation (3.32), the relation $\mathbf{A}' = \mathbf{C}'$ holds for the Schottky – Manakov case. This integrability case is unique in the class of systems with a quadratic integral, for which $\mathbf{A}' = \mathbf{C}'$. Indeed, the following assertion holds (see, e.g., [598]):

If $\mathbf{A}' = \mathbf{C}'$, *the eigenvalues of the matrix* $\mathbf{A}'$ *are different,and the matrix* $\mathbf{B}'$ *is nonsingular, then the system* (3.32) *admits a quadratic integral iff the following condition holds:*

$$\begin{gathered}\mathbf{A}' = \operatorname{diag}(a_1', a_2', a_3'), \quad \mathbf{B}' = \operatorname{diag}(b_1', b_2', b_3'),\\ b_1'^2(a_2' - a_3') + b_2'^2(a_3' - a_1') + b_3'^2(a_1' - a_2')\\ +(a_1' - a_2')(a_2' - a_3')(a_3' - a_1')k^2 = 0, \quad k^2 = 1.\end{gathered} \tag{3.39}$$

The same condition appears in the physical works [402, 546] devoted to the dynamics of the two-spin model (3.38). The authors of [402] independently discovered the Schottky case, which had been known more than a century earlier.

REMARK. A study of the algebraic integrability of the system (3.26), (3.31) can be found in [252]. By using the Kovalevskaya method, this system was studied by M. Adler and P. van Moerbeke [3, 2, 1], where the integrability was also discussed. A new integrable case with a fourth-degree integral was found in [1]. Conditions for meromorphy of the solution in the complex plane of time were given in [84, 83]. These imply that k in equality (3.39) should be an odd integer. Necessary conditions for algebraic integrability were also obtained in [188]; in this case k should be rational.

Table 3.2. Integrability cases of Poincaré – Zhukovskii equations.

	Author	Conditions for parameters and the first integral
1	Poincaré (1910) [472]	$\mathbf{A}' = \mathrm{diag}(a_1', a_2', a_3'), \quad \mathbf{B}' = \mathrm{diag}(b_1', b_2', b_3'), \quad \mathbf{C}' = \mathrm{diag}(c_1', c_2', c_3'),$ $a_1' = a_2', \quad b_1' = b_2', \quad c_1' = c_2',$ $F = M_3$
2	Schottky (1891) [522] Manakov (1976) [404]	$\mathbf{A} = \mathrm{diag}\Big(\frac{1}{\lambda_2+\lambda_3}, \frac{1}{\lambda_3+\lambda_1}, \frac{1}{\lambda_1+\lambda_2}\Big),$ $\mathbf{B} = 0, \quad \mathbf{C} = \mathrm{diag}\Big(\frac{1}{\lambda_0+\lambda_1}, \frac{1}{\lambda_0+\lambda_2}, \frac{1}{\lambda_0+\lambda_3}\Big),$ $F = M_1^2 + M_2^2 + M_3^2 + \Big(\frac{\lambda_1^2}{\lambda_1^2-\lambda_0^2}p_1^2 + \frac{\lambda_2^2}{\lambda_2^2-\lambda_0^2}p_2^2 + \frac{\lambda_3^2}{\lambda_3^2-\lambda_0^2}p_3^2\Big)$
3	Stekloff (1909) [550]	$\mathbf{A}' = \mathrm{diag}(\lambda_2^2\lambda_3^2, \lambda_1^2\lambda_3^2, \lambda_1^2\lambda_2^2),$ $\mathbf{B}' = \mathrm{diag}(\lambda_2\lambda_3(\lambda_2^2+\lambda_3^2), \lambda_1\lambda_3(\lambda_1^2+\lambda_3^2), \lambda_1\lambda_2(\lambda_1^2+\lambda_2^2)),$ $\mathbf{C}' = \mathrm{diag}(\lambda_2^4+\lambda_3^4+\lambda_2^2\lambda_3^2, \lambda_1^4+\lambda_3^4+\lambda_1^2\lambda_3^2, \lambda_1^4+\lambda_2^4+\lambda_1^2\lambda_2^2),$ $F = S_1((\lambda_2^2+\lambda_3^2)S_1 + 2\lambda_2\lambda_3K_1)$ $+ S_2((\lambda_1^2+\lambda_3^2)S_2 + 2\lambda_1\lambda_3K_2) + S_3((\lambda_1^2+\lambda_2^2)S_3 + 2\lambda_1\lambda_2K_3)$

Continued

4	Adler, van Moerbeke, 1982, [3]	$\mathbf{A}' = \operatorname{diag}(a_1', a_2', a_3'),\ \mathbf{B}' = \operatorname{diag}(b_1', b_2', b_3'),\ \mathbf{C}' = \operatorname{diag}(c_1', c_2', c_3'),\ \boldsymbol{\lambda} = (\lambda_1, \lambda_2, \lambda_3),$ $a_i' = -3\lambda_j^2\lambda_k^2,\ b_i' = (\lambda_i^4 - \lambda_j^4 - 2\lambda_j^3\lambda_k - \lambda_j^2\lambda_k^2 - 2\lambda_j\lambda_k^3 - \lambda_k^4),$ $c_i' = \frac{4}{3}(\lambda_i^4 - \lambda_j\lambda_k(\lambda_j^2 + \lambda_k^2 + \frac{5}{4}\lambda_j\lambda_k)), \quad \lambda_1 + \lambda_2 + \lambda_3 = 0,$ $F = \frac{1}{2}\boldsymbol{K}^2\Big((\boldsymbol{\lambda}, \boldsymbol{\lambda})(\boldsymbol{K}, \boldsymbol{S}) + \sum_k \Big(\big(\lambda_i\lambda_j - \frac{1}{3}\lambda_k^2\big)S_k^2 + (\lambda_i\lambda_j - 2\lambda_k^2)K_kS_k\Big)\Big)$ $+ \frac{1}{18}\boldsymbol{S}^2\Big(3(\boldsymbol{\lambda}, \boldsymbol{\lambda})(\boldsymbol{K}, \boldsymbol{S}) + \sum_k \Big(\frac{5}{3}(\lambda_i\lambda_j - \lambda_k^2)S_k^2 + (7\lambda_i\lambda_j - 4\lambda_k^2)K_kS_k\Big)$ $- \frac{1}{9}\sum_{i<j}(\lambda_i - \lambda_j)^2(K_iS_j + K_jS_i)S_iS_j$
5	Bogoyavlenskii 1983, [56, 61]	$\mathbf{A} = \operatorname{diag}(\alpha_2, \alpha_1, \alpha_1 + \alpha_2),\ \alpha_1 = 1 - a_1,\ \alpha_2 = 1 - a_2,$ $\mathbf{B} = 0, \quad \mathbf{C} = \operatorname{diag}(a_2 - a_1, a_1 - a_2, 0),$ $F = \big(\alpha_1M_1^2 - \alpha_2M_2^2 - (a_1 - a_2)p_3^2\big)^2 + 4\alpha_1\alpha_2M_1^2M_2^2, \quad (\boldsymbol{M}, \boldsymbol{p}) = 0,$
6	Borisov, Mamaev, 2000	$\mathbf{A} = \operatorname{diag}(a_1, a_2, a_3),\ \mathbf{B} = \lVert b_{ij}\rVert,\ \mathbf{C} = \lVert c_{ij}\rVert,$ $b_{13}\sqrt{a_2 - a_1} \mp (b_{22} - b_{11})\sqrt{a_3 - a_2} = 0, \quad b_{12} = 0,$ $b_{13}\sqrt{a_3 - a_2} \pm (b_{33} - b_{22})\sqrt{a_2 - a_1} = 0, \quad b_{23} = 0,$ $c_{13}\sqrt{a_2 - a_1} \mp (c_{22} - c_{11})\sqrt{a_3 - a_2} = 0, \quad c_{12} = 0,$ $c_{13}\sqrt{a_3 - a_2} \pm (c_{33} - c_{22})\sqrt{a_2 - a_1} = 0, \quad c_{23} = 0,$ $F = M_1\sqrt{a_2 - a_1} \mp M_3\sqrt{a_3 - a_2} = 0$
7	Borisov, Mamaev, Sokolov, 2001	$H = \frac{1}{2}\big(M_1^2 + M_2^2 + 2M_3^2\big) + \alpha M_3p_1 - \frac{1}{2}\alpha^2p_3^2, \quad \alpha = \text{const.}$ $F = M_3\big[M_3\big(\boldsymbol{M}^2 + \alpha^2M_2^2 + 2\alpha(M_3p_1 - M_1p_3) + \alpha^2(p_1^2 + p_3^2)\big) + 2\alpha(M_1 - \alpha p_3)(\boldsymbol{M}, \boldsymbol{p})\big]$
8	Tsiganov, Sokolov, 2002 [582, 538]	$H = aM_1^2 - \frac{1}{a}M_2^2 + M_1p_2 - M_2p_1 = aM_1^2 - \frac{1}{a}M_2^2 + Q_3,$ $F = -a^2(Q_2^2 + Q_3^2) + aQ_3(\boldsymbol{M}^2 - \boldsymbol{p}^2) + (Q_1^2 + Q_3^2), \quad \text{where } \boldsymbol{Q} = \boldsymbol{M} \times \boldsymbol{p}$

The following symmetric representation of integrals in involution is known (see, e.g., [465]). By using the skew-symmetric matrix $\mathbf{X} = ||X_{\mu\nu}||$ from (3.36), they can be written in the form

$$G_\mu = \sum_{\nu=0}^{3} \frac{X_{\mu\nu}^2}{\lambda_\mu^2 - \lambda_\nu^2}, \qquad \mu = 0, \dots, 3. \tag{3.40}$$

The following equalities hold:

$$\sum_{\mu=0}^{3} G_\mu = 0, \quad \sum_{\mu=0}^{3} \lambda_\mu G_\mu = 2H, \quad \sum_{\mu=0}^{3} \lambda_\mu^2 G_\mu = F_1, \tag{3.41}$$

where $F_1 = \boldsymbol{M}^2 + \boldsymbol{p}^2$ is a Casimir function and H is the Schottky Hamiltonian (3.34).

Thus, the set of linear combinations of the four integrals in (3.40) *defines a five-parameter family of integrable quadratic Hamiltonians on each level surface of the Casimir functions (three parameters are the differences $\lambda_\mu^2 - \lambda_\nu^2$ and two parameters are the coefficients of the linear combination obtained after dropping the Casimir functions). The Hamiltonian of the Schottky – Manakov case is a member of this family* (3.40).

In a different form containing an arbitrary parameter, the quadratic family (3.40) was written by F. Schottky [522]. It has the form

$$Q(s) = \sum_{\text{cycle } i,j,k} \left(\sqrt{(s - a_i)(s - a_4)} M_i + \sqrt{(s - a_j)(s - a_k)} p_i \right)^2 .$$

This family was rediscovered by L. Haine in [252].

A Lax representation with a rational spectral parameter (for the general n-dimensional case) was found by S. V. Manakov, see [404]. *The equations of motion* (3.26) *with Hamiltonian* (3.34) *admit a representation in Lax form with matrices depending linearly on the arbitrary parameter λ:*

$$\frac{d}{dt}(\mathbf{X} + \lambda \mathbf{J}^2) = [\mathbf{X} + \lambda \mathbf{J}^2, \ \boldsymbol{\Omega} + \lambda \mathbf{J}],$$

where $\mathbf{X}$ *and* $\boldsymbol{\Omega}$ *are defined by formula* (3.36) *and* $\mathbf{J}$ *is the inertia tensor (a symmetric positive definite matrix).*

The involutive integrals of motion are obtained by the series expansion in λ of the functions $P_k(\lambda) = \operatorname{tr}(\mathbf{X} + \lambda \mathbf{J}^2)^k$. This family is complete, i.e., the number of integrals is sufficient for the integrability. Other Lax representations with a spectral parameter (and also their relation to the bi-Hamiltonian property) are considered in [67, 95, 190].

By using the contraction, the algebra $so(4)$ passes to the algebra $e(3)$, and in this case the Schottky – Manakov case passes to the Clebsch case (S. P. Novikov [447]). Indeed, let us perform the following changes of the variables and parameters:

$$\boldsymbol{p} \to \frac{\boldsymbol{\gamma}}{\sqrt{x}}, \quad \lambda_0 \to -\frac{1}{\sqrt{cx}}, \tag{3.42}$$

for some constant c. Taking the limit as $x \to 0$ in the commutator relations (3.27) leads to the $e(3)$ Poisson bracket and taking the same limit in the integrals (3.40) yields the integrals (3.12).

Excluding the infinite summand (as $x \to 0$) proportional to the Casimir function from the Hamiltonian (presented in Table **??**) according to the rule $H \to H + \frac{\sqrt{c}}{2\sqrt{x}}(\boldsymbol{\gamma}^2 + x\boldsymbol{M}^2)$, we obtain the Hamiltonian function for the Clebsch case (see Sec. 3.1 of Ch. 3) on $e(3)$ in the form

$$H = \frac{1}{2}\left(\frac{M_1^2}{\lambda_2+\lambda_3} + \frac{M_2^2}{\lambda_3+\lambda_1} + \frac{M_3^2}{\lambda_1+\lambda_2}\right) + \frac{c}{2}(\lambda_1\gamma_1 + \lambda_2\gamma_2 + \lambda_3\gamma_3)\,.$$

In this case, the Clebsch integral can be represented as

$$K = M_1^2 + M_2^2 + M_3^2 + c\left(\lambda_1^2\gamma_1^2 + \lambda_2^2\gamma_2^2 + \lambda_3^2\gamma_3^2\right).$$

REMARK. Under the change of variables (3.42), we obtain a family of integrals $G_\mu(x)$, $\mu = 0, \dots, 3$ depending on a parameter x and remaining involutive for all x, i.e., on the whole pencil $\mathscr{L}_x$ given by (3.27).

It turns out that the Schottky – Manakov and Clebsch cases are not only related by a contraction but are also linearly isomorphic (A. I. Bobenko [50]). The appropriate transformation has the form

$$\begin{gathered} M_1 = \sqrt{(\lambda_2^2-\lambda_0^2)(\lambda_3^2-\lambda_0^2)}L_1, \quad M_2 = \sqrt{(\lambda_1^2-\lambda_0^2)(\lambda_3^2-\lambda_0^2)}L_2, \\ M_3 = \sqrt{(\lambda_1^2-\lambda_0^2)(\lambda_2^2-\lambda_0^2)}L_3, \\ p_1 = \sqrt{\lambda_1^2-\lambda_0^2}\gamma_1, \quad p_2 = \sqrt{\lambda_2^2-\lambda_0^2}\gamma_2, \quad p_3 = \sqrt{\lambda_3^2-\lambda_0^2}\gamma_3. \end{gathered} \tag{3.43}$$

The equations of motion in the variables $(\boldsymbol{L}, \boldsymbol{\gamma})$ are the same as those of the Clebsch case on the algebra $e(3)$ with Hamiltonian

$$\begin{gathered} H = \frac{1}{2}(\boldsymbol{L}, \mathbf{A}\boldsymbol{L}) + \frac{1}{2}(\boldsymbol{\gamma}, \mathbf{C}\boldsymbol{\gamma}), \\ \mathbf{A} = \operatorname{diag}\left(-\frac{(\lambda_2-\lambda_0)(\lambda_3-\lambda_0)}{\lambda_2+\lambda_3}, -\frac{(\lambda_1-\lambda_0)(\lambda_3-\lambda_0)}{\lambda_1+\lambda_3}, -\frac{(\lambda_1-\lambda_0)(\lambda_2-\lambda_0)}{\lambda_1+\lambda_2}\right), \\ \mathbf{C} = \operatorname{diag}\left(\frac{1}{\lambda_1+\lambda_0}, \frac{1}{\lambda_2+\lambda_0}, \frac{1}{\lambda_3+\lambda_0}\right). \end{gathered}$$

It was shown by A. V. Bolsinov [65] that a linear isomorphism also exists for higher-dimensional analogs of the systems considered. Although the explicit transformation (3.43) was presented in [50], it was already contained in implicit form in the work of F. Schottky (1891) [522]. The topological analysis and bifurcation diagrams are presented in the paper of A. A. Oshemkov [456] (see also [66]).

5 Steklov case

Another integrable case for which cross terms are present in the Hamiltonian (3.31) (i.e., $\mathbf{B} \not\equiv 0$) is found in the work of V. A. Steklov [550]. Necessary and sufficient conditions for its quadratic integrability were announced by A. P. Veselov in [598] and are described in more detail in [276].

Let the Hamiltonian H be given in the representation (3.33). *Suppose that the eigenvalues of the matrix $\mathbf{A}'$, as well as those of the matrix $\mathbf{C}'$, are different, and that the matrix $\mathbf{B}'$ is nonsingular,* ($\det \mathbf{B}' \neq 0$). *Then an additional quadratic integral of motion independent of H, F_1, and F_2 exists iff the following condition holds*:

$$\begin{gathered}\mathbf{A}' = \operatorname{diag}(a_1', a_2', a_3'), \quad \mathbf{B}' = \operatorname{diag}(b_1', b_2', b_3'), \quad \mathbf{C}' = \operatorname{diag}(c_1', c_2', c_3'),\\ b_1'^2(a_2' - a_3') + b_2'^2(a_3' - a_1') + b_3'^2(a_1' - a_2')\\ +(a_1' - a_2')(a_2' - a_3')(a_3' - a_1') = 0,\\ b_1'^2(c_2' - c_3') + b_2'^2(c_3' - c_1') + b_3'^2(c_1' - c_2')\\ +(c_1' - c_2')(c_2' - c_3')(c_3' - c_1') = 0.\end{gathered} \tag{3.44}$$

To describe an involutive family of integrals analogous to that given by (3.40), we introduce the skew-symmetric matrices **K**, **S**, **M**, and **P** corresponding to the vectors $\boldsymbol{K}, \boldsymbol{S}, \boldsymbol{M}$, and $\boldsymbol{P}$ whose entries are defined by the formulas

$$\begin{aligned} K_{ij} &= -\varepsilon_{ijk}K_k, & S_{ij} &= -\varepsilon_{ijk}S_k,\\ M_{ij} &= -\varepsilon_{ijk}M_k, & P_{ij} &= -\varepsilon_{ijk}p_k. \end{aligned} \tag{3.45}$$

The integrals can be represented in terms of these matrices in the form [67, 95]:

$$G_i = \sum_{j\neq i}^{3} \frac{\left(\lambda_j K_{ij} + \lambda_i S_{ij}\right)^2}{\lambda_i^2 - \lambda_j^2} \qquad i = 1, 2, 3; \tag{3.46}$$

for example, G_1 has the following form (the others are obtained by cyclic permutation of subscripts):

$$G_1 = \frac{(\lambda_2 K_3 + \lambda_1 S_3)^2}{\lambda_1^2 - \lambda_2^2} + \frac{(\lambda_3 K_2 + \lambda_1 S_2)^2}{\lambda_1^2 - \lambda_3^2}.$$

The following relation holds for the functions G_i:

$$\sum_{i=1}^{3} G_i = -\boldsymbol{K}^2 + \boldsymbol{S}^2, \tag{3.47}$$

which is a Casimir, see equation (3.29). Hence, the set of linear combinations of the integrals (3.46) defines a four-parameter family of integrable quadratic Hamiltonians (two parameters are the ratios $\dfrac{\lambda_1}{\lambda_3}, \dfrac{\lambda_2}{\lambda_3}$ and two parameters are the coefficients of the linear combination obtained after dropping the Casimir functions).

One usually relates the Steklov case with the Hamiltonian of the following form (by analogy with the Kirchhoff equations; see Sec. 3.1 of Ch. 3):

$$H = \frac{1}{2}\sum_{i=1}^{3}\lambda_i^4 G_i = \frac{1}{2}\sum_{\substack{\text{cyclic}\\ \text{permutation}\\ \text{of } i,j,k}} \left[\left(\lambda_i\lambda_j K_k + (\lambda_i^2+\lambda_j^2)S_k\right)^2 - \lambda_i^2\lambda_j^2 S_k^2\right]$$

which takes the standard form after appropriate redefinitions of parameters.

By analogy with the Schottky – Manakov case, for the *Steklov case, we can perform a contraction to obtain the integrable Steklov – Lyapunov family for the Kirchhoff equations* (3.1). To show this, let us perform the following change of variables, parameters, and integrals (3.46):

$$\begin{gathered} \boldsymbol{p} \to \frac{\boldsymbol{\gamma}}{\sqrt{x}}, \\ \lambda_i \to 1 + \sqrt{x}\lambda_i, \quad G_i \to \sqrt{x}G_i, \quad i = 1, 2, 3. \end{gathered} \tag{3.48}$$

The family of the integrals $\{G_i(x)\}$ thus obtained, depending on the parameter x remains involutive on the whole pencil (3.27), *and for $x \to 0$, they take the form* (3.13) *presented in the previous section.*

REMARK. In [50], the contraction of the integrable cases was carried out in a slightly different way. There elliptic functions are used to give a symmetric form to the parameterization of the Steklov cases on $so(4)$.

The linear change of variables

$$\begin{gathered} \mathbf{M} = \mathbf{JKJ} + \frac{1}{2}\left(\mathbf{J}^2\mathbf{S} + \mathbf{S}\mathbf{J}^2\right), \quad \mathbf{P} = \mathbf{S}, \\ \mathbf{J} = \operatorname{diag}(\lambda_1, \lambda_2, \lambda_3) \end{gathered}$$

transforms the family (3.46) into the Steklov – Lyapunov family (3.13) of the Kirchhoff equations [67, 95], i.e., as in the Clebsch and Schottky – Manakov case, they are linearly isomorphic.

Another, less symmetric, representation of the family of involutive integrals in the Steklov case was found by O. I. Bogoyavlenskii [61] (see also [59, 60]):

$$\begin{aligned} I_1 &= \left(\sqrt{\frac{\alpha_1-\alpha_3}{\alpha_1-\alpha_2}}K_1 + \sqrt{\frac{\beta_1-\beta_3}{\beta_1-\beta_2}}S_1\right)^2 + \left(\sqrt{\frac{\alpha_2-\alpha_3}{\alpha_1-\alpha_2}}K_2 + \sqrt{\frac{\beta_2-\beta_3}{\beta_1-\beta_2}}S_2\right)^2, \\ I_2 &= \left(\sqrt{\frac{\alpha_1-\alpha_2}{\alpha_1-\alpha_3}}K_1 + \sqrt{\frac{\beta_1-\beta_2}{\beta_1-\beta_3}}S_1\right)^2 - \left(\sqrt{\frac{\alpha_2-\alpha_3}{\alpha_1-\alpha_3}}K_3 + \sqrt{\frac{\beta_2-\beta_3}{\beta_1-\beta_3}}S_3\right)^2, \\ I_3 &= \left(\sqrt{\frac{\alpha_1-\alpha_2}{\alpha_2-\alpha_3}}K_2 + \sqrt{\frac{\beta_1-\beta_2}{\beta_2-\beta_3}}S_2\right)^2 + \left(\sqrt{\frac{\alpha_1-\alpha_3}{\alpha_2-\alpha_3}}K_3 + \sqrt{\frac{\beta_1-\beta_3}{\beta_2-\beta_3}}S_3\right)^2, \end{aligned} \tag{3.49}$$

where α_i and β_i are arbitrary parameters.

6 An integrable case with a fourth-degree integral (M. Adler and P. van Moerbeke)

The general integrable case found by M. Adler and P. van Moerbeke [1] is still the most complicated and least studied case in rigid body dynamics. It has no analog amongst the family of Kirchhoff equations, and under a contraction, the Hamiltonian degenerates into the Casimir function of the algebra $e(3)$.

In the original paper [1], the additional fourth-degree integral was presented in a rather cumbersome and asymmetric form. Slightly later on, A. Reyman and M. Semenov-Tian-Shansky presented the spectral Lax representation for this case using the exceptional algebra $\mathfrak{g}_2$ for its construction [484]. In [67], an analogous $L-A$ pair was obtained in a more natural way; the corresponding construction is also related to the algebra $\mathfrak{g}_2$ and to the existence of a compatible Poisson structure. However, the integral obtained from the $L - A$ pair requires additional and nontrivial simplifications, which were carried out by the authors; after that, the form presented in Table **??** is obtained.

Note that in [426], A. S. Mishchenko and A. T. Fomenko had all the machinery to find this case by using the argument shift method developed by them; but perhaps their over-formalized and general arguments prevented them from finding this concrete result. It is interesting that in his subsequent books (see, e.g., [577]), when presenting this case, A. T. Fomenko refers to [484] without ever making note of the relation to his construction.

For the Adler – van Moerbeke system, separation variables are still unknown and no topological analysis has yet been performed. In many respects, its existence is related to a special symmetry of the group $so(4)$, which admits a real representation in the form of direct sum $so(3) \oplus so(3)$; it is absent on $so(3,1)$ and does not admit generalization to higher dimensions.

When representing the Adler – van Moerbeke system on the direct sum $(\boldsymbol{K}, \boldsymbol{S}) \in so(3) \oplus so(3)$, it can be written in the following slightly different symmetric form, found by V. V. Sokolov, which is different from that presented in Table **??**:

$$H = \sum_{i=1}^{3} \left[-9\alpha_j^2\alpha_k^2 K_i^2 + 6\alpha_j\alpha_k(\alpha_i - \alpha_j)(\alpha_i - \alpha_k)K_i S_i + \alpha_j\alpha_k(4\alpha_i^2 - \alpha_j\alpha_k)S_i^2 \right],$$

$$F = 3\sum_{i,j} \alpha_j(\alpha_i - \alpha_j)K_i S_i S_j^2 - 9(\boldsymbol{K}, \boldsymbol{K})\sum_i \alpha_j\alpha_k K_i S_i$$
$$+ \sum_i (\alpha_i - \alpha_j)(\alpha_i - \alpha_k)K_i S_i^3 + \frac{3}{2}(\boldsymbol{S}, \boldsymbol{S})\sum_i (3\alpha_i^2 - \alpha_j^2 - \alpha_k^2)S_i^2,$$

where $\alpha_1 + \alpha_2 + \alpha_3 = 0$ and $\alpha_i = \text{const}$.

7 Borisov – Mamaev – Sokolov case [75]

Under a contraction, this case passes to the Sokolov case of the Kirchhoff equations, and, therefore, we immediately write it for the bracket pencil $\mathscr{L}_x$ (3.27):

$$H = \frac{1}{2}(M_1^2 + M_2^2 + 2M_3^2) + M_3(\alpha p_1 + \beta p_2) - \frac{1}{2}(\alpha^2 + \beta^2)p_3^2,$$
$$\alpha, \beta = \text{const};$$
$$\begin{aligned} F &= k_1 k_2, \quad k_1 = M_3, \\ k_2 &= M_3(M_1^2 + M_2^2 + M_3^2 + x(\beta M_1 - \alpha M_2)^2) \\ &\quad + 2(\alpha M_1 + \beta M_2)(M_1 p_1 + M_2 p_2) + 2M_3^2(\alpha p_1 + \beta p_2) \\ &\quad + M_3(\alpha p_1 + \beta p_2)^2 - (\alpha^2 + \beta^2)(2M_1 p_1 + 2M_2 p_2 + M_3 p_3)p_3. \end{aligned} \tag{3.50}$$

The functions k_1 and k_2 satisfy the relation

$$\dot{k}_1 = -2(\beta p_1 - \alpha p_2)k_1, \quad \dot{k}_2 = 2(\beta p_1 - \alpha p_2)k_2,$$

i.e., $k_1 = 0$ and $k_2 = 0$ are invariant relations. On the whole bracket pencil $\mathscr{L}_x$ (3.27), this case admits an integrable addition of linear terms of the form $c_1 M_3 + c_2 p_2$, $c_1, c_2 =$ const. The additional integral is rather lengthy to write down, but it can be easily reproduced, since its unknown coefficients are found in a straightforward manner from a linear system of algebraic equations [535, 538].

Case (3.50) has not yet been integrated by quadratures and neither is a Lax representation known, etc.

8 Tsiganov – Sokolov case

Under contraction, this case is mapped to the integrable case of the Kirchhoff equations mentioned in Sec. 3.1 of Ch. 3. We present the integrable family on the whole pencil $\mathscr{L}_x$:

$$H = aM_1^2 - \frac{x}{a}M_2^2 + M_1 p_2 - M_2 p_1 = aM_1^2 - \frac{x}{a}M_2^2 + Q_3,$$
$$F = -a^2(Q_2^2 + Q_3^2) + aQ_3(x\boldsymbol{M}^2 - \boldsymbol{p}^2) + x(Q_1^2 + Q_3^2),$$

where $\boldsymbol{Q} = \boldsymbol{M} \times \boldsymbol{p}$. This general case has not yet been integrated by quadratures and has been little studied up to now.

9 Bogoyavlenskii particular case for $(\boldsymbol{M}, \boldsymbol{p}) = 0$

O. I. Bogoyavlenskii found two particular integrable cases of the Poincaré – Zhukovskii equations with a fourth-degree integral [56]. It was observed by A. V. Tsiganov that the second case is mapped to the first by a linear group transformation preserving the Poisson brackets.

Bogoyavlenskii "first" case. Under a contraction, this case passes to the Chaplygin case (I) of the Kirchhoff equations. To make this connection more obvious, let us write the Hamiltonian H and integral F for the pencil of brackets $\mathscr{L}_x$, (3.27):

$$\begin{gathered} H = \frac{1}{2}\left(\alpha_2 M_1^2 + \alpha_1 M_2^2 + (\alpha_1+\alpha_2)M_3^2\right) - \frac{1}{2}(a_1 - a_2)\left(p_1^2 - p_2^2\right), \\ F = \left(\alpha_1 M_1^2 - \alpha_2 M_2^2 - (a_1 - a_2)p_3^2\right)^2 + 4\alpha_1\alpha_2 M_1^2 M_2^2, \\ \alpha_1 = 1 - x a_1, \quad \alpha_2 = 1 - x a_2, \quad a_1, a_2 = \text{const}. \end{gathered} \tag{3.51}$$

For $x = 1$, we obtain the integrable case presented in Table 3.2. Note that for $x \neq 0$ the Hamiltonian (3.51) describes the motion of a dynamically asymmetric body whose moments of inertia satisfy the relation $I_1^{-1} + I_2^{-1} = I_3^{-1}$.

The following transformation (defined for $x \neq 0$) preserves the Poisson bracket on $\mathscr{L}_x$:

$$\bar{M}_1 = \frac{p_1}{\sqrt{x}}, \quad \bar{M}_2 = \frac{p_2}{\sqrt{x}}, \quad \bar{M}_3 = M_3, \quad \bar{p}_1 = \sqrt{x}M_1, \quad \bar{p}_2 = \sqrt{x}M_2, \quad \bar{p}_3 = p_3. \tag{3.52}$$

Let us substitute the new variables in (3.51) and make the change of parameters

$$a_1 = -\frac{b_3 - b_2 - 2}{2x}, \quad a_2 = -\frac{b_3 - b_1 - 2}{2x}.$$

We find that the function

$$\begin{aligned} H_2 = H + \frac{b_1 + b_2}{4x}(x\bar{\boldsymbol{M}}^2 + \bar{\boldsymbol{p}}^2) &= \frac{1}{2}(b_1\bar{M}_1^2 + b_2\bar{M}_2^2 + b_3\bar{M}_3^2) \\ &+ \frac{1}{4x}\left((b_2+b_3)\bar{p}_1^2 + (b_1+b_3)\bar{p}_2^2 + (b_1+b_2)\bar{p}_3^2\right) \end{aligned} \tag{3.53}$$

exactly coincides with the Hamiltonian of the Bogoyavlenskii "second" case. Moreover, although for $x < 0$, $\sqrt{x} = i\sqrt{|x|}$ and so the transformation (3.52) is complex, the Hamiltonian (3.53) is still real in this case.

In Sec. 5.1 of Ch. 5, we present a more general family of particular integrable cases on the pencil of brackets $\mathscr{L}_x$ which admits the following as special cases: the Kovalevskaya case of the Euler – Poisson equations, the Chaplygin case (I) of the Kirchhoff equations, the Bogoyavlenskii case(I) of the Poincaré – Zhukovskii equations, and also various gyrostatic generalizations.

10 Generalization of the Hess case

As noted above (see Subsec. 2), since the equations governing the evolution of the vector $\dot{\boldsymbol{M}}$ in the Poincaré – Zhukovskii system coincide with an analogous equation in the Kirchhoff equations, there also exists the following invariant relation of Hess type:

$$\sqrt{a_3 - a_2}M_3 \pm \sqrt{a_2 - a_1}M_1 = 0, \quad a_1 < a_2 < a_3. \tag{3.54}$$

As for the Kirchhoff equations, in this case the terms in the Hamiltonian (3.31) containing $\boldsymbol{p}$ should be invariant under rotations about an axis perpendicular to the circular section of the gyration ellipsoid. The invariant relation (3.54) defines a distinguished torus in the phase space on which the solution can be obtained in terms of quadratures. The integration procedure can be performed by using the results of Ch. 4. This generalization of the Hess case was probably unknown previously, despite its being a rather natural one.

11 Integrable generalizations with linear terms in the Hamiltonian

Analogously to what was said for the Kirchhoff equations, we can give a mechanical meaning to Eqs. (3.26) and (3.30) if, along with quadratic terms, the Hamiltonian H contains linear terms. Depending on the physical setup modelled by the system of interest, the interpretation of the additional linear terms will be different. For the dynamics of a rigid body with fluid – filled cavities, additional linear terms arise when the cavities are multiply connected. For the four-dimensional Euler top, they arise when a balanced four-dimensional gyroscope is attached to the top. For a rigid body in the space of constant curvature, they arise when a balanced three-dimensional gyroscope is attached to the top (for the derivation of this system, see Appendix A). For a rigid body on S^3 in a fluid, they arise when the body is multiply connected. For a spin chain, they arise when the spin chain is placed in a constant external magnetic field.

Analogously to what was done for the Kirchhoff equations, we can also find here integrable cases generalizing those presented in Table 3.2. Cases 1 and 6 related to the presence of rotational symmetry are generalized in a very simple way: we add a gyrostat along the symmetry axis (see Ch. 4 for more details). No such generalizations have been found for the Schottky, Adler – van Moerbeke, and Bogoyavlenskii (II) cases.

The generalization of the Steklov case leads to an integrable case analogous to the Rubanovskii case for the Kirchhoff equations. It was found originally by O. I.Bogoyavlenskii [61] where it appeared in a form which is difficult to read. A nicer representation was subsequently found by H. Yehia [628]. Here, we present the most symmetric expression.

Analog of the Rubanovskii case on $so(4)$. In the symmetric form, the appropriate involutive family of integrals can be represented as follows:

$$J_S = G_S + \frac{1}{(\lambda_s^2 - \lambda_m^2)(\lambda_s^2 - \lambda_n^2)} \sum_{\substack{\text{cyclic}\\ \text{permutation}\\ \text{of } i,j,k}} r_k(\lambda_i \lambda_j K_k + \lambda_s^2 S_k), \tag{3.55}$$

$$s, m, n = 1, 2, 3,$$

where r_i, $i = 1, 2, 3$ are three arbitrary additional parameters and G_S are defined by (3.46). For example, the integral J_1 has the following explicit expression (the others are obtained by cyclic permutations):

$$J_1 = G_1 + \frac{r_1(\lambda_2\lambda_3 K_1 + \lambda_1^2 S_1) + r_2(\lambda_3\lambda_1 K_2 + \lambda_1^2 S_2) + r_3(\lambda_1\lambda_2 K_3 + \lambda_1^2 S_3)}{(\lambda_1^2 - \lambda_2^2)(\lambda_1^2 - \lambda_3^2)}.$$

The relation (3.47) also holds for these integrals, i.e., $\sum J_i = -\boldsymbol{K}^2 + \boldsymbol{S}^2$. The Hamiltonian H and the additional integral F can be represented as

$$\begin{aligned} H = \sum_i \lambda_i J_i = \sum_{\text{cycl.}} \Big(&(\lambda_i + \lambda_j) S_k^2 - \frac{\lambda_i \lambda_j}{\lambda_i + \lambda_j}(S_k + K_k)^2 \\ &- \lambda_i \lambda_j r_k (S_k + K_k) + (\lambda_i + \lambda_j)^2 r_k S_k \Big), \end{aligned}$$

$$F = \sum_i \lambda_i^2 J_i = \sum_{\text{cycl.}} \Big((\lambda_i^2 + \lambda_j^2) S_k^2 + 2\lambda_i \lambda_j S_k K_k \Big) + (\boldsymbol{r}, \boldsymbol{S}).$$

A Lax representation for this case is presented in [191].

Generalization of the Bogoyavlenskii case. The integrable system (3.51) also admits a generalization in which a constant gyrostatic momentum λ along the axis OM_3 is added, although this axis is not a symmetry axis for this case. The Hamiltonian and the integral have the following form (for clarity, we use their representation on the pencil $\mathscr{L}_x$):

$$\begin{aligned} H &= \frac{1}{2}\left(\alpha_2 M_1^2 + \alpha_1 M_2^2 + (\alpha_1 + \alpha_2) M_3^2\right) - \lambda M_3 - \frac{1}{2}(a_1 - a_2)(p_1^2 - p_2^2), \\ F &= \left(\alpha_1 M_1^2 - \alpha_2 M_2^2 - (a_1 - a_2)\gamma_3^2\right)^2 + 4\alpha_1\alpha_2 M_1^2 M_2^2 \\ &\quad + 4\lambda\big(M_3(\alpha_1 M_1^2 + \alpha_2 M_2^2) + (a_1 - a_2) p_3 (M_1 p_1 - M_2 p_2)\big) - 4\lambda^2 (M_1^2 + M_2^2), \end{aligned}$$

where $\alpha_1 = 1 - x a_1$, $\alpha_2 = 1 - x a_2$, and $a_1, a_2 = \text{const}$. This generalization was given in the above form by the authors in [82, 85]. In Sec. 5.1 of Ch. 5 this case is included into a richer family (in the sense that there is a greater number of free parameters) defined on the Poisson pencil $\mathscr{L}_x$.

3.3 A Remarkable Limit Case of the Poincaré – Zhukovskii Equations. A Countable Family of First Integrals

1 In this section, we consider the equations of motion of a rigid body in the case where a passage to the limit is carried out in the Poincaré – Zhukovskii equations (3.26). It differs from the limiting case used in the contraction and leads to the loss of the Hamiltonian property. Indeed, if in Eqs. (3.26) with Hamiltonian (3.31), we make the change of variables $\boldsymbol{\gamma} \to \mu\boldsymbol{\gamma}$ and let μ tend to zero, then a singularity arises in the commutator relations (3.24) However, on the level of the equations of motion, we obtain

$$\begin{gathered}\dot{\boldsymbol{M}} = \boldsymbol{M} \times \mathbf{A}\boldsymbol{M}, \quad \dot{\boldsymbol{\gamma}} = \boldsymbol{\gamma} \times \mathbf{B}\boldsymbol{M}, \\ \mathbf{A} = \operatorname{diag}(a_1,\, a_2,\, a_3), \quad \mathbf{B} = \operatorname{diag}(b_1,\, b_2,\, b_3).\end{gathered} \tag{3.56}$$

The system (3.56) describes the rotation of a body with a fluid – filled cavity in the case where the intensity of the fluid vortex in the cavity is small in comparison to the angular momentum (or vice versa). It is possible to give other interpretations of this passage to the limit, corresponding to all the various physical settings for the system (see Sec. 3.2) modelled by Eqs. (3.26).

The first vector equation in (3.56) can be independently integrated. It is the usual Euler case with the integrals

$$I_1 = \frac{1}{2}(\boldsymbol{M},\, \mathbf{A}\boldsymbol{M}), \quad I_2 = (\boldsymbol{M},\, \boldsymbol{M}).$$

The solution $\boldsymbol{M}(t)$ is a periodic function of t. See Ch. 2, Sec. 2. After substituting the known function $\boldsymbol{M}(t)$, the second equation in (3.56), related to the kinematic Poisson equations, is the following linear Hamiltonian system on $so(3)$:

$$\dot{\boldsymbol{\gamma}} = \boldsymbol{\gamma} \times \mathbf{B}\boldsymbol{M}(t), \quad \{\gamma_i, \gamma_j\} = \varepsilon_{ijk}\gamma_k \tag{3.57}$$

with periodic coefficients and linear Hamiltonian $H = (\mathbf{B}\boldsymbol{M}(t),\, \boldsymbol{\gamma})$. Equation (3.57), as well as system (3.56), has the geometric integral $I_3 = (\boldsymbol{\gamma},\, \boldsymbol{\gamma})$. For the integrability of (3.57), we need one more integral which depends periodically on t, say $I_4^{\star}(\boldsymbol{\gamma},\, t)$. Therefore, for integrability of the system (3.56), we need one more first integral $I_4(\boldsymbol{M},\, \boldsymbol{\gamma})$. This also follows from the last multiplier theory.

2 The additional integral of systems (3.56) and (3.57) always exists in the class of real-analytic functions. This is due to the fact that Eq. (3.57) is linear and defines a measure-preserving map of a two-dimensional sphere per period. All such mappings are integrable.

As is easily seen, the additional integral must be linear in $\boldsymbol{\gamma}$:

$$\begin{aligned}&I_4 = I_4(\boldsymbol{M},\, \boldsymbol{\gamma}) = (\boldsymbol{\Omega}(\boldsymbol{M}),\, \boldsymbol{\gamma}),\\ &I_4^{\star} = I_4^{\star}(\boldsymbol{\gamma},\, t) = (\boldsymbol{\Omega}^{\star}(t),\, \boldsymbol{\gamma}), \quad \boldsymbol{\Omega}^{\star}(t) = \boldsymbol{\Omega}(\boldsymbol{M}(t)),\end{aligned}$$

and, moreover, we obtain equations for the function $\boldsymbol{\Omega}^*(t)$ that are analogous to (3.57). Therefore, to find the first integral I_4 or I_4^*, it suffices to find at least one particular solution of Eqs. (3.57).

In general, however, we can obtain neither particular solutions of (3.57) nor the corresponding integrals (3.56) and (3.57) in closed algebraic form. Such a solution is possible only in the form of an infinite series, and it is not single-valued in the complex sense (and is not algebraic) [335].

REMARK. In [84] the Kovalevskaya exponents of system (3.56) are calculated. They are

$$\begin{gathered}\rho_1 = -1, \quad \rho_2 = \rho_3 = 2, \quad \rho_4 = 1, \quad \rho_{5,6} = 1 \pm n,\\ n = \left[\frac{b_1^2 a_{32} + b_2^2 a_{13} + b_3^2 a_{21}}{a_{23}a_{21}a_{13}}\right],\\ a_{ij} = a_i - a_j.\end{gathered} \tag{3.58}$$

The general solution of (3.56) has a finite-sheeted branching on the complex plane of time provided that n is rational. This condition is also necessary for the existence of an additional algebraic integral [188].

Physically, the absence of a "sufficiently good" (algebraic or polynomial) additional integral of system (3.56) in the general case is due to the loss of symmetries coming from the Hamiltonian property (the Poisson structure is a tensor invariant). However, the behavior of trajectories of (3.56), (3.57) is always regular, the Lyapunov exponents are equal to zero, and the real-analytic integral exists formally.

3 Let us find the conditions under which it is possible to explicitly find the additional integral. Otherwise it seems that analogous constructions are impossible. For this purpose it is necessary that $n = 2k + 1$, $k \in \mathbb{Z}$, in (3.58). For various k, the additional integrals related to a certain iteration process were first presented by A. V. Borisov and A. V. Tsygvintsev in [84, 83] without detailed proof. Here a natural procedure is given for constructing these integrals, whose existence seems to be slightly mysterious.

4 With the rescaled time $d\tau = M_1M_2M_3\,dt$ and in terms of new variables $u_i = M_i^2$, $s_i = \frac{\gamma_i}{M_i}$, the system (3.56) becomes

$$\begin{gathered}\boldsymbol{u}' = \mathbf{c}, \quad \mathbf{s}' = \mathbf{U}^{-1}\mathbf{A}_b\mathbf{s},\\ \boldsymbol{u} = (u_1, u_2, u_3), \quad \mathbf{s} = (s_1, s_2, s_3), \quad \mathbf{c} = (-2a_{23}, -2a_{31}, -2a_{12}),\\ a_{ij} = a_i - a_j, \quad \mathbf{U} = \operatorname{diag}(u_1, u_2, u_3),\\ \mathbf{A}_b = \begin{pmatrix} a_{23} & b_3 & -b_2\\ -b_3 & a_{31} & b_1\\ b_2 & -b_1 & a_{12}\end{pmatrix},\end{gathered} \tag{3.59}$$

where the prime denotes differentiation with respect to τ. Let us look for an integral of the system (3.59) in the form

$$F = \sum_{i=1,2,3} s_i u_i f^{(i)}(\boldsymbol{u}). \tag{3.60}$$

This leads to the following system of first-order partial differential equations for the vector $\boldsymbol{f} = (f^{(1)}, f^{(2)}, f^{(3)})$:

$$\mathbf{U}\widehat{D}\boldsymbol{f} = -\mathbf{A}_b\boldsymbol{f}, \tag{3.61}$$

where the scalar differential operator $\widehat{D}$ has the form

$$\widehat{D} = -2a_{23}\frac{\partial}{\partial u_1} - 2a_{31}\frac{\partial}{\partial u_2} - 2a_{12}\frac{\partial}{\partial u_3}. \tag{3.62}$$

To find the integral, it is necessary to find a particular solution of (3.61).

Theorem. *Suppose there exists an odd integer $n = 2k + 1$ for which the parameters a_i and b_j satisfy*

$$n^2 a_{23}a_{31}a_{12} + b_1^2 a_{23} + b_2^2 a_{31} + b_3^2 a_{12} = 0. \tag{3.63}$$

Then Eq. (3.61) *admits a polynomial particular solution* (*which can be written in explicit form*).

Proof Let us start by looking at $n = 1$ $(k = 0)$. Then we find a particular solution without difficulty. It corresponds to the constant eigenvector $\boldsymbol{f}_1$ of the matrix $\mathbf{A}_b$ corresponding to zero eigenvalue, since $\det \mathbf{A}_b = a_{23}a_{31}a_{12} + b_1^2 a_{23} + b_2^2 a_{31} + b_3^2 a_{12} = 0$, which exactly corresponds to conditions (3.63) for $n = 1$. We may use this idea to generalize to $n \neq 1$.

For $n \neq 1$, let us look for a solution of (3.61) in the form

$$\boldsymbol{f} = \mathbf{B}_1\mathbf{U}\widetilde{\boldsymbol{f}}, \quad \text{where} \quad \mathbf{B}_1 = \mathbf{A}_b^{-1} \tag{3.64}$$

– which we are free to do, as $n \neq 1$ implies $\det \mathbf{A}_b \neq 0$. Taking into account that $\widehat{D}\mathbf{U} = \mathbf{C}$, $\mathbf{C} = \operatorname{diag}(-2a_{23}, -2a_{31}, -2a_{12})$, we obtain from (3.61) the vector algebraic equation

$$\mathbf{U}\mathbf{B}_1\mathbf{C}\widetilde{\boldsymbol{f}} + \mathbf{U}\mathbf{B}_1\mathbf{U}\widehat{D}\widetilde{\boldsymbol{f}} = -\mathbf{U}\widetilde{\boldsymbol{f}}. \tag{3.65}$$

Let us introduce the matrices

$$\mathbf{A}_{b,n} = \begin{pmatrix} na_{23} & b_3 & -b_2 \\ -b_3 & na_{31} & b_1 \\ b_2 & -b_1 & na_{12} \end{pmatrix}. \tag{3.66}$$

After multiplying on the left by the matrix $\mathbf{A}_b\mathbf{U}^{-1}$ and collecting similar terms, we can write (3.65) in the same form as (3.61):

$$\mathbf{U}\widehat{D}\widetilde{\boldsymbol{f}} = -(\mathbf{A}_b - \mathbf{C})\widetilde{\boldsymbol{f}} = -\mathbf{A}_{b,3}\widetilde{\boldsymbol{f}}.$$

Since $\det \mathbf{A}_{b,n} = n(n^2 a_{23}a_{31}a_{12} + b_1^2 a_{23} + b_2^2 a_{31} + b_3^2 a_{12})$, we can define an iterative process by representing $\widetilde{\boldsymbol{f}}$ in the same form (3.64) and obtain the solution

$$\boldsymbol{f} = \mathbf{A}_{b,1}^{-1}\mathbf{U}\mathbf{A}_{b,3}^{-1}\mathbf{U}\cdots\mathbf{A}_{b,n-2}^{-1}\mathbf{U}\boldsymbol{f}_n, \tag{3.67}$$

where $\boldsymbol{f}_n$ is the eigenvector of the matrix $\mathbf{A}_{b,n}$ corresponding to zero eigenvalue, i.e., $\mathbf{A}_{b,n}\boldsymbol{f}_n = 0$. □

5 Let us consider the particular case $\mathbf{B} = n\mathbf{A}$, $n = 2k+1$, $k \in \mathbb{Z}$, in more detail. For $n > 0$, we have

$$\mathbf{A}_{b,n} = n\mathbf{A}_a = n\begin{pmatrix} a_{23} & a_3 & -a_2 \\ -a_3 & a_{31} & a_1 \\ a_2 & -a_1 & a_{12} \end{pmatrix},$$

while the eigenvector $\boldsymbol{f}_n$ is the same for all n and is equal to $\boldsymbol{f}_n = \boldsymbol{v}_+ = (1, 1, 1)$.

For $n < 0$, we have

$$\mathbf{A}_{b,n} = |n|\mathbf{A}_{-a} = |n|\begin{pmatrix} a_{23} & -a_3 & a_2 \\ a_3 & a_{31} & -a_1 \\ -a_2 & a_1 & a_{12} \end{pmatrix};$$

the eigenvector $\boldsymbol{f}_n$ is also the same for all n and can be written as

$$\boldsymbol{f}_n = \boldsymbol{v}_- = (a_1^{-1} - a_2^{-1} - a_3^{-1},\ -a_1^{-1} + a_2^{-1} - a_3^{-1},\ -a_1^{-1} - a_2^{-1} + a_3^{-1}).$$

We can use these formulae now to give explicit expressions for the integral for particular values of n.

1) $n = 1$.
   $I_4 = (\boldsymbol{M}, \boldsymbol{\gamma})$; this is the usual area integral for the Euler case (see Sec. 2.2 of Ch. 2).
2) $n = -1$.

$$I_4 = (\boldsymbol{\Omega}\boldsymbol{M}, \boldsymbol{\gamma}), \quad \boldsymbol{\Omega} = \mathbf{E} - \frac{2\mathbf{A}^{-1}}{\operatorname{Tr}\mathbf{A}^{-1}}; \tag{3.68}$$

   this is an analog of the area integral. As is shown in what follows (see Subsec. 6), for $n = -1$, Eqs. (3.56) can be reduced to the case $n = 1$ by a linear transformation.
3) $n = \pm 3$.
   In this case, the solution of Eq. (3.67) can be represented as

$$\boldsymbol{f} = (\mathbf{B}^s \pm \mathbf{B}^a)\mathbf{U}\boldsymbol{v}_\pm,$$

   where $\mathbf{B}^s$ and $\mathbf{B}^a$ are, respectively, symmetric and skew-symmetric matrices with the entries

$$b_{ij}^s = \begin{cases} 9a_i a_j, & i \neq j, \\ 9a_i^2 - \varepsilon_{ilm} a_{il} a_{im}, & i = j, \end{cases}$$

$$b_{ij}^a = -3a_{ij}a_k, \quad i \neq j \neq k,$$

   while the vectors $\boldsymbol{v}_\pm$ are defined in Subsec. 5. The additional integral I_4 is of degree four in the variables $\boldsymbol{M}$ and $\boldsymbol{\gamma}$; in the general case, this degree is equal to $2|k| + 2$.

6 It turns out [94] that for $n = -1$ and $\mathbf{B} = -\mathbf{A}$, the integral I_4 (3.68) remains unchanged for the more general system:

$$\begin{cases} \dot{\boldsymbol{M}} = \boldsymbol{M} \times \mathbf{A}\boldsymbol{M} + \varepsilon\boldsymbol{\gamma} \times \mathbf{A}^{-1}\boldsymbol{\gamma}, \\ \dot{\boldsymbol{\gamma}} = n\boldsymbol{\gamma} \times \mathbf{A}\boldsymbol{M}, \end{cases} \tag{3.69}$$

corresponding to the addition of the external Brun field to (3.56) (see Sec. 2.1 of Ch. 2) with the potential $V = \frac{1}{2}\varepsilon(\boldsymbol{\gamma}, \mathbf{A}^{-1}\boldsymbol{\gamma})$ in (3.56). The invariant measure of Eqs. (3.69) remains standard. In this case, we need to slightly modify the integrals I_1 and I_2:

$$\begin{aligned} I_1 &= \frac{1}{2}(\boldsymbol{M}, \mathbf{A}\boldsymbol{M}) - \frac{1}{2}\varepsilon(\boldsymbol{\gamma}, \mathbf{A}^{-1}\boldsymbol{\gamma}), \\ I_2 &= (\boldsymbol{M}, \boldsymbol{M}) + \varepsilon \det \mathbf{A}^{-1}(\boldsymbol{\gamma}, \mathbf{A}\boldsymbol{\gamma}). \end{aligned} \tag{3.70}$$

It is easy to show that integrals of the system (3.69) analogous to I_1 and I_2 do exist for $\mathbf{B} = n\mathbf{A}$ and any n, whereas the integral of type I_4 exists only for $n = \pm 1$. It seems that it is impossible to generalize it for $\varepsilon \neq 0$ and other $n = 2k + 1 = \pm 3, \pm 5, \dots$.

REMARK. In the variables $\boldsymbol{u}$ and $\boldsymbol{s}$, the system (3.69) can be represented as

$$u_i' = c_i + \varepsilon\lambda_k b_{kj} s_k s_j, \quad \boldsymbol{s}' = \mathbf{U}^{-1}\mathbf{A}_b\boldsymbol{s},$$

where $\lambda_k = a_k^{-1}$ and $b_{kj} = b_k - b_j$ and where the prime stands for rescaled time τ as in (3.59), but in repeating the arguments of Subsec. 3, it turns out that the procedure of iteration breaks down for $n \neq \pm 1$.

As is noted in [94], the case $n = -1$ in (3.69) is transformed to the case $n = 1$, corresponding to the Brun problem (or the Clebsch case), by using the linear transformation

$$\boldsymbol{W} = \boldsymbol{\Omega}\boldsymbol{M}, \quad \boldsymbol{\Omega} = \mathbf{E} - \frac{2\mathbf{A}^{-1}}{\operatorname{Tr}\mathbf{A}^{-1}}, \tag{3.71}$$

in terms of which (3.69) becomes

$$\begin{cases} \dot{\boldsymbol{W}} = \mathbf{J}^{-1}\boldsymbol{W} \times \boldsymbol{W} - \varepsilon\boldsymbol{\gamma} \times \mathbf{J}\boldsymbol{\gamma}, \\ \dot{\boldsymbol{\gamma}} = \mathbf{J}^{-1}\boldsymbol{W} \times \boldsymbol{\gamma}, \quad \mathbf{J} = \boldsymbol{\Omega}\mathbf{A}^{-1}, \end{cases} \tag{3.72}$$

which is the same as (3.69) for $n = 1$ up to time reversal $t \to -t$.

Note that the addition of a constant gyrostatic momentum to the system (3.56), i.e., the construction of a generalization of the Zhukovskii – Volterra problem, does not lead to a new integrable problem even for $n = -1$. In general, the problem of finding other possible generalizations of the countable family of integrals I_4 (for example, on $so(4)$, the gyrostat, etc.) remains unsolved for now. It is possible that they simply do not exist.

REMARK 1. In the general case, for arbitrary matrices **A** and **B** in (3.56), the general integral is not single-valued and is branching on the complex plane of time. Under the conditions $b_1 = b_2 = 0$ and $b_3 \neq 0$, it can be written explicitly:

$$\begin{aligned} I_4 &= \gamma_1 \sin\varphi + \gamma_2 \cos\varphi, \\ \varphi &= \frac{b_3}{\sqrt{a_{13}a_{32}}} \ln(\sqrt{a_{13}}M_1 + \sqrt{a_{32}}M_2). \end{aligned} \tag{3.73}$$

The existence of such complicated integrals of the system (3.56) is probably related to likely loss of the Hamiltonian property.

REMARK 2. Except for the cases $n = \pm 1$, the general solution of system (3.56) with integrals (3.60) has still not been obtained in quadratures; also, it is not clear whether or not it can be expressed in terms of elliptic functions. The topology of the level sets of the integrals has not been studied.

7 The system (3.56) can be also obtained in studying the nonholonomic problem of a dynamically asymmetric balanced ball (*Chaplygin ball*) rolling without slipping on the surface of the sphere (Fig. 3.2).

When there is no force field, the equations of motion have the form [94]

$$\begin{cases} \dot{\boldsymbol{M}} = \boldsymbol{M} \times \boldsymbol{\omega}, \\ \dot{\boldsymbol{\gamma}} = \dfrac{R}{R-a} \boldsymbol{\gamma} \times \boldsymbol{\omega}, \end{cases} \tag{3.74}$$

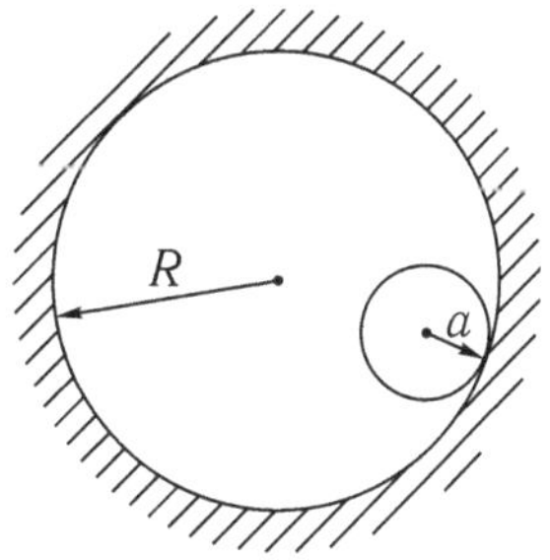

Fig. 3.2.

$$\boldsymbol{M} = \mathbf{I}\boldsymbol{\omega} + D\boldsymbol{\gamma} \times (\boldsymbol{\omega} \times \boldsymbol{\gamma}), \quad D = ma^2,$$

where m is the mass of the ball and $\mathbf{I}$ is the tensor of inertia with respect to the geometric center.

We shall not dwell here on studying the integrability of the system (3.74) but only note that as the nonholonomy parameter D tends to zero, the system (3.56) with the matrix $\mathbf{B} = \lambda \mathbf{A}$, $\lambda = \frac{R}{R-a}$ is obtained. This once more shows the necessity of studying Eqs. (3.56) and also allows us to generalize the integrals I_1, I_2, I_3, and I_4 to Eqs. (3.74). At present, such a generalization found in [94] is known only for $\lambda = \pm 1$; in both cases, it is possible to consider the more general situation (3.69) corresponding to the addition of the Brun field.

The case where $\lambda = 1$ and $R = \infty$ reduces to the classical Chaplygin problem for the rolling of a ball over the horizontal plane [130].

The case where $\lambda = -1$ and $a = 2R$ corresponds to the so-called *spherically suspended rigid body* (a model for a "ball-and-socket joint") in which a dynamically asymmetric body with a spherical cavity of radius $2R$ moves without slipping around a fixed ball of radius R. This integrable problem (and its generalizations with addition of the Brun field) was discovered by A. V. Borisov and Yu. N. Fedorov [94].

8 Suppose that $\mathbf{B} = \lambda \mathbf{A}$ and that two eigenvalues of the matrix $\mathbf{A}$ coincide, for example, $a_1 = a_2$. Then the system (3.56) and the following more general system

$$\begin{cases} \dot{\boldsymbol{M}} = \boldsymbol{M} \times \mathbf{A}\boldsymbol{M} + \boldsymbol{\gamma} \times \dfrac{\partial V}{\partial \boldsymbol{\gamma}}, \\ \dot{\boldsymbol{\gamma}} = \lambda \boldsymbol{\gamma} \times \mathbf{A}\boldsymbol{M}, \quad V = V(\gamma_3), \end{cases} \tag{3.75}$$

are algebraically integrable. A complete set of integrals of (3.75) is

$$
\begin{aligned}
I_1 &= \frac{1}{2}(M_1^2 + M_2^2) + (\lambda a_1)^{-1} V(\gamma_3), \\
I_2 &= M_3, \quad I_3 = \boldsymbol{\gamma}^2, \\
I_4 &= M_1\gamma_1 + M_2\gamma_2 + \frac{a_1 - a_3 + \lambda a_3}{\lambda a_1} M_3\gamma_3,
\end{aligned}
\tag{3.76}
$$

which ensures the Euler – Jacobi integrability as the system (3.75) has also standard invariant measure. The integrals (3.76) are analogous to the integrals of the Lagrange case of the Euler-Poisson system and admit nonholonomic generalizations [98].

3.4 Rigid Body in an Arbitrary Potential Field

As is shown in Sec. 1.4 of Ch. 1, the dynamics of a rigid body with a fixed point in an arbitrary potential field with potential V is described by a Hamiltonian system (1.60) (or (1.67)) with three degrees of freedom. The Hamiltonian function has the form

$$
\begin{gathered}
H = \frac{1}{2}(\boldsymbol{M}, \mathbf{A}\boldsymbol{M}) + V, \quad \mathbf{A} = \mathbf{I}^{-1}, \\
V \equiv V(\boldsymbol{\alpha}, \boldsymbol{\beta}, \boldsymbol{\gamma}) \equiv V(\lambda_0, \lambda_1, \lambda_2, \lambda_3) \equiv V(\theta, \varphi, \psi),
\end{gathered}
\tag{3.77}
$$

and for its integrability (in the Liouville sense), we need two more independent involutive integrals.

We may write $V(\boldsymbol{\alpha}, \boldsymbol{\beta}, \boldsymbol{\gamma}) \equiv V(\lambda_0, \lambda_1, \lambda_2, \lambda_3)$; here, $\boldsymbol{\alpha}, \boldsymbol{\beta}$, and $\boldsymbol{\gamma}$ are direction cosines and λ are the quaternionic Rodrigues – Hamilton parameters. Integrable cases for system (3.77) are known for the following three types of potentials:

1) The potential V is a linear function of $\boldsymbol{\alpha}$, $\boldsymbol{\beta}$, and $\boldsymbol{\gamma}$ (and is quadratic in λ). For a particular form of V, in the case of axial symmetry of the force field, we obtain the Euler – Poisson equations and therefore, in the general case, this system is called the *generalized Euler – Poisson equations* .
2) The potential V is quadratic in $\boldsymbol{\alpha}$, $\boldsymbol{\beta}$, and $\boldsymbol{\gamma}$, i.e., it is of degree four in quaternions. This problem was considered by Brun and Goryachev.
3) The potential V is linear in the quaternions λ. Although this case is in some sense simpler than the first two cases, it is placed in the last position because of the fact that it was not considered earlier. Probably this is related to the absence of its reasonable mechanical interpretation. We have called it the *quaternionic Euler – Poisson equations.*

Let us consider each of these three cases in turn and present all the known integrability cases characterized by necessary additional restrictions on free parameters. In this case the motion is regular, and for the generic initial conditions the trajectories are quasi-periodic orbits of three-dimensional tori, which are the common level surfaces of first integrals.

1 Generalized Euler – Poisson equations

First of all we note that any number of linear force fields reduces to three mutually perpendicular force fields of unit intensity, whose *force centers* (analogs of the center of mass for the gravity field) have fixed positions in the body [95].

The Hamiltonian function has the form

$$H = \frac{1}{2}(\mathbf{A}\boldsymbol{M},\ \boldsymbol{M}) + (\boldsymbol{r}_1,\ \boldsymbol{\alpha}) + (\boldsymbol{r}_2,\ \boldsymbol{\beta}) + (\boldsymbol{r}_3,\ \boldsymbol{\gamma}), \tag{3.78}$$

where $\boldsymbol{r}_1$, $\boldsymbol{r}_2$, and $\boldsymbol{r}_3$ are radius vectors of various force centers (electric, gravitational, etc.). They are called the *application centers*. In the case of one field, they reduce to the usual center of gravity.

Below we summarize the main results on reduction of the potential energy of the system (3.78) to the simplest form for various locations of the force centers $\boldsymbol{r}_1, \boldsymbol{r}_2$, and $\boldsymbol{r}_3$, which also take into account the body geometry (see [95] for details):

1) *The application centers of all fields lie on the same axis.*
   Using an appropriate choice of fixed axes in the space, we can reduce the potential energy to the form
   $$V = \sqrt{a^2 + b^2 + c^2}\,\alpha_1,$$
   where a, b, and c are the distances from the application centers to the fixed point. Therefore, this case reduces to a single force field, and, moreover, its application center $\boldsymbol{r}_1$ lies on the above-mentioned axis.
2) *The application centers of all fields lie in the same plane.*
   By a suitable choice of the axes fixed in space, the potential reduces to the form
   $$V = u\alpha_1 + v\alpha_2 + w\beta_2.$$
   Therefore, the system of forces reduces to two mutually orthogonal fields such that the radius vectors of their application centers $\boldsymbol{r}_1 = (u, v, 0)$ and $\boldsymbol{r}_2 = (0, w, 0)$ are not orthogonal in the general case.
3) *The application centers are arbitrary, but the tensor of inertia of the body is spherical* $(a_1 = a_2 = a_3)$.
   In this case, using the additional arbitrariness in choosing the moveable principal axes, we can reduce the potential energy to the form
   $$U = a\alpha_1 + b\beta_2 + c\gamma_3.$$

Depending on the locations of the application centers in the rigid body and the restrictions on the moments of inertia, the following integrability cases generalizing the corresponding cases in the Euler – Poisson – Equations are possible.

Euler case. The application centers are all located at the fixed point, i.e., $\boldsymbol{r}_1 = \boldsymbol{r}_2 = \boldsymbol{r}_3 = 0$ in the Hamiltonian (3.78). The additional integrals are the projections of the angular momentum on fixed axes, which form the vector-valued integral $\boldsymbol{N} = (N_1, N_2, N_3)$:

$$N_1 = (\boldsymbol{M}, \boldsymbol{\alpha}), \qquad N_2 = (\boldsymbol{M}, \boldsymbol{\beta}), \qquad N_3 = (\boldsymbol{M}, \boldsymbol{\gamma}). \tag{3.79}$$

Their Poisson brackets make up the algebra $so(3)$: $\{N_i, N_j\} = \varepsilon_{ijk}N_k$; hence the integrability is noncommutative (see Sec. 2.2 of Ch. 2 for more details).

Generalized Lagrange case. In this case, the body considered is dynamically symmetric and all three application centers lie on the dynamic symmetry axis, so the Hamiltonian may be written in the form

$$H = \frac{1}{2}(M_1^2 + M_2^2 + aM_3^2) + z_1\alpha_3 + z_2\beta_3 + z_3\gamma_3.$$

According to the reduction results, this case reduces to the usual Lagrange top in one field with the corresponding additional integrals $F_1 = (\boldsymbol{M}, \boldsymbol{\gamma})$, and $F_2 = M_3$ (Sec. 2.3 of Ch. 2).

Generalized Kovalevskaya case. The inertia ellipsoid is an ellipsoid of revolution, the relations $a_1 = a_2 = \frac{1}{2}a_3$, $(a_i = I_i^{-1})$ hold for the moments of inertia, and three application centers are arbitrarily located in the equatorial plane of the inertia ellipsoid. As shown above, we can restrict ourselves here to considering only two application centers.

The corresponding set of independent involutive integrals was found by A. G. Reyman and M. A. Semenov-Tian-Shansky [483, 485, 52]: one integral is quadratic in momenta, and the other one (analog of the Kovalevskaya integral) is of degree four in momenta.

As in the usual Kovalevskaya case, for this case the system considered admits a generalization in which a constant gyrostatic momentum along the dynamic symmetry axis is added. The Hamiltonians and the integrals have the following form [95, 485]:

$$\begin{aligned}
H &= \frac{1}{2}(M_1^2 + M_2^2 + 2(M_3 + \lambda)^2) - (\boldsymbol{r}_1, \boldsymbol{\alpha}) - (\boldsymbol{r}_2, \boldsymbol{\beta}),\\
F_1 &= (N_1\boldsymbol{r}_1 + N_2\boldsymbol{r}_2)^2 + 2N_3(\boldsymbol{r}_1 \times \boldsymbol{r}_2, \boldsymbol{M})\\
&\quad + 2(\boldsymbol{r}_1 \times \boldsymbol{r}_2, \boldsymbol{r}_2 \times \boldsymbol{\alpha} - \boldsymbol{r}_1 \times \boldsymbol{\beta}) + 4\lambda(\boldsymbol{r}_1 \times \boldsymbol{r}_2, \boldsymbol{N}),\\
F_2 &= \left(\frac{M_1^2 - M_2^2}{2} + g_\alpha\alpha_1 - h_\alpha\alpha_2 + g_\beta\beta_1 - h_\beta\beta_2\right)^2\\
&\quad + (M_1M_2 + g_\alpha\alpha_2 + h_\alpha\alpha_1 + g_\beta\beta_2 + h_\beta\beta_1)^2\\
&\quad - 2\lambda(M_3 + 2\lambda)(M_1^2 + M_2^2) - 4\lambda\left(\alpha_3(\boldsymbol{M}, \boldsymbol{r}_1) + \beta_3(\boldsymbol{M}, \boldsymbol{r}_2)\right),
\end{aligned} \tag{3.80}$$

where $\boldsymbol{r}_1 = (g_\alpha, h_\alpha, 0)$, $\boldsymbol{r}_2 = (g_\beta, h_\beta, 0)$, $\lambda = \text{const.}$ is the gyrostatic momentum, and N_i is defined by (3.79).

Neither the explicit integration of this case nor its qualitative and topological analysis has yet been carried out.

REMARK 1. A. G. Reyman and M. A. Semenov-Tian-Shansky found this case in the n-dimensional situation but under the following additional restrictions: the reduction centers $\boldsymbol{r}_1$ and $\boldsymbol{r}_2$ of two mutually perpendicular centers lie at the same distances from the fixed point, not necessarily at a right angle (or it can be assumed that $\boldsymbol{r}_1 \perp \boldsymbol{r}_2$

and the fields $\boldsymbol{\alpha}$ and $\boldsymbol{\beta}$ are nonperpendicular). As the reduction results show, these restrictions are not necessary.

REMARK 2. V. V. Sokolov and A. V. Tsiganov [540] found a generalization of this integrable case and the corresponding **L-A** pairs when pairs of terms bilinear in M_i, α_i and bilinear in M_i, β_i are added to Hamiltonian (3.80). More precisely,

$$\begin{gathered} H = \frac{1}{2}(M_1^2 + M_2^2 + 2M_3^2) - c_1\alpha_2 - b_1\beta_1 + \lambda M_3 \\ +c_2(M_3\alpha_1 - M_1\alpha_3) - b_2(M_3\beta_2 - M_2\beta_3), \\ c_i, b_i, \lambda = \text{const}, \end{gathered}$$

where the relation $c_1 b_2 - c_2 b_1 = 0$ holds for the constants c_i and b_i. Here, the additional integrals F_1 and F_2 are simplified further. They are obtained from the **L-A** pair [540], which generalizes that found by A. G. Reyman and M. A. Semenov-Tian-Shansky (this generalization also holds for higher dimensions).

For $\boldsymbol{r}_2 = 0$ (or for $\boldsymbol{r}_1 = 0$) the integral F_1 (3.80) turns into the area integral $(\boldsymbol{M}, \boldsymbol{\alpha})$ (or into $(\boldsymbol{M}, \boldsymbol{\beta})$, respectively). In this case the precession angle ψ is a cyclic variable, and the elimination of this variable leads to the usual Kovalevskaya case of the Euler–Poisson equations (Sec. 2.4 of Ch. 2). An analogous reduction is possible in the case where $\boldsymbol{r}_1 \parallel \boldsymbol{r}_2$.

For $\boldsymbol{r}_1 \perp \boldsymbol{r}_2$, we can choose, e.g., $g_\alpha = h_\beta$, $h_\alpha = g_\beta = 0$ or $h_\alpha = g_\beta$, $g_\alpha = h_\beta = 0$; then, instead of F_1, one has the linear integral $M_3 \pm N_3 = M_3 \pm (\boldsymbol{M}, \boldsymbol{\gamma})$, and $\varphi \mp \psi$ is a cyclic variable. The corresponding reduction and the resulting isomorphism to the Chaplygin integrable case in the Kirchhoff equations are considered in detail in Sec. 4.1 of Ch. 4. This integrable case was found by H. Yehia [619] before the appearance of [483, 485].

Generalization of the Delone case (1st Appelrot class)[1]. The *generalized Delone case (O. I. Bogoyavlenskii [59]) arises by imposing $\lambda = 0$ and $F_2 = z_1^2 + z_2^2 = 0$ in (3.80). It is then possible to use Dirac reduction to reduce to a system having two degrees of freedom* [95]. *The system restricts in the Dirac sense to the invariant relations*

$$\begin{aligned} z_1 &= \frac{M_1^2 - M_2^2}{2} + g_\alpha\alpha_1 - h_\alpha\alpha_2 + g_\beta\beta_1 - h_\beta\beta_2 = 0, \\ z_2 &= M_1M_2 + g_\alpha\alpha_2 + h_\alpha\alpha_1 + g_\beta\beta_2 + h_\beta\beta_1 = 0, \end{aligned} \tag{3.81}$$

which are central functions of the Dirac structure [95]. On the four-dimensional symplectic leaf of the Dirac bracket, there exist two integrals (3.80), which allows us to completely integrate the system.

1 See Ch. 2.

On the level surface of the invariant relations (3.81), there is the following additional integral of degree three:

$$F_3 = \{z_1, z_2\} = -M_3(M_1^2 + M_2^2) + 2\alpha_3(M_1 g_\alpha + M_2 h_\alpha) \\ + 2\beta_3(M_1 g_\beta + M_2 h_\beta). \tag{3.82}$$

Indeed,

$$\{F_3, H\} = 2z_1(-g_\alpha\alpha_2 - h_\alpha\alpha_1 - g_\beta\beta_2 - h_\beta\beta_1) \\ - 2z_2(-g_\alpha\alpha_1 + h_\alpha\alpha_2 - g\beta\beta_1 + h_\beta\beta_2), \tag{3.83}$$

although the Jacobi theorem, saying that the commutator of two integrals is also an integral, as a rule, cannot be generalized to the invariant relations.

By using (3.81) and (3.83), the integrability of the generalized Kovalevskaya top in the Delone case can be also proved without using the integral F_1 (3.80). It turns out that the full set of integrals, including F_1, z_1, z_2, and F_3, becomes dependent. Moreover, it is interesting to note that in the case of a single force field ($g_\alpha = g_\beta = h_\beta = 0$) the structure of the integral F_3 (3.82), cubic in momenta, is almost the same as that of the partial Goryachev – Chaplygin integral for the Euler – Poisson equations (see Sec. 2.5 of Ch. 2). The topological analysis of this case and the calculation of the Fomenko – Tsishang invariants were carried out by D. V. Zotev in [655].

Generalization of the 2nd and 3rd Appelrot classes (Kharlamov case). One more case where it is possible to reduce the generalized Kovalevskaya top to a system with two degrees of freedom was found by M. P. Kharlamov in [284]. The corresponding relations can be written in the simplest form if we do not assume that the vectors $\boldsymbol{\alpha}$ and $\boldsymbol{\beta}$ are normalized and if we set

$$\boldsymbol{\alpha}^2 = b^2, \quad \boldsymbol{\beta}^2 = b^2, \quad (\boldsymbol{\alpha}, \boldsymbol{\beta}) = c, \\ \boldsymbol{r}_1 = (1, 0, 0), \quad \boldsymbol{r}_2 = (0, 1, 0). \tag{3.84}$$

(As pointed out in [95], the simplest linear change of variables reduces this case to (4.4).), so that the invariant relations have the form

$$\begin{gathered} (\xi_1^2 + \xi_2^2)M_3 - \big(\alpha_3(\xi_1 M_1 + \xi_2 M_2) + \beta_3(\xi_2 M_1 - \xi_1 M_2)\big) = 0, \\ \left(\frac{M_1^2 - M_2^2}{2} + \xi_1\right)\left(c(\xi_1^2 - \xi_2^2) - (a^2 - b^2)\xi_1\xi_2\right) \\ +(M_1 M_2 + \xi_2)\left((a^2 - b^2)(\xi_1^2 - \xi_2^2) + 4c\xi_1\xi_2\right) = 0, \\ \xi_1 = \alpha_1 - \beta_2, \quad \xi_2 = \alpha + \beta_1. \end{gathered} \tag{3.85}$$

Setting $\beta_i = 0$ (or $\alpha_i = 0$), we arrive at the invariant relations defining the classical 2nd and 3rd Appelrot classes (see Sec. 2.4 of Ch. 2).

On the four-dimensional invariant manifold defined by the invariant relations (3.85), there exist two independent first integrals, for example, H and F_2 defined in (3.80). The separation of variables and the reduction to (degenerate) Abel – Jacobi equations are considered in Ch. 5. The topological analysis of the Kharlamov case was carried out in [289, 282].

Generalized spherical top. In this case, $a_1 = a_2 = a_3$, and for any location of the reduction centers $\boldsymbol{r}_1$, $\boldsymbol{r}_2$, and $\boldsymbol{r}_3$ in the body considered, the system remains integrable. Moreover, since the kinetic energy is invariant under changes of body-fixed axes, we can reduce the potential energy to the form

$$V = x\alpha_1 + y\beta_2 + z\gamma_3.$$

In the quaternion representation, it can be considered as an arbitrary quadratic form $V = \sum_{i=0}^{3} b_{ij}\lambda_i\lambda_j$. Hence, according to the analogy discussed in [95] (see also Appendix B), this case is isomorphic to the Neumann problem describing the motion of a point on a three-dimensional sphere S^3. A set of involutive (quadratic) integrals can be extracted from the work of J. Moser [432], in which the separation of variables for the Neumann system on S^n is described; this was carried out in the 19th century by Rossochatius [495], who also added interesting singular terms whose mechanical meaning is discussed in Sec. 4.1 of Ch. 4. Let us express the integrals in terms of the variables necessary for us in the most symmetric form:

$$\begin{aligned}
H &= 4\boldsymbol{M}^2 + \frac{1}{4}(a_0^2 + a_1^2 - a_2^2 - a_3^2)\alpha_1 \\
&\quad + \frac{1}{4}(a_0^2 - a_1^2 + a_2^2 - a_3^2)\beta_2 + \frac{1}{4}(a_0^2 - a_1^2 - a_2^2 + a_3^2)\gamma_3, \\
F_1 &= (\boldsymbol{M} + \boldsymbol{N}, \mathbf{A}(\boldsymbol{M} + \boldsymbol{N})) + (\boldsymbol{M} - \boldsymbol{N}, \mathbf{B}(\boldsymbol{M} - \boldsymbol{N})) \\
&\quad + \frac{1}{4}(a_0{+}a_1{-}a_2{-}a_3)\alpha_1 + \frac{1}{4}(a_0{-}a_1{+}a_2{-}a_3)\beta_3 + \frac{1}{4}(a_0{-}a_1{-}a_2{+}a_3)\gamma_3, \\
\mathbf{A} &= \operatorname{diag}\left(\frac{1}{a_0{+}a_1}, \frac{1}{a_0{+}a_2}, \frac{1}{a_0{+}a_3}\right), \quad \mathbf{B} = \operatorname{diag}\left(\frac{1}{a_2{+}a_3}, \frac{1}{a_1{+}a_3}, \frac{1}{a_1{+}a_2}\right), \\
F_2 &= a^2(\boldsymbol{M} + \boldsymbol{N})^2 - 4(\mathbf{C}\boldsymbol{M}, \boldsymbol{N}) + \frac{1}{4}(a_0^4 + a_1^4 - a_2^4 - a_3^4)\alpha_1 \\
&\quad + \frac{1}{4}(a_0^4 - a_1^4 + a_2^4 - a_3^4)\beta_2 + \frac{1}{4}(a_0^4 - a_1^4 - a_2^4 + a_3^4)\gamma_3, \\
\mathbf{C} &= \operatorname{diag}\left(a_2^2 + a_3^2, a_1^2 + a_3^2, a_1^2 + a_2^2\right),
\end{aligned}$$

where $\boldsymbol{N}$ is defined by the formulas (3.79) and $a^2 = a_0^2 + a_1^2 + a_2^2 + a_3^2$.

Remark. Considering F_1 as the Hamiltonian and using the quaternion representation (see Appendix B), we obtain an integrable problem of the motion of a four-dimensional rigid body with quadratic potential of a special form. This system can also be considered as a generalization of the Clebsch case (Sec. 3.1 of Ch. 3).

Analog of the Hess case. If the inertia ellipsoid of the rigid body considered is not symmetric with respect to the fixed point and the reduction centers of all three fields $\boldsymbol{r}_1$, $\boldsymbol{r}_2$, and $\boldsymbol{r}_3$ lie on a line perpendicular to its circular section (which passes through the middle axis), then, as was said above, the potential reduces to the case of a single field with the reduction center on the same axis. In this way we arrive at the usual Hess case for the motion of a rigid body in a gravitational field (Sec. 2.6 of Ch. 2) whose invariant relation has the form

$$\sqrt{a_2 - a_1}M_1 \pm \sqrt{a_3 - a_2}M_3 = 0, \qquad a_1 < a_2 < a_3.$$

As was noted by N. E. Zhukovskii [646], the center of gravity in this case moves according to the spherical pendulum law.

A detailed study of the Hess case for the linear field and for fields of a more general form is contained in Sec. 4.3 of Ch. 4, which also shows its connection with the existence of a cyclic variable and the Lagrange case.

2 The Brun system

Let us consider the case where the potential $V(\boldsymbol{\alpha}, \boldsymbol{\beta}, \boldsymbol{\gamma})$ depends quadratically on the direction cosines. This problem was studied by F. Brun as long ago as the 19th century [108]. Brun found two additional independent integrals of motion, but could not establish the integrability. For this purpose, it is necessary to use the Hamiltonian structure of the equations of motion and the Liouville theorem (instead of the last multiplier theory, which was usually used for integration of rigid body dynamics in the 19th century), and the fact that Brun's two integrals are in involution with each other. Later the integrability of an n-dimensional top with quadratic potential was studied in [482] (A. G. Reyman), but the most complete results are presented in the work of O.I.Bogoyavlenskii [58, 59, 60, 61]. These works also contain various physical interpretations of the problem.

Remark. In a small book [233], D. N. Goryachev studied systems with quadratic potential. He obtained general conditions for the existence of additional linear and quadratic integrals for such a system. Independently of Brun, he presented the integrable case in the presence of a single field, and in one particular case he found the second necessary quadratic integral for two force fields. All these integrals can be obtained from a more general system considered below.

Lax representation and first integrals [61, 95]. Let us consider the Hamiltonian system in the variables $\boldsymbol{M}$, $\boldsymbol{\alpha}$, $\boldsymbol{\beta}$, and $\boldsymbol{\gamma}$ defined by Eqs. (1.60), the Poisson bracket relations (1.59) of Ch. 1, and the Hamiltonian

$$H = \frac{1}{2}(\mathbf{I}^{-1}\boldsymbol{M}, \boldsymbol{M}) - x(\mathbf{I}\boldsymbol{\alpha}, \boldsymbol{\alpha}) - y(\mathbf{I}\boldsymbol{\beta}, \boldsymbol{\beta}) - z(\mathbf{I}\boldsymbol{\gamma}, \boldsymbol{\gamma}), \tag{3.86}$$

where $x, y, z \in \mathbb{R}$ and $\mathbf{I} = \operatorname{diag}(I_1, I_2, I_3)$ is the tensor of inertia of the body. The Hamiltonian (3.86) is obtained from (1.87) of Ch. 1 for $x = 0$ and therefore, physically such a potential can be regarded as an approximation of the Newton potential near a gravitating body.

Let us identify the three-dimensional vectors $\boldsymbol{M}, \boldsymbol{\alpha}, \boldsymbol{\beta}$, and $\boldsymbol{\gamma}$ with the skew-symmetric matrices $\mathbf{M}, \widetilde{\boldsymbol{\alpha}}, \widetilde{\boldsymbol{\beta}}$, and $\widetilde{\boldsymbol{\gamma}}$ by the formulas

$$M_{ij} = \varepsilon_{ijk}M_k, \quad \widetilde{\alpha}_{ij} = \varepsilon_{ijk}\alpha_k, \quad \widetilde{\beta}_{ij} = \varepsilon_{ijk}\beta_k, \quad \widetilde{\gamma}_{ij} = \varepsilon_{ijk}\gamma_k \tag{3.87}$$

and define the symmetric matrix

$$\mathbf{u} = x\widetilde{\boldsymbol{\alpha}}^2 + y\widetilde{\boldsymbol{\beta}}^2 + z\widetilde{\boldsymbol{\gamma}}^2, \tag{3.88}$$

where x, y, and z are defined in (3.86). Formulas (3.87) and (3.88) define the embedding of the phase space of the system in the space L^9 of 3×3 matrices, since any matrix $\mathbf{l}$ can be represented in the form $\mathbf{l} = \mathbf{M} + \mathbf{u}$. In this space, the commutator relations (1.80) of Ch. 1 define a Lie algebra structure corresponding to the semidirect sum $L^9 = so(3) \oplus_S \mathbb{R}^6$, where $so(3)$ is the algebra of matrices $\mathbf{M}$ and $\mathbb{R}^6$ is the space of symmetric matrices $\mathbf{u}$ whose commutator should be assumed to be zero. In the matrix form, the commutator relations for $\mathbf{l}_1 = \mathbf{M}_1 + \mathbf{u}_1$ and $\mathbf{l}_2 = \mathbf{M}_2 + \mathbf{u}_2$ can be written as

$$\begin{gathered}[\mathbf{M}, \mathbf{u}] = \mathbf{M}\mathbf{u} - \mathbf{u}\mathbf{M} \in \mathbb{R}^6, \quad [\mathbf{M}_1, \mathbf{M}_2] = \mathbf{M}_1\mathbf{M}_2 - \mathbf{M}_2\mathbf{M}_1 \in so(3), \\ [\mathbf{u}_1, \mathbf{u}_2] = 0.\end{gathered} \tag{3.89}$$

Remark. The standard matrix commutator for $gl(3)$ defines commutator relations that differ from (3.89) by the term $[\boldsymbol{u}_1, \boldsymbol{u}_2] \neq 0$. These two sets of commutator relations are compatible and so define a Poisson bracket pencil (see [95] for more details).

The Poisson structure (3.89) corresponding to the algebra L^9 has the Casimir functions

$$F_1 = \operatorname{Tr}(\mathbf{u}), \quad F_2 = \operatorname{Tr}(\mathbf{u}^2), \quad F_3 = \operatorname{Tr}(\mathbf{u}^3),$$

and when restricted to the six-dimensional manifold M^6 defined by these Casimir functions it is nondegenerate. For the Liouville integrability of the system, we need two additional involutive integrals defining three-dimensional tori on which the system follows quasi-periodic motions.

In the variables $\mathbf{M}$ and $\mathbf{u}$ the Hamiltonian (3.86) has the form

$$H = -\operatorname{Tr}\left(\frac{1}{4}\mathbf{M}\boldsymbol{\omega} + \mathbf{u}\mathbf{I}\right),$$

and the equations themselves can be written in the compact form:

$$\dot{\mathbf{M}} = [\mathbf{M}, \boldsymbol{\omega}] + \left[\mathbf{u}, \frac{\partial H}{\partial \mathbf{u}}\right], \quad \dot{\mathbf{u}} = [\mathbf{u}, \boldsymbol{\omega}], \tag{3.90}$$

where $\boldsymbol{\omega} = \|\omega_{ij}\|$ is the skew-symmetric matrix corresponding to the angular velocity, whose entries are $\omega_{ij} = \frac{\partial H}{\partial M_{ij}} = I_k^{-1} M_{ij}$, and $\frac{\partial H}{\partial \mathbf{u}} = \left\|\frac{\partial H}{\partial u_{ij}}\right\| = -\mathbf{I}$.

We can also represent Eqs. (3.90) in the form of a Lax pair with a rational dependence on a spectral parameter λ:

$$\begin{gathered}\dot{\mathbf{L}} = [\mathbf{L}, \mathbf{A}], \\ \mathbf{L} = \lambda\mathbf{M} + \mathbf{u} + \lambda^2\mathbf{B}, \quad \mathbf{A} = \boldsymbol{\omega} - \lambda\mathbf{I},\end{gathered} \tag{3.91}$$

where $\mathbf{B} = (\det \mathbf{I})\mathbf{I}^{-1}$.

We obtain two necessary independent involutive integrals of motion as the coefficients of λ^k in the traces of powers of the matrix $\mathbf{L}$:

$$G_1 = \mathrm{Tr}\left(\frac{1}{2}\mathbf{M}^2 + \mathbf{Bu}\right), \qquad G_2 = \mathrm{Tr}(\mathbf{M}^2\mathbf{u} + \mathbf{Bu}^2).$$

Modulo the Casimir functions, they can be represented in the following vector form:

$$G_1 = \frac{1}{2}\boldsymbol{M}^2 + \det\mathbf{I}\left(x(\boldsymbol{\alpha}, \mathbf{I}^{-1}\boldsymbol{\alpha}) + y(\boldsymbol{\beta}, \mathbf{I}^{-1}\boldsymbol{\beta}) + z(\boldsymbol{\gamma}, \mathbf{I}^{-1}\boldsymbol{\gamma})\right),$$
$$G_2 = (x+y+z)\boldsymbol{M}^2 + x(\boldsymbol{M}, \boldsymbol{\alpha})^2 + y(\boldsymbol{M}, \boldsymbol{\beta})^2 + z(\boldsymbol{M}, \boldsymbol{\gamma})^2 + V,$$
$$V = \det\mathbf{I}\left[I_1^{-1}(\boldsymbol{p}, \mathbf{C}\boldsymbol{p}) + I_2^{-1}(\boldsymbol{q}, \mathbf{C}\boldsymbol{q}) + I_3^{-1}(\boldsymbol{r}, \mathbf{C}\boldsymbol{r})\right],$$

where $\mathbf{C} = \mathrm{diag}(2yz - x^2,\ 2xz - y^2,\ 2xy - z^2)$, $\boldsymbol{p} = (\alpha_1, \beta_1, \gamma_1)$, $\boldsymbol{q} = (\alpha_2, \beta_2, \gamma_2)$, and $\boldsymbol{r} = (\alpha_3, \beta_3, \gamma_3)$.

The integrable system with Hamiltonian $H = G_1$ can be considered as the problem of motion of a spherical top or a material point on S^3 in a force field with potential of degree four in Rodrigues – Hamilton quaternionic variables [58, 340] (See Appendixes A and B).

After introducing the vector $\boldsymbol{N} = (N_1, N_2, N_3)$, (3.79), comprising the projections of the angular momentum on the fixed axis, we can consider the integrable system with the Hamiltonian $H = G_2$ as a system on the Lie algebra $e(4)$ (see Appendix B) integrable on the singular orbit defined by the variables $\boldsymbol{N} = (N_1, N_2, N_3)$, $\boldsymbol{p} = (\alpha_1, \beta_1, \gamma_1)$, $\boldsymbol{q} = (\alpha_2, \beta_2, \gamma_2)$, and $\boldsymbol{r} = (\alpha_3, \beta_3, \gamma_3)$. Indeed, as is easily seen, the algebra of variables $\boldsymbol{N}, \boldsymbol{p}, \boldsymbol{q}$, and $\boldsymbol{r}$ is isomorphic to the algebra of variables $\boldsymbol{M}, \boldsymbol{\alpha}, \boldsymbol{\beta}$, and $\boldsymbol{\gamma}$. But since $\boldsymbol{M}^2 = \boldsymbol{N}^2$, it follows that the integral G_2 on the algebra of $\boldsymbol{N}, \boldsymbol{p}, \boldsymbol{q}$, and $\boldsymbol{r}$ is similar to the Hamiltonian H (3.86) on the algebra of $\boldsymbol{M}, \boldsymbol{\alpha}, \boldsymbol{\beta}$, and $\boldsymbol{\gamma}$. In this sense, the integrals H and G_2 are reciprocal. *As Hamiltonians they define the same integrable system in different systems of variables corresponding to moving and fixed reference frames.*

REMARK 3. In [57], integrable cases for special systems of linked rigid bodies are obtained using the generalized Brun integrable system. However, they are not new examples, since their dynamics reduce to Eqs. (3.90).

REMARK 4. In [116], a hydrodynamic interpretation of system (3.90) is given: a free rigid body is linearly magnetized and moves in a homogeneous magnetic field (or a non-conducting linearly polarized rigid body moves freely in a homogeneous electric field). The existence conditions for two additional integrals presented in [116], as well as the integrals themselves, also exist in the general *Brun system*. Other physical interpretations of the general Brun system are presented in [61].

Dynamical symmetry case. Let us consider system (3.86) under the dynamical symmetry condition ($I_1 = I_2 = 1$). It turns out that it reduces to two degrees of freedom and to the Neumann system. In this case the Hamiltonian (3.86) of the system can be represented as

$$H = \frac{1}{2}(M_1^2 + M_2^2 + aM_3^2) - \frac{a-1}{a}(x\alpha_3^2 + y\beta_3^2 + z\gamma_3^2), \tag{3.92}$$

where $x, y, z, a = I_3^{-1} \in \mathbb{R}$. It follows from the equations of motion (3.90) that the component M_3 is an integral of the motion.

Let $\boldsymbol{N}$ be the projection of the momentum on the axes fixed relative to the absolute space (3.79). Let $\boldsymbol{p}$ be the same projection of the unit vector directed along the dynamical symmetry axis (in components $(p_1, p_2, p_3) = (\alpha_3, \beta_3, \gamma_3)$). It follows from direct calculations that the Poisson algebra generated by $\boldsymbol{N}$ and $\boldsymbol{p}$ is the Lie algebra $e(3)$:

$$\{N_i, N_j\} = \varepsilon_{ijk}N_k, \quad \{N_i, p_j\} = \varepsilon_{ijk}p_k, \quad \{p_i, p_j\} = 0. \tag{3.93}$$

These commutator relations were already mentioned in Sec. 1.4 of Ch. 1. Eliminating the integral $M_3 = \text{const}$, which is a Casimir function of the structure (3.93), we can write Hamiltonian (3.92) in the variables N_i and p_j (also using the fact that $\boldsymbol{M}^2 = \boldsymbol{N}^2$):

$$H = \frac{1}{2}\boldsymbol{N}^2 - \frac{a-1}{a}(xp_1^2 + yp_2^2 + zp_3^2). \tag{3.94}$$

The equations of motion with Hamiltonian (3.94) coincide with the equations of a point over the two-dimensional sphere in a force field with quadratic potential (Neumann problem). This relation was noted in [58] without using the equations of the bracket algebra (3.93) (see [95]).

The Brun problem in a single field is very well known. In this case the equations of motion have the form of a Hamiltonian system on $e(3)$ with the Hamiltonian

$$H = \frac{1}{2}(\mathbf{A}\boldsymbol{M}, \boldsymbol{M}) + \frac{1}{2}\mu(\mathbf{A}^{-1}\boldsymbol{\gamma}, \boldsymbol{\gamma})$$

and with the additional integral

$$F = \frac{1}{2}(\boldsymbol{M}, \boldsymbol{M}) - \frac{\mu}{2\det\mathbf{A}}(\mathbf{A}\boldsymbol{\gamma}, \boldsymbol{\gamma}).$$

This problem turns out to be equivalent to many other integrable dynamical systems arising in various areas of mechanics and physics, for example, the Clebsch case in the Kirchhoff equations (Sec. 3.1 of Ch. 3).

3 Quaternion Euler – Poisson equations

Let us consider the following last and least natural case of equations of motion of a rigid body with the potential linear in the Rodrigues – Hamilton quaternionic variables:

$$H = \frac{1}{2}(\mathbf{A}\boldsymbol{M}, \boldsymbol{M}) + \sum_{i=0}^{3} r_i\lambda_i, \quad r_i = \text{const}, \tag{3.95}$$

assuming that the equations of motion have the form (1.67) of Ch. 1. As was already mentioned, there are no such potentials in real (physical) rigid body mechanics, since their dependence on the position of the body is not single-valued (it is two-valued). As a justification for the consideration of such equations, we can refer to the quantum mechanics problems, the problems of point mass dynamics in a curved space S^3 [95], and also to some formal construction methods of **L**-**A** pairs [95] (see Appendix C). In fact, it turns out that in reducing the order of system (3.95), we obtain the usual Euler – Poisson equations with additional terms having various physical interpretations (Sec. 4.1 of Ch. 4).

An interesting feature of system (3.95) is the fact that by transformations linear in λ_i, we can reduce the general form of the potential

$$V = \sum_{i=0}^{3} r_i \lambda_i \tag{3.96}$$

to the form

$$V = r_0 \lambda_0. \tag{3.97}$$

That is, a linear transformation of the quaternion space λ_i (leaving the quaternion norm invariant) of the form

$$\begin{aligned} \widetilde{\lambda}_0 &= R^{-1}(r_0\lambda_0 + r_1\lambda_1 + r_2\lambda_2 + r_3\lambda_3), \\ \widetilde{\lambda}_1 &= R^{-1}(r_0\lambda_1 - r_1\lambda_0 - r_2\lambda_3 + r_3\lambda_2), \\ \widetilde{\lambda}_2 &= R^{-1}(r_0\lambda_2 + r_1\lambda_3 - r_2\lambda_0 - r_3\lambda_1), \\ \widetilde{\lambda}_3 &= R^{-1}(r_0\lambda_3 - r_1\lambda_2 + r_2\lambda_1 - r_3\lambda_0), \\ R^2 &= r_0^2 + r_1^2 + r_2^2 + r_3^2 \end{aligned} \tag{3.98}$$

reduces the potential (3.96) to the form (3.97). In this case the form of commutator relations for the variables M and $\widetilde{\lambda}$ remains the same as for M and λ. The existence of such linear transformations is a remarkable feature of the quaternion variables and the bracket (1.65) of Ch. 1; it has no analog for Poisson brackets of the algebras $e(3)$ and $so(4)$.

Probably, in the general asymmetric case where $a_1 \neq a_2 \neq a_3 \neq a_1$, the system (3.95) is not integrable, and there is not even one of the two necessary additional integrals. However, this is not proved anywhere, and for many reasons, the proof would not be natural. Note that even the application of the Kovalevskaya method to system (3.95) is not completely analogous to the classical Euler – Poisson problem.

For $a_1 = a_2$, there always exists the linear integral

$$\begin{aligned} F_1 = {} & M_3(r_0^2 + r_1^2 + r_2^2 + r_3^2) + N_3(r_1^2 + r_2^2 - r_0^2 - r_3^2) \\ & + 2N_2(r_1 r_0 - r_3 r_2) - 2N_1(r_1 r_2 - r_0 r_3), \end{aligned} \tag{3.99}$$

where N_i are the projections of the angular momentum on the fixed axes. As explained above, it may be assumed without loss of generality that $r_1 = r_2 = r_3 = 0$ and hence that this integral takes the natural form

$$F_1 = M_3 - N_3. \tag{3.100}$$

It turns out that this (linear) integral corresponds to the cyclic variable $\varphi + \psi$. This is discussed in detail in Sec. 4.1 of Ch. 4. Routh reduction in this cyclic variable leads to a Hamiltonian system on the algebra $e(3)$ with zero area constant $(\boldsymbol{M}, \boldsymbol{\gamma}) = 0$ and with Hamiltonian

$$H = \frac{1}{2}(M_1^2 + M_2^2 + a_3 M_3^2) + c(a_3 - 1)M_3 + r_0\gamma_2 + \frac{1}{2}\frac{c^2}{\gamma_3^2}, \tag{3.101}$$

where c is the constant value of the integral F_1 (3.100). The Hamiltonian (3.101) corresponds to the addition of a gyrostatic term linear in $\boldsymbol{M}$ and of the singular term $\frac{c^2}{2\gamma_3^2}$ to the usual Euler – Poisson equations; the physical meaning of the latter term is discussed in Ch. 4.

Let us present here the integrable cases of system (3.95), which turn out to be equivalent to the integrable cases of system (3.101).

Spherical top ($a_1 = a_2 = a_3$). The Hamiltonian has the form

$$H = \frac{1}{2}\boldsymbol{M}^2 + r_0\lambda_0,$$

and, as is shown in [95], the system is equivalent to the problem of motion of a material point over the three-dimensional sphere S^3. Since the potential depends only on λ_0, we can assume that the material point moves in a field whose center is fixed at either the north or south pole and that the potential depends only on the distance between the material point and the center (this is an analog of the problem of motion in a central field for $\mathbb{R}^3$). As in the plane case, the vector of the particle's angular momentum is preserved:

$$\boldsymbol{L} = \frac{1}{2}(\boldsymbol{N} - \boldsymbol{M}) = \text{const.} \tag{3.102}$$

where $\boldsymbol{N}$ stands for the projection of the angular momentum vector on the fixed axes.

The components of the vector $\boldsymbol{L}$ satisfy the Poisson bracket relations of $so(3)$: $\{L_i, L_j\} = \varepsilon_{ijk}L_k$, and *the integrability is not commutative*. Such a system is said to be *superintegrable*, and its three-dimensional tori are foliated by two-dimensional tori.

Kovalevskaya case. The Hamiltonian and the additional (fourth-degree) integral, both in involution with F_1, have the form

$$\begin{aligned} H &= \frac{1}{2}(M_1^2 + M_2^2 + 2M_3^2) + r_0\lambda_0, \\ F_2 &= (M_1N_1 + M_2N_2 + 2r_0\lambda_0)^2 + (N_1M_2 - N_2M_1 - 2r_0\lambda_3)^2 \\ &\quad + (N_3 - M_3)\big(M_3(\boldsymbol{M}^2 - M_3N_3) + 2r_0(M_2\lambda_1 - M_1\lambda_2 + \frac{\lambda_0}{2}(M_3 - N_3))\big). \end{aligned} \tag{3.103}$$

After reduction to the system (3.101) we obtain an integrable case that is embedded in the generalized Kovalevskaya family found by Goryachev and by Yehia (see Sec. 5.1 of Ch. 5 and Sec. 4.1 of Ch. 4).

Goryachev – Chaplygin case. The Hamiltonian and the additional integral have the form

$$\begin{aligned} H &= \frac{1}{2}(M_1^2 + M_2^2 + 4M_3^2) + r_0\lambda_0, \\ F_2 &= M_3(M_1^2 + M_2^2) + r_0(M_2\lambda_1 - M_1\lambda_2). \end{aligned} \tag{3.104}$$

After reduction to the system (3.101), this case is embedded in the generalized family presented in Sec. 4.1 of Ch. 4.

REMARK 5. The fact that the Lagrange and Hess cases do not generalize to system (3.95) is slightly surprising.

REMARK 6. With the addition to (3.103) and (3.104) of a constant gyrostatic momentum along the dynamical symmetry axis, we obtain the integrable cases corresponding to the generalized Yehia and Sretenskii cases of the Euler – Poisson equations; their integrals can be easily obtained from (3.101) by the reconstruction procedure described in Ch. 4.

In conclusion, we note that for the quaternion Euler – Poisson equations, the Kovalevskaya case, as well as the Goryachev – Chaplygin case, are general integrable cases. This allows one to use them for certain algebraic constructions (for constructing **L**-**A** pairs, etc.) and to establish various nontrivial interrelations between the corresponding integrable cases of the classical Euler – Poisson equations (Sec. 5.1 of Ch. 5).

4 Linear Integrals and Reduction

4.1 Linear Integrals in Rigid Body Dynamics

In this chapter we consider, for various forms of the equations of rigid body motion presented in Sec. 1.4 of Ch. 1, consequences of the existence of first integrals which are linear in angular momentum $\boldsymbol{M}$ (or, which is equivalent, in angular velocities $\boldsymbol{\omega}$ or generalized momenta $p_\theta, p_\varphi, p_\psi$, etc.). As is well known, in Hamiltonian mechanics [18, 19], a linear integral implies the existence of a cyclic variable and the possibility of reducing the order of the system. For a system given in Lagrangian form, the method for reduction of order was elaborated by E. Routh (it is often called *Routh reduction*). Owing to numerous studies initiated by the Marsden school, the reduction of dynamical systems of various types has become a full-fledged area of research. We shall not discuss this subject further, but leave the reader to consult the book [418]. In our book [95], we gave a specialized reduction algorithm applicable to systems admitting linear integrals: this does not require one to pass to the canonical form first and the algebraic form of the equations of motion is preserved. For the reduced system, not only the Hamiltonian but also the Poisson bracket changes form, and the latter can become nonlinear. In some cases, the reduced system turns out to be equivalent to some entirely different system; hence we have a method for finding isomorphic dynamical problems, which also extends to integrable problems.

In this section, we formulate several theorems on order reduction. We shall consider in turn three kinds of linear integrals – each of which has a natural physical interpretation – with their corresponding cyclic variables. Then we look at the reverse procedure, which allows one to extend results for the reduced system to the general (unreduced) equations. Using this scheme, from integrable families for the reduced system (with two degrees of freedom), we can obtain integrable cases of more general equations of rigid body motion in a potential field (see Sec. 3.4 of Ch. 3), i.e., for systems with three degrees of freedom. In this way we succeed in understanding the meaning of various additional singular terms of the type $\frac{a}{\gamma_3^2}$, $a = \text{const}$, in generalizations of integrable cases. Terms of this kind were introduced by D. N. Goryachev in the study and generalization of the cases of Goryachev – Chaplygin and of Kovalevskaya. The proper mechanical meaning of such additional terms remained unclear for a long time, although some "quantum mechanical" explanations were given. In [95], they are interpreted as being a consequence of a reduction procedure. Apart from the examples in this chapter, the reader can also look at some of the related problems in Ch. 3 (Sec. 3.4) and Ch. 5 (Sec. 5.1).

Linear integrals in the most general equations of rigid body dynamics around a fixed point were studied by D. N. Goryachev in [233]. In his work, he presented three typical possibilities considered below, which in some sense are unique (the proof of

DOI 10.1515/9783110544442-004

the latter fact is not expected to be easy). In Sec. 4.3 of this chapter, we apply the corresponding reductions to the linear invariant relations whose systematic introduction into dynamics is due to Levi-Civita ; he also tried to use them in the study of rigid body dynamics, as well as the study of celestial mechanics) [377]. In fact, the clearest and most important manifestation of the ideas of Levi-Civita is to be found in considering invariant relations of Hess type, which turns out to exist in many related problems of rigid body dynamics. In such cases, there also exist cyclic variables, order reduction is possible, with the result analogous to the Lagrange case and its generalizations. In particular, the possibility of reduction implies a number of qualitative features of the motion for the generalized Hess cases (the observation by N. E. Zhukovskii that in the Hess case the center of mass moves according to the spherical pendulum law is an example of this) typical for the motion of a heavy symmetric gyroscope.

It is well known that the existence of a first integral is due to the existence of a symmetry field and the possibility of order reduction, at least locally. This is the well-known *Noether theorem*, which is used for Hamiltonian systems with integrals linear in momenta and leads to some simplifications. For simplicity, we consider the canonical situation, although the arguments also easily extend to the general Poincaré – Chetaev equations, in particular, to the equations of rigid body dynamics in matrix realizations of Lie groups (defining the configuration spaces).

Indeed, for a system defined on the cotangent bundle TM with the canonical structure $\{q_i, p_j\} = \delta_{ij}$, the existence of an integral linear in the momenta

$$F = \sum_i v_i(\boldsymbol{q})p_i, \qquad \{F, H\} = 0, \tag{4.1}$$

leads to the phase flow

$$\frac{d\boldsymbol{q}}{ds} = \frac{\partial F}{\partial \boldsymbol{p}} = v(\boldsymbol{q}), \qquad \frac{d\boldsymbol{p}}{ds} = -\frac{\partial F}{\partial \boldsymbol{q}} \tag{4.2}$$

given by the Hamiltonian F and determining the action of a one-parameter symmetry group of the Hamiltonian H. In this case, we obtain a separate system of equations on the configurational space M:

$$\frac{d\boldsymbol{q}}{ds} = \boldsymbol{v}(\boldsymbol{q}). \tag{4.3}$$

Near a nonsingular point, we can rectify the field (4.3) and so represent it in new coordinates $Q_1, \dots, Q_{n-1}, Q_n$ in the form

$$\frac{dQ_1}{ds} = \dots = \frac{dQ_{n-1}}{ds} = 0, \quad \frac{dQ_n}{ds} = 1.$$

Obviously, the canonical momentum P_n corresponding to the coordinate Q_n coincides with the integral (4.1): $F = P_n$, and in view of the relation $\{H, P_n\} = \frac{\partial H}{\partial Q_n} = 0$, the coordinate Q_n is cyclic: in other words, we have reduction of order. □

In the order reduction procedure described below, which holds globally and is implemented in an algebraic way, we proceed according to a very similar scheme. Using the linear integral, we write systems (4.2) and (4.3). Since the system (4.3) decouples, it is easy to find its first integrals and also the integrals of the general system (4.3).

In what follows, the basic idea is to use this set of (as a rule, redundant) integrals as new variables for the initial system. If the algebra (nonlinear in general) of new variables with respect to the Poisson bracket is closed and the Hamiltonian is expressed only in terms of these variables, then we obtain a new Hamiltonian system for which the cyclic integral (4.1) is a Casimir function and the rank of the Poisson structure is lowered by two, i.e., the system is reduced. The advantages of the described reduction procedure preserving the algebraic character of the system and its various dynamic applications are discussed in our book [95]. Here, we consider only its use in the study of rigid body dynamics for three different variants described by the theorems presented below.

Let us consider the rigid body motion around a fixed point in a vector potential field, i.e., along with the potential forces, there also exist gyroscopic forces described by a vector potential and leading to additional terms linear in $\boldsymbol{M}$ in the Hamiltonian

$$H = \frac{1}{2}(\mathbf{A}\boldsymbol{M},\ \boldsymbol{M}) + (\boldsymbol{M},\ \boldsymbol{W}) + U, \tag{4.4}$$

where the functions U and $W = (W_1,\ W_2,\ W_3)$ defining the potential and the vector potential are assumed to be dependent on all the variables $\boldsymbol{q}$ defining the rigid body position. These variables might be, for example: the Euler angles θ, φ, and ψ; the direction cosines $\boldsymbol{\alpha}$, $\boldsymbol{\beta}$, and $\boldsymbol{\gamma}$; the Rodriguez parameters λ_0, λ_1, λ_2, and λ_3. Depending on the system of variables, we use the corresponding system of equations describing the motion (see Sec. 1.3 of Ch. 1). For greater generality we shall also assume that $\mathbf{A} = \mathbf{A}(\boldsymbol{q})$; this is necessary for modelling a body sliding over a plane or for modeling a gyroscope suspended in the gimbals. Here and in what follows, $\boldsymbol{N} = (N_1, N_2, N_3)$ are the projections of the angular momentum on the axes fixed in space.

1 Classical area integral $N_3 = (\boldsymbol{M},\ \boldsymbol{\gamma}) = c = \text{const}$

The symmetries leading to an integral of this kind are natural ones: they arise due to the invariance of the potential and the vector potential under rotations about an axis fixed in space. Homogeneous fields, in particular, the gravitational field, are obviously axially symmetric.

In this case, the precession angle ψ is a cyclic variable and the equations of motion can be presented on the algebra $e(3)$. That is, we can rewrite the Hamiltonian (4.4) in terms of the variables M_1, M_2,, M_3 and γ_1, γ_2,, γ_3 defining the symmetry unit vector in the fixed space:

$$\begin{aligned} H &= \frac{1}{2}(\mathbf{A}\boldsymbol{M},\ \boldsymbol{M}) + (\boldsymbol{M},\ \boldsymbol{W}(\theta,\ \varphi)) + U(\theta,\ \varphi) \\ &= \frac{1}{2}(\mathbf{A}\boldsymbol{M},\ \boldsymbol{M}) + (\boldsymbol{M},\ \boldsymbol{W}(\boldsymbol{\gamma})) + U(\boldsymbol{\gamma}), \end{aligned} \tag{4.5}$$

where

$$\gamma_1 = 2(\lambda_1\lambda_3 - \lambda_0\lambda_2), \quad \gamma_2 = 2(\lambda_0\lambda_1 + \lambda_2\lambda_3), \quad \gamma_3 = \lambda_0^2 - \lambda_1^2 - \lambda_2^2 + \lambda_3^2.$$

In the variables $(\boldsymbol{M}, \boldsymbol{\gamma})$, the Poisson bracket is determined by the algebra $e(3)$ (see Sec. 2.1 of Ch. 2). The symplectic leaf $\{\boldsymbol{\gamma}^2 = 1, (\boldsymbol{M}, \boldsymbol{\gamma}) = c\}$ of the algebra $e(3)$ is diffeomorphic to the cotangent bundle to the two-dimensional sphere S^2. This sphere is called the *Poisson sphere*. It is the configuration space of the reduced systems (where ψ is ignored).

For $c \neq 0$, reduction of order gives rise to additional gyroscopic forces having a singularity; this singularity can be interpreted as a *monopole*. In [447], the introduction of a monopole is considered as a noncanonical deformation of the Poisson bracket.

Here, we perform a reduction to the zero level set of the area integral in algebraic form without changing the bracket itself; in this case, the singularity corresponding to the monopole arises only in the Hamiltonian.

Theorem 1. *The equations of motion of the body with Hamiltonian* (4.5) *on the level set* $N_3 = c$ *of the integral are equivalent to the Hamiltonian equations on* $e(3)$ *on the zero level set of the area integral* $(\boldsymbol{M}, \boldsymbol{\gamma}) = 0$ *with the Hamiltonian*

$$\begin{aligned} H = {} & \frac{1}{2}(\mathbf{A}\boldsymbol{M}, \boldsymbol{M}) + (\boldsymbol{M}, \boldsymbol{W}) + c\frac{a_1\gamma_1 M_1 + a_2\gamma_2 M_2}{\gamma_1^2 + \gamma_2^2} \\ & + \frac{c^2}{2}\frac{a_1\gamma_1^2 + a_2\gamma_2^2}{(\gamma_1^2 + \gamma_2^2)^2} + \frac{c(W_1\gamma_1 + W_2\gamma_2)}{\gamma_1^2 + \gamma_2^2} + U(\boldsymbol{\gamma}). \end{aligned} \tag{4.6}$$

Proof It suffices to make the transformation $(\boldsymbol{M}, \boldsymbol{\gamma}) \to (\boldsymbol{M}, \boldsymbol{\gamma})$, which *preserves the structure of the algebra* $e(3)$ and transforms the integral $(\boldsymbol{M}, \boldsymbol{\gamma}) = c$ into $(\boldsymbol{M}, \boldsymbol{\gamma}) = 0$:

$$\begin{gathered} M_1 \to M_1 - c\frac{\gamma_1}{\gamma_1^2 + \gamma_2^2}, \quad M_2 \to M_2 - c\frac{\gamma_2}{\gamma_1^2 + \gamma_2^2}, \quad M_3 \to M_3, \\ \boldsymbol{\gamma} \to \boldsymbol{\gamma}. \end{gathered} \tag{4.7}$$

□

REMARK. For $c = 0$ and $W \equiv 0$, the reduced system is natural and there is no monopole.

The dynamical significance of Theorem 1, by contrast to the classical procedure of order reduction in the precession angle by means of the Routh method, is that all additional terms appearing in the reduced Hamiltonian are obtained explicitly in algebraic form. Moreover, since $(\boldsymbol{M}, \boldsymbol{\gamma}) = 0$, the reduced system with two degrees of freedom describes the motion of a point (defined by the vertical unit vector) on the (Poisson) sphere in a potential field with an additional magnetic term (even for $\boldsymbol{W} \equiv 0$, as long as $c \neq 0$) in the metric determined by the kinetic energy. We may express the reduced Hamiltonian (4.6) in terms of canonical variables if we use the natural variables on the cotangent bundle to the Poisson sphere $(\boldsymbol{M}, \boldsymbol{\gamma}) = 0$. This is equivalent to making the substitution

$$M_1 = -p_\varphi \cot\theta \sin\varphi + p_\theta \cos\varphi, \quad M_2 = -p_\varphi \cot\theta \cos\varphi - p_\theta \sin\varphi, \quad M_3 = p_\varphi,$$
$$\gamma_1 = \sin\theta \sin\varphi, \quad \gamma_2 = \sin\theta \cos\varphi, \quad \gamma_3 = \cos\theta.$$

Here θ and φ are spherical coordinates on the Poisson sphere and p_θ and p_φ are their corresponding canonically conjugate momenta. In contrast to this resulting canonical form of writing the Hamiltonian, the algebraic form of (4.6), as well as its analogs for other cyclic variables, helps in finding various analogies between problems, reveals connections between integrable cases, and provides a deeper insight into the algebraic nature of corresponding first integrals.

REMARK. For $a_1 = a_2$, the Hamiltonian (4.6) can be simplified using $(\boldsymbol{M}, \boldsymbol{\gamma}) = 0$ to

$$H = \frac{1}{2}(\mathbf{A}\boldsymbol{M}, \boldsymbol{M}) + (\boldsymbol{M}, \boldsymbol{W}) + U(\boldsymbol{\gamma}) - \frac{cM_3\gamma_3}{\gamma_1^2 + \gamma_2^2} + \frac{c^2}{2(\gamma_1^2 + \gamma_2^2)} + \frac{c(W_1\gamma_1 + W_2\gamma_2)}{\gamma_1^2 + \gamma_2^2}. \tag{4.8}$$

2 Integral $N_3 - M_3 = (M, \gamma) - M_3 = c = \mathrm{const}$

This integral corresponds to the cyclic variable $\psi - \varphi$ ($N_3 + M_3$ with $\psi + \varphi$ may be treated in an analogous fashion). It was studied for the first time by D. N. Goryachev [233]. The corresponding symmetries do not have an obvious physical interpretation; they not only arise due to the properties of the force field, but also are determined by the dynamical properties of the body. The body is dynamically symmetric, and in terms of the Rodrigues – Hamilton variables the Hamiltonian (4.4) takes the form

$$\begin{aligned} H = {} & \frac{1}{2}(M_1^2 + M_2^2 + aM_3^2) + 2\frac{M_1\lambda_1 + M_2\lambda_2}{\sqrt{\lambda_1^2 + \lambda_2^2}} W_1(\lambda_0, \lambda_3) \\ & + 2\frac{M_2\lambda_1 - M_1\lambda_2}{\sqrt{\lambda_1^2 + \lambda_2^2}} W_2(\lambda_0, \lambda_3) - 2M_3 W_3(\lambda_0, \lambda_3) + U(\theta, \varphi + \psi), \end{aligned} \tag{4.9}$$

where a is an arbitrary constant. Let us introduce the following new systems of variables:

$$\begin{gathered} K_1 = 2\frac{M_1\lambda_1 + M_2\lambda_2}{\sqrt{\lambda_1^2 + \lambda_2^2}}, \quad K_2 = 2\frac{M_2\lambda_1 - M_1\lambda_2}{\sqrt{\lambda_1^2 + \lambda_2^2}}, \quad K_3 = -2M_3, \\ s_1 = \lambda_3, \quad s_2 = \lambda_0, \quad s_3 = \sqrt{\lambda_1^2 + \lambda_2^2}, \end{gathered} \tag{4.10}$$

which have the commutation relations

$$\begin{gathered} \{K_3, K_1\} = K_2, \quad \{K_2, K_3\} = K_1, \quad \{K_1, K_2\} = K_3 + \frac{s_3(\boldsymbol{s}, \boldsymbol{K})}{s_3^2}, \\ \{K_i, s_j\} = \varepsilon_{ijk}s_k, \quad \{s_i, s_j\} = 0, \end{gathered} \tag{4.11}$$

so forming a closed algebra with respect to the nonlinear Poisson bracket generated by relations (4.11). This Poisson bracket is degenerate, having the Casimir functions

$$F_1 = s_3(\boldsymbol{s}, \boldsymbol{K}) = (\boldsymbol{M}, \boldsymbol{\gamma}) - M_3 = c, \quad F_2 = (\boldsymbol{s}, \boldsymbol{s}) = 1.$$

The one-parameter transformation

$$L = K - \alpha \frac{s}{s_3}, \qquad \alpha = \text{const.}, \tag{4.12}$$

preserves the bracket (4.11), and, moreover,

$$s_3(L,\ s) = s_3(K,\ s) - \alpha. \tag{4.13}$$

REMARK. We have presented the transformation (4.10) in the book [95] and in [92].

If we fix the integral $s_3(K,\ s) = c$ and choose $\alpha = c$, then since $s_3(L,\ s) = 0$, this leads to the vanishing of the nonlinear terms in the bracket (4.11), which becomes that of the algebra $e(3)$. The Hamiltonian (4.11) now takes the form

$$\begin{aligned} H = {} & \frac{1}{8}(L_1^2 + L_2^2 + aL_3^2) + (L,\ W(s)) + U(s) \\ & + c\frac{(s,\ W(s))}{s_3^2} + \frac{1}{2}\frac{c^2}{s_3^2} + c(a-1)L_3. \end{aligned} \tag{4.14}$$

Thus, we have proved the following analog of Theorem 1.

Theorem 2. *On the level set $N_3 - M_3 = c$ of the cyclic integral, under reduction of order, the system (4.9) with three degrees of freedom is mapped to the system on $e(3)$ with $(L,\ s) = 0$ and Hamiltonian function (4.14).*

Note that additional terms depending on c appear in (4.14). One of them can be interpreted as a gyroscopic momentum directed along the dynamical symmetry axis. The singular term $\frac{c}{s_3^2}$ was introduced into dynamics by D. N. Goryachev [235, 236].

Whereas the symmetrical origin of the integral $F = N_3 \pm M_3$ in rigid body dynamics is not obvious, its meaning can be easily understood by making an analogy with celestial mechanics in curved space [76], more precisely, with the motion of a material point on the spheres S^2 and S^3. This integral just corresponds to the projection of the angular momentum of the particle on an axis fixed in space with respect to which the potential is axially symmetric.

3 Integral $M_3 = c = \text{const}$ (Lagrange integral)

In this case the rigid body is dynamically symmetric, while the force field is invariant under rotations about the dynamical symmetry axis. The corresponding cyclic variable is the angle φ of proper rotation. It is convenient to write the Hamiltonian in terms of the direction cosines $\boldsymbol{\alpha}$, $\boldsymbol{\beta}$, and $\boldsymbol{\gamma}$:

$$\begin{aligned} H = {} & \frac{1}{2}(M_1^2 + M_2^2 + aM_3^2) + (M_1\alpha_1 + M_2\alpha_2)W_1^{(\alpha)}(\theta,\ \psi) \\ & + (M_1\alpha_2 - M_2\alpha_1)W_2^{(\alpha)}(\theta,\ \psi) + \cdots + M_3 W_3(\theta,\ \psi) + U(\theta,\ \psi), \end{aligned} \tag{4.15}$$

where the ellipsis stands for omitted terms which are linear in M_1 and M_2 and contain β and γ. Let us introduce new variables commuting with M_3 and determining the reduced system:

$$\begin{gathered} N_1 = (\boldsymbol{M}, \boldsymbol{\alpha}), \qquad N_2 = (\boldsymbol{M}, \boldsymbol{\beta}), \qquad N_3 = (\boldsymbol{M}, \boldsymbol{\gamma}), \\ \boldsymbol{p} = (\alpha_3, \beta_3, \gamma_3). \end{gathered} \tag{4.16}$$

The geometrical meaning of these variables is clear: the vector $\boldsymbol{N}$ is composed of the components of the angular momentum in the fixed coordinate system, and $\boldsymbol{p}$ are the components of the symmetry axis vector in the same system.

The commutation relations for the generators $(\boldsymbol{N}, \boldsymbol{p})$, (4.16), correspond to the coalgebra $e(3)$ (see Sec. 1.3 of Ch. 1). Its Casimir functions have the form

$$F_1 = \boldsymbol{p}^2 = 1, \qquad F_2 = (\boldsymbol{N}, \boldsymbol{p}) = \boldsymbol{p}^2 M_3 = c.$$

The terms linear in $\boldsymbol{M}$ in the Hamiltonian (4.15) can be found from the relations

$$\begin{aligned} M_1\alpha_1 + M_2\alpha_2 &= N_1 - p_1 M_3, & M_1\alpha_2 - M_2\alpha_1 &= p_2 N_3 - p_3 N_2, \\ M_1\beta_1 + M_2\beta_2 &= N_2 - p_2 M_3, & M_1\beta_2 - M_2\beta_1 &= p_3 N_1 - p_1 N_3, \\ M_1\gamma_1 + M_2\gamma_2 &= N_3 - p_3 M_3, & M_1\gamma_2 - M_2\gamma_1 &= p_1 N_2 - p_2 N_1 \end{aligned}$$

and, dropping constant terms, we obtain the Hamiltonian of the reduced system:

$$\begin{gathered} H = \frac{1}{2}\boldsymbol{N}^2 + \left(\boldsymbol{N}, \boldsymbol{W}^{(1)}\right) + \left(\boldsymbol{p} \times \boldsymbol{N}, \boldsymbol{W}^{(2)}\right) \\ + c\left(W_3(\boldsymbol{p}) - \left(\boldsymbol{p}, \boldsymbol{W}^{(1)}\right)\right) + U(\boldsymbol{p}), \\ \boldsymbol{W}^{(1)}(\boldsymbol{p}) = \left(W_1^{(\alpha)}, W_1^{(\beta)}, W_1^{(\gamma)}\right), \quad \boldsymbol{W}^{(2)}(\boldsymbol{p}) = \left(W_2^{(\alpha)}, W_2^{(\beta)}, W_2^{(\gamma)}\right). \end{gathered} \tag{4.17}$$

We have already performed a reduction corresponding to the integral $M_3 = \text{const}$ and used the variables (4.16) in Sec. 3.4 of Ch. 3 to establish a connection between the Brun problem (under the dynamical symmetry condition) and the Clebsch integrable case of the Kirchhoff equations.

4 Reconstruction of integrable systems

The most interesting is the inverse problem – that of obtaining new integrable cases of the three-degree-of-freedom system (4.4) from existing integrable cases of Hamiltonians defining a two-degree-of-freedom system on $e(3)$. Also, we show here in which way the integrable systems on the zero level set of the area integral $(\boldsymbol{L}, \boldsymbol{s}) = 0$ of the algebra $e(3)$ can be reconstructed up to the general integrable system (4.4) having the linear integral $M_3 \pm N_3$. First, let us formulate one general result whose proof is by direct verification.

Theorem 3. 1. *Suppose that on the algebra $e(3)$ there is given a Hamiltonian system with Hamiltonian function*

$$H(\boldsymbol{L},\boldsymbol{s}) = \frac{1}{2}(L_1^2 + L_2^2 + aL_3^2) + (\boldsymbol{L},\ \boldsymbol{W}(\boldsymbol{s})) + U(\boldsymbol{s})$$

and that the system is integrable when $(\boldsymbol{L},\boldsymbol{s}) = 0$, *(i.e., there exists an additional (partial) integral).*

Then by means of the transformation

$$\boldsymbol{L} = \boldsymbol{K} - \alpha\frac{\boldsymbol{s}}{s_3} \tag{4.18}$$

and using (4.10), *we obtain a system on the quaternion algebra in the variables* $\boldsymbol{M}, \lambda_0, \lambda_1, \lambda_2, \lambda_3$ (Sec. 1.3 of Ch. 1, formula (1.41)) *with Hamiltonian function*

$$H'(\boldsymbol{M},\lambda) = H - \alpha\frac{\lambda_3 W_1 + \lambda_0 W_2 + \sqrt{\lambda_1^2 + \lambda_2^2}\,W_3}{\lambda_1^2 + \lambda_2^2} - \frac{1}{2}\frac{\alpha^2}{\lambda_1^2 + \lambda_2^2} + \alpha(a-1)M_3, \tag{4.19}$$

integrable on the level surface $N_3 - M_3 = \alpha$.

2. *If we can choose the constants in the Hamiltonian H such that* H' *is independent of* α, *then the system* (4.19) *is integrable for an arbitrary value of the linear integral* $N_3 - M_3$. *Moreover, in the additional integral, after the transformations* (4.18) *and* (4.10), *it is necessary to set* $\alpha = (M_1\lambda_1 + M_2\lambda_2 + M_1\lambda_2 - M_2\lambda_1 - \sqrt{\lambda_1^2 + \lambda_2^2 M_3})$.

REMARK. An analogous reconstruction of integrable cases can be performed by using the linear integrals M_3 = const and N_3 = const. However, the generalizations of the integrable cases obtained in this process contain terms linear in velocities, which have no obvious mechanical interpretation.

Let us illustrate Theorem 3 by examining two examples.

Generalization of the Yehia – Kovalevskaya case. In [627], Yehia presents the particular case of integrability $(\boldsymbol{L},\ \boldsymbol{s}) = 0$, which generalizes the Kovalevskaya case with the Hamiltonian

$$\begin{aligned} H = \frac{1}{2}\left(L_1^2 + L_2^2 + 2(L_3 + \xi)^2\right) \\ +a\frac{\boldsymbol{s}^2}{s_3^2} + c_1 s_1 + c_2 s_2 + 2b_1 s_1 s_2 + b_2(s_2^2 - s_1^2) \end{aligned} \tag{4.20}$$

and the additional integral presented in Sec. 5.1 of Ch. 5, (5.4). After the transformation

$$\boldsymbol{L} = \boldsymbol{K} - \alpha\frac{\boldsymbol{s}}{s_3},$$

we obtain

$$\begin{aligned} H = \frac{1}{2}(K_1^2 + K_2^2 + 2K_3^2) + c_1 s_1 + c_2 s_2 + 2b_1 s_1 s_1 + b_2(s_2^2 - s_1^2) \\ + (2\xi - \alpha)K_3 + \frac{1}{2}\frac{2a\boldsymbol{s}^2 - 2\alpha s_3(\boldsymbol{K},\ \boldsymbol{s}) + \alpha^2\boldsymbol{s}^2}{s_3^2}, \end{aligned} \tag{4.21}$$

which defines an integrable system with the nonlinear bracket (4.11) on the symplectic leaf defined by

$$s_3(\boldsymbol{K}, \boldsymbol{s}) = \alpha, \qquad \boldsymbol{s}^2 = 1.$$

Moreover, since the structure of the Hamiltonian (4.21) does not change (the numerator of the last term is a Casimir function and so is equivalent to a constant), we may conclude that the Hamiltonian (4.20) defines the general integrability case for the nonlinear bracket (4.11). To obtain the integral on an arbitrary leaf, we redefine the constants in the Hamiltonian according to the rule

$$\xi \to \xi + \frac{\alpha}{2}, \qquad a \to a + \frac{\alpha^2}{2} \tag{4.22}$$

and set

$$\alpha = \frac{s_3(\boldsymbol{K}, \boldsymbol{s})}{\boldsymbol{s}^2}.$$

As a result, we obtain an additional integral, which is now general and has the form

$$\begin{aligned} F_2 = {} & \Big(K_1^2 - K_2^2 - 2\frac{K_1 s_1 - K_2 s_2}{s_3} F_0 - 2a\frac{s_1^2 - s_2^2}{s_3^2} - 2c_1 s_1 + 2c_2 s_2 - 2b_2 s_3^2\Big)^2 \\ & + 4\Big(K_1 K_2 - F_0\frac{M_1 s_2 + M_2 s_1}{s_3} - 2a\frac{s_1 s_2}{s_3^2} - c_1 s_2 - c_2 s_1 + b_1 s_3^2\Big)^2 \\ & + 4(\xi + F_0)\Big[-(K_3 + \xi)\Big(K_1^2 + K_2^2 + 2F_0 K_3 + 2a\frac{s_1^2 + s_2^2 + 2s_3^2}{s_3^2}\Big) \\ & \qquad - 2F_0(c_1 s_1 + c_2 s_2 + 2b_1 s_1 s_2 - b_2(s_1^2 - s_2^2)) \\ & \qquad + 2s_3(c_1 K_1 + c_2 K_2 + b_1(K_1 s_2 + K_2 s_1) \\ & \qquad - b_2(K_1 s_1 - K_2 s_2))\Big], \end{aligned} \tag{4.23}$$

$$F_0 = \frac{s_3(\boldsymbol{K}, \boldsymbol{s})}{\boldsymbol{s}^2}.$$

Substituting the expressions of $(\boldsymbol{K}, \boldsymbol{s})$ in terms of the momenta $\boldsymbol{M}$ and the quaternion parameters λ (4.10) into (4.20) and (4.23), we obtain the general integrable system with quaternion bracket in the variables $\boldsymbol{M}$ and λ, the linear integral $F_1 = (\boldsymbol{M}, \boldsymbol{\gamma}) - M_3$, and the fourth-degree integral (4.23).

Using the system (4.20) for $c_1 = c_2 = 0$, we can present a new generalization of the Kovalevskaya case in the direction cosines. In this case, the Hamiltonian and two additional integrals have the form

$$\begin{aligned} H &= 2(M_1^2 + M_2^2 + 2M_3^2 - \xi M_3) + \frac{1}{2}b_1(-\alpha_2 + \beta_1) \\ &\quad + \frac{1}{2}b_2(\alpha_1 + \beta_2) + \frac{2a}{1 - \gamma_3}, \qquad \xi, b_1, b_2, a = \text{const}, \\ F_1 &= M_3 - (\boldsymbol{M}, \boldsymbol{\gamma}), \\ F_2 &= k_1^2 + k_2^2 + 4k_3(k_4 + k_5 + k_6), \\ k_1 &= 4(M_1 N_1 + M_2 N_2) - b_2(1 - \gamma_3) + \frac{2a(\alpha_1 + \beta_2)}{1 - \gamma_3}, \end{aligned}$$

$$
\begin{aligned}
k_2 &= 4(M_2N_1 - M_1N_2) + b_1(1-\gamma_3) + \frac{2a(\alpha_2-\beta_1)}{1-\gamma_3},\\
k_3 &= \xi + N_3 - M_3,\\
k_4 &= (2M_3 - \xi)\left(4(\boldsymbol{M}^2 - M_3N_3) + \frac{2a(3-\gamma_3)}{1-\gamma_3}\right),\\
k_5 &= (M_3 - N_3)\left(b_1(-\alpha_2+\beta_1) + b_2(\alpha_1+\beta_2)\right),\\
k_6 &= 2b_1(M_2\alpha_3 - M_1\beta_3) - 2b_2(M_1\alpha_3 + M_2\beta_3).
\end{aligned}
\tag{4.24}
$$

Under the condition $a = 0$, we obtain the Yehia case [619] (1987) of the generalized Kovalevskaya system (see Sec. 3.4 of Ch. 3). Our arguments show that it is isomorphic to the family (1.20) discovered by Goryachev and Chaplygin [235, 128]. This family is discussed in Sec. 5.1 of Ch. 5. We presented this isomorphism in [92] (1997), although a little earlier Yehia [623] (1988) had used a restricted version (for $M_3 \pm N_3 = 0$). In [624] (2000), this analogy was rediscovered by using the classical Lagrangian representation of the equations of motion.

In the general case where $a \neq 0$, an integrable Hamiltonian in the quaternion representation was found by us in the first edition of this book (2001); in terms of the direction cosines, it was independently presented in [625] (2003).

Remark. The integral (4.23) can be written in the direction cosines by using the relations

$$
\begin{gathered}
\frac{K_1s_1 - K_2s_2}{s_3} = 2\frac{M_1\alpha_3 + M_2\beta_3}{1-\gamma_3}, \quad \frac{K_1s_2 + K_2s_1}{s_3} = 2\frac{M_1\beta_3 - M_2\alpha_3}{1-\gamma_3},\\
K_1^2 - K_2^2 = \frac{(\alpha_1-\beta_2)(M_1^2 - M_2^2) + 2(\alpha_2+\beta_1)M_1M_2}{1-\gamma_3},\\
2K_1K_2 = \frac{(\alpha_2+\beta_1)(M_1^2 - M_2^2) - 2(\alpha_1-\beta_2)M_1M_2}{1-\gamma_3}.
\end{gathered}
\tag{4.25}
$$

Generalized Goryachev – Chaplygin family. Consider an analogous generalization of the Goryachev – Chaplygin integrable case on the zero level set of the area integral $(\boldsymbol{L}, \boldsymbol{s})$ with the addition of singular terms [235] (see Sec. 5.1 of Ch. 5). The Hamiltonian has the form

$$
H = \frac{1}{2}(L_1^2 + L_2^2 + 4L_3^2) + \xi L_3 + \frac{a\boldsymbol{s}^2}{s_3^2} + b_1s_1 + b_2s_2. \tag{4.26}
$$

After the change of variable $\boldsymbol{L} = \boldsymbol{K} - \alpha\frac{\boldsymbol{s}}{s_3}$, $\alpha = \text{const}$, neglecting the inessential constants, the Hamiltonian takes the form

$$
\begin{aligned}
H = {}& \frac{1}{2}(K_1^2 + K_2^2 + 4K_3^2) + b_1s_1 + b_2s_2\\
&+(\xi - 3\alpha)K_3 + \frac{1}{2}\frac{2a\boldsymbol{s}^2 - 2\alpha s_3(\boldsymbol{K}, \boldsymbol{s}) + \alpha^2\boldsymbol{s}^2}{s_3^2}.
\end{aligned}
\tag{4.27}
$$

By analogy with the previous case, we see that the system (4.26) represents the general integrability case with nonlinear bracket (and hence quaternion bracket in the variables $\boldsymbol{M}$, λ) with the third-degree integral of the form

$$F = (K_3 + \frac{1}{2}\xi)\left(K_1^2 + K_2^2 + 2a\frac{\boldsymbol{s}^2}{s_3^2}\right) - s_3(b_1K_1 + b_2K_2).$$

It is interesting to note that the integral does not change its form as compared with its form for the algebra $e(3)$ (see Sec. 2.5 of Ch. 2).

In conclusion, we note that the reduction and reconstruction procedures presented in this section are used in Sec. 3.4 of Ch. 3 and Appendix C to analyze the quaternion Euler – Poisson equations and their integrable cases.

4.2 Dynamical Symmetry and Lagrange Integral

In this section, taking a unified point of view, we consider dynamical problems for which there exists an analog of the Lagrange integral existing in the Euler – Poisson equations. Recall that it is related to the precession angle ψ and the angle φ of proper rotation, which are cyclic coordinates.

The latter coordinate gives rise to the Lagrange integral M_3 = const., ω_3 = const., and makes possible the preservation of the projections of the angular velocity and angular momentum on the dynamical symmetry axis. This integral arises due to the invariance of the system under rotations about the dynamical symmetry axis.

It turns out that the Lagrange type integral exists for almost all problems of rigid body dynamics, which are of theoretical interest, and its existence leads to integrable cases that are, as a rule, important for applications. For example, the analog of the Lagrange integral for the Kirchhoff equations was found by Kirchhoff himself, who integrated it and found the most simple motions. For the Poincaré – Zhukovskii equations (on $so(4)$), the analog of the Lagrange integral was used by Poincaré in order to justify his theoretical conclusions about the precession of the Earth's rotation axis. In both of the last two cases, as for the classical Lagrange problem, we can obtain an explicit (elliptic) quadrature for the nutation angle θ defined by the gyroscopic function and also use all the results of the qualitative analysis of motion presented in Sec. 2.3 of Ch. 2.

1 Explicit quadrature for the generalized Lagrange case. Conditions for existence of the integral

We present here the explicit integration by quadratures for the Lagrange case in the most general form, assuming that the rigid body motion is described by Hamiltonian

$$H = \frac{1}{2}(\mathbf{A}\boldsymbol{M}, \boldsymbol{M}) + (\boldsymbol{M}, \boldsymbol{W}(\boldsymbol{y})) + U(\boldsymbol{y}), \tag{4.28}$$

where $\mathbf{A}$ is a constant (not necessary diagonal) matrix and the following Poisson structure defined by the pencil $\mathscr{L}_x$ is given:

$$\{M_i, M_j\} = -\varepsilon_{ijk}M_k, \quad \{M_i, y_j\} = -\varepsilon_{ijk}y_k, \quad \{y_i, y_j\} = -\varepsilon_{ijk}xM_k, \tag{4.29}$$

where x is the parameter of the pencil. In what follows, we consider the corresponding conditions for the more general situation $\mathbf{A} = \mathbf{A}(\boldsymbol{y})$ occurring in the model for a rigid body sliding on the plane in a gimbal suspension.

The following assertion may be proved by explicit calculations.

Theorem 4. *The system* (4.28) *with the bracket* (4.29) *admits a linear integral of the form*

$$F = M_3 = c, \quad c = \text{const.}, \tag{4.30}$$

whenever the following conditions hold:

$$\begin{gathered}\mathbf{A} = \operatorname{diag}(a_1, a_1, a_3),\\ U(\boldsymbol{y}) = U\left(y_3, \sqrt{y_1^2 + y_2^2}\right), \quad W_3(\boldsymbol{y}) = W_3\left(y_3, \sqrt{y_1^2 + y_2^2}\right),\\ y_1\frac{\partial W_1}{\partial y_2} - y_2\frac{\partial W_1}{\partial y_1} + W_2 = 0, \quad y_1\frac{\partial W_2}{\partial y_2} - y_2\frac{\partial W_2}{\partial y_1} - W_1 = 0.\end{gathered} \tag{4.31}$$

Subject to the conditions of Theorem 4, the Hamiltonian (4.28) takes the form

$$\begin{aligned}H =& \frac{1}{2}(M_1^2 + M_2^2 + aM_3^2) + M_3 W_3\left(y_3, \sqrt{y_1^2+y_2^2}\right)\\ &+ U\left(y_3, \sqrt{y_1^2+y_2^2}\right) + \frac{M_1y_1 + M_2y_2}{\sqrt{y_1^2 + y_2^2}}\widetilde{W}_1\left(y_3, \sqrt{y_1^2+y_2^2}\right)\\ &+ \frac{M_1y_2 - M_2y_1}{\sqrt{y_1^2 + y_2^2}}\widetilde{W}_2\left(y_3, \sqrt{y_1^2+y_2^2}\right).\end{aligned} \tag{4.32}$$

On the level surface $M_3 = c$, we can reduce the system (4.32) to a system with one degree of freedom. This is achieved as follows.

The corresponding reduced variables have the form

$$\begin{gathered}K_1 = \frac{M_1y_1 + M_2y_2}{\sqrt{y_1^2 + y_2^2}}, \quad K_2 = \frac{M_1y_2 - M_2y_1}{\sqrt{y_1^2 + y_2^2}},\\ \sigma_1 = \sqrt{y_1^2 + y_2^2}, \quad \sigma_2 = y_3;\end{gathered} \tag{4.33}$$

in this case it is obvious that $M_1^2 + M_2^2 = K_1^2 + K_2^2$. Note that an analogous system of variables was used by Poincaré in integrating the integrable case (found by him) in the Poincaré – Zhukovskii equations.

The Poisson structure in the variables (4.33) is given by

$$\begin{aligned}
&\{K_1, K_2\} = -c + K_1\frac{\sigma_2}{\sigma_1}, && \{\sigma_1, \sigma_2\} = xK_2,\\
&\{K_1, \sigma_1\} = xc\frac{K_2}{\sigma_1}, && \{K_1, \sigma_2\} = -x\frac{K_1K_2}{\sigma_1},\\
&\{K_2, \sigma_1\} = -\sigma_2 - xc\frac{K_1}{\sigma_1}, && \{K_2, \sigma_2\} = -\sigma_1 + x\frac{K_1^2}{\sigma_1}
\end{aligned} \tag{4.34}$$

and the other brackets vanish. The rank of the bracket (4.34) is equal to 2. Its Casimir functions are

$$F_1 = \boldsymbol{\sigma}^2 + x\boldsymbol{K}^2 + xc = c_1, \qquad F_2 = K_1\sigma_1 + c\sigma_2 = c_2,$$

where $\boldsymbol{K} = (K_1, K_2)$, $\boldsymbol{\sigma} = (\sigma_1, \sigma_2)$, and $\widetilde{\boldsymbol{W}} = (\widetilde{W}_1, \widetilde{W}_2)$. The reduced Hamiltonian is

$$H = \frac{1}{2}\boldsymbol{K}^2 + (\boldsymbol{K}, \widetilde{\boldsymbol{W}}(\boldsymbol{\sigma})) + U(\boldsymbol{\sigma}) + cW_3(\boldsymbol{\sigma}). \tag{4.35}$$

This system with one degree of freedom easily reduces to quadratures.

For example, for $\widetilde{\boldsymbol{W}} = 0$, after fixing the Casimir functions $F_1 = c_1$, $F_2 = c_2$, on the level set $H = h$ of the energy integral, we obtain

$$\begin{aligned}
\dot{\sigma}_2 = K_2\left(\sigma_1 - x\frac{\partial U_\star}{\partial \sigma_1}\right), \qquad & U_\star(\boldsymbol{\sigma}) = U(\boldsymbol{\sigma}) + cW_3(\boldsymbol{\sigma}),\\
K_2^2 = 2(h - U_\star) - \left(\frac{c_2 - c\sigma_2}{\sigma_1}\right)^2 &.
\end{aligned} \tag{4.36}$$

For the Euler – Poisson equations, the geometric meaning of the variable σ_2 is obvious: it is the cosine of the nutation angle. For the equations on $so(4)$, the angle has no simple interpretation.

For the homogeneous quadratic potential energy $U_\star = r_1\sigma_1^2 + r_2\sigma_2^2$, from (4.36) we obtain an elliptic quadrature of the form

$$\begin{aligned}
\dot{\sigma}_2^2 = 2\left((h - r_2\sigma_2^2)(1 - 2xr_1) - a_1(c' - (1 - 2xr_2)\sigma_2^2)\right)(c' - (1 - 2xr_2)\sigma_2^2)&\\
- (1 - 2xr_1)^2(c_2 - c\sigma_2)^2 = f(\sigma_2)&,\\
c' = c_1 - x(2h + c)&.
\end{aligned} \tag{4.37}$$

Expressions (4.36) and (4.37) generalize the known quadrature for the Lagrange case of rigid body dynamics [401]. The function $f(\sigma_2)$ is said to be *gyroscopic*.

For $x = 0$, from (4.37) we obtain the gyroscopic function of the Kirchhoff case, and for $x = 1$ that of the Poincaré case. For the classical Lagrange case corresponding to $x = 0$, $W_3 = 0$, and $U = -r\sigma_2$, the equation for σ_2 has the form

$$\dot{\sigma_2}^2 = -2r\sigma_2^3 - (2h + c)\sigma_2^2 + (2cc_2 - 2rc_1)\sigma_2 - c_2^2 + 2hc_1. \tag{4.38}$$

To obtain the absolute motion of the dynamical symmetry axis, we need to perform the quadrature for the precession angle ψ. For $x = 0$, it has the form $\dot{\psi} = a_1 \frac{M_1\gamma_1+M_2\gamma_2}{\gamma_1^2+\gamma_2^2} = a_1 \frac{K_1}{\sigma_1}$, i.e., it is determined by the evolution of variables of the reduced system. An analogous conclusion holds for the quadrature of the proper rotation angle φ.

We shall not dwell here on obtaining the general solution in absolute space; the majority of the results presented in Sec. 2.3 of Ch. 2 hold for this also. We only note that the solution of system (4.36) in elliptic functions is obtained only for the linear and quadratic dependence of the potential on the components $\boldsymbol{\gamma}$ (resp. $\boldsymbol{M}$, $\boldsymbol{\gamma}$). In the other cases, the gyroscopic function is a polynomial of degree greater than four, and now the solution is branching on the complex plane of time. However, the methods of qualitative analysis presented in Ch. 2 can be used to describe the motion. This once more highlights that an explicit integration of such systems in terms of theta functions (including the classical Lagrange top) would not give any further insight into the nature of real motions.

2 A top on a smooth plane in a gravitational field

This top differs from the previously presented systems in that the matrix $\mathbf{A}$ depends on the positional variables (see Sec. 1.4 of Ch. 1). If the body considered is dynamically symmetric, $I_1 = I_2$, is bounded by an axially symmetric surface, and, moreover, the dynamical and geometry symmetry axes coincide, then we can represent the Hamiltonian in the form (see Sec. 1.6 of Ch. 1)

$$\begin{aligned} H = &\frac{1}{2} a_1 f \left(M_1^2 + M_2^2 + m a_1 (\gamma_3 g_1 - g_2)^2 (M_1\gamma_1 + M_2\gamma_2)^2 \right) \\ &+ \frac{1}{2} a_3 M_3^2 + \mu \left((\gamma_1^2 + \gamma_2^2) g_1 + \gamma_3 g_2 \right), \\ f^{-1} &= 1 + m a_1 (\gamma_1^2 + \gamma_2^2)(\gamma_3 g_1 - g_2)^2, \quad \mathbf{I}^{-1} = \operatorname{diag}(a_1, a_1, a_3); \end{aligned} \tag{4.39}$$

where $\mathbf{I}$ is the body tensor of inertia with respect to the center of mass, μ is the body weight, and $g_1 = g_1(\gamma_3)$ and $g_2 = g_2(\gamma_3)$ are functions depending on the body surface geometry and defined by the equations

$$\boldsymbol{\gamma} = -\frac{\operatorname{grad} F(\boldsymbol{r})}{|\operatorname{grad} F(\boldsymbol{r})|}, \quad \boldsymbol{r} = (g_1(\gamma_3)\gamma_1,\ g_1(\gamma_3)\gamma_2,\ g_2(\gamma_3)). \tag{4.40}$$

In formula (4.40), the equation $F(\boldsymbol{r}) = 0$ defines the body surface, while owing to axial symmetry, $F = F(r_1^2 + r_2^2, r_3)$. The system (4.39) also reduces to one degree of freedom by using the variables (4.33), and the quadrature for $\gamma_3 = \cos\theta$ can be obtained in the form

$$\begin{aligned} \dot{\gamma}_3^2 = a_1 f (1 - \gamma_3^2) \Big(2\big(h - \mu((1-\gamma_3^2) g_1 + \gamma_3 g_2)\big) - \frac{a_1 (c - M_3\gamma_3)}{1 - \gamma_3^2} - a_3 M_3^2 \Big), \\ M_3 = \text{const.}, \quad (\boldsymbol{M}, \boldsymbol{\gamma}) = c = \text{const.} \end{aligned} \tag{4.41}$$

Comments. First of all, we call attention to the comments already made (Sec. 1.6 of Ch. 1). The most studied situations are those in which an axially symmetric body makes contact with the plane at a single point (apex) or along a circle (of a hoop disk or coin type). In the first case called the *Lagrange top on a smooth plane*, or *toy top*, analysis of the motion can be performed analogously to Sec. 2.3 of Ch. 2. In explicit integration of (4.41) we obtain a hyperelliptic quadrature (which was studied already by Klein [304, 305]). However, after time rescaling, which removes the denominator in (4.41), it is easy to show that all the bifurcation diagrams presented in Sec. 2.3 of Ch. 2 remain practically unchanged. Moreover, the apex of the top draws curves on the plane that are analogous to those which are drawn by the apex of the Lagrange top on the fixed sphere. They are contained, e.g., in the book of Grammel [239].

Owing to the friction of the apex of the top with the plane, its general evolution reduces to one for which the dynamical symmetry axis (under a suitable twisting) rapidly becomes vertical, and the top appears to be "sleeping" for some time. Various generalizations of this effect are presented in [610, 239, 280, 411, 498].

Most of the results obtained for the case of the disk motion are those concerning the regular precessions and their stability [411]. The author of [411] also studies the stability of vertical plane motions of a heavy elliptic disk whose equations are not integrable in general. In [5], the nonintegrability of the equations of motion of a circular but unbalanced disk on an absolutely smooth surface is studied.

Also, we note that in the classical formulation of the nonholonomic problem where there is no slipping, the equations describing the rolling of a circular disk are also integrable (Chaplygin – Appel – Korteweg problem) [10, 411]), but the dynamics described by them is essentially more complicated.

The nonintegrability of the problem of rigid body motion on a smooth plane is studied in [114] by using the separatrix splitting method. However, the results obtained in [114] are limited and do not allow one to find any nontrivial integrability cases for now.

3 A gyroscope in gimbal suspension in an axially-symmetric field

Using the variables (4.33), we obtain the following quadrature for the cosine of the nutation angle:

$$\begin{gathered}
\dot{\sigma_2}^2 = a_1 f(1 + a_1 g\sigma_1^2)\left(2(\widehat{h} - U(\sigma_2))\sigma_1^2 - a_1 f(c_2 - c\sigma_2)^2(1 + a_1 I_1^i\sigma_1^2)\right), \\
f^{-1} = (1 + a_1 I_1^i)\left(1 + a_1\left(I^e + (I_3^i - I_2^i)\frac{\sigma_2^2}{\sigma_1^2}\right)\right), \\
g = I^e + (I_3^i - I_2)\frac{\sigma_2^2}{\sigma_1^2}, \quad \sigma_1^2 = 1 - \sigma_2^2,
\end{gathered} \tag{4.42}$$

where $\widehat{h} = h - \frac{1}{2}a_3c^2$; the meaning of the parameters I^e, I_k^i, and I_k is explained in Sec. 1.4 of Ch. 1. As for analysis of the motion for system (4.42), we can refer the reader to the book [401]. Due to the exterior annulus the angular momentum vector experiences a secular drift in space even when there are no exterior forces. This drift, called the *Magnus effect*, is explained by the appearance of reaction moments of the exterior annulus that are perpendicular to its rotation axis. In the general case, the equations of an asymmetric gyroscope in a gimbal suspension are not integrable [117].

4 Case of axial symmetry in Chaplygin equations

As was shown in Sec. 1.7 of Ch. 1, the dynamics of a rigid body in a fluid in a gravitational field without initial impulse can be described by the Hamiltonian system on $e(3)$ with Hamiltonian

$$H = \frac{1}{2}(\boldsymbol{M}, \mathbf{A}\boldsymbol{M}) + \frac{1}{2}\mu^2 t^2(\boldsymbol{\gamma}, \mathbf{C}\boldsymbol{\gamma}). \tag{4.43}$$

Under the condition of axial symmetry, we can represent the Hamiltonian (4.43) in the form

$$H = \frac{1}{2}(M_1^2 + M_2^2 + aM_3^2) + \frac{1}{2}\mu^2 t^2 \gamma_3^2. \tag{4.44}$$

The additional integral has the form $F = M_3$.

For the reduction, we can use the system of variables (4.33), however, it is more convenient to write the second-order equation for the nutation angle: thus for $\gamma_3 = \cos\theta$, taking into account the relations $(M_1^2+M_2^2)(\gamma_1^2+\gamma_2^2) = (M_1\gamma_2+M_2\gamma_2)^2+(M_1\gamma_2-M_2\gamma_1)^2$, $\dot{\gamma}_3 = M_2\gamma_1 - M_1\gamma_2$, we find that

$$-\sin\theta\ddot{\theta} = \frac{(M_3 - c\cos\theta)(M_3\cos\theta - c)}{\sin^2\theta} - \mu^2 t^2 \sin^2\theta\cos\theta, \quad c = (\boldsymbol{M}, \boldsymbol{\gamma}). \tag{4.45}$$

If the body falls from rest, then $M_3 = 0$ and $c = 0$. We then obtain for the nutation angle the pendulum type nonautonomous equation [127]

$$\ddot{\theta} = \mu^2 t^2 \sin\theta\cos\theta. \tag{4.46}$$

The evolution of the other angle variables is given by

$$\dot{\varphi} = (a-1)M_3 + \frac{M_3 - c\cos\theta}{\sin^2\theta}, \quad \dot{\psi} = \frac{c - M_3\cos\theta}{\sin^2\theta}. \tag{4.47}$$

The equations for describing the rigid body fall under the axial symmetry conditions, which are more general than(4.45), are considered in Sec. 6.7 of Ch. 6.

5 An analogy between the Lagrange top and the Leggett system

We have described the reduction methods (and the appropriate systems of variables) for those problems of rigid body dynamics that admit one linear integral. At the same time, there exists a number of systems such that in the problem there exists a redundant set of linear integrals not commuting with each other. In the latter case, a consistent application of the described reduction is not always possible, since, as a rule, the involutive set composed from the linear integrals contains nonlinear integrals. In this case, following the scheme described in Sec. 4.1, we can immediately reduce the order by two degrees of freedom, which is attained by the choice of the corresponding reduced (algebraic) variables.

In [403], the authors consider the explicit integration of a variant of the Leggett system describing the dynamics of the atomic spin of superfluid ^{3}He-B. If we consider the quaternion equations of the dynamics (see Sec. 1.4 of Ch. 1), then the Hamiltonian of the system can be written in the form

$$H = \frac{1}{2}(M_1^2 + M_2^2 + M_3^2) + bM_3 + U(\lambda_0), \tag{4.48}$$

where $U = c\left(4\lambda_0^2 - \frac{3}{2}\right)^2$, $b, c = \text{const}$. Such a form of the Hamiltonian also describes the motion of a material point in curved space S^3 (see Appendix A).

A system of the form (4.48) always has the cyclic integral

$$F = (\boldsymbol{M}, \boldsymbol{\gamma}) - M_3 = \text{const}.$$

One more additional integral arises under the condition $b = 0$ (absence of a magnetic field). In the variables (4.10) it has the form

$$F = K_2^2(s_1^2 + s_3^2) + (K_1 s_1 + K_2 s_3)^2.$$

The integration of this system, as described in [403], is very complicated. On the other hand, as was shown in [95], it is a generalization of the Lagrange case (after a suitable reduction). Indeed, in this case, Eqs. (4.48) have the vector integral

$$\boldsymbol{L} = \big((\boldsymbol{M}, \boldsymbol{\alpha}) - M_1, (\boldsymbol{M}, \boldsymbol{\beta}) - M_2, (\boldsymbol{M}, \boldsymbol{\gamma}) - M_3\big)$$

whose components compose the algebra $so(3)$. Let us choose the following new variables, which commute simultaneously with all components of the vector $\boldsymbol{L}$:

$$\begin{gathered}
K_1 = \frac{\sqrt{(\boldsymbol{M} \times \boldsymbol{\lambda})^2}}{\sqrt{\boldsymbol{\lambda}^2}}, \quad K_2 = \frac{(\boldsymbol{M}, \boldsymbol{\lambda})}{\sqrt{\boldsymbol{\lambda}^2}}, \\
\sigma_1 = \sqrt{\boldsymbol{\lambda}^2}, \quad \sigma_2 = \lambda_0, \\
\boldsymbol{\lambda}^2 = \lambda_1^2 + \lambda_2^2 + \lambda_3^2.
\end{gathered} \tag{4.49}$$

They satisfy a nonlinear algebra with

$$\begin{aligned}
&\{K_2, K_1\} = \frac{p_1 \sigma_2}{2\sigma_1}, \quad &&\{K_2, \sigma_1\} = -\frac{\sigma_2}{2}, \\
&\{K_2, \sigma_2\} = \frac{\sigma_1}{2}, \quad &&\{K_1, \sigma_1\} = \{K_1, \sigma_2\} = 0
\end{aligned} \tag{4.50}$$

and the Casimir functions

$$F_1 = \boldsymbol{\sigma}^2, \qquad F_2 = K_1\sigma_1 = \text{const},$$

where $\boldsymbol{\sigma} = (\sigma_1\sigma_2)$ and $\boldsymbol{K} = (K_1, K_2)$. (A less convenient generators are used in [447, 223].)

The rank of the bracket (4.50) is equal to 2, and, therefore, any Hamiltonian system is integrable, in particular, the system (4.48) for $b = 0$ whose Hamiltonian in the new variables becomes

$$H = \frac{1}{2}\boldsymbol{K}^2 + U(\sigma_2).$$

An analogous representation is obtained for the generalization of the Lagrange case (see (4.34) and (4.35)) for $c = 0$, $\boldsymbol{W} = 0$, and $x = 0$. This analogy can be established directly (for example, in the Euler angles), but the algebraic approach to the reduction and order reduction problems as developed in [95] is particularly clear and simple.

4.3 Generalizations of the Hess Case

Following the general scheme suggested by analyzing analogs of the Lagrange case, we first give the general dynamic conditions leading to the existence of a Hess type invariant relation and then illustrate them by examining various concrete mechanical systems [102].

First let us formulate a general assertion on the existence of the Hess integral

$$M_3 = c = \text{const}. \tag{4.51}$$

for some given value of c for the system of the most general form having Hamiltonian

$$\begin{gathered} H = \frac{1}{2}(\boldsymbol{M}, \mathbf{A}'\boldsymbol{M}) + (\boldsymbol{M}, \boldsymbol{W}(\boldsymbol{q})) + U(\boldsymbol{q}), \\ \boldsymbol{q} \approx (\boldsymbol{\alpha}, \boldsymbol{\beta}, \boldsymbol{\gamma}) \approx (\theta, \varphi, \psi) \approx (\lambda_0, \lambda_1, \lambda_2, \lambda_3), \end{gathered} \tag{4.52}$$

where $\mathbf{A}'$ is a constant, not necessarily diagonal, matrix (we assume the use of non-principal axes). For an arbitrary c this gives conditions for the existence of a Lagrange type integral; we show them in Sec. 4.1, formula (4.13). The Hamiltonian (4.52) also describes the motion of a rigid body with a fixed point in several force fields of special forms (see Sec. 3.4 of Ch. 3, where we examine several generalizations of the Lagrange and Hess cases to the potential $U(\boldsymbol{q})$, $W(\boldsymbol{q}) \equiv 0$). Using direct calculations, we can show that the conditions for the relation (4.51) to hold have the form

$$\begin{gathered} a'_{11} = a'_{22}, \quad a'_{12} = 0, \\ (\widehat{L}_\alpha + \widehat{L}_\beta + \widehat{L}_\gamma)(U + cW_3) = 0, \\ (\widehat{L}_\alpha + \widehat{L}_\beta + \widehat{L}_\gamma)W_1 + W_2 + ca'_{23} = 0, \\ (\widehat{L}_\alpha + \widehat{L}_\beta + \widehat{L}_\gamma)W_2 - W_1 - ca'_{13} = 0, \end{gathered} \tag{4.53}$$

where $\widehat{L}_\alpha$, $\widehat{L}_\beta$, and $\widehat{L}_\gamma$ are the differential operators

$$\hat{L}_\alpha = \alpha_1 \frac{\partial}{\partial \alpha_2} - \alpha_2 \frac{\partial}{\partial \alpha_1}, \quad \hat{L}_\beta = \beta_1 \frac{\partial}{\partial \beta_2} - \beta_2 \frac{\partial}{\partial \beta_1}, \quad \hat{L}_\gamma = \gamma_1 \frac{\partial}{\partial \gamma_2} - \gamma_2 \frac{\partial}{\partial \gamma_1}.$$

REMARK. In a slightly different form, conditions (4.53) were given by Yehia in [638].

Linear and quadratic potentials. Let us us apply the conditions (4.53) to vector and scalar potentials $\boldsymbol{W}$ and U of particular form. Suppose that

$$\begin{aligned} \boldsymbol{W} &= \boldsymbol{K} + \sum_{i=1}^{3} \mathbf{B}^{(i)} \boldsymbol{\alpha}_i, \\ U &= \sum_{i=1}^{3} (\boldsymbol{r}^{(i)}, \boldsymbol{\alpha}_i) + \frac{1}{2} \sum_{i=1}^{3} (\boldsymbol{\alpha}_i, \mathbf{C}^{(i)} \boldsymbol{\alpha}_i), \end{aligned} \tag{4.54}$$

where $\boldsymbol{\alpha}_1 = \boldsymbol{\alpha}$, $\boldsymbol{\alpha}_2 = \boldsymbol{\beta}$, $\boldsymbol{\alpha}_3 = \boldsymbol{\gamma}$, $\boldsymbol{K}$, and $\boldsymbol{r}_i$ are constant vectors, $\mathbf{C}^{(i)}$ are symmetric 3×3 matrices, and $\mathbf{B}^{(i)}$ are arbitrary 3×3 matrices ($i = 1, 2, 3$).

The conditions for existence of the Hess integral for some (more particular) cases of system (4.54) are presented in [638, 350, 116] (see below).

Using (4.51) and (4.53), we find that

$$\boldsymbol{K} = (-ca_{13}', -ca_{23}', k_3 a_{33}'),$$

where k_3 is an arbitrary constant, and

$$\begin{gathered} b_{11}^{(i)} = b_{22}^{(i)}, \quad b_{12}^{(i)} = -b_{21}^{(i)}, \quad b_{13}^{(i)} = b_{23}^{(i)} = 0, \\ r_1^{(i)} = cb_{31}^{(i)}, \quad r_2^{(i)} = cb_{32}^{(i)}, \\ \mathbf{C}^{(i)} = \operatorname{diag}(c_{11}^{(i)}, c_{11}^{(i)}, c_{33}^{(i)}). \end{gathered}$$

Moreover, we can represent the Hamiltonian in the form

$$\begin{aligned} H = {} & \frac{1}{2}(a_{11}'(M_1^2 + M_2^2) + a_{33}'(M_3 + k_3)^2) \\ & + (M_3 - c)(a_{13}' M_1 + a_{23}' M_2) + b_{11}^{(1)}(M_1 \alpha_1 + M_2 \alpha_2) \\ & + b_{12}^{(1)}(M_1 \alpha_2 - M_2 \alpha_1) + b_{33}^{(1)} M_3 \alpha_3 + (M_3 - c)(b_{31}^{(1)} \alpha_1 + b_{32}^{(1)} \alpha_2) \\ & + \frac{1}{2}(c_{11}^{(1)}(\alpha_1^2 + \alpha_2^2) + c_{33}^{(1)} \alpha_3^2) + r_3^{(1)} \alpha_3 + \cdots, \end{aligned} \tag{4.55}$$

where analogous terms containing $\boldsymbol{\beta}$ and $\boldsymbol{\gamma}$ are omitted.

Let us consider the following reciprocal system of variables: the projection of the angular momentum vector on the fixed axes $N = (N_1, N_2, N_3) = ((\boldsymbol{M}, \boldsymbol{\alpha}), (\boldsymbol{M}, \boldsymbol{\beta}), (\boldsymbol{M}, \boldsymbol{\gamma}))$ and the vector $\boldsymbol{p} = (\alpha_3, \beta_3, \gamma_3)$. Their Poisson brackets make up the algebra $e(3)$ (Sec. 1.4 of Ch. 1). In the variables $(\boldsymbol{N}, \boldsymbol{p})$ under conditions (4.53) and (4.54), Hamiltonian (4.52) has the form

$$\begin{aligned} H = {} & \frac{1}{2} a_{11}' \boldsymbol{N}^2 + (\boldsymbol{b}_1, \boldsymbol{N}) + (\boldsymbol{b}_2 \times \boldsymbol{p}, \boldsymbol{N}) + (\boldsymbol{r} + c\boldsymbol{b}_3 - c\boldsymbol{b}_1, \boldsymbol{p}) + \frac{1}{2}(\boldsymbol{p}, \mathbf{C}\boldsymbol{p}) \\ & + (M_3 - c) f(\boldsymbol{M}, \boldsymbol{\alpha}, \boldsymbol{\beta}, \boldsymbol{\gamma}), \end{aligned} \tag{4.56}$$

where $\boldsymbol{b}_1 = (b_{11}^{(1)}, b_{11}^{(2)}, b_{11}^{(3)})$, $\boldsymbol{b}_2 = (b_{12}^{(1)}, b_{12}^{(2)}, b_{12}^{(3)})$, $\boldsymbol{b}_3 = (b_{33}^{(1)}, b_{33}^{(2)}, b_{33}^{(3)})$, $\boldsymbol{r} = (r_3^{(1)}, r_3^{(2)}, r_3^{(3)})$, $\mathbf{C} = \operatorname{diag}(c_{33}^{(1)}-c_{11}^{(1)}, c_{33}^{(2)}-c_{11}^{(2)}, c_{33}^{(3)}-c_{11}^{(3)})$, and, moreover, the function $f(\boldsymbol{M}, \boldsymbol{\alpha}, \boldsymbol{\beta}, \boldsymbol{\gamma})$ cannot be expressed in terms of the variables $\boldsymbol{N}$ and $\boldsymbol{p}$. (otherwise, we obtain a top of Lagrange type.)

Since $\boldsymbol{N}$ and $\boldsymbol{p}$ commute with M_3, the equations of motion for them on the level set $M_3 = c$ decouple and are described by the Hamiltonian system on $e(3)$ with Hamiltonian (4.56) taken under the condition $M_3 - c = 0$, i.e., by a system with two degrees of freedom.

Thus, the reduced system is isomorphic to the equations of the spherical top motion in a potential field on a fixed level set $(\boldsymbol{N}, \boldsymbol{p}) = M_3 = c$ of the area constant.

In general the system (4.56) is not integrable, i.e., existence of an invariant Hess type relation for system (4.52) does not mean its complete integrability. As was seen in Sec. 4.1, an analogous remark applies for the existence of the general Lagrange integral, which ensures only the possibility of order reduction by one degree of freedom.

Let us find additional conditions under which the system (4.56) is completely integrable. Indeed, for $b_1 = b_2 = b_3 = r = 0$ we have the Clebsch integrable case (Neumann system for $c = 0$), and for $b_1 = b_2 = b_3 = 0$ and $c = 0$ we have the Lagrange case for a single field.

REMARK 1. Up to now, we have been using a special coordinate system in which the axes do not coincide with the body principal axes and $\mathbf{A}$ is not diagonal. In coordinates for which the tensor of inertia is diagonal, $\mathbf{A} = \operatorname{diag}(a_1, a_2, a_3)$, the Hess type integral (4.51) has the form

$$\begin{aligned} F = \sqrt{a_2 - a_1}\sqrt{a_3 - a_2}(M_1\sqrt{a_2 - a_1} \pm M_3\sqrt{a_3 - a_2}) \\ -(K_1\sqrt{a_3 - a_2} \pm K_3\sqrt{a_2 - a_1}) = 0, \end{aligned} \tag{4.57}$$

where $\boldsymbol{K}$ is the constant vector in (4.54). The transformation to body coordinates is performed by using the matrix $\mathbf{U}$, (2.55).

REMARK 2. The Hess integral, in common with the Lagrange integral, exists for a more complicated system with five degrees of freedom [118] describing a body suspended on a weightless *rigid rod* (*string*), which moves in a gravitational field [509]. Even given the existence of the mentioned two integrals, three more involutive integrals are needed for integrability of this system. In general they are unknown, and the only integrable case is related to the full separation of motion, when the body's attachment point to the string coincides with the center of mass of the system.

Known integrable cases. In the case of a single force field that is axially symmetric in the space, i.e., for $U = U(\boldsymbol{\gamma})$, $\boldsymbol{W} = \boldsymbol{W}(\boldsymbol{\gamma})$, various authors found the following analogs of the Hess integral, which are sufficient for complete integrability in each case:

1. $U(\boldsymbol{\gamma}) = \mu\gamma_3$, $\boldsymbol{W}(\boldsymbol{\gamma}) = 0$, the classical Hess case of the Euler – Poisson equations [258];
2. $U(\boldsymbol{\gamma}) = \mu\gamma_3$, $\boldsymbol{W} = (ca_{13}', ca_{23}', k_3)$, $k_3 = \text{const}$, the gyrostatic generalization of the Hess case shown by L. N. Sretenskii [542];

3. $U(\gamma) = (\gamma, \mathbf{C}\gamma)$, $\mathbf{C} = \operatorname{diag}(c_1, c_1, c_3)$, $\boldsymbol{W} = (b_{11}\gamma_1 + b_{12}\gamma_2, -b_{12}\gamma_1 + b_{11}\gamma_2, b_{31}\gamma_1 + b_{32}\gamma_2 + b_{33}\gamma_3)$, a particular integrable case (S. A. Chaplygin [131], V. V. Kozlov and D. A. Onishchenko [350]) of the Kirchhoff equations.

Obviously, cases 1, 2, and 3 satisfy the general conditions (4.53) and (4.55).

REMARK. All generalizations of the Hess case found on the algebra $e(3)$ can be naturally extended to the case of bracket pencil $\mathscr{L}_x$, since the equations for $\boldsymbol{M}$ are the same on the whole pencil. In this case, the Hess invariant relation is independent of the pencil parameter.

For two force fields, the system (4.54) was considered in [116] (A. A. Burov, G. I. Subkhamkulov), although potentials (4.54) in it have a hydrodynamic interpretation. In [116], two particular cases were considered in which the Hess integral $M_3 = 0$ was found to exist for the system, but the problem of complete integrability was not discussed. It turns out that the system is integrable in one case and nonintegrable in the other case.

1st case:

$$U = \frac{1}{2}\left(c_{11}^{(1)}(\alpha_1^2 + \alpha_2^2) + c_{33}^{(1)}\alpha_3^2\right) + \frac{1}{2}\left(c_{11}^{(2)}(\beta_1^2 + \beta_2^2) + c_{33}^{(2)}\beta_3^2\right).$$

The Hamiltonian of the reduced system (4.56) in this case can be represented in the form

$$H = \frac{1}{2}a_{11}'\boldsymbol{N}^2 + \frac{1}{2}(c_{33}^{(1)} - c_{11}^{(1)})p_1^2 + \frac{1}{2}(c_{33}^{(2)} - c_{11}^{(2)})p_2^2.$$

Because of the relation $(\boldsymbol{N}, \boldsymbol{p}) = M_3 = 0$, this case *is isomorphic to the Neumann system*, which is integrable.

2nd case:

$$U = r_3\alpha_3 + \frac{1}{2}\left(c_{11}(\beta_1^2 + \beta_2^2) + c_{33}\beta_3^2\right).$$

The Hamiltonian for the reduced system is

$$H = \frac{1}{2}a_{11}'\boldsymbol{N}^2 + r_3 p_1 + \frac{1}{2}(c_{33} - c_{11})p_2^2, \quad (\boldsymbol{N}, \boldsymbol{p}) = 0.$$

and this is the same as that of a spherical pendulum in a gravitational field and the Brun field perpendicular to it. Such a system is unlikely to be integrable. As above, we consider the application of the above-mentioned general conditions to three additional problems of rigid body dynamics.

A rigid body on a smooth plane. In [119], A. A. Burov finds the Hess invariant relation for the dynamics of a rigid body with a gyrostat on a smooth horizontal plane. Since the angular velocities $\boldsymbol{\omega}$ and the system of principal axes are used in this case, this relation looks slightly surprising. We present here conditions for the existence of the Hess integral for equations in Hamiltonian form on the algebra $e(3)$ in the case where the potential of the force field is symmetric with respect to rotation around the vertical.

As was shown in Ch. 1, the equations of motion can be written in Hamiltonian form on the algebra $e(3)$ with Hamiltonian function

$$H = \frac{1}{2}(\mathbf{A}(\boldsymbol{M} - \boldsymbol{K}),\ \mathbf{IA}(\boldsymbol{M} - \boldsymbol{K})) + \frac{1}{2}m(\boldsymbol{a},\ \mathbf{A}(\boldsymbol{M} - \boldsymbol{K})) + U(\boldsymbol{\gamma}),$$
$$\boldsymbol{a} = \boldsymbol{r} \times \boldsymbol{\gamma}, \qquad \mathbf{A} = (\mathbf{I} + m\boldsymbol{a} \otimes \boldsymbol{a})^{-1}, \tag{4.58}$$

where $\boldsymbol{K}$ is the gyrostatic momentum vector, which is constant in body coordinates as used here, $\boldsymbol{\gamma}$ is the normal vector to the plane, and $\boldsymbol{M}$ is the angular momentum vector with respect to the contact point, which is related to the angular velocity by the formula

$$\boldsymbol{M} = \mathbf{I}\boldsymbol{\omega} + m\boldsymbol{a}(\boldsymbol{a},\ \boldsymbol{\omega}). \tag{4.59}$$

Here, $\mathbf{I}$ is the tensor of inertia with respect to the center of mass of the body and m is the mass of the body.

The vector $\boldsymbol{r}(\boldsymbol{\gamma})$ is found from the condition

$$\boldsymbol{\gamma} = -\frac{\operatorname{grad} F(\boldsymbol{r})}{|\operatorname{grad} F(\boldsymbol{r})|},$$

where $F(\boldsymbol{r}) = 0$ defines the equation of the (everywhere convex) body surface.

Let a body be bounded by an axially symmetric surface whose symmetry axis is perpendicular to the circular section of the gyration ellipsoid of the form

$$(\boldsymbol{M},\ \mathbf{I}^{-1}\boldsymbol{M}) = \text{const}.$$

Choosing the coordinate system such that one of its axes Ox_3 is perpendicular to the circular section and the other Ox_2 is directed along the middle axis of inertia; then if U depends only on γ_3 and the relation

$$K_2 = 0, \qquad a_{11}^{(0)}K_1 + a_{13}^{(0)}K_3 = ca_{13}^{(0)},$$

holds, where $\mathbf{A}^{(0)} = \|a_{ij}^{(0)}\| = \mathbf{I}^{-1}$, it follows that the invariant Hess relation takes the form

$$M_3 - c = 0.$$

Proof. The proof of this assertion is a direct verification, which is convenient to perform by using any tool of analytical computations.

In the Hess case, the Hamiltonian differs from Hamiltonian (4.39) in the Lagrange case on a smooth plane by the addition of a term of the form $(M_3 - c)f(\boldsymbol{M}, \boldsymbol{\gamma})$. It vanishes on the level surface of the Hess integral, where the passage to the reduced system defined by the variables (4.33) is also possible.

REMARK 3. In the chosen coordinate system,

$$\mathbf{A}^{(0)} = \begin{pmatrix} a_{11}^{(0)} & 0 & a_{13}^{(0)} \\ 0 & a_{22}^{(0)} & 0 \\ a_{13}^{(0)} & 0 & a_{33}^{(0)} \end{pmatrix},$$

the equation of the body surface has the form

$$F = F(x_1^2 + x_2^2, x_3) = 0,$$

whereas the vector $\boldsymbol{a}$ in (4.58) can be represented in the form

$$\boldsymbol{a} = (-f(\gamma_3)\gamma_2, f(\gamma_3)\gamma_1, 0),$$

where f depends only on γ_3.

REMARK 4. An analog of the Hess case for rigid body motion on an absolutely rough plane (nonholonomic system) has not been found yet. However, the generalization of the Lagrange problem of the axially symmetric rigid body rolling without slipping on the plane exists and was integrated by S. A. Chaplygin [411].

A gyroscope in gimbal suspension. In this problem, the kinetic energy also depends on the positional variables, which raises additional difficulties. Here again it is convenient to use a Hamiltonian representation of the system (for more details, see Sec. 1.4 of Ch. 1). Since the expressions obtained are rather cumbersome, we present here only the final result in the absence of gyrostatic momentum.

Let a dynamically asymmetric rigid body be clamped in a gimbal suspension in such a way that the axis of rotation of the rigid body relative to the gimbal suspension (see Fig. 1.10 of Ch. 1) *coincides with the normal to the circular section of the gyration ellipsoid and the body's potential energy in the exterior field is invariant under rotations of the body about the axis of rotation. Then there exists an invariant Hess type relation*

$$M_3 = 0 \tag{4.60}$$

in the coordinate system one of whose axes (Ox_3) *coincides with the perpendicular to the circular section.*

This case was (probably for the first time) shown by the authors in [102, 77].

When gravity is taken into account, the center of mass of the system must lie on the axis of rotation of the body relative to the gimbal suspension.

REMARK. It is easy to generalize this result to the case of a gyrostat. In this case the relation (4.60) takes the form $M_3 = c$, where c is a fixed constant depending on the gyrostatic momentum.

The Hess integral in Chaplygin equations. Let us present one more case [77, 102] of existence of the Hess invariant relation for the nonautonomous system describing a rigid body falling in a fluid without initial shock (Sec. 1.7 of Ch. 1). In this case, the surface bounding the body is axially symmetric, whereas the symmetry axis is perpendicular to the circular section of the gyration ellipsoid. The Hamiltonian can be represented as

$$H = \frac{1}{2}(M_1^2 + M_2^2 + aM_3^2 + 2a_{13}M_1M_3) + \frac{1}{2}\mu^2 t^2 \gamma_3^2. \tag{4.61}$$

In the chosen coordinate system, the invariant relation has the form

$$M_3 = 0, \tag{4.62}$$

and we can use it to obtain the following equation for the nutation angle, which does not differ from that in the axially symmetric case (see Sec. 4.2):

$$-\sin\theta\ddot{\theta} = -\frac{c^2\cos^2\theta}{\sin^2\theta} - \mu^2 t^2 \sin^2\theta\cos\theta, \quad c = (\boldsymbol{M}, \boldsymbol{\gamma}). \tag{4.63}$$

In addition, the angle of proper rotation is given by the system

$$\dot{\varphi} = -\frac{c\cos\theta}{\sin^2\theta} + a_{13}M_1, \quad \dot{M}_1 = a_3 M_1 M_2 + \mu t^2 \gamma_2\gamma_3,$$

where $\gamma_3 = \cos\theta$, $\gamma_1 = \sin\theta\sin\varphi$, and $\gamma_2 = \sin\theta\cos\varphi$, and M_2 can be found from the relation

$$M_1^2 + M_2^2 = \frac{c^2 + \dot{\gamma}_3^2}{1 - \gamma_3^2}.$$

For $c = 0$, Eq. (4.63) coincides with the Chaplygin equation. In this case, we can use the qualitative analysis methods presented in [345] (see Sec. 6.5 of Ch. 6).

5 Generalizations of Integrability Cases. Explicit Integration

5.1 Various Generalizations of the Kovalevskaya and Goryachev – Chaplygin Cases

Because of the special significance of the Euler-Poisson equations, both for mathematical reasons and for their important practical applications, the study of their integrable cases has led to the appearance of a number of works in which these cases were generalized in various directions. One of them corresponds to inclusion of a gyrostatic momentum in the physical system which involves additional potential terms in the Hamiltonian. Another possible generalization is to look for integrable cases defined on a pencil of Poisson brackets.[1] Finally, we can consider the most general case where additional parameters arise in the Hamiltonian, as well as in the Poisson brackets. We stress, as a justification for considering such a problem, i.e., fitting some integrable case into a more general integrable family, that it helps us to understand its dynamical origin deeply and also to highlight specific properties (for example, the possibility of integrating in terms of elliptic functions), which may not be generic for all representatives of the family.

A certain class of generalizations related to reductions and reconstructions of various mechanical systems has already been considered in Ch. 4.

1 Generalization of Kovalevskaya's case

The first generalization is due to S. A. Chaplygin [131], who considered a composition of two special cases of the Kirchhoff equations – the Kovalevskaya case and his own – on the zero level set of the area integral $(\boldsymbol{M}, \boldsymbol{\gamma})$. D. N. Goryachev [236] studied the same family with the addition of a singular term of the form $\frac{a}{\gamma_3^2}$. Goryachev did not refer to the earlier work of Chaplygin. Probably this is due to the fact that they looked at different applications (Kirchhoff's equations and the motion of a rigid body with a fixed point in potential fields, respectively) and the interconnection between these problems was not clearly understood at that time. In [235], D. N. Goryachev also added an analogous singular term to the Goryachev – Chaplygin case. In both cases, Goryachev's approach involved looking for all possible potentials admitting an additional integral of degree either three or four.

1 Note that the notion of a pencil of Poisson brackets is commonly applied to find integrable systems by use of the so called bi-Hamiltonian technique, due originally to Magri. Although the same notion is intended here, it is used in a *different* way.

DOI 10.1515/9783110544442-005

1. The first generalization relates the Kovalevskaya and Chaplygin cases to a just one integrable family on the zero level set $(\boldsymbol{M}, \boldsymbol{\gamma}) = 0$ of the area constant. The most general Hamiltonian has the form

$$\begin{aligned} H &= \frac{1}{2}(M_1^2 + M_2^2 + 2M_3^2) + \lambda M_3 \\ &\quad + r_1\gamma_1 + r_2\gamma_2 + 2b_1\gamma_1\gamma_2 + b_2(\gamma_2^2 - \gamma_1^2) + \frac{1}{2}\frac{a}{\gamma_3^2}, \end{aligned} \tag{5.1}$$

$$\lambda, r_1, r_2, b_1, b_2, a = \text{const.}$$

The corresponding first integral of degree four is

$$\begin{aligned} F &= \left(M_1^2 - M_2^2 - a\frac{\gamma_1^2 - \gamma_2^2}{\gamma_3^2} - 2r_1\gamma_1 + 2r_2\gamma_2 - 2b_2\gamma_3^2\right)^2 \\ &\quad + 4\left(M_1^2M_2^2 - a\frac{\gamma_1\gamma_2}{\gamma_3^2} - r_1\gamma_2 - r_2\gamma_1 + b_1\gamma_3^2\right)^2 \\ &\quad - 4\lambda(M_3 + \lambda)\left(M_1^2 + M_2^2 + a\left(1 + \frac{1}{\gamma_3^2}\right)\right) \\ &\quad + 8\lambda\gamma_3\big(M_1(r_1 + b_1\gamma_2 - b_2\gamma_1) + M_2(r_2 + b_1\gamma_1 + b_2\gamma_2)\big); \end{aligned} \tag{5.2}$$

by means of a rotation around the axis Ox_3 we can choose to exclude one of the parameters r_1, r_2, b_1, b_2 (traditionally one sets $b_1 = 0$).

The most general family of Hamiltonians (5.1) was found by Yehia in [631] (1996). (In a previous work [630], he had obtained the integrable family (5.1) without the addition of the singular term $\frac{a}{\gamma_3^2}$.) As mentioned previously, earlier results, namely, some special cases of (5.1), are due to D. N. Goryachev ($\lambda = 0$) [235] (1915) and S. A. Chaplygin ($a = 0, \lambda = 0$) [128] (1903).

2. In another paper, Yehia [625] (2003) considered a system with the Hamiltonian

$$H = \frac{1}{2}(M_1^2 + M_2^2 + 2M_3^2) + c(\gamma_1^2 - \gamma_2^2) + 2d\gamma_1\gamma_2 + \frac{\lambda}{\gamma_3^2} + \varepsilon\left(\frac{1}{\gamma_3^4} - \frac{1}{\gamma_3^6}\right),$$

$$c, d, \lambda, \varepsilon = \text{const.}$$

This has the additional fourth-degree partial integral (under the condition that $(\boldsymbol{M}, \boldsymbol{\gamma}) = 0$)

$$\begin{aligned} F &= \left[M_1^2 - M_2^2 + c\gamma_3^2 - \frac{\lambda(\gamma_1^2 - \gamma_2^2)}{\gamma_3^2}\right]^2 + \left(M_1M_2 + d\gamma_3^2 - \frac{2\lambda\gamma_1\gamma_2}{\gamma_3^2}\right) \\ &\quad + 2\varepsilon(M_1^2 + M_2^2)\left(\frac{1}{\gamma_3^4} - \frac{1}{\gamma_3^6}\right) + \varepsilon\frac{(\gamma_1^2 + \gamma_2^2)^2}{\gamma_3^{12}}(\varepsilon - 2\lambda\gamma_3^4) \\ &\quad + \frac{2\varepsilon}{\gamma_3^4}\left[c(\gamma_1^2 - \gamma_2^2) + 2d\gamma_1\gamma_2\right]. \end{aligned}$$

For $\varepsilon = 0$, this is the same as the classical result of D. N. Goryachev [235]. In [625], H.Yehia announced an integrable system with the Hamiltonian

$$H = \frac{1}{2}(M_1^2 + M_2^2 + 2M_3^2) + \frac{a\gamma_3}{(\gamma_1^2 + \gamma_2^2)^{3/4}} + \frac{b}{\sqrt{\gamma_1^2 + \gamma_2^2}} + \frac{\gamma_3\sqrt{c\gamma_1 + d\gamma_2 + \sqrt{(c^2 + d^2)(\gamma_1^2 + \gamma_2^2)}}}{\sqrt{\gamma_1^2 + \gamma_2^2}},$$

but an additional integral was not explicitly given.

3. Another partially integrable family (on the zero level set $(\boldsymbol{M}, \boldsymbol{\gamma}) = 0$ of the area integral) was found by Yehia and Elmandouh. It also involves the addition of the singular Goryachev term $\lambda\gamma_3^{-2}$ to the Hamiltonian of the general case found by V. V. Sokolov [535, Ch. 3, Sec. 1]:

$$H = \frac{1}{2}(M_1^2 + M_2^2 + 2M_3^2) + b\gamma_2 - aM_3\gamma_1 + \frac{a^2}{2}\gamma_3^2 - kM_3 + ak\gamma_1 + \frac{1}{2}\frac{\lambda}{\gamma_3^2},$$

$$\begin{aligned}F = {} & \left(M_1^2 - M_2^2 + 2a\gamma_1\left(M_3 - \frac{a}{2}\gamma_1 - k\right) + 2b\gamma_2 + a^2\gamma_2^2 - \frac{\lambda(\gamma_1^2 - \gamma_2^2)}{\gamma_3^2}\right)^2 \\ & + 4\left(M_1M_2 + a\gamma_2\left(M_3 - \frac{a}{2}\gamma_1 - k\right) - b\gamma_1 - \frac{a^2}{2}\gamma_1\gamma_2 - \frac{\lambda\gamma_1\gamma_2}{\gamma_3^2}\right)^2 \\ & + 4k(M_3 - k)\left(M_1^2 + M_2^2 + 2aM_1\gamma_3 + a^2(\gamma_1^2 + \gamma_2^2 + 2\gamma_3^2) + \frac{\lambda(1 + \gamma_3^2)}{\gamma_3^2}\right) \\ & - 8bkM_2\gamma_2\gamma_3 + 4ka^2\gamma_3\left(k\gamma_3 - 2(\boldsymbol{M}, \boldsymbol{\gamma})\right),\end{aligned}$$

where a, b, k, λ are arbitrary constants. As mentioned above, for $\lambda = 0$ we get the general, unrestricted integrable case (i.e., without the zero condition on $(\boldsymbol{M}, \boldsymbol{\gamma})$).

4. Yet another partially integrable family on the zero level set $(\boldsymbol{M}, \boldsymbol{\gamma}) = 0$ of the area integral was again found by Yehia and Elmandouh (2010) [642]. The Hamiltonian is

$$H = \frac{1}{2}\left(M_1^2 + M_2^2 + 2\left(M_3 - \frac{a}{2}\gamma_1\right)^2\right) - \frac{a^2}{4}(\gamma_1^2 + 2\gamma_3^2) + b\gamma_2 + \frac{\varepsilon}{\sqrt{\gamma_1^2 + \gamma_2^2}} + \frac{1}{2}\frac{\mu}{\gamma_3^2},$$

and the additional integral is

$$F = \left(M_1^2 - M_2^2 + 2a\gamma_1\left(M_3 - \frac{a}{2}\gamma_1\right) + a^2\gamma_2^2 + 2b\gamma_2 - \frac{\mu(\gamma_1^2 - \gamma_2^2)}{\gamma_3^2}\right)^2$$
$$+ 4\left(M_1M_2 + a\gamma_2\left(M_3 - \frac{a}{2}\gamma_1\right) - \frac{a^2}{2}\gamma_1\gamma_2 - b\gamma_1 - \frac{\mu\gamma_1\gamma_2}{\gamma_3^2}\right)^2$$
$$+ 4\varepsilon\left(\frac{M_1^2 + M_2^2}{\sqrt{\gamma_1^2 + \gamma_2^2}} + \frac{\varepsilon}{\gamma_1 + \gamma_2} + \sqrt{\gamma_1^2 + \gamma_2^2}\left(a^2 + \frac{\mu}{\gamma_3^2}\right)\right),$$

where a, b, ε, μ are any constants. For $\mu = 0$ this family reduces to the general integrable case (for arbitrary values of the integral $(\boldsymbol{M}, \boldsymbol{\gamma})$) which was found by Sokolov [535] (2002).

A special case of this family, with $a = 0$, appeared in earlier work of Yehia and Bedweihy [641] (1987) (see also [634]).

In [579], A. V. Tsiganov looked at integrable systems on the two-dimensional sphere S^2 having an additional integral of degree four in momenta. In contrast to [525], the additional integrals are explicitly given in this work, and, moreover, in a particular case, the Chaplygin integral [128] for the rigid body motion in a fluid is obtained (for $(\boldsymbol{M}, \boldsymbol{\gamma}) = 0$).

Other integrable systems involving the addition of various potential terms to the Kovalevskaya and Goryachev – Chaplygin cases were recently found by H. Yehia [640]; in many respects, this illustrates the increasing possibilities of packages for analytical computations. Unfortunately, such studies lack transparency and, moreover, lose any physical significance. Attempts to represent such results in the form of tables can be found in [626].

5. *Another direction for generalizations* is to change the Poisson bracket structure, possibly with a corresponding change in the Hamiltonian (see Sec. 3.2 of Ch. 3).

Let us consider the following one-parameter family of Poisson brackets, the pencil $\mathscr{L}_x$:

$$\{M_i, M_j\} = -\varepsilon_{ijk}M_k, \quad \{M_i, \gamma_j\} = -\varepsilon_{ijk}\gamma_k, \quad \{\gamma_i, \gamma_j\} = -x\varepsilon_{ijk}M_k, \tag{5.3}$$

where $x = \text{const}$ is a parameter. For $x = 1$, the bracket (5.3) corresponds to the algebra $so(4)$, for $x = 0$, to the algebra $e(3)$, and for $x = -1$, to the algebra $so(1, 3)$.

Setting $x = \pm R^2$, $\boldsymbol{\gamma} \to R\boldsymbol{\gamma}$ and letting R tend to infinity, we obtain a contraction of the algebras $so(4)$ and $so(1, 3)$ to $e(3)$.

The bracket (5.3) has two quadratic Casimir functions

$$F_1 = (\boldsymbol{M}, \boldsymbol{\gamma}), \qquad F_2 = x(\boldsymbol{M}, \boldsymbol{M}) + (\boldsymbol{\gamma}, \boldsymbol{\gamma}). \tag{5.4}$$

In what follows, the level surfaces of these functions are denoted $F_1 = c_1$ and $F_2 = c_2$.

The classical Kovalevskaya case (Sec. 2.4 of Ch. 2) admits a generalization to the bracket pencil and, moreover, the possibility of adding a gyrostatic term remains:

$$H = \frac{1}{2}(M_1^2 + M_2^2 + 2M_3^2) + \lambda M_3 + \mu\gamma_1. \tag{5.5}$$

The additional integral changes somewhat, containing, as it does, the parameter x:

$$\begin{aligned} F &= k_1^2 + k_2^2 - \lambda\{k_1, k_2\} - 4\lambda^2(M_1^2 + M_2^2) \\ &= (M_1^2 - M_2^2 - 2\mu\gamma_1 + \mu^2 x)^2 + 4(M_1 M_2 - \mu\gamma_2)^2 \\ &\quad - 4\lambda\big(M_3(M_1^2 + M_2^2 + \mu^2 x) - 2\mu M_1\gamma_3\big) - 4\lambda^2(M_1^2 + M_2^2). \end{aligned} \tag{5.6}$$

Here, $k_1 = M_1^2 - M_2^2 - 2\mu\gamma_1 + \mu^2 x$ and $k_2 = 2(M_1 M_2 - \mu\gamma_2)$.

This generalization was presented in the work [307] of I. V. Komarov. The explicit integration for $\lambda = 0$ was carried out in [311, 82, 85] by using a generalization of the Kovalevskaya method (see also Sec. 5.2 of Ch. 5). For $\lambda \neq 0$, the integration of system (5.6) reduces to that of the Clebsch case at a nonzero value of the area constant. This reduction uses the generalized Haine – Horozov transformation presented in Sec. 5.2 of Ch. 5.

For $x \neq 0$, no bifurcation analysis of (1.5) has ever been carried out (either for $\lambda = 0$ or for $\lambda \neq 0$), and the existence or non-existence of an isomorphism (either orbital or topological) between the systems with $x = 0$ and with $x \neq 0$ remains open. All that we know is that they cannot be transformed to one another by means of inhomogeneous real linear transformations. The Lax spectral representation for system (5.6) is constructed in [540].

Note that the commutator term $\{k_1, k_2\}$ in the integral (5.6) also has a form similar to the Goryachev – Chaplygin integral:

$$\{k_1, k_2\} = 4\big(M_3(M_1^2 + M_2^2 + \mu^2 x) - 2\mu M_1\gamma_3\big). \tag{5.7}$$

This fact still has no reasonable explanation. We only note that there is also a curious interconnection on the Lax pair level between the Kovalevskaya and Goryachev – Chaplygin tops, which was shown in [51].

For $\lambda = 0$, there is an analog of the particular (periodic) Delone solution of system (5.5); in this case, on the level of the invariant relations $k_1 = k_2 = 0$, the commutator $\{k_1, k_2\}$ also defines a third-degree integral (see Sec. 2.4 of Ch. 2).

6. *G. K. Suslov's interpretation of the Kovalevskaya top integrability.* In the well-known textbook [562], G. K. Suslov presented a system of three new variables for the Kovalevskaya top, which very simply vary in a certain new time: their trajectory is the ellipse obtained as an intersection of a cylinder and a plane. The arguments are generalized to the case of system (5.5) for $\lambda = 0$ and $x \neq 0$.

Indeed, if we set

$$k_1 = M_1^2 - M_2^2 - 2\mu\gamma_1 + \mu^2 x, \quad k_2 = 2(M_1 M_2 - \mu\gamma_2), \\ z = M_1^2 + M_3^2,$$

then the following simple equations are obtained for them:

$$\dot{k}_1 = 2M_3 k_2, \quad \dot{k}_2 = -2M_3 k_1, \quad \dot{z} = M_3 k_2;$$

after rescaling time as $d\tau = M_3\,dt$ they reduce to linear equations. These equations possess the integrals

$$h = z - \frac{1}{2}k_1 + \mu^2 x = \text{const.}, \quad k^2 = k_1^2 + k_2^2 = \text{const.},$$

which define the mentioned ellipse in the space of variables (k_1, k_2, z).

7. *A "composition" of the Kovalevskaya and Chaplygin cases.* There is a generalization of the Kovalevskaya and Chaplygin integrable cases (with gyrostat), which embeds both of them in a larger integrable family on the whole pencil $\mathscr{L}_x$. In this general case, the analog of the area constant is also set equal to zero: $(\boldsymbol{M}, \boldsymbol{\gamma}) = 0$, i.e., the generalization is an example of a partial integrable case. It is convenient to represent the Hamiltonian in the form [61]:

$$\begin{aligned} H = {} & \frac{1}{2}\left(\alpha_2 M_1^2 + \alpha_1 M_2^2 + (\alpha_1 + \alpha_2) M_3^2\right) - \lambda M_3 \\ & + r_1\gamma_1 + r_2\gamma_2 - \frac{1}{2}(a_1 - a_2)(\gamma_1^2 - \gamma_2^2), \end{aligned} \tag{5.8}$$

where $\alpha_1 = 1 - xa_1$, $\alpha_2 = 1 - xa_2$, and a_1, a_2 are both constant.

In this case, the integral has the form

$$\begin{aligned} F &= k_1^2 + \alpha_1\alpha_2 k_1^2 - \lambda\{k_1, k_2\} - 4\lambda^2(M_1^2 + M_2^2), \\ k_1 &= \alpha_1 M_1^2 - \alpha_2 M_2^2 - (a_1 - a_2)\gamma_3^2 - 2(\gamma_1 r_1 - \gamma_2 r_2) + x\frac{\alpha_1 r_1^2 - \alpha_2 r_2^2}{\alpha_1\alpha_2}, \\ k_2 &= M_1 M_2 - 2\frac{\alpha_1 r_1\gamma_2 + \alpha_2 r_2\gamma_1}{\alpha_1\alpha_2} + 2x\frac{r_1 r_2}{\alpha_1\alpha_2}, \\ \{k_1, k_2\} &= 4M_3\left(\alpha_1 M_1^2 - \alpha_2 M_2^2 - x\frac{\alpha_1 r_1^2 + \alpha_2 r_2^2}{\alpha_1\alpha_2}\right) \\ &\quad - 4\gamma_3((a_1 - a_2)(M_1\gamma_1 - M_2\gamma_2) - 2(M_1 r_1 + M_2 r_2)). \end{aligned} \tag{5.9}$$

Remark. The equations of motion for F can be represented as

$$\dot{F} = \{F, H\} = (a_1 - a_2)(\boldsymbol{M}, \boldsymbol{\gamma}) f(\boldsymbol{M}, \boldsymbol{\gamma}),$$

where $f(\boldsymbol{M}, \boldsymbol{\gamma})$ is some (polynomial) function of $\boldsymbol{M}$ and $\boldsymbol{\gamma}$. Thus, for $a_1 = a_2$, this becomes the general integrability case corresponding to the gyrostatic generalization of the Kovalevskaya case.

2 Tsiganov – Sokolov case with fourth-degree integral

Let us present an integrable case that has recently been found by Tsiganov [582] and Sokolov [538] on the bracket pencil (5.3) with a homogeneous quadratic Hamiltonian, as well as with a more general one with the addition of terms linear in $\boldsymbol{M}$ and $\boldsymbol{\gamma}$. In this case, the additional integral is a fourth-degree polynomial in the variables $\boldsymbol{M}$ and $\boldsymbol{\gamma}$.

1. As mentioned previously, in the homogeneous case, the Hamiltonian H_0 and the additional integral F_0 have the form

$$\begin{aligned} H_0 &= aM_1^2 - \frac{x}{a}M_2^2 + M_1\gamma_2 - M_2\gamma_1 = aM_1^2 - \frac{x}{a}M_2^2 + Q_3, \\ F_0 &= -a^2(Q_2^2 + Q_3^2) + aQ_3(x\boldsymbol{M}^2 - \boldsymbol{\gamma}^2) + x(Q_1^2 + Q_3^2), \end{aligned} \tag{5.10}$$

where $a = \text{const}$, $\boldsymbol{Q} = \boldsymbol{M} \times \boldsymbol{\gamma}$, $\boldsymbol{M}^2 = (\boldsymbol{M}, \boldsymbol{M})$, $\boldsymbol{\gamma}^2 = (\boldsymbol{\gamma}, \boldsymbol{\gamma})$. For $x > 0$ (i.e., for $so(4)$), the integral F_0 can be written as a product of the linear and cubic functions:

$$F_0 = -f_1 \cdot f_2, \quad f_1 = a\left(M_1 + \frac{\gamma_1}{\sqrt{x}}\right) - \sqrt{x}\left(M_2 + \frac{\gamma_2}{\sqrt{x}}\right),$$
$$f_2 = \boldsymbol{M}^2\left(a\left(M_1 - \frac{\gamma_1}{\sqrt{x}}\right) - \sqrt{x}\left(M_2 - \frac{\gamma_2}{\sqrt{x}}\right)\right) + \frac{2(\boldsymbol{M}, \boldsymbol{\gamma})}{\sqrt{x}}\left(aM_1 - \sqrt{x}M\right),$$

which evolve according to

$$\dot{f}_1 = -\left(M_3\sqrt{x} + \gamma_3\right)f_1, \quad \dot{f}_2 = \left(M_3\sqrt{x} + \gamma_3\right)f_2.$$

According to these equations, the relations $f_1 = 0$ and $f_2 = 0$ define invariant manifolds.

In the limiting case as $x \to 0$ (i.e., for $e(3)$), the integral can also be written in terms of invariant relations:

$$\widetilde{F}_0 = \gamma_1\left(\gamma_1\boldsymbol{M}^2 - 2M_1(\boldsymbol{M}, \boldsymbol{\gamma})\right).$$

It is interesting that on the algebra $e(3)$, i.e., for $x = 0$, the case (5.10) can be reduced to a simple integrable system on $so(3, 1)$ with an additional quadratic integral. Indeed, for $x = 0$, the components of the vectors $\boldsymbol{M}$ and $\boldsymbol{Q}$ have commutation relations

$$\{M_i, M_j\} = \varepsilon_{ijk}M_k, \quad \{M_i, Q_j\} = \varepsilon_{ijk}Q_k, \quad \{Q_i, Q_j\} = -\boldsymbol{\gamma}^2\varepsilon_{ijk}M_k, \tag{5.11}$$

i.e., they form a representation of the algebra $so(3, 1)$. In this case, the integral F_0 can be expressed in terms of the variables M and Q:

$$F_0 = -a^2(Q_2^2 + Q_3^2) - a\boldsymbol{\gamma}^2Q_3 \quad (\boldsymbol{\gamma}^2 = c = \text{const.}),$$

i.e., it contains only terms quadratic and linear in Q.

2. The integrable case (5.10) can be generalized by the addition of terms linear in $\boldsymbol{M}$ and $\boldsymbol{\gamma}$ (for all x). Let us write the corresponding Hamiltonian H and the additional integral F:

$$\begin{aligned} H &= H_0 + (\boldsymbol{\lambda}, \boldsymbol{M}) + \delta\gamma_3, \quad \text{where } \boldsymbol{\lambda} = (\lambda_1, \lambda_2, \lambda_3) = \text{const.}, \ \delta = \text{const.}, \\ F &= F_0 + F_1 + F_2 + F_3, \\ F_3 &= 2a^2\left(\lambda_2(M_3Q_2 - M_2Q_3) - Q_1(\lambda_1M_1 - \delta\gamma_1)\right) \\ &\quad + a(x\boldsymbol{M}^2 - \boldsymbol{\gamma}^2)(\lambda_1M_1 + \lambda_2M_2 - (\lambda_3M_3 - \delta\gamma_3)) \\ &\quad + 2x\left(\lambda_1(M_1Q_3 - M_3Q_1) + Q_2(M_2\lambda_3 - \delta\gamma_2)\right), \end{aligned}$$

$$\begin{aligned} F_2 &= a^2\big((2\lambda_3^2 - x\delta^2)M_1^2 - (\lambda_2^2 + \lambda_3^2 - x\delta^2)M^2 - 2\delta\lambda_2 Q_1 \\ &\quad - \delta\gamma_1(2\lambda_3 M_1 - \delta\gamma_1)\big) - 2a\lambda_3(\lambda_1 Q_1 + \lambda_2 Q_2) \\ &\quad - x\big((2\lambda_3^2 - x\delta^2)M_2^2 - (\lambda_1^2 + \lambda_3^2 - x\delta^2)\boldsymbol{M}^2 \\ &\quad + 2\delta\lambda_1 Q_2 - \delta\gamma_2(2\lambda_3 M_2 - \delta\gamma_2)\big), \\ F_1 &= 2a\lambda_3\big(\lambda_1(\lambda_3 M - \delta\gamma_1) + \lambda_2(\lambda_3 M_2 - \delta\gamma_2) + (\lambda_3^2 - x\delta^2)M_3\big). \end{aligned} \tag{5.12}$$

Commentary. The story of how the above family was discovered is as follows. In [582], A. V. Tsiganov found the most general integrable family on $so(4)$ with the Hamiltonian being a sum of terms quadratic and linear in $\boldsymbol{M}$ and in $\boldsymbol{\gamma}$. The rather cumbersome form of the Hamiltonian and the additional integral did not allow him to identify the homogeneous quadratic case (5.10) in [582] (see also [87]).

Note that in [582], to find the mentioned integrable cases, the author used the r-matrix method in which the Lax matrix for the two-node XXX model of the Heisenberg magnetic was generalized (this method allows one to write the system directly in separating variables using symplectic coordinates on the second Sklyanin bracket (see [66])).

On the other hand, the case (5.10) was obtained by V. V. Sokolov in [538] by the method of indeterminate coefficients and using software for analytic computations (not only on $so(4)$ but on the whole pencil). He also showed the possibility of adding the linear terms (5.12), although he did not explicitly write the corresponding integrals (we have presented them here in a form that is probably not the simplest one). It seems correct, therefore, to call the whole family in (5.12) the *Tsiganov – Sokolov family*. Here, we are crediting the $so(4)$ case (for which A. V. Tsiganov also found the separation of variables in [582]) to both A. V. Tsiganov and V. V. Sokolov and the cases on $e(3)$ and $so(3,1)$ to V. V. Sokolov alone.

3 Generalization of the Goryachev – Chaplygin case

In this section, we present generalizations of the Goryachev – Chaplygin case analogous to those of the Kovalevskaya case. Such a generalization to the pencil (5.3) is related to the extension of the Andoyer variables to the pencil, which turn out to be separating for all representatives of the pencil. This idea of finding analogs of the Goryachev – Chaplygin case is due to [82, 85], and, as is shown in Sec. 5.3, it is related to a nonlinear transformation of the Poisson brackets to a standard form.

1. We first consider a generalization with the addition of a gyrostatic momentum term and a singular term to the Hamiltonian:

$$H = \frac{1}{2}(M_1^2 + M_2^2 + 4M_3^2) + \lambda M_3 + \mu\gamma_1 + \frac{1}{2}\frac{a}{\gamma_3^2}, \tag{5.13}$$

where λ, μ, a = const. In this case, the additional integral is of degree three in the momenta:

$$F = \left(M_3 + \frac{\lambda}{2}\right)\left(M_1^2 + M_2^2 + \frac{a}{\gamma_3^2}\right) - \mu M_1 \gamma_3.$$

In [235], D. N. Goryachev himself presented the system (5.13) for zero gyrostatic momentum $\lambda = 0$; for $\lambda \neq 0$ and $a = 0$, it was presented by L. N. Sretenskii [542]. The generalization (5.13) was proposed in complete form by I. V. Komarov and V. B. Kuznetsov [315] (1987); moreover, they gave a certain quantum mechanical interpretation for the singular term $\frac{1}{2}a\gamma_3{}^{-2}$. The family (5.13) was also presented by H. Yehia [633], albeit essentially later on (1996). This is explained by the fact that the work [315] was published only in Russian.

2. V. V. Sokolov and A. V. Tsiganov obtained a more general form of case (5.13) by including quadratic cross terms in $\boldsymbol{M}$ and $\boldsymbol{\gamma}$ into the Hamiltonian. The Hamiltonian and the additional cubic (partial) integral for this family (valid on the level set $(\boldsymbol{M}, \boldsymbol{\gamma}) = 0$) have the following form [540]:

$$\begin{aligned} H &= \frac{1}{2}\left(M_1^2 + M_2^2 + 4M_3^2 + \frac{\varepsilon}{\gamma_3^2}\right) + \lambda M_3 + \mu_1\gamma_1 + \mu_2\gamma_2 \\ &\quad + a_1(2M_3\gamma_1 - M_1\gamma_3) + a_2(2M_3\gamma_2 - M_2\gamma_3), \\ F &= \left(M_3 + a_1\gamma_1 + a_2\gamma_2 + \frac{\lambda}{2}\right)\left(M_1^2 + M_2^2 + \frac{\varepsilon}{\gamma_3^2}\right) - (\mu_1 M_1 + \mu_2 M_2)\gamma_3. \end{aligned} \tag{5.14}$$

For $\varepsilon = 0$, reduction to the Abel – Jacobi equations is achieved by using the Andoyer variables. If there is no gravity field, i.e., if $\mu_1 = \mu_2 = 0$, then the integral F can be represented as the following product of two terms, one linear and the other quadratic:

$$F = k_1k_2, \quad k_1 = M_3 + a_1\gamma_1 + a_2\gamma_2 + \frac{\lambda}{2}, \quad k_2 = M_1^2 + M_2^2 + \frac{\varepsilon}{\gamma_3^2},$$

which evolve (on the level set $(\boldsymbol{M}, \boldsymbol{\gamma}) = 0$) according to

$$\dot{k}_1 = 2(a_1\gamma_2 - a_2\gamma_1)k_1, \quad \dot{k}_2 = -2(a_1\gamma_2 - a_2\gamma_1)k_2$$

i.e., the relations $k_1 = 0$ and $k_2 = 0$ define invariant manifolds. For $\varepsilon \neq 0$, integration of system (5.14) was carried out in [539].

3. To obtain a generalization of the Goryachev – Chaplygin case to the bracket pencil (5.3), let us construct an analog of the Andoyer variables for the pencil. These new variables will be denoted (l, L, g, G, h, H).

To start with, set the component M_3 to be equal to one of the momentum variables:

$$L = M_3. \tag{5.15}$$

The coordinate l ($\{l, L\} = 1$) canonically conjugate to L on the subalgebra of $so(3)$ with generators M_1, M_2, and M_3 can be found by integrating the Hamiltonian flow with the

Hamiltonian function $\mathscr{H} = L$:

$$\frac{dM_1}{dl} = \{M_1, L\} = M_2, \quad \frac{dM_2}{dl} = \{M_2, L\} = -M_1, \quad \frac{dM_3}{dl} = \{M_3, L\} = 0. \tag{5.16}$$

Taking account of the commutation relation $\{M_1, M_2\} = -M_3$, we find that

$$M_1 = \sqrt{G^2 - L^2} \sin l, \qquad M_2 = \sqrt{G^2 - L^2} \cos l, \tag{5.17}$$

where $G^2 = M_1^2 + M_2^2 + M_3^2$ is the Casimir function of this subalgebra of $so(3)$.

Next, take G as the second momentum variable and choose $\mathscr{H} = G$ as the Hamiltonian. Then, with g being the variable canonically conjugate to G, the flow corresponding to $\mathscr{H}$ on the whole pencil $\mathscr{L}_x$ has the form

$$\frac{d\boldsymbol{M}}{dg} = 0, \qquad \frac{d\boldsymbol{\gamma}}{dg} = \frac{1}{G}\boldsymbol{\gamma} \times \boldsymbol{M}. \tag{5.18}$$

According to (5.18), $\boldsymbol{M}$ is independent of g, and for $\boldsymbol{\gamma}$, using Eqs. (5.18) and the Casimir functions (5.4), we find that

$$\boldsymbol{\gamma} = \frac{H}{G^2}\boldsymbol{M} + \frac{\alpha}{G}(\boldsymbol{M} \times \boldsymbol{e}_3 \sin g + G\boldsymbol{M} \times (\boldsymbol{M} \times \boldsymbol{e}_3) \cos g), \quad \text{where} \quad \alpha^2 = \frac{c_2 - xG^2 - \frac{H^2}{G^2}}{G^2 - L^2} \quad \text{and} \quad \boldsymbol{e}_3 = (0, 0, 1). \tag{5.19}$$

Here, $c_2 = x\boldsymbol{M}^2 + \boldsymbol{\gamma}^2$ and H traditionally denotes the area constant $c_1 = (\boldsymbol{M}, \boldsymbol{\gamma})$.

We thus obtain symplectic coordinates on the whole pencil $\mathscr{L}_x$, defined by (5.15), (5.17) and (5.19). For $x = 0$ and $c = 1$, they become the standard Andoyer variables of rigid body dynamics.

4. We may use (5.15) and (5.19) to find a generalization of the Goryachev – Chaplygin partially integrable case for the pencil $\mathscr{L}_x$. Choose the Hamiltonian in the form

$$\mathscr{H} = \frac{1}{2}(G^2 + 3L^2) + \lambda L + a\left(\cos l \cos g + \frac{L}{G} \sin l \sin g\right), \tag{5.20}$$

where a and λ are constants.

As compared with [333], (5.20) has an additional term linear in L, which is interpreted as a component of the gyrostatic momentum on the algebra $e(3)$.

The system (5.20) *admits a separation of variables.* Indeed, let us perform the canonical change of variables

$$L = p_1 + p_2, \quad G = p_1 - p_2, \qquad q_1 = l + g, \quad q_2 = l - g. \tag{5.21}$$

In this case, the Hamiltonian (5.20) is

$$\mathscr{H} = \frac{1}{2}\frac{p_1^3 - p_2^3}{p_1 - p_2} - \lambda\frac{p_1^2 - p_2^2}{p_1 - p_2} + \frac{a}{p_1 - p_2}(p_1 \sin q_1 + p_2 \sin q_2), \tag{5.22}$$

and clearly the system is now in separated form. Using (5.15), (5.17), and (5.19), let us express the Hamiltonian (5.20) in terms of the variables $\boldsymbol{M}$ and $\boldsymbol{\gamma}$. We find that for zero value of the area constant, $(\boldsymbol{M}, \boldsymbol{\gamma}) = H = 0$,

$$\mathscr{H} = \frac{1}{2}(M_1^2 + M_2^2 + 4M_3^2) + \lambda M_3 + \mu\frac{\gamma_1}{|\boldsymbol{\gamma}|}, \quad |\boldsymbol{\gamma}| = \sqrt{\gamma_1^2 + \gamma_2^2 + \gamma_3^2}. \tag{5.23}$$

The additional (partial) integral is

$$F = \left(M_3 + \frac{\lambda}{2}\right)(M_1^2 + M_2^2) - \mu M_1\frac{\gamma_3}{|\boldsymbol{\gamma}|}. \tag{5.24}$$

For the algebra $e(3)$, we have $|\boldsymbol{\gamma}| = 1$ and obtain the classical Goryachev – Chaplygin case; for the latter, the above separation-of-variables method was suggested by V. V. Kozlov [333]. For $\varkappa \neq 0$, the family (5.23), (5.24) was found in [82, 85]. The generalization presented here is due to the fact that for $(\boldsymbol{M}, \boldsymbol{\gamma}) = 0$, the change of variables $\gamma_i \to \frac{\gamma_i}{|\gamma|}$ reduces the bracket of the pencil (5.3) to the bracket of the algebra $e(3)$. Using the above transformation, we can also generalize other systems. Problems related to nonlinear transformations of the Poisson brackets and the isomorphisms between various integrable systems generated by these transformations are considered in Sec. 5.3 of Ch. 5.

4 Three integrable families on $so(3, 1)$

Three exotic integrable families on the Poisson bracket algebra $so(3, 1)$ with a quadratic Hamiltonian of the form

$$H = \frac{1}{2}(\boldsymbol{M}, \boldsymbol{A}\boldsymbol{M}) + \frac{1}{2}(\boldsymbol{b}, \boldsymbol{M} \times \boldsymbol{\gamma}) \tag{5.25}$$

were presented by Sokolov [535, 538, 536] and by Tsiganov and Goremykin [585]. First of all, we note that it is easy to generalize the quadratic integrability cases of Shottky – Manakov, Steklov – Lyapunov, and Rubanovskii to $so(3, 1)$. They are presented, for example, in [60, 61] and may be obtained by solving the linear system for indeterminate coefficients. They are not described here because of the absence of any realistic mechanical interpretation for them; the best that can be said is that they are the Kirchhoff equations describing the rigid body motion in a fluid that fills the Lobachevskii space. The next collection of cases must be looked at not so much as physical systems, but as being important nontrivial examples which have figured in the development of the theory of integrable systems. This will be a collection of integrable or partially integrable cases which, as it turns out, exist only for the algebra $so(3, 1)$ and cannot be generalized to the whole pencil $\mathscr{L}_\varkappa$ (The same is true, for example, of the Adler – van Moerbeke case on $so(4)$); moreover, there are integrals not only of the third and fourth degrees but of the sixth degree in the phase variables.

First of all, we note (and this is essential in the course of V. V. Sokolov's arguments) that the Hamiltonians on $so(3,1)$ having an invariant relation linear in $\boldsymbol{M}$, $F = (\boldsymbol{\lambda}, \boldsymbol{M}) = 0$ ($\boldsymbol{\lambda}$ is a constant vector), can all be represented as

$$H = c_1(\boldsymbol{\lambda}, \boldsymbol{a})\boldsymbol{M}^2 + c_2(\boldsymbol{\lambda}, \boldsymbol{M})(\boldsymbol{a}, \boldsymbol{M}) + (\boldsymbol{\lambda}, \boldsymbol{M} \times \boldsymbol{\gamma}), \tag{5.26}$$

where c_1 and c_2 are constant parameters and $\boldsymbol{a}$ is an arbitrary constant vector. In what follows, we set $\boldsymbol{\lambda} = \boldsymbol{b}$ from (5.25). Let us present three of Sokolov's cases with integrals of third, fourth, and sixth degrees, respectively; in this case, the condition $x = -(a_1^2 + a_2^2 + a_3^2)$ always holds. That is, these cases are restricted to $\mathscr{L}_{x=-\boldsymbol{a}^2}$, or, in other words, to $so(3,1)$.

1. $c_1 = 1$, $c_2 = -2$:

$$F = (\boldsymbol{b}, \boldsymbol{M})\left(2(\boldsymbol{a}, \boldsymbol{M} \times \boldsymbol{\gamma}) + \boldsymbol{\gamma}^2 - x\boldsymbol{M}^2\right). \tag{5.27}$$

2. $c_1 = 1$, $c_2 = -1$:

$$F = (\boldsymbol{b}, \boldsymbol{M})^2\left(2(\boldsymbol{a}, \boldsymbol{M} \times \boldsymbol{\gamma}) - (\boldsymbol{a}, \boldsymbol{M})^2 + \boldsymbol{\gamma}^2 - x\boldsymbol{M}^2\right). \tag{5.28}$$

The case (5.27) was presented by A. V. Tsiganov and O. V. Goremykin in [585] and the case (5.28) by V. V. Sokolov in [535, 538] who also presented the classification theorem which asserts that cases (5.27) and (5.28) exhaust all possibilities for the existence of third- and fourth-degree integrals for systems on $so(3,1)$ with Hamiltonians of the form (5.25).

The third integrable case found by V. V. Sokolov in collaboration with T. Wolf [536] is of the sixth degree.

3. $c_1 = 1$, $c_2 = -\frac{1}{2}$:

$$\begin{aligned} F = (\boldsymbol{b}, \boldsymbol{M})\Big\{&\boldsymbol{M}^2(\boldsymbol{d}, \boldsymbol{M} \times \boldsymbol{a})^2 + 2(\boldsymbol{d}, \boldsymbol{M} \times \boldsymbol{a})\left(\boldsymbol{d}, \boldsymbol{M} \times (\boldsymbol{M} \times \boldsymbol{\gamma})\right) \\ &-\boldsymbol{\gamma}^2(\boldsymbol{M}, \boldsymbol{d})^2 - (\boldsymbol{d}, \boldsymbol{M} \times \boldsymbol{\gamma})^2 - \left((\boldsymbol{a}, \boldsymbol{b})^2 + x\boldsymbol{b}^2\right)\boldsymbol{M}^2\boldsymbol{\gamma}^2\Big\}, \end{aligned} \tag{5.29}$$

where $\boldsymbol{d} = \boldsymbol{b} \times \boldsymbol{a}$.

The integrals (5.27), (5.28), and (5.29) can be generalized by adding certain linear terms to the Hamiltonian (5.25) [536]. These terms have the following form, respectively:

1. $H_1 = (\boldsymbol{k}, \boldsymbol{M}) + \mu_1(\boldsymbol{b}, \boldsymbol{\gamma})$, $\boldsymbol{k} = (k_1, k_2, k_3)$, $k_i, \mu = \text{const.}$;
2. $H_1 = (\mu_1\boldsymbol{a} + \mu_2\boldsymbol{a} \times \boldsymbol{b}, \boldsymbol{M}) + \mu_3(\boldsymbol{b}, \boldsymbol{\gamma})$, $\mu_1, \mu_2, \mu_3 = \text{const.}$;
3. $H_1 = \mu_1(\boldsymbol{a} \times \boldsymbol{b}, \boldsymbol{M})$, $\mu_1 = \text{const.}$.

The integrals corresponding to these terms are cumbersome, so we do not present them here; however, they can be easily obtained by solving some linear systems.

5 Goryachev case

In conclusion, we present one more exotic partially integrable case, which has no direct relation to the Euler – Poisson equations but has a third-degree integral close to the

Goryachev – Chaplygin integral. It was presented by D. N. Goryachev [235]: for $(\boldsymbol{M}, \boldsymbol{\gamma}) = 0$, the Hamiltonian and the additional integral have the form

$$H = \frac{1}{2}\left(M_1^2 + M_2^2 + \frac{4}{3}M_3^2\right) - \frac{3}{4}\left(\frac{b_1}{3}\gamma_1 + b_2\right)\gamma_3^{-2/3}, \tag{5.30}$$
$$F = \frac{8}{3}(M_1^2 + M_2^2)M_3 + \frac{64}{27}M_3^3 + \left(\left(2M_1 - \frac{4}{3}\frac{M_3\gamma_1}{\gamma_3}\right)b_1 - 4\frac{b_2M_3}{\gamma_3}\right)\gamma_3^{1/3},$$
$$b_1, b_2 = \text{const.}$$

Some systems on the sphere having a third-degree integral and even more strange than (5.30) are presented in [581].

Some generalizations of system (5.30) having a third-degree integral in momenta were constructed in [581, 174]. They also have no direct mechanical interpretation.

5.2 Separation of Variables

1 Separating transformations in integrable problems of rigid body dynamics

In this section, we collect the most important methods for finding explicit solutions for the integrable cases of rigid body dynamics. We shall content ourselves with the presentation of the separating transformations (which are in general defined not just on the configuration space but on the whole phase space) and do not present the whole integration procedure, which is related to the inversion of Abelian integrals and the use of theta-functions of various kinds.

Separating variables are important, because by using them, it sometimes becomes possible to simplify the solutions of a problem. One of such problems involves topological analysis: having the separating variables, one can reduce the study of bifurcations of the level surfaces of first integrals to the analysis of the multiplicity of roots of the corresponding characteristic polynomial. Another problem is that of constructing action-angle variables, which are necessary for perturbation theory and for various quantization procedures (in particular, the WKB (quasi-classical) quantization).

Unfortunately, for many problems of rigid body dynamics, despite having sufficiently many first integrals to prove their integrability separating transformations have not been found. This lack does not prevent topological analysis, for which it is possible to obtain bifurcation sets directly by studying the critical level sets of the first integrals. Perhaps one should admit the possibility that for such systems to be “separable”, it will require a reformulation of basic notions.

The known separating transformations, which we have tried to present here in their most natural form, comprise a precious heritage of treasures for the study of dynamical systems in general and help to show the peculiarity of dynamical problems related to the dynamics of tops in particular. We have already presented some explicit solutions for the Euler, Lagrange, and Clebsch cases in Chs. 2 and 3, and they will not be discussed in detail here. Note that upon restricting to zero value for the area constant $(\boldsymbol{M}, \boldsymbol{\gamma}) = 0$, the Clebsch case becomes equivalent to the Neumann problem, which is integrated by using sphero-conic coordinates on the Poisson sphere (see Sec. 1.7 of Ch. 1). For the Zhukovskii – Volterra problem, in addition to the two solutions which have already been presented in Sec. 2.7 of Ch. 2, we present here another solution, which is more geometric and allows one to appreciate the difficulties related to explicit expressions in terms of elliptic functions.

The separation of variables for the Goryachev – Chaplygin case presented in the previous section can be generalized to the case of the algebra $so(4)$. This is related to a certain universal system of phase variables, which are separating for a whole family of integrable systems. Such an approach is close to the initial suggestion of Jacobi who recommended "to go in the reverse direction", and having found a certain remarkable substitution, to seek for problems for which it can be successfully applied [274]. Jacobi considered only transformations of the configuration space, which is very restrictive. Jacobi's idea was developed in the recent works [362, 583, 588, 539], whose results are described in [87, 587]; in these works, the explicit integration procedure (e.g., for the Kovalevskaya gyrostat and the generalized Goryachev – Chaplygin case, see below) reduces to representation of the system in the form of a Hamiltonian system with a quadratic Poisson bracket (Sklyanin bracket) for which there exist universal symplectic coordinates. They are also separating for a whole class of integrable problems. With this approach, finding an embedding of a given system in one of the known quadratic algebras becomes the main problem for its explicit integration. This requires experience and is a kind of art.

Zhukovskii – Volterra system. Let us look for an explicit solution for the Zhukovskii – Volterra case based on a method suggested by A. Wangerin in 1889 [607] and developed in [615]. As compared with the original analytical solution given by V. Volterra, which is discussed in Sec. 2.7 of Ch. 2, this method is more illustrative and geometric.

We use the equations of motion for the Zhukovskii – Volterra case in the form

$$\dot{\boldsymbol{M}} = \boldsymbol{M} \times \mathbf{A}\boldsymbol{M} + \boldsymbol{M} \times \boldsymbol{K}, \tag{5.31}$$

which possess the energy and momentum integrals

$$\begin{gathered}(\boldsymbol{M}, \mathbf{A}\boldsymbol{M}) + (\boldsymbol{M}, 2\boldsymbol{K}) = 2h = \text{const.},\\ \boldsymbol{M}^2 = f = \text{const.},\end{gathered} \tag{5.32}$$

where $\mathbf{A} = \operatorname{diag}(a_1, a_2, a_3)$ and $\boldsymbol{K} = (k_1, k_2, k_3) \in \mathbb{R}^3$. Now we make use of the following geometric observation.

The intersection line of two generic quadrics of the form

$$(\mathbf{A}_i\boldsymbol{M},\ \boldsymbol{M}) + (\boldsymbol{M},\ \boldsymbol{K}_i) = c_i, \qquad i = 1,\ 2 \tag{5.33}$$

(where $\boldsymbol{M},\ \boldsymbol{K}_i \in \mathbb{R}^3$, $c_i = \text{const.}$, and $\mathbf{A}_i$ are 3×3 symmetric matrices), lies on a cone in $\mathbb{R}^3$.

Indeed, assuming that it is possible to find a vector $\boldsymbol{\xi}$ and a pair of constants λ and μ satisfying the system of equations

$$\begin{gathered}(\lambda\mathbf{A}_1+\mu\mathbf{A}_2)\boldsymbol{\xi}=-\frac{1}{2}(\lambda\boldsymbol{K}_1+\mu\boldsymbol{K}_2),\\ ((\lambda\mathbf{A}_1+\mu\mathbf{A}_2)\boldsymbol{\xi},\ \boldsymbol{\xi})+(\boldsymbol{\xi},\ \lambda\boldsymbol{K}_1+\mu\boldsymbol{K}_2)=\lambda c_1+\mu c_2,\end{gathered}\tag{5.34}$$

then, defining the transformed variable $\widetilde{\boldsymbol{M}}$ by the shift $\widetilde{\boldsymbol{M}}=\boldsymbol{M}-\boldsymbol{\xi}$, we find the equation of the cone in the form

$$((\lambda\mathbf{A}_1+\mu\mathbf{A}_2)\widetilde{\boldsymbol{M}},\ \widetilde{\boldsymbol{M}})=0.\tag{5.35}$$

□

Let us apply this shift transformation to the integrals in (5.32). Using (5.34), we find that the vector $\boldsymbol{\xi}$ is

$$\boldsymbol{\xi}=-(\mathbf{A}+x\mathbf{E})^{-1}\boldsymbol{K},\qquad \text{with}\quad \mu/\lambda=x,$$

where $\mathbf{E}$ is the identity matrix, and x has to satisfy the fourth-order equation

$$-((\mathbf{A}+x\mathbf{E})^{-1}\boldsymbol{K},\ \boldsymbol{K})=2h+xf.\tag{5.36}$$

Since Eq. (5.36) has poles on the real line, it admits at least one real solution x_0, for which we also find the vector $\boldsymbol{\xi}(x_0)$. Writing now Eq. (5.35) in the form

$$\frac{\widetilde{M}_1^2}{b_1^2}+\frac{\widetilde{M}_2^2}{b_2^2}=\widetilde{M}_3^2,\qquad \widetilde{\boldsymbol{M}}=\boldsymbol{M}-\boldsymbol{\xi}(x_0),\tag{5.37}$$

with

$$b_1^2=\frac{x_0+a_3}{x_0+a_1},\qquad b_2^2=\frac{x_0+a_3}{x_0+a_2},$$

we may parameterize $\widetilde{M}_1$ and $\widetilde{M}_2$ as follows:

$$\widetilde{M}_1=b_1\widetilde{M}_3\sin\varphi,\qquad \widetilde{M}_2=b_2\widetilde{M}_3\cos\varphi.\tag{5.38}$$

Substituting now (5.38) into the energy integral (5.32), we obtain the following quadratic equation for M_3:

$$a(\varphi)\widetilde{M}_3^2+b(\varphi)\widetilde{M}_3+c=0,\tag{5.39}$$

where the coefficients $a(\varphi)$, $b(\varphi)$ and c have the form

$$\begin{aligned}a(\varphi)&=\frac{1}{2}(a_1b_1^2\sin^2\varphi+a_2b_2^2\cos^2\varphi+a_3),\\ b(\varphi)&=x_0\Big(\frac{k_1b_1}{x_0+a_1}\sin\varphi+\frac{k_2b_2}{x_0+a_2}\cos\varphi+\frac{k_3}{x_0+a_3}\Big),\\ c&=-h+(\boldsymbol{K},\ (\mathbf{A}+x_0\mathbf{E})^{-1}\boldsymbol{K})\\ &\quad+\frac{1}{2}((\mathbf{A}+x_0\mathbf{E})^{-1}\boldsymbol{K},\ \mathbf{A}(\mathbf{A}+x_0\mathbf{E})^{-1}\boldsymbol{K}).\end{aligned}\tag{5.40}$$

Taking (5.37) into account, we obtain from (5.31) the equation for the evolution of $\widetilde{M}_3$,

$$\dot{\widetilde{M}}_3 = (a_2 - a_1)\widetilde{M}_1\widetilde{M}_2 + \frac{b_2^2}{b_1^2}\widetilde{M}_1 k_2 - \frac{b_1^2}{b_2^2}\widetilde{M}_2 k_1$$

and we may use this to obtain an equation describing the evolution of the angle φ. Using (5.37), (5.38), and (5.39) and performing direct calculations, we obtain

$$\dot{\varphi}^2 = \frac{(x_0 + a_1)(x_0 + a_2)}{x_0^2}(b(\varphi)^2 - 4a(\varphi)c) = P_2(\cos\varphi, \sin\varphi), \tag{5.41}$$

where $a(\varphi)$, $b(\varphi)$, and c are defined according to (5.40) and $P_2(x, y)$ is a polynomial of degree two in (x, y). The quadrature of Eq. (5.41) is performed by using elliptic functions. It can be shown in this case [615] that with a suitable choice of the signs and the root x_0, the projections of the angular velocity are real-valued functions of time. Finally, we note the implicit character of the fourth-degree solution (coming not least from the fact that we need to solve Eq. (5.36) for x), which does not allow us to draw any conclusions about the dynamics. Moreover, the transformations into elliptic functions from (5.41) are not related to the Poisson structure of the problem.

The Kovalevskaya case was historically the first case for which separation of variables cannot be obtained by the Hamilton – Jacobi method, i.e., the separating variables are not found on the configurational space. However, in this case, there exists a nontrivial phase variable transformation that contains both the generalized coordinates and the momenta, which lead to the Abel – Jacobi equations and to separating variables on the plane. The resulting integrable system describes the motion of a material point on the Euclidean plane under the action of a certain potential, which is what the so-called Kolosov analogy implies (see below).

Let us have a look at the general Kovalevskaya case with the Hamiltonian

$$H = \frac{1}{2}(M_1^2 + M_2^2 + 2M_3^2) + \gamma_1, \tag{5.42}$$

defined on the pencil of brackets

$$\{M_i, M_j\} = -\varepsilon_{ijk}M_k, \ \{M_i, \gamma_j\} = -\varepsilon_{ijk}\gamma_k, \ \{\gamma_i, \gamma_j\} = -x\varepsilon_{ijk}M_k. \tag{5.43}$$

The explicit integration of this system was carried out for the first time in [311] by using the Haine – Horozov analogy (see below). The reduction to the Abel – Jacobi equations (see [82, 85]) presented here for $x \neq 0$ uses an argument of G. K. Suslov who suggested his own original method for integrating the Kovalevskaya case [562].

The bracket (5.43) has two Casimir functions

$$\begin{aligned} l &= (\boldsymbol{M}, \boldsymbol{\gamma}), \\ c &= x(\boldsymbol{M}, \boldsymbol{M}) + (\boldsymbol{\gamma}, \boldsymbol{\gamma}). \end{aligned} \tag{5.44}$$

The additional integral (the generalized Kovalevskaya integral presented in [309]; see also Sec. 5.1) has the form

$$\begin{gathered} k^2 = k_1^2 + k_2^2, \\ \text{with} \quad k_1 = M_1^2 - M_2^2 - 2\gamma_1 + x, \quad k_2 = 2M_1M_2 - 2\gamma_2. \end{gathered} \tag{5.45}$$

Defining the new variables:

$$\begin{aligned} z_1 &= M_1 + iM_2, \qquad & z_2 &= M_1 - iM_2, \\ \zeta_1 &= k_1 + ik_2, \qquad & \zeta_2 &= k_1 - ik_2, \\ \zeta_1\zeta_2 &= k^2, \end{aligned}$$

making use of the following equations of motion for z_1 and z_2:

$$i\dot{z}_1 = M_3 z_1 - \gamma_3, \quad -i\dot{z}_2 = M_3 z_2 - \gamma_3$$

together with the relations in (5.45), we may express M_3, γ_1, γ_2, and γ_3 by the formulas

$$\begin{aligned} \gamma_1 &= \frac{1}{4}\left(z_1^2 + z_2^2\right) - \frac{1}{4}(\zeta_1 + \zeta_2) + \frac{x}{2}, \\ \gamma_2 &= -\frac{i}{4}\left(z_1^2 - z_2^2\right) + \frac{i}{4}(\zeta_1 - \zeta_2), \\ \gamma_3 &= i\frac{\dot{z}_1 z_2 + \dot{z}_2 z_1}{z_1 - z_2}, \qquad M_3 = i\frac{\dot{z}_1 + \dot{z}_2}{z_1 - z_2}. \end{aligned} \tag{5.46}$$

Now let us substitute (5.46) into the integrals, whose values are fixed to be l and c as in (5.44), and into the Hamiltonian (5.42), whose value is fixed to be h say. Solving the resulting equations for z_1 and z_2, we find that

$$\begin{gathered} \dot{z}_1^2 = R_1 - \frac{\zeta_1}{4}(z_1 - z_2)^2, \quad \dot{z}_2^2 = R_2 - \frac{\zeta_2}{4}(z_1 - z_2)^2, \\ \dot{z}_1\dot{z}_2 = -R - \frac{1}{4}(2h - x)(z_1 - z_2)^2, \end{gathered} \tag{5.47}$$

where

$$\begin{gathered} R = R(z_1, z_2) = \frac{1}{4}z_1^2 z_2^2 - \frac{h}{2}\left(z_1^2 + z_2^2\right) + l(z_1 + z_2) + \frac{k^2}{4} - c + xh - \frac{x^2}{4}, \\ R_1 = R(z_1, z_1), \qquad R_2 = R(z_2, z_2). \end{gathered} \tag{5.48}$$

It remains to eliminate ζ_1 and ζ_2 from (5.47) and (5.48). For this purpose, we use the Kovalevskaya integral (5.45):

$$\left(R_1 - \dot{z}_1^2\right)\left(R_2 - \dot{z}_2^2\right) = \frac{\zeta_1\zeta_2}{16}(z_1 - z_2)^4 = \frac{k^2}{16}(z_1 - z_2)^4.$$

Regrouping the terms, the last equation gives us

$$\begin{aligned} \left(\frac{\dot{z}_1}{\sqrt{R_1}} + \frac{\dot{z}_2}{\sqrt{R_2}}\right)^2 &= \left(\frac{\dot{z}_1\dot{z}_2}{\sqrt{R_1R_2}} + 1\right)^2 - \frac{k^2(z_1 - z_2)^4}{16R_1R_2} = f_1, \\ \left(\frac{\dot{z}_1}{\sqrt{R_1}} - \frac{\dot{z}_2}{\sqrt{R_2}}\right)^2 &= \left(\frac{\dot{z}_1\dot{z}_2}{\sqrt{R_1R_2}} - 1\right)^2 - \frac{k^2(z_1 - z_2)^4}{16R_1R_2} = f_2, \end{aligned} \tag{5.49}$$

where the functions f_1 and f_2 can be expressed in terms of z_1 and z_2 if the product $\dot{z}_1\dot{z}_2$ is substituted from Eqs. (5.47).

The final step is the introduction of the Kovalevskaya variables by the formulas

$$s_1 = \frac{R - \sqrt{R_1R_2}}{2(z_1 - z_2)^2}, \qquad s_2 = \frac{R + \sqrt{R_1R_2}}{2(z_1 - z_2)^2}. \tag{5.50}$$

We may express $R = (s_1 + s_2)(z_1 - z_2)^2$ and $\sqrt{R_1R_2} = (s_2 - s_1)(z_1 - z_2)^2$ from (5.50) and substitute in the right-hand sides of (5.49). As a result, we find that

$$\begin{gathered} f_1 = \frac{f(s_1)}{(s_1 - s_2)^2}, \quad f_2 = \frac{f(s_2)}{(s_1 - s_2)^2}, \\ f(s) = \left(2s + \frac{1}{4}(2h - x) + \frac{1}{4}k\right)\left(2s + \frac{1}{4}(2h - x) - \frac{1}{4}k\right). \end{gathered} \tag{5.51}$$

On the other hand, from the left-hand sides of (5.49) we obtain relations that define the passage to certain curvilinear coordinates

$$\frac{dz_1}{\sqrt{R_1}} + \frac{dz_2}{\sqrt{R_2}} = \frac{ds_1}{\sqrt{\varphi(s_1)}}, \qquad \frac{dz_1}{\sqrt{R_1}} - \frac{dz_2}{\sqrt{R_2}} = \frac{ds_2}{\sqrt{\varphi(s_2)}}, \tag{5.52}$$

where $\varphi(s)$ is the third-degree polynomial

$$\varphi(s) = 4s^3 + 2hs^2 + \left(\frac{1}{16}(2h - x)^2 - \frac{k^2}{16} + \frac{c}{4}\right)s + \frac{l^2}{16}.$$

Substituting (5.52) into Eqs. (5.49) and taking account of (5.51), we find the equations of motion for the variables s_1 and s_2:

$$\dot{s}_1 = \frac{\sqrt{f(s_1)\varphi(s_1)}}{s_1 - s_2}, \quad \dot{s}_2 = \frac{\sqrt{f(s_2)\varphi(s_2)}}{s_2 - s_1}. \tag{5.53}$$

We can find the polynomial $\varphi(s)$ using standard methods for reducing elliptic integrals to standard form. We next present an alternative method similar to that presented by Suslov [562].

It follows from (5.50) that the Kovalevskaya variables (s_1, s_2) are solutions of the quadratic equation

$$Q(z_1, z_2, s) = (z_1 - z_2)^2 s^2 - Rs + G = 0, \tag{5.54}$$

where

$$\begin{aligned} G = \frac{R^2 - R_1R_2}{4(z_1 - z_2)^2} &= -\frac{l}{8}z_1z_2(z_1 + z_2) \\ &+ \frac{1}{64}\left((2h - x)^2 - k^2 + 4c\right)(z_1 + z_2)^2 - \frac{lh}{4}(z_1 + z_2) + \frac{l^2}{4}. \end{aligned}$$

Let us calculate the squares of derivatives $\left(\frac{\partial Q}{\partial s}\right)^2$, $\left(\frac{\partial Q}{\partial z_1}\right)^2$, and $\left(\frac{\partial Q}{\partial z_2}\right)^2$ and then use Eq. (5.54) to eliminate, respectively, s, z_1, and z_2 from each of them. Then we find by direct calculation that

$$\left(\frac{\partial Q}{\partial s}\right)^2 = R_1R_2, \quad \left(\frac{\partial Q}{\partial z_1}\right)^2 = \varphi(s)R_2, \quad \left(\frac{\partial Q}{\partial z_2}\right)^2 = \varphi(s)R_1.$$

It follows from (5.54) that the total differential of the function Q is zero. Thus:

$$dQ = \frac{\partial Q}{\partial z_1}dz_1 + \frac{\partial Q}{\partial z_2}dz_2 + \frac{\partial Q}{\partial s}ds = 0.$$

The two equations of (5.52) follow from the condition $dQ = 0$ upon division by the product $\sqrt{\varphi(s)R_1R_2}$ and taking into consideration the possibility of extracting square roots with different signs.

Isomorphism between the Kovalevskaya – (Komarov, Yehia) gyrostat on the pencil $\mathscr{L}_x$ and the Clebsch systems on $e(3)$ The Hamiltonian and the additional integral of the Kovalevskaya gyrostat on the pencil $\mathscr{L}_x$ can be written in the form

$$\begin{aligned} H &= \frac{1}{2}(M_1^2 + M_2^2 + 2M_3^2 - 2M_3\lambda) + \mu\gamma_1, \\ K &= \left(\frac{M_1^2 - M_2^2}{2} - \mu\gamma_1 + x\frac{\mu^2}{2}\right)^2 + (M_1M_2 - \mu\gamma_2)^2 \\ &\quad + \lambda\left((M_3 - \lambda)(M_1^2 + M_2^2 + x\mu^2) - 2\mu M_1\gamma_3\right), \quad \lambda, \mu = \text{const.} \end{aligned} \tag{5.55}$$

In the case where $\lambda = 0$ and $x = 0$, it is known (Haine and Horozov [253]) that a nonlinear complex transformation transforms the Kovalevskaya system into the Neumann system. This transformation has the form

$$\begin{aligned} q_1 &= -\frac{iM_1}{M_2}, \quad q_2 = \frac{M_1^2 + M_2^2 + 1}{2M_2}, \quad q_3 = \frac{i(M_1^2 + M_2^2 - 1)}{2M_2}, \\ L_1 &= -\frac{i\mu\gamma_3}{M_2}, \quad L_2 = \frac{2\mu M_1\gamma_3 - M_3(M_1^2 + M_2^2 - 1)}{2M_2}, \\ L_3 &= \frac{i\left(2\mu M_1\gamma_3 - M_3(M_1^2 + M_2^2 + 1)\right)}{2M_2}, \end{aligned} \tag{5.56}$$

and, moreover, the relations $\boldsymbol{q}^2 = 1$ and $(\boldsymbol{L}, \boldsymbol{q}) = 0$ hold for $\boldsymbol{q}$ and $\boldsymbol{L}$. Fixing the level set of the integrals $H = h$ and $K = k$ (5.55) and making use of the invariants (5.44) for $x \neq 0$, we can write the equations of motion in terms of the variables $\boldsymbol{L}$ and $\boldsymbol{q}$ as

$$\dot{\boldsymbol{q}} = \boldsymbol{q} \times \boldsymbol{L}, \quad \dot{\boldsymbol{L}} = \boldsymbol{q} \times \mathbf{Q}\boldsymbol{q}, \tag{5.57}$$

where the symmetric matrix $\mathbf{Q}$ is expressed in terms of the parameters and constants of the integrals by the formula

$$\mathbf{Q} = \begin{pmatrix} -h & -i\mu l & \mu l \\ -i\mu l & -\frac{1}{4} + m & i\left(\frac{1}{4} + m\right) \\ \mu l & i\left(\frac{1}{4} + m\right) & \frac{1}{4} - m \end{pmatrix}, \quad m = \mu^2(\boldsymbol{\gamma}^2 + x\boldsymbol{M}^2) - \mu^2 x\left(h - \frac{\mu^2 x}{4}\right) - k. \tag{5.58}$$

Thus, for $\lambda = 0$, the transformation (5.56) reduces the Kovalevskaya top to the Neumann problem, which is separated in the spherical coordinates. This allows us to understand a little more deeply the nature of the strange manipulations of Kovalevskaya. The Kovalevskaya variables are in essence the sphero-conic coordinates for system (5.57); the transformation (5.56) defines one of the methods for explicit integration of the Kovalevskaya case. It is interesting to note that the transformation (5.56) is necessarily complex. For $\lambda \neq 0$ it can be extended to connect two systems, which are not completely explicitly integrated: the problem of the Kovalevskaya gyrostat and the Clebsch case at nonzero value of the area constant. In what follows, we also treat $\lambda \neq 0$. (As mentioned previously, the transformation (5.56) was used for the first time in [311] for integration of the Kovalevskaya case $\lambda = 0$ on $so(4)$ and $so(3,1)$.)

Let us consider a generalization of the transformation (5.56) where the vectors $\boldsymbol{q}$ and $\boldsymbol{L}$ are modified as follows:

$$\boldsymbol{q} \to \boldsymbol{q}, \quad \boldsymbol{L} \to \boldsymbol{L} + \lambda(\boldsymbol{q} + i\boldsymbol{e}_1 \times \boldsymbol{q}),$$

where $\boldsymbol{e}_1 = (1, 0, 0)$. Here, we have

$$\boldsymbol{q}^2 = 1, \quad (\boldsymbol{q}, \boldsymbol{L}) = \lambda. \tag{5.59}$$

For $\lambda \neq 0$, the parameter m in the matrix $\mathbf{Q}$ must be modified as follows:

$$m = \mu^2(\boldsymbol{y}^2 + x\boldsymbol{M}^2) - \mu^2 x\left(h + \lambda^2 - \frac{\mu^2 x}{4}\right) - k.$$

Remark. For the case $x = 0$ and $\lambda \neq 0$, the analogy considered is due to I. V. Komarov and A. V. Tsiganov in [313, 314].

The equations of motion (5.57) are a system on the symplectic leaf of the algebra $e(3)$ defined by relations (5.59) with a Hamiltonian and an integral quadratic in $\boldsymbol{L}$ and $\boldsymbol{q}$:

$$\begin{aligned} \widetilde{H} &= (\boldsymbol{L}, \boldsymbol{L}) + (\boldsymbol{q}, \mathbf{Q}\boldsymbol{q}) = \lambda^2 + \mu^2 x - h, \\ \widetilde{K} &= (\mathbf{Q}\boldsymbol{L}, \boldsymbol{L}) - \det \mathbf{Q}(\boldsymbol{q}, \mathbf{Q}^{-1}\boldsymbol{q}) = k - \lambda^2 h + \mu^2 x(\lambda^2 - \mu^2 x), \end{aligned}$$

i.e., it defines the Clebsch system at a nonzero value of the area constant.

Kolosov's analogy and its generalizations. In [319], G. V. Kolosov presented a transformation involving time as well as phase variables that reduces the Kovalevskaya problem on $e(3)$ to point dynamics on the Euclidean plane in a certain potential field for which the elliptic coordinates are separating variables. This is the Kolosov analogy, which allows one to use some results from celestial mechanics in the study of rigid body dynamics. Let us look for an analogous procedure for the Kovalevskaya problem on the pencil (5.43). In this case, *the analog of the Kolosov transformation leads to a particle dynamics system on a certain axisymmetric surface of nonconstant curvature.*

Following [310], we start by using the equations of motion to rewrite the Hamiltonian function. For this purpose, let us make the following change of variables in the equations of motion (5.53), (here we use Hamiltonian (5.42) and the integrals in the forms (5.44) and (5.45)):

$$s_i \to s_i - \frac{1}{2}\left(h + \frac{x}{4}\right), \qquad i = 1,\ 2, \tag{5.60}$$

and let us represent them in the form

$$\frac{(s_1 - s_2)^2 \dot{s}_i^2}{f(s_i)} = g(s_i), \qquad i = 1,\ 2$$

$$\begin{aligned} f(s) &= 4\left(s - \frac{x}{4} + \frac{1}{2}k\right)\left(s - \frac{x}{4} - \frac{1}{2}k\right), \\ g(s) &= 4s^3 - \left(2h + \frac{3}{2}x\right)s^2 + \left(c - k^2 + \frac{x^2}{4}\right)s + \varkappa, \\ \varkappa &= \frac{1}{2}(k^2 h + 2l^2 - ch) + \frac{x}{4}(h^2 + k^2 - c) - \left(\frac{x}{4}\right)^3 . \end{aligned} \tag{5.61}$$

Remark. The form of the change of variables (5.60) is determined by the requirement that in the polynomial $g(s)$ only the coefficients of even powers of s depend on the energy constant h.

Using the two equations in (5.61), we subtract the first from the second and then eliminate the constant $\varkappa$ to obtain the expression for the energy:

$$\begin{aligned} H &= \frac{(s_1 - s_2)^2}{2(s_1^2 - s_2^2)}\left(\frac{\dot{s}_1^2}{f(s_1)} - \frac{\dot{s}_2^2}{f(s_2)}\right) + U(s_1,\ s_2), \\ U(s_1,\ s_2) &= \frac{2(s_1^2 + s_1 s_2 + s_2^2) + \frac{1}{2}\left(c - k^2 + \frac{x^2}{4}\right)}{s_1 + s_2} . \end{aligned} \tag{5.62}$$

After rescaling time as $d\tau = 2(s_1 - s_2)\,dt$ and passing to canonical momenta $p_i = \frac{\partial H}{\partial s_i'}$, for $s_i' = \frac{ds_i}{d\tau}$, $i = 1,\ 2$, we obtain a system with separating variables.

Now let us consider the variables $s_1 - \frac{x}{4}$ and $s_2 - \frac{x}{4}$ as elliptic coordinates on the plane (u, v):

$$\begin{aligned} u &= \frac{2}{k}\left(s_1 - \frac{x}{4}\right)\left(s_2 - \frac{x}{4}\right) + \frac{k}{2}, \\ v &= \pm\frac{2}{k}\sqrt{\left(s_1^2 - \frac{k^2}{4}\right)\left(\frac{k^2}{4} - s_2^2\right)} . \end{aligned}$$

Writing the energy (5.62) in terms of elliptic coordinates (u, v), we obtain the expressions

$$\begin{aligned} H &= T + U, \\ T &= \frac{1}{2}\left(1 + \frac{x}{2\rho}\right)(u'^2 + v'^2), \\ U &= \frac{4\rho^2 - 2uk + c + 3x\rho + x^2}{2\rho + x} = \frac{2(\rho^2 + \rho_1^2) - k^2 + c + 3x\rho + x^2}{2\rho + x}, \\ \rho &= \sqrt{u^2 + v^2}, \qquad \rho_1 = \sqrt{(u - k)^2 + v^2} . \end{aligned} \tag{5.63}$$

The system (5.63) describes the motion of a point particle on a curved surface in a potential field. The *Gaussian curvature* of the surface, which is not constant, can be written in terms of the metric defined by the kinetic energy T:

$$K = -\frac{4x}{(2\rho + x)^3}.$$

As follows from (5.63), the curvature can change its sign only at points at which the potential energy is infinite. Hence, the motion is restricted to domains in which the curvature has the same sign. For $x = 0$ we have $K = 0$, which coincides with the classical result of Kolosov giving an analogy of the Kovalevskaya case with particle motion in flat space.

Historical commentary. In [318], G. V. Kolosov extended the Hamilton – Jacobi method for the integration of various problems of rigid body dynamics. In particular, he considered the cases of Goryachev – Chaplygin, Clebsch, and Bobylev – Steklov and also the 1st Chaplygin case for the Kirchhoff equations. Analogously to what was known for material point motion, writing the equations in the canonical Hamiltonian form, G. V. Kolosov looked for canonical transformations in the phase space that separate the variables, thus trying to generalize his results about the Kovalevskaya case. In [318], he also presented a particular periodic solution for the Clebsch case that is characterized by the property of having two independent invariant relations of the form $F_1 = aM_3 + b\gamma_3$ and $F_2 = (\alpha M_1\gamma_1 + \beta M_2\gamma_2)/(\varepsilon M_1^2 + \delta\gamma_2^2)$ for some $a, b, \alpha, \beta, \varepsilon, \delta = \text{const}$. These relations arise under restriction on constants of the first integrals obtained from bifurcation conditions (i.e., when the integrals become dependent).

In essence, the technique of introducing canonical action-angle variables described above (on the basis of [310]), using the Abel – Jacobi equations, is an extension of the observations of Kolosov, who had tried to give a reasonable geometric meaning to the nontrivial algebraic transformations suggested by Kovalevskaya and Chaplygin in the context of canonical transformations of phase space.

In connection with the problem of separating the variables for the Kovalevskaya case, we mention an interesting piece of work [361] in which the author constructed a transformation that *simultaneously* separates the Kovalevskaya and Goryachev – Chaplygin cases. The separation of variables for the Kovalevskaya case by using an approach of embedding it in the so-called second Sklyanin algebra, construction of action variables, and a Lax representation in the form of 2×2 **L**-**A** pairs were all carried out in [360] (see also [87]).

Chaplygin case (I). We consider the explicit integration of the Chaplygin case on the bracket pencil (5.43) for zero value of the area constant. Introduce the following notation for the Casimir functions:

$$(\boldsymbol{M}, \boldsymbol{\gamma}) = 0, \quad x\boldsymbol{M}^2 + \boldsymbol{\gamma}^2 = c. \tag{5.64}$$

The Hamiltonian and the additional integral have the form (see Sec. 3.2 of Ch. 3):

$$H = \frac{1}{2}(\alpha_2 M_1^2 + \alpha_1 M_2^2 + (\alpha_1 + \alpha_2) M_3^2) - \frac{1}{2}(a_1 - a_2)(\gamma_1^2 - \gamma_2^2) = \frac{h}{2},$$
$$F = (\alpha_1 M_1^2 - \alpha_2 M_2^2 - (a_1 - a_2)\gamma_3^2)^2 + 4\alpha_1\alpha_2 M_1^2 M_2^2 = k^2,$$

where

$$\alpha_1 = 1 - xa_1 \quad \text{and} \quad \alpha_2 = 1 - xa_2 \quad (k > 0).$$

For this problem, the variables were separated by S. A. Chaplygin [128] on $e(3)$ (i.e., when $x = 0$) and by O. I. Bogoyavlenskii [61] on $so(4)$ (i.e., when $x = 1$), and the separation of variables on the whole pencil (i.e., when x is arbitrary) was performed in [85, 82]. In the generic case the separating variables s_1 and s_2 are defined by the formulas [61, 128]

$$s_1 = \frac{u+k}{v}, \quad s_2 = \frac{u-k}{v}, \qquad u = \alpha_1 M_1^2 + \alpha_2 M_2^2, \quad v = (a_2 - a_1)\gamma_3^2. \tag{5.65}$$

The evolution of s_1 and s_2 is governed by

$$\dot{s}_1 = -\sqrt{(1 - s_1^2)(\delta_1 - \beta_1 s_1)}, \quad \dot{s}_2 = -\sqrt{(1 - s_2^2)(\delta_2 - \beta_2 s_2)}, \tag{5.66}$$

where

$$\delta_{1,2} = 2(h \pm k) - x\big((h \pm k)(a_1 + a_2) + (a_1 - a_2)^2 c\big),$$
$$\beta_{1,2} = 2(a_1 - a_2)\big(c - \frac{1}{2}x(h \pm k + c(a_1 + a_2))\big),$$

i.e., the equations of motion can be integrated in terms of elliptic functions of time.

Kharlamov's case for the generalized Kovalevskaya top. In [284], M. P. Kharlamov presented a system with two degrees of freedom resulting from reduction of the generalized Kovalevskaya top in two fields (see Ch. 3, Sec. 3.4). Following [284], we set

$$H = \omega_1^2 + \omega_2^2 + \frac{1}{2}\omega_3^2 - \alpha_1 - \beta_2,$$
$$\boldsymbol{\alpha}^2 = a^2, \quad \boldsymbol{\beta}^2 = b^2, \quad (\boldsymbol{\alpha}, \boldsymbol{\beta}) = 0,$$

i.e., the vectors $\boldsymbol{\alpha}$ and $\boldsymbol{\beta}$ are not normalized.

The separating variables are given by the relations (M. P. Kharlamov and A. Yu. Savushkin [287, 288])

$$s_1 = \frac{x^2 + z^2 + a^2 - b^2}{2x}, \quad s_2 = \frac{x^2 + z^2 - a^2 + b^2}{2x},$$
$$x^2 = (\alpha_1 - \beta_2)^2 + (\alpha_2 + \beta_1)^2, \quad z^2 = \alpha_3^2 + \beta_3^2$$

and satisfy the inequalities $s_1^2 \geqslant ya^2$ and $s_2^2 \leqslant b^2$. On the common level set of invariant relations and first integrals, we obtain the (degenerate) Abel – Jacobi equations

$$\dot{s}_1 = \sqrt{-(s_1^2 - a^2)\Phi(s_1)}, \quad \dot{s}_2 = \sqrt{(b^2 - s_2^2)\Phi(s_2)},$$
$$\Phi(s) = ms^2 - ls + \frac{l^2 - 1}{4m},$$

where the constants m and l are connected with the constant values of the integrals of motion by relations of the form

$$M = \frac{2G - (a^2 + b^2)H}{(a^2 - b^2)^2} = m, \quad L^2 = 2(a^2 + b^2)M^2 + 2HM + 1 = l^2,$$

where G is the quadratic integral

$$G = \left(\alpha_1\omega_1 + \alpha_2\omega_2 + \frac{1}{2}\alpha_3\omega_3\right)^2 + \left(\beta_1\omega_1 + \beta_2\omega_2 + \frac{1}{2}\beta_2\omega_3\right)^2$$
$$+\omega_3\left(\gamma_1\omega_1 + \gamma_2\omega_2 + \frac{1}{2}\gamma_3\omega_3\right) - b^2\alpha_1 - a^2\beta_2 + c(\alpha_2 + \beta_1).$$

Separation of variables for the Goryachev – Chaplygin and Kovalevskaya – Goryachev – Chaplygin tops. In accordance with [539, 588], we use these names to refer, respectively, to the systems on $e(3)$ having the form

1. (GCh) $H = \frac{1}{2}(M_1^2 + M_2^2 + 4M_3^2) + \lambda M_3 + \mu\gamma_1 + \frac{1}{2}\frac{\delta}{\gamma_3^2}$,

where $\lambda, \mu, \delta = \text{const.}$ and $(\boldsymbol{M}, \boldsymbol{\gamma}) = 0$;

2. (KGCh) $H = \frac{1}{2}(M_1^2 + M_2^2 + 2M_3^2) + \lambda M_3 + \mu\gamma_1 + 2b_1\gamma_1\gamma_2$
$+b_2(\gamma_2^2 - \gamma_1^2) + \frac{1}{2}\frac{\delta}{\gamma_2^2}$,

where $\lambda, \mu, b_1, b_2 = \text{const.}$ and $(\boldsymbol{M}, \boldsymbol{\gamma}) = 0$.

For these two systems, the integrals are, respectively, of degree three and four in $\boldsymbol{M}$ and are presented in Sec. 5.1 (see cases (5.1) and (5.13)). For their explicit integration, the main difficulties in both cases are given by the gyrostat term $\lambda \neq 0$ and the singular term $\delta \neq 0$. These were partially overcome in [539, 588] (see also [87]), where the authors presented the separation of variables obtained by embedding the system in a Sklyanin quadratic bracket algebra for which a universal system of separating symmetric coordinates is known. This approach is close to the approach of Jacobi, who introduced a general system of elliptic coordinates that allowed him to integrate a number of meaningful mechanical problems. In contrast to the Jacobi method, the separating variables in the above-mentioned works have been defined not only on the configuration space, but on the whole phase space. A defect of the methods of [539, 588] is the complex form taken by the proposed transformations, and from the mechanical point of view the problem cannot be regarded as solved. The results of [539, 588] are presented with some commentaries in our book [87]. We have nothing more to say about them here, since such a separation has not found application to the study of dynamics so far.

2 Action variables and separating variables

Separating variables can be used to construct bifurcation diagrams and to carry out topological analysis [286]. They are also useful for obtaining action-angle variables, which are necessary for studying perturbation problems and also for quantization purposes.

A natural algorithm for introducing action-angle variables was proposed in [310] for the Kovalevskaya case. It uses the usual method of introducing angle-action variables for systems with variables separated by means of the Hamilton – Jacobi construction. This procedure was formalized and then generalized by A. P. Veselov and S. P. Novikov [601] to the general theory of algebro-geometric Poisson brackets. However, we use the initial algorithm of [310] and introduce the action variables for the Kovalevskaya, Goryachev – Chaplygin, and Chaplygin cases on the whole bracket pencil (5.43).

The above-mentioned algorithm consists of several items.

1. Finding the Abel variables s_1 and s_2, which commute in the initial Poisson structure: $\{s_1, s_2\} = 0$.

   REMARK. The existence of the commuting pair of Abel variables s_1 and s_2 can be also proved by using arguments presented in the book [333], which are related to the reduction of equations of the form (5.53) to a standard form on the torus.

2. Using the energy equation in the variables s_i and $\dot{s}_i$, one introduces the canonical momenta p_i satisfying the additional requirement that the system p_j, s_i be separating.
3. Having the separated variables, one introduces the action-angle variables according to the well-known algorithm [410].

Kovalevskaya case. One can verify that the Kovalevskaya variables s_1 and s_2 (5.50) commute [310, 308]. Moreover, they satisfy Eqs. (5.53):

$$\dot{s}_1 = \frac{\sqrt{f(s_1)\varphi(s_1)}}{s_1 - s_2}, \quad \dot{s}_2 = -\frac{\sqrt{f(s_2)\varphi(s_2)}}{s_1 - s_2}, \tag{5.67}$$

where

$$\begin{aligned} f(s) &= \left(2s + \frac{1}{4}(2h - x) + \frac{1}{4}k\right)\left(2s + \frac{1}{4}(2h - x) - \frac{1}{4}k\right), \\ \varphi(s) &= 4s^3 + 2hs^2 + \left(\frac{1}{16}(2h - x)^2 - \frac{k^2}{16} + \frac{c}{4}\right)s + \frac{l^2}{16} \end{aligned} \tag{5.68}$$

(The change of variables from (z_1, z_2) to (s_1, s_2) is a complicated analog of the change from Euclidean to polar coordinates: the variable s_1 varies from 0 to ∞, whereas s_2 is a parametrization of the circle).

Let us isolate the motion integral $\varkappa = \frac{1}{16}\left((2h - x)^2 - k^2\right)$ in (5.68):

$$\begin{aligned} f(s) &= 4s^2 + (2h - x)s + \varkappa, \\ \varphi(s) &= s\left(4s^2 + 2hs + \varkappa + \frac{c}{4}\right) + \frac{l^2}{16} \end{aligned} \tag{5.69}$$

and then eliminate the constant $\varkappa$ from (5.69). From the resulting equation, the energy h can be expressed as a function of s_i and $\dot{s}_i$:

$$h = -2(s_1 + s_2) + \frac{l^2}{64 s_1 s_2} + \frac{x}{4} + \frac{\sqrt{a_1^2 + x_1^2} - \sqrt{a_2^2 + x_2^2}}{s_1 - s_2},$$
$$a_i = \frac{16 s_i^2 x + 4 s_i c + l^2}{64 s_i}, \quad x_i = \frac{(s_1 - s_2)\dot{s}_i}{2\sqrt{s_i}}. \tag{5.70}$$

Instead of the velocities $\dot{s}_i$, let us introduce the generalized momenta p_i by the formula

$$p_i = \int \frac{\partial h}{\partial \dot{s}_i} \frac{d\dot{s}_i}{\dot{s}_i} + F(s_i), \tag{5.71}$$

where $F(s_i)$ is an arbitrary function of s_i; the addition of $F(s_i)$ does not change the equations of motion (since it defines a canonical transformation).

Integration (for $F = 0$) yields

$$p_i = \frac{1}{2\sqrt{s_i}} \ln \frac{x_i + \sqrt{x_i^2 + a_i^2}}{a_i}. \tag{5.72}$$

Writing the Hamiltonian of the Kovalevskaya system (5.70) in terms of the variables s_i and p_i, we obtain

$$h = -s_1 - s_2 + \frac{l^2}{8 s_1 s_2} + \frac{x}{8} + \frac{a_1 \cosh(2p_1\sqrt{s_1}) - a_2 \cos(2p_2\sqrt{-s_2})}{s_1 - s_2}. \tag{5.73}$$

The variables s_i are separating for the Hamiltonian (5.73). Introducing the separation constant $\varkappa_1$, we obtain the following two equations, which can be integrated independently of one another:

$$2s_1^2 + s_1\left(h - \frac{x}{4}\right) + \frac{l^2}{64 s_1} + \varkappa_1 = a_1 \cosh(2p_1\sqrt{s_1}),$$
$$2s_2^2 + s_2\left(h - \frac{x}{4}\right) + \frac{l^2}{64 s_2} + \varkappa_1 = a_2 \cos(2p_2\sqrt{-s_2}). \tag{5.74}$$

Substituting expressions (5.72) into (5.74), we obtain the Kovalevskaya equations (5.67); the separation constant $\varkappa_1$ is related to the constant $\varkappa$ by the formula $\varkappa = 2\varkappa_1 - \frac{c}{8}$.

Finally, the action variables are found by the formula

$$I_i = \frac{1}{2\pi} \oint p(s_i)\, ds_i,$$

where the dependence of p_i on s_i is assumed to be expressed from (5.74), whereas the integration domain depends on the system parameters and the values of the integrals.

Goryachev – Chaplygin case. Now the variables q_1 and q_2 from (5.21) are separating. Moreover, it is easy to verify that on the level set $(\boldsymbol{M}, \boldsymbol{\gamma}) = 0$ they commute with one another.

Denoting by $\varkappa$ the separation constant, we find from (5.22) that

$$\begin{aligned} \frac{p_1^3}{2} - \lambda p_1^2 + a p_1 \sin q_1 - h p_1 &= \varkappa, \\ \frac{p_2^3}{2} - \lambda p_2^2 - a p_2 \sin q_2 - h p_2 &= \varkappa. \end{aligned} \tag{5.75}$$

The action variables can be obtained from the formula

$$I_i = -\frac{1}{2\pi} \oint q_i(p_i)\, dp_i,$$

where the dependence of q_i on p_i is expressed by using (5.75).

REMARK. Since

$$\dot{p}_i = -\frac{\partial H}{\partial q_i} = -\frac{a p_i \cos q_i}{p_1 - p_2}, \tag{5.76}$$

we obtain, by using (5.75), the Abel equations in the form

$$\dot{p}_i = -\frac{\sqrt{\Phi(p_i)}}{2(p_1 - p_2)}, \quad \Phi(z) = 4a^2 z^2 - (2\varkappa - z^3 + 2\lambda z^2 + 2hz)^2. \tag{5.77}$$

Chaplygin case (I). At zero value of the area integral $(\boldsymbol{M}, \boldsymbol{\gamma}) = 0$, the variables s_1 and s_2 (5.65) commute. From the system of equations (5.66) on the pencil, we find the energy $E = h + \text{const.}$ as a function of s_1, s_2, $\dot{s}_1$, and $\dot{s}_2$:

$$E = \frac{1}{2}\left(\frac{\dot{s}_1^2}{g(s_1)} + \frac{\dot{s}_2^2}{g(s_2)}\right) - \frac{2c}{x}(1 - x a_1)(1 - x a_2)\big(U(s_1) + U(s_2)\big), \tag{5.78}$$

where

$$\begin{aligned} g(s) &= (1 - s^2)\big(x(a_2 - a_1)s + 2 - x(a_1 + a_2)\big), \\ U(s) &= \big(x(a_2 - a_1)s + 2 - x(a_1 + a_2)\big)^{-1}. \end{aligned}$$

Defining the conjugate momenta by the formula (5.71),

$$p_i = \frac{\dot{s}_i}{(1 - s_i^2)(x(a_1 - a_2)s_i + 2 - x(a_1 + a_2))},$$

the Hamiltonian function is written in terms of the separated variables:

$$H = \frac{1}{2}\big(g(s_1)p_1^2 + g(s_2)p_2^2\big) - \frac{2c}{x}(1 - x a_1)(1 - x a_2)\big(U(s_1) + U(s_2)\big). \tag{5.79}$$

Now, introducing the separation constant $\varkappa$, we may write

$$\begin{aligned} \frac{1}{2}g(s_1)p_1^2 - \frac{2c}{x}(1 - x a_1)(1 - x a_2)U(s_1) &= \varkappa, \\ \frac{1}{2}g(s_2)p_2^2 - \frac{2c}{x}(1 - x a_1)(1 - x a_2)U(s_2) &= E - \varkappa \end{aligned} \tag{5.80}$$

and so calculate the action variables:

$$I_i = \frac{1}{2\pi} \oint p_i(s_i)\, ds_i.$$

The action variables were recently constructed for the Clebsch and Schottky–Manakov cases in [406] (V. G. Marikhin and V. V. Sokolov) by finding the complete Hamilton–Jacobi equation for these problems. The action variables for the Steklov case are easily obtained by using its isomorphism (presented in Sec. 5.3 of Ch. 5) to a system on the sphere S^2 with separated variables.

5.3 Algebraic Transformations of Poisson Brackets. Isomorphisms and Explicit Integration

In Hamiltonian mechanics, the canonical transformations preserving the (canonical) Poisson bracket are well known; as a rule, they are applied in order to simplify the equations of motion (for example, in perturbation theory). They can be defined by using a generating function.

In the noncanonical case where the Poisson bracket has a certain algebraic form (for example, when it is linear or quadratic), we can formulate the problem of searching for transformations that either preserve the initial algebraic structure of the bracket or transform it into some other bracket of interest. In the case of a degenerate Poisson bracket this problem might be posed not simultaneously for all symplectic leaves, but for a particular distinguished leaf having some additional properties (in particular, for singular orbits of Lie coalgebras).

In the case of Lie–Poisson brackets, even classical researchers [556] considered linear (i.e., group) transformations that allowed them to attain simplifications in a Hamiltonian.

The first examples of nonlinear transformations preserving a linear bracket structure (and, therefore, having singularities) were presented in our book [95]. In the same book, we also gave a transformation of a quadratic bracket into a linear bracket defined on a particular distinguished symplectic leaf that allowed us to find an isomorphism between the generalized Kovalevskaya case and the Chaplygin case and to explain the singular terms of the Goryachev case [92, 95].

In [80] we gave an algebraic transformation, defined on distinguished symplectic leaves of the algebras $e(3)$ and $so(4)$, that transforms one algebraic structure into the other. This allowed us to extend the classical Goryachev–Chaplygin case from the algebra $e(3)$ to the whole Poisson pencil $\mathscr{L}_x$ and so to obtain the $so(4)$ analog of Goryachev-Chaplygin as the special case $x = 1$ of this extension. In [583] the same transformation was used for the generalized Kovalevskaya case on the level set $(\boldsymbol{M}, \boldsymbol{\gamma}) = 0$.

In [312], the problem of finding transformations between the algebras $e(3)$, $so(4)$, and $so(3,1)$ is discussed in more detail. In particular, new types of transformations are presented that allow one to generalize various integrable cases and to find interconnections between systems on these different algebras (this can be used for explicit integration of various systems). Generally speaking, we can say that the study of algebraic transformations of Poisson brackets has turned out to have an importance for dynamics comparable with the problem of finding first integrals.

Here, we present all the results known to us which have been obtained in this direction, and also present some techniques that allow us to introduce a global and universal system of symplectic coordinates on families of Poisson brackets.

1 Group transformations of the pencil

Let us consider the pencil of Lie – Poisson brackets

$$\{M_i, M_j\} = \varepsilon_{ijk} M_k, \quad \{M_i, \gamma_j\} = \varepsilon_{ijk}\gamma_k, \quad \{\gamma_i, \gamma_j\} = x\varepsilon_{ijk} M_k, \tag{5.81}$$

which has been used repeatedly already. As is easily shown, for a fixed x, the Lie group G corresponding to this Lie – Poisson bracket is the set of of matrices preserving the metric $\mathbf{g} = \operatorname{diag}(1, 1, 1, x)$, i.e.,

$$G = \{\mathbf{U} | \mathbf{U}^T \mathbf{g} \mathbf{U} = \mathbf{g}\} \tag{5.82}$$

(for $x > 0$, $G \approx SO(4)$, for $x = 0$, $G \approx E(3)$, and for $x < 0$, $G \approx SO(3,1)$).

Hence, the bracket (5.81) is invariant under similarity transformations by elements of the group G. In matrix form, they can be represented as follows. Define the following matrix (identifying the algebra and the coalgebra by the Killing metric $\operatorname{Tr}(\mathbf{XY})$):

$$\mathbf{X} = \begin{pmatrix} 0 & M_3 & -M_2 & \gamma_1 \\ -M_3 & 0 & M_1 & \gamma_2 \\ M_2 & -M_1 & 0 & \gamma_3 \\ -\frac{1}{x}\gamma_1 & -\frac{1}{x}\gamma_2 & -\frac{1}{x}\gamma_3 & 0 \end{pmatrix}. \tag{5.83}$$

Then the similarity transformations preserving the bracket (5.81) can be written as

$$\mathbf{X}' = \mathbf{U}^{-1}\mathbf{X}\mathbf{U}, \quad \mathbf{U} \in G. \tag{5.84}$$

Using (5.82), we find that $\mathbf{U}^{-1} = \mathbf{g}^{-1}\mathbf{U}^T\mathbf{g}$.

REMARK. One can show that the matrix $\mathbf{gX}$ is skew-symmetric.

For $x = 0$ and $x > 0$, the transformations (5.84) can be represented in a simpler form.

For $x = 0$,

$$\boldsymbol{M}' = \mathbf{Q}\boldsymbol{M} + \boldsymbol{a} \times \mathbf{Q}\boldsymbol{\gamma}, \quad \boldsymbol{\gamma}' = \mathbf{Q}\boldsymbol{\gamma}, \quad \mathbf{Q} \in SO(3), \quad \boldsymbol{a} \in \mathbb{R}^3. \tag{5.85}$$

For $x > 0$, we may use the fact that the group G in (5.82) is isomorphic to $SO(4)$ whose Lie algebra admits the real decomposition $so(4) = so(3) \oplus so(3)$. Introducing the new variables $\boldsymbol{K}$ and $\boldsymbol{S}$, we obtain a Poisson bracket algebra as follows:

$$\begin{gathered} \boldsymbol{K} = \frac{1}{2}\left(\boldsymbol{M} + \frac{1}{\sqrt{x}}\boldsymbol{\gamma}\right), \quad \boldsymbol{S} = \frac{1}{2}\left(\boldsymbol{M} - \frac{1}{\sqrt{x}}\boldsymbol{\gamma}\right), \\ \{K_i, K_j\} = \varepsilon_{ijk}K_k, \quad \{S_i, S_j\} = \varepsilon_{ijk}S_k, \quad \{K_i, S_j\} = 0. \end{gathered} \tag{5.86}$$

The transformation (5.84) is represented in the form

$$\boldsymbol{K}' = \mathbf{Q}_1\boldsymbol{K}, \quad \boldsymbol{S}' = \mathbf{Q}_2\boldsymbol{K}, \quad \mathbf{Q}_1, \mathbf{Q}_2 \in so(3),$$

or, in the initial variables,

$$\begin{aligned} \boldsymbol{M}' &= \frac{\mathbf{Q}_1 + \mathbf{Q}_2}{2}\boldsymbol{M} + \frac{1}{\sqrt{x}}\frac{\mathbf{Q}_1 - \mathbf{Q}_2}{2}\boldsymbol{\gamma}, \\ \boldsymbol{\gamma}' &= \sqrt{x}\frac{\mathbf{Q}_1 - \mathbf{Q}_2}{2}\boldsymbol{M} + \frac{\mathbf{Q}_1 + \mathbf{Q}_2}{2}\boldsymbol{\gamma}. \end{aligned} \tag{5.87}$$

2 Transformations related to symplectic Andoyer variables

According to the Darboux theorem (Ch. 1) we can construct, on each symplectic leaf, the symplectic (canonical) variables (Darboux coordinates). For the Poisson structure pencil (5.81) whose symplectic leaves are parameterized by the constants of the Casimir functions

$$(\boldsymbol{M}, \boldsymbol{\gamma}) = l_x, \quad \boldsymbol{\gamma}^2 + x\boldsymbol{M}^2 = c_x, \tag{5.88}$$

for generic fixed l_x and c_x, only two systems of symplectic variables are known; these are the Andoyer-type and Kovalevskaya-type variables and their conjugates [310, 82]. Apart from this, on the orbit $l_x = 0$ (using the isomorphisms presented below), we can construct various canonical variables using the fact that the orbit of the algebra $e(3)$ with $l_0 = 0$ is symplectomorphic to the cotangent bundle to the sphere T^*S^2. The algorithm for constructing the Andoyer variables (L, l, G, g, H, h) on the pencil (5.81) is described in detail between equations (5.15) and (5.18) of the present chapter.

1. The simplest transformation preserving the structure (5.81) and leaving the symplectic leaves (5.88) fixed has the form

$$l \to l + \varphi(L), \quad L, G, g \to L, G, g,$$

where $\varphi(L)$ is an arbitrary function. The characteristic feature of this transformation is that in the initial variables it has the following very simple form:

$$\begin{gathered} \boldsymbol{M}' = \mathbf{Q}\boldsymbol{M}, \quad \boldsymbol{\gamma}' = \mathbf{Q}\boldsymbol{\gamma}, \\ \mathbf{Q} = \begin{pmatrix} \cos\varphi(M_3) & \sin\varphi(M_3) & 0 \\ -\sin\varphi(M_3) & \cos\varphi(M_3) & 0 \\ 0 & 0 & 1 \end{pmatrix}. \end{gathered} \tag{5.89}$$

In contrast to transformations (5.85), the angle of rotation in this case can depend on M_3. In [312], such transformations are called *generalized rotations*.

It is interesting that the simple transformation (5.89) can produce nontrivial integrable cases. For example, from the Kovalevskaya case we obtain a family (parameterized by the arbitrary function φ) of integrable Hamiltonians of the form

$$H' = \frac{1}{2}(M_1'^2 + M_2'^2 + 2M_3'^2) + a\big(\gamma_1' \cos\varphi(M_3') - \gamma_2' \sin\varphi(M_3')\big). \qquad (5.90)$$

2. We can naturally define reduction of the algebra $e(3)$ to the zero leaf ($l_0 = 0$), i.e., to the cotangent bundle to the sphere T^*S^2. Indeed, the transformation

$$\boldsymbol{M} \to \boldsymbol{M}' = \boldsymbol{M}, \quad \boldsymbol{\gamma} \to \boldsymbol{\gamma}' = \frac{\boldsymbol{M} \times (\boldsymbol{\gamma} \times \boldsymbol{M})}{|\boldsymbol{M} \times (\boldsymbol{\gamma} \times \boldsymbol{M})|}$$

transforms the brackets of the pencil (5.81) into the brackets of the algebra $e(3)$, and, moreover, all symplectic leaves of the pencil are mapped onto the leaf $e(3)$ given by the relations $\boldsymbol{\gamma}'^2 = 1$, $(\boldsymbol{M}', \boldsymbol{\gamma}') = 0$, which is symplectomorphic to T^*S^2.

For $(\boldsymbol{M}, \boldsymbol{\gamma}) = l_x = H = 0$, we obtain the following proposition.

Proposition 1. *The mapping*

$$\boldsymbol{M} \to \boldsymbol{M}, \quad \boldsymbol{\gamma} \to \frac{\boldsymbol{\gamma}}{|\boldsymbol{\gamma}|} \qquad (5.91)$$

is a symplectic mapping of orbits $l_x = 0$ of the pencil onto the orbit of the algebra $e(3)$ for which $l_0 = 0$ and $c_0 = 1$ (i.e., the tangent bundle to the sphere).

The mapping (5.91) allows us to "lift" (extend) the integrable cases from $e(3)$ to the whole pencil.

The generalization of the Goryachev – Chaplygin case to the pencil [80]:

$$H = \frac{1}{2}(M_1^2 + M_2^2 + 4M_3^2) + \mu\frac{\gamma_1}{|\boldsymbol{\gamma}|}. \qquad (5.92)$$

The generalization of the Kovalevskaya – Goryachev – Chaplygin case to the pencil [583]:

$$H = M_1^2 + M_2^2 + 2M_3^2 + \delta\frac{\boldsymbol{\gamma}^2}{\gamma_3^2} + 2a_0M_3 + \alpha_1\frac{\gamma_1}{|\boldsymbol{\gamma}|} + \beta_1\frac{\gamma_2}{|\boldsymbol{\gamma}|}$$
$$-\alpha_2\beta_2\frac{\gamma_1\gamma_2}{\boldsymbol{\gamma}^2} - \frac{1}{4}(\alpha_2^2 - \beta_2^2)\frac{\gamma_1^2 - \gamma_2^2}{\boldsymbol{\gamma}^2}. \qquad (5.93)$$

3. In conclusion, we also present the following nonlinear transformation which preserves the brackets of the (sub) algebra $so(3)$

$$M_1' = \frac{M_1M_2}{\sqrt{M_1^2 + M_2^2}}, \quad M_2' = \frac{M_2^2 - M_1^2}{2\sqrt{M_1^2 + M_2^2}}, \quad M_3' = \frac{M_3}{2}.$$

This is equivalent to the canonical transformation which in the Andoyer variables takes the simple form

$$(L, l) \to \left(\frac{L}{2}, 2l\right).$$

3 Orbit isomorphism for algebras $e(3)$ and $so(3,1)$

Consider the mapping presented in [312]:

$$\boldsymbol{M} \to \boldsymbol{M}, \quad \boldsymbol{\gamma} \to \boldsymbol{p} = b\boldsymbol{\gamma} + c\boldsymbol{\gamma} \times \boldsymbol{M}. \tag{5.94}$$

One can verify by direct calculation that if the brackets for $\boldsymbol{M}$ and $\boldsymbol{\gamma}$ are given by (5.81) with $x = 0$, the brackets for $\boldsymbol{M}$ and $\boldsymbol{p}$ have the form

$$\{M_i, M_j\} = \varepsilon_{ijk}M_k, \quad \{M_i, p_j\} = \varepsilon_{ijk}p_k, \quad \{p_i, p_j\} = -c^2\boldsymbol{\gamma}^2\varepsilon_{ijk}M_k.$$

Thus, the orbit of the algebra $e(3)$ with the values c_0 and l_0 of the Casimir functions is mapped onto the orbit of the algebra $so(3,1)$ with the pencil parameter $x = -c^2c_0^2$ and the Casimir functions in (5.88) taking the values

$$c_x = b^2c_0 - c^2l_0, \quad l_x = bl_0. \tag{5.95}$$

On the other hand, for the mapping inverse to (5.94), we have

$$\boldsymbol{M} \to \boldsymbol{M}, \quad \boldsymbol{\gamma} \to \boldsymbol{y} = \frac{b^2\boldsymbol{y} + c^2(\boldsymbol{y}, \boldsymbol{M})\boldsymbol{M} - bc(\boldsymbol{y} \times \boldsymbol{M})}{b(b^2 + c^2\boldsymbol{M}^2)}, \tag{5.96}$$

where it is assumed that $\boldsymbol{M}$ and $\boldsymbol{\gamma}$ satisfy relations (5.81) for $x \neq 0$. The brackets for the new variables have the form

$$\begin{gathered}\{M_i, M_j\} = \varepsilon_{ijk}M_k, \quad \{M_i, y_j\} = \varepsilon_{ijk}y_k, \\ \{y_i, y_j\} = \frac{xb^4 + b^2c^2(\boldsymbol{y}^2 + x\boldsymbol{M}^2) + c^4(\boldsymbol{y}, \boldsymbol{M})^2}{b^2(b^2 + c^2\boldsymbol{M}^2)^2}\varepsilon_{ijk}M_k.\end{gathered} \tag{5.97}$$

If b and c satisfy the equation

$$xb^4 + b^2c^2c_x + c^4l_x^2 = 0,$$

then relations (5.97) correspond to the algebra $e(3)$. In the case where b and c are real, this is possible only for $x < 0$; in this case, the mapping (5.96) defines a mapping of the orbit of the algebra $so(3,1)$ given by the values c_x and l_x of the Casimir functions onto the orbit of the algebra $e(3)$ with the following values of the Casimir functions:

$$c_0 = \frac{c_x^2 \pm \sqrt{c_x^2 - 4\sqrt{-x}\,l_x^2}}{2b^2}, \quad l_0 = \frac{l_x}{b}. \tag{5.98}$$

This orbit isomorphism defines an isomorphism between the Kovalevskaya integrable case on $so(3,1)$ and the Kovalevskaya – Sokolov case on $e(3)$. Indeed, the Kovalevskaya – Sokolov Hamiltonian on $e(3)$

$$H = \frac{1}{2}(M_1^2 + M_2^2 + 2M_3^2) + a\big(c(\gamma_3M_2 - \gamma_2M_3) + b\gamma_1\big) \tag{5.99}$$

is mapped by (5.94) to the Kovalevskaya Hamiltonian

$$H' = \frac{1}{2}(M_1^2 + M_2^2 + 2M_3) + ap_1 \tag{5.100}$$

on $so(3,1)$. Thus, we also have a way of explicitly integrating the Kovalevskaya–Sokolov case on $e(3)$, since the Kovalevskaya case on the whole pencil has already been integrated by various methods [311, 82].

Using the transformations (5.94) and (5.96), we can also obtain analogs on $so(3,1)$ of the Kovalevskaya and Goryachev – Chaplygin cases, which are different from those obtained from (5.92) and (5.93). Their dynamical significance is still not clear.

4 A transformation related to the Euler angles

In the book [95], we presented a transformation that preserves the algebra $e(3)$ and is related to the Euler angles parameterizing the orbits of $e(3)$ by the formulas

$$\begin{gathered}
\gamma_1 = |\boldsymbol{\gamma}|\sin\theta\sin\varphi, \quad \gamma_2 = |\boldsymbol{\gamma}|\sin\theta\cos\varphi, \quad \gamma_3 = |\boldsymbol{\gamma}|\cos\theta, \\
M_1 = \frac{\sin\varphi}{\sin\theta}(p_\psi - p_\varphi\cos\theta) + p_\theta\cos\varphi, \\
M_2 = \frac{\cos\varphi}{\sin\theta}(p_\psi - p_\varphi\cos\theta) - p_\theta\sin\varphi, \\
M_3 = p_\varphi.
\end{gathered} \tag{5.101}$$

Setting $p_\psi = -l = \text{const.}$ and taking (5.101) into account, we obtain the following proposition.

Proposition. *For* $(\boldsymbol{M}, \boldsymbol{\gamma}) = 0$, *the mapping*

$$\boldsymbol{\gamma} \to \boldsymbol{\gamma}' = \frac{\boldsymbol{\gamma}}{|\boldsymbol{\gamma}|}, \quad \boldsymbol{M} \to \boldsymbol{M}' = \begin{pmatrix} M_1 + l\dfrac{\gamma_1|\boldsymbol{\gamma}|}{\gamma_1^2 + \gamma_2^2} \\ M_2 + l\dfrac{\gamma_2|\boldsymbol{\gamma}|}{\gamma_1^2 + \gamma_2^2} \\ M_3 \end{pmatrix} \tag{5.102}$$

is a symplectic transformation of the orbit of the pencil (5.81) *given by the relations* $l_x = 0$ *onto the orbit of the algebra* $e(3)$ *for which* $\boldsymbol{\gamma}'^2 = 1$ *and* $(\boldsymbol{M}, \boldsymbol{\gamma}') = l$.

The mapping (5.102) gives rise to singular terms in the Hamiltonian, at the poles of the Poisson sphere ($\gamma_1 = \gamma_2 = 0$ and $\gamma_3 = \pm 1$), and to gyroscopic terms.

Generalizations of transformations (5.102) of the form

$$\boldsymbol{\gamma} \to \boldsymbol{\gamma}' = \boldsymbol{\gamma}, \quad \boldsymbol{M} \to \boldsymbol{M}' = \boldsymbol{M} + \mathbf{g}(\boldsymbol{\gamma}, \boldsymbol{M}),$$

for some vector-valued function $\mathbf{g}$, and also of the form

$$\boldsymbol{\gamma} \to \boldsymbol{\gamma}' = \boldsymbol{\gamma} + \mathbf{Q}(\boldsymbol{\gamma})\boldsymbol{M}, \quad \boldsymbol{M} \to \boldsymbol{M}' = \boldsymbol{M},$$

for some vector-valued function $\boldsymbol{Q}$, are discussed in [312].

5 A transformation for explicit integration of the Steklov – Lyapunov case

Let us consider another map from the whole algebra $e(3)$ onto the "zero leaf" of the same algebra that preserves the Poisson brackets:

$$\boldsymbol{q} = a\frac{(\boldsymbol{M} - \mathbf{B}\boldsymbol{\gamma}) \times \boldsymbol{\gamma}}{|(\boldsymbol{M} - \mathbf{B}\boldsymbol{\gamma}) \times \boldsymbol{\gamma}|}, \quad \boldsymbol{J} = \boldsymbol{M} - a^{-2}\mathbf{C}[\boldsymbol{q}, \boldsymbol{q} \times \boldsymbol{\gamma}]_+, \tag{5.103}$$

where $\mathbf{B} = \operatorname{diag}(b_1, b_2, b_3)$, $\mathbf{C} = \operatorname{diag}(b_2 - b_3, b_3 - b_1, b_1 - b_2)$, and $[\cdot, \cdot]_+$ is the anti-commutator given by the relation

$$[\boldsymbol{u}, \boldsymbol{v}]_{+i} = \sum_{j,k} |\varepsilon_{ijk}| u_j u_k.$$

By direct calculation (for $x = 0$ in (5.81)) we can show that

$$\begin{gathered} \{J_i, J_j\} = \varepsilon_{ijk} J_k, \quad \{J_i, q_j\} = \varepsilon_{ijk} q_k, \quad \{q_i, q_j\} = 0, \\ \boldsymbol{q}^2 = a^2, \quad (\boldsymbol{J}, \boldsymbol{q}) = 0. \end{gathered} \tag{5.104}$$

Thus, the map (5.103) defines a symplectic map (with singularities in general) of each orbit of the algebra $e(3)$ onto the orbit defined by relations (5.104), i.e., onto the cotangent bundle to the sphere T^*S^2. In what follows, without loss of generality, we set $a = 1$.

As explained by A. V. Tsiganov in [584], the transformation (5.103) provides us with a natural way to separate the variables in the Steklov – Lyapunov system on $e(3)$. Indeed, on T^*S^2, let us introduce the parameters μ and ν and define the Hamiltonian

$$H = \boldsymbol{J}^2 - 4(\boldsymbol{q}, \mathbf{B}\boldsymbol{q})\left(\nu + \mu(\boldsymbol{q}, \mathbf{B}\boldsymbol{q})\right) - \mu\left(\boldsymbol{q}, (\mathbf{C}^2 - \mathbf{B}^2)\boldsymbol{q}\right). \tag{5.105}$$

This Hamiltonian describes an integrable system on the sphere $q_1^2 + q_2^2 + q_3^2 = 1$, with fourth-degree potential, and is known to be separable in the sphero-conic coordinates (λ_1, λ_2)[58]:

$$q_1^2 = \frac{(b_1 - \lambda_1)(b_1 - \lambda_2)}{(b_2 - b_1)(b_3 - b_1)}, \quad q_2^2 = \frac{(b_2 - \lambda_1)(b_2 - \lambda_2)}{(b_1 - b_2)(b_3 - b_2)}, \quad q_3^2 = \frac{(b_3 - \lambda_1)(b_3 - \lambda_2)}{(b_1 - b_3)(b_2 - b_3)}. \tag{5.106}$$

After eliminating an inessential constant, the Hamiltonian (5.105) takes the form

$$H' = \frac{1}{\lambda_1 - \lambda_2}\left(g(\lambda_1)p_1^2 + \upsilon(\lambda_1) - g(\lambda_2)p_2^2 - \upsilon(\lambda_2)\right), \tag{5.107}$$

where $g(\lambda) = (\lambda - b_1)(\lambda - b_2)(\lambda - b_3)$ and $\upsilon(\lambda) = -4\left[\mu\lambda^3 - \left(\nu + \frac{3}{2}\mu \operatorname{tr}\mathbf{B}\right)\lambda^2\right]$.

Applying the transformation (5.103) and making the substitution

$$\mu = \boldsymbol{\gamma}^2, \quad \nu = (\boldsymbol{M}, \boldsymbol{\gamma}) - \boldsymbol{\gamma}^2 \operatorname{tr}\mathbf{A},$$

in (5.105), we obtain the Hamiltonian

$$H = \boldsymbol{M}^2 - 2(\boldsymbol{M}, \mathbf{A}\boldsymbol{\gamma}) + (\boldsymbol{\gamma}, \mathbf{C}^2\boldsymbol{\gamma}) - (\boldsymbol{M}, \boldsymbol{\gamma}) \operatorname{tr}\mathbf{B} - \boldsymbol{\gamma}^2 \operatorname{tr}(\mathbf{C}^2 - \mathbf{B}^2),$$

where $\mathbf{A} = \frac{1}{2}\operatorname{tr}\mathbf{BE} - \mathbf{B}$, defined on the algebra $e(3)$, which coincides (up to addition of Casimir functions) with the Hamiltonian of the Lyapunov integrable case.

Using (5.107), it is easy to obtain the equations of motion in separated form (Abel – Jacobi equations):

$$\frac{d\lambda_1}{\sqrt{R(\lambda_1)}} = \frac{4dt}{\lambda_1 - \lambda_2}, \qquad \frac{d\lambda_2}{\sqrt{R(\lambda_2)}} = \frac{4dt}{\lambda_1 - \lambda_2},$$

$$R(\lambda) = g(\lambda)\left(\boldsymbol{\gamma}^2\lambda^3 - \left((\boldsymbol{M}, \boldsymbol{\gamma}) + 2\boldsymbol{\gamma}^2 \operatorname{tr}\mathbf{A}\right)\lambda^2 + \alpha_1\lambda + \alpha_0\right),$$

where $\alpha_1 = \frac{1}{4}h$ and α_0 are separation constants and h is a fixed value of the Hamiltonian (5.107). (These separation variables differ from the Kötter variables [87] by the shift $\lambda_i \to \lambda_i + \operatorname{tr}\mathbf{A}$.) Note that in [87], we presented another method, due to Kötter [324], for integrating the Steklov – Lyapunov case. Compared to the procedure presented here, based on study of the transformation (5.103), Kötter's method is considerably more complicated but leads to the same results.

6 Periodic Solutions, Nonintegrability, and Transition to Chaos

6.1 Nonintegrability of Rigid Body Dynamics Equations. Chaotic Motions. A Survey of Results and Unsolved Problems

Until now, no other integrable cases have been known for the Euler – Poisson equations except for the classical cases of Euler, Lagrange, and Kovalevskaya, and no other partially integrable cases for zero value of the area constant have been found except for those of Goryachev and Chaplygin.

It turns out that these equations, as well as the majority of other equations of rigid body dynamics, are nonintegrable in the general case. By nonintegrability we mean the absence of an additional first integral (required by the Liouville – Arnold theorem) having some specified algebraic or analytical structure. The simplest integrals are linear or quadratic (or even polynomial) in the Euler – Poisson variables $(\boldsymbol{M}, \boldsymbol{\gamma})$. For yet more general systems the integrals may be algebraic, meromorphic (in the complex sense), or real-analytic functions.

1 Nonintegrability of Euler – Poisson equations

Single-valued and meromorphic integrals. Already in the 19th century, there were a number of studies devoted to the proof of nonintegrability of the Euler – Poisson equations. In essence, the first work was the study of S. V. Kowalewsky [329, 330], who showed that the cases of Euler, Lagrange and her own exhaust all the cases for which the general solution of the equations of motion is represented in the form of a Laurent series in the complex plane of time and has no singularities except for poles. The branching of solutions on the complex plane of time leads to the accumulation of singularities and to the impossibility of integrating these equations in the class of known special functions (for example, in θ-functions). This reasoning is the basis for the Kovalevskaya method, which is also called the Kovalevskaya – Lyapunov method or the Painlevé test.

However, the connection between the nonrigorous argument of the Kovalevskaya method with the presence or absence of additional integrals (the so-called Painlevé – Golubev conjecture) was revealed for the Euler – Poisson equations not so long ago.

Using methods based on analysis of branching of solutions on the complex plane of time, Kozlov and Ziglin proved results on the nonexistence of integrals of, respectively, single-valued [333] and meromorphic [650] type. In one of the methods for proving the absence of single-valued integrals proposed by V. V. Kozlov, only the dynamically asymmetric case is considered and the structure of the “secular set” determined by the series expansion of a perturbing function (whose role is played by the potential

DOI 10.1515/9783110544442-006

energy) in the action-angle variables of the integrable Euler – Poinsot problem (see below) is studied. In this case, the absence of single-valued integrals is related to the branching of the action variables under the perturbation.

The idea of the second proof goes back to Lyapunov and uses analysis of the monodromy of the variational equations for some particular solutions. Using his method, A. M. Lyapunov [394] showed that the solution is branched on the complex plane of time in all cases except for the Euler, Lagrange, and Kovalevskaya cases. Lyapunov's proof is essentially the method of the small parameter. S. L. Ziglin used Lyapunov's arguments in order to prove the absence of meromorphic first integrals. In [650], it is shown that an additional meromorphic general integral of the Euler – Poisson equation can exist only in the Euler, Lagrange, and Kovalevskaya cases, whereas an additional partial integral (on the level set $(\boldsymbol{M}, \boldsymbol{\gamma}) = 0$) can also exist in the Goryachev – Chaplygin case.

In the recent article [652] of S. L. Ziglin, some assumptions are removed concerning the complex character of the additional integral used in [650]. In another work of S. L. Ziglin ([651]), he shows nonintegrability by quadratures of variational equations for a particular meromorphic solution of the Euler – Poisson equations and the Euler, Lagrange, Kovalevskaya, and Goryachev – Chaplygin cases.

A recent proof of the absence of meromorphic integrals [398] for the Euler – Poisson system was obtained by using techniques based on differential Galois theory.

Algebraic integrals. The Husson theorem [263] asserts that there is no additional algebraic integral in the variables $\boldsymbol{M}, \boldsymbol{\gamma}$ except for the Euler, Lagrange, and Kovalevskaya cases. Arguments of Bruns are the foundation of the Husson method; Bruns proved the absence of additional algebraic integrals in the three-body problem – apart from known ones related to obvious symmetries.

Let us briefly consider the formulation of the Husson theorem. The subject of algebraic nonintegrablity is discussed in the survey paper [570] of P. Ya. Kochina. Note that the proof itself starts with an elementary lemma, which says that each algebraic integral of the Euler – Poisson equations is equivalent to some integral of rational form. This is followed by three nontrivial assertions, two of which were proved by Husson, and one by R. Liouville. The first Husson assertion is that for the existence of an additional rational integral, except for the Euler – Poinsot case, it is necessary that the body have a dynamical symmetry. R.Liouville showed that when the body possesses a dynamical symmetry, the existence of each rational integral is equivalent to the existence of some polynomial integral. The second of Husson's assertions shows that a general polynomial integral can exist only in the three known cases of Euler, Lagrange, and Kovalevskaya. Sometimes, the second of Husson's assertions is linked to the name of Burgatti [112], who proposed a shorter proof immediately after the appearance of Husson's paper, but Burgatti's proof turned out to be wrong! Improved and simplified proofs of the Husson and Liouville assertions were given by A. I. Dokshevich (see [162]).

In [614, Sec. 129], A. Wintner wrote that the elegant negative results of Bruns about the algebraic nonintegrability of the three-body problem "have no significance for the dynamics." He related this assertion to the fact that the algebraic form of the integral is not invariant under arbitrary (analytic) changes of variables. This criticism also applies to the Husson theorem, and the only justification of such results is the fact that in almost all known integrable dynamics problems, the additional integral is polynomial in momenta and algebraic (for an appropriate algebraization of the equations of motion) in coordinates.

Real-analytic integrals. Stronger results based on the small parameter method of Poincaré [474, 333] were obtained in several works, all of which are discussed in detail in the book [335] of V. V. Kozlov. They are related to the proof of the absence of additional analytic integrals, analysis of long-period resonances (the birth of a large number of isolated periodic solutions), and to the separatrix-splitting effect. We next have a look at these two methods, which go back to Poincaré. We first present the general analytical and geometrical ideas of such kinds of proofs, which turn out to be related to dynamical effects which lead to complicated (chaotic) behavior of trajectories. Then we point out the works in which these general methods are applied to concrete problems in the study of rigid body dynamics.

I. The Poincaré method and perturbation theory. In the Poincaré method, which we present for the case of two degrees of freedom, one considers a system to be close to integrable if it can be written in the form of a canonical Hamiltonian system with the Hamiltonian function

$$H(I, \varphi, \mu) = H_0(I_1, I_2) + \mu H_1(I_1, I_2, \varphi_1, \varphi_2) + \cdots, \tag{6.1}$$

where $I = (I_1, I_2) \in D$, $\varphi = (\varphi_1, \varphi_2) \mod 2\pi \in \mathbb{T}^2$ (D is a domain in $\mathbb{R}^2$) are the action-angle variables for the unperturbed problem (for example, for the Euler – Poinsot case), and μ is a small parameter (for the Euler – Poisson equations it is usually the body mass). The function H_1 is the first-order perturbation. When the Euler – Poisson equations are viewed as a small perturbation of the Euler – Poinsot system, H_1 is equal to $(\boldsymbol{r}, \boldsymbol{\gamma})$, where $\boldsymbol{r}$ is the radius vector of the center of mass in the coordinate system attached to the body, μ is a physical constant measuring the strength of the gravitational field, and γ is a unit vector in the direction of the field. The study of the Hamiltonian system (6.1) was called by Poincaré the *main problem of dynamics*. In this case, it is assumed that the function $H(I, \varphi, \mu)$ is analytic not only in the phase variables I and φ but also in the small parameter $\mu \in (-\mu_0, \mu_0)$, $\mu_0 \ll 1$.

The additional integral is also assumed to be analytic in (I, φ, μ) and to be representable as a formal series

$$F(I, \varphi, \mu) = \sum_{k=0}^{\infty} \mu^k F_k(I, \varphi). \tag{6.2}$$

The unperturbed system with the Hamiltonian H_0 is immediately integrated to give

$$I = I(0), \quad \varphi = \omega(I(0))t + \varphi(0), \quad \omega = (\omega_1, \omega_2), \quad \omega_i = \frac{\partial H_0}{\partial I_i}, \quad i = 1, 2. \tag{6.3}$$

The perturbing function H_1 can be expanded in a convergent Fourier series:

$$H_1(I, \varphi) = \sum_{k\in\mathbb{Z}^2} H_1^k(I)e^{i(k,\varphi)}, \quad k = (k_1, k_2) \in \mathbb{Z}^2. \tag{6.4}$$

Definition. The (Poincaré) *secular set* $\mathscr{B}$ is the set of variables $I \in D$ such that there exists a nonzero integer-valued vector $k = (k_1, k_2) \in \mathbb{Z}^2$ for which the following conditions are satisfied:

1. $k_1\omega_1(I) + k_2\omega_2(I) = 0$;
2. $H_1^k(I) \neq 0$.

The main Poincaré assertion on the nonintegrability of the perturbed system consists in the following.

Theorem 1 ([474, 333]). *Let the following conditions hold:*
1. *H is an analytic function;*
2. *the unperturbed system is nondegenerate, i.e.,* $\det\left\|\frac{\partial^2 H_0}{\partial} I_i \partial I_j\right\| \neq 0$ *almost everywhere in D;*
3. *the secular set $\mathscr{B}$ has the key property, i.e., a real-analytic function vanishes on the whole of D if it vanishes on $\mathscr{B}$.*

Then for small $\mu \neq 0$ the perturbed system (6.1) *has no additional analytic integral* (6.2) *independent of H.*

Note that the secular set contains nondegenerate resonance tori of the unperturbed system (6.3) on which $k_1\omega_1 + k_2\omega_2 = 0$ and $\det\left\|\frac{\partial^2 H_0}{\partial I_i \partial I_j}\right\| \neq 0$. These tori are entirely filled with periodic trajectories, and under a perturbation they blow up, giving birth to finitely many nondegenerate (isolated) periodic solutions. The existence of a large number of such nondegenerate resonances constitutes an obstruction to foliation of the phase space by invariant tori and to Liouville integrability.

The simplest and most natural way to introduce a small parameter into the Euler – Poisson system and so to subject it to analysis by the secular set technique is to view it as a perturbation of the Euler – Poinsot system. In other words, the perturbation is the potential energy of the system. The Euler – Poisson Hamiltonian function is written in the form $\mathscr{H} = \mathscr{H}_0 + \mu\mathscr{H}_1$, where $\mathscr{H}_0$ is the Euler – Poinsot Hamiltonian.

For the Euler – Poisson equations, the expansion of the perturbation term $\mathscr{H}_1$ (potential energy of the gravity field) into a double Fourier series in angular harmonics φ_1, φ_2 of the action-angle variables of the unperturbed Euler – Poinsot problem $\mathscr{H}_0$

has the form

$$\mathscr{H}_1 = \sum_{-\infty}^{\infty} H_{m,1} e^{i(m\varphi_1+\varphi_2)} + \sum_{-\infty}^{\infty} H_{m,-1} e^{i(m\varphi_1-\varphi_2)} + \sum_{-\infty}^{\infty} H_{m,0} e^{im\varphi_1},$$

where the coefficients $H_{m,i}$ of the expansion depend on I_1, I_2 and also on the area constant $I_3 = (\boldsymbol{M}, \boldsymbol{\gamma})$. Their expressions in terms of complete elliptic integrals can be found in [515, 16, 333].

The secular set $\mathscr{B}$ of the system $\mathscr{H}_0 + \mu\mathscr{H}_1$ is the union $\bigcup_{m\in\mathbb{Z}} \mathscr{B}_m$ of the sets $\mathscr{B}_m$ consisting of points $I = (I_1, I_2)$ satisfying the following conditions:

1°. $m\omega_1(I) \pm \omega_2(I) = 0$, $\omega_i = \dfrac{\partial \mathscr{H}_0}{\partial I_i}$, $i = 1, 2$;

2°. $H_{m,\pm 1}(I) \neq 0$.

It turns out [333] that the set $\mathscr{B}$ consists of infinitely many lines $\mathscr{B}_m$ passing through the origin and accumulating near one of the two straight lines $2\mathscr{H}_0(I_1, I_2) = I_2^2/B$ (B is the mean moment of inertia). It is easy to show that in the usual phase space (for example, in the Andoyer variables), the secular set consists of resonant tori that accumulate near the singular (separatrix) manifold. By the condition 2°, under a perturbation, these tori disappear and are replaced by trajectories of nondegenerate periodic solutions. By the Poincaré theorem [474, 333], the birth of a large number of nondegenerate solutions means that integrability of the problem is excluded.

In *New Methods of Celestial Mechanics* [474], H. Poincaré made the observation that the Euler – Poisson equations are nonintegrable, but he did not give a rigorous proof. The difficulties of his proof stem from the fact that on the separatrix (composed of doubly asymptotic solutions for unstable permanent rotations about the middle axis of the ellipsoid of inertia for the unperturbed Euler – Poinsot problem) the unperturbed Hamiltonian fails to be analytic, and a correct analysis of the structure of the secular set was carried out by V. V. Kozlov [333] under the condition of dynamical asymmetry. Thus, a theorem on nonexistence of an additional *analytic* integral of the Euler – Poisson equations that is independent of the classical energy, area, and geometric integrals was proved in the absence of dynamical symmetry: $A \neq B \neq C \neq A$, where A, B, and C are the body's principal moments of inertia. This result of V. V. Kozlov essentially strengthens the Husson theorem on the absence of a new *algebraic* integral.

Also, V. V. Kozlov showed that in the dynamically asymmetric case, the separatrices of unstable permanent rotations of the Euler – Poinsot problem split under the perturbation. In general, splitting of the separatrices leads to nonintegrability. Let us have a look in more detail at problems of this kind.

II. Separatrix splitting, homoclinic structure, and nonintegrability. Once again, let us consider the Hamiltonian system with two degrees of freedom and the Hamiltonian

$$H(p, q, \mu) = H_0(p, q) + \mu H_1(p, q) + \cdots, \quad (p, q) \in \mathbb{R}^4, \quad \mu \in (-\mu_0, \mu_0), \tag{6.5}$$

where the unperturbed Hamiltonian H_0 is assumed to define an integrable system, H_1 is a perturbing function, and μ is a small parameter. It is assumed that the function H is analytic in all its arguments.

Assume that on some fixed energy level, the unperturbed system has hyperbolic periodic trajectories. Due to the property of integrability, the asymptotic surfaces corresponding to these hyperbolic trajectories coincide and define singular invariant manifolds of the unperturbed system. They can be heteroclinic (when the asymptotic solutions approach different periodic trajectories as $t \to \pm\infty$) or homoclinic (when the asymptotic solutions tend to the same periodic trajectory as $t \to \pm\infty$). As was noted by Poincaré [474], under a perturbation, the arriving and departing separatrices for the corresponding periodic solutions are no longer coincident. This phenomenon, which is closely related to nonintegrability of Hamiltonian systems, is called *separatrix splitting*. Poincaré also obtained a necessary condition for separatrix splitting in the perturbed problem, which reduces to the condition that the integral along the separatrix of a certain function be different from zero. Here, we present one of the most often used forms, which is usually called the *Poincaré – Mel'nikov integral*.

Assume that the Hamiltonian system has two periodic solutions (which may coincide):

$$z_{pi}(t,\mu) = z_{pi}^{(0)}(t) + \mu z_{pi}^{(1)}(t) + \cdots, \quad z = (p_1, p_2, q_1, q_2), \quad i = 1, 2, \tag{6.6}$$

and a one-parameter family

$$z_{ai}(t,\alpha,\mu) = z_{ai}^{(0)}(t,\alpha) + \mu z_{ai}^{(1)}(t,\alpha) + \cdots, \quad i = 1, 2, \tag{6.7}$$

of asymptotic solutions. $\alpha \in \mathbb{R}$ parameterizes the asymptotic trajectories on the separatrix surface. The asymptotic solutions z_{a1} and z_{a2} satisfy

$$z_{a1} \xrightarrow{t\to-\infty} z_{p1}, \quad z_{a2} \xrightarrow{t\to+\infty} z_{p2},$$

and, moreover, $z_{ai}^{(0)} = z_a^{(0)}$ for $i = 1, 2$ (i.e., $z_a^{(0)}$ is a doubly asymptotic solution of the unperturbed system).

Assume that the unperturbed system has a first integral G_0 (i.e., $\{G_0, H_0\} = 0$) satisfying

$$\operatorname{grad} G_0\Big|_{z_{pi}^{(0)}} = 0, \quad i = 1, 2. \tag{6.8}$$

The integral

$$J(\alpha) = \int_{-\infty}^{\infty} \{G_0, H_1\}(z_a^{(0)}(t,\alpha))dt,$$

which is convergent because of (6.8), is called the *Poincaré – Mel'nikov integral*. The following assertions hold for $J(\alpha)$[653, 335].

(I) *If $J(\alpha) \not\equiv 0$, then for small $\mu \neq 0$, the perturbed separatrices defined by the asymptotic solutions* (6.7) *are split.*

(II) *If $J(\alpha)$ has a simple root α^*, i.e., $J(\alpha^*) = 0$ and $\frac{dJ}{d\alpha}(\alpha^*) \neq 0$, then the perturbed separatrices intersect transversally, and there exists a nondegenerate doubly asymptotic solution of the perturbed equations.*

Under certain additional conditions, splitting of the separatrix for a Hamiltonian system (with n degrees of freedom in general) excludes the existence of a complete set of analytic integrals. This subject is discussed in detail in the book [335]. For two degrees of freedom, the proofs simplify, and we can simply use the theorem on the nonintegrability in a neighborhood of a hyperbolic equilibrium state of a two-dimensional diffeomorphism [143].

Therefore, for two degrees of freedom the following assertion holds.

(III) *If $J(\alpha^*) = 0$ and $\frac{dJ}{d\alpha}(\alpha^*) \neq 0$, then for small values of μ, the Hamiltonian system* (6.5) *has no additional analytic first integral.*

In [344], an analogous theorem on the absence of a nontrivial analytic symmetry field was proved under the same conditions. Also, we note that the nonintegrability of the equations of motion, which is established by verifying the above conditions for $J(\alpha)$, is due to the fact that near the nondegenerate homoclinic (heteroclinic) trajectory there exist quasi-random motions on a set of zero measure and also a large (but finite) number of long-period, periodic solutions of the perturbed system [335].

In many important problems, the improper integral $J(\alpha)$ can be calculated by using residues.

In [653], S. L. Ziglin proved the absence of an additional integral for the Euler – Poisson equations, viewed as a perturbation of the dynamically asymmetric Euler – Poinsot case, by using the separatrix splitting property and, in addition, removed the condition of analytic dependence on the small parameter μ, which had been necessary in [333]. The analysis of the separatrix splitting made in [333, 653] leads to the Hess conditions under which one pair of separatrices splits, whilst the other pair remains doubled and defines the invariant manifold. Different variants of intersections of separatrices and of corresponding dynamical effects were investigated in detail by S. A. Dovbysh [167, 164] in relation to different values of the parameters of the system. The results of [167, 164] were obtained via a detailed analysis of the Poincaré – Mel'nikov integrals, which were trigonometrical polynomials in this case. In [165], the separatrix splitting was studied under the Hess conditions. For finite perturbations transversal splitting of the separatrix was established numerically in [88].

III. Separatrices of unstable equilibria. Perturbation of the Lagrange case. The case of dynamical symmetry needs a different approach, since in this case there are no separatrices to an unstable periodic solution for the Euler – Poinsot system, whose splitting would lead to nonintegrability of the perturbed system. We must look for other ways of introducing the small parameter into the general Euler – Poisson system.

One such way yields homoclinic trajectories leading to the restricted problem of rigid body rotation whose integrability is discussed in Sec. 6.5 of Ch. 6. This, however, proves nonintegrability of the axisymmetric case only in a certain limit variant. Another introduction of the small parameter μ, for which $\mu = 0$ is the Lagrange case, was studied by the Poincaré method and via the separatrix splitting.

The paper [168] studies the separatrices not of a hyperbolic periodic solution but of an equilibrium point (the vertical uniform rotation of a top) with corresponding eigenvalues $\pm(\alpha \pm i\beta)$, $\alpha\beta \neq 0$ $(\alpha, \beta \in \mathbb{R})$. This saddle-focus type of equilibrium arises if the rotation of a "sleeping" top is unstable and if the area constant is different from zero. In this case, it is possible to apply the Devaney theorem, which gives conditions for the existence of a set of quasi-random solutions; this leads to the absence of analytic integrals and of symmetry fields. Indeed, we can show that for small perturbations of the Lagrange problem, the equilibrium corresponding to the top's vertical rotation does not disappear and remains of saddle-focus type, but there arise transversal homoclinic trajectories γ. A Poincaré mapping can be defined on a suitably small section Π transversal to γ, and it is seen to be conjugate to the Bernoulli shifts of N symbols on a compact invariant hyperbolic subset of Π.

The proof of the Devaney theorem is based on the fact that the Poincaré mapping, constructed with respect to some small section Π, is a mapping of Smale horseshoe type. A proof of nonintegrability for a family of perturbations of the Lagrange problem for $(\boldsymbol{M}, \boldsymbol{\gamma}) = c \neq 0$ was obtained by S. A. Dovbysh who considered perturbations due to deviation of the center of mass from the dynamical symmetry axis [168] (The degenerate case of a spherical inertia tensor is excluded from this analysis). Previously, on the basis of an analog of the Mel'nikov method, Holmes and Marsden [262] analyzed a particular perturbation of the Lagrange problem for which one of the two equal moments of inertia changes. For $(\boldsymbol{M}, \boldsymbol{\gamma}) = c \neq 0$, transversality of the intersection of separatrices gives rise to a Smale horseshoe on the critical level set of the energy integral and hence to nonintegrability.

For zero value of the area constant, $c = 0$, the characteristic exponents are real, and the Devaney theorem is not applicable. In the work [168], S. A. Dovbysh uses the Turaev – Shil'nikov theorem for proving nonintegrability of the perturbed Lagrange problem.

In [63], S. V. Bolotin obtained sufficient conditions (in terms of the Birkhoff normal form) for nonintegrability of a Hamiltonian system in the presence of real characteristic exponents and proved the nonintegrability of the perturbed Lagrange top for zero area constant.

We also mention the works of T. V. Salnikova [518, 517]. In [518], the nonintegrability of the perturbed Lagrange top was proved by using the Poincaré method in a weaker sense (the additional integral is assumed to be analytic in the small parameter). In the paper [517] the branching of solutions and the nonexistence of single-valued integrals in the perturbed Lagrange top was studied.

Under the simultaneous conditions of a) dynamical symmetry, b) the center of masses being situated in the equatorial plane of the ellipsoid of inertia, and c) zero value of the area constant, the nonintegrability of the Euler – Poisson equations was studied by S. V. Bolotin using variational methods (Morse theory) [64]. These conditions are satisfied by most of the known integrable cases of the Euler – Poisson equations: the spherical top, the Kovalevskaya case, and the Goryachev – Chaplygin case. The system's behavior depends on a single parameter a, namely, the ratio between the equatorial and axial moments of inertia (by the triangle inequality $\frac{1}{2} \leqslant a \leqslant \infty$). Bolotin [64] showed that for $a > 4$ there are additional transversal homoclinic trajectories, which leads to the nonintegrability and chaotic behavior.

Computer experiments carried out in the interval $\frac{1}{2} \leqslant a \leqslant 4$ (the lower boundary is determined by the triangle inequality) show that, with high reliability, integrability is possible only for $a = 1$ (spherical top), $a = 2$ (Kovalevskaya case), and $a = 4$ (Goryachev – Chaplygin case). The present authors have studied numerically the value of the area between intersecting separatrices. This value is invariant under canonical transformations (in contrast to the angle between the separatrices at their intersection point), and it has been proposed as a measure of nonintegrability (or stochasticity) by S. V. Bolotin. In Fig. 6.1, we plot the dependence on the parameter a of this area. It is equal to zero only for the known integrable cases. It is interesting to note that since the case of the spherical top is strongly degenerate, the graph at the point $a = 1$ almost coincides with the horizontal axis. In this case, the calculation errors become commensurable with the area value, and the computer computations become unreliable.

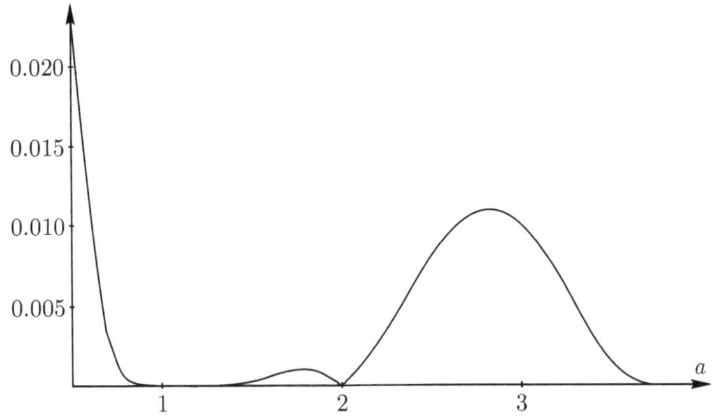

Fig. 6.1. Dependence of the separatrix splitting area A on the parameter a. $E = 5$, $H = 0$, $r = (1, 1, 0)$, $M = 1.2$, and $I = \mathrm{diag}(a, a, 1)$.

Despite a large number of studies on nonintegrability of the Euler – Poisson equations, the problem remains unsolved in the general form. Up to now sufficient conditions for the existence of an additional real-analytic integral have not been found; although, probably, they are the Euler, Lagrange and Kovalevskaya conditions (and the Goryachev – Chaplygin conditions for zero value of the area constant). At present, it is clear that in many respects the studies devoted to this problem are of interest not so much for rigid body dynamics itself, but rather contribute to the development of general analytical methods for proving nonintegrability.

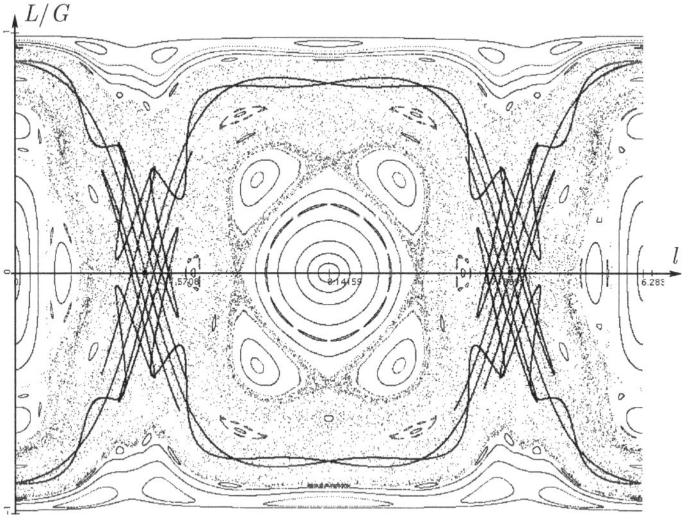

Fig. 6.2. $E = 1, \mathbf{I} = \mathrm{diag}(1, 6/5, 3/2), \mu\boldsymbol{r} = (0,\ 0.5,\ 0), c = 0.$

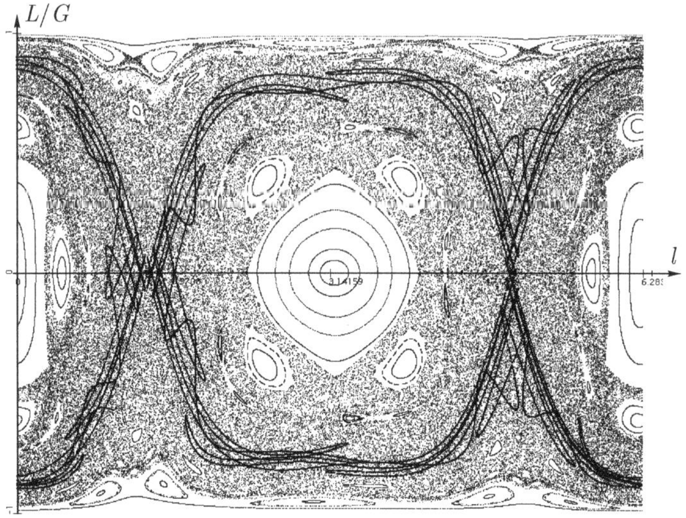

Fig. 6.3. $E = 1, \mathbf{I} = \mathrm{diag}(1, 6/5, 3/2), \mu\boldsymbol{r} = (0,\ 0.5,\ 0), c = 0.04.$

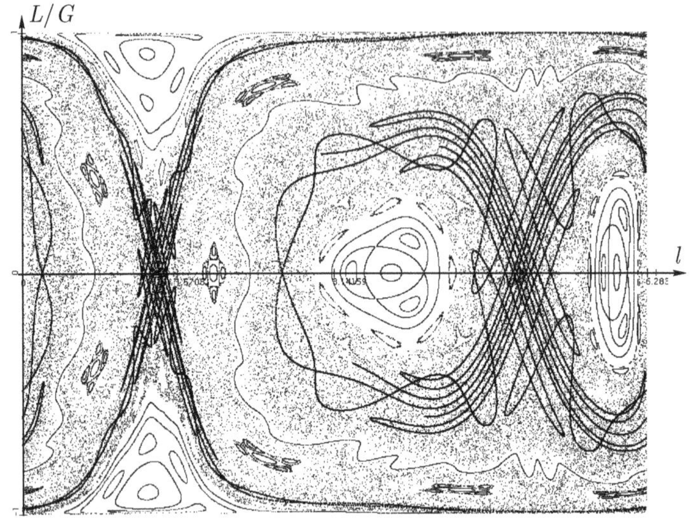

Fig. 6.4. $E = 1, \mathbf{I} = \mathrm{diag}(1, 6/5, 3/2), \mu\boldsymbol{r} = (0,\ 0.5,\ 0), c = 0.5.$

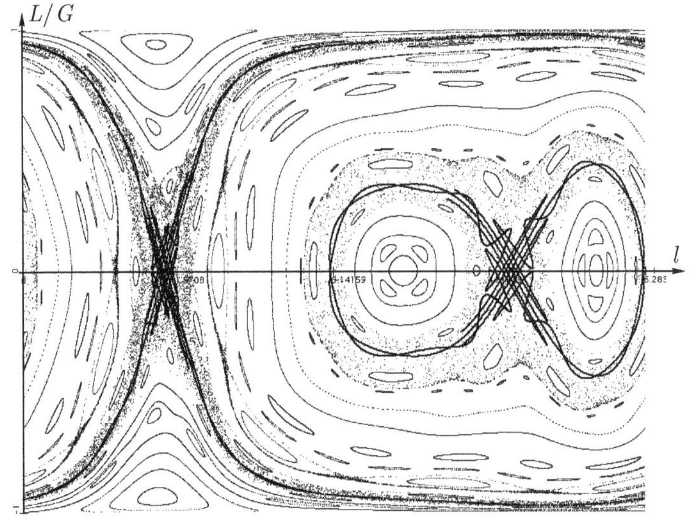

Fig. 6.5. $E = 1, \mathbf{I} = \mathrm{diag}(1, 6/5, 3/2), \mu\boldsymbol{r} = (0,\ 0.5,\ 0), c = 1.$

Numerical studies. The first numerical studies of the Poincaré mapping and the separatrix splitting are contained in [205, 88]. These works study the dependence of the separatrix splitting on the body mass μ. By rescaling $\boldsymbol{M}$ and t,

$$\boldsymbol{M} = \sqrt{\mu}\boldsymbol{M}', \quad dt = \frac{1}{\sqrt{\mu}}d\tau,$$

the parameter μ can be eliminated from the equations of motion, but this results in rescalings of the values of the integrals of the system. Thus, we may equivalently decide to regard the body mass as fixed and look at various values of the energy and area integral. Figs. 6.2 – 6.5 show possible variants of the transversal separatrix splittings and also the stochastic layer surrounding them for various values of the area integral c and a fixed energy. Fig. 6.4 shows the invariant torus separating the separatrices of two hyperbolic points. Its existence for small values of μ was predicted in [164] (S. A. Dovbysh) by the use of techniques based on the Moser theorem on invariant curves. Since the separatrices intersect each other transversally in this case, the problem is nonintegrable. Thus, we have a “computer” proof of the nonintegrability for finite values of μ.

2 Nonintegrability of Kirchhoff equations

The second class of important systems, after the Euler – Poisson equations, to be studied from the viewpoint of their integrability and nonintegrability, are the Kirchhoff equations describing the behavior of a rigid body in an ideal fluid.

Since the Kirchhoff equations can be written in the form (see Ch. 3)

$$\begin{gathered}\dot{\boldsymbol{M}} = \boldsymbol{M} \times \frac{\partial \mathscr{H}}{\partial \boldsymbol{M}} + \boldsymbol{p} \times \frac{\partial \mathscr{H}}{\partial \boldsymbol{p}}, \quad \dot{\boldsymbol{p}} = \boldsymbol{p} \times \frac{\partial \mathscr{H}}{\partial \boldsymbol{M}}, \\ \mathscr{H} = \frac{1}{2}(\mathbf{A}\boldsymbol{M}, \boldsymbol{M}) + (\mathbf{B}\boldsymbol{M}, \boldsymbol{p}) + \frac{1}{2}(\mathbf{C}\boldsymbol{p}, \boldsymbol{p});\end{gathered} \tag{6.9}$$

they reduce to a Hamiltonian system with two degrees of freedom on the common level set of the integrals $(\boldsymbol{M}, \boldsymbol{p}) = c_1$ and $(\boldsymbol{p}, \boldsymbol{p}) = c_2$.

Introducing a small parameter into Eqs. (6.9) by rescaling $\boldsymbol{p} \to \mu\boldsymbol{p}$, we write the Hamiltonian $\mathscr{H}$ in the form

$$\mathscr{H} = \mathscr{H}_0 + \mu\mathscr{H}_1 + \mu^2\mathscr{H}_2,$$

where $\mathscr{H}_0 = \frac{1}{2}(\mathbf{A}\boldsymbol{M}, \boldsymbol{M})$ is the Hamiltonian of the Euler – Poinsot case and $\mathscr{H}_1 = (\mathbf{B}\boldsymbol{M}, \boldsymbol{p})$ and $\mathscr{H}_2 = \frac{1}{2}(\mathbf{C}\boldsymbol{p}, \boldsymbol{p})$ are perturbation terms. The Fourier series expansions of $\mathscr{H}_1$ and $\mathscr{H}_2$ in the action-angle variables I, φ of the Euler – Poinsot problem are obtained in [26, 91] and have the form

$$\begin{aligned}\mathscr{H}_1 &= \sum_{-\infty}^{\infty} H_{m,0}e^{im\varphi_1} + \sum_{-\infty}^{\infty} H_{m,\pm1}e^{i(m\varphi_1 \pm \varphi_2)}, \\ \mathscr{H}_2 &= \sum_{-\infty}^{\infty} H^{\star}_{m,0}e^{im\varphi_1} + \sum_{-\infty}^{\infty} H^{\star}_{m,\pm1}e^{i(m\varphi_1 \pm \varphi_2)} + \sum_{-\infty}^{\infty} H^{\star}_{m,\pm2}e^{i(m\varphi_1 \pm \varphi_2)}.\end{aligned}$$

The dependence of the coefficients $H_{m,0}$, $H_{m,\pm1}$, $H^{\star}_{m,0}$, $H^{\star}_{m,\pm1}$, $H^{\star}_{m,\pm2}$ on I_1 and I_2 expressed in terms of complete elliptic integrals is also presented in [26, 91], where the resulting expansions for $\mathscr{H}_1$ and $\mathscr{H}_2$ are applied in the proof of the "complicated" structure of the secular set and the absence of any of analytic, single-valued, or algebraic integrals in the situation of dynamical asymmetry $a_1 \neq a_2 \neq a_3 \neq a_1$, where a_i are the eigenvalues of the matrix **A**.

REMARK. Fourier series expansions in terms of action-angle variables for the Euler – Poinsot case of the direction cosines $\boldsymbol{\alpha}$, $\boldsymbol{\beta}$, and $\boldsymbol{\gamma}$ and quadratic functions of them was considered by several authors (M. Arribas, Y. V. Barkin, A. V. Borisov, H. Kinoshita, Yu. A. Sadov, and A. Elipe) [26, 91, 24, 300, 515]. From the modern viewpoint, one needs a Fourier series form which allows one to perform the computer calculations in the simplest way by using systems of analytical (symbolic) calculations. These results are particularly relevant in celestial mechanics in studying various effects related to the Newton interaction between rigid bodies.

First let us consider the conditions for the existence of real-analytic integrals and then meromorphic and algebraic integrals.

Real-analytic integrals. Using the Poincaré method involving the analysis of the secular set and a large number of nondegenerate long-periodic orbits and the separatrix splitting in the case of dynamical asymmetry $a_1 \neq a_2 \neq a_3 \neq a_1$, one obtains the following necessary integrability conditions for the Kirchhoff equations [350, 26, 91].

Theorem 2. *Under the condition $a_1 \neq a_2 \neq a_3 \neq a_1$ and for μ sufficiently small, there is no additional first integral of Eqs. (6.9), independent of the energy, area, and geometric integrals, unless one of the following conditions holds for the coefficients:*

$$\begin{aligned} &\textit{either} \quad a) \quad \mathbf{B} \neq 0: \ \textit{then } \mathbf{B} \textit{ has to be diagonal and} \\ &\qquad\qquad\quad \frac{b_1 - b_3}{a_2} + \frac{b_2 - b_1}{a_3} + \frac{b_3 - b_2}{a_1} = 0; \\ &\textit{or} \quad b) \quad \mathbf{B} \equiv 0: \ \textit{then } \mathbf{C} \textit{ has to be diagonal and} \\ &\qquad\qquad\quad \frac{c_1 - c_3}{a_2} + \frac{c_2 - c_1}{a_3} + \frac{c_3 - c_2}{a_1} = 0. \end{aligned}$$

REMARK. Applying the Poincaré method to prove this theorem requires the additional integral to have analytic dependence on the small parameter μ. For the separatrix splitting method this condition is not required, although μ is expected to be small.

Condition b) is sufficient as well as necessary and gives rise to the integrable Clebsch case (Sec. 3.2 of Ch. 3). As numerical experiments show, condition a) is not sufficient for integrability. If condition a) holds, then the separatrix splitting has in general a higher order of smallness and cannot be analyzed by using the Poincaré – Mel'nikov integral.

In the proof of the theorem based on the study of the splitting of separatrices to unstable permanent rotations for perturbations of the Euler – Poinsot case, we can obtain the conditions shown in Table 3.1 leading to the Hess type integral. This integral defines an invariant manifold composed of a pair of separatrices that do not split under the perturbation but remain doubled. Numerical studies of chaotic motions and of the separatrix construction can be found in [91, 88, 15].

The study of the integrability of Eqs. (6.9) in the dynamically symmetric case, $a_1 = a_2$, is more complicated. In [519] (see also [334]), it is shown that for $\mathbf{B} = 0$, the additional integral can exist only under the conditions $c_{ij} = 0, \forall i \neq j$ and $c_1 = c_2$. Thus, the small parameter μ may be introduced in the Kirchhoff system in such a way that $\mu = 0$ results in the dynamically symmetric case by setting $c_1 = c_2 + \mu$, $c_{ij} = \mu c'_{ij}$ $(i \neq j)$. In the case $\mu = 0$, there exists the additional integral $M_3 = c$.

Branching of solutions. Single-valued and algebraic integrals. The conditions for meromorphicity, for $a_1 \neq a_2 \neq a_3 \neq a_1$, were studied by means of the *Lyapunov method* in [503]. It turns out that the general solution to Eqs. (6.9) is meromorphic only under the Clebsch or Steklov conditions under which there exists an additional integral, and the integration can be performed in θ-functions. In the case of dynamical symmetry, the meromorphicity of general solutions was not studied systematically.

The study of conditions for the existence of an additional *single-valued* integral is a much more complicated problem in this case (it was carried out in [26, 91] and requires a detailed analysis of the structure of the secular set). It turns out that in the situation where $a_1 \neq a_2 \neq a_3 \neq a_1$, the branching of solutions indeed leads to the absence of single-valued integrals. The following assertion holds for Eqs. (6.9).

Theorem 3. [26, 91] *For $a_1 \neq a_2 \neq a_3 \neq a_1$, necessary (and also sufficient) conditions for the existence of an additional and single-valued integral are either the Clebsch or the Steklov conditions.*

REMARK. Because of the homogeneity of the right-hand sides in (6.9), the Bruns lemma holds for this system, according to which the existence of an algebraic first integral implies the existence of a rational integral. But a rational integral must be single-valued, and, therefore, its absence follows from Theorem 3. In other words, the Clebsch and Steklov conditions exhaust all cases allowing the existence of an additional algebraic integral in the case of dynamical asymmetry. This is an analog of the Husson result obtained for the Euler – Poisson equations.

For $a_1 = a_2$ and for diagonal matrices $\mathbf{A}$, $\mathbf{B}$, and $\mathbf{C}$, the algebraic integrability of Eqs. (6.9) was considered in [513] by using the Husson method (note that because of the homogeneity of the equations, instead of algebraic integrals, we can also consider integrals meromorphic in a neighborhood of the origin). The following assertion was proved.

Theorem 4. [513, 512] *Suppose that $a_1 = a_2$ and $b_1 = b_2$. Then no additional general or partial integral of the system* (6.9) *exists except for the Clebsch case* ($b_1 = b_2 = b_3$,

$\frac{c_2 - c_3}{a_1} + \frac{c_3 - c_1}{a_2} + \frac{c_1 - c_2}{a_3} = 0$), *the Kirchhoff case* ($a_1 = a_2, b_1 = b_2, c_1 = c_2$), *and the Chaplygin case* ($b_i = 0, a_1 = \frac{1}{2}a_3, c_1 + c_2 - c_3 = 0$), *all of which are integrable.*

Note that in this theorem, the case of off-diagonal entries in the matrices **B** and **C** is not considered. Precisely in the nondiagonal case, the additional first integral was found unexpectedly by V. V. Sokolov [537].

3 Related problems of rigid body dynamics

Amongst other works dealing with the subject of the nonintegrability of equations of rigid body dynamics, we mention [114], which treated the problem of the triaxial ellipsoid sliding along a plane (see also [520]) and [268], in which the problem of the inertial rotation of a rigid body with an asymmetric rotor about a fixed point was considered. The latter work refines the results of [262], where some terms had been neglected in the Hamltonian. The paper [74] treats the transition to chaos in the Liouville equations with an adiabatic perturbation, i.e., it depends periodically on a slow time $\tau = \mu t$. In this work the conditions for adiabatic chaos and also the uncertainty curves describing the domains of chaotic motions are obtained. Appendix H contains an analysis of separatrix splitting and the nonintegrability for the Chaplygin equations describing the plane motion of a heavy rigid body in a fluid in the presence of circulation.

It is interesting that for the Poincaré – Zhukovskii equations (i.e., the Kirchhoff equations on $so(4)$), no theorems on nonintegrability have been obtained, other than the theorem on the absence of first integrals up to degree four in momentum.

6.2 Periodic and Asymptotic Solutions in Euler – Poisson Equations and Related Problems

1 Various types of partial solutions of the equations of rigid body dynamics

As mentioned previously, various types of periodic solutions may be considered for the equations of rigid body motion around a fixed point, in the case where the potential $V(\gamma)$ is axisymmetric. Most commonly, one looks at periodic solutions of the reduced system (i.e., closed cycles on the Poisson sphere, typically corresponding to quasi-periodic two-frequency solutions for the complete system). For a heavy rigid body, these are periodic solutions of the Euler – Poisson equations. Alternatively, but essentially more rarely, one considers "true" periodic solutions of the rigid body in absolute space under which all three apexes of the trihedron related to the body describe closed curves in absolute space. For a heavy rigid body, these are periodic solutions of the general equations for momentum and direction cosines ($\boldsymbol{M}, \boldsymbol{\alpha}, \boldsymbol{\beta}, \boldsymbol{\gamma}$) or for momentum and quaternion equations ($\boldsymbol{M}, \lambda_0, \boldsymbol{\lambda}$) (see Ch. 1). In the case of unstable periodic solutions (or equilibrium states), one studies the motions asymptotic to them. Here, we shall not discuss results concerning quasi-periodic motions, which are studied, for example, in [255, 27]. Neither shall we deal with the general theory of periodic solutions, which is contained in many classical and modern books [474, 45, 333]. However, we give an overview of the classification, which is related to various methods for finding such solutions and bibliographical hints related to applications to top dynamics.

All the periodic and asymptotic solutions of rigid body dynamics, known at present, can be divided into different classes according to the type of methods required for their study.

1. *Exact (analytical) solutions* having an explicit analytical representation. The majority of these solutions to the Euler – Poisson equations are collected in the books [224, 228, 373], and many of them are periodic or asymptotic solutions. We recall that particular solutions of the Kirchhoff equations are less studied and mention here only the classical results of S. A. Chaplygin [131] and V. A. Steklov [558].
2. *Lyapunov solutions* obtained by small parameter methods in a neighborhood of known equilibrium states (for example, the Staude rotations) applied to small variations in the initial conditions or the values of the first integrals and fixed parameters of the system. Such solutions were obtained and studied using the Lyapunov theorem concerning the birth of periodic solutions near equilibrium states. The first detailed study of periodic and asymptotic solutions to the equations of motion of a heavy rigid body which are expressed as power series in a formal small parameter was carried out by E. Mettler [423]. Assuming that the center of mass lies in the principal plane of the ellipsoid of inertia, V. MacMillan obtained periodic solutions close to the vertical rotations [400]. The Lyapunov method was used for analyzing periodic solutions, e.g., in [407, 408, 526]. The series for asymptotic solutions of the Euler – Poisson equations were constructed in [593] (see also [594, 227]). These solutions were found for an arbitrary mass distribution; they depend on two arbitrary constants and describe the motions asymptotic to an unstable uniform rotation.
3. *Poincaré solutions* which are born from resonant tori of the Euler – Poinsot problem under small perturbations. The proof of existence and the stability analysis of these periodic solutions is based on the Poincaré theorem concerning the birth of pairs of nondegenerate periodic solutions due to blow-up of resonant tori of integrable systems. For the Euler – Poisson equations, in both the dynamically asymmetric and dynamically symmetric cases, these solutions were presented by V. V. Kozlov [333]. Here, the simplest result is the theorem saying that permanent rotations about the principal axes of inertia of the Euler – Poinsot problem do not disappear following the addition of a perturbation. A more complicated result underlying the proof of nonintegrability of the perturbed system is the theorem saying that from any resonant torus of the dynamically asymmetric Euler – Poinsot problem belonging to the secular set, at least two isolated periodic solutions are

born under a perturbation. They exist for a small perturbation, depending analytically on the perturbation parameter: one of them is linearly stable, whereas the other is unstable. For a potential quadratic in $\boldsymbol{\gamma}$, for example, for the Newtonian potential (and also for the generalized quadratic potential corresponding to the Grioli problem for motion in a magnetic field), the Poincaré solutions were presented in [26, 149]. The periodic solutions that are born from resonant tori in the Lagrange and Kovalevskaya cases are also known [518].

4. *Periodic solutions of a rapidly rotating rigid body.* These solutions and results of their analysis are collected in the book [17]; their study is based on a combination of the small parameter method and analysis of quasi-linear autonomous systems. In studying the Goryachev – Chaplygin case L. N. Sretenskii [545] was one of the first scientists to make use of these kinds of solutions. In [545], it is assumed that the top has a large initial angular velocity around the principal axis of inertia passing through the center of gravity, in which case the solution is expressed in terms of almost periodic functions of time. The quasi-periodic solutions of a rapidly rotating body are also studied in [54].
5. *Periodic solutions obtained by using topological methods* based on the calculus of variations in the large, on topology, and on geodesic theory [333, 349, 292, 447, 448, 563]. For the zero value of the area constant, these solutions are conveniently presented by using the well-known geometric fact that on the two-dimensional sphere there are at least three closed non-self-intersecting geodesics. The Euler – Poisson equations (and more generally, the equations of rigid body motion in any axisymmetric potential force field) have six distinct periodic solutions, which are perturbations of uniform rotations about the principal axes of the ellipsoid of inertia (see V. V. Kozlov [349, 333] and M. P. Kharlamov [292]). For a nonzero value of the area constant there are large obstructions to constructing the corresponding solutions due to gyroscopic terms in the Routh function. These solutions may be obtained only by imposing strong additional restrictions; for a more detailed discussion of the problem, see Sec. 6.3 of Ch. 6.
6. *Pendulum-like solutions.* These solutions (which can be oscillatory, asymptotic or rotational) for the Euler – Poisson equations were studied, for example, in [409, 414, 415, 618, 248, 166] (see also references therein).

We note that the above-mentioned methods of investigation cover quite a narrow range of periodic solutions in rigid body dynamics. It is therefore difficult to detect solutions of the first type, and moreover, the resulting explicit analytical formulae expressed in terms of special functions do not always allow one to perform a useful qualitative analysis of the motion. Typically, solutions of the second and third types are restricted by the requirement that the perturbations of initial parameters or initial conditions be sufficiently small: this is due to the necessity of restricting to within the radius of convergence of series in the small parameter. Solutions of the fourth type arise only for sufficiently high angular velocities of rotation (for example, near the Euler – Poinsot problem). The fifth method is not constructive and is an existence theorem for periodic solutions. Therefore, it is not possible even to present concrete periodic solutions related to this method. The problem of finding periodic solutions far from integrable cases remains practically unstudied. These solutions should be found by the use of computer methods. Here, we mention an interesting book [34], which is mainly devoted to "plane" problems of the rigid body dynamics; in this book, various periodic orbits are described numerically, and whilst some of them have a somewhat exotic character, others may be useful in applications.

Computer methods and Genealogy. A fairly complete solution to the problem of finding periodic and asymptotic solutions can be provided by an analysis based on global study of the Poincaré mapping depending on the system parameters. This kind of analysis uses algorithms to continue periodic solutions with respect to a parameter, the theory of bifurcations of periodic solutions, and numerical constructions of invariant manifolds – the homoclinic structure and splitting of separatrices.

The effective application of computers uses two complementary tools. On the one hand, the Poincaré mapping helps in making convenient choices of initial conditions or of values of first integrals. On the other hand, computer visualization of the behavior of a solution can help in making certain conclusions about its properties. The latter can be achieved very effectively by means of computer visualization of motions of the reduced system (on the Poisson sphere), as well as in absolute space. As a rule, for rigid body dynamics, special algorithms are necessary: for computation of rotation numbers, for computation of the separatrix splitting value, and so on. All algorithms of this kind are discussed in Sections 6.3 and 6.5 of this chapter, which are, respectively, devoted to the global analysis of the restricted problem of heavy rigid body dynamics and to the search for new classes of periodic solutions (in absolute space) having interesting geometric properties.

2 Methods for studying mappings

Search for fixed points of a mapping. Here and in what follows, we consider area-preserving mappings $T : \boldsymbol{x} \to T(\boldsymbol{x})$, $\boldsymbol{x} \in \mathbb{R}^2$ of the plane to itself that arise as the period mapping for a nonautonomous system with one degree of freedom or as the Poincaré mapping for an autonomous Hamiltonian system with two degrees of freedom on the level surface of energy. All the problems considered are presented in more detail in [399]. The search for a fixed point of a mapping T amounts to numerically solving the equation

$$T(\boldsymbol{x}) = \boldsymbol{x}. \tag{6.10}$$

If the linearization (also known as monodromy matrix) $L = \frac{\partial T}{\partial x}$ of the mapping in the neighborhood of the fixed point has all eigenvalues different from 1, then the fixed point is isolated and can be found by using the Newton method. This gives rise to the following iterative formula for finding the fixed point

$$(L - E)(\boldsymbol{x}_{i+1} - \boldsymbol{x}_i) = -(T(\boldsymbol{x}_i) - \boldsymbol{x}_i). \tag{6.11}$$

If the mapping is induced by the phase flow $\dot{\boldsymbol{x}} = \boldsymbol{v}(\boldsymbol{x})$, then the monodromy matrix can be found by solving the variational equation

$$\dot{\delta \boldsymbol{x}} = \frac{\partial \boldsymbol{v}}{\partial \boldsymbol{x}}(\boldsymbol{x})\delta \boldsymbol{x}. \tag{6.12}$$

For this purpose, as initial conditions for $\delta \boldsymbol{x}$, we choose equal small displacements along the coordinate axes ($\delta \boldsymbol{x}_{1,2}^0 = (\varepsilon, 0), (0, \varepsilon)$). Let us simultaneously integrate the equations of motions and (6.12) from the point $\boldsymbol{x}_0$ up to the next intersection with the section plane. The resulting vectors $\delta \boldsymbol{x}_i$ divided by ε are then the columns of the monodromy matrix to leading order $L = \frac{1}{\varepsilon}(\delta \boldsymbol{x}_1, \delta \boldsymbol{x}_2) + O(\varepsilon)$.

The linear stability of a fixed point of the mapping is determined by *multipliers*, which are the eigenvalues $\lambda_{1,2}$ of the monodromy matrix. Since the mapping $T(\boldsymbol{x})$ is area-preserving, it follows that $\det L = \lambda_1\lambda_2 = 1$. Hence the eigenvalues are either complex conjugates of one another and lie on the unit circle or they are real and inverse to one another. In the case of complex conjugate eigenvalues, the fixed point is stable (of elliptic type). If the eigenvalues are real, the fixed point is unstable (of hyperbolic type or of hyperbolic type with inversion if $\lambda_i < 0$). It is not possible to draw a conclusion about the stability of parabolic ($\lambda_1 = \lambda_2 = -1$) and degenerate ($\lambda_1 = \lambda_2 = 1$) fixed points using linear approximation.

Extension of the fixed points of a mapping with continuous variation of a parameter. Let us consider a self-mapping $T(\boldsymbol{x}, \mu)$ of the plane dependent on a parameter. Let $\boldsymbol{x}_0$ be a fixed point of T for the value $\mu = \mu_0$ of the parameter. For a small variation $\mu = \mu_0 + \delta\mu$ of the parameter, T has a fixed point close to $\boldsymbol{x}_0$; the first approximation of the latter $\boldsymbol{x}_1 = \boldsymbol{x}_0 + \delta \boldsymbol{x}$ is calculated from the equation

$$(L - E)\delta \boldsymbol{x} = -\frac{\partial T}{\partial \mu}(\boldsymbol{x}_0, \mu_0)\delta\mu. \tag{6.13}$$

Further corrections of the fixed point position are found by applying Newton's method (6.11). A sequential application of the procedure just described allows us to construct a family of fixed points of the mapping under the variation of a parameter.

3 Bifurcation of fixed points of an area-preserving mapping

There exist some restrictions for possible scenarios of fixed point bifurcations imposed by the condition that the Poincaré index be preserved [193]. The Poincaré index of a closed curve C not passing through any fixed points of the mapping considered is defined to be the number of complete turns of the vector $\theta(\boldsymbol{x}) = T(\boldsymbol{x}) - \boldsymbol{x}$ as $\boldsymbol{x}$ moves along the curve. The Poincaré index of a fixed point is defined to be the index of any closed curve in a neighborhood of the fixed point that goes around this point and around no other fixed points. It is easy to show that the Poincaré index of an elliptic point is equal to $+1$, that of a hyperbolic point is equal to -1, and that of a hyperbolic point with inversion is equal to $+1$. The index of an arbitrary contour is equal to the sum of the indices of all fixed points enclosed by the contour. By definition, the Poincaré index is integer-valued, and it must be preserved under variation of system parameters, since the mapping is analytic as a function of the parameters. Therefore, the sum of fixed point indices is preserved in a given domain under a continuous variation of the system parameters if none of the fixed points cross the boundary of the domain. This analysis results in a rule which strongly restricts the possible fixed point bifurcations. Consider some of them.

1. *Fork-type bifurcation.* As the parameter changes, the multipliers of the fixed point pass through the value $\lambda_1 = \lambda_2 = 1$ in such a way that the point type changes from elliptic to hyperbolic. In this case the Poincaré index changes from $+1$ to -1. As a result, near the initial point, two stable (unstable) fixed points with indices $+1$ (-1) are born (disappear).
2. *Period doubling bifurcation.* As the parameter changes, the multipliers of the fixed point pass through the value $\lambda_1 = \lambda_2 = -1$ in such a way that the point type changes from elliptic to hyperbolic with inversion. In this case the Poincaré index is preserved. However, for the mapping $T^2(\boldsymbol{x})$, the Poincaré index changes from $+1$ to -1. As a result, near to the initial fixed point two stable (unstable) second-order fixed points with indices $+1$ (-1) are born (disappear).

Note that when moving away from the integrable case, the most common way for loss of fixed point stability to occur is via the birth of a pair of stable fixed points of doubled period. Moreover, such bifurcations often form infinite sequences (cascades of period doubling bifurcations) described above.

6.3 Absolute and Relative Choreographies in Rigid Body Dynamics

In this section, we present a new family of periodic solutions at zero value of the area constant. This family includes a number of already known analytic periodic solutions; it is related with both Poincaré and Lyapunov solutions. In addition, we are able, by construction, to connect all these solutions for all intermediate values of parameters and, in particular, to cases far from integrable. A remarkable fact that distinguishes this family from other families is that the solutions found are periodic in absolute space. This property holds only exceptionally: known solutions that are periodic on the Poisson sphere (in the reduced system) are not periodic in absolute space.

We recall that, to define the absolute motion of a body corresponding to a known solution $t \mapsto (\boldsymbol{\omega}(t), \boldsymbol{\gamma}(t))$ of the reduced system, one must perform an additional quadrature (see Ch. 1)

$$\dot{\psi} = \frac{\omega_1\gamma_1 + \omega_2\gamma_2}{\gamma_1^2 + \gamma_2^2}. \tag{6.14}$$

Hence, if $\boldsymbol{\omega}(t)$ and $\boldsymbol{\gamma}(t)$ are T-periodic, then the periodicity condition in absolute space is

$$\Delta\psi = \int_{t_0}^{t_0+T} \frac{\omega_1\gamma_1 + \omega_2\gamma_2}{\gamma_1^2 + \gamma_2^2}\,dt = \frac{m_1}{m_2}\pi, \quad m_1, m_2 \in \mathbb{Z}. \tag{6.15}$$

The number of known solutions that are periodic in absolute space is extremely small: examples are (i) solutions of the Delaunay class, (ii)Steklov solutions, (iii)some particular Goryachev – Chaplygin solutions. As results stated below show, all these solutions are included in the same class of particular solutions, namely, the class of *absolute choreographies for zero value of the area constant.*

1 Birth of Absolute Choreographies for $(\boldsymbol{M}, \boldsymbol{\gamma}) = 0$

The Euler – Poinsot case. Recall that in the integrable Euler – Poinsot case, the body performs inertial motion ($\boldsymbol{r} = 0$) and the Hamiltonian and the additional integral have the form

$$\mathscr{H}_0 = \frac{1}{2}(\boldsymbol{M}, \mathbf{A}\boldsymbol{M}), \quad \mathscr{F}_3 = \boldsymbol{M}^2 = \text{const.}, \tag{6.16}$$

where $\mathbf{A} = \operatorname{diag}(a_1, a_2, a_3) = \mathbf{I}^{-1}$ and $\mathbf{I} = \operatorname{diag}(I_1, I_2, I_3)$ is the tensor of inertia of the body. In what follows, we assume that $I_1 > I_2 > I_3$.

For the construction of the Poincaré mapping on the level set of first integrals $\boldsymbol{\gamma}^2 = 1$ and $(\boldsymbol{M}, \boldsymbol{\gamma}) = 0$ we use the Andoyer variables l, L, g, G, H (Ch. 1). The Poincaré section is defined by fixing the cyclic variable of the Euler-Poinsot case, $g = \text{const}$. As has been noted (see Ch. 2), permanent rotations about the principal axes of inertia correspond to nondegenerate fixed points (rotations about the maximal and minimal axes correspond to stable points and rotations about the middle axis correspond to unstable points). More complicated periodic solutions lie on resonance tori. As is well known, frequencies of motions on resonance tori become commensurable, i.e., there exist $p, q \in \mathbb{Z} \setminus \{0\}$ such that $p\omega_g + q\omega_l = 0$, where ω_l and ω_g are frequencies corresponding to the angular Andoyer variables l and g. Each resonance torus is defined by the pair p, q and and filled by periodic orbits. On the Poincaré section, each periodic orbit defines fixed points of order q. In what follows, we speak of solutions of order q; see Fig. 6.6. One of the tools useful for searching for resonance tori is the study of the *rotation number*, equal to the ratio of frequencies $n = \omega_l/\omega_g$. The rotation number depends only on the torus considered, and for a resonance torus it is an irreducible fraction $n = p/q$, $p, q \in \mathbb{Z} \setminus \{0\}$. For a detailed exposition of rotation numbers and their properties for the Euler – Poinsot problem, see [333].

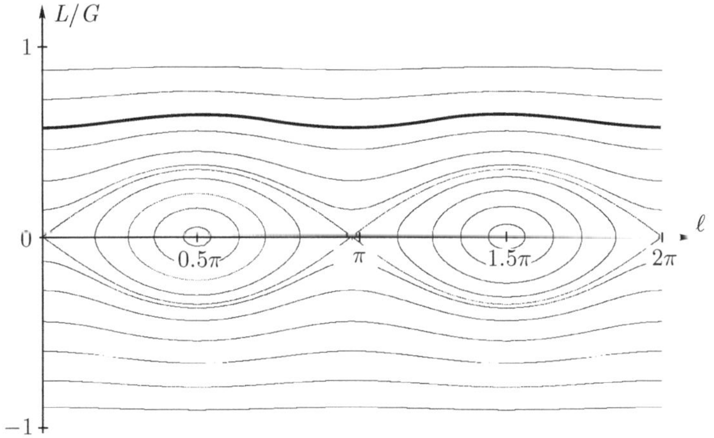

Fig. 6.6. The Poincaré mapping in the Euler – Poisson problem for $E = 2$, $c = 0$, $\mathbf{A} = \mathrm{diag}(1,\ 1.25,\ 3)$, $\mu = 0$. The bold line denotes the resonance torus with the rotation number $n = 1$, the section $g = \pi/2$.

On a given energy level $\mathscr{H}(l, L, G) = E$, the invariant tori can be parameterized by one variable; in the present case, $\varepsilon = \frac{2E}{G^2}$. Thus, the rotation number is a function of one variable and can be written in the form (see [333])

$$n(\varepsilon)^{-1} = \begin{cases} \dfrac{1}{\pi} \displaystyle\int\limits_{-\pi/2}^{\pi/2} \dfrac{(a_1 + (a_2 - a_1)\sin^2 y)dy}{\sqrt{(a_2 - a_1 - (\varepsilon - a_1)\sin^2 y)(a_3 - a_1 - (\varepsilon - a_1)\sin^2 y)}}, & \varepsilon \in [a_1, a_2], \\ \dfrac{1}{2\pi} \displaystyle\int\limits_{0}^{2\pi} \dfrac{(a_1 \sin^2 l + a_2 \cos^2 l)\,dl}{\sqrt{(\varepsilon - a_1 \sin^2 l - a_2 \cos^2 l)(a_3 - a_1 \sin^2 l - a_2 \cos^2 l)}}, & \varepsilon \in [a_2, a_3], \end{cases} \tag{6.17}$$

and $\varepsilon = a_1$ corresponds to rotation of the body about the maximal axis of inertia, $\varepsilon = a_3$ to rotation about the minimal axis of inertia, $\varepsilon = a_2$ to rotation about the middle axis of inertia. A graph of the rotation number plotted against the variable that parameterizes tori for fixed moments of inertia is presented in Fig. 6.7.

REMARK. In general, a vector field without singular points on a two-dimensional torus has infinitely many rotation numbers. These are related by

$$n' = \frac{pn + l}{qn + m}, \tag{6.18}$$

where $\begin{pmatrix} p & l \\ q & m \end{pmatrix}$ is an element of $SL_2(\mathbb{Z})$ (its entries are integers and its determinant is equal to ± 1). In our case, the rotation number is fixed by the choice of the Andoyer variables and the frequencies ω_g and ω_l.

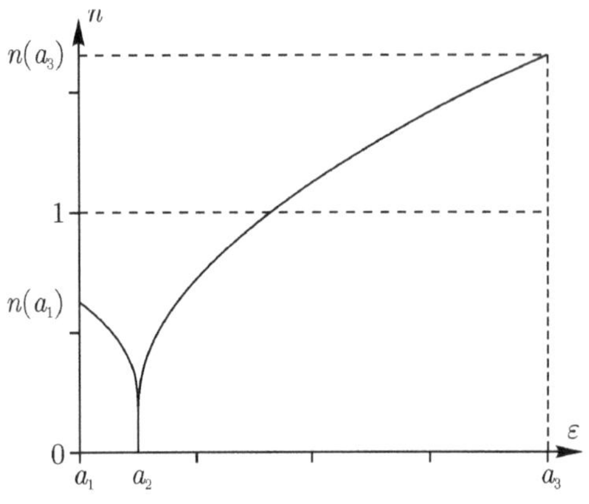

Fig. 6.7. The graph of the rotation number against the variable parameterizing tori on the given energy level $E = 2$ for $\boldsymbol{a} = (1,\ 1.25,\ 3)$, $\mu = 0$.

Consider the most interesting case where the frequencies are equal, $\omega_l = \omega_g$, and the rotation number becomes unity. As is shown in [333], the qualitative behavior of the graph in Fig. 6.7 is preserved for any moments of inertia; therefore, for the existence of the corresponding torus it is necessary that at least one of the following inequalities hold:

$$\lim_{\varepsilon \to a_1} n(\varepsilon) > 1, \quad \lim_{\varepsilon \to a_3} n(\varepsilon) > 1, \tag{6.19}$$

where, as is shown in [333],

$$\begin{aligned} &\lim_{\varepsilon \to a_1} n(\varepsilon) = ((I_1/I_3 - 1)(I_1/I_2 - 1))^{1/2}, \\ &\lim_{\varepsilon \to a_3} n(\varepsilon) = (((1 - I_3/I_2)(1 - I_3/I_1))^{-1/2} - 1)^{-1}. \end{aligned} \tag{6.20}$$

Note that the first of inequalities (6.19) implies $I_2 + I_3 < I_1$, which contradicts the triangle inequality. Therefore, on the interval $\varepsilon \in (a_1,\ a_2)$ (i.e., inside the separatrix on the phase portrait, see Fig. 6.6), there is no torus with rotation number $n = 1$. The second of inequalities (6.19) defines a domain on the plane of parameters $(I_2/I_1,\ I_3/I_1)$. The part of this domain for which the triangle inequality $I_2 + I_3 > I_1$ holds is the existence domain for the torus with rotation number $n = 1$ (see Fig. 6.8). The Poincaré mapping in Fig. 6.6 is constructed for the parameters $(I_2/I_1,\ I_3/I_1)$ lying in the existence domain of the torus found with $n = 1$. The invariant curve consisting of first-order fixed points and corresponding to this torus is shown as a bold line.

It turns out that on the Poisson sphere, solutions lying on the resonance torus have the form of closed curves with self-intersections of figure-eight type and the points of self-intersection lie in the equatorial plane (see Fig. 6.9(a)). Moreover, all these solutions are also periodic in absolute space. This fact may be proved as a consequence of the following proposition.

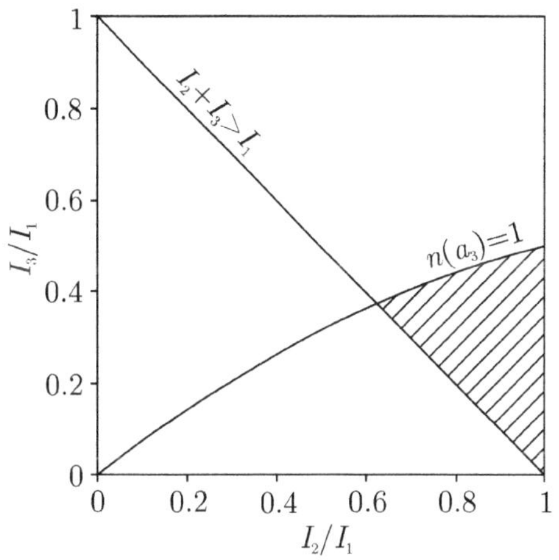

Fig. 6.8. The existence domain of the resonance torus $n = 1$.

Proposition 1. *All periodic solutions of the reduced system in the Euler – Poinsot problem are also periodic in absolute space.*

The proof of this proposition can be easily established via explicit integration of the absolute motion in the Euler – Poinsot problem.

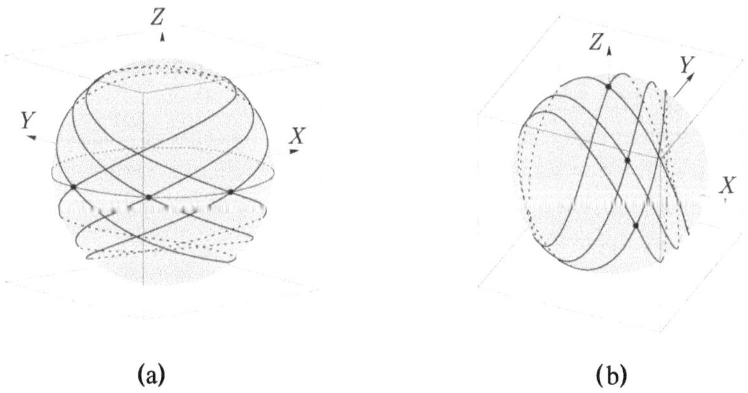

Fig. 6.9. Periodic solutions lying on the resonance torus $n = 1$. (a) The motion on the Poisson sphere. (b) The trace of the apex e_1 in absolute space.

Figure 6.9(b) shows several traces of the unit vector e_1 (apex) attached to the axis Ox for periodic solutions lying on the torus considered. All these traces also have the form of a figure eight and their self-intersection points lie in a vertical plane. In what follows, in all figures, we show the trace of the apex e_1 to illustrate the motion in absolute space.

Perturbation of the Euler – Poinsot case. Let us ask the question: do there exist periodic solutions in absolute space under perturbations of the Euler – Poinsot problem $\mathscr{H} = \mathscr{H}_0 + \mu\mathscr{H}_1$? First, we consider perturbations at the zero value of the area integral. As small parameter μ we take the strength of the gravitational field, so that if the displacement of the center of mass of the body relative to the fixed point is $\boldsymbol{r}$, then $\mathscr{H}_1 = (\boldsymbol{r}, \boldsymbol{\gamma})$. The following proposition holds.

Proposition 2. *Perturbation of the system by displacement of the center of mass within the plane perpendicular to the minimal axis of inertia of the body* ($r_1 \neq 0$, $r_2 \neq 0$, $r_3 = 0$) *gives rise to the birth of a pair of solutions (stable and unstable), periodic in absolute space, from the resonance torus of the Euler – Poinsot problem with the rotation number* $n = 1$ *at the zero value of the area integral.*

For small displacements of the center of mass, the existence of periodic solutions of the reduced system can be proved by the Poincaré theorem on the appearance of periodic solutions from resonance tori of the unperturbed problem (see [333]). The proof of the periodicity of solutions in absolute space and the generalization of this result to the case of finite displacements of the center of mass were made by the authors by using computer methods. The generalization to the case of finite displacements of the center of mass was made by the method of continuation by a parameter and proof of the periodicity in absolute space follows from explicit integration of the absolute motion of the rigid body using Everhart's method of different orders. The minimal value of the gap in the trajectory (which can be defined, for example, as the displacement of the apex $\boldsymbol{e}_1$ during a period of relative motion) obtained by using the Everhart method of order 27 is 10^{-13}. In addition, as the accuracy of the method increases, the obvious convergence of the gap of the trajectory to zero occurs.

Note that if the system is perturbed in such a way that the center of mass of the body is displaced in a direction transverse to the plane perpendicular to the minimal axis of inertia of the body, so that $r_3 \neq 0$, then periodic solutions of the reduced system still emerge, but they are not periodic in absolute space.

The phase portrait for a perturbation where the center of mass is displaced along the maximal inertia axis is shown in Fig. 6.10. It is remarkable that almost all of the phase portrait is determined just by specifying the periodic solutions emerging from the resonance torus $n = 1$, together with the permanent rotations. We list the most interesting properties of the solutions indicated in Proposition 2.

1. The vector $\boldsymbol{\gamma}$ on the Poisson sphere traces out a figure-eight curve, with the self-intersection point lying in the equatorial plane ($\gamma_3 = 0$) and symmetric with respect to the equatorial plane. In the case where the center of mass is displaced along one of the principal axes, the figure of eight possesses the symmetry of the group C_{2v} (i.e., the figure of eight is centrally symmetric). The trajectories on the Poisson sphere corresponding to various displacements of the center of mass are shown in Fig. 6.9(a).

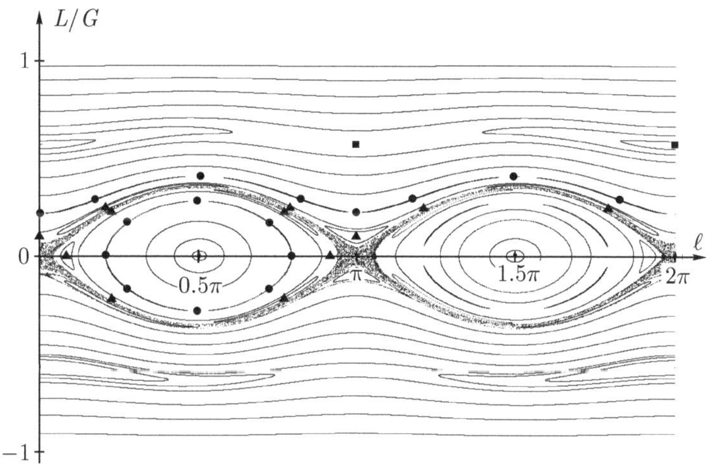

Fig. 6.10. The Poincaré mapping of the Euler – Poinsot problem for $E = 2$, $c = 0$, $\mathbf{A} = \mathrm{diag}(1,\ 1.25,\ 3)$, $\boldsymbol{r} = (1,\ 0,\ 0)$, $\mu = 0.01$, and $g = \pi/2$. Square, round, and triangle markers denote, respectively, first-, second-, and third-order fixed points corresponding to nontrivial absolute choreographies.

2. The solution indicated is symmetric with respect to the transformation $\sigma(\gamma_1, \gamma_2, \gamma_3, \omega_1, \omega_2, \omega_3) = (\gamma_1, \gamma_2, -\gamma_3, -\omega_1, -\omega_2, \omega_3)$. Using this property in the calculation of the integral (6.15), one can prove that the solutions are periodic in absolute space.
3. The apexes of the principal axes of inertia $\boldsymbol{e}_1$ and $\boldsymbol{e}_2$ trace out figure-eight curves on a sphere with the same symmetry group as on the Poisson sphere and $\boldsymbol{e}_3$ traces out a convex, closed curve without self-intersections. The traces of the apex $\boldsymbol{e}_1$ corresponding to various displacements of the center of mass are shown in Fig. 6.9(b).

By analogy with the figure-eight periodic trajectories in the three-body problem of celestial mechanics discovered by Moore, Chenciner, Montgomery, and Simó (see [133, 531]), in what follows, we call such solutions *absolute choreographies*. The term *choreography* is used to refer to the fact that several particles move along some fixed closed curves. The use of this term in the present situation is justified by the fact that for the absolute motion of the rigid body, all three apexes (similarly to three particles) move along closed curves (the so-called triply connected choreography), two of which are figures of eight.

The absolute choreographies considered form a family depending on the parameters of the system. By an appropriate choice of units and by rescaling, we can reduce the number of independent parameters on which the family depends to four: $(E,\ I_2/I_1,\ I_3/I_1,\ r_2/r_1)$. In this case, increase in the perturbation of the Euler – Poinsot problem corresponds to decrease in energy E from infinity under the constant displacement of the center of mass and constant mass of the body. In what follows, we restrict ourselves to the study of a three-parameter family with a generic fixed direction r_2/r_1=const. of displacement of the center of mass.

Scenarios for the birth of absolute choreographies. As was noted above, absolute choreographies in the problem of motion of a heavy rigid body with a fixed point appear from resonance tori of the Euler – Poinsot problem with rotation number equal to one. If such a torus does not exist (see Fig. 6.8), absolute choreographies may still appear. Moreover, in this case, an absolute choreography appears via a subcritical period doubling bifurcation from the permanent rotation about the minimal inertia axis (of the Euler – Poinsot problem) when energy E decreases. Note that period doubling bifurcations of permanent rotations were studied in [96]. Moreover, a cascade of doublings of the period appearing from the permanent rotations in the Euler – Poinsot problem was found in [96], but whereas they studied periodicity of solutions of the reduced system, they did not notice that such solutions also correspond to periodic motion in absolute space.

Therefore, in the Euler – Poisson problem, if the energy decreases from infinity, two scenarios for the appearance of absolute choreographies exist:

1. They emerge from resonance tori of the Euler – Poinsot problem. The values of the moments of inertia must lie in the existence domain of the torus $n = 1$ in Fig. 6.8.
2. They emerge from a permanent rotation of the Euler – Poinsot problem by period doubling bifurcation. The values of the moments of inertia lie outside the existence domain of the torus $n = 1$ in Fig. 6.8.

These scenarios for appearance of choreographies can be illustrated by Figs. 6.11(a) and 6.11(b).

Note that while in the first case choreographies appear in pairs (stable and unstable), in the second case only a stable choreography appears. The link between these cases is made by continuous variation of the moments of inertia; moreover, during this variation another period doubling bifurcation occurs, causing the appearance or disappearance of an unstable choreography.

2 Genealogy of Choreographies for $(\boldsymbol{M}, \boldsymbol{\gamma}) = 0$

We shall not give a complete bifurcation analysis for all possible values of parameters of the family of absolute choreographies having $(\boldsymbol{M}, \boldsymbol{\gamma}) = 0$. Here we just describe relations between this family of solutions and known families of partial periodic solutions and of critical periodic solutions for integrable cases. We also describe various typical situations for the emergence of choreographies and for their bifurcations.

Relationship with integrable cases. We consider the evolution of solutions obtained for a dynamically symmetric rigid body ($I_1 = I_2$) depending on the ratio I_3/I_1 (due to the triangle inequality, $I_3/I_1 \in (0, 2]$ in this case) for a finite value of energy and for displacements of the center of mass within the equatorial plane ($r_3 = 0$). The point $I_3/I_1 = 1/2$ corresponds to the Kovalevskaya case whose phase portrait is shown in Fig. 6.12. In this case, the phase portrait contains only one absolute choreography, which is stable and which appears from the permanent rotation by a period doubling bifurcation. It turns out that it coincides with well-known *Delaunay solutions* (see Ch. 2) whose periodicity in absolute space was proved analytically in [230].

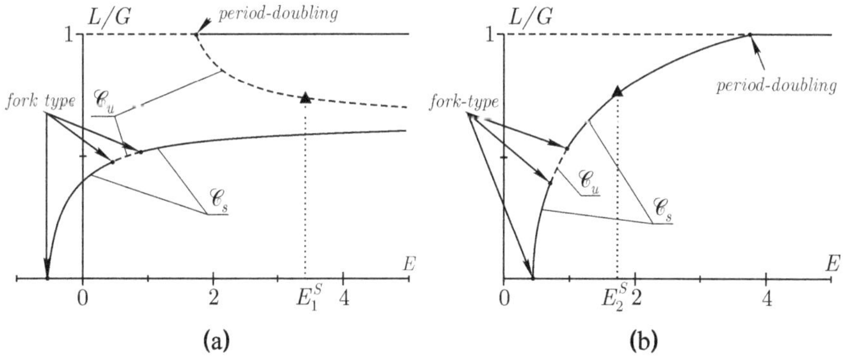

Fig. 6.11. Bifurcation diagrams for absolute choreographies: (a) for choreographies appearing according to the first scenario, $\mathbf{A} = \text{diag}(1, 1.25, 3)$, $\mu\boldsymbol{r} = (0, 1, 0)$; (b) for choreographies appearing according to the second scenario, $\mathbf{A} = \text{diag}(1, 1.5, 2.25)$, $\mu\boldsymbol{r} = (1, 0, 0)$. $\mathscr{C}_s$ are stable and $\mathscr{C}_u$ are unstable choreographies; E_1^s and E_2^s are values of energy corresponding to the Steklov solutions.

If the ratio I_3/I_1 decreases, the phase portrait becomes partially chaotic but the stable choreography is preserved. Moreover, for some value of I_3/I_1 (which, in general, depends on the energy), the permanent rotation about the minimal axis of inertia gives rise to an unstable absolute choreography (again by a period doubling bifurcation). If $I_3/I_1 = 1/4$, we arrive at the Goryachev – Chaplygin integrable case whose phase portrait is shown in Fig. 6.13. In this case, both solutions (stable and unstable) corresponding to absolute choreographies are seen on the phase portrait. Their periodicity in absolute space was pointed out in [77]; an analytic proof of this fact for some particular cases can be found in [231, 229]. If the parameter I_3/I_1 decreases further, the phase portrait again becomes partially chaotic but periodic solutions corresponding to absolute choreographies are preserved, up to $I_3/I_1 \to 0$.

Continuous variation of energy and the relationship with Staude solutions. Consider the evolution of absolute choreographies when energy varies and moments of inertia are fixed. The ratio L/G (in Andoyer coordinates) may be viewed as a coordinate on the Poincaré section and the evolution can be visualized by the graph of L/G against the energy of the system, see Fig. 6.11.

Scenario 1. Figure 6.11(a) shows bifurcations of the stable and unstable choreographies appearing in the first scenario (see item 2). Stable choreographies $\mathscr{C}_s$ are

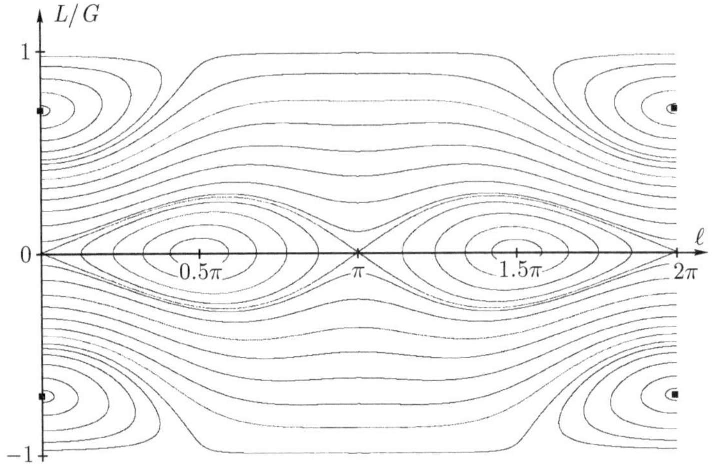

Fig. 6.12. The phase portrait for the Kovalevskaya case with zero value of the area constant ($E = 2$, $c = 0$, $\mathbf{A} = \mathrm{diag}(1, 1, 2)$, $\mu \boldsymbol{r} = (1, 0, 0)$). The squares denote fixed points corresponding to absolute choreographies.

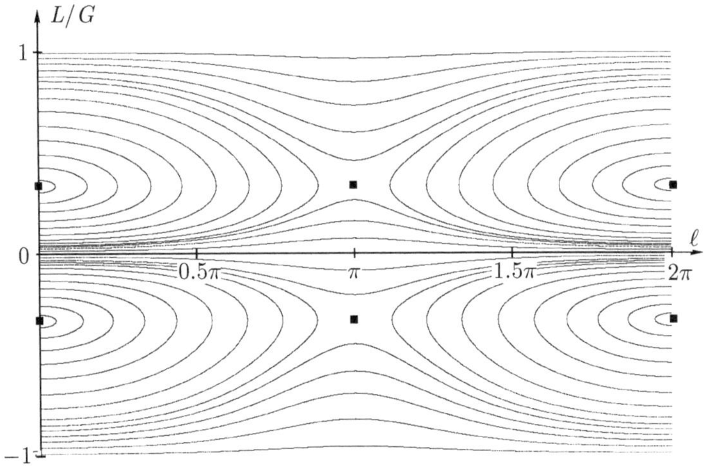

Fig. 6.13. The phase portrait for the Goryachev – Chaplygin case with zero value of the area constant ($E = 2$, $c = 0$, $\mathbf{A} = \mathrm{diag}(1, 1, 4)$, $\mu \boldsymbol{r} = (1, 0, 0)$). The squares denote fixed points corresponding to absolute choreographies.

represented by the continuous line and unstable choreographies $\mathscr{C}_u$ by the dotted line. We see that for large energies, both choreographies on the phase portrait tend to the torus with the rotation number $n = 1$. As the energy decreases, the unstable choreography merges with the rotation about the minimal axis of inertia (which lies on the line $L/G = 1$) by the period doubling bifurcation. As the energy decreases, the stable choreography changes stability type two times, passing through two fork-type bifurcations, and then merges with the rotation about the maximal axis of inertia (which lies on the line $L/G = 0$) via another fork-type bifurcation. Critical energies at which bifurcations occur depend in general on the moments of inertia of the body and form a three-dimensional bifurcation diagram in the parameter space $(E, I_2/I_1, I_3/I_1)$.

Remark 1. Note that some of solutions appearing from the stable choreography via the fork-type bifurcation are also absolute choreographies. For them, the corresponding trajectories of apexes in absolute space possess less symmetry and one of the loops of the figure of eight becomes larger than the other.

If the center of mass is displaced along the middle axis of inertia, then a cascade of period doublings occurs for these nonsymmetric choreographies under continuous variation of energy and, moreover, all solutions which appear are also absolute choreographies.

Scenario 2. Figure 6.11(b) shows bifurcations in the case of the second variant for appearance of choreographies (see item 2). In this case, the unstable solution is absent and the stable solution appears from the rotation about the minimal axis of inertia (which lies on the line $L/G = 1$) by a period doubling bifurcation. The further evolution of the stable choreography does not differ qualitatively from the previous case.

Thus, on the one hand, the choreographies found are Poincaré solutions appearing from the resonance tori when the energy decreases from infinity. On the other hand, when the energy increases, the same choreographies appear from the rotations about the principal axes of inertia, which are Lyapunov solutions for the Staude permanent rotations.

Connection with Steklov solutions It is known that solutions found by V. A. Steklov [555, 358] are periodic in absolute space. Their periodicity follows from the explicit analysis of quadratures, which are expressed in elliptic functions [290]. The following remarkable fact holds: the motion of the apexes in absolute space for the Steklov solutions traces out a figure of eight [290]. It turns out that the Steklov solution is a particular case of the family of absolute choreographies found by the authors.

Indeed, for the Steklov solution, in addition to the triangle inequality and the assumptions $I_1 > I_2 > I_3$, the moments of inertia are restricted by the following conditions:

$$\begin{aligned} &I_2 > 2I_3, \ \boldsymbol{r} = (0,1,0) \quad \text{for solutions of the first type,} \\ &I_1 > 2I_3, \ \boldsymbol{r} = (1,0,0) \quad \text{for solutions of the second type.} \end{aligned} \tag{6.21}$$

For the moments of inertia satisfying conditions (6.21), one can calculate specific values of the energy $E_{1,2}^S(I_1, I_2, I_3)$ for which Steklov solutions of the first and second type exist. Therefore, the families of Steklov solutions are parameterized by two parameters are in the family of absolute choreographies found by the authors. In Fig. 6.11, the choreographies corresponding to Steklov solutions are marked by triangles.

Figure 6.14 shows the existence domains for Steklov solutions of the first and second types on the plane of the parameters $(I_2/I_1,\ I_3/I_1)$. One can observe from these

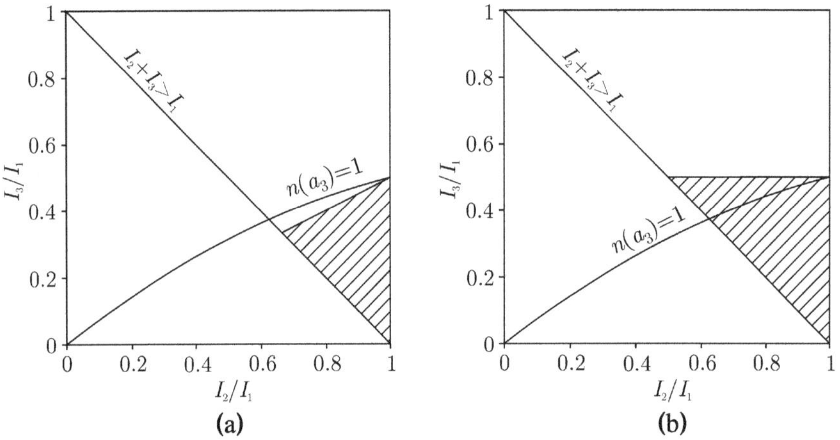

Fig. 6.14. The existence domains for Steklov solutions on the plane $(I_2/I_1,\ I_3/I_1)$: (a) of the first type; (b) of the second type.

figures that all solutions of the first type and some solutions of the second type lie in the existence domain of the torus $n = 1$ of the Euler – Poinsot problem. *Under continuous variation of energy, these solutions turn into Poincaré solutions for the torus* $n = 1$ *(according to the first scenario for the birth of choreographies).*

Steklov solutions of the second type and lying outside the existence domain of the torus $n = 1$*, merge under continuous variation of energy with the rotation about the principal axis of inertia via a period doubling bifurcation (according to the second scenario of the birth of choreographies).*

3 More Complicated Choreographies for $(M, \gamma) = 0$

Consider the problem of the existence of more complicated periodic motions in absolute space. Again, we start with the Euler – Poinsot case. Obviously, except for the torus with rotation number $n = 1$, in the Euler – Poinsot problem there exists a countable set of resonance tori on the fixed energy level that are filled with periodic solutions of different orders. By Proposition 1, all these solutions are absolute choreographies. The corresponding traces of apexes in absolute space have the form of closed curves with number of self-intersections not less than the order of the solution.

For all resonance tori, by using inequalities similar to (6.19), one can find the existence domains on the plane $(I_2/I_1, I_3/I_1)$. The same inequalities imply that tori with $n(\varepsilon) \geqslant 1$ lie only outside the separatrix ($\varepsilon \in (a_2, a_3)$) and tori with $n < 1$ can lie inside or outside the separatrix. We recall that $\varepsilon = \frac{2E}{G^2}$ and $n(\varepsilon)$ is given by (6.17).

Now we consider a perturbation of the Euler – Poinsot problem (as above, on the zero level set of the area integral). By the Poincaré theorem, under a perturbation, all resonance tori split into an even number of periodic solutions (with an equal number of stable and unstable ones). Some of these solutions can also be absolute choreographies. Moreover, as computer studies demonstrate, the following proposition holds.

Proposition 3.

1. *If the center of mass is displaced in a plane perpendicular to the minimal axis of inertia ($r_3 = 0$), then resonance tori of the Euler – Poinsot problem lying outside the separatrix ($\varepsilon \in (a_2, a_3)$) give rise to solutions periodic in absolute space.*
2. *If the center of mass is displaced in the plane perpendicular to the maximal axis of inertia ($r_1 = 0$), then resonance tori of the Euler – Poinsot problem lying inside the separatrix ($\varepsilon \in (a_1, a_2)$) give rise to solutions periodic in absolute space.*

This proposition implies that if the center of mass is displaced in the direction of the middle axis of inertia, then all periodic solutions of the reduced system appearing from resonance tori are periodic for the complete system. The corresponding Poincaré section, on which absolute choreographies appearing from resonance tori with rotation numbers $n = 1/2$ and $n = 1/3$ are marked, is shown in Fig. 6.15. The choreographies themselves (the traces of the apex e_1) are shown in Fig. 6.16.

4 Relative choreographies

We now consider perturbations of the Euler – Poinsot problem on a nonzero level set of the area integral. In the reduced system, periodic Poincaré solutions [333] appear from resonance tori in this case as well. However, these solutions are not periodic in absolute space.

We state some general properties of the absolute motion that are related with periodic solutions of the reduced system.

Proposition 4. *There always exists a coordinate frame uniformly rotating about the vertical axis in which the absolute motion corresponding to a periodic solution of the reduced system is also periodic with the same period, i.e., all apexes of the rigid body trace out closed curves in this frame.*

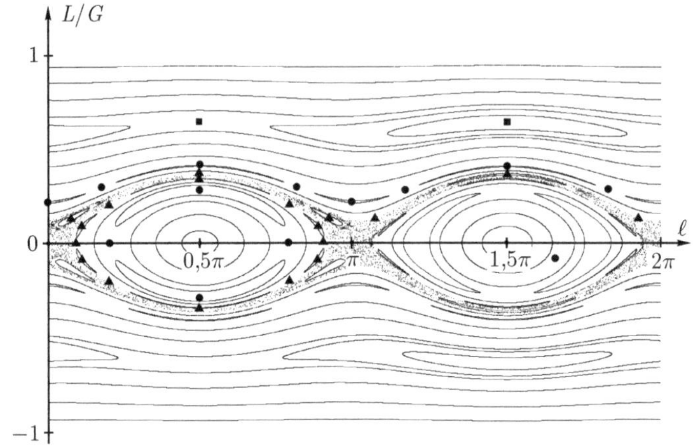

Fig. 6.15. The Poincaré mapping ($g = \pi/2$) of the Euler – Poisson problem for $E = 2$, $c = 0$, $\boldsymbol{a} = (1, 1.25, 3)$, $\boldsymbol{r} = (0, 1, 0)$, $\mu = 0.02$. The square, round, and triangle markers denote first-, second-, and third-order fixed points, respectively.

Proof. The proof is based on the method proposed in [333]. Since for periodic solutions of the reduced system, $\boldsymbol{\omega}$ and $\boldsymbol{\gamma}$ are periodic functions of time, the right-hand side of (6.14) may be expanded in a convergent Fourier series

$$\dot{\psi} = \sum_{n=-\infty}^{+\infty} a_n e^{i\frac{2\pi}{T}nt}.$$

Integrating this relation gives

$$\psi(t) = a_0 t + \sum_{\substack{n=-\infty \\ n\neq 0}}^{+\infty} \frac{a_n T}{2\pi i n} e^{i\frac{2\pi}{T}nt} + \text{const.} = a_0 t + \Psi(t),$$

where $\Psi(t)$ is a T-periodic function of time. It is clear that in the coordinate frame rotating about the vertical with angular speed $\Omega = a_0$, for a given solution, all Euler angles θ, φ, and ψ are T-periodic functions of time.

Hence, the absolute motion of the body corresponding to periodic solutions that appear from resonance tori, in the case of a nonzero value of the area integral, can be interpreted as a *relative choreography* (this term is taken from [133, 531]), i.e., as a periodic motion, in the coordinate frame which rotates uniformly about the vertical with angular speed Ω. Periodic solutions that appear at the zero value of the area integral if the displacement of the center of mass does not satisfy the conditions of Proposition 2 (i.e., if $r_3 \neq 0$) can be interpreted similarly.

Generally speaking, to each periodic solution there corresponds a countable number of relative choreographies. Indeed, the shift of angular velocity

$$\Omega' = \Omega + \frac{p}{q}\frac{2\pi}{T}, \quad p \in \mathbb{Z},\ q \in \mathbb{N}, \tag{6.22}$$

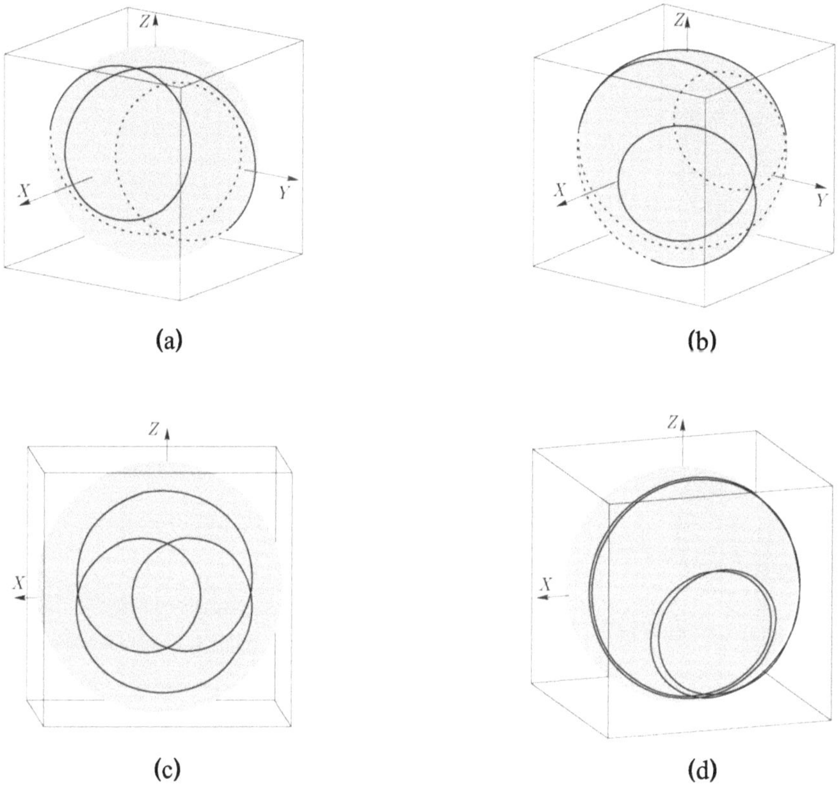

Fig. 6.16. Examples of absolute choreographies corresponding to periodic solutions of second and third orders that appear from tori $n = 1/2$ and $n = 1/3$ when the center of mass is displaced along the middle axis of inertia ($\boldsymbol{r} = (0, 1, 0)$). (a) and (b) outside the separatrix ($\varepsilon \in (a_2, a_3)$); (c) and (d) inside the separatrix ($\varepsilon \in (a_1, a_2)$). The trace of the maximal axis is shown. All figures correspond to $E = 2$, $\mathbf{A} = \mathrm{diag}(1, 1.25, 3)$, $\mu = 0.01$.

where T is the period of the relative motion, preserves the property of closedness of trajectories in the rotating coordinate frame.

Examples of different relative choreographies all corresponding to the same periodic solution that appears from the torus with rotation number $n = 1$ are shown in Fig. 6.17.

5 Open problems

In conclusion, we list some open problems concerning the new solutions described in this section that, in our opinion, deserve attention in the further study of choreographies.

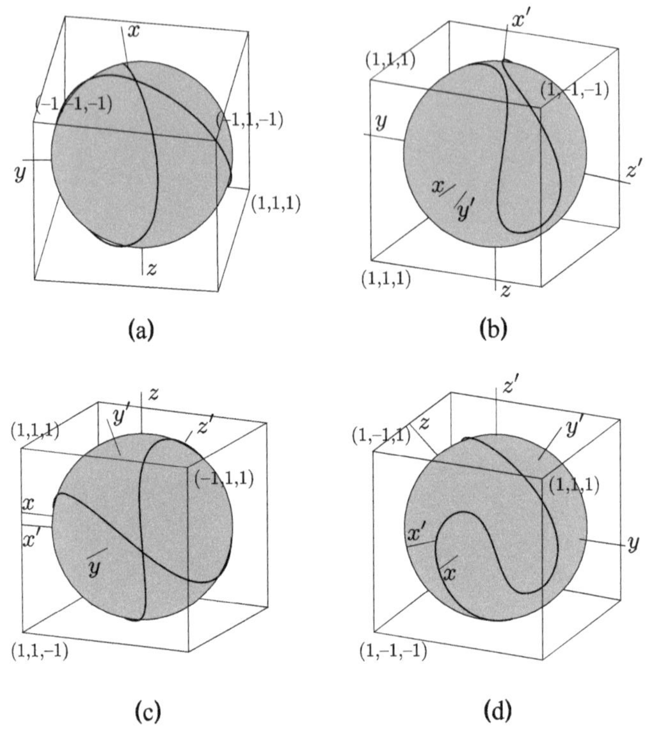

Fig. 6.17. The traces of the apex in the rotating coordinate system that correspond to the periodic solution appeared from the torus $n = 1$. For displacements of the center of mass along the minimal axis ($\boldsymbol{r} = (0, 0, 1)$) on the zero constant of the area integral ($c = 0$): (a) $\Omega = -0.3334243$; (b) $\Omega = 2.0623605$. For displacements of the center of mass along the middle axis ($\boldsymbol{r} = (0, 1, 0)$) and a nonzero constant of the area integral ($c = 1$): (c) $\Omega = 0.0969602$; (d) $\Omega = 2.4865794$. All figures correspond to $\boldsymbol{a} = (1, 1.25, 3)$, $E = 4$, and $\mu = 1$.

1. Find an analytic proof of the periodicity of choreographies in absolute space. A similar proof based on variational and topological principles was found by A. Chenciner and R. Montgomery [133] for choreographies in the three-body problem.
2. Do a complete bifurcation analysis of the new solutions for all parameters of the system, E, I_2/I_1, I_3/I_1, and r_2/r_1. Obviously, this analysis will mostly have to be performed by using computer methods because of the large number of parameters in the problem and the absence of analytic representations of solutions.
3. Study the stability of the new solutions in terms of the parameters of the system. This analysis can also be performed by numerical simulations for the reasons stated above.

6.4 Chaotic Motions. Genealogy of Periodic Orbits

Period doubling, cascades, and separatrix splitting. The absence of analytical integrals of the equations of motion is closely related to complicated (chaotic) behavior of the system trajectories. In this case, analytical methods based on the use of a small perturbation parameter enable exact results only near to integrable situations and, in some sense, provide insight only into the very first stage of the motion stochastization under a perturbation.

A real system can be significantly far (with respect to the small perturbation parameter μ) from an integrable system, and for such a system, the analytical methods cannot be applied. Computer methods are an effective tool for these kinds of studies; with their help, we can continue to vary the parameter μ, which is no longer small, for some invariant objects (periodic solutions, invariant tori, and separatrices) and also compute various characteristics of chaotic motions (the Lyapunov exponents, the entropy, etc.). Using the well-known Poincaré theorem for periodic motions and KAM-theory for quasi-periodic motions, we can obtain a justification for such a continuous variation in nondegenerate situations. For example, nondegenerate periodic orbits (corresponding to fixed points of the Poincaré mapping in the case of two degrees of freedom) can be numerically traced up to a bifurcation point at which various reconstructions can occur. All this is described in detail in the book [247].

For example, the following situation may take place: an initial stable solution remains stable as μ changes until at a certain value $\mu = \mu_1$, called a *bifurcation point*, it loses stability, at which point a new stable periodic solution is born, but this solution now has double the period of the old one. As μ continues to change, there occurs another period doubling bifurcation (loss of stability accompanied by birth of a new solution with double the period) at the value $\mu = \mu_2$. This process then continues with the system passing through a sequence of bifurcation points $\mu = \mu_n$ as μ varies. Under certain conditions, such period doubling bifurcations may have an accumulation point, i.e., the sequence of bifurcation values $\mu = \mu_n$ is a convergent one, similar to a geometric progression. As was shown by M. Feigenbaum, this sequence has universal properties. In fact, as $n \to \infty$, the value δ_n,

$$\delta_n = \frac{\mu_{n-1} - \mu_{n-2}}{\mu_n - \mu_{n-1}},$$

tends to a certain universal constant δ (the *Feigenbaum constant*) independent of the form of the two-dimensional area-preserving mapping. It is equal to $\delta = 8.721\ldots$ Universality can be proved by the use of renormalization group analysis [399].

Apart from the universality of δ, there exist two other universal numbers for two-dimensional area-preserving mappings. It turns out that in a suitably chosen coordinate system, the structure of the periodic trajectories copies itself with scaling factors $\alpha = 4.018\ldots$ and $\beta = 16.363\ldots$ along the coordinate axes.

Obviously, the birth of a large number of stable and unstable periodic orbits near the limiting value μ_∞ indicates a complicated behavior of the system, and the avalanchelike period doubling process itself can be considered as a scenario of transition to chaos. Note only that as μ varies, the period doubling scenario is not the only route to chaos in the system, since bifurcations can also occur with unstable (hyperbolic) orbits and their separatrices. Methods of perturbation theory allow us to estimate only a measure of the separatrix splitting and the width of the stochastic layer for small values of μ (if the technique of Poincaré – Melnikov integral is applicable). As a rule, for finite and increasing μ, the width of the stochastic layer and the degree of separatrix splitting, see Sec. 5.2 of Ch. 5, increase up to a certain (maximal) value, which can also be used as a global characteristic of chaos. For the general case, the stochastic layer is determined by the intersection of separatrices of various resonances, in particular, by long-periodic resonances. Such a situation is described in more detail in Sec. 6.5 of Ch. 6.

For a number of dynamical systems, an important role is played by the continuations and bifurcations of quasi-periodic solutions (invariant tori). For example, according to Green's conjecture, the global stochastization of the standard map (the Chirikov map) is related to the breakdown of the invariant torus whose rotation number is the most poorly approximated by rational numbers, i.e., equal to the golden section [399]. Physically, chaos develops according to several scenarios, which coexist with each of other in any real dynamical system.

The paper [96] considers the continuous variation of the parameter μ (body weight) for permanent stable rotations about the large and small axes of inertia in the Euler – Poinsot problem. When the displacement of the center of mass from the fixed point lies in the principal plane of the ellipsoid of inertia, for certain values of the moments of inertia, a cascade of period doublings was detected, which leads to global chaotization of the motion for which no domains of the Poincaré section exhibit regular dynamics (large stable resonances). Analogous cascades can occur as a result of continuous variation of the energy.

Global section. Genealogy of periodic orbits. This section deals with an important aspect of analysis of stochastic effects in the Euler – Poisson equations (and, in general, in autonomous Hamiltonian systems with two degrees of freedom). The illustrations of chaotic behavior appearing in this book are mostly based on the use of the Poincaré map and, therefore, on the choice of a certain section of the three-dimensional isoenergetic level surface $H = E$. The topological structure of this level surface can be complicated; the topological classification of isoenergetic manifolds is contained in several works (e.g., S. B. Katok [281], R. P. Kuzmina [359], Ya. V. Tatarinov [566], I. N. Gashenenko [213]) in which additional assumptions about the center of mass location are made.

In order that the Poincaré section completely describes the entire phase flow on a fixed energy level, every trajectory must intersect it transversally infinitely many times during the motion. In such a case, we speak about the *global transversal Poincaré section*. The topology of the isoenergetic manifold is an obstruction to the existence of such sections.

For the Euler – Poisson equations, even if the center of mass lies on the principal axis of inertia, the nonsingular integrable manifolds Q^3 (whose complete list was obtained by S. B. Katok in [281]) are not in general connected and consist of several (two or three) connected components each of which can only be one of the following four types: $\mathbb{R}P^3$, S^3, $S^1 \times S^2$, and $K^3 = (S^1 \times S^2)\#(S^1 \times S^2)$, where the symbol # denotes the connected sum. The manifold K^3 represented as a connected sum can be also obtained from the three-dimensional sphere S^3 by gluing on two handles. On the other hand, from results of [69], it follows that the global transversal section can exist only if the isoenergetic manifold is $S^1 \times S^2$ and the corresponding global Poincaré section is diffeomorphic to S^2.

In [213, 211], it was proposed to use an almost everywhere transversal global section to study particular solutions and routes to chaos in the Euler – Poisson equations. For such a global section, the author of [213, 211] proposed the two-dimensional surface

$$P^2_{h,c} = \{H = h,\ (\boldsymbol{M}, \boldsymbol{\gamma}) = c,\ (\boldsymbol{\gamma}, \boldsymbol{\gamma}) = 1,\ F = 0\} \subset Q^3, \tag{6.23}$$

where

$$F = \mu(\boldsymbol{M} \times \boldsymbol{\gamma}, \boldsymbol{r}) = \frac{1}{2}\frac{d}{dt}(\boldsymbol{M}, \boldsymbol{M}).$$

This almost everywhere transversal, global section is characterized by the property that the magnitude of the angular momentum $|\boldsymbol{M}|$ attains its extremal value there. The study of this section is natural from a physical point of view, since by compactness of the isoenergetic manifold, the extremum is permanently attained for almost every motion, and the dependence of $|\boldsymbol{M}|$ on time is oscillatory (but not regular). In [213], it is proved that for regular values of (h, c), each section $P^2_{h,c}$ is a closed smooth manifold in Q^3 and that it separates every component of Q^3 into two closed manifolds with common boundary. Every trajectory on Q^3 repeatedly intersects the section $P^2_{h,c}$, while the transversality requirement is violated (i.e., the trajectory is tangent to the section) only on some curve ∂N, which separates $P^2_{h,c}$ into domains $N^{\pm}$: within each domain the flows intersect in the same direction and this direction is opposite for N^+ and N^-. Moreover, on $P^2_{h,c}$, the minimum (maximum) value of $|\boldsymbol{M}|$ is attained in the domain N^+ (N^-).

The example of the Poincaré mapping on the section $P^2_{h,c}$ and also separately on the domains N^+ and N^- in the space of the variables (M_1, M_2, M_3) is presented in Fig. 6.18. For typical values of the parameters, the section $P^2_{h,c}$ consists of two surfaces with one inside the other, both of which are isomorphic to the sphere S^2 and glued with one another at two points. The general classification of the topological types of the section $P^2_{h,c}$ depending on the parameters (h, c) can be found in [212, 209] (and in [213, 210] for the heavy gyrostat).

REMARK. Sometimes it is more convenient to use several local Poincaré sections instead of one complicated global section, because in that case we often obtain a very simple and illustrative presentation of the flow.

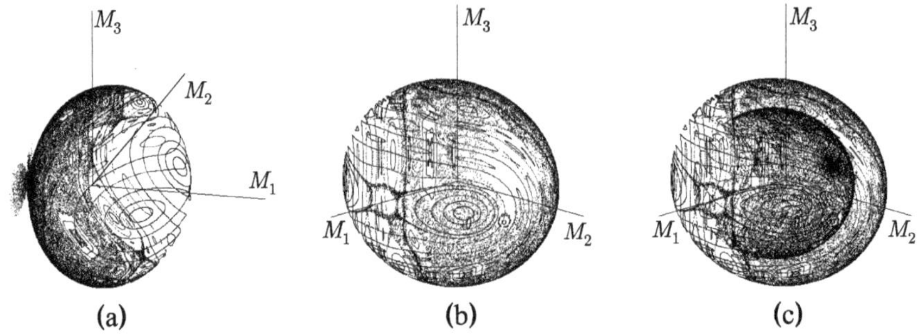

Fig. 6.18. Typical Poincaré mappings on the section $P^2_{h,c}$: (a) involving intersections with the surface in the positive direction (N^+); (b) involving intersections in the negative direction (N^-); (c) involving intersections in both directions.

At the minimum value of the energy h, the section $P^2_{h,c}$ degenerates to the point corresponding to the Staude permanent rotation. As h increases from its minimum value, the evolution of this rotation leads to an interesting (nonanalytical) class of periodic motions of the reduced system. The birth of new periodic solutions takes place when h attains other critical values. The corresponding Staude solutions can also be continued in h, and we obtain a new class of periodic orbits. Along with the described way of continuing the Staude periodic solutions, other periodic solutions "are born and die" in other domains of the phase space; these are resonances, and the largest of them determine, as a rule, the whole phase portrait structure. Under a change of parameters h and c, the interconnection of various classes of periodic solutions defines their *genealogy*. This genealogy is almost unstudied for the Euler – Poisson equations when the number of free parameters is large.

The study of the global picture of stochastization for the Hill problem from celestial mechanics having only one essential parameter was carried out by C. Simó and T. Stuchi in [533]. In this system they detected the meandering tori originating as a consequence of the fact that a plane mapping loses the "twisting" property. We have detected analogous tori in the restricted problem of rigid body dynamics for the Hess conditions (see below). It would be interesting to extend the methods of [533] to the Euler – Poisson equations.

The complete analysis of global dynamics of the Euler – Poisson equations in general is a very broad program in which computer and analytical methods are combined. It is sensible to investigate more concrete problems, which may be expected to be solved with less effort.

Novikov's conjecture and periodic solutions on the Poisson sphere. In [349, 292], by means of variational methods, it was proved that at zero value of the area constant $c = 0$, there exist no less than six different periodic solutions of the Euler – Poisson equations under the condition $h > \max V$, $V = \mu(\boldsymbol{r}, \boldsymbol{\gamma})$, i.e., when the domain of the possible motion coincides with the whole Poisson sphere. The projections of these periodic solutions on the Poisson sphere define three different closed non-self-intersecting curves.

For $c \neq 0$, it was asserted in [447] (S. P. Novikov) that for $h > \max V_c$ $\left(V_c = V + c^2/2(I\gamma, \gamma)\right)$, the equations of rigid body rotation with a fixed point in an axisymmetric potential field with potential V have at least one periodic solution. But because of inaccuracies in the extension of Morse theory to multivalued functionals developed in [447, 448], this assertion cannot be regarded as proved. In [563], I. A. Taimanov proved the existence of a closed geodesic for a sufficiently large value of the area constant (corresponding to a "strong magnetic field"). The assertion in [447] that such periodic solutions were already known to classical researchers and can be found in terms of elliptic functions appears inaccurate.

For specific parameter values, computer experiments allow us to draw conclusions on the existence of several (non-self-intersecting) periodic orbits and also to study their stability, but it is not clear how using computer methods, we can examine, case by case, all possible parameter values.

One more problem, in which we can attain some progress using computer simulation, is related to the search for solutions periodic in absolute space. Such solutions were considered in the previous section.

Remark. The study of periodic orbits for a classical system is very important for its quantization and for the study of the system on the quantum (or quasi-classical) level [560]. The probability density of a quantum system is distributed extremely nonuniformly, and there exist domains called *scars* in which the amplitude is very large. These domains are concentrated near to unstable periodic orbits, and, as Berry and Mount would have it, quasi-classical quantum mechanics "sews the wave flesh on the classical bones" [41].

6.5 Chaos Evolution in the Restricted Problem of Heavy Rigid Body Rotation

In this section, we study the process of phase portrait chaotization in the restricted problem of rotation of a heavy rigid body with a fixed point. We present two mechanisms of chaotization that complement one another: homoclinic structure growth and the development of cascades of period doubling bifurcations. We describe the adiabatic behavior on the zero level set of the area integral as the energy tends to zero. We find meandering tori arising from the violation of the twist condition of the Poincaré mapping.

1 The restricted problem of rigid body dynamics

The Euler – Poisson equations describing the rigid body motion around a fixed point in a homogeneous gravity field have the form (see Ch. 2)

$$\mathbf{I}\dot{\boldsymbol{\omega}} + \boldsymbol{\omega} \times \mathbf{I}\boldsymbol{\omega} = \mu \boldsymbol{r} \times \boldsymbol{\gamma}, \quad \dot{\boldsymbol{\gamma}} = \boldsymbol{\gamma} \times \boldsymbol{\omega}, \tag{6.24}$$

where $\boldsymbol{\omega} = (\omega_1, \omega_2, \omega_3)$, $\boldsymbol{r} = (r_1, r_2, r_3)$, and $\boldsymbol{\gamma} = (\gamma_1, \gamma_2, \gamma_3)$ are components of the angular velocity vector, of the radius vector of the center of mass, and of the unit vector of the vertical in the system of principal axes of inertia $(\boldsymbol{e}_1, \boldsymbol{e}_2, \boldsymbol{e}_3)$, which is rigidly fixed in the body and passes through the fixed point, $\mathbf{I} = \operatorname{diag}(I_1, I_2, I_3)$ is the tensor of inertia with respect to the fixed point in the same axes, and $\mu = mg$ is the body weight.

Let us consider the problem of motion of a dynamically symmetric body under the following assumptions: $I_1 = I_2 = 1$, $I_3 = \delta$, and $\boldsymbol{r} = (0, \delta, 0)$. The triangle inequality holds for the moments of inertia as long as $\delta < 2$, and hence, to the chosen configuration there corresponds some real mass distribution. The angular velocity part of the equations of motion takes the form

$$\dot{\omega}_1 = (1 - \delta)\omega_2\omega_3 - \delta\gamma_3, \quad \dot{\omega}_2 = (\delta - 1)\omega_1\omega_3, \quad \dot{\omega}_3 = \gamma_1. \tag{6.25}$$

Let us consider the limiting case $\delta \to 0$ of this problem, where the body degenerates into a line segment. Such a passage to the limit, producing a restricted problem from the general one, is completely analogous to what happens in the three-body problem of celestial mechanics, and so we call this limiting model *the restricted problem of heavy rigid body dynamics*. In the limit, the moment of inertia and the moment of the gravity force with respect to the dynamical symmetry axis simultaneously tend to zero. Therefore, in the limit, we obtain a nontrivial equation for the proper rotation. This passage to the limit was suggested for the first time by V. V. Kozlov and D. V. Treshchev in [353]. An analogous passage to the limit for a more general case of a dynamically asymmetric body was considered by A. A. Burov in [113].

After passing to the limit as $\delta \to 0$, Equations (6.25), we arrive at the equations of the restricted problem of heavy rigid body dynamics,

$$\dot{\omega}_1 = \omega_2\omega_3, \quad \dot{\omega}_2 = -\omega_1\omega_3, \quad \dot{\omega}_3 = \gamma_1, \quad \dot{\boldsymbol{\gamma}} = \boldsymbol{\gamma} \times \boldsymbol{\omega} \tag{6.26}$$

and the integrals of system (6.26), which are obtained from the energy and area integrals of the initial problem take the form

$$\omega_1^2 + \omega_2^2 = 2h, \quad \omega_1\gamma_1 + \omega_2\gamma_2 = c, \quad \gamma_1^2 + \gamma_2^2 + \gamma_3^2 = 1. \tag{6.27}$$

Let us reduce Eqs. (6.26) to the common level surfaces of the integrals (6.27) [353]. For this purpose, we choose ξ and η as variables which parameterize ω_1, ω_2, and ω_3 as follows:

$$\omega_1 = \sqrt{2h}\sin\xi, \quad \omega_2 = \sqrt{2h}\cos\xi, \quad \omega_3 = \eta. \tag{6.28}$$

Using (6.26) we have

$$\ddot{y}_3 = (\omega_1^2 + \omega_2^2) y_3$$

whence we obtain, taking into account (6.27), $y_3 = A\cos(\sqrt{2h}t)$, where A depends on the values h and c of the integrals (6.27). Substituting now for y_3 into Eqs. (6.26) and into the expressions for the integrals (6.27), we find that

$$\begin{aligned} y_1 &= \frac{c}{\sqrt{2h}} \sin\xi - \sqrt{1 - \frac{c^2}{2h}} \sin(\sqrt{2h}t)\cos\xi, \\ y_2 &= \frac{c}{\sqrt{2h}} \cos\xi + \sqrt{1 - \frac{c^2}{2h}} \sin(\sqrt{2h}t)\sin\xi, \\ y_3 &= \sqrt{1 - \frac{c^2}{2h}} \cos(\sqrt{2h}t). \end{aligned} \tag{6.29}$$

It is easy to show that for all rigid body motions, the arguments of the square roots in (6.29) are nonnegative. After substituting from (6.28) and (6.29) into the equations of motion (6.26), we obtain the following equations for the remaining two variables:

$$\dot{\xi} = \eta, \quad \dot{\eta} = \frac{c}{\sqrt{2h}} \sin\xi - \sqrt{1 - \frac{c^2}{2h}} \sin(\sqrt{2h}t)\cos\xi. \tag{6.30}$$

We can represent Eqs. (6.30) in the Hamiltonian form

$$\begin{aligned} &\dot{\xi} = \frac{\partial \mathscr{H}}{\partial \eta}, \quad \dot{\eta} = -\frac{\partial \mathscr{H}}{\partial \xi}, \\ &\mathscr{H} = \frac{\eta^2}{2} + \frac{c}{\sqrt{2h}} \cos\xi + \sqrt{1 - \frac{c^2}{2h}} \sin(\sqrt{2h}t)\sin\xi. \end{aligned} \tag{6.31}$$

Thus, the limiting case of the problem of rigid body rotation about a fixed point reduces to a Hamiltonian system with one and a half degrees of freedom.

We now consider some properties of the absolute motion in the restricted rigid body problem. For this purpose, we write the quadrature for the precession angle

$$\dot{\psi} = \frac{\omega_1 y_1 + \omega_2 y_2}{y_1^2 + y_2^2}. \tag{6.32}$$

Substituting (6.27) and (6.29) into (6.32) and integrating the resulting expression with respect to time, we obtain

$$\psi(t) = \begin{cases} \arctan\left(\dfrac{\sqrt{2h}}{c} \tan(\sqrt{2h}t)\right) + \psi(0), & c \neq 0, \\ \psi(0), & c = 0. \end{cases} \tag{6.33}$$

In the limiting case $\delta \to 0$, the body degenerates into a line segment, and, therefore, in the fixed coordinate system, its motions are defined only by the apex $\boldsymbol{e}_3$, which is parallel to the third axis of inertia. The dependence of $\boldsymbol{e}_3$ on the Euler angles is given by

$$\boldsymbol{e}_3 = (\sin\theta\sin\psi, \ -\sin\theta\cos\psi, \ \cos\theta). \tag{6.34}$$

Taking into account the relation $\cos\theta = \gamma_3$ and using (6.33), we conclude that for any values of the integrals and initial conditions of Eqs. (6.31), the motion of the rod is periodic with frequency $\sqrt{2h}$. In particular, for $c = 0$, there is no precession, and any motion of the body is a uniform rotation in the vertical plane. Thus, in the limit the equations of motion are divided into two parts. The first part is related to the absolute motion of the body and can be integrated explicitly. The second part is related to proper rotation of the body and, generally speaking, it has no relation to the real body motion (since, in fact, it describes the rotation of an infinitely thin rod about its axis). An interesting fact is that even in the case of chaotic proper rotation, the motion of the body in the fixed coordinate system is regular and periodic. Besides, by making the passage to the limit of the restricted problem, we avoid problems related to the topology of the nontrivial energy levels and the choice of the global Poincaré transversal section. Also, we note that the main properties of the growth of chaos described here for the restricted problem also manifest themselves in the general Euler – Poisson system [96].

Note that system (6.31) can be also viewed as a mathematical pendulum with a time-periodic perturbation of a special form. Indeed, setting $\nu = \sqrt{1 - \frac{c^2}{2h}}$ and treating ν as a small parameter, we obtain

$$\mathscr{H} = \mathscr{H}_0 + \nu\mathscr{H}_1 + o(\nu), \quad \mathscr{H}_0 = \frac{\eta^2}{2} + \cos\xi, \quad \mathscr{H}_1 = \sin(\sqrt{2h}t)\sin\xi. \tag{6.35}$$

For $\nu = 0$, the system (6.31) is integrable. Its analytical nonintegrability for $\nu \neq 0$ was shown in [353] for the case $c \neq 0$ and in [169] for the case $c = 0$ by using a numerical construction of splitting separatrices to the unstable periodic solution. The Kovalevskaya exponents were calculated in [86] and hence the algebraic nonintegrability of the problem considered was shown. We first consider the process of transition to chaos in this problem for $c = 0$.

2 Transition to chaos for $c = 0$

Let us consider how the phase flow of system (6.31) for $c = 0$ changes as the total body energy h decreases from $+\infty$ to 0, see (6.27). For this purpose, we study the Poincaré mapping on the cylinder ($\xi \bmod 2\pi$, η) for Eqs. (6.30), corresponding to a period of the Hamiltonian (6.31).

This map is symmetric with respect to the involutions ($\eta \to -\eta$, $\xi \to \pi - \xi$), and ($\xi \to -\xi$, $t \to -t$). This allows us to restrict ourselves to consideration of the system dynamics in the domain $\eta \geqslant 0$, $\xi \in (0, \pi)$; however, for clarity, we present phase portraits in the entire domain in what follows. We may also note that our interest is mainly in how the system behaves for very small values of η, since for large η, we can neglect the terms depending on ξ and t in Hamiltonian (6.31), and the motion becomes close to an integrable one. In the phase portrait, the response to the value of η is evident by the fact that as η increases, the invariant curves differ less and less from horizontal lines.

Note that there exist the following two phase portrait chaotization mechanisms, which complement one another:

- *Transversal intersections of stable and unstable invariant manifolds (which are separatrices)* and the homoclinic structure of the mapping and the Smale horseshoe arising in this case. This mechanism results in the formation of a chaotic layer near to unstable periodic solutions.
- *Cascades of period doubling bifurcations*. After passing the cascade of period doubling bifurcations, the phase portrait is characterized by the existence of trajectories with an arbitrarily large period. We can say that the chaotic behavior corresponding to the emergence of such trajectories is local, since it is related to a particular set of periodic solutions. The global chaotization of the phase portrait near to the cascade is due to the fact that the regularity regions are "partitioned" into smaller regions around which chaos emerges according to the first chaotization mechanism.

As is seen below, in the problem considered, the joint action of both mechanisms leads to an almost complete chaotization of the phase portrait.

Now let us consider the stages of the phase portrait chaotization of this system under the energy decrease in more detail.

Splitting of resonant tori. For $h = +\infty$, the system (6.31) is integrable, and the corresponding phase portrait on the cylinder ($\xi \mod 2\pi,\ \eta$) has the form of horizontal circles $\eta = \text{const.}$, and, moreover, to each circle there corresponds a different rotation frequency. As the energy h decreases, we start to observe splitting of the invariant curves (the so-called resonant tori) for which the rotation frequency is commensurable with the forcing force frequency $\sqrt{2h}$. In Fig. 6.19, we present the corresponding phase portrait of the system for a large but not infinite energy value. As is seen from the figure, there is a considerable splitting of the resonant tori for this value of h, but still, the phase portrait is close to the integrable case. For this case, the resonances (periodic solutions) of first order play a determining role for the phase portrait. All periodic solutions that develop following the addition of a perturbation (energy decrease) are enumerated in Fig. 6.19. For this value of the energy, resonances of higher orders do not materialize, i.e., they split only weakly and are little different from usual invariant tori.

Remark. Permanent body rotations about the axes Ox and Oy are associated to the fixed points labelled 1 and 2 in Fig. 6.19 and the same rotations in the opposite directions are associated with different points on the diagram having the same labels. Periodic solutions in the absolute space are associated to fixed points with labels 3 and 4; these are called choreographies after [71]. In the course of motion along the periodic trajectories, the apexes Ox and Oy trace out closed trajectories in the form of figure-eight curves, while the third apex Oz moves along a circle in a vertical plane.

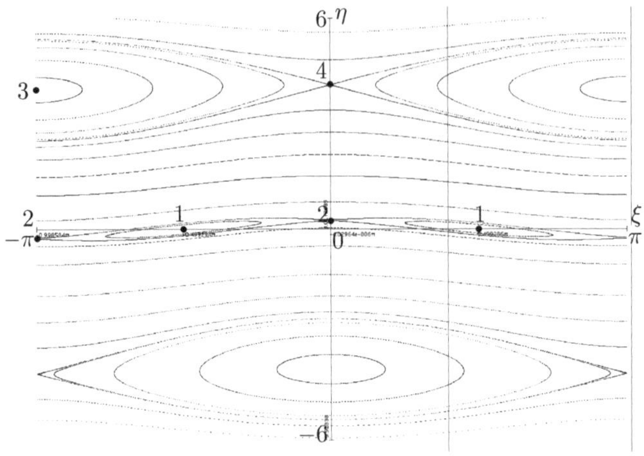

Fig. 6.19. Phase portrait for $c = 0$, $h = 8$.

Homoclinic structure of the mapping. As the energy decreases, separatrix splitting occurs near to unstable periodic solutions. In this case, narrow chaotic layers are born around stable periodic solutions of different orders (see Fig. 6.20). The method for constructing separatrices can be found in [459].

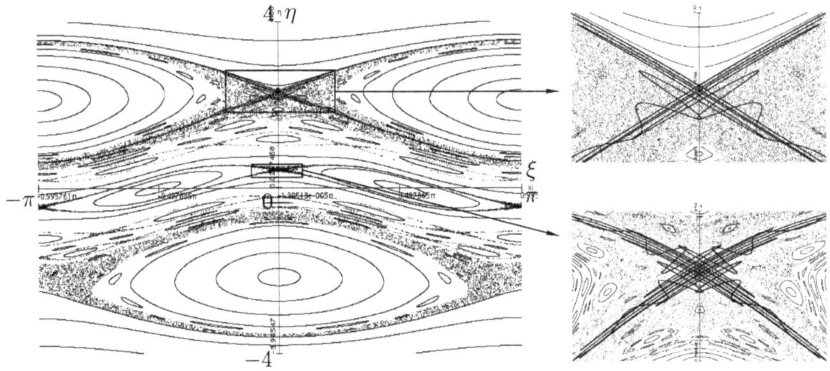

Fig. 6.20. Phase portrait for $c = 0$ and $h = 2.5$.

Remark. It is interesting that the upper pair of periodic solution separatrices corresponding to the unstable choreography hardly splits at all compared with the lower pair (see Fig. 6.20, to the right). We note that such a situation is observed across almost the whole range of h (except for very small values).

When the energy decreases yet further, the invariant tori separating the chaotic layers break down, and the chaotic layers merge with each other. In this case, the separatrices corresponding to different fixed points intersect each other transversally. After all chaotic layers have merged, a single chaotic layer arises, consisting of a net of intersections of the separatrices to resonances of different orders. The corresponding phase portrait and intersections of separatrices of fixed points of 1st, 2nd, 3rd, 4th, 5th, and 7th orders are presented in Fig. 6.21.

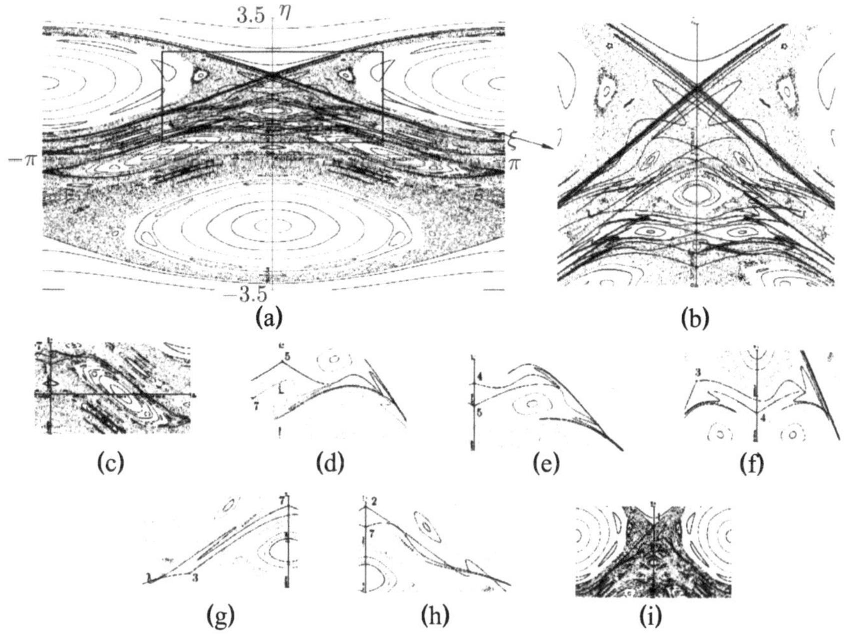

Fig. 6.21. Phase portrait and intersections of separatrices of the 1st, 2nd, 3rd, 4th, 5th, and 7th orders for $c = 0$ and $h = 2$. (a) and (b) are the phase portrait and its zoomed-up fragment in the domain of separatrix construction. (c) – (i) are sequential separatrix intersections. The numbers near fixed points are their periods.

REMARK. It follows from the closure theorem [576] and the existence of intermediate intersections of higher-order separatrices that the separatrices corresponding to the upper and lower fixed points of the first order in Fig. 6.21 also intersect, and their closures coincide. However, near a critical value of h at which the last invariant torus separating the chaotic layers breaks down, this intersection can occur at very large distances (in the sense of the number of iterations) from fixed points. Therefore, for finding the first intersection of separatrices, we need an exponentially large time.

Cascades of period doubling bifurcations. At the next chaotization stage, as the energy continues to decrease, the remaining regions of regularity break down via a cascade of period doubling bifurcations. The cascade is an infinite sequence of period doubling bifurcations (or multiple period increase bifurcations) under variations of system parameters within finite limits. After passing through the cascade, an almost completely homogeneous chaotic layer is formed near the origin (Fig. 6.22).

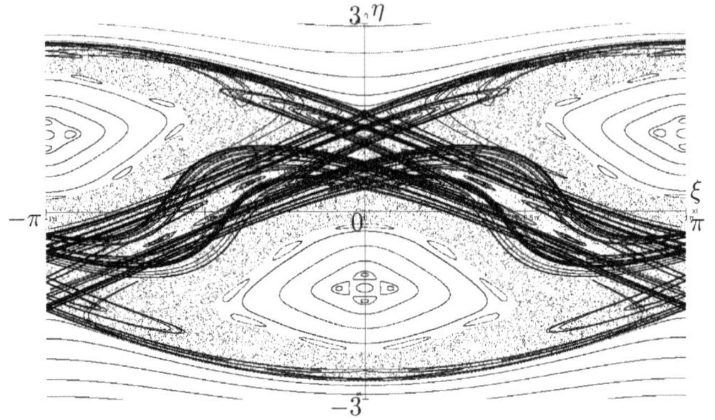

Fig. 6.22. Phase portrait and separatrices to first-order fixed points with $c = 0$ and $h = 1$.

Note that the cascade stage is, as a rule, preceded by some preliminary bifurcations. For illustration, we present the bifurcation scheme for decreasing energy for the two most important stable first-order periodic solutions

- *Periodic solution* 1 (Fig. 6.19).
  1) a period doubling bifurcation;
  2) a fork-type bifurcation;
  3) a cascade of period doubling bifurcations.
- *Periodic solution* 3 (Fig. 6.19).
  1) a double period doubling bifurcation for which the stability type of a fixed point is preserved, and two pairs of stable and two pairs of unstable solutions of doubled period are born from it;
  2) a cascade of period doubling bifurcations of the already born pair of second-order stable periodic solutions;
  3) a fork-type bifurcation of the main solution;
  4) a cascade of period doubling bifurcations.

The third periodic solution lies on the symmetry plane $\xi = \pi$, see Fig. 6.16. In [488], it was shown that for a reversible map, under a bifurcation of a periodic solution lying on the symmetry plane, new periodic solutions obtained by this process intersect this plane at two points. Moreover, under continuous variation of a parameter, these solutions do not go away from the symmetry plane. Therefore, in studying the period doubling bifurcations of the third periodic solution, it suffices to trace the births of periodic solutions that lie on the axis $\xi = \pi$. The corresponding projection of the tree of period doubling bifurcations on the plane (h, η) is shown in Fig. 6.23. The rescaling factors for the parameter h are presented in Table 6.1 for the first six period doubling bifurcations. As is seen from the table, the values of the rescaling factors converge quite rapidly to the Feigenbaum constant $\delta \approx 8.721$.

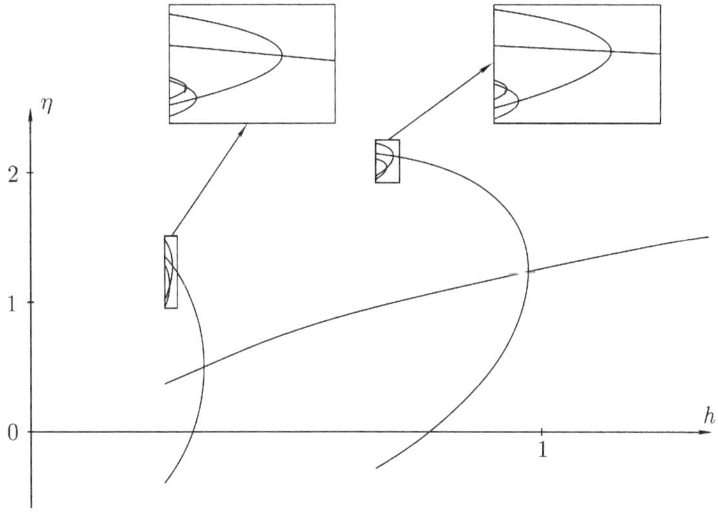

Fig. 6.23. Projection of the tree of period doubling bifurcations starting from absolute choreographies on the plane (h, η).

Table 6.1. Rescaling factors for the bifurcation tree (Fig. 6.23).

Number of bifurcation	h_n	δ_n	Number of bifurcation	h_n	δ_n
2	0.287733199		2	0.734936403	
3	0.282333473		3	0.721745207	
4	0.281711468	8.681161784	4	0.720016837	7.632159557
5	0.281640344	8.745399858	5	0.719817826	8.684798818
6	0.28163219	8.721983486	6	0.719795025	8.728028583

As was explained above, after passing through cascades of period doubling bifurcations, for small η, the system phase portrait is almost completely chaotic. However, for a certain (sufficiently small) value of the energy a stable first-order periodic solution is born again on the axis $\xi = \pi$. As the energy decreases further, this solution undergoes a fork-type bifurcation, and then the newly-formed regularity domain becomes chaotic via another cascade of period doubling bifurcations.

The case with $c = 0$ and $h \to 0$. Adiabatic behavior. Let us consider the way the phase portrait changes as the energy h tends to zero. For this purpose, we recall that if h is fixed while η tends to infinity, the system (6.31) approaches an integrable system in which all its solutions are uniform rotations with respect to ξ for constant η. Now let us consider Eqs. (6.31) for η large as a time-periodic perturbation of this integrable system. Thus, the second and third terms in the Hamiltonian (6.31) are considered to be perturbations, and the increase in these perturbations corresponds to decrease in η.

Let us define the rotation number as follows:

$$n = \frac{1}{2\pi} \int_0^T \dot{\xi} dt \quad \in \mathbb{R}, \tag{6.36}$$

where $T = \frac{2\pi}{\sqrt{2h}}$ is the period of the perturbation. As usual, the addition of a perturbation leads to a breakdown of the resonant tori for which the relation $n = \frac{p}{q}$, $p, q \in \mathbb{N}$, holds. By the Poincaré theorem [335], an even number of periodic solutions of order q are born from resonant tori for which the denominator of the rotation number is q. Substituting from (6.30) in (6.36), for large η, we obtain

$$n = \frac{\eta}{\sqrt{2h}} + o(\eta). \tag{6.37}$$

It follows from (6.37) that the rotation number grows without bound as η increases. Therefore, for a fixed h in system (6.31), there are infinitely many first-order periodic solutions born out of the splitting of resonant tori with integer rotation numbers, i.e., when $q = 1$.

Let us consider the variation of those periodic solutions having a given rotation number n. It follows from (6.37) that as h decreases, the periodic solutions with a given rotation number move towards the origin. At a certain stage η will become too small for the last two terms in the Hamiltonian (6.31) to be considered any longer as a small perturbation. As a result, the further evolution of periodic solutions can be studied only by using computer methods, in particular, by the method of continuous variation of a parameter. It turns out that, as h decreases further, the periodic solutions born from resonant tori continue to approach the chaotic layer, which was formed after cascades of period doubling bifurcations. For a certain critical value of h, this layer merges with the narrow chaotic layer near unstable periodic solutions with rotation number n. After that, the remaining regularity domains near stable periodic solutions with rotation number n become chaotic via cascades of period doubling bifurcations. The evolution of the phase portrait described here is observed for periodic solutions with integer rotation numbers $n \geqslant 3$, to which there correspond first-order fixed points on the section.

Therefore, as the energy h tends to zero, we observe infinitely many cascades of period doublings of first-order fixed points, while these points approach the chaotic layer in the domain of small $|\eta|$.

For $c = 0$ and $h = 0$, we also obtain an integrable system, but for $c = 0$ and for any arbitrarily small $h > 0$ a chaotic layer of finite width (not tending to zero as $h \to 0$) forms in the phase portrait. In some respects, such a behavior is analogous to that of a system with adiabatic chaos. Indeed, for $c = 0$, the equations of motion with Hamiltonian

$\mathcal{H} = \frac{\eta^2}{2} + \sin(\sqrt{2h}t)\sin\xi$ with h small can be treated as an adiabatic system with slowly varying parameter $\tau = \sqrt{2h}t$, $\dot{\tau} = \sqrt{2h} \to 0$. The methods for studying such systems were developed in [441] (see also [19, 613]). Note that these methods are closely related to studying jumps of the adiabatic invariant in passing through separatrices to hyperbolic fixed points (periodic solutions) and also to obtaining explicit criteria for the splitting of these separatrices. A specific feature of our problem is the change in the stability of the main periodic solutions during the perturbation period. As a result of this stability change, the technique developed in [441] cannot be applied straightforwardly. However, experiments show that as $h \to 0$, a certain region of chaos is not removable. The problem of finding the limit value of η bounding this chaotic layer as $h \to 0$ seems to be of interest.

REMARK. Note that for any arbitrarily small value of $c \neq 0$ of the area constant $(\boldsymbol{M}, \boldsymbol{\gamma})$ and for h at its minimum value $h = h_{\text{min}} = c^2/2$, the system (6.30) reduces to the mathematical pendulum equation and is also integrable. But in the corresponding perturbed system with $h > h_{min}$ the area of the chaotic layer tends to zero as $h \to h_{min}$.

3 Case $c \neq 0$

The system (6.31) has two parameters, and therefore, in the general case it is necessary to study the dynamics of the system on the plane of parameters (c, h) in the domain $h > c^2/2$. Here, we restrict ourselves to the case with $c = 1$, although, as the simulations for other values of c show, the general laws in the evolution of chaos are preserved.

Let us briefly describe the main stages of the phase portrait evolution as h decreases.

1. For $h = +\infty$ (as for the case $c = 0$) the system (6.31) is close to an integrable one, and the phase portrait is foliated by parallel circles.
2. As h decreases, the resonance tori split and the resulting chaotic layers are enlarged (see Fig. 6.24). Eqs. (6.31) are now invariant only under the involution $(\xi \to -\xi, t \to -t)$, and therefore, the phase portrait is symmetric only with respect to the plane $\xi = 0$. This leads to the conclusion that for sufficiently large energies, there are no solutions 1 (Fig. 6.19), whereas one of the solutions 2 is stable.
3. As h decreases further, the chaotic layers merge into one layer (see Fig. 6.25). Qualitatively, there is no difference between this case and the case $c = 0$ considered previously.
4. Next, almost all regularity domains within the single chaotic layer (as in the case $c = 0$) break down through a cascade of period doubling bifurcations. For $c = 1$, the regularity domain around the solution 2 and some small higher-order resonances are exceptions (see Fig. 6.26). Note that in the general case (for $c \neq 0$), they can also become chaotic. For a sufficiently small c, for example, the solution 2 loses

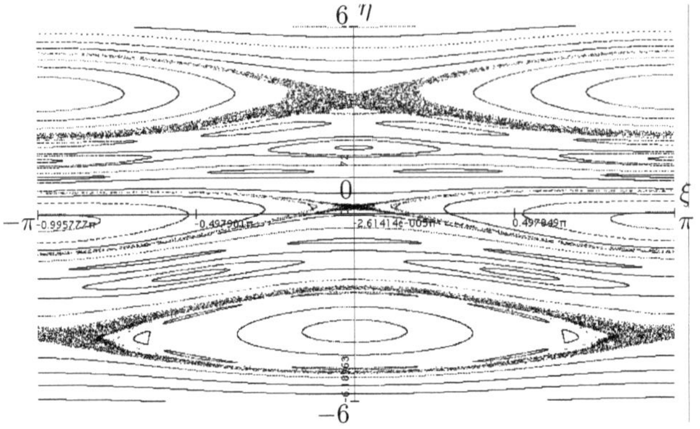

Fig. 6.24. Phase portrait for $c = 1$ and $h = 8$.

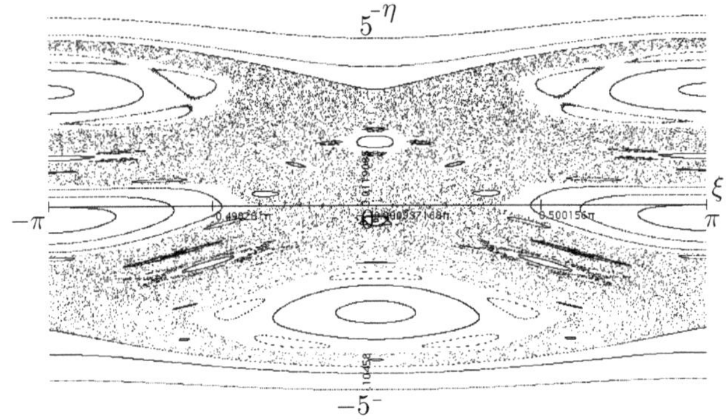

Fig. 6.25. Phase portrait for $c = 1$ and $h = 4.5$.

its stability through a fork-type bifurcation. However, at later stages, it bifurcates again and becomes stable.

5. Yet further decrease in h leads to enlargement of the regularity domain around the solution 2 and to the diminishing of the chaotic layer (see Fig. 6.27). For $h = h_{\min} = c^2/2$, this process finishes by arriving at an integrable system (the mathematical pendulum). During this final stage, there arise the so-called meandering tori to whose description we now turn.

4 Meandering tori

Let us look at the phase portrait evolution near to the minimum energy value $h_{\min} = c^2/2$, i.e., near to the mathematical pendulum, in more detail. In the phase portrait, there exists a pair of first-order periodic solutions (stable and unstable). It turns out

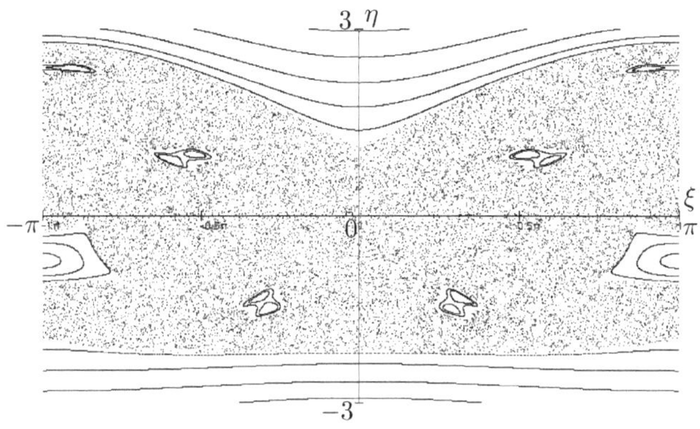

Fig. 6.26. Phase portrait for $c = 1$ and $h = 0.935$.

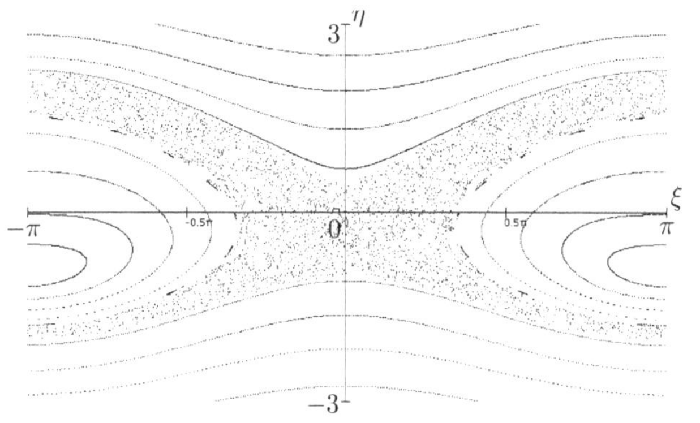

Fig. 6.27. Phase portrait for $c = 1$ and $h = 0.510152$.

that as the energy h increases (i.e., as a perturbation is added), the twisting property of the Poincaré map becomes invalid in a small neighborhood of the stable periodic solution. As a result of the twisting property violation, there arise *meandering invariant tori* [532] for which the radius of the corresponding invariant curves on the section cannot be represented as a single-valued function of the angle (see Figs. 6.28 and 6.29). The meandering tori in Hill's problem of celestial mechanics were found in [533].

Let us briefly describe the mechanism of formation of meandering tori. Since the twisting property is no longer valid, the extremum (a maximum in this case) of the rotation number does not coincide with the main periodic solution but lies on a certain invariant curve (torus). As a result, on different sides of this curve, there exist resonant tori (called *resonances* in what follows) with the same rotation numbers $n = p/q$, $p, q \in \mathbb{N}$. Under the increase in perturbation (growth of h), the maximum value of the rotation number approaches p/q; as a result of this, the resonances approach each other. Moreover, as h grows, their splitting enlarges. For a certain critical value of h^*,

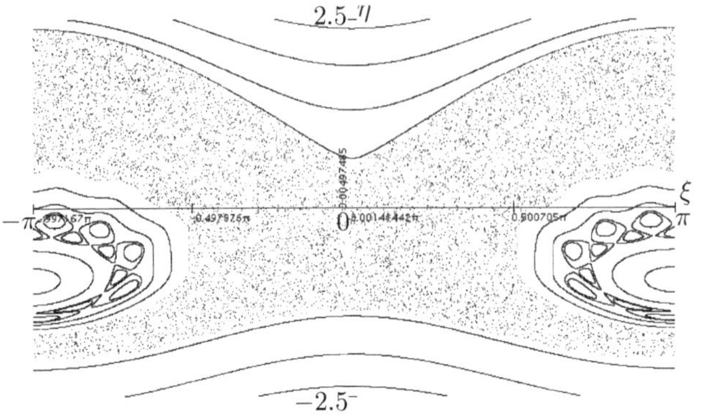

Fig. 6.28. Meandering tori near resonances with rotation number $n = 2/9$ for $c = 1$ and $h = 0.5539$.

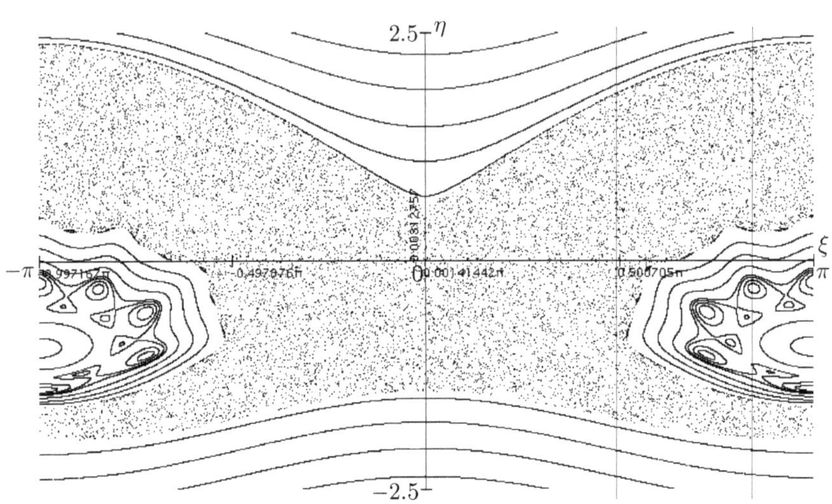

Fig. 6.29. Meandering tori near resonances with rotation number $n = 1/4$ for $c = 1$ and $h = 0.57362$.

the resonances merge, whereas the separatrices belonging to different resonances intersect each other. After that, there arise tori enveloping stable periodic solutions of both resonances; they are said to be *meandering tori*. With further increase in h, the stable periodic solutions of one of the resonances approach unstable solutions of the other and then mutually annihilate. The annihilation occurs at the moment when the maximum value of the rotation number becomes equal to p/q.

Note that meandering tori exist only within sufficiently narrow energy intervals $h \in (h^*, h(n_{\max} = p/q))$. For the meandering tori depicted in Figs. 6.28 and 6.29, this interval is of order 10^{-4}. Moreover, the width of a meandering torus width strongly depends on the value of splitting of the initial resonant tori. Therefore, in the phase portrait, the meandering tori born out of resonances of small order being sufficiently distant from the integrable case are seen the most clearly.

6.6 Adiabatic Chaos in the Liouville Equations

Let us consider the Liouville equations (Sec. 2.8 of Ch. 2) with parameters varying periodically and slowly in time. It turns out that in the general case, such a system exhibits chaotic behavior, which has its own specific features [74]. We first present some general theoretical methods based on analysis of jumps in the value of the adiabatic invariant and the separatrix splitting and then compare them with results of computer experiments.

1 Hamiltonian systems with one and a half degrees of freedom. Adiabatic invariant jumps and adiabatic chaos

Let us consider a nonautonomous Hamiltonian system with one degree of freedom and slowly varying time dependence, i.e., the system depends explicitly on $\tau = \varepsilon t$, $\varepsilon \ll 1$. Note that all the arguments presented below for such nonautonomous systems hold also for autonomous Hamiltonian systems with two degrees of freedom whose evolution can be decoupled into slow and rapid components. Some such systems were considered in [440, 74].

For a system

$$\dot{q} = \frac{\partial H}{\partial p}, \quad \dot{p} = -\frac{\partial H}{\partial q} \tag{6.38}$$

with one and a half degrees of freedom, we assume that the Hamiltonian $H = H(p, q, \tau)$ is an analytic function of the variables (p, q, τ) and depends periodically on the phase q with period 2π and the slow time $\tau = \tau_0 + \varepsilon t$ with period 1 ($0 < \varepsilon \ll 1$ is a small parameter, $0 \leqslant \tau_0 < 1$). The periodic dependence of the Hamiltonian on time allows us to define the *"slow" Poincaré mapping* as the period map for the period τ mod 1 $\approx \left(t \bmod \frac{1}{\varepsilon}\right)$. Also, it is assumed that for each fixed value of the parameter $\tau = \tau_0$, the "frozen" integrable system with Hamiltonian $H = H(p, q, \tau_0)$ has separatrices of the same (homoclinic or heteroclinic) type. For example, the pendulum with slowly varying length and some physical problems arising in plasma theory [19, 612] are described by equations of the form (6.38). To describe the system's evolution far from the separatrices, one uses adiabatic and improved adiabatic approximations corresponding to the first two steps of standard perturbation theory [441]. If in the system (6.38) we pass to the action-angle variables (I, φ) for each fixed τ_0 ("frozen" system), then the equations of motion take the form

$$\dot{I} = -\varepsilon\frac{\partial H_1}{\partial \varphi}, \quad \dot{\varphi} = \omega_0 + \varepsilon\frac{\partial H_1}{\partial I}, \tag{6.39}$$

and the Hamiltonian can be represented as

$$H = H_0(I, \tau) + \varepsilon H_1(I, \varphi, \tau).$$

The theorem of averaging in one-frequency systems [19] states that if the action variable I is far away from separatrices, it undergoes a variation of order ε over times $\frac{1}{\varepsilon}$, i.e., it is an adiabatic invariant (AI) (according to KAM-theory, under certain nonlinearity conditions, this is true for arbitrary times).

A more precise description of the behavior of system (6.39) can be obtained at the next step of perturbation theory. We can define the improved adiabatic invariant (IAI) J in the form

$$J = I + \varepsilon u(p, q, \tau), \quad u = -\frac{1}{\omega_0}(H_1(p, q, \tau) - \langle H_1 \rangle),$$

where the brackets $\langle \dots \rangle$ denote averaging over φ. The evolution equation for IAI is

$$\dot{J} = -\varepsilon^2 \frac{\partial H_2}{\partial J}. \tag{6.40}$$

Therefore, far away from the separatrices, the improved adiabatic invariant undergoes oscillations of order ε^2.

Near the separatrix, the adiabatic description of the system no longer holds and must be modified. Depending on the initial conditions, two types of behavior of the adiabatic invariant are possible in a domain close to the separatrix.

In the first case, which corresponds to passage through a resonance, the trajectory goes through the resonance and does not stick, whereas the IAI undergoes a jump of order ε [441, 569]. At the same time, in domains far away from the separatrices, the variation of the IAI is of order ε^2.

In the second case, which corresponds to a capture into resonance [19], a point in a neighborhood of the separatrix moves in such a way that the commensurability is almost preserved, which leads to the IAI variation by a quantity of order 1 over times of order $\frac{1}{\varepsilon}$. However, as is shown in [19], the measure of such trajectories behaves like $\sqrt{\varepsilon}$ as $\varepsilon \to 0$.

Moreover, as noted in [437], if the IAI and the phase vary together in a coherent fashion, then stable periodic trajectories may appear. The measure of the regions of stability near these stable periodic trajectories is of order 1 as $\varepsilon \to 0$, although it is very small (typically $\sim 2\,\%$).

Here, we consider the trajectories that are neither trapped in a resonance nor lie in a neighborhood of periodic trajectories near separatrices. The measure of such trajectories is close to being a full measure, and a probabilistic description is applicable to them.

In [441], formulas for the IAI jump value ΔJ for passage through the separatrix were obtained. It was shown that the jump value can be considered as a function of a random variable with a given distribution. For multiple passages through the separatrix, the IAI variation has the character of a random walk with a random step size of order ε, and there is a nonzero probability of detecting an IAI value different from the initial one by a quantity of order 1 after $\sim \frac{1}{\varepsilon}$ steps of the Poincaré mapping. A stochastic process of such a kind in dynamical systems is usually called *adiabatic chaos*. However, we note that the appearance of this kind of chaos is accompanied by the usual effects: splitting of separatrices and the appearance of quasi-random oscillations [612, 279]. It is a characteristic feature of adiabatic chaos that *the chaotic domain does not diminish as* $\varepsilon \to 0$ in contrast to the standard situation [19, 506].

The general formulas for the adiabatic invariant jump in Hamiltonian systems and examples of their applications to various problems of mechanics and physics are presented in [441, 569, 438, 573]. In Sec. 2, we consider a system with the property that for fixed $\tau = \tau_0$ its separatrices lie in the shaded domains depicted in Fig. 6.30. If the phase portrait of the "frozen" system remains symmetric with respect to the axis Op (Fig. 6.31), then for the trajectories intersecting the separatrix, the IAI jump is calculated by the formula

$$\Delta J = -\frac{\varepsilon\Theta(\tau^*)a}{2\pi}\ln(2\sin\pi\xi) + \mathcal{O}\left(\varepsilon^{3/2}\left(|\ln\varepsilon| + \frac{1}{1-\xi}\right)\right), \tag{6.41}$$

where $\Theta(\tau) = \frac{dS}{dt}$ is the rate of variation of the area S of the domain bounded by one of the separatrices and τ^* is the moment when the trajectory crosses the separatrix, which is found from the equation $I(\tau^*) = \frac{S(\tau^*)}{2\pi}$, $a = \left[-\det\left\|\frac{\partial^2 H}{\partial p\partial q}\right\|\right]^{-1/2}$. For a small variation ($\sim\varepsilon$) of the initial condition, the change in the value $\xi \in (0, 1)$ is of order 1, and hence we can consider ξ as a random variable whose distribution is uniform, since the system is Hamiltonian [441]. For asymmetric domains, formula (6.41) becomes more complicated [441].

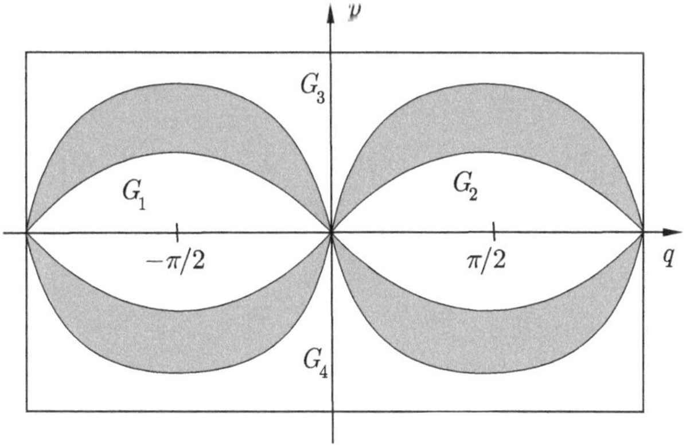

Fig. 6.30. Separatrices of system (6.45) for $\delta = \delta_{\min} = \frac{1}{4}$ and $\delta = \delta_{\max} = \frac{3}{4}$.

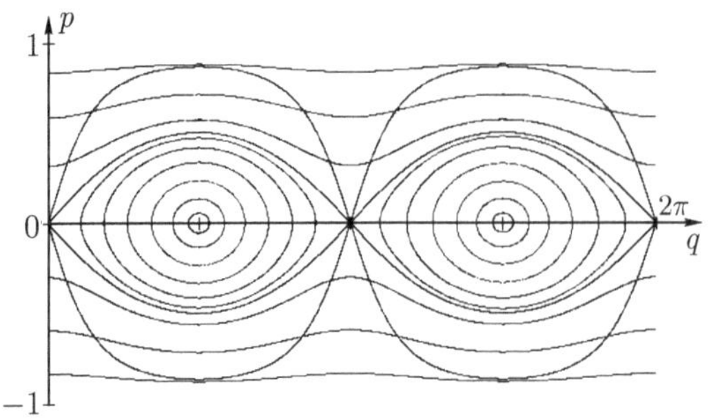

Fig. 6.31. Phase portrait of the "frozen" system (6.45) for $\tau = 0.5$ when $\delta(\tau)$ is defined by (6.46).

2 Rigid body dynamics with slowly varying parameters

The Liouville equations (see Sec. 2.8 of Ch. 2) have the form

$$\dot{\boldsymbol{M}} = \boldsymbol{M} \times \frac{\partial H}{\partial \boldsymbol{M}}, \quad H = \frac{1}{2}(\boldsymbol{M}, \mathbf{A}\boldsymbol{M}) - (\boldsymbol{M}, \boldsymbol{K}), \tag{6.42}$$

where $\mathbf{A} = \operatorname{diag}(a_1, a_2, a_3)$ is the matrix with entries $a_i = 1/I_i$ (I_i are the principal moments of inertia) and the vector $\boldsymbol{K} = (K_1, K_2, K_3)$ describes the gyrostatic momentum in the body.

REMARK. Recall that if $\mathbf{A}$ and $\boldsymbol{K}$ are independent of time, then we obtain the Zhukovskii – Volterra system (Sec. 2.7 of Ch. 2). For explicit dependence of $\mathbf{A}(t)$ and $\boldsymbol{K}(t)$ on time, Eqs. (6.42) were studied by J. Liouville who observed several simple integrable cases. A more complete study of the integrability and proof of nonintegrability of the system in the case of periodic dependence of $\mathbf{A}(t)$ and $\boldsymbol{K}(t)$ on time were carried out in [93, 74] by using the separatrix splitting method (and later in [179]). In [93], the dependence on time is not slow, and the system is close to being an integrable system.

Here, we consider the situation where $\mathbf{A}$ and $\boldsymbol{K}$ vary slowly and periodically in time with period $\frac{1}{\varepsilon}$. Equations (6.42) can be written in canonical form on the level surface of the integral $\boldsymbol{M}^2 = G^2$ after making the change of variables

$$M_1 = G\sqrt{1-p^2}\sin q, \quad M_2 = G\sqrt{1-p^2}\cos q, \quad M_3 = Gp. \tag{6.43}$$

Moreover, if we rescale time and the system parameters by $dt' = (a_3 - a_1)dt$, $\delta = (a_2 - a_1)/(a_3 - a_1)$, $v_i = a_i K_i (a_3 - a_1)$, then a canonical system is obtained with the Hamiltonian

$$H = \frac{1}{2}p^2 + \frac{1}{2}\delta(1-p^2)\cos^2 q - v_1\sqrt{1-p^2}\sin q - v_2\sqrt{1-p^2}\cos q - v_3 p. \tag{6.44}$$

In the case where there is no gyrostatic momentum $\boldsymbol{K} = 0$ and the Hamiltonian (6.44) simplifies to

$$H = \frac{1}{2}p^2 + \frac{1}{2}\delta(1-p^2)\cos^2 q, \quad \delta = \delta(\tau) = \delta(\tau_0 + \varepsilon t). \tag{6.45}$$

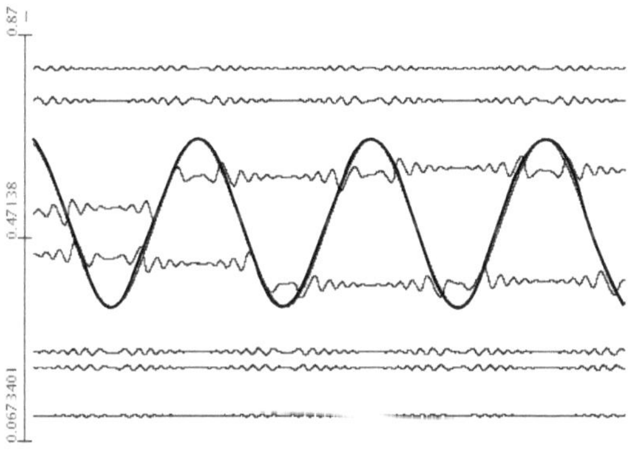

Fig. 6.32. Dependence of the action on time for separate trajectories from different domains. (The periodic trajectory corresponds to the action for the separatrix of the "frozen" system.)

Now suppose that in (6.45) the parameter δ varies according to the law

$$\delta = \frac{1}{2}\left(1 - \frac{1}{2}\cos 2\pi\tau\right). \tag{6.46}$$

For any fixed $\tau = \tau_0$, the phase portrait of the "frozen" system contains symmetric separatrices (Fig. 6.31). The characteristic dependence of AI on the slow time τ is presented in Fig. 6.32 (by symmetry, the action variables coincide in the domains G_1 and G_2 and also in G_3 and G_4; see Fig. 6.30). The variation in the AI for the separatrix of the "frozen" system is indicated by the boldface line in Fig. 2.7: the motion is regular far away from the separatrix, and I = const. (up to $O(\varepsilon)$). The chaotic region corresponds to the segment $[I_{min}, I_{max}]$ (in this case, $I_{min} = 1/3$ and $I_{max} = 2/3$) and the AI has jump-like behavior with random step of order ε. On the other hand, for the Poincaré mapping T during the period τ mod 1, in the domain bounded by the curves $I = I_{min}$ and $I = I_{max}$, we observe stochastic behavior,

The value of the IAI jump for one intersection of the separatrix is given by the expression

$$\Delta J = -\frac{1}{2\pi}\varepsilon a\Theta(\tau^*)\ln(2\sin\pi\xi) + \mathcal{O}\left(\varepsilon^{3/2}\left(|\ln\varepsilon| + (1-\varepsilon)^{-1}\right)\right), \tag{6.47}$$

where $\Theta(t) = \frac{2\delta'}{\sqrt{\delta(1-\delta)}}$ is the growth rate of the area under the separatrix for the Hamiltonian (6.45) and $a = \frac{1}{\sqrt{\delta(1-\delta)}}$.

Since the trajectory intersects the separatrix two times during one period in τ, and, moreover, at the second intersection, the quantity $\Theta(\tau^*)$ has the same magnitude as at the first intersection with the opposite sign, the total IAI jump up to order $O(\varepsilon)$ is equal to

$$\Delta J = -\frac{\varepsilon a\Theta(\tau^*)}{2\pi}\ln\frac{\sin\pi\xi_1}{\sin\pi\xi_2} = -\varepsilon\frac{\delta'(\tau^*)}{2\pi(1-\delta)\delta}\ln\frac{\sin\pi\xi_1}{\sin\pi\xi_2}, \tag{6.48}$$

and averaging over the random variables ξ_1 and ξ_2 gives

$$\langle \Delta J \rangle = 0,$$
$$\langle (\Delta J)^2 \rangle = \varepsilon^2 \frac{a^2 \Theta^2(\tau^*)}{2\pi^2} \int_0^1 \ln^2(2 \sin \pi\xi) d\xi = \frac{a^2 \Theta^2(\tau^*)}{24} \varepsilon^2. \tag{6.49}$$

Below we present the results of studying the Poincaré mapping and the IAI jumps numerically for the case where $\delta(\tau)$ is defined by (6.46), i.e., by ($\mathbf{A} = \mathbf{A}(t)$ and $\boldsymbol{K} = 0$). The phase portraits for $\tau_0 = 0.5$ and for various values of ε are presented in Fig. 6.33. It is seen that as ε decreases, the chaotic region diminishes but remains finite and bounded by the curves $I(p, q, \tau_0) = I_{\min}$ and $I(p, q, \tau_0) = I_{\max}$. The appearance of chaos is caused by a random IAI variation at the moment of intersection with the separatrix. The IAI variation over the interval $\sim 1/\varepsilon$ is plotted versus initial conditions in Fig. 6.34. In this figure, the initial IAI value J_0 is represented along the horizontal axis and the jump $\left(\frac{\Delta J}{\varepsilon}\right)^2$ is represented along the vertical axis. When the trajectory crosses the separatrix (unless a resonant capture occurs), the IAI changes by a quantity of order $\sim \varepsilon$. No higher peaks correspond to these values of J_0 in the dependence $\left(\frac{\Delta J}{\varepsilon}\right)^2 (J_0)$. If the trajectory is trapped into resonance, then $\Delta J \sim 1$, and we observe high peaks in the dependence of $\left(\frac{\Delta J}{\varepsilon}\right)^2 (J_0)$ on J_0. The heights of the peaks and the widths between them are determined by the parameters and the initial conditions of system (6.45). In some cases, for trajectories that are almost identical at the initial moment, the IAI values can differ considerably after the separatrix is crossed (see Fig. 6.35).

In Fig. 6.34, we also present the analytical function (6.49) that illustrates the IAI variation averaged over φ [441, 436]. As may be seen, it is smooth and does not completely reflect the real behavior of the system. At the same time, the use of numerical methods allows us to study the system's evolution in detail depending on the whole set of initial conditions for fixed parameter values. In Fig. 6.36, we present the characteristic form of the surface describing the IAI variation after a time interval of order $\sim 1/\varepsilon$ as a function of the initial value J_0 and the initial phase φ_0. As may be seen, this surface has a very complicated character and cannot be described analytically. Also, we may note that for the majority of real physical systems, even small values of ε are, nevertheless, so large that formulas (6.49) can still be used. The example of the IAI jump study presented here is one of the possible methods for the numerical description of systems with slowly varying parameters. Another method is based on analysis of the separatrix splitting conditions [74].

3 Separatrix splitting and adiabatic chaos conditions

Let us consider the case where $\boldsymbol{K} \neq 0$. For simplicity, we assume that $K_1 = K_3 = 0$ and $K_2 = K(\tau) = K(\varepsilon t)$. For the canonical system (6.44) in this case we have Hamiltonian

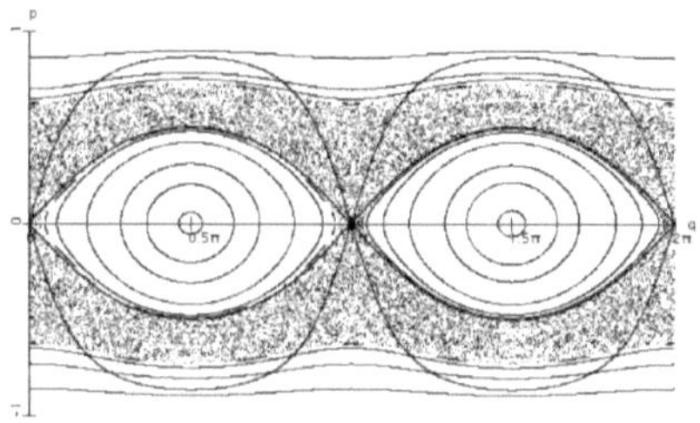

(a) τ_0=0.5, ε=0.01

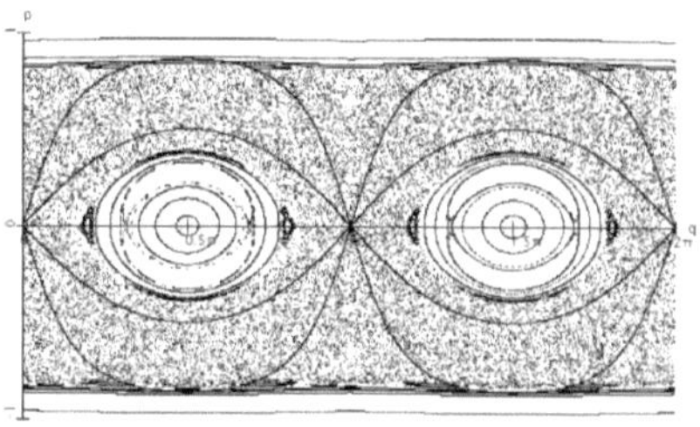

(b) τ_0=0.5, ε=0.05

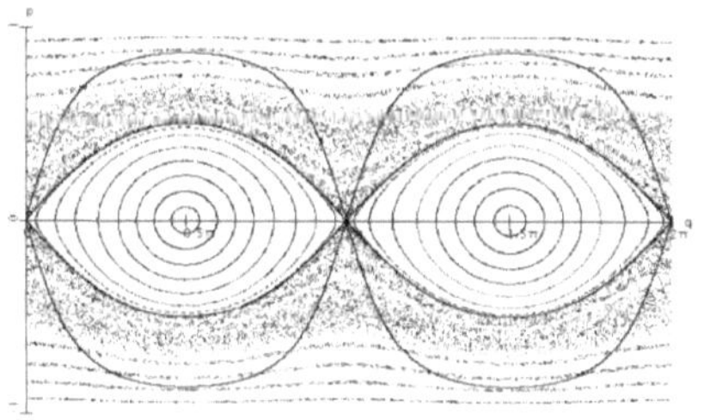

(c) τ_0=0.5, ε=0.001

Fig. 6.33. Phase portrait for the "slow" Poincaré mapping of system (6.45) for $\varepsilon = 0.01$ (a), $\varepsilon = 0.05$ (b), and $\varepsilon = 0.001$ (c).

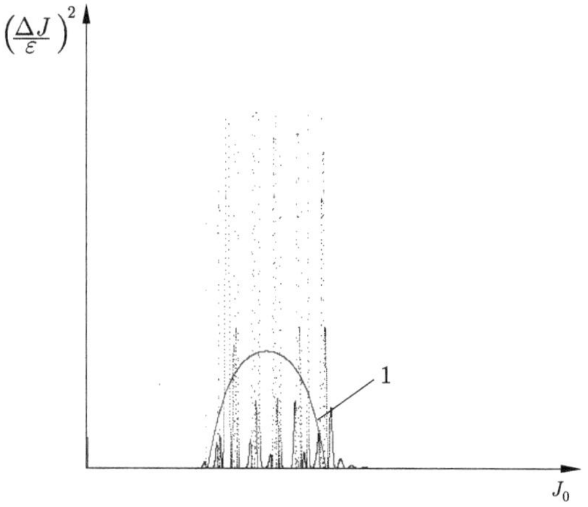

Fig. 6.34. IAI variation over the interval $\frac{1}{\varepsilon}$ versus the initial value J_0; the number 1 denotes the analytic dependence (6.49).

$$H = \frac{1}{2}p^2 + \frac{1}{2}\delta(1 - p^2)\cos^2 q - \nu\sqrt{1 - p^2}\cos q. \tag{6.50}$$

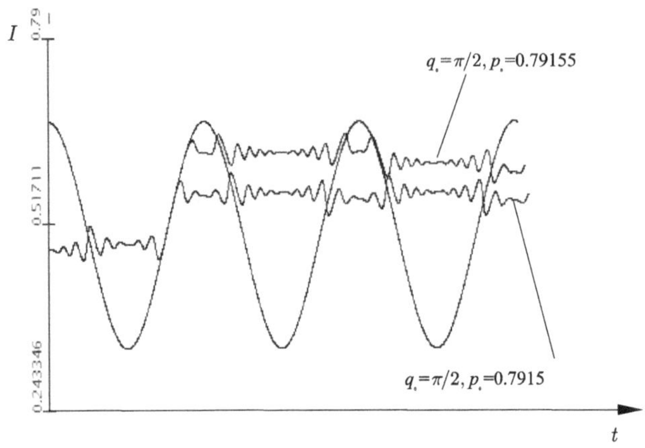

Fig. 6.35. Divergence of two trajectories during a resonant capture with very close initial conditions (q_0, p).

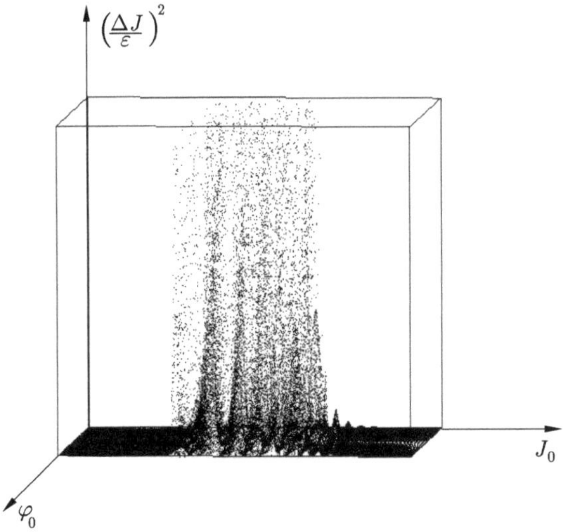

Fig. 6.36.

The phase portrait of the "frozen" system (the integrable Zhukovskii – Volterra case) has the form depicted in Fig. 6.37. For the system (6.50) conditions for the emergence of adiabatic chaos may be specified; some of these are obtained from analysis of AI jumps and others are obtained via the separatrix splitting method. In Fig. 6.38, we show the characteristic phase portrait of system (6.50) when parameters are slowly varying; analogously to the above case, the chaotic regions bounded by the curves $I(p, q, \tau_0) = I_{\min}$ and $I(p, q, \tau_0) = I_{\max}$ are clearly seen.

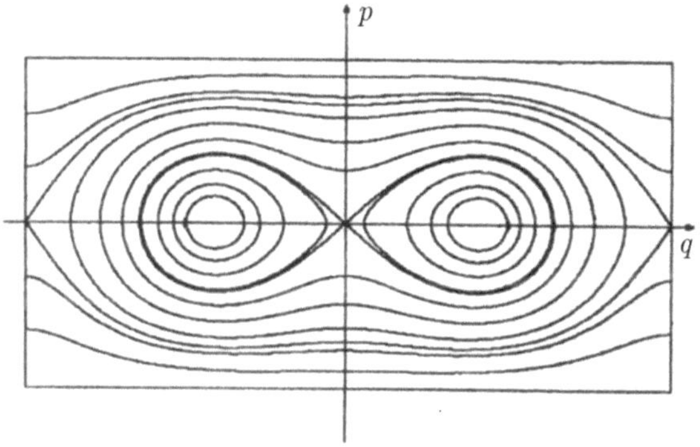

Fig. 6.37. Phase portrait of the "frozen" system (6.51) for $\nu = 0.1$ and $\tau_0 = 0.5$.

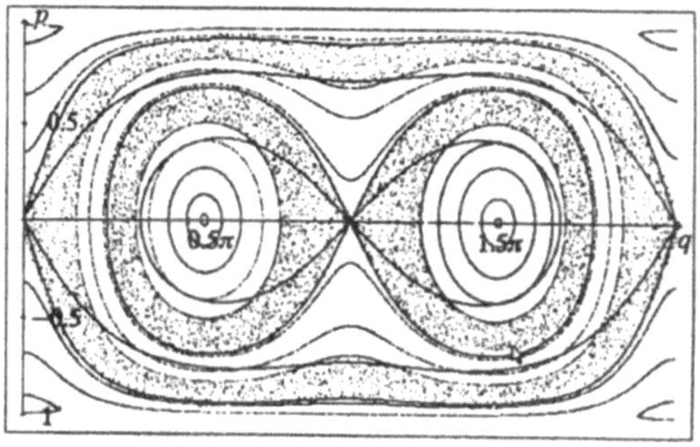

Fig. 6.38. Phase portrait of the "slow" Poincaré mapping of system (6.51) for $\nu = 0.1$, $\delta = 0.5(1 + 0.3\cos 2\pi\tau)$, $\varepsilon = 0.05$, and $\tau = 0$.

As is shown in [612, 279, 439], necessary conditions for the separatrix splitting of the Poincaré mapping of the system (6.50) can be obtained to first order in ε from analysis of the "adiabatic" Poincaré – Mel'nikov integral. If the equations of motion of system (6.50) are written in the form

$$\dot{q} = \frac{\partial H}{\partial p}(p, q, \delta(z), \nu(z)), \quad \dot{p} = -\frac{\partial H}{\partial q}(p, q, \delta(z), \nu(z)), \qquad \dot{z} = \varepsilon, \tag{6.51}$$

where $\delta(z)$ and $\nu(z)$ are periodic functions of the same period $2\pi n$, then in the extended phase space the Poincaré mapping is given by the intersection of trajectories with the planes $z = z_0 + 2\pi n$, $n \in \mathbb{Z}$. According to the results of [279], to first order in ε, the splitting separatrix value on the plane of the Poincaré section is the same along almost the entire length of the separatrix of the "frozen" system and depends on the parameter z_0, which determines the Poincaré section. This value is proportional to the adiabatic

Poincaré – Mel'nikov function

$$M_A(z_0) = \int_{-\infty}^{+\infty} \left[\frac{\partial H}{\partial z}(p_0(t, z_0), q_0(t, z_0), z_0) - \frac{\partial H}{\partial z}(P(z_0), Q(z_0), z_0) \right] dt, \tag{6.52}$$

where $p_0(t, z_0)$, $q_0(t, z_0)$ is the solution for the separatrix of the "frozen" system to the corresponding hyperbolic fixed point $(P(z_0), Q(z_0))$ (in the case of the heteroclinic separatrix the second term of integrand has the same value for both hyperbolic points). The geometric meaning of the function M_A in (6.52) is that [279, 439]

$$M_A(z) = \frac{dA(z)}{dz}, \tag{6.53}$$

where $A(z)$ is the area under the separatrix of the "frozen" system. For the separatrix to be nonsplitting up to first order in ε, it is necessary that $A(z) = \text{const}$ for all z.

Let us consider the system with Hamiltonian (6.50) and with $v = 0$ (i.e., $\boldsymbol{K} = 0$). For this system the separatrices of the "frozen" system have the form shown in Fig. 6.30, whereas the area $A(z)$ is easily calculated:

$$A(z) = \int p(H(P(z), Q(z), z))dq = \arcsin \sqrt{\delta(z)}. \tag{6.54}$$

The condition $A(z) = \text{const.}$ leads to the condition $\delta(z) = \text{const.}$ determining the usual Euler – Poinsot case. Equation (6.54) together with (6.16) yields the expression

$$M_A(z) = \frac{\delta'(z)}{2\sqrt{\delta(1-\delta)}}, \tag{6.55}$$

which shows that when δ is periodic and slowly varying, the separatrices of the perturbed problem are always split and intersect transversally. This leads to the nonexistence of an additional analytic integral in this problem and to the appearance of quasi-random oscillations and a stochastic layer.

The picture of the split separatrices for various values of the parameter $z_0 = \tau_0$ is presented in Figs. 6.39, 6.40, and 6.41; it is seen from these figures that along almost the whole length of the separatrices the distance between them is almost constant and depends on z_0, as was predicted by formula (6.52).

For $v_1 = v_3 = 0$ and $v_2 = v$ $(0 < v < \delta < 1)$, the phase portrait of the "frozen" system has the form shown in Fig. 6.37 (phase portrait of the Zhukovskii – Volterra system). The calculation of the area $A(z)$ for one pair of separatrices (to the point $p = 0, q = \pi$) leads to the expression

$$A_1(z) = 4 \left[\arcsin b - \frac{v}{1-\delta} \frac{1}{2a} \ln \left| \frac{a + \frac{b}{\sqrt{1-b^2}}}{a - \frac{b}{\sqrt{1-b^2}}} \right| \right], \tag{6.56}$$

where $b = \sqrt{\delta - v}$ and $a = \frac{\delta}{1-\delta}$. The area $A_2(z)$ for the other pair of separatrices (to the point $p = 0, q = 0$) is obtained from $A_1(z)$ by the substitution $v \to -v$.

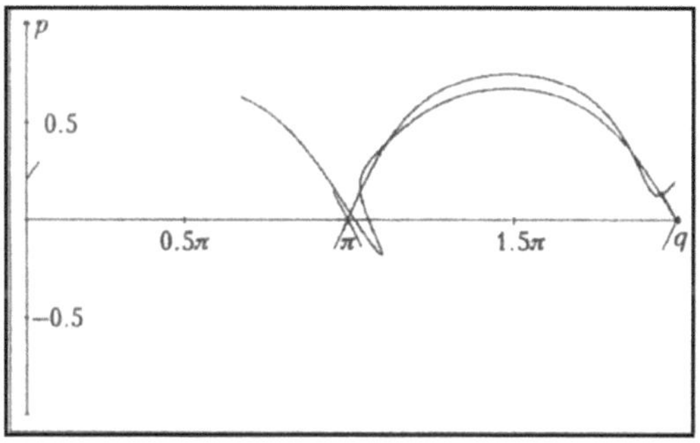

Fig. 6.39. Separatrices of system (6.51) for $\tau_0 = -\frac{1}{4}$ ($\varepsilon = 0.05$).

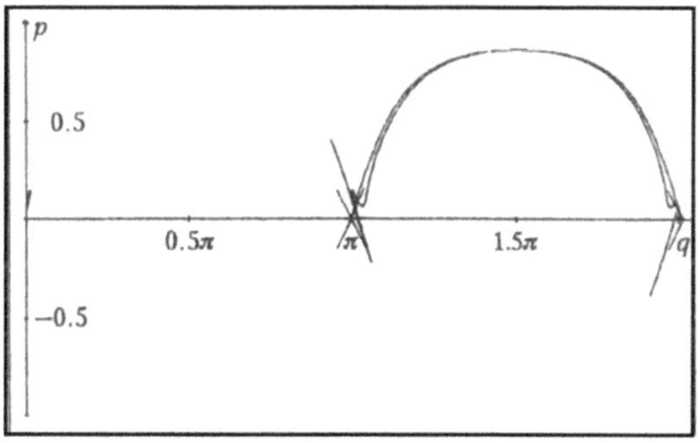

Fig. 6.40. Separatrices of system (6.51) for $\tau_0 = 0$ ($\varepsilon = 0.05$).

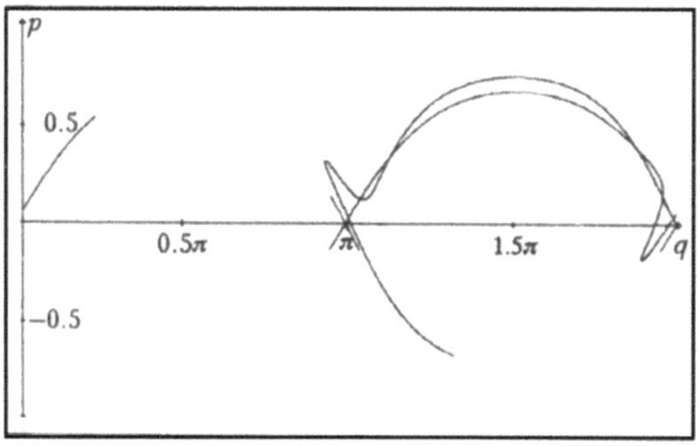

Fig. 6.41. Separatrices of system (6.51) for $\tau_0 = \frac{1}{2}$ ($\varepsilon = 0.05$).

Note that the splitting conditions of different pairs of separatrices do not coincide, therefore if system parameters vary slowly and periodically in time, then at least one pair of separatrices splits. In Fig. 6.42, we present the separatrices of the perturbed problem under the condition $A_1(z) = \text{const}$; it is seen that one pair of separatrices "does not split": the distance between the separatrices of this pair is not constant along the length and is proportional to ε^2, whereas the other pair remains split and the distance

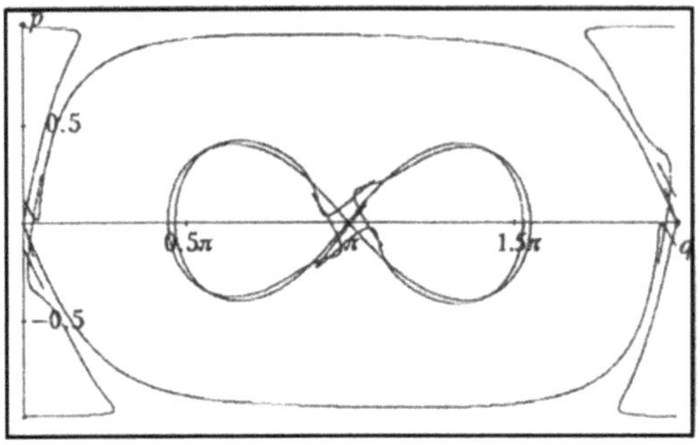

Fig. 6.42. Separatrices of system (6.51) for $\tau_0 = \frac{1}{4}$ in the case where the condition $A(z) = \text{const}$ ($\varepsilon = 0.05$) holds.

between the separatrices is proportional to ε. Under the "nonsplitting" condition, the chaos picture is shown in Fig. 6.43; it is clearly seen from this pricture that the thickness of the stochastic layer near to the "nonsplit" separatrices is of order ε. Note that $\Theta \equiv 0$ for the "nonsplitting" separatrices, and so the formulas (6.41) and (6.47) for the IAI jump are not applicable.

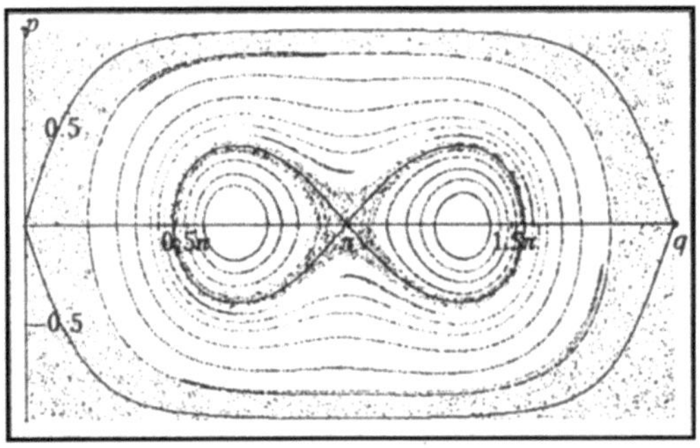

Fig. 6.43. Phase portrait of the "slow" Poincaré mapping of system (6.51) for $A(z) = \text{const.}$, $\tau_0 = \frac{1}{4}$, $\varepsilon = 0.05$.

Let us compare the conditions for the appearance of chaos obtained from analysis of separatrix splitting with the scenario of the development of adiabatic chaos resulting from random AI jumps. If the separatrices split, then the AI exhibits the behavior shown in Fig. 6.44 in the variables I and τ; if $A_1(z) \equiv \text{const.}$ it has the behavior shown in Fig. 6.45. These figures show that the condition $A(z) \equiv \text{const.}$ yields an obstruction to the appearance of adiabatic chaos in domains lying outside an ε-neighborhood of the separatrix. In this neighborhood, the characteristic length scales of the stochastic layer are of order ε^2.

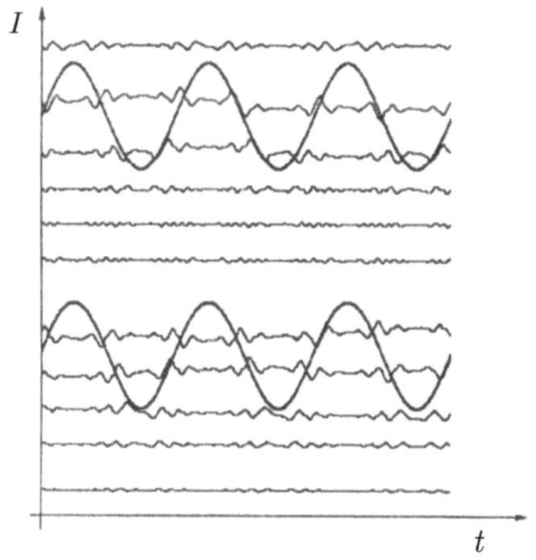

Fig. 6.44. Dependence of the variable I for various trajectories of system (6.51) for $\nu = 0.1$ and $\delta = 0.5(1 + 0.3\cos 2\pi\tau)$. The bold line indicates the action for the separatrices of the "frozen" system.

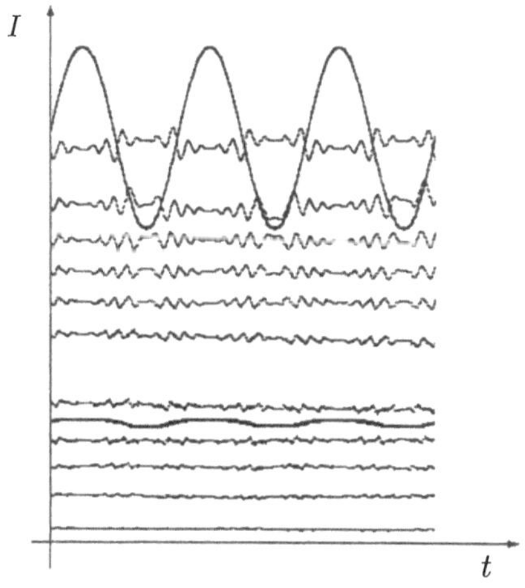

Fig. 6.45. Dependence of the variable I for various trajectories of system (6.51) for $A_1(z) = 0.8$.

6.7 Falling Heavy Rigid Body in an Ideal Fluid. Probabilistic Effects and Attracting Sets

1 Equations of motion and particular cases

In Sec. 1.6 of Ch. 1, we presented the equations describing a heavy rigid body moving in an infinite volume of vortex-free incompressible fluid. For an initial horizontal impulse P_1, they can be represented in the form of the Poincaré – Chetaev equations on the algebra $so(3) \oplus_s \mathbb{R}^9 = \{\boldsymbol{M}, \boldsymbol{\alpha}, \boldsymbol{\beta}, \boldsymbol{\gamma}\}$:

$$\begin{gathered}\dot{\boldsymbol{M}} = \boldsymbol{M} \times \frac{\partial H}{\partial \boldsymbol{M}} + \boldsymbol{\alpha} \times \frac{\partial H}{\partial \boldsymbol{\alpha}} + \boldsymbol{\beta} \times \frac{\partial H}{\partial \boldsymbol{\beta}} + \boldsymbol{\gamma} \times \frac{\partial H}{\partial \boldsymbol{\gamma}}, \\ \dot{\boldsymbol{\alpha}} = \boldsymbol{\alpha} \times \frac{\partial H}{\partial \boldsymbol{M}}, \quad \dot{\boldsymbol{\beta}} = \boldsymbol{\beta} \times \frac{\partial H}{\partial \boldsymbol{M}}, \quad \dot{\boldsymbol{\gamma}} = \boldsymbol{\gamma} \times \frac{\partial H}{\partial \boldsymbol{M}},\end{gathered} \tag{6.57}$$

where

$$H = \frac{1}{2}(\mathbf{A}\boldsymbol{M}, \boldsymbol{M}) + (\mathbf{B}\boldsymbol{M}, P_1\boldsymbol{\alpha} - \mu t\boldsymbol{\gamma}) + \frac{1}{2}\big(\mathbf{C}(P_1\boldsymbol{\alpha} - \mu t\boldsymbol{\gamma}), P_1\boldsymbol{\alpha} - \mu t\boldsymbol{\gamma}\big) + \mu(\boldsymbol{r}, \boldsymbol{\gamma}); \tag{6.58}$$

here $\mathbf{A} = \operatorname{diag}(a_1, a_2, a_3)$, the matrices $\mathbf{B}$ and $\mathbf{C}$ are symmetric, $\boldsymbol{r} = (r_1, r_2, r_3)$ is the radius vector of the center of mass, μ is the body weight in fluid, and P_1 is the initial impulse (impact) in the horizontal direction. Let us consider in more detail particular cases of the system (6.57)—(6.58), which are different from those of Ch. 1.

Plane-parallel motion. The plane-parallel rigid body motion is given by the invariant relations $M_1 = M_2 = 0$ and $\alpha_3 = \gamma_3 = 0$. It is easy to show that a necessary condition for the existence of such motions is that the body possess dynamical symmetry with respect to the given (invariant) plane, which leads to the relations

$$b_{11} = b_{22} = b_{33} = b_{12} = 0, \quad c_{13} = c_{23} = 0.$$

Moreover, it can be shown that by a suitable translation and rotation of the axes attached to the body we can achieve that $\mathbf{B} = 0$ and that $\mathbf{C}$ is diagonal. Let the angle of rotation of the moving axes be measured in the way shown in Fig. 6.46; then for the projections of the fixed unit vectors on the moving axes we have

$$\alpha_1 = \sin\varphi, \quad \alpha_2 = -\cos\varphi, \quad \gamma_1 = \cos\varphi, \quad \gamma_2 = \sin\varphi.$$

Using (6.57) and (6.58) and noting that $\mathbf{A} = \operatorname{diag}(a_1, a_2, a_3)$, $\mathbf{C} = \operatorname{diag}(c_1, c_2, c_3)$ and $\mathbf{r} = (x, y, 0)$, we obtain the following nonautonomous second-order equation for the angle of rotation φ:

$$\begin{aligned}a_3\ddot{\varphi} = {} & (c_1 - c_3)\left(\mu^2 t^2 \sin\varphi\cos\varphi + P_1\mu t\cos 2\varphi - P_1^2\sin\varphi\cos\varphi\right) \\ & + \mu(x\sin\varphi - y\cos\varphi),\end{aligned} \tag{6.59}$$

For the balanced body ($x = y = 0$) without initial impulse ($P_1 = 0$) we obtain a remarkably simple equation

$$\ddot{\varphi} = kt^2 \sin\varphi \cos\varphi, \quad k = \frac{\mu^2(c_1 - c_3)}{a_3}. \tag{6.60}$$

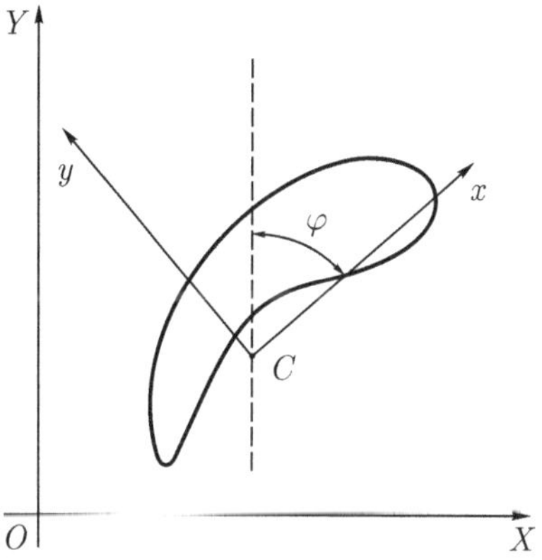

Fig. 6.46.

Remark. In [345, 347, 478], this equation is called the *Chaplygin equation*, but one should keep in mind that although it was obtained, along with other interesting results, by S. A. Chaplygin in 1890 in his student work, he did not publish it first. This was probably due to the fact that he could not explicitly integrate it. However, S. A. Chaplygin did include this study later in the first collection of his work published in his lifetime (1933) [129].

Equations (6.60) were also obtained independently by D. N. Goryachev (1893) [232] and by V. A. Steklov (1894) [556, 554]. In particular, V. A. Steklov showed that when a body falls, the amplitude of its oscillations with respect to the horizontal axis decreases, whereas the frequency of its oscillations grows. This conclusion was drawn by V. A. Steklov in the supplement to his book [556], in which he made a number of errors in analyzing the asymptotic behavior of the body. The Steklov problem of the asymptotic description of solutions to Eq. [556, 554] was solved by V. V. Kozlov [345], who showed that for almost all initial conditions, the body motion tends to uniformly accelerated fall with the wider side of the body perpendicular to the gravitational field with oscillation around the horizontal axis at increasing frequency of order t and decreasing amplitude of order $1/\sqrt{t}$. Numerical analysis of the asymptotic motions with different numbers of half-turns is contained in [153]. Analytic expressions for asymptotic descriptions of a falling body in various forms were obtained in [478, 153].

In [156], the buoyancy effects were described and studied. With vortex-free flow around the body, it is assumed that at the initial moment the wider side of the body is aligned with the horizontal and the body acquires a horizontal velocity. Initially, the body starts moving downwards. However, if the body's added mass in the lateral direction is sufficiently large, it will abruptly move upward with its narrower side up and subsequently rise higher than its initial elevation.

Experimental and numerical results on a plate falling in a fluid are contained in [35, 564] (see also references therein).

Motion of an axially symmetric body (round disk). There an an important special case for which the system (6.57)–(6.58) has the additional (autonomous) linear so-called "Lagrange integral"

$$M_3 = \text{const}.$$

A necessary condition for M_3 to be an integral is for the body to have axial symmetry and in this case the moving body-fixed axes can be chosen in such a way that

$$\mathbf{A} = \operatorname{diag}(a_1, a_1, a_3), \quad \mathbf{B} = \operatorname{diag}(b_1, b_1, b_3), \quad \mathbf{C} = \operatorname{diag}(c_1, c_1, c_3),$$
$$\boldsymbol{r} = (0, 0, z).$$

As was shown in Sec. 4.1 of Ch. 4, the evolution of the projections $\boldsymbol{N} = \big((\boldsymbol{M}, \boldsymbol{\alpha}), (\boldsymbol{M}, \boldsymbol{\beta}), (\boldsymbol{M}, \boldsymbol{\gamma})\big)$ of the angular momentum on the fixed axes and the evolution of the vector $\boldsymbol{n} = (\alpha_3, \beta_3, \gamma_3)$ of the symmetry axis is then described by the following nonautonomous Hamiltonian system on $e(3)$:

$$\dot{\boldsymbol{N}} = \frac{\partial \bar{H}}{\partial \boldsymbol{N}} \times \boldsymbol{N} + \frac{\partial \bar{H}}{\partial \boldsymbol{n}} \times \boldsymbol{n}, \quad \dot{\boldsymbol{n}} = \frac{\partial \bar{H}}{\partial \boldsymbol{N}} \times \boldsymbol{n}, \tag{6.61}$$

where, ignoring inessential terms, one can write the Hamiltonian (1.108) as

$$\begin{aligned}\bar{H}(t) = {} & \frac{1}{2} a_1 \boldsymbol{N}^2 + b_1 (P_1 N_1 - \mu t N_3) + M_3 (b_3 - b_1)(P_1 n_1 - \mu t n_3) \\ & + \frac{1}{2}(c_3 - c_1)(P_1 n_1 - \mu t n_3)^2 + \mu z n_3 .\end{aligned} \tag{6.62}$$

The trajectory of the origin of the moving coordinate system C can be obtained from the equations

$$\begin{aligned}\dot{x}_1 &= b_1 N_1 + (b_3 - b_1) M_3 n_1 + P_1\big(c_1 + (c_3 - c_1) n_1^2\big) - \mu t (c_3 - c_1) n_1 n_3, \\ \dot{x}_2 &= b_1 N_2 + (b_3 - b_1) M_3 n_2 + P_1 (c_3 - c_1) n_1 n_2 - \mu t (c_3 - c_1) n_2 n_3, \\ \dot{x}_3 &= b_1 N_3 + (b_3 - b_1) M_3 n_3 + P_1 (c_3 - c_1) n_1 n_3 - \mu t \big(c_1 + (c_3 - c_1) n_3^2\big).\end{aligned} \tag{6.63}$$

In the absence of an initial (horizontal) impulse, i.e., when $P_1 = 0$, there exists a further additional integral

$$N_3 = (\boldsymbol{M}, \boldsymbol{\gamma}) = \text{const}.$$

In this case we obtain, for the nutation angle θ ($\gamma_3 = \cos\theta$), from (6.61) and (6.62), as in the case of plane-parallel motion, the following nonautonomous second-order equation [129] (due to S. A. Chaplygin):

$$\begin{aligned}a_1^{-1} \ddot{\theta} = {} & \frac{(M_3 \cos\theta - N_3)(M_3 - N_3 \cos\theta)}{\sin^3\theta} \\ & + (c_3 - c_1)\mu^2 t^2 \sin\theta \cos\theta - M_3 (b_3 - b_1) \mu t \sin\theta + \mu z \sin\theta .\end{aligned} \tag{6.64}$$

If the body starts from rest, then $M_3 = N_3 = 0$, and we obtain

$$a_1^{-1} \ddot{\theta} = (c_3 - c_1)\mu^2 t^2 \sin\theta \cos\theta + \mu z \sin\theta .$$

An analog of the Hess case. Along with the two special cases of the system just considered, there may also occur a situation where the system (6.57)-(6.58) admits an invariant relation analogous to the Hess case in the Euler – Poisson equations. For its existence, it is necessary that the surface bounding the body be axially symmetric and the symmetry axis be perpendicular to the circular cross-section of the ellipsoid of gyration (recall that the ellipsoid of gyration is defined by the equation $(\boldsymbol{x}, \mathbf{A}\boldsymbol{x}) = 1$). Let us take one of the body-fixed axes to be directed along the symmetry axis, and let the other two axes be such that $a_{23} = 0$; then the parameters in the Hamiltonian (6.58) have the form

$$\mathbf{A} = \begin{pmatrix} a_1 & 0 & a_{13} \\ 0 & a_1 & 0 \\ a_{13} & 0 & a_3 \end{pmatrix}, \quad \mathbf{B} = \operatorname{diag}(b_1, b_1, b_3), \quad \mathbf{C} = \operatorname{diag}(c_1, c_1, c_3), \quad \boldsymbol{r} = (0, 0, z).$$

With such a choice of the (moving) coordinate system, the invariant relation takes the simplest form (see Sec. 2.6 of Ch. 2)

$$M_3 = 0. \tag{6.65}$$

The equation for the nutation angle θ is the same as (6.64) with $M_3 = 0$, i.e.,

$$a_1^{-1}\ddot{\theta} = \frac{N_3^2 \cos^2\theta}{\sin^3\theta} + (c_3 - c_1)\mu^2 t^2 \sin\theta\cos\theta + \mu z \sin\theta. \tag{6.66}$$

The distinction between the case of an axially symmetric body and this analog of the Hess case manifests itself in the equations governing the evolutions of the precession and proper rotation angles (see Sec. 2.6 of Ch. 2). The analysis of the Hess case for Eqs. (6.57) and (6.58) was first carried out by the authors in [102].

2 Isotropic body motion

Let us look at the simplest special case of system (6.58) found by Steklov [556, 558] for which the equations may be integrated by quadratures. In this case,

$$\mathbf{A} = \operatorname{diag}(a_1, a_2, a_3), \quad \mathbf{B} = b\mathbf{E}, \quad \mathbf{C} = c\mathbf{E}, \quad \boldsymbol{r} = 0,$$

i.e., the added mass tensor is spherical, but if $\mathbf{B} \neq \mathbf{0}$, the body does not have three symmetry planes. If $\mathbf{B} = 0$, the motion is trivial: the center of mass moves along a parabola, whereas the motions of the apexes $\boldsymbol{\alpha}$, $\boldsymbol{\beta}$, and $\boldsymbol{\gamma}$ are the same as in the Euler – Poinsot case.

The equations governing the evolution of angular momentum in the moving frame decouple and are identical to the Euler – Poinsot case:

$$\dot{\boldsymbol{M}} = \boldsymbol{M} \times \mathbf{A}\boldsymbol{M}.$$

To find the trajectory of the center of mass, it is convenient to rewrite the equations of motion in the fixed frame:

$$\begin{gathered} \dot{N}_1 = b\mu t N_2, \quad \dot{N}_2 = -b\mu t N_1 - bP_1 N_3, \quad \dot{N}_3 = bP_1 N_2, \\ \dot{x}_1 = bN_1 + cP_1, \quad \dot{x}_2 = bN_2, \quad \dot{x}_3 = bN_3 - c\mu t, \end{gathered} \tag{6.67}$$

where $\boldsymbol{N} = \big((\boldsymbol{\alpha}, \boldsymbol{M}), (\boldsymbol{\beta}, \boldsymbol{M}), (\boldsymbol{\gamma}, \boldsymbol{M})\big)$ is the angular momentum in the fixed frame. Obviously, the squared angular momentum is an integral of the motion: $\boldsymbol{M}^2 = \boldsymbol{N}^2 = \text{const}$.

If the initial impulse is equal to zero, $P_1 = 0$, then the first three equations in (6.67) are integrated in terms of elementary functions:

$$N_1 = A\sin(b\mu t^2/2 + \varphi_0), \quad N_2 = A\cos(b\mu t^2/2 + \varphi_0), \quad N_3 = \text{const.},$$

where A and φ_0 are arbitrary constants. The motion of the body along the vertical axis is uniformly accelerated: $x_3 = -c\mu t^2/2$, whereas the projection of the trajectory on the plane x_1, x_2 is a spiral described by the Fresnel integrals and converging to a fixed point on the plane.

For large time t, the following asymptotic representation holds:

$$x_1 = x_1^0 - \frac{A}{\mu}\frac{\cos(b\mu t^2/2 + \varphi_0)}{t} + O(t^3),$$
$$x_2 = x_2^0 + \frac{A}{\mu}\frac{\sin(b\mu t^2/2 + \varphi_0)}{t} + O(t^3).$$

If $P_1 \neq 0$, then the equations for $\boldsymbol{N}$ cannot be integrated in terms of elementary functions, and, moreover, the center of mass undergoes a drift with rate cP_1 along the axis Ox_1.

3 Qualitative analysis of plane-parallel motion of a balanced body without initial impulse

We have shown that for a special choice of the moving axes (for which the kinetic energy is diagonal), the rotation angle of the body measured from to the vertical (Fig. 6.46) is described by Eq. (6.59), and the motion of the origin C of the moving system is described by the equations

$$\begin{aligned} \dot{X} &= (\boldsymbol{\alpha}, \mathbf{C}\boldsymbol{p}) = P_1(c_1\sin^2\varphi + c_2\cos^2\varphi) - \mu t(c_1 - c_2)\sin\varphi\cos\varphi, \\ \dot{Y} &= (\boldsymbol{\gamma}, \mathbf{C}\boldsymbol{p}) = P_1(c_1 - c_2)\sin\varphi\cos\varphi - \mu t(c_1\cos^2\varphi + c_2\sin^2\varphi). \end{aligned} \tag{6.68}$$

REMARK. Equation (6.59) corresponds to a *nonautonomous* Hamiltonian system with one degree of freedom. Such systems have been studied in greatest detail for the case in which the Hamiltonian is a periodic function of time. In the general case, they exhibit chaotic behavior. In this system the Hamiltonian is nonperiodic in time and, as is shown below, the dependence of the angle $\varphi(t)$ is asymptotic.

First consider the “simplest” case where a balanced body ($x = y = 0$) falls without initial impulse ($P_1 = 0$). Then after the change of variables $2\varphi = \theta$, Eq. (6.59) takes the form

$$\ddot{\theta} = kt^2\sin\theta, \quad k = \frac{\mu^2(c_1 - c_2)}{a_3}. \tag{6.69}$$

In what follows, we assume that $c_1 > c_2$, i.e., $k > 0$, and $0 \leqslant \theta < 2\pi$.

Stationary (equilibrium) solutions. Small oscillations. Doubly asymptotic solutions. The simplest "equilibrium" solutions $\theta(t) = \text{const.}$ of equation (6.69) are of the form

$$1)\ \theta = 0, \qquad 2)\ \theta = \pi. \tag{6.70}$$

The first solution corresponds to falling with the narrow side downward ($X = X_0$, $Y = Y_0 - \mu c_1 t^2/2$), and the second solution corresponds to falling with the wide side downward ($X = X_0$, $Y = Y_0 - \mu c_2 t^2/2$). Indeed, since the added mass satisfies $c_1^{-1} < c_2^{-1}$, the rotation angle of the body measured from the vertical satisfies $\varphi = \pi n$ when the axis Cx is vertical, and $\varphi = \pi/2 + \pi n$ when the axis Cy is vertical.

Linearizing Eq. (6.69) near fixed points (6.70), we obtain

$$\begin{aligned} &1)\quad \ddot{\xi} = kt^2\xi, &&\theta = \xi,\\ &2)\quad \ddot{\xi} = -kt^2\xi, &&\theta = \pi - \xi. \end{aligned}$$

The general solution of these equations is expressed in terms of the Bessel functions:

$$\begin{aligned} &1)\quad \xi(t) = \sqrt{t}\left(C_1 I_{1/4}\left(\sqrt{k}t^2/2\right) + C_2 K_{1/4}\left(\sqrt{k}t^2/2\right)\right),\\ &2)\quad \xi(t) = \sqrt{t}\left(C_1 J_{1/4}\left(\sqrt{k}t^2/2\right) + C_2 Y_{1/4}\left(\sqrt{k}t^2/2\right)\right), \end{aligned} \tag{6.71}$$

where $I_\nu(x)$ and $K_\nu(x)$ are Bessel functions of the second kind and $J_\nu(x)$ and $Y_\nu(x)$ are Bessel functions of the first kind. Thus, the first solution is linearly unstable, while the second is linearly (asymptotically) stable. Indeed, using the asymptotics of the Bessel functions J_ν and Y_ν for large values of the argument, we find that

$$\xi(t) = \frac{A\sin\left(\sqrt{k}t^2/2 + \alpha_0\right)}{\sqrt{t}} + O\left(t^{-5/2}\right), \quad A = \text{const.}$$

Hence, the amplitude of oscillations decays as $t^{-1/2}$, and their frequency grows infinitely as t^2.

As was shown in [345], using variational methods, we can prove that there exist two solutions that are asymptotic to the unstable equilibrium ($\theta = 0$) and approach it from different sides. Moreover, by the invariance of Eq. (6.69) under the involution $t \to -t$ there exists a solution $\theta_*(t)$ with initial data $\theta_*(0) = \pi$ for which [345]

$$\theta_*(t) + \theta_*(-t) = 2\pi, \quad \lim_{t\to-\infty}\theta_*(t) = 0, \quad \lim_{t\to+\infty}\theta_*(t) = 2\pi.$$

Hence, the solution $\theta_*(t)$ is doubly asymptotic (there also exists an analogous doubly asymptotic solution that goes around the circle $\theta \bmod 2\pi$ in the other direction). In this case, the body executes one half-turn, and the trajectory of its center C described by Eqs. (6.68) is depicted in Fig. 6.47a. Note that the top point of the trajectory is a cusp: close to this point, the equation of the trajectory has the form $Y = \lambda X^{2/3}$, $\lambda = \text{const.}$ In Fig. 6.47b, we show the change in the angle θ for this doubly asymptotic solution.

The existence of doubly asymptotic trajectories with an arbitrary number of half-turns was proved in [42].

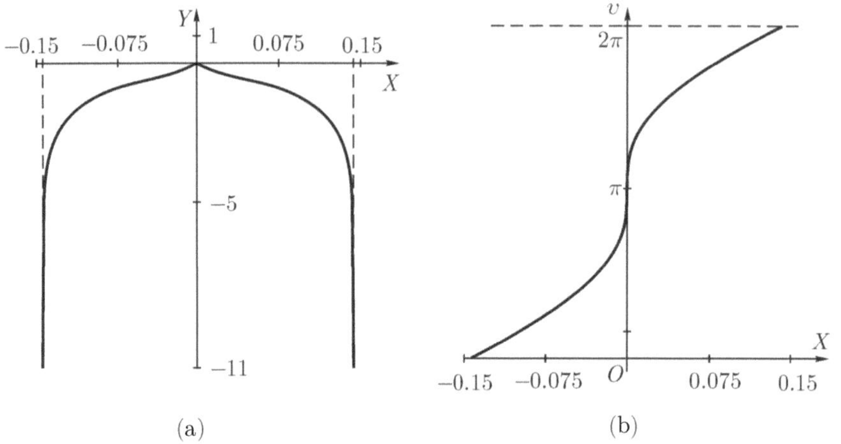

Fig. 6.47. The trajectory of the body center C and the angle φ versus the coordinate x for the doubly asymptotic solution for $k = 1$, $a_3/\mu = 0.1$ $(P_1 = 0)$.

Asymptotic behavior of solutions of the Chaplygin equation. Now we prove the following theorem.

Theorem 5 ([345]). *For any solution $\theta(t)$ of Eq. (6.69), we have*

$$\lim_{t\to\infty}\theta(t) = 0 \quad or \quad \lim_{t\to\infty}\theta(t) = \pi.$$

Proof. Rescale time as $\frac{1}{2}t^2 = \tau$ so that Eq. (6.69) takes the form

$$\theta'' + \frac{1}{2\tau}\theta' - k\sin\theta = 0, \qquad \frac{d\theta}{d\tau} = \theta', \tag{6.72}$$

i.e., we have obtained the equation describing the plane pendulum "with dissipation decaying in time".

Consider the function $\mathscr{E}(\theta, \theta')$, the energy of the "unperturbed system":

$$\mathscr{E} = \frac{1}{2}(\theta')^2 + k\cos\theta. \tag{6.73}$$

Its derivative along the solutions of system (6.72) is negative:

$$\frac{d\mathscr{E}}{d\tau} = -\frac{(\theta')^2}{2\tau} \leqslant 0.$$

Integrating this relation, we find that

$$\mathscr{E}(\tau) = \mathscr{E}_0 - \int_{\tau_0}^{\tau} \frac{\left(\theta'(s)\right)^2}{2s} ds.$$

As the function (6.73) is bounded below, we conclude that for all initial conditions not coinciding with the fixed points $\theta = 0$ and $\theta' = 0$ of Eq. (6.72),

1°. There exists $\lim_{\tau\to\infty} \mathscr{E}(\tau) = \mathscr{E}_\star$.

2°. The integral $I = \int_{\tau_0}^{\infty} \frac{(\theta')^2}{2\tau} d\tau$ converges.

Now let us show that $\mathscr{E}_\star$ coincides with a critical value of the function $\mathscr{E}$, i.e., $\mathscr{E}_\star = \pm k$.

Assume the contrary and consider first the case $\mathscr{E}_\star > k$. Clearly, for sufficiently large τ, according to (6.72) and (6.73), the system stays close to the phase curve of the plane pendulum corresponding to rotational motion, and hence $\frac{1}{2}(\theta')^2 \geqslant \varkappa > 0$, $\varkappa = \text{const.}$ (i.e., the kinetic energy is bounded away from zero). Therefore, the integral $I = \int_{\tau_0}^{\infty} \frac{(\theta')^2}{2\tau} d\tau$ diverges (as $\ln \tau$), which contradicts 2° above.

Now consider the case $-k < \mathscr{E}_\star < k$. For sufficiently large τ, on the phase plane (θ, θ'), the point stays in a small neighborhood of the closed curve of the "unperturbed system", $\frac{1}{2}(\theta')^2 + k\cos\theta = \mathscr{E}_\star = \text{const.}$ In this case, since $\mathscr{E} > \mathscr{E}_\star$, this point always remains outside. Moreover, the time taken to go along half of this curve, $\Delta\tau = \tau_{n+1} - \tau_n$, measured between two consecutive moments for which $\theta' = 0$ is equal to approximately half the period of the unperturbed motion of the plane pendulum and certainly does not exceed the whole period:

$$\Delta\tau \leqslant T = \int_{\theta_1}^{\theta_2} \frac{2d\theta}{\sqrt{2(\mathscr{E}_\star - k\cos\theta)}}, \tag{6.74}$$

where θ_i are *consecutive* roots of the equation $\mathscr{E}_\star - k\cos\theta = 0$. Obviously, there exists a positive constant $\varkappa$ (depending only on $\mathscr{E}_\star$ and independent of τ_n and τ_{n+1}) such that

$$\int_{\tau_n}^{\tau_{n+1}} \frac{\left(\theta'(\tau)\right)^2}{2\tau} d\tau \geqslant \frac{\varkappa}{\tau_{n+1}}. \tag{6.75}$$

Using (6.74), we also obtain the inequality $\tau_{n+1} \leqslant nT + \tau_1$, $\tau_1 \geqslant \tau_0$. This yields

$$I \geqslant \int_{\tau_1}^{\infty} \frac{(\theta')^2}{2\tau} d\tau \geqslant \varkappa \sum_{n=1}^{\infty} \frac{1}{nT + \tau_1}.$$

But this series, and hence the integral, diverges, which again contradicts 2°; therefore, $\mathscr{E}_\star = k$ or $\mathscr{E}_\star = -k$. □

A conjecture due to V. V. Kozlov [345] says that $\mathscr{E}_\star = -k$ *for almost all solutions of Eq.* (6.69).

This is equivalent to saying that the set of all trajectories tending, as $t \to \pm\infty$, to the unstable equilibrium state $\theta = 0$ (2π) has measure zero.

Numerical analysis. One can perform a numerical analysis of Eq. (6.60) based on Theorem 5 [153]. For this purpose, on the phase plane $(\theta, \dot{\theta})$ (more precisely, on the cylinder θ mod 2π, $\dot{\theta} \in (-\infty, +\infty)$), at the initial instant of time $t = t_0$, we construct domains within which the body executes the same number of half-turns as $t \to +\infty$ (or as $t \to -\infty$) before it "is attracted" to the solution $\theta = \pi$. As is seen from Fig. 6.48, these domains are spaced regularly, and, moreover, their width decreases when $|\dot{\theta}|$ grows, so that for a large initial $|\dot{\theta}|$, we can speak only about the probability of the body falling with its "upper" or "lower" side down as $t \to \infty$. The boundaries of the domains are filled with initial conditions corresponding to motions that asymptotically approach the unstable equilibrium state $\theta = 0\ (2\pi)$. It is seen from the figure that the domains corresponding to $t \to +\infty$ and $t \to -\infty$ turn out to be mirror reflections of one another with respect to the line $\theta = \pi$. When these domains overlap, their boundaries intersect at points lying on the line $\theta = \pi$. These intersection points correspond to initial conditions for the doubly asymptotic solutions of Eq. (6.69) with different numbers of body half-turns.

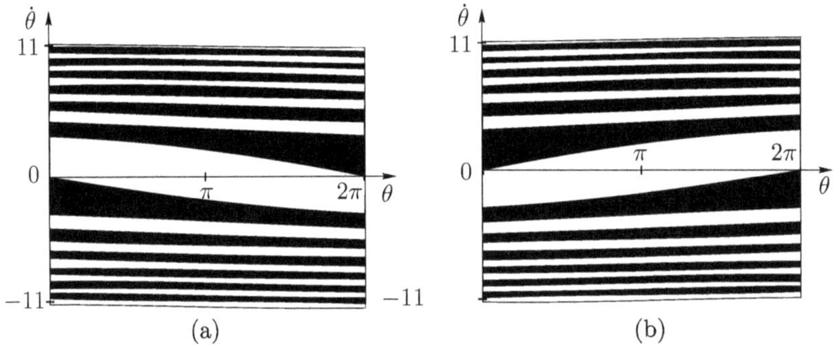

Fig. 6.48. Domains on the phase plane corresponding to initial conditions with $t_0 = 0$ when the body executes the same number of half-turns under the variation of t from 0 to $+\infty$ (in the case a) and under the variation of t from $-\infty$ to 0 (in the case *b*) ($k = 1$).

REMARK. On the cylinder $\dot{\theta}$, θ mod 2π, all boundaries of the domains are glued into one smooth curve, which is analogous to the helix whose pitch decreases as $|\dot{\theta}|$ increases. To one side of this curve lie the domains with an even number of half-turns, and to the other side lie the domains with an odd number of half-turns.

Thus, numerical calculations corroborate the conjecture that $\theta(t) \xrightarrow[t\to\pm\infty]{} \pi$ for almost all solutions, and, moreover, in the three-dimensional space t, θ, θ', the solutions asymptotically approaching the unstable equilibrium state $\theta = \pi$ fill two-dimensional surfaces. Moreover, there exists a countable set of doubly asymptotic solutions which differ from each other by the number of half-turns executed by them when t varies from $-\infty$ to $+\infty$.

In Fig. 6.49, we present the trajectories of the center C of the body executing doubly asymptotic motions with one and with three half-turns.

Body trajectory. Substituting the asymptotic expansion for small oscillations (6.71) into Eqs. (6.68) and integrating, we obtain an asymptotic representation for the trajectory of the motion in the form

$$X(t) = A\frac{\cos\left(\sqrt{k}t^2/2 + \theta_0\right)}{\sqrt{t}} + O\left(t^{-3/2}\right), \quad Y(t) = -\mu c_2 t^2 + O\left(t^{-1/2}\right),$$

where A and θ_0 are some constants. Hence, for large times, the trajectory is close to a sinusoid with constant step $\Delta y = \pi\mu c_2/\sqrt{k}$ and decaying amplitude [345]. (The step ΔY is calculated between two consecutive zeros of the function $X(t)$.) A typical trajectory is shown in Fig. 6.50.

4 Plane-parallel motion for $P_1 \neq 0$

Now let us present the main qualitative features of the behavior of the system of equations (6.59) and (6.68) in the general case ($P_1 \neq 0$). Only the main ideas of proofs are given here, as they are analogous to the ones in the previous section.

If $P_1 \neq 0$, there exist no equilibrium solutions similar to (6.70). On the other hand, the theorem on the asymptotic behavior of solutions holds in the general case also.

The following assertion generalizing Theorem 5 holds.

Theorem 6. [345] *For any solution $\varphi(t)$ of Eq. (6.59), we have*

$$1.\ \lim_{t\to+\infty}\varphi(t) = \pi n \quad \text{or} \quad 2.\ \lim_{t\to\infty}\varphi(t) = \frac{\pi}{2} + \pi n, \quad n \in \mathbb{Z}.$$

The proof again uses the new time variable $\tau = \frac{1}{2}t^2$. We introduce the function

$$\mathscr{E} = \frac{1}{2}(\varphi')^2 + \frac{1}{2}k^2\cos\varphi + \frac{F(\varphi)}{\sqrt{2\tau}} + \frac{G(\varphi)}{2\tau}, \quad k = \frac{\mu^2(c_1 - c_2)}{a_3},$$

$$F(\varphi) = \frac{P_1\mu(c_1 - c_2)}{2a_3}\sin 2\varphi, \quad G(\varphi) = \frac{P_1^2(c_1 - c_2)}{2a_3}\sin^2\varphi + \frac{\mu}{a_3}(x\sin\varphi + y\cos\varphi).$$

The derivative of $\mathscr{E}$ along solutions of the system satisfies the equation

$$\frac{d\mathscr{E}}{d\tau} + \frac{F(\varphi)}{(2\tau)^{3/2}} + \frac{G(\varphi)}{2\tau^2} = -\frac{(\varphi')^2/2}{\tau}.$$

Since the functions $F(\varphi)$ and $G(\varphi)$ are bounded, we conclude that $\mathscr{E}(\tau) \xrightarrow[\tau\to\infty]{} \mathscr{E}_\star$, $\mathscr{E}_\star =$ const, and we find that the integral $\int_{\tau_0}^{\infty} \frac{(\varphi')^2/2}{\tau}d\tau$ converges. Modifying the proof of the theorem of the previous section, we can show that $\mathscr{E}_\star$ is a critical value of the function $\frac{1}{2}(\varphi)^2 + \frac{1}{2}k^2\cos^2\varphi$.

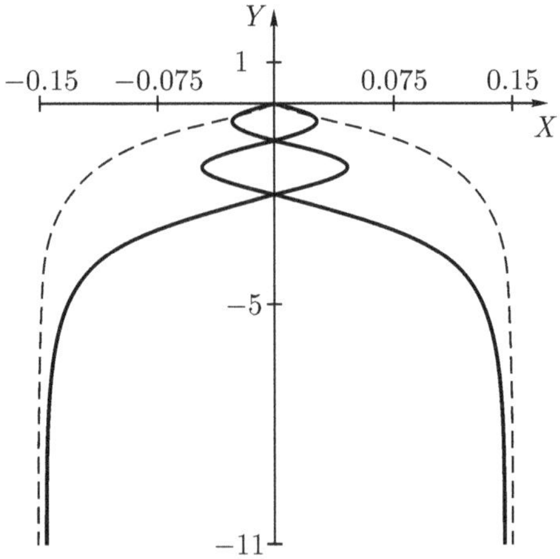

Fig. 6.49. Trajectory of the body's center C in the case of doubly asymptotic solutions with one (dotted line) and three (solid line) half-turns for $k = 1$ and $a_3/\mu = 0.1$.

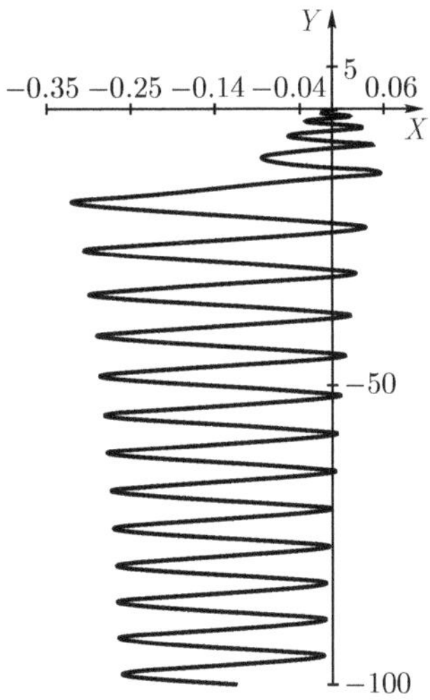

Fig. 6.50. Typical trajectory of a body falling without initial impulse.

Probably, *almost all solutions of the equation tend to one of the solutions* $\varphi(t) = \frac{\pi}{2} + \pi n$ (i.e., the body tends to fall with its wide side downward for the set of initial conditions of a complete measure). This is confirmed by numerical experiments.

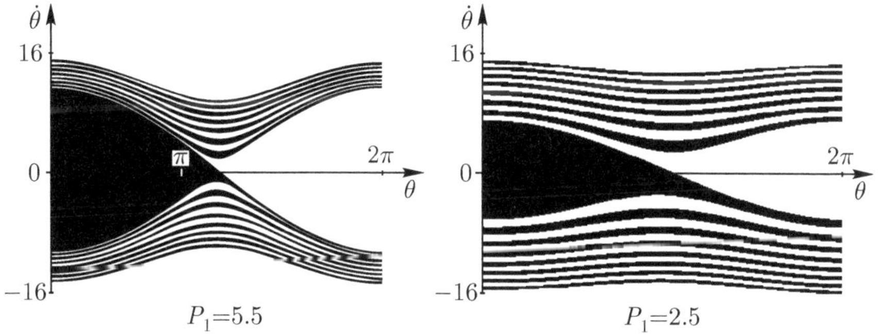

Fig. 6.51.

However, there exist solutions tending to the unstable state $\varphi = 0$. As is shown in [345], they can be obtained in the form of a formal power series

$$\varphi(t) = \frac{\varphi_1}{t} + \frac{\varphi_2}{t^2} + \frac{\varphi_3}{t^3} + \ldots, \qquad (6.76)$$

where $\varphi_1 = \frac{P_1}{\mu}$, and the other φ_k are found using an appropriate iteration procedure. Although the series (6.76) diverges, there exists a solution $\varphi_a(t)$ for which it is an asymptotic approximation:

$$\varphi_a(t) - \sum_{k=1}^{n} \frac{\varphi_k}{t^k} = O\left(t^{-n-1}\right).$$

Remark. Analogously, we can find an asymptotic power series corresponding to solutions which tend to $\varphi = \pi/2$ as t tends to ∞.

With the use of a computer, we may search for domains in the phase plane at the initial instant of time $t = t_0$ in each of which the same number of half-turns is executed by the body before the trajectory is attracted to the solution $\varphi = \frac{\pi}{2}$ as $t \to +\infty$. The boundaries of these domains are filled with asymptotic solutions. As in the case $P_1 = 0$, each domain for $t_0 = 0$ assigned to a certain number of half-turns as $t \to -\infty$ turns out to be the symmetric reflection with respect to the line $\varphi = \pi/2$ of the domain corresponding to the same number of half-turns as $t \to +\infty$. The points of intersection with domain boundaries as $t \to +\infty$ and $t \to -\infty$ correspond to doubly asymptotic solutions.

The typical trajectory of the center C of a body thrown at an angle to the horizontal is presented in Fig. 6.52. In Fig. 6.53, we present the trajectories in the case of doubly asymptotic motion with one and with three half-turns.

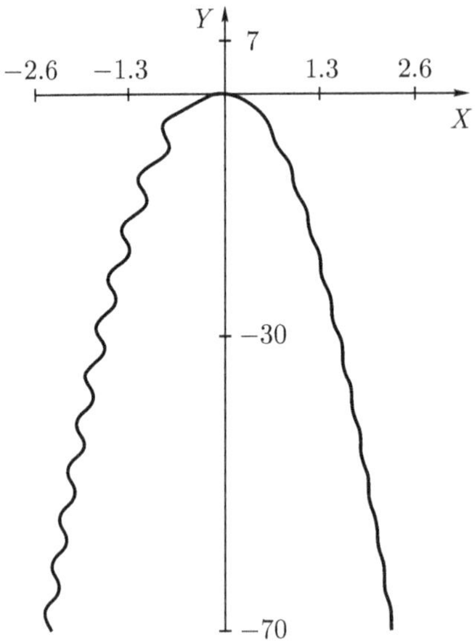

Fig. 6.52. Typical trajectory of a rigid body thrown at an angle to the horizontal.

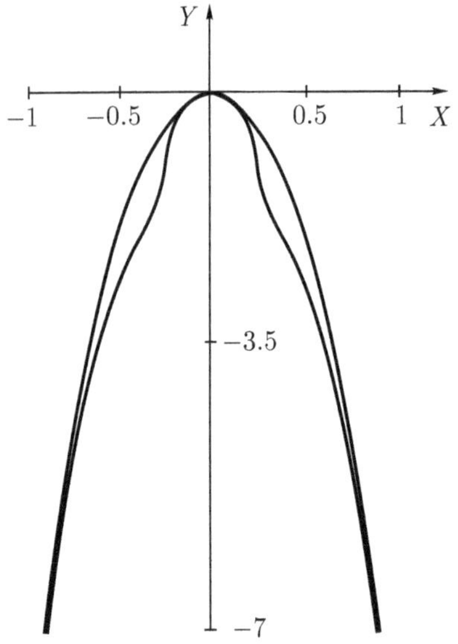

Fig. 6.53. Body trajectories for doubly asymptotic motions with one (upper curve) and three half-turns.

As is shown in [345], in the general case, the body trajectory is asymptotically a parabola:

$$X(t) = -P_1 t + o(t), \quad Y(t) = -\frac{\mu t^2}{2c_3} + o(t^2).$$

5 A body with three symmetry planes

As above (for the plane-parallel motion), before studying the general system (6.57)–(6.58), we consider in detail the particular case of motion without initial impulse ($P_1 = 0$) under the additional restrictions

$$\mathbf{B} = 0, \quad \boldsymbol{r} = 0. \tag{6.77}$$

In this case, we obtain a nonautonomous Hamiltonian system (on $e(3)$) for $\boldsymbol{M}$ and $\boldsymbol{\gamma}$ with Hamiltonian

$$\bar{H} = \frac{1}{2}(\mathbf{A}\boldsymbol{M}, \boldsymbol{M}) + \frac{1}{2}\mu^2 t^2 (\mathbf{C}\boldsymbol{\gamma}, \boldsymbol{\gamma}). \tag{6.78}$$

In the general case, we can assume that $\mathbf{A}$ is a diagonal matrix and $\mathbf{C}$ is an arbitrary symmetric matrix.

Equilibrium solutions and normal oscillations. The equations of motion of system (6.78) have the form

$$\dot{\boldsymbol{M}} = \boldsymbol{M} \times \mathbf{A}\boldsymbol{M} + \mu^2 t^2 \boldsymbol{\gamma} \times \mathbf{C}\boldsymbol{\gamma}, \quad \dot{\boldsymbol{\gamma}} = \boldsymbol{\gamma} \times \mathbf{A}\boldsymbol{M} \tag{6.79}$$

and the simplest solution is

$$\boldsymbol{M} = 0, \quad \boldsymbol{\gamma} = \pm\boldsymbol{\xi}_i, \quad i = 1, 2, 3, \tag{6.80}$$

where $\boldsymbol{\xi}_i$ are eigenvectors of the matrix $\mathbf{C}$ (if any of the eigenvalues of $\mathbf{C}$ coincide, there are infinitely many eigenvectors $\boldsymbol{\xi}_i$).

Let us linearize the system (6.79) in a neighborhood of the equilibrium solution (6.80). By linear coordinate transformations we can reduce the equations of motion to the form of "normal oscillations":

$$\ddot{x}_k + t^2 \omega_k x_k = 0, \quad k = 1, 2, \tag{6.81}$$

where x_k are appropriate local coordinates near the fixed points $\boldsymbol{\gamma} = \boldsymbol{\xi}_i$. The solutions of system (6.81) are conveniently expressed in terms of Bessel functions (see (6.71)). It is easy to show that if all eigenvalues of $\mathbf{C}$ are different, then the (asymptotically) stable solution of system (6.81) which has the form (6.71) for large t corresponds to a local minimum of the function $V(\boldsymbol{\gamma}) = \frac{1}{2}(\boldsymbol{\gamma}, \mathbf{C}\boldsymbol{\gamma})$. All the other eigenvectors produce linearly unstable solutions.

Asymptotic behavior of solutions. Now we shall show that analogously to the plane-parallel case, for arbitrary initial conditions, the vector $\boldsymbol{\gamma}$ tends to one of the eigenvectors of the matrix $\mathbf{C}$. The assertion formulated below is the most general version of Theorems 5 and 6 on the asymptotic behavior of a rigid body. Its proof is rather complicated, and, therefore, we first have presented the arguments for the particular case of plane-parallel motion.

Theorem 7. [348] *For any solution* $\boldsymbol{\gamma}(t)$ *of Eqs.* (6.79), *we have*

$$\lim_{t\to\infty} V\big(\boldsymbol{\gamma}(t)\big) = \mathcal{E}_c,$$

where $V(\boldsymbol{\gamma}) = \frac{1}{2}(\boldsymbol{\gamma}, \mathbf{C}\boldsymbol{\gamma})$ *and* $\mathcal{E}_c$ *is a critical value of the function* $V(\boldsymbol{\gamma})$.

Proof. We give here only the main stages of the proof and leave the details to the reader. Let us make the transformation

$$\frac{1}{2}t^2 = \tau, \qquad t\boldsymbol{M} = \boldsymbol{m}.$$

We obtain the equations of motion in the form

$$\frac{d\boldsymbol{m}}{d\tau} = -\frac{1}{2\tau}\boldsymbol{m} + \boldsymbol{m}\times\mathbf{A}\boldsymbol{m} + \mu^2\boldsymbol{\gamma}\times\mathbf{C}\boldsymbol{\gamma}, \quad \frac{d\boldsymbol{\gamma}}{d\tau} = \boldsymbol{\gamma}\times\mathbf{A}\boldsymbol{m}. \tag{6.82}$$

It is easy to show that $\operatorname{div}\boldsymbol{v} = -\frac{1}{2\tau}$, so that the system (6.82) is the same as the Kirchhoff equations with dissipation decaying in time. Let us consider the energy

$$\mathcal{E} = \frac{1}{2}(\boldsymbol{m}, \mathbf{A}\boldsymbol{m}) + \frac{1}{2}\mu^2(\boldsymbol{\gamma}, \mathbf{C}\boldsymbol{\gamma}) \tag{6.83}$$

of the "unperturbed" system. Calculating the derivative $\mathcal{E}$ along the solutions of (6.82), we find that

$$\frac{d\mathcal{E}}{d\tau} = -\frac{(\boldsymbol{m}, \mathbf{A}\boldsymbol{m})}{2\tau}.$$

This yields

1°. $\mathcal{E} \xrightarrow[t\to\infty]{} \mathcal{E}_\star = \text{const}$;

2°. the integral $I = \displaystyle\int_{\tau_0}^{\infty} \frac{\big(\boldsymbol{m}(\tau), \mathbf{A}\boldsymbol{m}(\tau)\big)}{2\tau}\,d\tau$ converges.

Now it remains to show that $\mathcal{E}_\star = \mathcal{E}_c$, where $\mathcal{E}_c$ is a critical value of the function (6.83) and hence of the function $V(\boldsymbol{\gamma})$. The proof of Theorem 1 was essentially based on the fact that the "unperturbed" system was a Hamiltonian system with one degree of freedom; now we have to apply a modification of that proof.

Denote by $K(\tau) = \frac{1}{2}\big(\boldsymbol{m}(\tau), \mathbf{A}\boldsymbol{m}(\tau)\big)$ the kinetic energy of the unperturbed system. Let us show that if $\mathcal{E}_\star \neq \mathcal{E}_c$, then for the function $K(\tau)$ there exist positive constants ε and $T_1 < T_2$ such that

1) if $K(\tau_\star) < \varepsilon$, then $\exists \Delta\tau_1 \leqslant T_1, \quad K(\tau_\star + \Delta\tau_1) \geqslant \varepsilon$;

2) if $K(\tau_\star) = \varepsilon$ and $K'(\tau_\star) > 0$, then $\forall \Delta\tau_2 \geqslant T_2, \quad K(\tau_\star + \Delta\tau_2) \geqslant \varepsilon$.

That is, the system departs rapidly from regions in which $K(\tau)$ is close to zero and stays for a long time inside regions in which $K(\tau)$ is bounded away from zero (see Fig. 6.54). We now show that this contradicts property 2°.

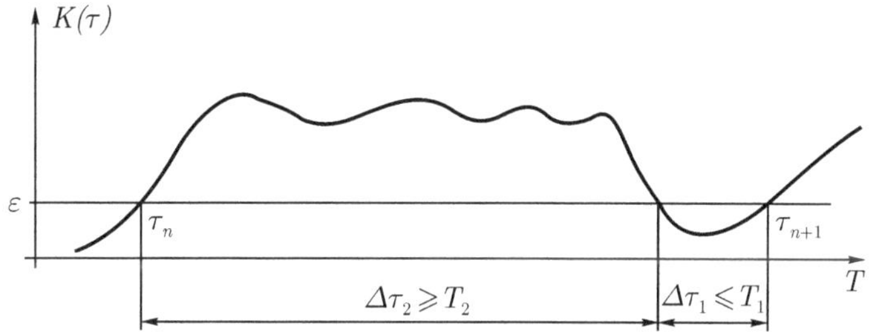

Fig. 6.54.

We will find a lower bound for the integral

$$I_n = \int_{\tau_0}^{\tau_n} \frac{K(\tau)}{\tau} d\tau,$$

where τ_n are consecutive moments of intersection with the line $K(\tau) = \varepsilon$ such that $K(\tau_n) > 0$ (see Fig. 6.54). We do this in two stages: first, we show that I_n is bounded below by an integral whose upper boundary increases as n grows slower than τ_n. Then, in turn, we show that this second integral is bounded below by the partial sum of a divergent series.

Since $\tau_n = \tau_{n-1} + \Delta\tau_1 + \Delta\tau_2 \geqslant \tau_{n-1} + \Delta\tau_2 \geqslant \tau_{n-1} + T_2 \geqslant nT_2 + \tau_0$, we obtain

$$I_n \geqslant \int_{\tau_0}^{\tau_0+nT_2} \frac{K(\tau)}{\tau} d\tau.$$

On the basis of properties 1 and 2, we conclude that the minimum value of the integral on the right-hand side is obtained if all the intervals of time during which $K(\tau) \leqslant \varepsilon$ are situated the closest to the origin τ_0 and are separated by the maximum possible interval $\Delta\tau_1 - T_1$; clearly this value will be greater than the sum:

$$I_n \geqslant \int_{\tau_0}^{\tau_0+nT_2} \frac{K(\tau)}{\tau} d\tau \geqslant \sum_{k=1}^{\overline{n}} \int_{\tau_0+kT_1+(k-1)T_2}^{\tau_0+kT_1+kT_2} \frac{\varepsilon}{\tau} d\tau \geqslant \sum_{k=1}^{\overline{n}} \frac{\varepsilon \cdot T_2}{\tau_0 + k(T_1 + T_2)},$$

where $\overline{n} = \min\big(k, k(T_1 + T_2) \geqslant nT_2\big)$. But the latter sum diverges as $\overline{n} \to \infty$.

To prove properties 1 and 2, we choose a time sufficiently large that $\mathscr{E} \approx \mathscr{E}_\star$ and choose ε sufficiently small that in the neighborhood $K(\tau) \leqslant \varepsilon$, the vector field has the form

$$\dot{\boldsymbol{m}} = \mu^2 \boldsymbol{\gamma} \times \mathbf{C}\boldsymbol{\gamma} + O(\varepsilon^2), \quad \dot{\boldsymbol{\gamma}} = O\big(\sqrt{\varepsilon}\big). \tag{6.84}$$

Since $\mathscr{E}_\star$ is distant from $\mathscr{E}_c$, the vector field (6.84) is large enough (i.e., $|\boldsymbol{\gamma} \times \mathbf{C}\boldsymbol{\gamma}|$ is not a small quantity). From this, we can obtain estimates for T_1 and T_2 if we use the following conditions:

$\Delta m = \sqrt{\varepsilon} \sim |\boldsymbol{\gamma} \times \mathbf{C}\boldsymbol{\gamma}| \cdot T_1$, i.e. T_1 is the time for which the system leaves the neighborhood $K(\tau) \leqslant \varepsilon$;

$\Delta y \sim 1 \sim \sqrt{\varepsilon} T_2$, i.e., for time T_2, the vector field (6.84) changes so that the point returns to the neighborhood $K(\tau) \leqslant \varepsilon$.
That is, $T_1 = O(\varepsilon^{1/2})$ and $T_2 = O(\varepsilon^{-1/2})$.

We thus have shown that the condition $\mathscr{E}_\star \neq \mathscr{E}_c$ contradicts the convergence of the integral. □

For the fall of an arbitrary body having three symmetry planes, there also exists the following conjecture due to V. V. Kozlov: $\mathscr{E}_\star = \mathscr{E}_c^{\min}$ *for almost all solutions* $\boldsymbol{\gamma}$ *of Eqs.* (6.79). This means that as $t \to \infty$, the body's orientation in space almost always tends to become such that the axis corresponding to the largest added mass is vertical.

Computer analysis. The theorem formulated above leads to the following natural question: what are the structures of the sets in the space of initial conditions which correspond to different asymptotic regimes as $t \to \pm\infty$ (i.e., the basins of attraction)? Let us choose $t_0 = 0$ and parameterize the common four-dimensional level surface of the integrals

$$(\boldsymbol{M}, \boldsymbol{\gamma}) = c = \text{const.}, \quad \boldsymbol{\gamma}^2 = 1$$

by the Andoyer variables (L, G, l, g), and fix the 2-dimensional surface of initial conditions for $t_0 = 0$ by the equations

$$g = g_0, \quad E = \frac{1}{2}(\boldsymbol{M}, \mathbf{A}\boldsymbol{M}) = \text{const.}$$

We color a point on this surface in the color corresponding to the side on which the body falls as $t \to \infty$. A typical picture is presented in Figs. 6.55, 6.56, and 6.57.

We see that the body falls in such a way that the axis corresponding to the maximal added mass is vertical, either with its wide side downwards or upwards, which confirms the conjecture formulated above. In this case, the structure of the boundaries of these domains is fractal: when enlarged, the pattern of the surface occurs repeatedly at increasingly smaller scales (i.e., it has the fractal properties).

Therefore, if we use the analogy with integrable and nonintegrable (regular and chaotic) systems, then the plane-parallel case can be called the integrable case, while the general case of the system (6.78)–(6.79) can be called the nonintegrable case. Indeed, in the plane-parallel case, the boundaries of the basins of attraction corresponding to various orientations of the body are regular, while for the system (6.78)–(6.79) they are fractal. As is seen from Fig. 6.57, if the system (6.79) has one more additional integral (in this case, the Lagrange integral), then the boundaries of the basins of attraction become regular.

The fractal structure of the boundaries separating the different types of behavior as $t \to \infty$ is closely related to probabilistic effects arising in the description of asymptotic motions. Indeed, for a complicated fractal pattern of distribution of the initial conditions corresponding to different types of asymptotic behavior and for specified given initial conditions (known only with finite degree of accuracy), the asymptotic

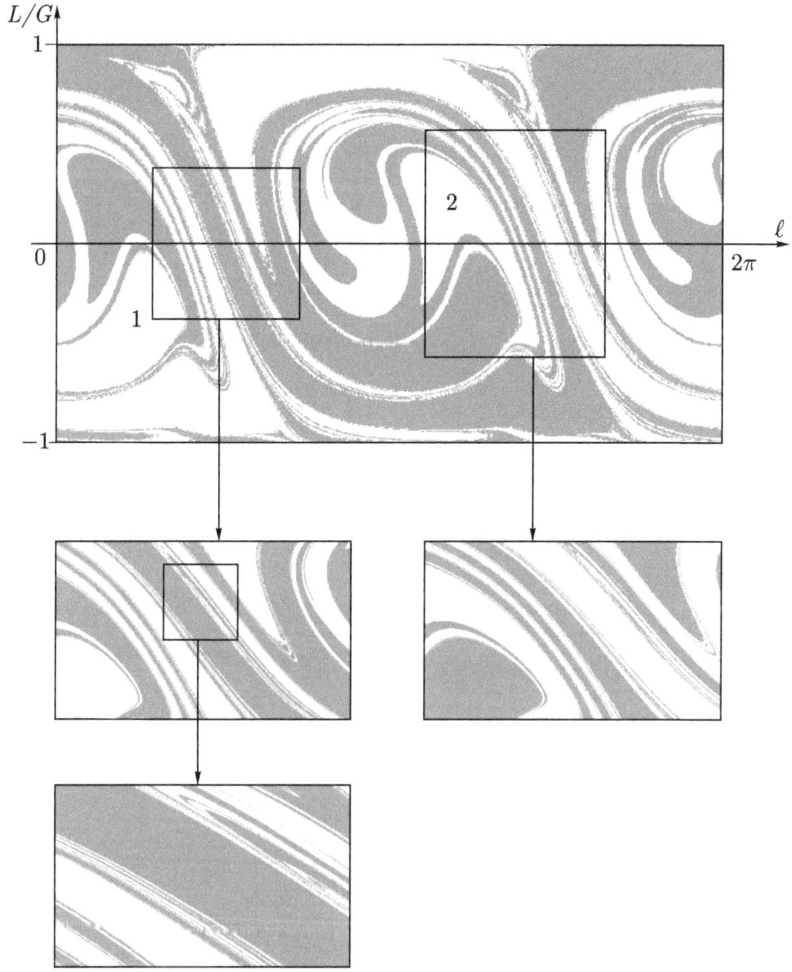

Fig. 6.55. Typical pattern of the domains corresponding to two different positions of the body, which are limiting as $t \to +\infty$ (in which the eigenvector corresponding to the maximal added mass is vertical; two colors correspond to its two possible directions). On the four-dimensional level surface of first integrals, the two-dimensional surface is defined by Eqs. (6.78). The values of the system parameters are as follows: $\mathbf{A} = \mathrm{diag}(1.8, 1.5, 2)$, $\mathbf{C} = \mathrm{diag}(0.5, 2.9, 1.4)$, $\mu = 1$, $(\boldsymbol{M}, \boldsymbol{\gamma}) = 1$, and $E_0 = 7$.

behavior becomes unpredictable, and we are only able to give a probabilistic description of the motion. This is a specific probabilistic chaos generated by the structure of the initial conditions. A similar probabilistic description was proposed by A. I. Neishtadt for studying the motion of a rigid body around a fixed point under the action of dissipative moments, both constant and linear in ω [442]. It turns out that for small values of these moments, the system dynamics is of probabilistic character, and explicit formulas for probabilities realizing the evolution of the system to one of the uniform

(a) E_0=20, t_0=0.6

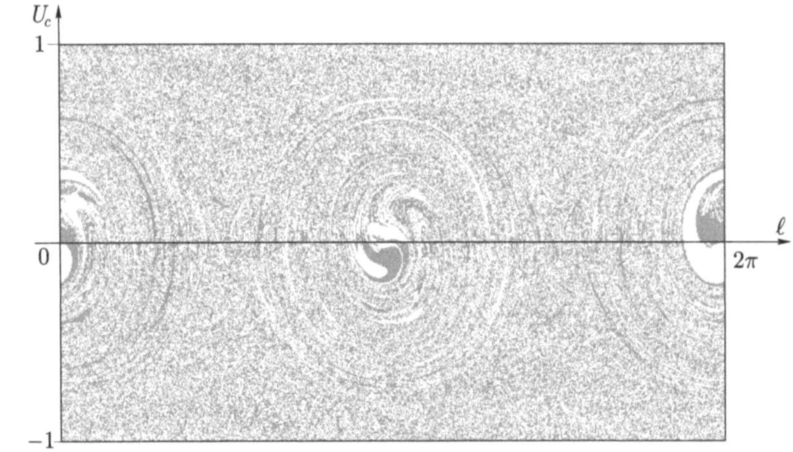

(b) E_0=70, t_0=0.3

Fig. 6.56. Typical partition of the surface of the initial conditions in accordance with the system behavior as $t \to \infty$ for increasing initial energy and initial time t_0. The values of the system parameters are as follows: $\mathbf{A} = \mathrm{diag}(1.8, 1.5, 2)$, $\mathbf{C} = \mathrm{diag}(0.5, 2.9, 1.4)$, $\mu = 1$, $(\boldsymbol{M}, \boldsymbol{\gamma}) = 1$, and $E_0 = 1$.

each rotations were obtained in [442]. A direct extension of the analytical results of [442] to the system of equations (6.79) and (6.82) involves considerable difficulties due to the higher dimension of the system and to the dependence of the “dissipation parameter” ε on time: $\varepsilon \sim \frac{1}{\tau}$.

The above-mentioned type of chaos can also be called *asymptotic* or *scattering-related* in accordance with the accepted terminology for such systems.

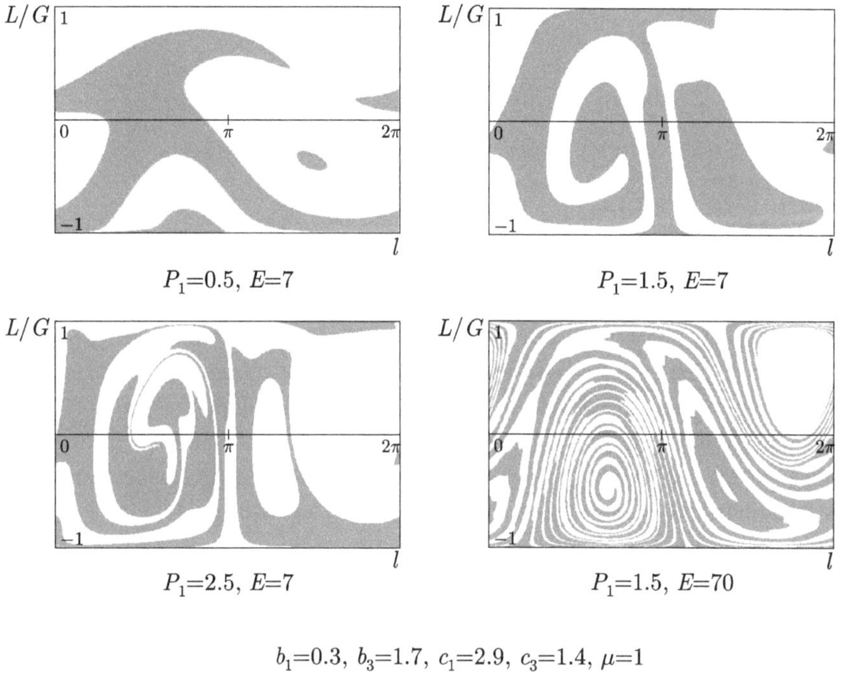

Fig. 6.57. Analog of the Lagrange case, i.e., the case of existence of the integral $M_3 = \text{const}$. The regular structure of the basin of attraction.

REMARK. The fall of a heavy body in a fluid differs essentially from its inertial motion described by the Kirchhoff equations. In the general case, the latter system is nonintegrable [350, 91] and exhibits typical chaotic behavior (Hamiltonian chaos) [89, 15].

6 Falling body with screw symmetry. Steklov solutions and their stability

For the general case $P_1 \neq 0, \mathbf{B} \neq 0$ of the system (6.57)–(6.58), after the transformation

$$\frac{1}{2}t^2 = \tau, \quad \boldsymbol{M} = t\boldsymbol{m},$$

we obtain the equations of motion in the form

$$\begin{gathered}
\frac{d\boldsymbol{m}}{d\tau} = -\frac{1}{2\tau}\boldsymbol{m} + \boldsymbol{m} \times \frac{\partial H}{\partial \boldsymbol{m}} + \boldsymbol{\alpha} \times \frac{\partial H}{\partial \boldsymbol{\alpha}} + \boldsymbol{\gamma} \times \frac{\partial H}{\partial \boldsymbol{\gamma}}, \\
\frac{d\boldsymbol{\alpha}}{d\tau} = \boldsymbol{\alpha} \times \frac{\partial H}{\partial \boldsymbol{m}}, \quad \frac{d\boldsymbol{\gamma}}{d\tau} = \boldsymbol{\gamma} \times \frac{\partial H}{\partial \boldsymbol{m}}, \\
H = H_0 + \frac{1}{\sqrt{2\tau}}H_1 + \frac{1}{2\tau}H_2,
\end{gathered}$$

$$H_0 = \frac{1}{2}(\boldsymbol{m}, \mathbf{A}\boldsymbol{m}) - \mu(\mathbf{B}\boldsymbol{m}, \boldsymbol{\gamma}) + \frac{\mu^2}{2}(\mathbf{C}\boldsymbol{\gamma}, \boldsymbol{\gamma}),$$
$$H_1 = P_1(\mathbf{B}\boldsymbol{m}, \boldsymbol{\alpha}) - P_1\mu(\mathbf{C}\boldsymbol{\alpha}, \boldsymbol{\alpha}), \quad H_2 = \frac{P_1}{2}(\boldsymbol{\alpha}, \mathbf{C}\boldsymbol{\alpha}) + \mu(\boldsymbol{r}, \boldsymbol{\gamma}). \tag{6.85}$$

Differentiating the energy with respect to time along the system yields

$$\frac{dH}{d\tau} = -\frac{(\boldsymbol{m}, \mathbf{A}\boldsymbol{m})}{2\tau} + \frac{1}{2\tau}\left(\mathbf{B}\boldsymbol{m}, \mu\boldsymbol{\gamma} - \frac{2P_1}{\sqrt{2\tau}}\boldsymbol{\alpha}\right) + \frac{W_1}{(2\tau)^{3/2}} + \frac{W_2}{(2\tau)^2},$$
$$W_1 = -P_1\mu(\mathbf{C}\boldsymbol{\alpha}, \boldsymbol{\gamma}), \quad W_2 = P_1^2(\mathbf{C}\boldsymbol{\alpha}, \boldsymbol{\alpha}) + 2\mu(\boldsymbol{r}, \boldsymbol{\gamma}). \tag{6.86}$$

For this system, the asymptotic laws of the motion formulated in the previous sections are no longer true. Moreover, there exist complicated attracting motion regimes as $t \to \infty$ different from translational motions (Fig. 6.58). First of all, we consider stability conditions (for $\mathbf{B} \neq 0$) of the classical uniformly accelerated rotations for Eqs. (6.85) and find a range of parameter values over which stability is lost (and more complicated regimes become stable). In what follows, we also consider the case of a nonzero initial impulse $P_1 = 0$.

Linear stability of Steklov solutions. Let $P_1 = 0$; then the equations for $\boldsymbol{m}$ and $\boldsymbol{\gamma}$ decouple and the area integral can be represented as

$$(\boldsymbol{m}, \boldsymbol{\gamma}) = \frac{\sigma}{\sqrt{2\tau}}, \quad \sigma = \text{const}, \tag{6.87}$$

i.e., $(\boldsymbol{M}, \boldsymbol{\gamma}) = \sigma$.

Moreover, if $\boldsymbol{r} = 0$ and **A**, **B**, and **C** *are simultaneously diagonalizable*, then Eqs. (6.85) admit particular solutions analogous to the equilibrium solutions in (6.80). In the basis of the common eigenvectors of the matrices **A**, **B**, and **C**, we have

$$\gamma_k = \pm 1, \quad \gamma_i = \gamma_j = 0, \quad m_k = \pm\frac{\sigma}{\sqrt{2\tau}}, \quad m_i = m_j = 0, \quad i \neq j \neq k \neq i, \tag{6.88}$$

i.e., if one takes into account the serial number k and possible signs $\pm$, there exist six particular solutions. In this case, the body falls in such a way that its axis Oe_k remains vertical and the angular velocity of rotation about Oe_k is determined by

$$\Omega^{(k)} = -\mu b_k t + \sigma a_k,$$

i.e., the body rotation accelerates uniformly. The velocity of the origin of the moving coordinate system referred to the moving coordinate axes is determined by the expression $\boldsymbol{v} = \frac{\partial H}{\partial \boldsymbol{p}} = (\sigma\mathbf{B} - \mu t\mathbf{C})\boldsymbol{\gamma}$, whence, using (1.104), we find that

$$x_i = \text{const}, \quad x_j = \text{const}, \quad x_k = -\mu c_k\frac{t^2}{2} + \sigma b_k t + \text{const},$$

i.e., the origin accelerates uniformly along the vertical axis, as in the case of a freely falling body. These uniformly accelerated motions were found by V. A. Steklov [558]

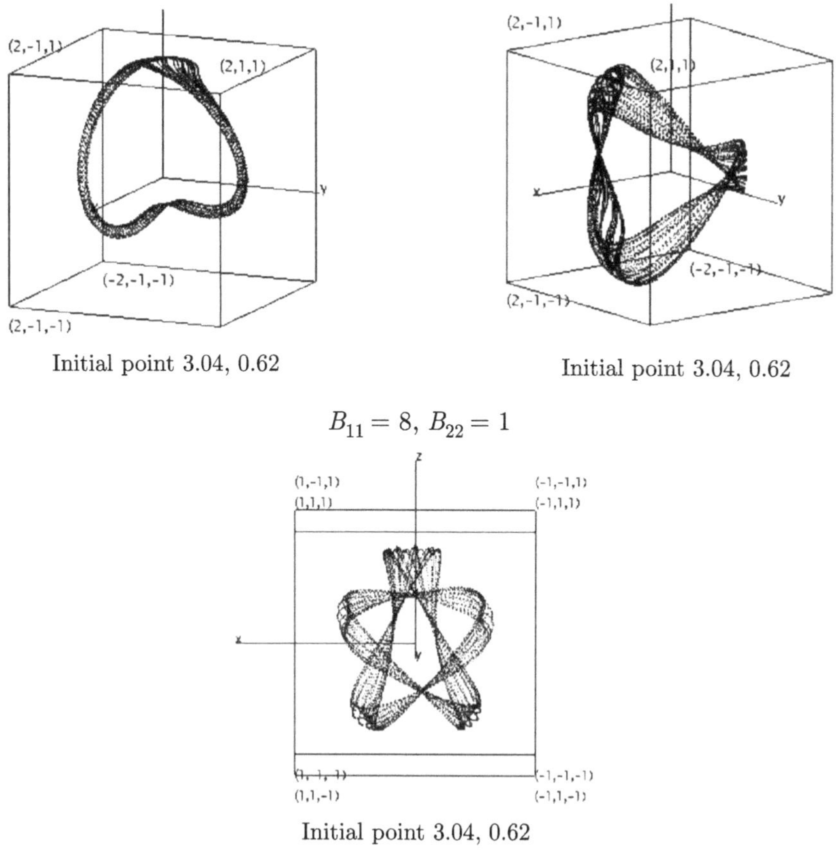

Fig. 6.58. Fall a body with screw symmetry. Limit invariant sets $A_{11} = 1$, $A_{22} = 1.2$, $A_{33} = 2$, $C_{11} = 1.6$, $C_{22} = 0.1$, $C_{33} = 0$, $E = 3$, $P_1 = 0$, $m = 1$, $x = y = z = 0$, $B_{11} = 8$, $B_{22} = 0$.

(1895) and S. A. Chaplygin [129] (1890). In what follows, they are called the *Steklov solutions*.

Analogously to solutions (6.80), solutions (6.88) are always unstable in the whole phase space (with respect to the variables $\boldsymbol{M}$ and $\boldsymbol{\gamma}$). This instability was shown by V. A. Steklov [558]. On the other hand, stability with respect to the position variables $\boldsymbol{\gamma}$ depends on the system parameters and requires separate consideration.

To study the "positional" stability of solutions (6.88), let us choose new variables

$$v_i = \frac{d\gamma_i}{d\tau}, \quad v_j = \frac{d\gamma_j}{d\tau}, \quad i \neq j \neq k \neq i, \tag{6.89}$$

and complement these equations with the area integral (6.87). We may then express the variables m_i, m_j, and m_k in terms of v_i, v_j, and σ. Using the relation $\gamma_k = \pm 1 \mp \frac{1}{2}(\gamma_i^2 + \gamma_j^2)$, near to solutions (6.88), we obtain the linearized equations for the new variables in the form

$$
\begin{aligned}
\frac{dy_i}{d\tau} &= v_i, \qquad \frac{dy_j}{d\tau} = v_j, \\
\frac{dv_i}{d\tau} &= -a_i^{-1} a_j \varkappa_i^{(k)} y_i + a_i^{-1} a_j \mu \big(b_i - b_k + a_i(b_j - b_k)\big) v_j \\
&\quad + \frac{\sigma}{\sqrt{2\tau}} a_i^{-1} \big(\mu a_j a_k (b_i - b_k) y_i + (a_i a_k + a_j a_k - a_i a_j) v_j\big) \\
&\quad - \frac{1}{2\tau} \big(v_i - \sigma^2 a_i^{-1} a_j a_k (a_k - a_i) y_j + \mu (b_j - b_k) y_j\big), \\
\frac{dv_j}{d\tau} &= \dots, \\
\varkappa_i^{(k)} &= \mu^2 \big(a_i(c_i - c_k) - (b_i - b_k)^2\big),
\end{aligned} \tag{6.90}
$$

where the expression for $\frac{dv_j}{d\tau}$ is obtained by the subscript interchange $i \leftrightarrow j$.

We recall a theorem from the theory of ordinary differential equations [140].

Let a linear system

$$x' = \big(\mathbf{A} + \mathbf{V}(t) + \mathbf{R}(t)\big)x \tag{6.91}$$

be given.

Theorem 8. *Suppose that* $\mathbf{A}$ *is a constant matrix with distinct characteristic roots* $\mu_j, j = 1, \dots, n$. *Assume that the matrix* $\mathbf{V}$ *is differentiable and satisfies the condition*

$$\int_0^\infty \left|\mathbf{V}'(\mathbf{t})\right| dt < \infty, \tag{6.92}$$

and suppose that $\mathbf{V}(\mathbf{t}) \to \mathbf{0}$ *as* $t \to \infty$. *Suppose that the matrix* R *is integrable and that*

$$\int_0^\infty \left|\mathbf{R}(\mathbf{t})\right| dt < \infty. \tag{6.93}$$

Denote by $\lambda_j(t), j = 1, \dots, n$, *the roots of the equation* $\det\big(\mathbf{A} + \mathbf{V}(t) - \lambda\mathbf{E}\big) = 0$. *Obviously, if necessary, we can choose* μ_j *so that* $\lim_{t\to\infty} \lambda_j(t) = \mu_j$. *For a given* k, *let us set*

$$D_{kj}(t) = \operatorname{Re}\big(\lambda_k(t) - \lambda_j(t)\big).$$

Assume that all j, $1 \leqslant j \leqslant n$, *lie in one of the classes* I_1 *and* I_2, *where*

$$j \in I_1 \ \mathit{if} \int_0^t D_{kj}(t) d\tau \to \infty \ \mathit{as} \ t \to \infty$$

and

$$\int_{t_1}^{t_2} D_{kj}(\tau)d\tau > -K \quad (t_2 \geqslant t_1 \geqslant 0), \tag{6.94}$$

$$j \in I_2 \text{ if } \int_{t_1}^{t_2} D_{kj}(\tau)d\tau < K \quad (t_2 \geqslant t_1 \geqslant 0); \tag{6.95}$$

here, k is fixed and K is a constant. Let $\boldsymbol{p}_k$ be the eigenvector of **A** *corresponding to the eigenvalue μ_k, so that*

$$\mathbf{A}\boldsymbol{p}_k = \mu_k' \boldsymbol{p}_k. \tag{6.96}$$

Then there exist a solution φ_k of the system (6.91) *and a number t_0, $0 \leqslant t_0 \leqslant \infty$, such that*

$$\lim_{t\to\infty} \varphi_k(t) \exp\left[-\int_{t_0}^{t} \lambda_k(\tau)d\tau\right] = \boldsymbol{p}_k. \tag{6.97}$$

In our case, for the solution (6.90) with serial number k assume $\boldsymbol{x} = (y_i, v_i, v_j)$, $i, j \neq k$ and according to (6.90) therefore, $\mathbf{R}(t) \equiv 0$ in the theorem, and the condition (6.92) holds for $\mathbf{V}(\tau)$. Since the eigenvalues of the linear system (6.90) are obviously expanded in powers of $\tau^{-1/2}$:

$$\lambda_k(\tau) = \lambda_k^{(0)} + \frac{\lambda_k^{(1)}}{\sqrt{\tau}} + \frac{\lambda_k^{(2)}}{\tau} + \dots,$$

it follows that the conditions $\operatorname{Re}\lambda_k^{(0)} \leqslant 0$ are necessary stability conditions for the system (6.90) (analogously, the conditions $\operatorname{Re}\lambda_k^{(0)} > 0$ are sufficient conditions for instability), where the $\lambda_k^{(0)}$ are the eigenvalues of the system for $\tau = \infty$. To find the $\lambda_k^{(0)}$, we can obtain the (biquadratic) characteristic polynomial

$$\lambda^4 - \lambda^2(\varkappa_i^{(k)} + \varkappa_j^{(k)} - \varkappa_k^{(k)}) + \varkappa_i^{(k)}\varkappa_k^{(k)} = 0. \tag{6.98}$$

A necessary stability condition for solutions of (6.88) is that there exists a pair of purely imaginary roots of the polynomial (6.98) (more precisely, this is the condition for the absence of exponential instability in τ). From this stability condition we obtain the corresponding restrictions on the parameters:

$$\begin{gathered}\varkappa_i^{(k)} \cdot \varkappa_j^{(k)} > 0, \quad \varkappa_i^{(k)} + \varkappa_j^{(k)} - \varkappa_k^{(k)} < 0,\\ D = (\varkappa_i^{(k)})^2 + (\varkappa_j^{(k)})^2 + (\varkappa_k^{(k)})^2 - 2\varkappa_i^{(k)}\varkappa_j^{(k)} - 2\varkappa_i^{(k)}\varkappa_k^{(k)} - 2\varkappa_j^{(k)}\varkappa_k^{(k)} > 0.\end{gathered} \tag{6.99}$$

Now let us study in detail the stability of each of the solutions depending on the parameters. Without loss of generality, we set $k = 3$, $\mu = 1$, $c_3 = 0$, and $b_3 = 0$ (the last two conditions may be fulfilled by using the integrals $\boldsymbol{\gamma}^2 = 1$ and $(\boldsymbol{M}, \boldsymbol{\gamma}) = \text{const}$). Let us fix a_1, a_2, a_3, c_1, and c_2, and on the plane of the parameters b_1, b_2, let us construct a domain in which inequalities (6.99) hold. We can show that the relations (6.99) take the following form in this case:

$$\begin{gathered}(a_1c_1 - b_1^2)(a_2c_2 - b_2^2) > 0, \quad \Phi = a_1c_2 + a_2c_1 + 2b_1b_2 > 0, \\ D = ((a_1c_2 - a_2c_1)^2 + 4(a_1b_2 + a_2b_1)(c_2b_1 + c_1b_2) > 0.\end{gathered} \tag{6.100}$$

It is straightforward to show that there exist the following three qualitatively different cases:

1° $c_3 = 0 > c_1 > c_2$ (i.e., $c_1 < 0$ and $c_2 < 0$); in this case, on the plane b_1, b_2, there are no domains in which inequalities (6.99) hold. We can show that the solutions (6.98) are partitioned either into a pair of real solutions or a quadruple of complex solutions;

2° $c_1 > c_3 = 0 > c_2$ (i.e., $c_1 > 0$ and $c_2 < 0$); in this case, the domain defined by (6.100) lies between the lines $b_1 = \pm\sqrt{a_1c_1}$ and the branches of the hyperbola defined by the relation $D = 0$ (see Fig. 6.59a);

3° $c_1 > c_2 > c_3 = 0$ (i.e., $c_1 > 0$ and $c_2 > 0$); in this case, the domains defined by (6.100) lie between the lines $b_1 = \pm\sqrt{a_1c_1}$ and $b_2 = \pm\sqrt{a_2c_2}$ and the branches of the hyperbola $D = 0$ (see Fig. 6.59b).

Remark. It is easy to show that the curves $\Phi = 0$ and $D = 0$ intersect each other at the points where they intersect some of the lines $b_i = \pm\sqrt{a_ic_i}$.

If $b_1 = b_2 = 0$, then the conditions (6.100) lead to the result presented previously in [348], namely, that only case 3°, i.e., the fall for which the axis corresponding to the largest added mass is vertical, turns out to be stable. Hence, addition of the matrix **B** allows us to stabilize (at least in the linear sense) the motion for which the "mean" axis is vertical and does not allow us to stabilize the motion for which the "small" axis is vertical.

Now, for definiteness, we set $c_1 > c_2 > c_3 = 0$, and on the plane of parameters b_1, b_2, we draw the regions of linear stability of the Steklov solutions corresponding to falling both with the "wide" side downwards and the "middle" side downwards (see Fig. 6.60; a fall with the "narrow" side downwards is always unstable). It is clearly seen in the figure that there are regions (white) in which all three Steklov solutions are unstable.

On Lyapunov stability. We are able to prove (Lyapunov) asymptotic stability for one of the Steklov solutions (6.88): for the case where the body falls with its "widest" side perpendicular to the gravitational field.

As in the foregoing discussion, we set, without loss of generality, $i = 1, j = 2, k = 3$ and $b_3 = 0$, $c_1 > c_2 > c_3 = 0$ in (6.88). We look for a Lyapunov function in the form

$$V = H_2 + \frac{1}{\tau}W.$$

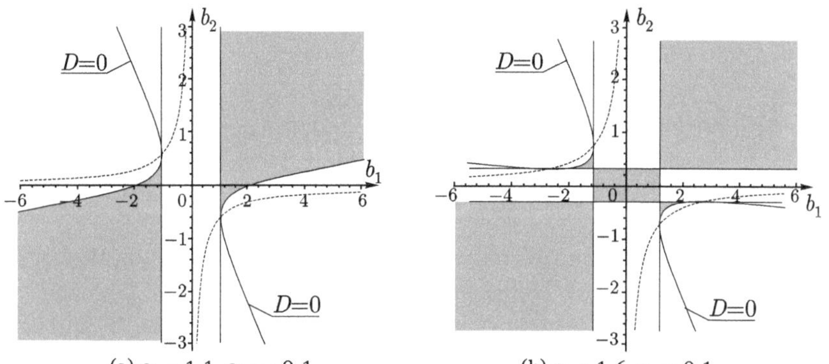

(a) $c_1 = 1.1, c_2 = -0.1$ (b) $c_1 = 1.6, c_2 = 0.1$

Fig. 6.59. Typical patterns of domains in the plane of parameters b_1, b_2 for which the necessary stability conditions (6.100) of Steklov solutions hold for various relations of parameters of the matrix **C**; in this case, $\mathbf{A} = \mathrm{diag}(1, 1.2, 2)$. (Gray denotes the domains in which the necessary stability conditions (6.100) hold.)

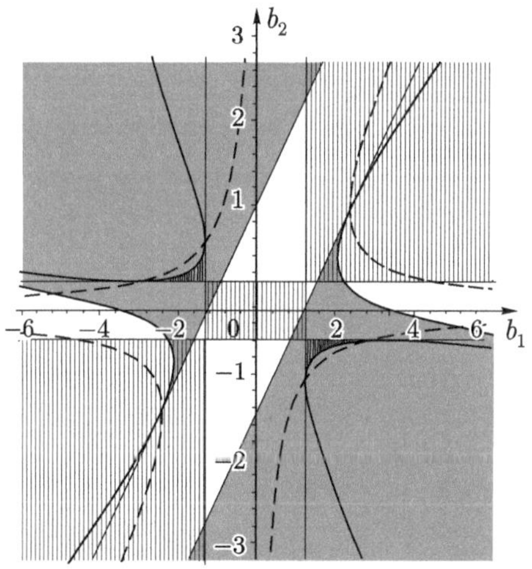

Fig. 6.60. Typical patterns of the domains of (linear) stability on the plane of parameters b_1, b_2 for the Steklov solutions corresponding to the fall with the “wide” side downward (i.e., the eigenvector in the direction of the maximal added mass is vertical) and with the “middle” side downward. $\mathbf{A} = \mathrm{diag}(1, 1.2, 2)$, $\mathbf{C} = \mathrm{diag}(1.6, 0.1, 0)$. (The hatched and gray regions indicate the domains in which conditions (6.100) hold for the above solutions.)

Here H_2 is the quadratic part of the Hamiltonian near to this solution. In the variables y_1, y_2, v_1, and v_2 H_2 has the form

$$H_2 = \frac{1}{2}(a_2^{-1}v_1^2 + a_1^{-1}v_2^2) + \frac{1}{2a_1}\left(a_1c_1 - b_1^2 - \frac{2\sigma}{\sqrt{2\tau}}a_3b_1 + \frac{\sigma^2}{4\tau^2}a_3(a_1 - a_3)\right)y_1^2$$
$$+ \frac{1}{2a_2}\left(a_2c_2 - b_2^2 - \frac{2\sigma}{\sqrt{2\tau}}a_3b_2 + \frac{\sigma^2}{4\tau^2}a_3(a_2 - a_3)\right)y_2^2. \tag{6.101}$$

We now look for the function W in the form of a homogeneous quadratic form in y_1, y_2, v_1, and v_2 with constant coefficients.

It is easy to see that near to the origin and for large τ, the function H_2 and hence V is positive definite under the conditions

$$a_1c_1 - b_1^2 > 0, \quad a_2c_2 - b_2^2 > 0. \tag{6.102}$$

As was shown earlier, these inequalities define one of the stability regions of this solution of the linearized system (see Fig. 6.59b). Thus, for those parameter values at which it will be possible to choose a function V whose derivative along solutions of the linear system is strictly negative (for sufficiently large τ) we can show the asymptotic stability in the region defined by inequalities (6.102)

The derivative of the function V along the solutions of system (6.90) has the form

$$\frac{dV}{d\tau} = -\frac{1}{\tau}G_1 + \frac{1}{\tau^{3/2}}G_2 + \frac{1}{\tau^2}G_3,$$

where G_1, G_2, and G_3 are homogeneous quadratic forms in the variables y_1, y_2, v_1, and v_2. Therefore, for large τ, the sign of the derivative $\frac{dV}{d\tau}$ is determined by the quadratic form G_1, which must be positive definite in the case of asymptotic stability.

By direct calculation, we can show that W must have the form

$$W = k_1v_1y_1 + k_2v_2y_2;$$

the addition of other terms leads to the appearance of strictly negative terms in the function G_1. With such a choice of W, G_2 and G_3 are independent of v_1 and v_2, and we have

$$\begin{aligned} G_1 &= 2k_1a_1^{-1}a_2(a_1c_1 - b_1^2)v_1^2 + 2k_2a_2^{-1}a_1(a_2c_2 - b_2^2)v_2^2 \\ &\quad + a_2^{-1}(1 - 2a_2k_1)y_1^2 + a_1^{-1}(1 - 2a_1k_2)y_2^2 \\ &\quad + \frac{1}{2}a_1^{-1}(b_1 - 2k_1(a_1b_2 + a_2b_1))y_1v_2 - \frac{1}{2}a_2^{-1}(b_2 - 2k_2(a_1b_2 + a_2b_1))y_2v_1. \end{aligned} \tag{6.103}$$

It is easy to obtain the conditions for positive definiteness of the form G_1:

$$\begin{aligned} &0 < k_1 < \frac{1}{2a_2}, \quad 0 < k_2 < \frac{1}{2a_1}, \\ &-4(a_1b_2 + a_2b_1)^2k_1^2 - 16a_1a_2(c_1a_1 - b_1^2)k_1k_2 - b_1^2 \\ &\qquad +4(2a_1a_2c_1 - a_2b_1^2 + a_1b_1b_2)k_1 > 0, \\ &-4(a_1b_2 + a_2b_1)^2k_2^2 - 16a_1a_2(c_2a_2 - b_2^2)k_1k_2 - b_2^2 \\ &\qquad +4(2a_1a_2c_2 - a_1b_2^2 + a_2b_1b_2)k_2 > 0. \end{aligned} \tag{6.104}$$

Let us highlight the following two cases:

1. $b_1 b_2 > 0$; then choosing $k_1 = \frac{1}{2} b_1 (a_1 b_2 + a_2 b_1)^{-1}$ and $k_2 = \frac{1}{2} b_2 (a_1 b_2 + a_2 b_1)^{-1}$, we obtain a diagonal and positive definite form (6.103);
2. $b_1 b_2 < 0$; in this case, sufficient conditions for the solvability of inequalities (6.104) are determined by solutions of a fourth-order equation (and have a very cumbersome form). At the same time, since only one term in the last two relations of (6.104) is positive, we can obtain necessary conditions for solvability of (6.104) in the form

$$\Phi_1 = 2a_1 a_2 c_1 - a_2 b_1^2 + a_1 b_1 b_2 > 0, \quad \Phi_2 = 2a_1 a_2 c_2 - a_1 b_2^2 + a_2 b_1 b_2 > 0,$$
$$b_1 b_2 > \max(-a_1 c_2, -a_2 c_1). \tag{6.105}$$

In Fig. 6.61, the regions colored in gray are those in which the solution considered is always stable and the hatched regions are those in which the additional necessary stability conditions (6.105) hold. As is seen from the figure, for $b_1 b_2 < 0$, the region of asymptotic stability does not coincide with the whole region of sign definiteness of the quadratic form (6.101).

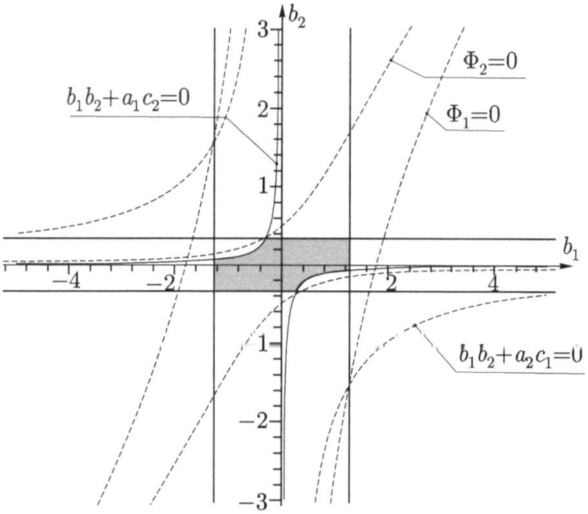

Fig. 6.61. Region of asymptotic stability of the Steklov solution corresponding to the fall with the "wide" side downward for $\mathbf{A} = \text{diag}(1, 1.2, 2)$ and $\mathbf{C} = \text{diag}(1.6, 0.1, 0)$.

REMARK. The linear and nonlinear stability analyses of Steklov solutions were carried out by M. V. Deryabin in [155, 154]. In particular, conditions (6.104) were obtained in the form of general inequalities for coefficients while ignoring the differences between the three types of Steklov solutions, for which the relation between the stability and the instability can be different. Here, we carried out a geometric analysis of possible parameter values for which the stability conditions (6.104) hold and drew conclusions on the existence of regions of parameter values for which all Steklov solutions are unstable. Numerical experiments show that in this case there exist more complicated invariant attracting sets in the phase space, such as the two-dimensional torus (see Fig. 6.58), to which the trajectories of the system (6.85) tend as $t \to +\infty$. The existence of such an invariant set has not yet been proved analytically, since the bifurcation theory and qualitative methods for systems of the form (6.85), for which the linear "dissipation" decays in time with respect to the parameter $\varepsilon \sim \frac{1}{T}$, have not yet been developed. In our analysis we also obtained simpler conditions for linear stability and Lyapunov stability by the systematic use of the Hamiltonian form for the equations of motion.

A Derivation of the Kirchhoff, Poincaré – Zhukovskii, and Four-Dimensional Top Equations

1 Rigid body motion in an ideal incompressible fluid

While the Euler – Poisson equations are well known and their derivation from the principles of dynamics is contained in many textbooks, a discussion of the physical origin of the Kirchhoff and Poincaré – Zhukovskii equations can be found only in the original classical works and in the treatise [365] of Lamb. We briefly present this derivation here using modern notation and the formalism of the Poincaré – Chetaev equations.

Kirchhoff equations. Let us consider the problem of rigid body motion in an infinite volume of ideal incompressible fluid. For this purpose, we assume that the body moving in the fluid is bounded by a simply connected surface and that the motion is inertial, i.e., it is executed only under the action of the hydrodynamic pressure of the fluid. We assume that the fluid has free boundaries and is stationary at infinity independently of the motion of the rigid body. Another condition is that the fluid flow has a single-valued velocity potential, i.e., it is vortex-free. This condition is reasonable in the sense that if the motion is vortex-free at the initial instant of time, then it remains so at all subsequent times. This assertion is a consequence of *Lagrange's theorem* [365] well known in hydrodynamics.

As will be shown, if all the above conditions hold, then the equations of motion of the rigid body, which are a system of six ordinary differential equations, decouple from the partial differential equations describing the fluid motion.

Let us consider a rigid body τ with surface Σ and volume V that moves in standard Euclidean space $\mathbb{R}^3 = \{x_1, x_2\, x_3\}$ filled with a homogeneous ideal incompressible fluid of constant density ρ. If the fluid flow is potential with potential φ, then the fluid velocity $\boldsymbol{v}_f$ is found in the form

$$\boldsymbol{v}_f = \frac{\partial \varphi}{\partial \boldsymbol{x}}, \qquad \boldsymbol{x} = (x_1, x_2, x_3) \in \mathbb{R}^3. \tag{A.1}$$

Moreover, $\boldsymbol{v}_f \to 0$ as $|\boldsymbol{x}| \to \infty$ due to the property of the fluid being stationary at infinity. It is assumed that the surface bounding the body is a smooth manifold.

We may calculate the kinetic energy of the moving fluid,

$$T_f = \frac{1}{2} \int\limits_{\mathbb{R}^3 \setminus \tau} \rho {v_f}^2 \, d^3x, \tag{A.2}$$

where d^3x is the volume element. Using the Gauss – Ostrogradskii formula, the volume integral (A.2) is transformed into a surface integral. For this purpose, let us define the vector field $\boldsymbol{u}$ by the formula

$$\boldsymbol{u} = \varphi \frac{\partial \varphi}{\partial \boldsymbol{x}} = \varphi \boldsymbol{v}_f, \tag{A.3}$$

DOI 10.1515/9783110544442-007

where the velocity potential φ satisfies Laplace's equation $\Delta\varphi = 0$, due to the incompressibility condition $\operatorname{div} \boldsymbol{v}_f = 0$. We have

$$\operatorname{div} \boldsymbol{u} = \operatorname{div}\left(\varphi \frac{\partial \varphi}{\partial \boldsymbol{x}}\right) = \varphi\Delta\varphi + \left(\frac{\partial \varphi}{\partial \boldsymbol{x}}\right)^2 = {v_f}^2. \tag{A.4}$$

Let us apply formula (A.4) to the fluid domain V, which is bounded by the surface Σ to one side and some sphere Σ' of a suitably large radius R to the other side. The kinetic energy of this fluid domain is equal to

$$T_f(V) = \frac{1}{2}\int\limits_V \operatorname{div} \boldsymbol{u}\, d^3x = \frac{1}{2}\int\limits_{\Sigma+\Sigma'} (\boldsymbol{u}, \boldsymbol{n})\, d\sigma = \frac{1}{2}\int\limits_{\Sigma+\Sigma'} \varphi\left(\frac{\partial \varphi}{\partial \boldsymbol{x}}, \boldsymbol{n}\right) d\sigma. \tag{A.5}$$

As can be shown from hydrodynamic arguments, the condition $\boldsymbol{v}_f \to 0$ implies that the potential φ has the order $o\left(\frac{1}{R}\right)$ at infinity, and, therefore, $\frac{\partial\varphi}{\partial x} = o\left(\frac{1}{R^2}\right)$. Hence, while $T_f(V) \sim T_f$ as $R \to \infty$, the contribution to $T_f(V)$ of the integral over the surface Σ' tends to zero as $R \to \infty$.

To calculate the integral (A.5) over the surface Σ, we choose a Cartesian coordinate system $Ox_1x_2x_3$ fixed in the body τ, with origin O at some point inside the body. Denote by $\boldsymbol{v} = (v_1, v_2, v_3)$ and $\boldsymbol{\omega} = (\omega_1, \omega_2, \omega_3)$ the projections of the velocity of the point O and the body's angular velocity on the axes of the chosen coordinate system, respectively (Fig. A.1).

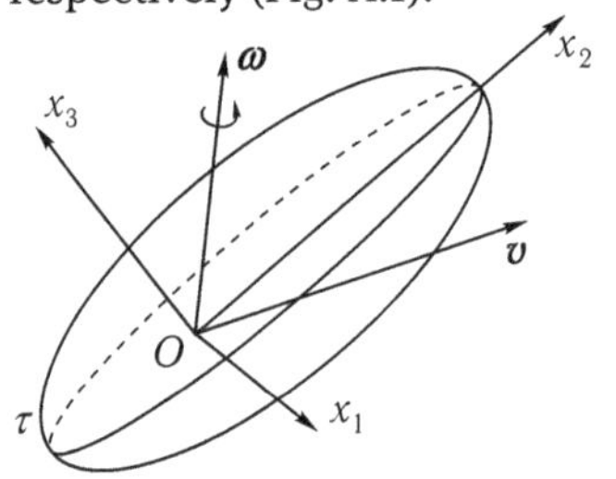

Fig. A.1.

Following Kirchhoff [302], we seek for the potential φ in the form

$$\varphi(t, \boldsymbol{x}) = \sum_{i=1}^{3} v_i\varphi_i + \sum_{i=1}^{3} \omega_i\chi_i, \tag{A.6}$$

where v_i and ω_i are assumed to be known functions of time determined by the rigid body motion, while the functions φ_i and χ_i are harmonic functions: $\Delta\varphi_i = \Delta\chi_i = 0$. By construction, $\Delta\varphi = 0$.

For a point A of the surface Σ of the body τ, the following Euler formula of velocity distribution in a rigid body holds:

$$\boldsymbol{v}_A = \boldsymbol{v} + \boldsymbol{\omega} \times \overline{OA}. \tag{A.7}$$

The boundary conditions for the potential (A.6) have the form

$$(\boldsymbol{v}_A, \boldsymbol{n}) = \left(\frac{\partial \varphi}{\partial \boldsymbol{x}}, \boldsymbol{n}\right), \tag{A.8}$$

where $\boldsymbol{n} = (n_1, n_2, n_3)$ is the normal unit vector. They express the nonpenetration condition: the normal component of the fluid relative velocity near to the surface of the body is equal to zero. We obtain the following equations for the functions φ_i and χ_i:

$$\left(\frac{\partial \varphi_i}{\partial \boldsymbol{x}}, \boldsymbol{n}\right) = n_i, \quad \left(\frac{\partial \chi_i}{\partial \boldsymbol{x}}, \boldsymbol{n}\right) = \sum_{j,k} \varepsilon_{ijk}x_jn_k, \quad i, j, k = 1, 2, 3. \tag{A.9}$$

The problem of finding a harmonic function satisfying conditions of the form (A.9) on the boundary is the *Neumann problem*. This problem is solvable, and its solution is unique and independent of v_i and ω_i.

Assume that the functions φ_i and χ_i have already been found. Then from formulas (A.5), (A.6), and (A.7), we obtain the following expression for the fluid kinetic energy:

$$T_f = \frac{1}{2}(\mathbf{A}'\boldsymbol{\omega},\ \boldsymbol{\omega}) + (\mathbf{B}'\boldsymbol{\omega},\ \boldsymbol{v}) + \frac{1}{2}(\mathbf{C}'\boldsymbol{v},\ \boldsymbol{v}), \tag{A.10}$$

where $\mathbf{A}'$, $\mathbf{B}'$, and $\mathbf{C}'$ are 3×3 matrices with constant entries determined by the body's geometry. Since the rigid body's kinetic energy is also a quadratic form in the variables $\boldsymbol{v}$ and $\boldsymbol{\omega}$, it follows that the total energy of the system has the form

$$T_{f+b} = \frac{1}{2}(\mathbf{A}\boldsymbol{\omega},\ \boldsymbol{\omega}) + (\mathbf{B}\boldsymbol{\omega},\ \boldsymbol{v}) + \frac{1}{2}(\mathbf{C}\boldsymbol{v},\ \boldsymbol{v}). \tag{A.11}$$

Thus, under the given assumptions, the kinetic energy of the whole system (A.11) is determined only by the values of $\boldsymbol{\omega}$ and $\boldsymbol{v}$, and the matrices **A**, **B**, and **C** are constant. This is a consequence of the fact that the fluid is ideal and incompressible.

Physically, the matrices **A**, **B**, and **C** generalize the concepts of "added masses" and "moments of inertia," which arise in elementary statements of the problem of rigid body motion in a fluid, for example, the motion of a sphere or of a small plate [46, 365]). The total number of parameters of the matrices **A**, **B**, and **C** is equal to 21 (since the matrices **A** and **C** can be assumed to be symmetric). However, by simple arguments, we can show that by suitable choices of the point O and the orientation of the axes of the frame $Ox_1x_2x_3$, it is possible to reduce the matrix **A** to diagonal form and **B** to symmetric form. In what follows, we assume that this reduction has been carried out, which allows us to reduce the total number of parameters to 15. If the body additionally admits a certain (discrete or continuous) symmetry group, then in the kinetic energy (A.11) we can exclude yet more parameters. In Table A.1 we list various symmetry groups by presenting a set of generating elements, the matrices **A**, **B**, and **C** for each group, and also examples of the corresponding bodies. Note that in all cases, the symmetry is admitted by the surface geometry, as well as by the mass distribution of the body.

To deduce the equations of rigid body motion, it is necessary to invoke Hamilton's principle $\delta \int_{t_1}^{t_2} T_{f+b}\, dt = 0$ as applied to the body-fluid system. The extremals of this action satisfy the Euler – Poincaré equations on the group $E(3)$ (see Sec. 1.1 of Ch. 1).

To write the Euler – Poincaré equations on the group $E(3)$, as quasi-velocities we use the angular velocity $\boldsymbol{\omega}$ and the velocity $\boldsymbol{v}$ of the distinguished body point (origin O) referred to the body-fixed coordinate system. In this case, according to Sec. 1.4 of Ch. 1, we can represent the equations in the vector form

$$\begin{cases} \dfrac{d}{dt}\dfrac{\partial T}{\partial \boldsymbol{\omega}} = \dfrac{\partial T}{\partial \boldsymbol{\omega}} \times \boldsymbol{\omega} + \dfrac{\partial T}{\partial \boldsymbol{v}} \times \boldsymbol{v}, \\ \dfrac{d}{dt}\dfrac{\partial T}{\partial \boldsymbol{v}} = \dfrac{\partial T}{\partial \boldsymbol{v}} \times \boldsymbol{\omega}. \end{cases} \tag{A.12}$$

Table A.1.

	Symmetry	Conditions for the parameters		Examples
1	Plane of symmetry *xy*	$\mathbf{A} = \operatorname{diag}(a_1, a_2, a_3)$ $\mathbf{B} = \begin{pmatrix} 0 & 0 & b_{13} \\ 0 & 0 & b_{23} \\ b_{13} & b_{23} & 0 \end{pmatrix}$ $\mathbf{C} = \begin{pmatrix} c_{11} & c_{12} & 0 \\ c_{12} & c_{22} & 0 \\ 0 & 0 & c_{33} \end{pmatrix}$		
2	Two mutually perpendicular planes of symmetry *xy* and *xz*	$\mathbf{A} = \operatorname{diag}(a_1, a_2, a_3)$ $\mathbf{C} = \operatorname{diag}(c_1, c_2, c_3)$ $\mathbf{B} = \begin{pmatrix} 0 & 0 & 0 \\ 0 & 0 & b_{23} \\ 0 & b_{23} & 0 \end{pmatrix}$		
3	Three mutually perpendicular planes of symmetry *xy*, *xz*, and *yz*	$\mathbf{A} = \operatorname{diag}(a_1, a_2, a_3)$ $\mathbf{C} = \operatorname{diag}(c_1, c_2, c_3)$ $\mathbf{B} = 0$		Triaxial ellipsoid, parallelepiped
4	Rotation through angle $\alpha = \pi$ about the axis Oz	$\mathbf{A} = \operatorname{diag}(a_1, a_2, a_3)$ $\mathbf{B} = \begin{pmatrix} b_{11} & b_{12} & 0 \\ b_{12} & b_{22} & 0 \\ 0 & 0 & b_{33} \end{pmatrix}$ $\mathbf{C} = \begin{pmatrix} c_{11} & c_{12} & 0 \\ c_{12} & c_{22} & 0 \\ 0 & 0 & c_{33} \end{pmatrix}$		Two-blade screw propeller
5	Rotation through angle $\alpha = \pi/2$ about the axis Oz	$\mathbf{A} = \operatorname{diag}(a_1, a_2, a_3)$, $\mathbf{B} = \operatorname{diag}(b_1, b_2, b_3)$, $\mathbf{C} = \operatorname{diag}(c_1, c_2, c_3)$,	$a_1 = a_2$ $b_1 = b_2$ $c_1 = c_2$	Four-blade screw propeller
6	Axis of symmetry Oz (analogously for rotation through angle $\alpha = \frac{2\pi}{n}$ $n \neq 2, 4$ about the axis Oz)	$\mathbf{A} = \operatorname{diag}(a_1, a_2, a_3)$, $\mathbf{C} = \operatorname{diag}(c_1, c_2, c_3)$, $\mathbf{B} = 0$	$a_1 = a_2$ $c_1 = c_2$	Ellipsoid of rotation, three-blade screw propeller

The equations of motion were obtained in this form by G. Kirchhoff [302] (see also [95]), and A. Clebsch gave them the Hamiltonian form in [139]. Indeed, performing the Legendre transform, we obtain the following equations on the algebra $e(3)$:

$$\begin{gathered} \boldsymbol{M}=\frac{\partial T}{\partial \boldsymbol{\omega}}, \quad \boldsymbol{p}=\frac{\partial T}{\partial \boldsymbol{v}}, \\ H=(\boldsymbol{M}, \boldsymbol{\omega})+(\vec{p}, \vec{v})-T|_{\boldsymbol{\omega}, \boldsymbol{v} \rightarrow \boldsymbol{M}, \boldsymbol{p}}, \\ \dot{\boldsymbol{M}}=\boldsymbol{M} \times \frac{\partial H}{\partial \boldsymbol{M}}+\boldsymbol{p} \times \frac{\partial H}{\partial \boldsymbol{p}}, \quad \dot{\boldsymbol{p}}=\boldsymbol{p} \times \frac{\partial H}{\partial \boldsymbol{M}}. \end{gathered} \tag{A.13}$$

The Hamiltonian function is the kinetic energy expressed in the variables $\boldsymbol{M}$ and $\boldsymbol{p}$:

$$H=\frac{1}{2}(\mathbf{A}\boldsymbol{M}, \boldsymbol{M})+(\mathbf{B}\boldsymbol{M}, \boldsymbol{p})+\frac{1}{2}(\mathbf{C}\boldsymbol{p}, \boldsymbol{p}), \tag{A.14}$$

where for the matrices $\mathbf{A}$, $\mathbf{B}$, and $\mathbf{C}$ we use the same notation as in (A.11), which, however, is obtained from the latter by a simple linear transformation. In hydromechanics, the vectors $\boldsymbol{M}$ and $\boldsymbol{p}$ are called the impulsive couple and the impulsive force, respectively.

Equations of motion for a multiply connected body. We now consider the inertial motion of a multiply connected body τ (a body with holes) in a fluid. The domain $\mathbb{R}^3 \setminus \tau$ surrounding the body is also multiply connected; suppose that it admits $n+1$ closed paths (contours) not homotopic to each other, and among them let there be n contours, denoted by $l_1, \dots, l_n$, which cannot be contracted to a point inside the domain $\mathbb{R}^3 \setminus \tau$ (the number n of closed nonfree contours is connected with the number m of holes in the body by the relation $n = C_{m+1}^2 = \frac{1}{2}m(m+1)$) (Fig. A.2).

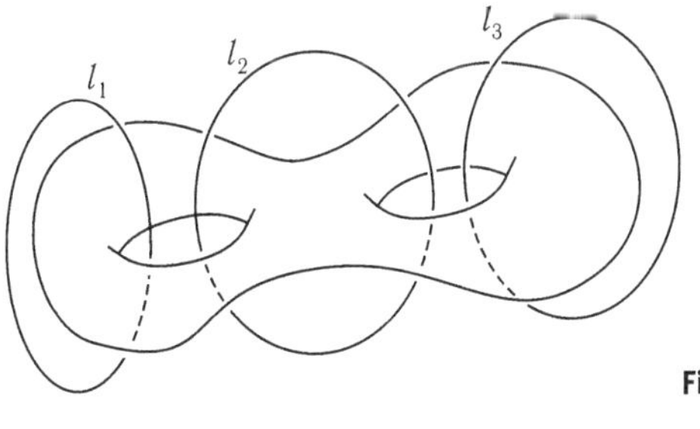

Fig. A.2.

As in the simply connected case, we assume that the fluid flow is potential, $\boldsymbol{v}_f = \frac{\partial \varphi}{\partial \boldsymbol{x}}$. However, since the domain is not simply connected, the velocity potential φ is not uniquely determined by the Laplace equation and the boundary conditions $\frac{\partial \varphi}{\partial \boldsymbol{n}}$. To uniquely solve Laplace's equation in this case, along with the boundary conditions, it is necessary to prescribe the circulations $\varkappa_i$ around all of the contours l_i, $i = 1, \dots, n$. (As follows from Green's theorem, the circulations along homotopic contours coincide.)

Following G. Lamb [365], let us represent the potential in the form

$$\varphi(t, \boldsymbol{x}) = \sum_{i=1}^{3} v_i(t)\varphi_i(\boldsymbol{x}) + \sum_{i=1}^{3} \omega_i(t)\chi_i(\boldsymbol{x}) + \varphi_0, \tag{A.15}$$

where v_i and ω_i, $i = 1, 2, 3$, are the translational and angular rigid body velocities, respectively, φ_i and χ_i are functions single-valued in the whole domain $\mathbb{R}^3 \setminus \tau$ and determined by the Laplace equation and the boundary conditions (A.9). The potential φ_0 defines a circular motion in the domain and is defined by the equation $\Delta\varphi_0 = 0$ with zero boundary conditions $\frac{\partial\varphi_0}{\partial \boldsymbol{n}} = 0$ and prescribed circulations $\varkappa_i$ along the contours l_i, $i = 1, \ldots, n$. Since φ_0 depends linearly on the circulations, we represent it in the form

$$\varphi_0(\boldsymbol{x}) = \sum_{i=1}^{n} \varkappa_i\psi_i(\boldsymbol{x}), \tag{A.16}$$

where each of the potentials ψ_i has unit circulation along the contour l_i and zero circulation along the other contours $l_j, j \neq i$. Thus, ψ_i, $i = 1, \ldots, n$ are completely defined by the body geometry and are independent of the velocities $\boldsymbol{v}, \boldsymbol{\omega}$ and the circulations $\varkappa_i$.

Remark. The potentials ψ_i and hence φ_0 are not single-valued functions in the domain $\mathbb{R}^3 \setminus \tau$.

In calculating the fluid kinetic energy and the pressure forces acting on the body, it is also necessary to take into account the circulation flow independent of time. As was shown in [365], when this is done, T_f differs from (A.10) only by a constant term consisting of a homogeneous quadratic form in the circulations $\varkappa_1, \ldots, \varkappa_n$. At the same time, terms linear in $\varkappa_i$ of the form

$$\rho\sum_{i=1}^{n}\int_{\Sigma} \varkappa_i\psi_i\frac{\partial\varphi_\mu}{\partial n}d\sigma, \quad \rho\sum_{i=1}^{n}\int_{\Sigma} \varkappa_i\psi_i\frac{\partial\chi_\mu}{\partial n}d\sigma, \quad \mu = 1, 2, 3, \tag{A.17}$$

are added, respectively, to the linear momentum $\frac{\partial T_{f+b}}{\partial \boldsymbol{v}}$ and the angular momentum $\frac{\partial T_{f+b}}{\partial \boldsymbol{\omega}}$, where the integration is over the body's surface and ρ is the constant fluid density.

Hence the existence of circulation flows is equivalent to the addition of constant gyroscopic forces acting on the body. For Eqs. (A.13) in the variables $\boldsymbol{M}$ and $\boldsymbol{p}$ there are additional linear terms in the Hamiltonian

$$H = \frac{1}{2}(\mathbf{A}\boldsymbol{M}, \boldsymbol{M}) + (\mathbf{B}\boldsymbol{M}, \boldsymbol{p}) + \frac{1}{2}(\mathbf{C}\boldsymbol{p}, \boldsymbol{p}) + (\boldsymbol{a}, \boldsymbol{M}) + (\boldsymbol{b}, \boldsymbol{p}), \tag{A.18}$$

where $\boldsymbol{a}$ and $\boldsymbol{b}$ are constant vectors linear in $\varkappa_i$.

Remark. The equations of motion for a rigid body in a fluid in a gravitational field are defined on the algebra $e(3) \oplus_s \mathbb{R}^{12}$ and are presented in Ch. 1, Sec. 1.4 (1.8). In this general case the Hamiltonian function (A.18) includes a potential energy term $U = U(\boldsymbol{\alpha}, \boldsymbol{\beta}, \boldsymbol{\gamma}, \boldsymbol{x})$, where $\boldsymbol{\alpha}$, $\boldsymbol{\beta}$, and $\boldsymbol{\gamma}$ are direction cosines and $\boldsymbol{x}$ is the radius vector of the center of mass. The two most important cases of rigid body dynamics in a fluid with a potential field, for which the equations of motion can be written on the algebra $e(3)$, were presented by S. A. Chaplygin [128, 129]. (For these cases, the system of equations for the angular momentum vector $\boldsymbol{M}$ and the vertical unit vector $\boldsymbol{\gamma}$ decouples):

1. *The body in the fluid falls under the influence of a gravitational field with zero initial shock.* In this case, we obtain a nonautonomous system (Sec. 1.6 of Ch. 1 and Sec. 6.7 of Ch. 6).
2. *The motion is in a gravitational field balanced by the buoyancy force.* In this case, we obtain an autonomous system with terms linear in $\boldsymbol{\gamma}$ (Sec. 3.1 of Ch. 3).

In both cases, the body must be bounded by a surface having three mutually perpendicular planes of symmetry.

2 Poincaré – Zhukovskii equations

So far we have considered the equations of motion for a rigid body in a fluid; now we pass to the consideration of another class of problems concerning the motion about a fixed point of a rigid body containing cavities which are filled with an ideal incompressible fluid. Here, the most interesting case is where the fluid executes motion with homogeneous vorticity [421, 429, 472]. In this case, the six-dimensional system of equations describing the evolution of the body's angular momentum $\boldsymbol{M}$ and the fluid vorticity $\boldsymbol{\xi}$ decouples. The case of potential fluid flow in a simply connected cavity leads only to a change in the body's moments of inertia and defines an invariant manifold $\boldsymbol{\xi} = 0$. For the potential flow in a multiply connected cavity we obtain the same equations of motion as those of a rigid body with a gyrostat; this system was studied by N. E. Zhukovskii in detail [647]. A system equivalent to a body with a gyrostat is said to be *Zhukovskii equivalent*. We can show that homogeneous vortex fluid motion is possible only for an ellipsoidal cavity [429].

Let us choose a moving coordinate system $O\boldsymbol{e}_1, \boldsymbol{e}_2, \boldsymbol{e}_3$ with origin at a point O fixed in space and with axes attached to the shell (see Fig. A.3). The vector $\boldsymbol{x}_c = (x_{c1}, x_{c2}, x_{c3})$ defines the coordinates of the center of the cavity in the system $O\boldsymbol{e}_1, \boldsymbol{e}_2, \boldsymbol{e}_3$.

The equation of the cavity boundary may be written as

$$F = (\boldsymbol{x} - \boldsymbol{x}_c, \mathbf{B}(\boldsymbol{x} - \boldsymbol{x}_c)) = 1, \qquad \text{(A.19)}$$

where $\mathbf{B}$ is a symmetric matrix whose eigenvalues coincide with the inverse squares $\lambda_i = \frac{1}{b_i^2}$, $i = 1, 2, 3$, of the principal semiaxes of the cavity.

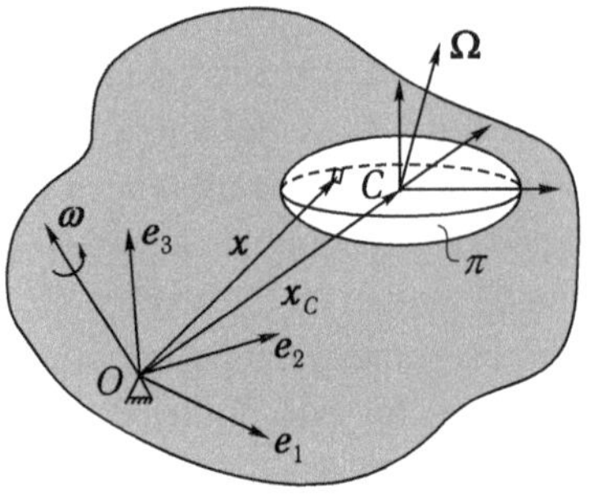

Fig. A.3.

An important feature of the ellipsoidal cavity is that in its interior there exists a particular solution of the Euler equations for the ideal fluid for which the velocities $\boldsymbol{v}(t, \boldsymbol{x})$ satisfying the hydrodynamics equations and the boundary conditions are linear in coordinates. It is for this reason that the vortex flow homogeneous at the initial instant remains homogeneous at all instants of time (H. Poincaré and P. L. Dirichlet). Let us present this solution in explicit form.

For this purpose, let us represent the fluid velocity field in the cavity as

$$\boldsymbol{v}(t, \boldsymbol{x}) = \boldsymbol{\omega}(t) \times \boldsymbol{x} + \mathbf{D}(t)(\boldsymbol{x} - \boldsymbol{x}_c), \tag{A.20}$$

where $\boldsymbol{\omega}(t)$ is the shell angular velocity and $\mathbf{D}(t)$ is a 3×3 matrix. The vorticity is the same for all points of the cavity and is equal to

$$\boldsymbol{\Omega}(t) = \frac{1}{2} \operatorname{rot} \boldsymbol{v}(t, \boldsymbol{x}) = \boldsymbol{\omega}(t) + \boldsymbol{d}^a(t); \tag{A.21}$$

here, $\boldsymbol{d}^a$ is the vector with components $\frac{1}{2}(d_{32} - d_{23},\ d_{13} - d_{31},\ d_{21} - d_{12})$ corresponding to the antisymmetric part of $\mathbf{D}(t)$.

The boundary nonpenetration conditions have the form

$$(\boldsymbol{v}(t, \boldsymbol{r}_A),\ \boldsymbol{n}) = (\boldsymbol{\omega} \times \boldsymbol{r}_A,\ \boldsymbol{n}), \tag{A.22}$$

where $\boldsymbol{r}_A$ is the radius vector of a point A on the boundary of the domain and $\boldsymbol{n}$ is a normal vector at the point A.

Let us choose the normal vector in the form $\boldsymbol{n} = \nabla F = \mathbf{B}(\boldsymbol{x} - \boldsymbol{x}_c)$. Equating the coefficients multiplying $(x_i - x_{ci})(x_j - x_{cj})$, we find from Eq. (A.22) that

$$\mathbf{BD} + (\mathbf{BD})^T = 0. \tag{A.23}$$

That is, $\mathbf{BD}$ is a skew-symmetric matrix. Following [429, 472], we set $\mathbf{BD} = \mathbf{B}^{1/2}\boldsymbol{\Xi}\mathbf{B}^{1/2}$, i.e.,

$$\boldsymbol{v} = \boldsymbol{\omega} \times \boldsymbol{x} + \mathbf{B}^{-1/2}\boldsymbol{\Xi}\mathbf{B}^{1/2}(\boldsymbol{x} - \boldsymbol{x}_c), \tag{A.24}$$

where $\boldsymbol{\Xi} = ||\xi_{ij}||$ [421] is an arbitrary skew-symmetric matrix depending on time.

Denote by the same letter $\boldsymbol{\xi}$ the vector with components $\boldsymbol{\xi}_k = \varepsilon_{ijk}\xi_{ij}$ corresponding to this matrix and analogously for $\boldsymbol{\Omega}$: $\Omega_{ij} = -\varepsilon_{ijk}\Omega_k$. Then in matrix form we have

$$\boldsymbol{\Omega}(t) = \boldsymbol{\omega} + \frac{1}{2}\left(\mathbf{B}^{-1/2}\boldsymbol{\Xi}(t)\mathbf{B}^{1/2} - \mathbf{B}^{1/2}\boldsymbol{\Xi}(t)\mathbf{B}^{-1/2}\right). \tag{A.25}$$

To explain the geometric meaning of the vector $\boldsymbol{\xi}$, let us make the change of variables [472]:

$$\boldsymbol{x}' = \mathbf{B}^{1/2}(\boldsymbol{x} - \boldsymbol{x}_c); \tag{A.26}$$

then the new velocities are

$$\boldsymbol{v}' = \mathbf{B}^{1/2}\boldsymbol{v}.$$

Since in the new variables the equation of the cavity is that of the unit sphere centered at the origin, the usual rotation of this sphere with angular velocity $\boldsymbol{\xi}$,

$$\boldsymbol{v}_v' = \boldsymbol{\xi} \times \boldsymbol{x}',$$

corresponds to the vortex flow. Making the inverse change of variables, we obtain the velocity field in the cavity in terms of the initial variables,

$$\boldsymbol{v}(t, \boldsymbol{x}) = \boldsymbol{\omega} \times \boldsymbol{x} + \mathbf{B}^{-1/2}\Xi(t)\mathbf{B}^{1/2}(\boldsymbol{x} - \boldsymbol{x}_c). \tag{A.27}$$

Therefore, $\boldsymbol{\xi}$ is the angular velocity of an imaginary sphere centered at the point. Moreover, to any rotational motion of this imaginary sphere there corresponds a motion of the fluid in the cavity. The map (A.26) defines an equivalence between this system and a system of connected tops. This equivalence admits a simple generalization to the case of n cavities [61].

The configuration space of the problem is the group $SO(3) \times SO(3) \approx SO(4)$, where the first factor corresponds to the shell's rotations with respect to axes fixed in space, and the second factor corresponds to rotations of the imaginary sphere with respect to the shell. Let us represent the equations of motion in the form of the Poincaré equations on this Lie group. Denote by $\boldsymbol{w}_1, \boldsymbol{w}_2, \boldsymbol{w}_3$ and $\boldsymbol{\zeta}_1, \boldsymbol{\zeta}_2, \boldsymbol{\zeta}_3$ the corresponding invariant vector fields on the two copies of $SO(3)$. Then

$$\boldsymbol{\omega} = \omega_1\boldsymbol{w}_1 + \omega_2\boldsymbol{w}_2 + \omega_3\boldsymbol{w}_3, \quad \boldsymbol{\xi} = \xi_1\boldsymbol{\zeta}_1 + \xi_2\boldsymbol{\zeta}_2 + \xi_3\boldsymbol{\zeta}_3.$$

Since the angular velocities are referred to the same coordinate axes $\boldsymbol{e}_1, \boldsymbol{e}_2$, and $\boldsymbol{e}_3$ attached to the shell, which play the role of fixed axes for the imaginary sphere (analogous to the fixed axes for the body), the vector fields $\boldsymbol{w}_1, \boldsymbol{w}_2$, and $\boldsymbol{w}_3$ are left-invariant and $\boldsymbol{\zeta}_1, \boldsymbol{\zeta}_2$, and $\boldsymbol{\zeta}_3$ are right-invariant. Their commutation relations differ by a sign:

$$[\boldsymbol{w}_i, \boldsymbol{w}_j] = \varepsilon_{ijk}\boldsymbol{w}_k, \quad [\boldsymbol{\zeta}_i, \boldsymbol{\zeta}_j] = -\varepsilon_{ijk}\boldsymbol{\zeta}_k.$$

The Poincaré equations governing the evolution of angular velocities are

$$\begin{aligned} \frac{d}{dt}\frac{\partial L}{\partial \omega_i} + \varepsilon_{ijk}\omega_j\frac{\partial L}{\partial \omega_k} &= \boldsymbol{w}_i(L), \\ \frac{d}{dt}\frac{\partial L}{\partial \xi_i} - \varepsilon_{ijk}\xi_j\frac{\partial L}{\partial \xi_k} &= \boldsymbol{\zeta}_i(L), \quad i, j, k = 1, 2, 3. \end{aligned} \tag{A.28}$$

When there are no external forces, the Lagrange function L coincides with the total kinetic energy, and, moreover, it is independent of the configuration variables on the group. In this case, Eqs. (A.28), written in vector form, are

$$\frac{d}{dt}\frac{\partial T}{\partial \boldsymbol{\omega}} + \boldsymbol{\omega} \times \frac{\partial T}{\partial \boldsymbol{\omega}} = 0, \quad \frac{d}{dt}\frac{\partial T}{\partial \boldsymbol{\xi}} - \boldsymbol{\xi} \times \frac{\partial T}{\partial \boldsymbol{\xi}} = 0. \tag{A.29}$$

Calculating the kinetic energy of the system using (A.20) and taking into account the relation $\int(\boldsymbol{x}-\boldsymbol{x}_c)\rho\, d^3x = 0$, we obtain

$$T = \frac{1}{2}(\boldsymbol{\omega}, \mathbf{I}\boldsymbol{\omega}) - \mathrm{Tr}(\widetilde{\boldsymbol{\omega}}\mathbf{D}\mathbf{J}) - \frac{1}{2}\mathrm{Tr}(\mathbf{D}^2\mathbf{J}), \tag{A.30}$$

where $\mathbf{I}$ is the tensor of inertia of the body-fluid system with respect to the fixed point O, $\widetilde{\boldsymbol{\omega}} = ||\widetilde{\omega}_{ij}|| = ||\varepsilon_{ijk}\omega_k||$, and $\mathbf{J}$ is the tensor of inertia of the cavity with respect to its geometric center,

$$J_{ij} = \rho \int_{\pi} (x_i - x_{ci})(x_j - x_{cj})\, d^3x$$

and here the integration extends over the cavity volume π. In the coordinate system whose axes are parallel to the principal axes of the ellipsoidal cavity, we have $\mathbf{J} = \mathrm{diag}\left(\frac{1}{5}m_f b_1^2, \frac{1}{5}m_f b_2^2, \frac{1}{5}m_f b_3^2\right)$, where m_f is the fluid mass. The kinetic energy (A.30) for such a system has the form

$$\begin{aligned} T = {} & \frac{1}{2}(I_1\omega_1^2 + I_2\omega_2^2 + I_3\omega_3^2) \\ & + \frac{2}{5}m_f(b_2b_3\omega_1\xi_1 + b_3b_1\omega_2\xi_2 + b_1b_2\omega_3\xi_3) \\ & + \frac{1}{10}m_f\left((b_2^2+b_3^2)\xi_1^2 + (b_3^2+b_1^2)\xi_2^2 + (b_1^2+b_2^2)\xi_3^2\right). \end{aligned}$$

In Hamiltonian form the equations of motion (Poincaré – Chetaev equations) are obtained via the Legendre transform:

$$\begin{gathered} \boldsymbol{M} = \frac{\partial T}{\partial \boldsymbol{\omega}} = \mathbf{I}\boldsymbol{\omega} + \mathbf{I}_1\boldsymbol{\xi}, \quad \boldsymbol{K} = \frac{\partial T}{\partial \boldsymbol{\xi}} = \mathbf{I}_1\boldsymbol{\omega} + \mathbf{I}_2\boldsymbol{\xi}, \\ H = (\boldsymbol{\omega}, \boldsymbol{M}) + (\boldsymbol{\xi}, \boldsymbol{K}) - T = \frac{1}{2}(\boldsymbol{M}, \mathbf{A}\boldsymbol{M}) + (\boldsymbol{M}, \mathbf{B}\boldsymbol{K}) + \frac{1}{2}(\boldsymbol{K}, \mathbf{C}\boldsymbol{K}), \\ \dot{\boldsymbol{M}} = \boldsymbol{M} \times \frac{\partial H}{\partial \boldsymbol{M}}, \quad \dot{\boldsymbol{K}} = -\boldsymbol{K} \times \frac{\partial H}{\partial \boldsymbol{K}}. \end{gathered} \tag{A.31}$$

The Poisson structure of Eqs. (A.31) corresponds to the algebra $so(4)$ (in the direct sum decomposition $so(4) = so(3) \oplus so(3)$):

$$\{M_i, M_j\} = -\varepsilon_{ijk}M_k, \quad \{K_i, K_j\} = \varepsilon_{ijk}K_k. \tag{A.32}$$

REMARK 1. Equations (A.21) governing the evolution of the angular velocity $\boldsymbol{\omega}$ of the rigid body and of the vorticity vector $\boldsymbol{\xi}$ of the fluid flow in the cavity can be obtained by another method, which was used by N. E. Zhukovskii. The first triple of Eqs. (A.28) and (A.31) is obtained from the balance of angular momentum $\boldsymbol{M}$ for the body-fluid system.

In this case, according to (A.27), the total angular momentum can be represented as

$$\boldsymbol{M} = \mathbf{I}\boldsymbol{\omega} + \rho\int_{\pi} \boldsymbol{x} \times \left(\mathbf{B}^{-1/2}\boldsymbol{\Xi}\mathbf{B}^{1/2}(\boldsymbol{x} - \boldsymbol{x}_c)\right) = \mathbf{I}\boldsymbol{\omega} + \mathbf{I}_1\boldsymbol{\xi},$$

and, in the moving coordinate system, its evolution is governed by the equation

$$\dot{\boldsymbol{M}} + \boldsymbol{\omega} \times \boldsymbol{M} = \boldsymbol{N},$$

where $\boldsymbol{N}$ is the moment of external forces ($\boldsymbol{N} = 0$ in our case).

To obtain the equation for $\boldsymbol{\xi}$, we may use the Helmholtz equation for the vorticity $\boldsymbol{\Omega}$ [647] in the moving coordinate system $O\boldsymbol{e}_1\boldsymbol{e}_2\boldsymbol{e}_3$:

$$\frac{\partial \boldsymbol{\Omega}}{\partial t} + (\boldsymbol{v}, \nabla)\boldsymbol{\Omega} - (\boldsymbol{\Omega}, \nabla)\boldsymbol{v} = \operatorname{rot} \boldsymbol{f}.$$

Thus, if the volume force $\boldsymbol{f}$ depends linearly on coordinates, we obtain the following equation for the vectors $\boldsymbol{\omega}$ and $\boldsymbol{\xi}$ defining the particular solution (A.20):

$$\dot{\boldsymbol{\Omega}} - \boldsymbol{\omega} \times \boldsymbol{\Omega} - \mathbf{D}\boldsymbol{\Omega} = \operatorname{rot} \boldsymbol{f}, \quad \mathbf{D} = \mathbf{B}^{-1/2}\Xi\mathbf{B}^{1/2},$$

where $\boldsymbol{\Omega}$ is given by (A.25).

REMARK 2. A generalization of the "Poincaré problem" – which consists in the study of oscillations of the absolutely rigid shell containing the core of ideal fluid – can be found in the two-volume geophysical work of P. Melchior [421]. There one can also find interesting historical references and analysis of double resonance and the effect of gyrostatic rigidity discovered by Poincaré [472].

3 Motion of a rigid body with a gyrostat in curved space – stationary motions

We now present a derivation of the equations of motion for a free rigid body in a curved space and for a free rigid body with a gyrostat attached to it. For these cases, the natural setting for the equations is the pencil of Poisson brackets $\mathfrak{L}_x$, (3.27) (Sec. 3.2 of Ch. 3). For simplicity, we confine ourselves to the case of the three-dimensional sphere S^3 and then just present the equations of motion in the Lobachevskii space without further explanations.

Free body motion in S^3. We first consider the equations of motion for a free rigid body on the three-dimensional sphere S^3. Note that the position of a two-dimensional body (small plate) on the surface of the usual two-dimensional sphere S^2 can be characterized by the element of the group $SO(3)$ that determines the body's position on the sphere and its orientation with respect to fixed axes (Fig. A.4).

This visual illustration turns out to be useful for understanding the interconnection between the motion of a free rigid body on S^3 and the rotation of a four dimensional rigid body about a fixed point (Euler equations on $SO(4)$).

We imagine the three-dimensional sphere as a surface in $\mathbb{R}^4$: $q_0^2 + \boldsymbol{q}^2 = 1$. The position and orientation of the body with respect to the coordinates q_μ are given by an element of the group $SO(4)$, and thus, the problem of the free rigid body motion in S^3 reduces to the problem of the four-dimensional rigid body with a fixed point in the flat space $\mathbb{R}^4$.

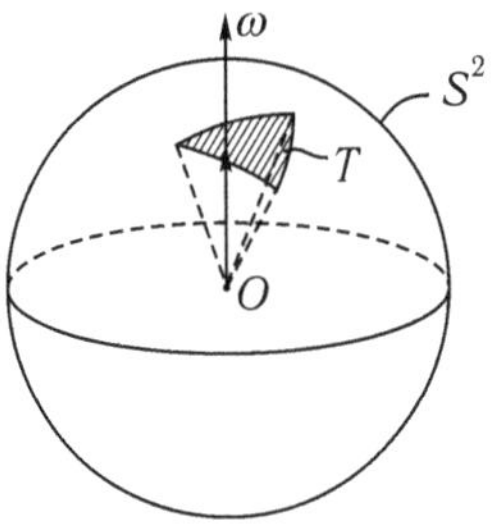

Fig. A.4. Rigid body on the sphere.

We may write the equations describing the rotation of the four-dimensional free rigid body in the form of Euler equations on the Lie algebra $so(4)$ as follows. Let us introduce a coordinate system attached to the body. The coordinates x_μ in this moving coordinate system are related to the coordinates q_μ of the fixed space by

$$q_\mu = \sum_{\nu=0}^{3} B_{\mu\nu} x_\nu, \tag{A.33}$$

where $B_{\mu\nu}$ are the entries of an orthogonal matrix from the group $SO(4)$.

The Lagrange function L of the free rigid body is equal to the sum of all the kinetic energies of the volumes that make up the body:

$$L = \frac{1}{2} \sum_{x \in \text{body}} m \sum_{\mu\nu\sigma} \dot{B}_{\mu\nu} \dot{B}_{\mu\sigma} x_\nu x_\sigma .$$

It can be represented as a quadratic functions of the quasi-velocities $\omega_{\mu\nu}$ [61]:

$$L = \frac{1}{2} \sum_{\mu\nu\sigma} J_{\mu\nu} \omega_{\mu\sigma} \omega_{\nu\sigma}, \tag{A.34}$$

where $\omega_{\mu\nu} = -\omega_{\nu\mu} = \sum_\sigma B^{-1}{}_{\mu\sigma} \dot{B}_{\sigma\nu}$ is an element of the algebra $so(4)$ that is the “angular velocity in the body” and $\|\mathbf{J}\|_{\mu\nu} = \left\| \sum_{x \in \text{body}} m x_\mu x_\nu \right\|$ is the tensor of moments of inertia. Writing the Poincaré – Chetaev equations on the group $SO(4)$ (Sec. 1.6 of Ch. 1), we obtain the following commutator equation:

$$\mathbf{X} = [\mathbf{X}, \boldsymbol{\omega}]; \tag{A.35}$$

here, $[\cdot, \cdot]$ is the matrix commutator and the entries of the matrix of the angular momentum $\mathbf{X} \in SO^*(4)$ (“in the body”) are defined by the formula

$$\mathbf{X} = \left\| \frac{\partial L}{\partial \omega_{\nu\mu}} \right\|, \qquad \mathbf{X} = \frac{1}{2} (\mathbf{J}\boldsymbol{\omega} + \boldsymbol{\omega}\mathbf{J}) .$$

Let us write the system of equations of motions using the vectors $\boldsymbol{M}$ and $\boldsymbol{p}$ whose components are related to the entries of the angular momentum matrix $\mathbf{X}$ by

$$M_i = \frac{1}{2} \varepsilon_{ijk} X_{jk}, \quad p_i = X_{0i}, \quad i, j, k = 1, 2, 3. \tag{A.36}$$

$$\begin{aligned} \dot{\boldsymbol{M}} &= \boldsymbol{M} \times \frac{\partial H}{\partial \boldsymbol{M}} + \boldsymbol{p} \times \frac{\partial H}{\partial \boldsymbol{p}}, \\ \dot{\boldsymbol{p}} &= \boldsymbol{p} \times \frac{\partial H}{\partial \boldsymbol{M}} + \boldsymbol{M} \times \frac{\partial H}{\partial \boldsymbol{p}} . \end{aligned} \tag{A.37}$$

Equations (A.37) are Hamiltonian equations on the algebra $so(4)$ in the standard matrix representation (see Sec. 1.2). Choosing the body-fixed coordinate system in such a way that $\mathbf{J} = \mathrm{diag}(\lambda_0, \lambda_1, \lambda_2, \lambda_3)$, we obtain the Hamiltonian function of the free rigid body in the variables $\boldsymbol{M}$ and $\boldsymbol{p}$ by using the Legendre transforms $H = \mathrm{Tr}(\mathbf{X}\boldsymbol{\omega}) - L\big|_{\boldsymbol{\omega}\to\boldsymbol{M},\boldsymbol{p}}$ in the form

$$H = \frac{1}{2}(\boldsymbol{M}, \mathbf{A}\boldsymbol{M}) + \frac{1}{2}(\boldsymbol{p}, \mathbf{B}\boldsymbol{p}), \tag{A.38}$$

where

$$\mathbf{A} = \mathrm{diag}\left(\frac{1}{\lambda_2+\lambda_3}, \frac{1}{\lambda_1+\lambda_3}, \frac{1}{\lambda_1+\lambda_2}\right),$$

$$\mathbf{B} = \mathrm{diag}\left(\frac{1}{\lambda_0+\lambda_1}, \frac{1}{\lambda_0+\lambda_2}, \frac{1}{\lambda_0+\lambda_3}\right).$$

These equation were studied in detail and integrated in the nineteenth century by W. Frahm and F. Schottky (see Sec. 3.2 of Ch. 3).

It should be noted that in contrast to the free rigid body motion in Euclidean space (Euler – Poinsot problem), integrability of the inertial motion on S^3 and L^3 is essentially more complicated from the viewpoint of the integration procedure, as well as from that of the qualitative (topological) analysis of the motion [456].

Roughly speaking, we can say that the distinction with respect to the flat space case is that the body motion on the sphere S^3 and its rotation are no longer separated, and, therefore, the body rotation influences the system motion as a whole. (For L^2, this effect was already noticed by N. E. Zhukovsky [644].) Analogs of permanent and screw motions for Eqs. (A.37) and (A.38) were presented in [95] (see also [104]).

REMARK. Analysis of motion of a small two-dimensional spherical plate over the sphere S^2 under the action of potential forces was carried out in [115], where an analog of the Lagrange case arising under the body dynamic symmetry was presented.

Motion of two linked bodies. A balanced gyrostat. Let us consider a balanced gyrostat in S^3, i.e., the mechanical system consisting of the following two elements: a “supporting body” T_1 and a “supported body” T_2, which are joined in such a way that the mass distribution of the system does not change in time. In many respects, the analysis presented below repeats the arguments of Sec. 2.8 of Ch. 2 applied to the description of linked bodies. Here, however, they are presented in a more invariant way that is applicable to the general n-dimensional case.

To each of the bodies let us associate its own coordinate system. Let $\mathbf{B}$ and $\mathbf{Q}$ be transition matrices from the absolute coordinate system (q) to the system (x) of the “supporting body” and from the system of the “suporting body” (x) to the system (y) of the “supported body”:

$$q_\mu = \sum_\nu B_{\mu\nu} x_\nu, \qquad x_i = \sum_\nu Q_{\mu\nu} y_\nu. \tag{A.39}$$

Let us introduce the following notation:

$J_{\mu\nu} = \sum\limits_{T_1+T_2} m x_\mu x_\nu$ are the entries of the inertia matrix of the first and second bodies in the coordinate system of the supporting body (here, $\sum\limits_{T_1+T_2}$ denotes the summation with respect to the elements of the first and second bodies);

$(\widetilde{J}_2)_{\mu\nu} = \sum\limits_{T_2} m y_\mu y_\nu$ is the inertia matrix of the supported body in the coordinate system (y);

$(J_1)_{\mu\nu} = \sum\limits_{T_1} m x_\mu x_\nu$ and $(J_2)_{\mu\nu} = \sum\limits_{T_2} m x_\mu x_\nu$ are the inertia matrices of the supporting body and of the supported body in the coordinate system (x);

$\widetilde{\boldsymbol{\omega}}_2 = \mathbf{Q}^{-1}\dot{\mathbf{Q}}$ is the matrix of angular velocities of the supported body in the coordinate system (y);

$\boldsymbol{\omega} = \mathbf{B}^{-1}\dot{\mathbf{B}}$ and $\boldsymbol{\omega}_2 = \mathbf{Q}\widetilde{\boldsymbol{\omega}}_2\mathbf{Q}^{-1} = \dot{\mathbf{Q}}\mathbf{Q}^{-1}$ are, respectively, the angular velocities of the supporting body and of the supported body in the supporting body coordinate system (x).

The constant mass distribution condition (gyrostat balance condition) $J_{\mu\nu} = \sum\limits_{T_1+T_2} m x_\mu x_\nu = \text{const.}$ (i.e., $\dot{J}_{\mu\nu} = 0$) is equivalent to the relation

$$\widetilde{\boldsymbol{\omega}}_2\widetilde{\mathbf{J}}_2 = \widetilde{\mathbf{J}}_2\widetilde{\boldsymbol{\omega}}_2. \tag{A.40}$$

Taking into account that the “supported body” has a fixed axis with respect to the “supporting body”, we write the solutions of Eq. (A.40) in the form

$$\widetilde{\boldsymbol{\omega}}_2 = \begin{pmatrix} 0 & 0 & 0 & 0 \\ 0 & 0 & 0 & 0 \\ 0 & 0 & 0 & a \\ 0 & 0 & -a & 0 \end{pmatrix}, \quad \widetilde{\mathbf{J}}_2 = \begin{pmatrix} \lambda_0 & 0 & 0 & 0 \\ 0 & \lambda_1 & 0 & 0 \\ 0 & 0 & \lambda_2 & 0 \\ 0 & 0 & 0 & \lambda_3 \end{pmatrix}. \tag{A.41}$$

In what follows, we assume that $a = a(t)$ is some given function of time. As a result of this condition there are no additional degrees of freedom corresponding to the supported body. The Lagrange function of the system is equal to the sum of the kinetic energies of both bodies. Taking into account relations (A.41), we can represent it in the form

$$L = -\frac{1}{2}\operatorname{Tr}\boldsymbol{\omega}^2(\mathbf{I}_1 + \mathbf{I}_2) + \operatorname{Tr}(\boldsymbol{\omega}\mathbf{K}), \tag{A.42}$$

where $\boldsymbol{\omega}$, as introduced above, is the angular velocity of the supporting body and the entries of the angular momentum matrix $\mathbf{K} = \mathbf{J}_2\boldsymbol{\omega}_2 + \boldsymbol{\omega}_2\mathbf{J}_2$ of the supported body are given functions of time. By contrast with flat space, $\mathbf{K}$ depends not only on the direction of the axis of rotation but on the point at which the bodies are fixed.

The equations of motion have the form (A.35), but now $\mathbf{X} = \left\| \frac{\partial L}{\partial \omega_{\mu\nu}} \right\| = \frac{1}{2}(\mathbf{I}\boldsymbol{\omega} + \boldsymbol{\omega}\mathbf{I}) + \mathbf{K}$, $\mathbf{I} = \mathbf{I}_1 + \mathbf{I}_2$. For $\mathbf{K}(t) = \text{const.}$ ($a(t) = \text{const.}$), we obtain the *balanced* gyrostat problem.

The equations of motion can be written as Hamiltonian equations in vector form (A.37). The Hamiltonian for the balanced gyrostat can be represented in the form

$$H = \frac{1}{2}(\boldsymbol{M} - \mathbf{P}, \mathbf{A}(\boldsymbol{M} - \mathbf{P})) + \frac{1}{2}(\boldsymbol{p} - \mathbf{S}, \mathbf{B}(\boldsymbol{p} - \mathbf{S})), \tag{A.43}$$

where the components of the vectors **P** and **S** are expressed in terms of the gyrostatic momentum matrix by the formulas

$$P_i = \frac{1}{2}\varepsilon_{ijk}K_{jk}, \quad S_i = K_{0i}, \quad i, j, k = 1, 2, 3.$$

Probably the system (A.37) with Hamiltonian (A.43) is not integrable for arbitrary parameter values in contrast to the flat space case. For this system there is no general additional quadratic integral and the behavior can be stochastic for certain choices of the parameters. An interesting problem is that of finding integrable cases of this system (and analogous equations for L^3) with some additional restrictions on the parameters of the Hamiltonian function (see also Sec. 3.2 of Ch. 3).

Kirchhoff equations on S^3 and L^3. If the Hamiltonian in Eqs. (A.37) is assumed to be an arbitrary (positive-definite) quadratic form in the variables $\boldsymbol{M}$ and $\boldsymbol{p}$, then we obtain the Kirchhoff equations describing inertial rigid body motion in an infinite volume of vortex-free ideal fluid in S^3 (respectively L^3), which are analogous to the Kirchhoff equations (Sec 3.1 of Ch. 3). *They coincide with the Zhukovskii – Poincaré equations on the algebra* $so(4)$ *(respectively* $so(3,1)$*)*; an overview of their integrable cases is contained in Sec. 3.2 of Ch. 3.

As for the physical significance of these equations, we cite the words of Garret Birkhoff from his well-known book [46]:

"The previous formulas have obvious analogs for imaginary rigid body motions in an ideal fluid in non-Euclidean spaces. Of course, it is questionable that these analogs of the classical formulas have even limited physical significance. Nevertheless, it would be interesting to establish some of the analogs of these (classical[1]) *formulas in order to illustrate the influence of the space curvature (if it exists) on the reaction of infinitely ranged ideal fluid on the body under the steady motion"*

Free body motion in Lobachevskii space. Let us present the equations of motion of a free body in the Lobachevskii space L^3 without giving a detailed derivation for them.

Denote by x^σ coordinates in a system attached to the body and by q^μ coordinates in the "absolute" space. These coordinates are related to each other by $q^\mu = B^\mu_\sigma(t)x^\sigma$, where the matrix $\mathbf{B} = \|B^\mu_\sigma\|$ belongs to the group $SO(1,3)$. Let us pass to the quasi-velocities $\boldsymbol{\omega} \in so(1,3)$ and the quasi-momenta $\mathbf{X} \in so^*(1,3)$, $\mathbf{X} \approx (\boldsymbol{M}, \boldsymbol{p})$ using the formulas

$$\boldsymbol{\omega} = \mathbf{B}^{-1}\dot{\mathbf{B}} \quad \text{(in the enties of } \omega^\sigma_\tau = (B^{-1})^\sigma_\mu \dot{B}^\mu_\tau\text{)},$$

$$\mathbf{X} = \frac{\partial L}{\partial \boldsymbol{\omega}} = \mathbf{g}\mathbf{J}\boldsymbol{\omega} + \boldsymbol{\omega}\mathbf{g}\mathbf{J},$$

$$M^i = -\frac{1}{2}\varepsilon^{ijk}g^{jl}X^k_l, \quad p^i = -X^i_0, \quad i, j, l, k = 1, 2, 3,$$

1 The authors addition.

where $\mathbf{g} = \operatorname{diag}(-1, 1, 1, 1)$ is the metric tensor of the Minkowski space $\mathbb{M}^4$ and $J^{\sigma\tau} = \sum_{p\in\text{body}} m_p x_p^\sigma x_p^\tau$ is the tensor of moments of inertia in the coordinate system attached to the body.

In the variables $\boldsymbol{M}$ and $\boldsymbol{\pi}$, the equations of motion have the form

$$\begin{aligned} \dot{\boldsymbol{M}} &= \boldsymbol{M} \times \frac{\partial H}{\partial \boldsymbol{M}} + \boldsymbol{p} \times \frac{\partial H}{\partial \boldsymbol{p}}, \\ \dot{\boldsymbol{p}} &= \boldsymbol{p} \times \frac{\partial H}{\partial \boldsymbol{M}} - \boldsymbol{M} \times \frac{\partial H}{\partial \boldsymbol{p}} \end{aligned} \tag{A.44}$$

and are a Hamiltonian system on the (co)algebra $so(1, 3)$. In the variables $\boldsymbol{M}$ and $\boldsymbol{p}$, the Hamiltonian function can be written in the form

$$\begin{aligned} H &= \frac{1}{2}(\boldsymbol{M}, \mathbf{A}\boldsymbol{M}) + \frac{1}{2}(\boldsymbol{p}, \mathbf{B}\boldsymbol{p}), \\ \mathbf{A} &= \operatorname{diag}((\lambda_2 + \lambda_3)^{-1}, (\lambda_3 + \lambda_1)^{-1}, (\lambda_1 + \lambda_2)^{-1}), \\ \mathbf{B} &= \operatorname{diag}((\lambda_0 - \lambda_1)^{-1}, \quad (\lambda_0 - \lambda_2)^{-1}, \quad (\lambda_0 - \lambda_3)^{-1}), \end{aligned} \tag{A.45}$$

where

$$\mathbf{J} = \operatorname{diag}(\lambda_0, \lambda_1, \lambda_2, \lambda_3).$$

Since the relation $(x^0)^2 - (x^1)^2 - (x^2)^2 - (x^3)^2 = R^2$ holds in the Lobachevskii space, the inequality $(x^0)^2 > (x^i)^2$, $i = 1, 2, 3$, holds, and, therefore, $\lambda_0 > \lambda_i$, $i = 1, 2, 3$. The system (A.45) is the integrable Schottky – Manakov case on the bracket pencil $\mathcal{L}_x$ (See Sec. 3.2 of Ch. 3). The problem of describing the motion of a two-dimensional pseudo-spherical plate on the Lobachevskii plane was first studied by N. E. Zhukovskiy [644].

Comment. The derivation of the equations of motion for the four-dimensional body (and gyrostat) presented here extends to the n-dimensional situation without changes. Here, it is also necessary to use the angular velocities and the angular momenta (which now are elements of the algebras $so(n)$ and $so(1, n-1)$ and the coalgebras $so^*(n)$ and $so^*(1, n-1)$, respectively), in the coordinate system attached to the body. The result is a commutator equation of the form (A.35) which is an analog of the integrable Euler equation for a free top [23, 404].

B The Lie Algebra $e(4)$ and Its Orbits

The quaternion equations of rigid body dynamics (Sec. 1.4 of Ch. 1) can be written in terms of the Poisson bracket defined by the algebra $e(4)$. More precisely, we speak about a singular orbit of $e(4)$. The higher-dimensional analog also plays an important role in higher-dimensional rigid body dynamics, for which it is necessary to consider singular orbits of $e(n)$. Indeed, the major part of the results on the analogy between material point dynamics on n-dimensional spheres S^n or ellipsoids E^n ($n \geqslant 2$) in potential fields and rigid body motion (n-dimensional body motion in a potential field in general) are obtained precisely by restricting the rigid body dynamics to this singular orbit. Note that celestial mechanics in spaces of constant curvature also deals with problems concerned with material point motion on S^n [95].

For the singular orbits of $O(n)$, $e(n)$, and $U(n)$, which also play important roles in dynamics, we refer the reader to our book [95]. Bearing in mind the application to three-dimensional rigid body motions, we only cover singular orbits of $e(4)$.

The Lie algebra $e(4)$ of the group $E(4)$ of rigid motions of four-dimensional space is a semidirect sum $so(4) \oplus_S R^4$ and can be realized by matrices of the form

$$\mathbf{X} = \begin{pmatrix} 0 & L_3 & -L_2 & \pi_1 & \lambda_1 \\ -L_3 & 0 & L_1 & \pi_2 & \lambda_2 \\ L_2 & -L_1 & 0 & \pi_3 & \lambda_3 \\ -\pi_1 & -\pi_2 & -\pi_3 & 0 & \lambda_0 \\ 0 & 0 & 0 & 0 & 0 \end{pmatrix}.$$

For its ten generators $\boldsymbol{\pi} = (\pi_1, \pi_2, \pi_3)$, $\boldsymbol{L} = (L_1, L_2, L_3)$, $\boldsymbol{\lambda} = (\lambda_1, \lambda_2, \lambda_3)$, and λ_0, the following commutation relations hold (see [170]):

$$\begin{aligned} &\{L_i, L_j\} = \varepsilon_{ijk} L_k, \quad \{L_i, \pi_j\} = \varepsilon_{ijk}\pi_k, \quad \{\pi_i, \pi_j\} = \varepsilon_{ijk} L_k, \\ &\{L_i, \lambda_j\} = \varepsilon_{ijk}\lambda_k, \quad \{L_i, \lambda_0\} = 0, \quad \{\pi_i, \lambda_j\} = -\delta_{ij}\lambda_0, \\ &\{\pi_i, \lambda_0\} = \lambda_i, \quad \{\lambda_i, \lambda_j\} = \{\lambda_i, \lambda_0\} = 0 \end{aligned} \tag{B.1}$$

for all i, j, and k assuming the values 1, 2, and 3.

The bracket (B.1) is degenerate and has two Casimir functions

$$F_1 = \sum_{\mu=0}^{3} \lambda_\mu^2, \quad F_2 = \sum_{\mu=0}^{3} W_\mu^2, \tag{B.2}$$

where W is the four-dimensional *Pauli – Lubanski vector* (more precisely, its Euclidean analog for $e(4)$, as, strictly speaking it is defined for the Poincaré group)

$$W_0 = -(\boldsymbol{\lambda}, \boldsymbol{L}), \boldsymbol{W} = \boldsymbol{L}\lambda_0 + \boldsymbol{\pi} \times \boldsymbol{\lambda}. \tag{B.3}$$

DOI 10.1515/9783110544442-008

Its commutation relations with the generators are the same as those of the four-dimensional vector λ:

$$\begin{aligned} &\{L_i, W_j\} = \varepsilon_{ijk} W_k, \quad \{L_i, W_0\} = 0, \quad \{\pi_i, W_j\} = -\delta_{ij} W_0, \\ &\{\pi_i, W_0\} = W_i, \quad \{\lambda_\mu, W_\nu\} = 0, \quad \{W_\mu, W_\nu\} = 0. \end{aligned} \tag{B.4}$$

Symplectic leaves of the phase space are coadjoint orbits of $E(4)$ and generic coadjoint orbits are level surfaces $\{F_1 = c_1,\ F_2 = c_2\}$ of the Casimir functions.

In the nondegenerate (regular) case, the dimension of a symplectic leaf is equal to 8. However, for $c_1 = 0$ or $c_2 = 0$, the leaf dimension reduces by two. All singular symplectic leaves (singular symplectic orbits) of the form $c_2 = 0$ and $c_1 \neq 0$ are diffeomorphic to the (co)tangent bundle TS^3 $\left(T^*S^3\right)$ of the three-dimensional sphere, and the following relations hold for the vectors $\boldsymbol{L}$ and $\boldsymbol{\pi}$:

$$(\boldsymbol{\lambda}, \boldsymbol{L}) = 0, \qquad \boldsymbol{L}\lambda_0 + \boldsymbol{\pi} \times \boldsymbol{\lambda} = 0. \tag{B.5}$$

The equations of motion of a rigid body with a fixed point presented in Sec. 1.4 of Ch. 1 are written for another representation of $e(4)$ that corresponds to the canonical subalgebra decomposition $so(4) \simeq so(3) \oplus so(3)$. To pass to this representation, we need to introduce the new variables $\boldsymbol{M}$ and $\boldsymbol{N}$ by the formulas

$$\boldsymbol{M} = \frac{1}{2}(\boldsymbol{\pi} - \boldsymbol{L}), \quad \boldsymbol{N} = \frac{1}{2}(\boldsymbol{\pi} + \boldsymbol{L}).$$

In this case, the algebra (B.1) decomposes into two seven-dimensional isomorphic intersecting subalgebras.

The commutation relations for the subalgebra $(\boldsymbol{M}, \lambda)$ are as follows:

$$\{M_i, M_j\} = -\varepsilon_{ijk} M_k,\ \{M_i, \lambda_0\} = \frac{1}{2}\lambda_i,\ \{M_i, \lambda_j\} = -\frac{1}{2}\left(\varepsilon_{ijk}\lambda_k + \delta_{ij}\lambda_0\right); \tag{B.6}$$

for the subalgebra $(\boldsymbol{N}, \lambda)$, they are

$$\{N_i, N_j\} = \varepsilon_{ijk} N_k,\ \{N_i, \lambda_0\} = \frac{1}{2}\lambda_i,\ \{N_i, \lambda_j\} = \frac{1}{2}\left(\varepsilon_{ijk}\lambda_k - \delta_{ij}\lambda_0\right). \tag{B.7}$$

The rank of the Poisson structure of each of the subalgebras is equal to 6, and they have the common Casimir $F_1 = \sum_{\mu=0}^{3} \lambda_\mu^2$. In these variables, the invariant relations (B.5) have the form

$$\begin{aligned} (\boldsymbol{N} - \boldsymbol{M})\lambda_0 + (\boldsymbol{N} + \boldsymbol{M}) \times \boldsymbol{\lambda} &= 0, \\ (\boldsymbol{N} - \boldsymbol{M}, \boldsymbol{\lambda}) &= 0. \end{aligned} \tag{B.8}$$

By using quaternionic multiplication, they are written in the shorter form $\boldsymbol{M} = \lambda^{-1}\boldsymbol{N}\lambda$ (the mechanical meaning of these relations is revealed in Sec. 1.4 of Ch. 1). On the orbit T^*S^3, the following additional relations hold:

$$\begin{aligned} \pi &= 2\left(\lambda_0 \boldsymbol{\lambda} \times \boldsymbol{M} + \boldsymbol{\lambda}(\boldsymbol{M}, \boldsymbol{\lambda}) + \lambda_0^2 \boldsymbol{M}\right), \\ \boldsymbol{L} &= 2\left(\lambda_0 \boldsymbol{\lambda} \times \boldsymbol{M} + \boldsymbol{\lambda}(\boldsymbol{M}, \boldsymbol{\lambda}) - \boldsymbol{\lambda}^2 \boldsymbol{M}\right). \end{aligned} \tag{B.9}$$

The Lie – Poisson bracket (B.1) defines the following Hamiltonian system in terms of the variables $x = (\boldsymbol{M}, \boldsymbol{N}, \boldsymbol{\lambda}, \lambda_0)$

$$\begin{aligned}
\dot{\boldsymbol{M}} &= \boldsymbol{M} \times \frac{\partial H}{\partial \boldsymbol{M}} + \frac{1}{2}\boldsymbol{\lambda} \times \frac{\partial H}{\partial \boldsymbol{\lambda}} + \frac{1}{2}\boldsymbol{\lambda}\frac{\partial H}{\partial \boldsymbol{\lambda}_0} - \frac{1}{2}\boldsymbol{\lambda}_0\frac{\partial H}{\partial \boldsymbol{\lambda}}, \\
\dot{\boldsymbol{N}} &= -\boldsymbol{N} \times \frac{\partial H}{\partial \boldsymbol{N}} - \frac{1}{2}\boldsymbol{\lambda} \times \frac{\partial H}{\partial \boldsymbol{\lambda}} + \frac{1}{2}\boldsymbol{\lambda}\frac{\partial H}{\partial \boldsymbol{\lambda}_0} - \frac{1}{2}\boldsymbol{\lambda}_0\frac{\partial H}{\partial \boldsymbol{\lambda}}, \\
\dot{\lambda}_0 &= -\frac{1}{2}\left(\boldsymbol{\lambda}, \frac{\partial H}{\partial \boldsymbol{M}}\right) - \frac{1}{2}\left(\boldsymbol{\lambda}, \frac{\partial H}{\partial \boldsymbol{N}}\right), \\
\dot{\boldsymbol{\lambda}}_0 &= \frac{1}{2}\boldsymbol{\lambda} \times \frac{\partial H}{\partial \boldsymbol{M}} - \frac{1}{2}\boldsymbol{\lambda} \times \frac{\partial H}{\partial \boldsymbol{N}} + \frac{1}{2}\boldsymbol{\lambda}_0\frac{\partial H}{\partial \boldsymbol{M}} + \frac{1}{2}\boldsymbol{\lambda}_0\frac{\partial H}{\partial \boldsymbol{N}}
\end{aligned} \tag{B.10}$$

for any Hamiltonian function $H = H(\boldsymbol{M}, \boldsymbol{N}, \boldsymbol{\lambda}, \lambda_0)$. On each of the orbits (singular or regular), Eqs. (B.10) (by Darboux's theorem) can be written in canonical form.

Mechanically, (B.10) is the most general and elegant Hamiltonian form for the equations of motion of a rigid body with a fixed point; this form contains the components of the angular momentum in both the moving and fixed coordinate systems.

C Quaternion Equations and L-A Pair for the Generalized Goryachev – Chaplygin Top

In Sec. 3.4 of Ch. 3, we presented the quaternionic Euler – Poisson equations (3.95) for which the potential was a linear function of the quaternion variables. Let us show how, using these equations and performing a reduction with respect to a linear integral (Sec. 4.1 of Ch. 4), we can obtain a Lax representation [80] for the Goryachev – Chaplygin top; this representation is different from that found by A. I. Bobenko and V. B. Kuznetsov [51].

Let us consider the space $\mathcal{L}$ of complex 3×3 matrices with basis

$$\mathbf{M}_1 = \left(\begin{array}{cc|c} 0 & \frac{1}{2}i & 0 \\ \frac{1}{2}i & 0 & 0 \\ \hline 0 & 0 & 0 \end{array}\right), \qquad \mathbf{M}_2 = \left(\begin{array}{cc|c} 0 & \frac{1}{2} & 0 \\ -\frac{1}{2} & 0 & 0 \\ \hline 0 & 0 & 0 \end{array}\right),$$

$$\mathbf{M}_3 = \left(\begin{array}{cc|c} -\frac{1}{2}i & 0 & 0 \\ 0 & \frac{1}{2}i & 0 \\ \hline 0 & 0 & 0 \end{array}\right), \qquad \mathbf{M}_4 = \left(\begin{array}{cc|c} -\frac{1}{6}i & 0 & 0 \\ 0 & -\frac{1}{6}i & 0 \\ \hline 0 & 0 & \frac{1}{3}i \end{array}\right),$$

$$\mathbf{P}_1 = \left(\begin{array}{cc|c} 0 & 0 & \frac{1}{2} \\ 0 & 0 & 0 \\ \hline -\frac{1}{2}x & 0 & 0 \end{array}\right), \qquad \mathbf{P}_2 = \left(\begin{array}{cc|c} 0 & 0 & -\frac{1}{2}i \\ 0 & 0 & 0 \\ \hline -\frac{1}{2}xi & 0 & 0 \end{array}\right), \tag{C.1}$$

$$\mathbf{P}_3 = \left(\begin{array}{cc|c} 0 & 0 & 0 \\ 0 & 0 & \frac{1}{2} \\ \hline \frac{1}{2}xi & 0 & 0 \end{array}\right), \qquad \mathbf{P}_4 = \left(\begin{array}{cc|c} 0 & 0 & 0 \\ 0 & 0 & -\frac{1}{2}i \\ \hline 0 & \frac{1}{2}xi & 0 \end{array}\right).$$

With respect to the standard matrix commutator $[\cdot, \cdot]$, they form a semisimple algebra for which the Cartan decomposition $\mathcal{L} = H + V$ holds; here, the subalgebra $H = su(2) \oplus su(1)$ is made up of the matrices M_i, and the subalgebra $V = \mathbb{C}^2$ of the matrices P_i.

Remark. Here, x is a parameter, so that (C.1) defines a pencil of Lie algebras. For $x > 0$ these algebras are isomorphic to the algebra $su(3)$, for $x < 0$ to the algebra $su(2, 1)$, and for $x = 0$ to the semidirect sum $(so(2) \oplus su(1)) \oplus_s \mathbb{C}^2$.

Because of the semisimplicity, we may identify the algebra $\mathcal{L}$ with the coalgebra $\mathcal{L}^*$ using the inner product (Killing form)

$$g = -\operatorname{Tr}(\mathbf{X} \cdot \mathbf{Y}), \qquad \mathbf{X}, \mathbf{Y} \in \mathcal{L}. \tag{C.2}$$

Denote by $m_1, m_2, m_3, m_4, p_1, p_2, p_3$, and p_4 the coordinates in the coalgebra; then, after the identification (of algebra elements), we obtain the matrix

$$\mathbf{X} = \begin{pmatrix} -i(m_3 + m_4) & im_1 + m_2 & \frac{1}{x}(p_1 - ip_2) \\ im_1 - m_2 & i(m_3 - m_4) & \frac{1}{x}(p_3 - ip_4) \\ -p_1 - ip_2 & -p_3 - ip_4 & 2im_4 \end{pmatrix}.$$

DOI 10.1515/9783110544442-009

REMARK. If we denote the above elements of the basis of the algebra $\mathcal{L}$ by $\mathbf{E}_i$, $i = 1, \dots, 8$, and introduce the vector $\boldsymbol{e}$ whose components are the coordinates on the coalgebra in the basis dual to $\mathbf{E}_i$, then the identification formula takes the form $\mathbf{X} = \sum_{i=1}^{8} \xi_i \mathbf{E}_i$, where $\boldsymbol{\xi} = g^{-1}\boldsymbol{e}$ and g is defined by formula (C.2).

The corresponding Lie – Poisson bracket (more precisely, the pencil of brackets linearly depending on the parameter x) for the coordinate functions of the coalgebra has the form

$$
\begin{aligned}
&\{m_i, m_j\} = \varepsilon_{ijk} m_k, && \{m_i, m_4\} = 0,\\
&\{m_i, p_j\} = \frac{1}{2}(\varepsilon_{ijk} p_k - \delta_{ij} p_4), && \{m_i, p_4\} = \frac{1}{2} p_i, \quad i, j, k - 1, 2, 3,\\
&\{p_i, p_j\} = \frac{1}{2} x(\varepsilon_{ijk} m_k + 3\varepsilon_{ij3} m_4), && \{p_i, p_4\} = -\frac{1}{2} x(m_i - 3\delta_{i3} m_4).
\end{aligned}
$$

For $x = 0$, the bracket $\{\,,\,\}_\theta$ for the variables $m_1, m_2, m_3, p_1, p_2, p_3$, and p_4 (which compose a subalgebra) coincides with the bracket for the angular momentum $\boldsymbol{M}$ and the quaternions λ_0 and $\boldsymbol{\lambda}$ in the rigid body dynamics (Sec. 1.3 of Ch. 1).

In accordance with the general method developed in [67, 95], let us construct an $\mathbf{L}$-matrix whose invariants define commuting functions for the whole family of brackets $\{\,\}_\theta + \lambda(\{\,\}_\lambda + \{\,\}_a)$, where $\mathbf{a} \in V$ is a shift of the argument. Restricting ourselves to problems of real dynamical significance, we set $x = 1$ and write the $\mathbf{L}$-matrix in the form

$$
\mathbf{L} = (\mathbf{h}\lambda + \mathbf{v} + \mathbf{a}\lambda^2),
$$

where

$$
\mathbf{h} = \left(\begin{array}{cc|c} -i(m_3+m_4) & im_1+m_2 & 0 \\ im_1-m_2 & i(m_3-m_4) & \\ \hline \multicolumn{2}{c|}{0} & 2im_4 \end{array}\right),
$$

$$
\mathbf{v} = \left(\begin{array}{cc|c} \multicolumn{2}{c|}{0} & p_1 - ip_2 \\ & & p_3 - ip_4 \\ \hline -p_1-ip_2 & -p_3-ip_4 & 0 \end{array}\right), \tag{C.3}
$$

$$
\mathbf{a} = \left(\begin{array}{cc|c} \multicolumn{2}{c|}{0} & a_1 - ia_2 \\ & & a_3 - ia_4 \\ \hline -a_1-ia_2 & -a_3-ia_4 & 0 \end{array}\right).
$$

For an arbitrary shift $\mathbf{a}$ there exists among the invariants of the matrix $\mathbf{L}$ a function linear in m, of the form

$$
F_1 = m_4 a^2 + (\boldsymbol{m}, \boldsymbol{\gamma}_a), \tag{C.4}
$$

where $a^2 = \sum_{i=1}^{4} a_i^2$, $\boldsymbol{m} = (m_1, m_2, m_3)$, and the components of the vector $\boldsymbol{\gamma}_a$ are expressed in terms of the a_i by the formulas

$$
\boldsymbol{\gamma}_a = (2(a_1a_3 + a_2a_4),\ 2(-a_2a_3 + a_1a_4),\ a_3^2 + a_4^2 - a_1^2 - a_2^2).
$$

Let us choose $a_1 = a_2 = 0$. Then the integral (C.4) takes the form

$$F_1 = m_3 + m_4. \tag{C.5}$$

Let us consider the quadratic invariant of the matrix **L** and take it as the Hamiltonian:

$$F_2 = \operatorname{Tr}(\mathbf{h}^2 + 2\mathbf{va}) = m_1^2 + m_2^2 + m_3^2 + 3m_4^2 + a_4p_4 + a_3p_3. \tag{C.6}$$

It is necessary to set

$$m_4 = -m_3 + c, \qquad c = \text{const}. \tag{C.7}$$

in the **L**-matrix (C.3) and in the Hamiltonian (C.6).

This Hamiltonian defines a certain (formal) integrable Hamiltonian system on the family of brackets $\{,\}_\theta + \lambda(\{\ \}_\lambda + \{\ \}_a)$ [95]. Finding the matrix **A** for this system is not difficult.

To pass from the found formal system to the generalization of the Goryachev – Chaplygin case (see Sec. 2.5 of Ch. 2), we perform reduction in the **L** and **A** matrices using the linear integral (C.7). For this purpose, in the matrix **L** and in the Hamiltonian (C.6) we set

$$m_4 = -m_3 + c, \qquad c = \text{const}. \tag{C.8}$$

We obtain the **L**-matrix and the Hamiltonian of an integrable system on the subalgebra m_1, m_2, m_3, p_1, p_2, p_3, p_4, which corresponds to a rigid body with an attached gyrostat having gyrostatic momentum equal to c:

$$\mathbf{L} = \begin{pmatrix} -i\lambda c & (im_1+m_2)\lambda & p_1-ip_2 \\ (im_1-m_2)\lambda & i\lambda(2m_3-c) & p_3-ip_4+(a_3+ia_4)\lambda^2 \\ -p_1-ip_2 & -p_3-ip_4-(a_3+a_4)\lambda^2 & -2i(m_3-c)\lambda \end{pmatrix}, \tag{C.9}$$

$$H = m_1^2 + m_2 + 4m_3^2 - 6m_3c + 2a_4p_4 + 2a_3p_3, \tag{C.10}$$

$$\mathbf{A} = dH = \begin{pmatrix} i(4m_3 - 3c) & -im_1 - m_2 & 0 \\ -im_1 + m_2 & -i(4m_3 - 3c) & -a_3 + ia_4 \\ 0 & a_3 + ia_4 & 0 \end{pmatrix}.$$

This system is an integrable case for the quaternion bracket $\{\ \}_\theta$. We can show that it admits an integral linear in m_i, of the form

$$\begin{gathered} F_3 = m_3 - (\mathbf{m}, \boldsymbol{\gamma}), \\ \boldsymbol{\gamma} = (2(p_2p_4 + p_1p_3),\ 2(p_2p_3 - p_1p_4),\ p_3^2 + p_4^2 - p_1^2 - p_2^2) \end{gathered} \tag{C.11}$$

in which the standard reduction with symmetry described in Sec. 4.1 of Ch. 4 is possible, leading to a system on a nonlinear bracket.

Reduction leads to the generators, which are integrals of the vector field $\mathbf{v} = \{\cdot, F_3\}$, of the following form:

$$\begin{gathered} K_1 = \frac{M_1p_1 + M_2p_2}{\sqrt{p_1^2 + p_2^2}}, \quad K_2 = \frac{M_2p_1 - M_1p_2}{\sqrt{p_1^2 + p_2^2}}, \quad K_3 = M_3, \\ s_1 = p_3, \qquad s_2 = p_4, \quad s_3 = \pm\sqrt{p_1^2 + p_2^2}. \end{gathered} \tag{C.12}$$

The bracket corresponding to them is nonlinear:

$$\{K_i, K_j\} = \varepsilon_{ijk}K_k + \varepsilon_{ij3}\frac{F_4}{s_3^2},$$
$$\{K_i, s_j\} = \varepsilon_{ijk}s_k, \quad \{s_i, s_j\} = 0, \tag{C.13}$$

where $F_3 = (\boldsymbol{K}, \boldsymbol{s})s_3$ is a Casimir function of the bracket (C.13). For zero value of the area constant, $(\boldsymbol{K}, \boldsymbol{s}) = 0$, (C.13) becomes the bracket of the algebra $e(3)$, whereas we can show that in the new variables (C.12) the Hamiltonian (C.10) coincides with the Goryachev – Chaplygin Hamiltonian:

$$H = (K_1^2 + K_2^2 + 4K_3^2) - 2a_4s_2 - 2a_3s_1. \tag{C.14}$$

For $F_3 = c \neq 0$, the **L-A** pair (C.9) and (C.10) corresponds to the Goryachev case with an additional singular term. According to the procedure described in Sec. 4.1 of Ch. 4, the nonlinearity in the bracket (C.13) can be removed by the transformation

$$\mathbf{L} = \mathbf{K} - c\frac{\boldsymbol{s}}{s_3}, \tag{C.15}$$

which reduces it to the form of the algebra $e(3)$, whereas it reduces the Hamiltonian (C.14) to the form

$$H^* = \frac{1}{2}(L_1^2 + L_2^2 + 4L_3^2) - a_4s_2 - a_3s_1 + 3cL_3 + \frac{1}{2}\frac{c^2}{s_3^2}. \tag{C.16}$$

This integrable Hamiltonian (C.16) can be interpreted as a generalization of the Goryachev – Chaplygin case $(\boldsymbol{L}, \boldsymbol{s}) = 0$ with two additional terms: one linear in L_3 (corresponding to a constant gyrostatic momentum), and the other singular term. The integrable generalization with just the additional gyrostatic momentum term was presented by L. N. Sretenskii [543], the generalization with just the addition of the singular potential by D. N. Goryachev himself [97], and the general case where both terms with arbitrary independent coefficients can be added to the Hamiltonian was presented in [315, 633] (see also Sec. 5.1 of Ch. 5).

Thus, the above **L-A** pair also represents integrable generalizations of the Goryachev – Chaplygin case. It differs from a somewhat mysterious **L-A** pair presented in [51], which is obtained by deleting a row and a column from the **L-A** pair of the Kovalevskaya case.

D The Hess Case and Quantization of the Rotation Number

This appendix is a somewhat revised version of the paper [49]. We present here a continuation of the qualitative analysis of rigid body dynamics in the Hess case which was started in Ch. 2, Sec. 2.6. In particular, we have constructed a bifurcation diagram for the Hess case and shown that the nonsingular level surface of first integrals is a two-dimensional torus. The vector field on the torus in the Hess case does not admit an invariant measure with density smoothly depending on phase variables. For such systems (see, e.g., [22]) the graph showing the dependence of the rotation number on parameters is, as a rule, a Cantor ladder whose rational values correspond to horizontal segments on which there are one or several limit cycles (for details, see [468]). A similar situation is also encountered in nonholonomic mechanics [47, 68]. Nevertheless, it turns out that there exists no Cantor ladder in the Hess case. This is due to the fact that in this case the system on the torus is reduced by central projection to a linear Hamiltonian system with periodic coefficients. In this case, limit cycles (horizontal segments) can arise only for integer values of the rotation number $\nu = n \in \mathbb{Z}$.

An analogous system on the torus occurs in explanations of the Josephson effect in superconductor physics [222, 111, 265] and in the dynamics of coupled oscillators [183]. In [222, 111, 265], the above-mentioned property is usually called the effect of *quantization of the rotation number*. In some works (see, e.g., [183]), proofs are given by using a reduction to the real Riccati equation [388] with periodic coefficients, for which the Poincaré map for a period is linear-fractional. For the Hess case such a reduction is used in [391].

It turns out that the rotation number in the Hess case in the Euler – Poisson equations does not exceed unity, therefore, one can observe only ont horizontal segment corresponding to $\nu = 0$. However, the Hess case admits a generalization for which the rotation number can be arbitrarily large, and limit cycles occur, as before, only for integer values of the rotation number. Various generalizations of the Hess case are also discussed in [102].

We conclude by giving an example of a Hamiltonian system which possesses an invariant submanifold analogous to the Hess case, but on which the dependence of the rotation number on parameters is a Cantor ladder.

1 Equations of motion and conservation laws

For completeness of exposition we recall here the Euler – Poisson equations (see Ch. 2) describing the motion of a heavy rigid body with a fixed point in a coordinate system

DOI 10.1515/9783110544442-010

attached to the body:

$$\dot{\boldsymbol{M}} = \boldsymbol{M} \times \frac{\partial H}{\partial \boldsymbol{M}} + \boldsymbol{\gamma} \times \frac{\partial H}{\partial \boldsymbol{\gamma}}, \quad \dot{\boldsymbol{\gamma}} = \boldsymbol{\gamma} \times \frac{\partial H}{\partial \boldsymbol{M}},$$
$$H = \frac{1}{2}(\boldsymbol{M}, \mathbf{I}^{-1}\boldsymbol{M}) - mg(\boldsymbol{r}, \boldsymbol{\gamma}), \tag{D.1}$$

where $\boldsymbol{\gamma}$ is the unit vector directed along the gravitional field, $\boldsymbol{M}$ is the angular momentum, $\mathbf{I}$ and m are the inertia tensor and the mass of the body, respectively, and g is the free-fall acceleration.

Equations (D.1) admit, in addition to the energy integral $H = \text{const}$, an area integral and a geometric integral of the form

$$F_1 = (\boldsymbol{M}, \boldsymbol{\gamma}), \quad \boldsymbol{\gamma}^2 = 1.$$

We recall that the conditions satisfied by the parameters of the system (D.1) in the Hess case are most naturally described by a *gyration ellipsoid*, the level surface of the kinetic energy in the space of momenta $\boldsymbol{M} = (M_1, M_2, M_3)$:

$$(\boldsymbol{M}, \mathbf{I}^{-1}\boldsymbol{M}) = \text{const}. \tag{D.2}$$

1°. All semiaxes of the ellipsoid (D.2) are different (i.e., the principal moments of inertia of the body do not coincide.
2°. The center of mass of the body $\boldsymbol{r}$ is perpendicular to the circular section of the gyration ellipsoid (see Fig. 2.45).

Let us choose a moving coordinate system $Ox_1x_2x_3$ with $Ox_3 \| \boldsymbol{r}$ and with the axes Ox_1 and Ox_2 directed so that the inertia tensor $\mathbf{I}$ of the body has the form

$$\mathbf{I} = \begin{pmatrix} I_{11} & 0 & I_{13} \\ 0 & I_{22} & 0 \\ I_{13} & 0 & I_{33} \end{pmatrix}.$$

Since in this coordinate system the section formed by the intersection of the gyration ellipsoid (D.2) with the plane $M_3 = 0$ is a circle, the relation $(\mathbf{I}^{-1})_{11} = (\mathbf{I}^{-1})_{22}$ is satisfied, whence we obtain the condition

$$I_{11} = I_{22} + \frac{I_{13}^2}{I_{33}}. \tag{D.3}$$

Moreover, in this case the following inequality holds for the rigid body [410, p. 149]:

$$-I_{33} < I_{13} < I_{33}. \tag{D.4}$$

We note that $Ox_1x_2x_3$ is not the principal coordinate system for the rigid body ($I_{13} \neq 0$). Nevertheless, the chosen coordinate system allows us to better illustrate the analogy between between the Hess case and the Lagrange case.

Proposition 1. *In the Hess case the system* (D.1) *possesses the invariant manifold*

$$M_3 = 0. \tag{D.5}$$

In [654, 336], the Hess case with a small weight mg is considered as a perturbation of the integrable Euler case. It turns out that the system (D.1) is nonintegrable and the invariant submanifold (D.5) is a pair of unsplit separatrices. In Sec. 2.6 (see also [48]), it is noted that in this case the manifold (D.5) separates two different stochastic layers (see Fig. **??**).

Let us decrease the number of parameters by passing to dimensionless variables:

$$\boldsymbol{M} \to \mu_0 \boldsymbol{M}, \qquad t \to \tau_0 t,$$
$$\mu_0 = g^{1/2} m^{1/4} I_{22}^{3/4}, \qquad \tau_0 = \frac{I_{22}^{1/4}}{g^{1/2} m^{1/4}}.$$

In this case, the Hamiltonian of the system can be represented as

$$H = \frac{1}{2}\left(M_1^2 + M_2^2 + 2aM_3M_1\right) - by_3,$$

and the equations of motion (D.3) on the invariant Hess submanifold have the form

$$\begin{gathered} \dot{M}_1 = -aM_1M_2 - by_2, \quad \dot{M}_2 = aM_1^2 + by_1, \\ \dot{y}_1 = -aM_1y_2 - M_2y_3, \quad \dot{y}_2 = aM_1y_1 + M_1y_3, \quad \dot{y}_3 = M_2y_1 - M_1y_2, \\ a = \frac{I_{13}}{I_{33}}, \quad b^2 = \frac{mr_3^2}{I_{22}}. \end{gathered} \tag{D.6}$$

We note that the system (D.6) admits two transformations Σ_1 and Σ_2 of the variables and parameters which preserve the form of the equations

$$\begin{aligned} \Sigma_1 : &\quad M_2 \to -M_2,\ y_1 \to -y_1,\ a \to -a, \\ \Sigma_2 : &\quad M_1 \to -M_1,\ y_1 \to -y_1,\ y_3 \to -y_3,\ b \to -b. \end{aligned}$$

Using (D.4), we obtain that without loss of generality in the system (D.6) we may assume $a \in [0, 1)$ and $b \in [0, \infty)$.

The energy, area and geometrical integrals can be represented as

$$E = \frac{1}{2}(M_1^2 + M_2^2) - by_3, \quad F_1 = M_1y_1 + M_2y_2, \quad \boldsymbol{y}^2 = 1. \tag{D.7}$$

Thus, the integral generic submanifold of the system (D.6) is two-dimensional and, as will be shown below, turns out to be a two-dimensional torus $\mathbb{T}^2$.

We note that, in addition to the first integrals (D.7), the system (D.6) possesses a generalized symmetry field (see [100])

$$\boldsymbol{u} = -M_2\frac{\partial}{\partial M_1} + M_1\frac{\partial}{\partial M_2} - y_2\frac{\partial}{\partial y_1} + y_1\frac{\partial}{\partial y_2},$$

which satisfies the equation

$$[\boldsymbol{u}, \boldsymbol{v}] = aM_2\boldsymbol{u},$$

where $\boldsymbol{v}$ is the vector field of the system (D.6).

An invariant measure for the system (D.6) has been found only in the following particular cases:

- in the absence of a gravitional field ($b = 0$): an invariant measure with singular density $\rho = M_2^{-1}$;
- in the case $a = 0$: an invariant measure with constant density $\rho = \text{const}$.

We note that the invariant submanifold (D.5) is isolated in the phase space of the Hamiltonian system, in contrast to the case of integrable Hamiltonian systems for which the invariant submanifolds are the level surface of first integrals and form families (which fill entirely some regions of the phase space). Therefore, it is impossible to restrict the initial standard invariant measure of the system (D.1); in particular, this leads to a situation where, as shown below, the invariant tori of the system (D.6) can contain limit cycles.

2 Bifurcation diagram

Consider the issue of conditions for degeneracy of the integrals (D.7). For this we note that they are exactly the same as the integrals of the Lagrange top with $M_3 = 0$. Therefore, we perform the analysis by using the same variables as in rigid body dynamics.

On the common level set of the geometric integral $\boldsymbol{\gamma}^2 = 1$ and the area integral $F_1 = p_\psi$ we choose new variables:

$$M_1 = p_\psi \frac{\sin\varphi}{\sin\theta} + p_\theta\cos\varphi, \quad M_2 = p_\psi\frac{\cos\varphi}{\sin\theta} - p_\theta\sin\varphi,$$
$$\gamma_1 = \sin\theta\sin\varphi, \quad \gamma_2 = \sin\theta\cos\varphi, \quad \gamma_3 = \cos\theta,$$

where $\varphi \mod 2\pi$ is the angle of proper rotation and $\theta \in (0, \pi)$ is the angle of nutation. As a result, the equations of motion take the form

$$\begin{gathered}\dot\theta = p_\theta, \quad \dot p_\theta = p_\psi^2\frac{\cos\theta}{\sin^3\theta} - b\sin\theta,\\ \dot\varphi = -p_\psi\frac{a\sin\theta\sin\varphi + \cos\theta}{\sin^2\theta} - ap_\theta\cos\varphi.\end{gathered} \tag{D.8}$$

Remark. As is seen, the equations for the variables (θ, p_θ) decouple and coincide with the equations of motion of the spherical pendulum [646].

Their only integral is

$$E = \frac{1}{2}p_\theta^2 + \frac{p_\psi^2}{2\sin^2\theta} - b\cos\theta.$$

In this case, the degeneracy conditions can be written as

$$\frac{\partial E}{\partial \theta} = -p_\psi^2 \frac{\cos\theta}{\sin^3\theta} + b\sin\theta = 0, \qquad \frac{\partial E}{\partial p_\theta} = p_\theta = 0.$$

This yields a result that is well known in rigid body dynamics.

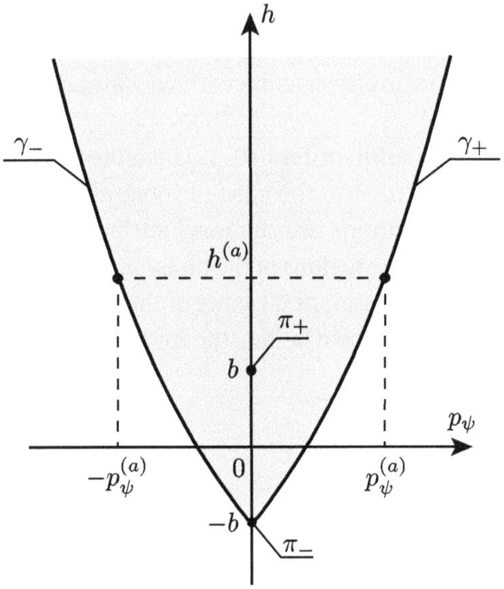

Fig. D.1. Bifurcation diagram (grey denotes the region of possible values of first integrals) and the point $(p_\psi^{(a)}, h^{(a)})$.

Proposition 2. *The first integrals become dependent if the values $E = h$, $F_1 = p_\psi$ on the plane (p_ψ, h) correspond*

- *either to points of the curves given by the equations*

$$\gamma_+ = \left\{ p_\psi(\theta_c) = \frac{\sqrt{b}\sin^2\theta_c}{\sqrt{\cos\theta_c}}, \quad h(\theta_c) = \frac{b(1-3\cos^2\theta_c)}{2\cos\theta_c} \right\}, \quad \theta_c \in \left(0, \frac{\pi}{2}\right)$$

$$\gamma_- = \left\{ p_\psi(\theta_c) = -\frac{\sqrt{b}\sin^2\theta_c}{\sqrt{\cos\theta_c}}, \quad h(\theta_c) = \frac{b(1-3\cos^2\theta_c)}{2\cos\theta_c} \right\}, \quad \theta_c \in \left(0, \frac{\pi}{2}\right) \tag{D.9}$$

- *or to the points*

$$\pi_+ = \{p_\psi = 0,\ h = b\} \quad \textit{and} \quad \pi_- = \{p_\psi = 0,\ h = -b\}. \tag{D.10}$$

On the plane of the values of first integrals (p_ψ, h) the curves γ_+, γ_- and the points π_+, π_- form a bifurcation diagram of the integral map of the system (D.6). We show on it the region of possible values of the integrals (see Fig. D.1). Now this diagram gives a full picture of rearrangements of the integral manifold of the system.

If the point (p_ψ, h) lies on one of the curves γ_+, γ_-, then the invariant submanifold given by the integrals (D.7) is a circle. According to (D.8), $\dot\theta = 0$, $\dot p_\theta = 0$, and the evolution of the angle φ is governed by

$$\dot\varphi = \mp\frac{a\sqrt{b}\sin\theta_c}{\sqrt{\cos\theta_c}}\sin\varphi \mp \sqrt{b\cos\theta_c}, \tag{D.11}$$

where the upper and lower signs correspond to the curves γ_+ and γ_-, respectively.

Equation (D.11) describes two possible types of behavior of the angle φ:

1) $|\tan\theta_c| > a$, then $\dot\varphi$ does not change sign anywhere and the angle φ increases monotonously for γ_- and decreases for γ_+;
2) $|\tan\theta_c| < a$, then two equilibria arise on the circle, one of them is unstable and the second is asymptotically stable.

The change of these regimes on the curves γ_+ and γ_- occurs at the points $(p_\psi^{(a)}, h^{(a)})$ and $(-p_\psi^{(a)}, h^{(a)})$, respectively, where

$$p_\psi^{(a)} = \frac{b^{1/2}}{a^{1/2}(1+a^2)^{3/4}}, \qquad h^{(a)} = \frac{b(1-2a^2)}{2a(1+a^2)^{1/2}}.$$

Specifically, if $h < h^{(a)}$, there are no equilibria on the circle, otherwise there are equilibria on it.

Remark. The bifurcation diagram does not depend on the parameter a, whereas the position of the point $(p_\psi^{(a)}, h^{(a)})$ does. Moreover, if $a \to \infty$, then $h^{(a)} \to -b$, i.e., at all points of the curves γ_+, γ_- equation (D.11) possesses equilibria.

It can be shown that for those values of the first integrals p_ψ and h which satisfy (D.9) or (D.10) the system (D.6) has no fixed points. *Consequently, for all possible values of the first integrals p_ψ and h which do not correspond to the points lying on the curves γ_+, γ_- and to the points π_+ and π_-, the level surface of the first integrals is diffeomorphic to the two-dimensional torus $\mathbb{T}^2$.*

3 Zero level set of the area integral

Let us consider separately the Hess case on the zero level set of the area integral $F_1 = p_\psi = 0$. In this case the solution is written in terms of quadratures. The equations of motion (D.8) and the energy integral with zero value of the area integral have the form

$$\begin{gathered}\dot\theta = p_\theta, \quad \dot p_\theta = -b\sin\theta, \quad \dot\varphi = -ap_\theta\cos\varphi, \\ E = \frac{1}{2}p_\theta^2 - b\cos\theta.\end{gathered} \tag{D.12}$$

The system (D.12) possesses an invariant measure with density

$$\rho = \frac{1}{\cos\varphi},$$

which has a singularity for $\varphi = \frac{\pi}{2}$ and $\varphi = \frac{3\pi}{2}$, which correspond to its partial periodic solutions (D.12).

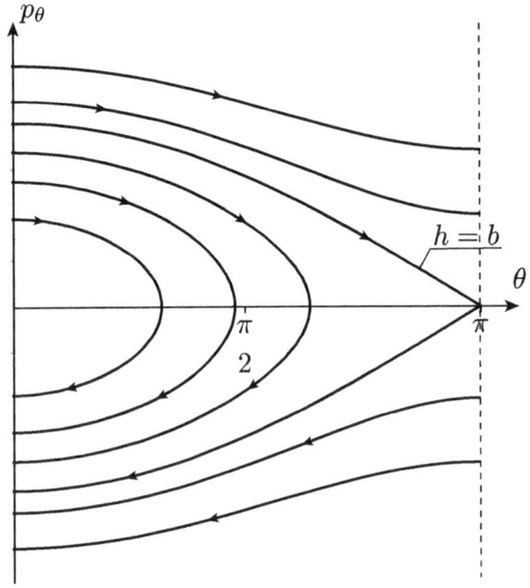

Fig. D.2. Projections of the system trajectories onto the plane (θ, p_θ).

The equations for (θ, p_θ) decouple, and the trajectories on this plane are shown in Fig. D.2. The explicit dependence $\theta(t)$ on the fixed level set of the energy integral $E = h$ is given by the quadrature

$$\dot{\theta}^2 = 2(h + b\cos\theta).$$

We see that there are two cases:

1) $-b < h < b$, in this case all trajectories pass through the pole $\theta = 0$ and reach the turning point $\theta_\star$ given by the equation (see Fig.D.2)

$$h + b\cos\theta_\star = 0;$$

2) $b < h$, in this case all trajectories pass through both poles of the Poisson sphere $\theta = 0$ and $\theta = \pi$.

In both cases the shape of the trajectories is the same, it does not depend on the value of h and, according to (D.12), is given by the equation

$$\frac{d\varphi}{d\theta} = -a\cos\varphi.$$

This equation has an explicit solution, so that the projections of the trajectories onto the Poisson sphere for $\varphi \neq \frac{\pi}{2}$ and $\varphi \neq \frac{3\pi}{2}$ are given by

$$\frac{1 - \sin\varphi}{\cos\varphi} = Ce^{-a\theta}, \tag{D.13}$$

where C is the constant of integration which parameterizes various trajectories.

Since the definition of the angular coordinates θ and φ has the singularity $\theta = 0$ and $\theta = \pi$, equation (D.13) defines only a segment of the system trajectories between the poles of the Poisson sphere. On the other hand, when $h = \pm b$, the system (D.6) has no singularities at the points $\gamma_1 = \gamma_2 = 0$ and $\gamma_3 = \pm 1$, therefore, the projection of the trajectory smoothly continues through the poles, the angle φ changes by a value equal to π, and $\dot{\theta}$ changes sign. Consequently, two segments of the trajectory are glued together at each of the poles and are given by equation (D.13) with different constants C_1 and C_2. Specifically, if $\theta = 0$, they satisfy the relation

$$C_1 C_2 = -1, \tag{D.14}$$

while if $\theta = \pi$, then

$$e^{2a\pi} C_1 C_2 = -1. \tag{D.15}$$

As stated above, when $-b < h < b$, the region of possible motions on the Poisson sphere is bounded by the inequality $0 < \theta < \theta_\star$. In this case, the motion occurs along the trajectory segment given by (D.13) at some $C = C_1$; as the turning point $\theta = \theta_\star$ is reached, a stop occurs and the point returns along the same trajectory until it reaches the pole $\theta = 0$. According to (D.14), after the pole the motion continues along the trajectory segment given by the same equation (D.13) at $C = -C_1^{-1}$ up to $\theta = \theta_\star$, where a U-turn occurs, see Fig.D.3a. This implies that the motion of the system (D.6) is periodic with period

$$T = 4 \int\limits_0^{\theta_\star} \frac{d\theta}{\sqrt{2(h + b\cos\theta)}}.$$

Thus, the invariant tori $\mathbb{T}^2$ given by (D.7) are foliated in this case by periodic trajectories.

When $b < h$, the situation is different: first, at each $h = \text{const}$ the phase space of the system (D.6) has a pair of periodic trajectories $\mathcal{C}_h^{(-)}$ and $\mathcal{C}_h^{(+)}$ which are projected onto the Poisson sphere into a special trajectory consisting of two meridians $\varphi = \frac{\pi}{2}$ and $\varphi = \frac{3\pi}{2}$, the directions of the trajectory are different for $\mathcal{C}_h^{(-)}$ and $\mathcal{C}_h^{(+)}$. The other trajectories on the torus $\mathbb{T}^2$ asymptotically approach $\mathcal{C}_h^{(+)}$ as $t \to +\infty$ and $\mathcal{C}_h^{(-)}$ as $t \to -\infty$, they are projected onto the Poisson sphere into trajectory segments (D.13) for which the constant C changes when the point passes through the poles according to the rules (D.14) and (D.15), see Fig.D.3(a). If we denote the sequence of the constants (D.13) on these segments by $C = C_m$, $m = 0, 1, \ldots$, then for $t \to +\infty$ we obtain

$$C_{2m} = C_0 (e^{-2a\pi})^m, \quad C_{2m+1} = C_0^{-1} (e^{2a\pi})^m.$$

Thus, we have

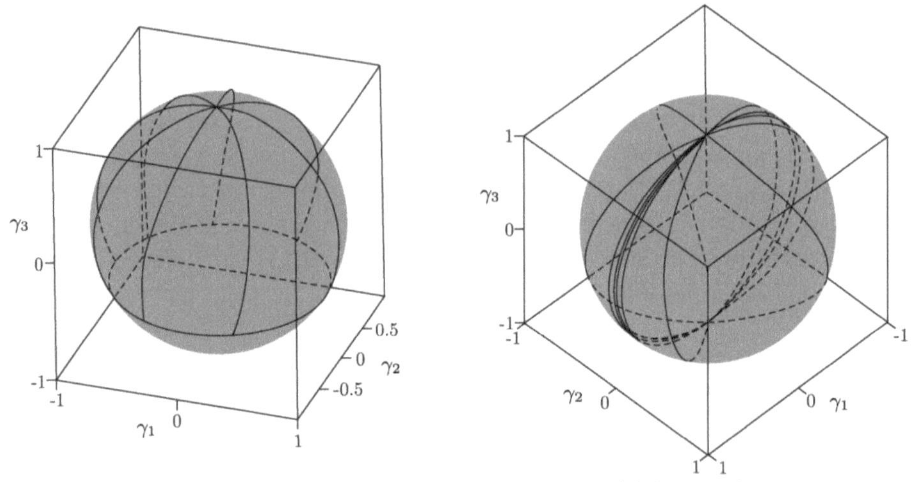

(a) $h = 0.5, b = 1$ (b) $h = 1.5, b = 1$

Fig. D.3. Trajectories on the Poisson sphere.

Proposition 3. *For the system* (D.6), *on the zero level set of the area integral the invariant tori* $\mathbb{T}^2$ *given by* (D.7) *are characterized by the following:*
- *when* $-b < h < b$*, they are foliated by periodic trajectories;*
- *when* $b < h$*, they contain a pair of periodic solutions (cycles) — a stable one and an unstable one, all the other solutions tend to a stable cycle as* $t \to +\infty$ *and to an unstable cycle as* $t \to -\infty$*.*

4 The rotation number and limit cycles

As stated above (see Sec. D2), if the values of the first integrals do not satisfy conditions (D.9) and (D.10), their common level surface is diffeomorphic to the two-dimensional torus $\mathbb{T}^2$. To analyze the flow on these tori, we parameterize them by suitable angular coordinates using the well-known Andoyer variables [8] (G, g, l), which on the level set of the area integral $F_1 = p_\psi$ are given as follows:

$$M_1 = G\sin l, \quad M_2 = -G\cos l,$$

$$\gamma_1 = \frac{p_\psi}{G}\sin l + \sqrt{1 - \frac{p_\psi^2}{G^2}}\sin g\cos l,$$

$$\gamma_2 = -\frac{p_\psi}{G}\cos l + \sqrt{1 - \frac{p_\psi^2}{G^2}}\sin g\sin l,$$

$$\gamma_3 = -\sqrt{1 - \frac{p_\psi^2}{G^2}}\cos g.$$

In terms of the new variables the energy integral E becomes

$$E = \frac{G^2}{2} - b\sqrt{1 - \frac{p_\psi^2}{G^2}}\cos g.$$

Let us fix the level set of the energy integral $E = h$ and eliminate the variable g. Finally we obtain the equations of motion (D.6) in the form

$$\begin{gathered} \dot{G}^2 = R(G), \quad \dot{l} = aG\sin l + \frac{b}{G^2}p_\psi, \\ R(G) = b^2\left(1 - \frac{p_\psi^2}{G^2}\right) - \frac{1}{4}(G^2 - 2h)^2, \end{gathered} \tag{D.16}$$

where $l \mod 2\pi$ and $G \in [G_1, G_2]$ with G_1 and G_2 being positive solutions of the equation $R(G) = 0$, which always exist for p_ψ and h belonging to the region of possible motions.

In the system (D.16), the equation for G decouples, and its solution is expressed in terms of the elliptic Jacobi function:

$$\begin{gathered} G(t) = 2\omega\sqrt{c - k^2\operatorname{sn}^2(\omega t, k)}, \\ h = \omega^2(3c - 1 + k^2), \quad p_\psi^2 = \frac{16c\omega^6}{b^2}(1 - c)(c - k^2), \\ b^2 = \omega^4\Big(\big(c + (1 + k)^2\big)\big(c + (1 - k)^2\big) - 4c^2\Big). \end{gathered}$$

Substituting the function $G(t)$ into the second equation (D.16) and eliminating the parameter ω by rescaling time by $t \to \omega^{-1}t$, we obtain

$$\frac{dl}{dt} = 2\alpha(t) + 2\beta(t)\sin l, \tag{D.17}$$

where the functions $\alpha(t)$ and $\beta(t)$ are periodic in time (with period T) and in this case are represented as

$$\begin{gathered} \alpha(t) = \frac{\sqrt{c(1 - c)(c - k^2)}}{2(c - k^2\operatorname{sn}^2(t, k))}, \quad \beta(t) = a\sqrt{c - k^2\operatorname{sn}^2(t, k)}, \\ T = \int_0^{2\pi} \frac{dx}{\sqrt{1 - k^2\sin^2 x}}, \quad k \in (0, 1), \quad c \in (k^2, 1). \end{gathered} \tag{D.18}$$

Thus, the problem reduces to investigating equation (D.17), which describes the vector field on the torus $\mathbb{T}^2 = \{(t, l),\ t \mod T,\ l \mod 2\pi\}$.

We recall that for the flow on the torus there exists an invariant — a rotation number, which in this case can be calculated by the formula

$$\nu = T\lim_{t\to\infty}\frac{l(t)}{2\pi t}.$$

As is well known (see, e.g., [22]), the graph showing the dependence of the rotation number on parameters is in the general case a Cantor ladder: rational rotation numbers correspond to horizontal segments, and on the tori there is one or several limit cycles (for details, see [468]). One of the exceptions is the case where the vector field on the family of tori admits an invariant measure smoothly depending on the parameters, and the rotation number ν is a smooth function of the parameters.

A numerical graph showing the dependence of the rotation number ν of the system (D.17) on c is presented in Fig. D.4. The graph consists of two segments: on the segment $(c_{\min}, c_\star)$ the rotation number ν smoothly depends on c, and on the segment $(c_\star, 1)$ we have the horizontal segment $\nu \equiv 0$. Thus, we see that, although tori with limit cycles occur in the system (D.17), it is not a generic system (since there is no Cantor ladder). Let us consider the explanation of such behavior of the rotation number.

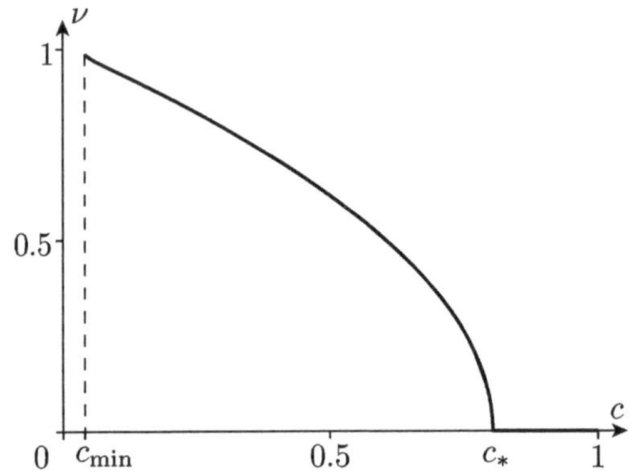

Fig. D.4. Graph showing the dependence of the rotation number ν on c for fixed $a = 0.25$, $k = 0.2$.

We first show that the horizontal segment corresponds to the torus with *two* limit cycles. To do so, we use a Poincaré section for a period, which defines the map of the circle onto itself:

$$\Theta(l) : S^1 \to S^1. \tag{D.19}$$

Figure D.5 shows the first iteration of this map for $c \in (c_\star, 1)$, we see that its graph intersects the diagonal (of a square), the intersection points correspond to the periodic solutions (D.17). As is seen, in this case there are two limit cycles.

Moreover, it turns out that in equation (D.17) with arbitrary periodic functions $\alpha(t)$ and $\beta(t)$ there can be no more than two limit cycles, and if they are present, the rotation number takes only an integer value. This is due to the following property of the system (D.17).

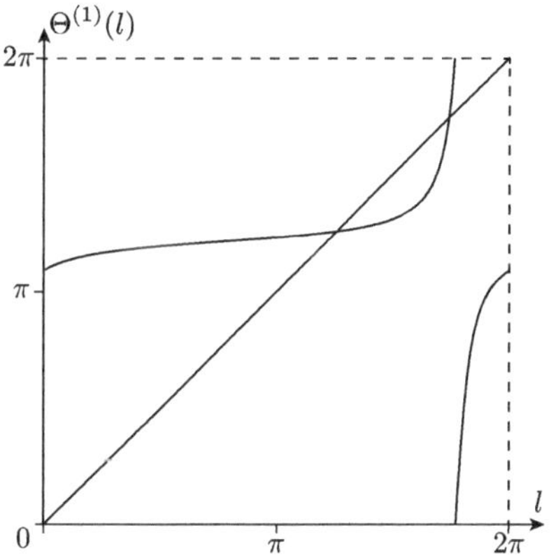

Fig. D.5. The first iteration of the map $\Theta(l)$ and the diagonal of a square for the values $a = 0.25$, $k = 0.2$, $c = 0.9$.

Proposition 4. *The general solution of equation* (D.17) *is expressed in terms of solutions of the linear Hamiltonian system with periodic coefficients:*

$$\dot{x} = \frac{\partial \mathcal{H}}{\partial y}, \quad \dot{y} = -\frac{\partial \mathcal{H}}{\partial x},$$
$$\mathcal{H} = \frac{1}{2}\left(\alpha(t)x^2 + 2\beta(t)xy + \alpha(t)y^2\right), \tag{D.20}$$
$$l(t) = 2\left(\arctan\frac{y(t)}{x(t)} + \pi N_T\right), \quad N_T \in \mathbb{Z},$$

where N_T is the number of complete half-turns (turns through the angle π) for period T about the origin of coordinates of the vector $\boldsymbol{r} = (x(t), y(t))$; when the turns occur counterclockwise, $N_T > 0$, while when they occur clockwise, $N_T < 0$.

Proof. Consider an equation for the half-angle $\phi = \frac{l}{2}$, which can be rewritten as

$$\dot{\phi} = \alpha(t)(\sin^2\phi + \cos^2\phi) + 2\beta(t)\sin\phi\cos\phi.$$

Using the Liouville method of doubling the number of variables, we recast it as a Hamiltonian system:

$$\dot{\phi} = \frac{\partial \mathcal{H}}{\partial r}, \quad \dot{r} = -\frac{\partial \mathcal{H}}{\partial \phi},$$
$$\mathcal{H} = r\left(\alpha(t)(\sin^2\phi + \cos^2\phi) + 2\beta(t)\sin\phi\cos\phi\right).$$

After a canonical change of variables $(r, \phi) \to (x, y)$ of the form

$$x = \sqrt{2r}\cos\phi, \quad y = \sqrt{2r}\sin\phi$$

we obtain (D.20). □

From a geometric point of view, the solution of equation (D.17) is obtained by *central projection* of the trajectory $\boldsymbol{r}(t)$ of the linear system (D.20) onto a unit circle (followed by doubling of the resulting angle).

We recall that the general solution of the linear system $\boldsymbol{r}(t) = (x(t), y(t))$ can be expressed in terms of the fundamental matrix $\mathbf{S}(t)$:

$$\boldsymbol{r}(t) = \mathbf{S}(t)\boldsymbol{C},$$

where $\boldsymbol{C} = (C_1, C_2) = \boldsymbol{r}(0)$ is some constant vector. The matrix $\mathbf{S}(t)$ satisfies the equations

$$\dot{\mathbf{S}} = \begin{pmatrix} \alpha(t) & -\beta(t) \\ \beta(t) & \alpha(t) \end{pmatrix} \mathbf{S}, \quad \mathbf{S}(0) = \mathbf{E},$$

and since the system is Hamiltonian, $\det \mathbf{S}(t) \equiv 1$, i.e., $\mathbf{S}$ is an element of the group $SL(2)$.

Let $t = nT + \Delta t$, $\Delta t \in [0, T)$, then due to periodicity of the coefficients α and β the following relation holds:

$$\mathbf{S}(t) = \mathbf{S}(\Delta t)\mathbf{P}^n,$$

where $\mathbf{P} = \mathbf{S}(T) \in SL(2)$ is a constant matrix, which is called a monodromy matrix. The first iteration of the Poincaré map (D.19) has the form

$$\Theta^{(1)}(l) = 2\arctan\left(\frac{P_{11}\tan\frac{l}{2} + P_{12}}{P_{21}\tan\frac{l}{2} + P_{22}}\right) + 2\pi N_T. \tag{D.21}$$

Every linear transformation

$$\boldsymbol{r} = \mathbf{Q}\tilde{\boldsymbol{r}}, \quad \det \mathbf{Q} = 1$$

preserves the canonical form of the Hamiltonian system (D.20), therefore the fundamental matrix $\mathbf{S}(t)$ and the monodromy matrix $\mathbf{P}$ are defined up to a similarity transformation:

$$\mathbf{S} \to \mathbf{Q}^{-1}\mathbf{S}\mathbf{Q}, \quad \mathbf{P} \to \mathbf{Q}^{-1}\mathbf{P}\mathbf{Q}, \quad \det \mathbf{Q} = 1.$$

As is well known, with regard to these transformations, the group $SL(2)$ is divided into three nonintersecting orbits, and, depending on to which of these orbits the monodromy matrix $\mathbf{P}$ belongs, the behavior of the system (D.17) is qualitatively different.

1) $|\mathrm{Tr}\mathbf{P}| > 2$, in this case the matrix $\mathbf{P}$ is brought by real transformation $\mathbf{Q}$ to the diagonal form

$$\begin{pmatrix} \lambda & 0 \\ 0 & \lambda^{-1} \end{pmatrix}, \quad \lambda \neq 1,$$

and the eigenvectors of the matrix $\mathbf{P}$ in the system on the torus (D.17) correspond to two limit cycles, one of which is stable and the other is unstable. The rotation number ν turns out to be integer. It is equal to the number of half-turns made by the eigenvector of the matrix $\mathbf{P}$ within period T.

2) $|\mathrm{Tr}\mathbf{P}| = 2$, the matrix $\mathbf{P}$ is brought to the form

$$\begin{pmatrix} 1 & 1 \\ 0 & 1 \end{pmatrix},$$

on the torus there exists one unstable limit cycle, and the other trajectories asymptotically tend to one of its sides as $t \to +\infty$ and tend to the other side as $t \to -\infty$. The rotation number is also integer and is equal to the number of half-turns within period T of the only eigenvector of the matrix $\mathbf{P}$. This case corresponds to a bifurcation in which a pair of cycles merges to form one cycle, which then disappears as $|\mathrm{Tr}\mathbf{P}|$ decreases;

3) $|\mathrm{Tr}\mathbf{P}| < 2$, the monodromy matrix is transformed to the orthogonal form

$$\begin{pmatrix} \cos\vartheta & -\sin\vartheta \\ \sin\vartheta & \cos\vartheta \end{pmatrix}, \quad \vartheta \in [0, 2\pi),$$

consequently, the Poincaré map $\Theta(l)$ of the system (D.17) is conjugate to rotation through the constant angle ϑ; if ϑ is rational, all trajectories are periodic, while if ϑ is irrational, they are quasi-periodic. The rotation number is

$$\nu = N_T + \frac{\vartheta}{2\pi},$$

where N_T is the number of complete half-turns of the abscissa axis. In addition, due to a smooth dependence of the vector field of the linear system (D.20) on the parameters, both the angle ϑ and the rotation number ν smoothly depend on the parameters, and this is why there is no Cantor ladder in Fig. D.4.

Thus, the following theorem holds:

Theorem. *Let* $\mathbf{P}$ *be a monodromy matrix of the linear system* (D.20) *corresponding to the initial equation* (D.17). *Then if*

1) $|\mathrm{Tr}\mathbf{P}| > 2$, *there are two limit cycles on the torus, one of which is stable and the other is unstable;*
2) $|\mathrm{Tr}\mathbf{P}| = 2$, *there is one unstable limit cycle on the torus;*
3) $|\mathrm{Tr}\mathbf{P}| < 2$, *all trajectories on the torus are either periodic or quasi-periodic.*

Moreover, numerical analysis shows that when the coefficients in (D.17) are given by (D.18), the rotation number ν does not exceed unity and the only horizontal segment is $\nu = 0$. This suggests that no trajectory on the torus $\mathbb{T}^2$ makes a complete turn by angle l within period T.

Figure D.6 shows for different fixed values of a a partition of the region of possible values of the parameters (c, k) and a corresponding partition of the region of possible values of the integrals (p_ψ, h) on the bifurcation diagram into zones in which

- all trajectories of the system (D.17) are periodic or quasi-periodic (light grey);
- there are two limit cycles and all trajectories tend from an unstable cycle to a stable one (dark grey).

The boundary between the regions corresponds to the case where the system has one limit cycle and the boundary lines on the bifurcation diagram meet at the point $h = b$.

REMARK. To analyze the system (D.17), the authors of some papers use a relationship with the Riccati equation

$$\dot{z} = R_2(t)z^2 + R_1(t)z + R_0(t), \tag{D.22}$$

which is satisfied by the function $z = \tan\frac{l}{2}$. In this case, the coefficients $R_0(t) = \beta(t)$, $R_1(t) = 2\alpha(t)$ and $R_2(t) = \beta(t)$ are periodic functions of time. We note that the reduction of the Hess case to integration of the Riccati equation was first pointed out by P. A. Nekrasov [445].

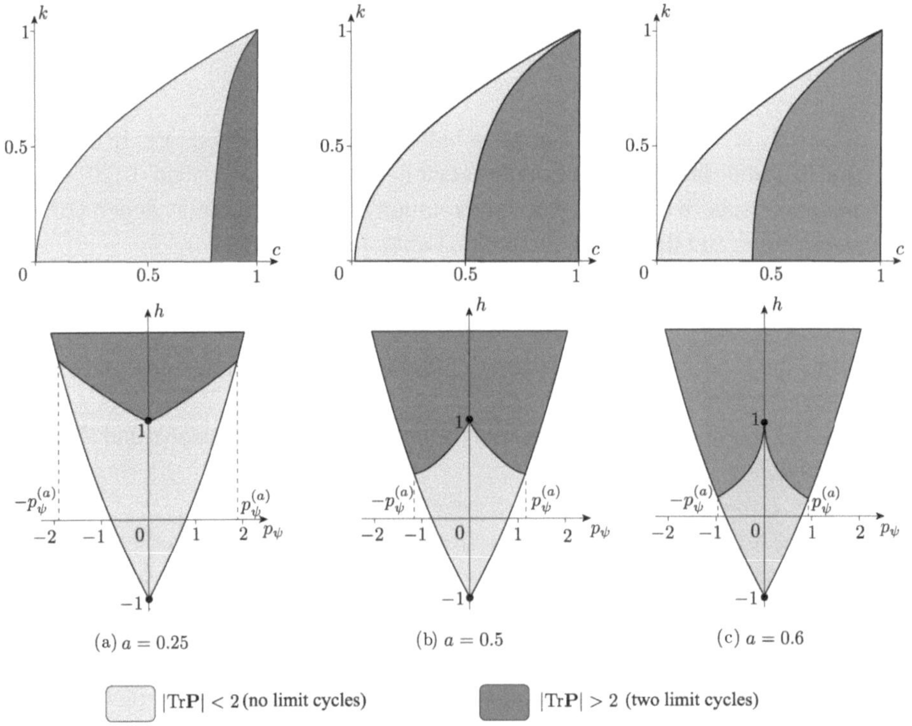

Fig. D.6. Regions with and without two limit cycles on the parameter plane (c, k) and on the bifurcation diagram for fixed values of a and $b = 1$.

5 Generalization of the Hess case

The Hess case on a pencil of brackets and quantization of the rotation number. We now consider a generalization of the Hamiltonian system (D.1) on a pencil of Poisson brackets that is defined by

$$\{M_i, M_j\} = -\varepsilon_{ijk}M_k, \quad \{M_i, \gamma_j\} = -\varepsilon_{ijk}\gamma_k, \quad \{\gamma_i, \gamma_j\} = -x\varepsilon_{ijk}M_k, \tag{D.23}$$

where x is a real-valued parameter. The Casimir functions of this bracket are

$$F_1 = (\boldsymbol{M}, \boldsymbol{\gamma}), \quad F_2 = \boldsymbol{\gamma}^2 + x\boldsymbol{M}^2.$$

Let us represent the equations of motion in the vector form

$$\dot{\boldsymbol{M}} = \boldsymbol{M} \times \frac{\partial H}{\partial \boldsymbol{M}} + \boldsymbol{\gamma} \times \frac{\partial H}{\partial \boldsymbol{\gamma}}, \quad \dot{\boldsymbol{\gamma}} = \boldsymbol{\gamma} \times \frac{\partial H}{\partial \boldsymbol{M}} + x\boldsymbol{M} \times \frac{\partial H}{\partial \boldsymbol{\gamma}}. \tag{D.24}$$

The general form of the Hamiltonian which is quadratic in $\boldsymbol{M}$, $\boldsymbol{\gamma}$ and for which equations (D.24) possess the invariant manifold $M_3 = 0$ looks like

$$H = \frac{1}{2}(\mathbf{A}\boldsymbol{M}, \boldsymbol{M}) + (\mathbf{B}\boldsymbol{M}, \boldsymbol{\gamma}) + \frac{1}{2}(\mathbf{C}\boldsymbol{\gamma}, \boldsymbol{\gamma}) + \mu\gamma_3 + \kappa M_3,$$

$$\mathbf{A} = \begin{pmatrix} a_{11} & 0 & a_{13} \\ 0 & a_{11} & 0 \\ a_{13} & 0 & a_{33} \end{pmatrix}, \quad \mathbf{B} = \operatorname{diag}(b_1, b_1, b_3), \quad \mathbf{C} = \operatorname{diag}(c_1, c_1, c_3).$$

- When $x = 0$, relations (D.23) define the algebra $e(3)$, and the system (D.24) reduces to the Kirchhoff equations describing the rigid body dynamics in an ideal fluid;
- When $x = 1$, relations (D.23) define the algebra $so(4)$, and the system (D.24) reduces to the Poincaré – Zhukovskii equations describing the dynamics of a rigid body with cavities filled with fluid;
- When $x = -1$, relations (D.23) define the algebra $so(3, 1)$, and the system (D.24) describes the motion of a rigid body in the Lobachevskii space.

In order to bring the system on the invariant manifold $M_3 = 0$ to a standard form similar to (D.17), we use a generalization of the Andoyer variables, as presented in [8].

Setting $M_3 = 0$ and fixing the level set of the Casimir functions:

$$F_1 = p_\psi, \quad F_2 = s,$$

we define the variables (G, g, l) as follows:

$$M_1 = G \sin l, \quad M_2 = G \cos l,$$

$$\gamma_1 = \frac{p_\psi}{G} \sin l - \delta_x \sin g \cos l, \quad \gamma_2 = \frac{p_\psi}{G} \cos l + \delta_x \sin g \sin l, \quad \gamma_3 = -\delta_x \cos g,$$

$$\delta_x = \sqrt{s - xG^2 - \frac{p_\psi^2}{G^2}}, \tag{D.25}$$

where $(l, g) \mod 2\pi$. The equations of motion for the variables (G, g) decouple and have the form

$$\dot{G} = \frac{\partial H}{\partial g}, \quad \dot{g} = -\frac{\partial H}{\partial G},$$
$$H = \frac{1}{2}(a_{11} - xc_1)G^2 + \frac{1}{2}(c_3 - c_1)\,\delta_x^2 \cos^2 g + \mu\delta_x \cos g. \tag{D.26}$$

Using the known solutions $G(t)$ and $g(t)$ of the system (D.26) for l, we can represent the equation in the form (D.17):

$$\dot{l} = \tilde{\alpha}(t) + \tilde{\beta}(t)\sin l,$$
$$\tilde{\alpha}(t) = \delta_x\left(b_1 - b_3 - (c_3 - c_1)\frac{p_\psi}{G^2(t)}\right)\cos g(t) + \frac{\mu s}{G^2(t)} + \kappa, \quad \tilde{\beta}(t) = a_{13}G(t). \tag{D.27}$$

Thus, the problem reduces to investigating the vector field on the torus $\mathbb{T}^2$, on which, as the previous section implies, there can be no more than two limit cycles and the graph of the rotation number cannot be a Cantor ladder, but contains horizontal segments corresponding only to integer rotation numbers.

As noted above, in the Hess case, for the Euler – Poisson equations the rotation number does not exceed unity. We show that in the general case the rotation number can be arbitrarily large and contain arbitrarily many horizontal segments $\nu = n \in \mathbb{Z}$. For this purpose we consider equations (D.24) with $x = 0$ and the Hamiltonian

$$H = \frac{1}{2}(M_1^2 + M_2^2 + a_{33}M_3^2) + (\kappa + b\gamma_3 + aM_1)M_3.$$

In this case, equation (D.27) can be represented as

$$\dot{l} = aG\sin l - \kappa_0\cos(Gt) + \kappa,$$
$$G = \text{const}, \quad \kappa_0 = b\sqrt{s - \frac{p_\psi^2}{G^2}}. \tag{D.28}$$

A typical dependence of the rotation number ν on κ for equation (D.28) is shown in Fig. D.7, in which horizontal segments corresponding to limit cycles are clearly seen.

Remark. An equation analogous to (D.28) simulates the Josephson effect in superconductor physics and arises in the dynamics of coupled oscillators. In this connection, it is considered in [222, 111, 265, 183], where it was also shown that, in equation (D.17), limit cycles (Arnold's languages) occur only for integer values of the rotation number.

Cubic perturbation and a Cantor ladder. To conclude, we consider a Hamiltonian system that also possesses the invariant manifold $M_3 = 0$, but on this manifold the dependence of the rotation number ν on one of the system parameters is a Cantor ladder. Let us set $x = 0$ and add into the Hamiltonian a term cubic in $\boldsymbol{\gamma}$ and $\boldsymbol{M}$:

$$\dot{\boldsymbol{M}} = \boldsymbol{M} \times \frac{\partial H}{\partial \boldsymbol{M}} + \boldsymbol{\gamma} \times \frac{\partial H}{\partial \boldsymbol{\gamma}}, \quad \dot{\boldsymbol{\gamma}} = \boldsymbol{\gamma} \times \frac{\partial H}{\partial \boldsymbol{M}},$$
$$H = \frac{1}{2}(\boldsymbol{M}, \boldsymbol{M}) + (\kappa + aM_1 + \varepsilon\gamma_1\gamma_2)M_3. \tag{D.29}$$

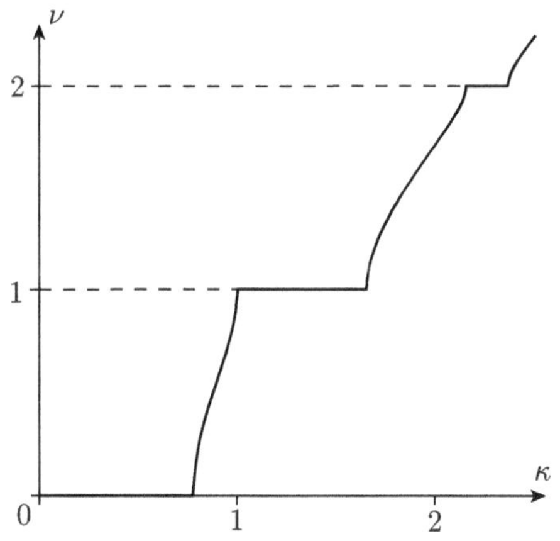

Fig. D.7. Graph showing the dependence of the rotation number ν on κ for fixed $a = 1$, $\kappa_0 = 1$, $G = 1$.

Using the variables (D.25) and setting $s = 1$, we obtain the following equations of motion on the invariant manifold $M_3 = 0$:

$$\begin{gathered} g = -Gt, \quad G = \text{const.}, \\ \dot{l} = \kappa + aG\sin l + \frac{\varepsilon}{2}\lambda_1(t)\sin(2l) + \varepsilon\lambda_2(t)\cos(2l), \\ \lambda_1(t) = 2\frac{p_\psi^2}{G^2} - 1 + \left(1 - \frac{p_\psi^2}{G^2}\right)\cos^2(Gt), \quad \lambda_2(t) = p_\psi\sin(Gt)\sqrt{1 - \frac{p_\psi^2}{G^2}}. \end{gathered} \tag{D.30}$$

In this case, the equation is seen to contain several harmonics in the variable l. A typical dependence of the rotation number ν on κ for equation (D.30) is shown in Fig. D.8.

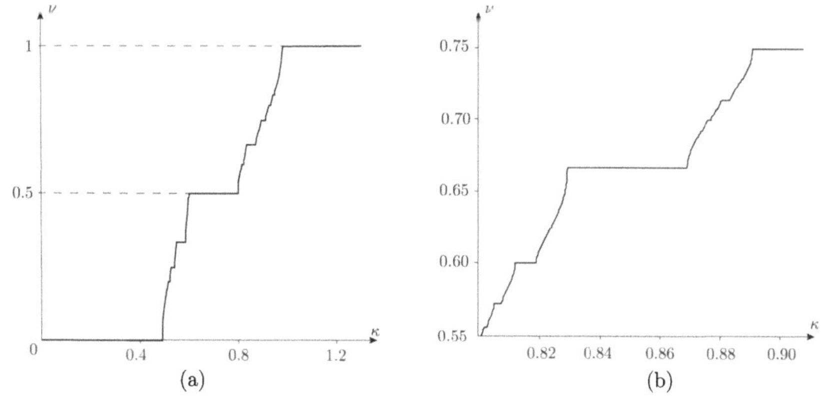

Fig. D.8. Graph showing the dependence of the rotation number ν on κ for fixed $G = 1$, $p_\psi = 0.5$, $\varepsilon = 0.5$, $a = 0.5$. Figure (b) is an enlarged fragment of graph (a).

E Ferromagnetic Dynamics in a Magnetic Field

A neutral ferromagnet becomes magnetized along the axis of rotation: this is the quantum mechanical *Barnett effect*. The magnetic moment $\boldsymbol{B}$ is related to the angular velocity $\boldsymbol{\omega}$ by $\boldsymbol{B} = \boldsymbol{\Lambda}_1 \boldsymbol{\omega}$, where $\boldsymbol{\Lambda}_1$ is a certain symmetric linear operator. An analogous phenomenon takes place for the rotation of a superconducting rigid body (the London effect). If the body rotates about a fixed point in a homogeneous magnetic field with strength $\boldsymbol{H}$, then it is acted on by magnetic forces of strength $\boldsymbol{B} \times \boldsymbol{H}$. Denoting $\boldsymbol{\gamma} = \boldsymbol{H}$, we can write the equations of motion in the following form [521]:

$$\begin{gathered}\dot{\boldsymbol{M}} = \boldsymbol{M} \times \mathbf{A}\boldsymbol{M} + \boldsymbol{\Lambda}\boldsymbol{M} \times \boldsymbol{\gamma}, \quad \dot{\boldsymbol{\gamma}} = \boldsymbol{\gamma} \times \mathbf{A}\boldsymbol{M}, \\ \boldsymbol{M} = \mathbf{I}\boldsymbol{\omega}, \quad \boldsymbol{\Lambda} = \boldsymbol{\Lambda}_1 \mathbf{A}, \quad \mathbf{A} = \mathbf{I}^{-1} = \operatorname{diag}(a_1, a_2, a_3).\end{gathered} \tag{E.1}$$

In the general case, Eqs. (E.1) are probably not Hamiltonian, and since they have the standard invariant measure and two trivial integrals

$$F_1 = (\boldsymbol{M}, \boldsymbol{\gamma}) = c_1 \text{ and } \quad F_2 = \boldsymbol{\gamma}^2 = 1, \tag{E.2}$$

we need two more independent integrals for integrability (by the Euler – Jacobi theorem).

As was shown in [341], Eqs. (E.1) are Hamiltonian in the following two cases: (i) $\boldsymbol{\Lambda}$ proportional to $\mathbf{A}$ and (ii) $\boldsymbol{\Lambda} = \operatorname{diag}(\lambda_1^2, \lambda_2^2, \lambda_3^2)$, with $\mathbf{A}$ the identity matrix. For both of these special cases Eqs. (E.1) reduce to the Kirchhoff equations on the algebra $e(3)$ (Sec. 3.1 of Ch. 3), while for case (ii) Eqs. (E.1) are integrable. Let us consider these cases in turn.

1. V. A. Samsonov [521]. Let $\boldsymbol{\Lambda} = \lambda\mathbf{A}$; then Eqs. (E.1) admit the additional integral

$$F_3 = (\mathbf{A}\boldsymbol{M}, \boldsymbol{M}). \tag{E.3}$$

It was shown by V. V. Kozlov in [341] that in this case Eqs. (E.1) reduce to the Kirchhoff equations on $e(3)$ (see Sec. 3.1 of Ch. 3) with Hamiltonian

$$H = \frac{1}{2}(\widetilde{\mathbf{A}}\widetilde{\boldsymbol{M}}, \widetilde{\boldsymbol{M}}) + (\widetilde{\mathbf{B}}\widetilde{\boldsymbol{M}}, \widetilde{\boldsymbol{\gamma}}) + \frac{1}{2}(\widetilde{\mathbf{C}}\widetilde{\boldsymbol{\gamma}}, \widetilde{\boldsymbol{\gamma}}),$$

where $\widetilde{\mathbf{A}} = \mathbf{A}$, $\widetilde{\mathbf{B}} = -\lambda\mathbf{A}$, and $\widetilde{\mathbf{C}} = \lambda^2\mathbf{A}$. The variables $\widetilde{\boldsymbol{M}}$ and $\widetilde{\boldsymbol{\gamma}}$ are obtained from $\boldsymbol{M}$ and $\boldsymbol{\gamma}$ by using the simple linear transformation:

$$\widetilde{\boldsymbol{M}} = \boldsymbol{M} + \lambda\boldsymbol{\gamma}, \quad \widetilde{\boldsymbol{\gamma}} = \boldsymbol{\gamma}. \tag{E.4}$$

With the help of the axial symmetry $a_1 = a_2$, Eqs. (E.1) may be completely integrated: a further additional integral was presented in [521]. It has the form

$$F_4 = \widetilde{M}_3 = M_3 + \lambda\gamma_3 = \text{const},$$

and the corresponding integrability case is the integrable Kirchhoff case (Sec. 3.1 of Ch. 3).

DOI 10.1515/9783110544442-011

2. V. V. Kozlov [341]. This nontrivial integrable case of (E.1) with two quadratic integrals was presented in [341]. It requires the conditions $a_1 = a_2 = a_3 = 1$ and $\boldsymbol{\Lambda} = \mathrm{diag}(\lambda_1^2, \lambda_2^2, \lambda_3^2)$. The additional integrals are

$$F_3 = (\boldsymbol{\Lambda M}, \boldsymbol{M}), \quad F_4 = \frac{1}{2}\boldsymbol{M}^2 - (\boldsymbol{\Lambda M}, \boldsymbol{\gamma}) + \frac{1}{2}\det\boldsymbol{\Lambda}(\boldsymbol{\Lambda}^{-1}\boldsymbol{\gamma}, \boldsymbol{\gamma}). \tag{E.5}$$

By an interesting linear change of variables suggested by L. E. Veselova [602], this integrable case turns into the Clebsch case. The change of variables has the form

$$\begin{aligned} L_i &= \frac{1}{2}\sqrt{\lambda_j^{-1} + \lambda_k^{-1}}\left(\lambda_i M_i + \det\boldsymbol{\Lambda}^{1/2}\gamma_i\right), \\ p_i &= \frac{1}{2}\frac{1}{\sqrt{\lambda_j^{-1} + \lambda_k^{-1}}}\left(\lambda_i M_i - \det\boldsymbol{\Lambda}^{1/2}\gamma_i\right), \\ &\{ijk\} \text{ is cyclic permutation of } \{123\} \end{aligned} \tag{E.6}$$

In these variables, the equations of motion take the form

$$\begin{gathered} \frac{d\boldsymbol{L}}{d\tau} = \boldsymbol{L} \times \mathbf{J}^{-1}\boldsymbol{L} + \boldsymbol{p} \times \mathbf{J}\boldsymbol{p}, \quad \frac{d\boldsymbol{p}}{d\tau} = \boldsymbol{p} \times \mathbf{J}^{-1}\boldsymbol{L}, \\ \mathbf{J} = \mathrm{diag}(\lambda_2^{-1} + \lambda_3^{-1}, \lambda_3^{-1} + \lambda_1^{-1}, \lambda_1^{-1} + \lambda_2^{-1}), \quad dt = \det\mathbf{J}^{1/2}d\tau. \end{gathered} \tag{E.7}$$

Thus, indeed, they define the Clebsch case of the Kirchhoff equations on $e(3)$ (see Sec. 3.1 of Ch. 3) with Hamiltonian and additional integrals

$$\begin{gathered} H = \frac{1}{2}(\boldsymbol{L}, \mathbf{J}^{-1}\boldsymbol{L}) + \frac{1}{2}(\boldsymbol{p}, \mathbf{J}\boldsymbol{p}), \\ G_1 = (\boldsymbol{L}, \boldsymbol{p}), \quad G_2 = \boldsymbol{p}^2, \quad G_3 = \frac{1}{2}\boldsymbol{L}^2 - \frac{1}{2}\det\mathbf{J}(\boldsymbol{p}, \mathbf{J}^{-1}\boldsymbol{p}). \end{gathered} \tag{E.8}$$

The connection between the integrals in (E.8), (E.2), and (E.5) is given by the relations

$$\begin{gathered} H = \frac{1}{4}(F_3 + \det\boldsymbol{\Lambda}F_2), \quad G_1 = \frac{1}{4}(F_3 - \det\boldsymbol{\Lambda}F_2), \\ G_2 = \frac{1}{4\det\mathbf{J}}\left(-\mathrm{Tr}\,\mathbf{J}\det\boldsymbol{\Lambda}^{1/2}F_1 + \det\boldsymbol{\Lambda}^{1/2}\,\mathrm{Tr}\,\boldsymbol{\Lambda}^{1/2}F_2 + \frac{\mathrm{Tr}\,\boldsymbol{\Lambda}^{1/2}}{\det\boldsymbol{\Lambda}^{1/2}}F_3 + 2F_4\right), \\ G_3 = \frac{1}{4\det\mathbf{J}}(\mu_1 F_1 + \mu_2 F_2 + \mu_3 F_3 + \mu_4 F_4), \\ \mu_1 = \det\boldsymbol{\Lambda}^{1/2}\,\mathrm{Tr}\,\boldsymbol{\Lambda}^{-3/2} + 2\,\mathrm{Tr}\,\boldsymbol{\Lambda}^{1/2}\,\mathrm{Tr}\,\boldsymbol{\Lambda}^{-1/2} - 1, \\ \mu_2 = \frac{1}{2}\left(\mathrm{Tr}\,\boldsymbol{\Lambda} + \det\boldsymbol{\Lambda}^{1/2}\,\mathrm{Tr}\,\boldsymbol{\Lambda}^{1/2}\,\mathrm{Tr}\,\boldsymbol{\Lambda}^{-1}\right), \\ \mu_3 = \frac{1}{2}\left(\frac{\mathrm{Tr}\,\boldsymbol{\Lambda}}{\det\boldsymbol{\Lambda}} + \frac{\mathrm{Tr}\,\boldsymbol{\Lambda}^{1/2}\,\mathrm{Tr}\,\boldsymbol{\Lambda}^{-1}}{\det\boldsymbol{\Lambda}^{1/2}}\right), \quad \mu_4 = -\mathrm{Tr}\,\boldsymbol{\Lambda}^{-1} - 3\frac{\mathrm{Tr}\,\boldsymbol{\Lambda}^{1/2}}{\det\boldsymbol{\Lambda}^{1/2}}. \end{gathered}$$

The zero level set $G_1 = (\boldsymbol{L}, \boldsymbol{p}) = 0$ of the area integral of system (E.7) corresponds (as is standard for the Clebsch system) to the Neumann problem of point motion on the sphere in a quadratic potential (Sec. 3.1 of Ch. 3) for which reduction to a natural two-degree-of-freedom Hamiltonian system with separated variables is possible (Sec. 1.7 of Ch. 1). In the initial variables of system (E.1), the condition $G_1 = 0$ is represented by the condition

$$F_3 = (\boldsymbol{\Lambda M}, \boldsymbol{M}) = \det \boldsymbol{\Lambda}.$$

REMARK. Transformation (E.6) is analogous to that used by S. A. Chaplygin in the nonholonomic problem of the rolling of a dynamically asymmetric ball [130] in order to reduce the system to the zero level set of the area integral for which the separation of variables is possible.

3. Let us present one more case in which there exists at least one integral. Indeed, under the condition $\boldsymbol{\Lambda} = \lambda \mathbf{E}$ one obtains the integral $F_3 = \boldsymbol{M}^2$. Under the additional condition $a_1 = a_2 = a$ there exists one more trivial linear integral $F_4 = aM_3 + \gamma_3$ of the Lagrange type integral.

F The Landau – Lifshitz Equation, Discrete Systems, and the Neumann Problem

1 Landau – Lifshitz Equation

In the theory of ferromagnetism, the Landau – Lifshitz equation

$$\frac{\partial \boldsymbol{S}}{\partial t} = \boldsymbol{S} \times \left(\frac{\partial^2 \boldsymbol{S}}{\partial x^2} + \mathbf{J}\boldsymbol{S}\right), \qquad \boldsymbol{S}^2 = 1, \tag{F.1}$$

which describes the evolution of the magnetization vector $\boldsymbol{S}(x, t) \in \mathbb{R}^3$ in space and time plays a fundamental role; here, $\mathbf{J} = \operatorname{diag}(J_1, J_2, J_3)$ is a diagonal matrix that characterizes the interaction anisotropy.

Traveling ("cnoidal") wave type solutions of (F.1) , i.e., $\boldsymbol{S}(x, t) = \boldsymbol{q}(x - at)$ (a = const is the wave speed), satisfy the equation

$$-a\dot{\boldsymbol{q}} = \boldsymbol{q} \times (\ddot{\boldsymbol{q}} + \mathbf{J}\boldsymbol{q}), \qquad \boldsymbol{q}^2 = 1, \tag{F.2}$$

which after taking the vector product of both sides with $\boldsymbol{q}$ takes the form

$$\ddot{\boldsymbol{q}} + a\dot{\boldsymbol{q}} \times \boldsymbol{q} = -\mathbf{J}\boldsymbol{q} + \lambda\boldsymbol{q}, \quad \boldsymbol{q}^2 = 1, \quad \lambda = (\boldsymbol{q}, \mathbf{J}\boldsymbol{q}) - \dot{\boldsymbol{q}}^2. \tag{F.3}$$

Let us introduce the total angular momentum of the system

$$\boldsymbol{M} = \dot{\boldsymbol{q}} \times \boldsymbol{q} - a\boldsymbol{q}, \tag{F.4}$$

which is analogous to that used by Wilson for the Dirac monopole quantization.

Remark. The angular momentum (F.4) was presented for the first time by Poincaré as a vector integral of the motion of a charged particle in a magnetic monopole field (see also [95]).

It is easy to verify that the commutator of $\boldsymbol{M}$ with $\boldsymbol{q}$ is defined by the algebra $e(3)$:

$$\{M_i, M_j\} = \varepsilon_{ijk}M_k, \quad \{M_i, q_j\} = \varepsilon_{ijk}q_k, \quad \{q_i, q_j\} = 0. \tag{F.5}$$

In this case, $(\boldsymbol{M}, \boldsymbol{q}) = -a$, whereas the Hamiltonian of the system has the form

$$H = \frac{1}{2}\boldsymbol{M}^2 + \frac{1}{2}(\boldsymbol{q}, \mathbf{J}\boldsymbol{q}). \tag{F.6}$$

The system (F.5) – (F.6) is the Clebsch integrable case of the Kirchhoff equations (see Sec. 3.1 of Ch. 3); for $(\boldsymbol{M}, \boldsymbol{q}) = -a = 0$ (i.e., for the stationary solution of standing wave type), it is isomorphic to the Neumann system. This analogy was found by A. P. Veselov in [600].

DOI 10.1515/9783110544442-012

REMARK. Along with the Landau – Lifshitz equations, there exists another system related to the asymmetric chiral O_3-field whose equations have the form $\frac{\partial \boldsymbol{u}}{\partial \xi} = \boldsymbol{u} \times \mathbf{K}\boldsymbol{u}$, $\frac{\partial \boldsymbol{v}}{\partial \eta} = \boldsymbol{v} \times \mathbf{K}\boldsymbol{v}$ ($\boldsymbol{u}, \boldsymbol{v} \in \mathbb{R}^3$ and $\mathbf{K}$ is a diagonal matrix); its self-similar solutions reduce to the integrable Schottky – Manakov system (see Sec. 3.2 of Ch. 3). Indeed, as is shown in [600], the solutions depending only on $t = \xi + \eta$ satisfy the system $\dot{\boldsymbol{u}} = \boldsymbol{u} \times \mathbf{K}\boldsymbol{v}$, $\dot{\boldsymbol{v}} = \boldsymbol{v} \times \mathbf{K}\boldsymbol{u}$, which is a particular case of the equations of a free top on $SO(4)$.

2 Anisotropic XYZ Heisenberg Model

Let us consider the Hamiltonian system with Hamiltonian function

$$H = -\sum_{n=1}^{N} \sum_{\alpha\beta} J_{\alpha\beta} S_n^\alpha S_{n+1}^\beta \tag{F.7}$$

on the direct sum of $so(3)$-algebras with $S_1, \ldots, S_N$, $S_i \in so(3)$, where the the $\boldsymbol{S}_n = (S_n^1, S_n^2, S_n^3)$ are all classical spin vectors (with commutation relations analogous to those for angular momentum $\boldsymbol{M}$) normalized by the condition $\boldsymbol{S}_n^2 = 1$. The components of $J_{\alpha\beta}$ define the interaction tensor, which is assumed to be diagonal and triaxial: $\mathbf{J} = \operatorname{diag}(J_1, J_2, J_3)$, $J_3 > J_2 > J_1 > 0$. The system (F.7) is called the anisotropic XYZ Heisenberg model (see Sec. 3.2 of Ch. 3).

Both classical and quantum versions of the system (F.7) have been considered. The problem of finding stationary solutions of the model was considered in [240, 596], and in particular, in the quantum case in order to calculate wave functions for the quantum Hamiltonian. Earlier studies of the latter problem go back to Bethe, who considered the anisotropic XXX model, and to Baxter, who (in principle) constructed all eigenfunctions for the completely anisotropic XYZ chain of quantum spins 1/2. For completely anisotropic chains with arbitrary spins, only a few isolated results have been obtained [240].

Let us consider stationary solutions of the system (F.7), which is in essence a system of interacting tops. These solutions satisfy the equation

$$\boldsymbol{S}_n \times \mathbf{J}(\boldsymbol{S}_n + \boldsymbol{S}_{n+1}) = 0, \; n = 1, \ldots, N \tag{F.8}$$

or

$$\boldsymbol{S}_{n-1} + \boldsymbol{S}_{n+1} = \lambda_n \mathbf{J}^{-1} \boldsymbol{S}_n, \tag{F.9}$$

where the factor λ_n is found from the condition $|\boldsymbol{S}_{n+1}| = 1$. Thus

$$1 = |\boldsymbol{S}_{n+1}|^2 = |-\boldsymbol{S}_{n-1} + \lambda_n \mathbf{J}^{-1}\boldsymbol{S}_n|^2 = \lambda_n^2 |\mathbf{J}^{-1}\boldsymbol{S}_n|^2 - 2\lambda_n(\boldsymbol{S}_n, \mathbf{J}^{-1}\boldsymbol{S}_n) + 1,$$

giving rise to the following two possibilities: either $\lambda_n = 0$ or

$$\lambda_n = \frac{2(\boldsymbol{S}_{n-1}, \mathbf{J}^{-1}\boldsymbol{S}_n)}{|\mathbf{J}^{-1}\boldsymbol{S}_n|^2}. \tag{F.10}$$

The first possibility $\lambda_n = 0$ is usually rejected for physical reasons, and the system (F.9) – (F.10) defines a discrete (two-dimensional) mapping whose study is of independent theoretical interest.

It is easy to verify that the system (F.10) has first integrals (i.e., functions invariant under the mapping (F.9) – (F.10)):

$$F_1 = (\boldsymbol{S}_n, \mathbf{J}^{-1}\boldsymbol{S}_{n+1}), \quad F_2 = |\mathbf{J}\boldsymbol{S}_n|^2 + |\mathbf{J}\boldsymbol{S}_{n+1}|^2 - (\boldsymbol{S}_n, \mathbf{J}\boldsymbol{S}_{n+1})^2. \tag{F.11}$$

If we introduce a change of variables analogous to the one in (F.4) (with $a = 0$): $\boldsymbol{M} = \boldsymbol{S}_n \times \mathbf{J}\boldsymbol{S}_{n+1}$, $\boldsymbol{\gamma} = \boldsymbol{S}_n$, then we have

$$\begin{aligned} \det \mathbf{J}^2 F_1^2 &= (\mathbf{J}^2\boldsymbol{M}, \boldsymbol{M}) - \det \mathbf{J}^2 (\mathbf{J}^{-2}\boldsymbol{\gamma}, \boldsymbol{\gamma}), \\ F_2 &= \boldsymbol{M}^2 + (\mathbf{J}^2\boldsymbol{\gamma}, \boldsymbol{\gamma}). \end{aligned} \tag{F.12}$$

In this case, as is directly apparent from the special form of the change of variables, we also have $F_3 = (\boldsymbol{M}, \boldsymbol{\gamma}) = 0$.

The integrals (F.12) coincide with the integrals of the Neumann system (see Sec. 3.2 of Ch. 3) written in the algebra $e(3)$ defined by the variables $\boldsymbol{M}$ and $\boldsymbol{\gamma}$ on the zero level set of the area integral $(\boldsymbol{M}, \boldsymbol{\gamma}) = 0$. They were presented by Y. I. Granovskii and A. S. Zhedanov [240]. Therefore we may consider the system (F.9) – (F.10) as being *a discrete analog of the Neumann system*, and the existence of the integrals (F.12) enables us to conclude that the system is integrable.

This name is also justified by the fact that Eq. (F.8), which is almost equivalent to (F.9) – (F.10), becomes the stationary Landau – Lifshitz equation

$$\boldsymbol{S} \times \left(\frac{d^2\boldsymbol{S}}{dx^2} + \mathbf{J}\boldsymbol{S}\right) = 0, \qquad \boldsymbol{S}^2 = 1, \tag{F.13}$$

in the continuous limit (see (F.1)); this equation can be also represented in the form

$$\frac{d^2\boldsymbol{S}}{dx^2} + \mathbf{J}\boldsymbol{S} - \lambda(x)\boldsymbol{S}, \qquad \boldsymbol{S}^2 = 1, \tag{F.14}$$

which, as was shown in the first subsection, is equivalent to the standard (continuous) Neumann problem.

We may note that the discrete Landau – Lifshitz equation is of physical interest when the continual approximation is not applicable, i,e., when the system changes considerably over distances of the same order as the lattice spacing.

Higher-dimensional generalizations. A. P. Veselov suggested a higher-dimensional analog of the mapping (F.9) – (F.10) in which the vectors $\boldsymbol{S}$ belong to the k-dimensional sphere $\boldsymbol{S}_n \in S^k \subset \mathbb{R}^{k+1}$ and $\mathbf{J} = \operatorname{diag}(J_0, \ldots, J_k)$. In this case, the integrals analogous to those in (F.12) can be written in the following convenient symmetric form if we use Sec. 3.1 of Ch. 3 (K. Uhlenbeck integrals [589]):

$$F_\alpha(x, y) = x_\alpha^2 + \sum_{\beta \neq \alpha} \frac{(\boldsymbol{x} \wedge \mathbf{J}\boldsymbol{y})_{\alpha\beta}^2}{J_\alpha^2 - J_\beta^2}, \qquad \sum_{\alpha=0}^{n} F_\alpha = 1, \tag{F.15}$$

where $\boldsymbol{x} = \boldsymbol{S}_n$, $\boldsymbol{y} = \boldsymbol{S}_{n+1}$, $(\boldsymbol{x}\wedge\boldsymbol{y})_{\alpha\beta} = x_\alpha y_\beta - x_\beta y_\alpha$, $\alpha = 0, 1, \ldots, n$. These integrals were also presented by A. P. Veselov in [596]. In [597], the discrete analog of the Liouville theorem was presented; for the integrals (F.15), it allows us to define the concept of involutivity and take the meaning of a group shift on Liouville tori to the mapping (F.9) – (F.10). Moreover, in [597], general formulas for the solutions were presented in terms of theta functions.

3 Ellipsoidal billiards and discrete tops

As was shown by G. Birkhoff [45], the Jacobi problem of finding the geodesics on an ellipsoid becomes a system of integrable billiards (elliptic billiards) as the length of one of the semiaxes of the ellipsoid tends to zero. In this case, a point moves inside the ellipse along a line, and a reflection is performed according to the ideal law: the angle of incidence is equal to the reflection angle, with no change in the magnitude of the velocity. In the n-dimensional case, explicit formulas for the analogs of the mapping (F.9) – (F.10) are as follows [596]:

$$\begin{gathered}\frac{\boldsymbol{q}_{k+1}-\boldsymbol{q}_k}{|\boldsymbol{q}_{k+1}-\boldsymbol{q}_k|} - \frac{\boldsymbol{q}_k-\boldsymbol{q}_{k-1}}{|\boldsymbol{q}_k-\boldsymbol{q}_{k-1}|} = \lambda_k \mathbf{A}\boldsymbol{q}_k, \\ \lambda_k = \frac{2(\mathbf{A}\boldsymbol{q}_k, \boldsymbol{q}_k-\boldsymbol{q}_{k-1})}{|\boldsymbol{q}_k-\boldsymbol{q}_{k-1}||\mathbf{A}\boldsymbol{q}_k|^2},\end{gathered} \tag{F.16}$$

and, moreover, $(\mathbf{A}\boldsymbol{q}_k, \boldsymbol{q}_k) = 1$, $\boldsymbol{q}_k \in \mathbb{R}^{n+1}$, $\mathbf{A} = \operatorname{diag}(a_1, \ldots, a_n)$, is the equation of the n-dimensional ellipsoid. The mapping (F.16) has a complete set of integrals in involution (in the sense of [597])

$$F_\alpha = p_\alpha^2 + \sum_{\beta\neq\alpha} \frac{(\boldsymbol{p}\wedge\boldsymbol{q})_{\alpha\beta}^2}{b_\beta - b_\alpha}, \qquad b_\alpha = a_\alpha^{-1}, \qquad \boldsymbol{p}_k = \frac{\boldsymbol{q}_{k+1}-\boldsymbol{q}_k}{|\boldsymbol{q}_{k+1}-\boldsymbol{q}_k|}. \tag{F.17}$$

In [597], explicit formulas were also obtained for the impact points expressed in theta functions.

The papers [597, 433] consider an n-dimensional discrete analog of the Schottky – Manakov equations (of a free top) on $SO(N)$:

$$\begin{gathered}\boldsymbol{M}_{k+1} = \boldsymbol{\omega}_k \boldsymbol{M}_k \boldsymbol{\omega}_k^{-1}, \quad \boldsymbol{M}_k = \boldsymbol{\omega}_k^{-1}\mathbf{I} - \mathbf{I}\boldsymbol{\omega}_k, \quad \boldsymbol{\omega}_k \in so(N), \\ \mathbf{I} = \operatorname{diag}(I_1, \ldots, I_N).\end{gathered} \tag{F.18}$$

For a four-dimensional top, the corresponding family of first integrals can be explicitly written by using the results of Sec. 3.2 of Ch. 3. The paper [433] also discusses the problems of obtaining general solutions in theta functions.

Comments. There has been quite a lot of published research recently on discretizations of the classical cases of rigid body dynamics (see, e.g., [189], where the necessary references can be found). Arguments from mechanics analogous to the Birkhoff passage to the limit in the problem of geodesics on an ellipsoid allow us to pass from the $(n + 1)$-dimensional top (on $SO(n + 1)$) to the n-dimensional one ($SO(n)$). However, examples that have a physical meaning and are described by such discrete systems are unknown so far.

Moreover, the real discretizations arising when numerical methods are applied do not fit in with this scheme, and, for example, the standard Chirikov mapping, used as a model example for studying chaotic dynamics, is obtained by discretizing the integrable problem of the mathematical pendulum. Possibly the study of such discretizations will be is useful to enhance numerical methods for integrating nonlinear systems.

G Dynamics of Tops and Material Points on Spheres and Ellipsoids

1 Point motion on the sphere and ellipsoid ($n = 2, 3$). An analogy with top dynamics

The equations of motion of a rigid body around a fixed point for the spherical tensor of inertia $\mathbf{A} = \lambda\mathbf{E}$, $\lambda = \text{const.}$, $\mathbf{E} = \|\delta_{ij}\|$, in a potential field are isomorphic to the equations of motion of a material point on the surface of the three-dimensional sphere S^3 in an analogous field. This analogy was found in [58, 340] (see also [95]). Moreover, in the axially symmetric field $V = V(\boldsymbol{\gamma})$, the spherical top dynamics on the zero level set of the area integral $(\boldsymbol{M}, \boldsymbol{\gamma}) = 0$ is equivalent to the motion of a material point on the two-dimensional sphere S^2. It turns out that this analogy also holds in the higher-dimensional case if we use singular orbits of $e(n)$; this is studied in detail in [95].

In what follows we consider only the real rigid body and two- and three-dimensional spheres, but we also present an even more general analogy by considering the motion of a point on ellipsoids E^2 and E^3 instead of on spheres S^2 and S^3. There exists an orbital isomorphism between the dynamics of a heavy rigid body (which is now dynamically completely asymmetric) and the dynamics of a point on an ellipsoid in a potential field. A particular example of this isomorphism is the analogy between the integrable Jacobi geodesics problem and the Clebsch case of the Kirchhoff equations presented in Sec. 3.1 of Ch. 3. As is shown in [95, 99], a similar analogy holds between the Clebsch case on $e(4)$ and the Jacobi problem on the three-dimensional ellipsoid E^3. We first consider the two-dimensional case.

Two-dimensional ellipsoid and sphere (E^2, S^2). Let a point move in Euclidean space with coordinates $\boldsymbol{q} = (q_1, q_2, q_3)$ on the surface of the ellipsoid E^2 given by the equation $(\boldsymbol{q}, \mathbf{B}\boldsymbol{q}) = 1$, $\mathbf{B} = \operatorname{diag}(b_1, b_2, b_3)$. The equations of motion with Lagrange multipliers can be represented as

$$\ddot{\boldsymbol{q}} = \lambda\mathbf{B}\boldsymbol{q} - \frac{\partial V}{\partial \boldsymbol{q}}, \quad \text{where } \lambda = \frac{\left(\mathbf{B}\boldsymbol{q}, \frac{\partial V}{\partial \boldsymbol{q}}\right) - (\dot{\boldsymbol{q}}, \mathbf{B}\dot{\boldsymbol{q}})}{(\mathbf{B}\boldsymbol{q}, \mathbf{B}\boldsymbol{q})} \tag{G.1}$$

(the point mass is assumed to be one). In the redundant canonical variables $(\boldsymbol{p}, \boldsymbol{q}) \in \mathbb{R}^6$, we have the Hamiltonian [19, 95]

$$H = \frac{1}{2}(\boldsymbol{p} \times \boldsymbol{n})^2 + V(\boldsymbol{q}), \quad \boldsymbol{n} = \frac{\mathbf{B}\boldsymbol{q}}{\sqrt{(\mathbf{B}\boldsymbol{q}, \mathbf{B}\boldsymbol{q})}}; \tag{G.2}$$

here, $\boldsymbol{n}$ is the normal to the surface. For the system (G.2), let us make the canonical transformation $\boldsymbol{p}, \boldsymbol{q} \to \boldsymbol{p}', \boldsymbol{q}'$ corresponding to a contraction along the principal axes of the ellipsoid, $\boldsymbol{q}' = \mathbf{B}^{1/2}\boldsymbol{q}$, $\boldsymbol{p}' = \mathbf{B}^{-1/2}\boldsymbol{p}$, and let us pass to new variables by the formulas

$$\boldsymbol{\gamma} = \boldsymbol{q}', \quad \boldsymbol{M} = \boldsymbol{p}' \times \boldsymbol{q}'. \tag{G.3}$$

DOI 10.1515/9783110544442-013

As a result, we obtain the Hamiltonian system defined by the bracket of $e(3)$, the zero level set of the Casimir function, $(\boldsymbol{M}, \boldsymbol{\gamma}) = 0$, and Hamiltonian

$$H = \frac{1}{2} \det \mathbf{B} \frac{(\boldsymbol{M}, \mathbf{A}\boldsymbol{M})}{(\boldsymbol{\gamma}, \mathbf{B}\boldsymbol{\gamma})} + V(\boldsymbol{\gamma}), \quad \mathbf{A} = \mathbf{B}^{-1}. \tag{G.4}$$

After rescaling time as $d\tau = \frac{\det \mathbf{B}}{(\boldsymbol{\gamma}, \mathbf{B}\boldsymbol{\gamma})} dt$, on the energy level $H = E = \frac{1}{2} c \det \mathbf{B}$, we obtain the Hamiltonian system on $e(3)$ with the Hamiltonian

$$H' = \frac{1}{2}(\boldsymbol{M}, \mathbf{A}\boldsymbol{M}) + \frac{1}{2}(\boldsymbol{\gamma}, \mathbf{B}\boldsymbol{\gamma})\left(\frac{2V(\boldsymbol{\gamma})}{\det \mathbf{B}} - c\right) \tag{G.5}$$

and zero energy level $H' = 0$.

REMARK. A more general assertion holds, namely, a Hamiltonian system on $e(3)$ of the form

$$H = \frac{F(\boldsymbol{M}, \boldsymbol{\gamma})}{G(\boldsymbol{\gamma})} + V(\boldsymbol{\gamma})$$

on the energy level $H = h$ is orbitally equivalent (after rescaling time as $dt = G(\boldsymbol{\gamma})\, d\tau$) to the system on $e(3)$ with the Hamiltonian

$$H' = F(\boldsymbol{M}, \boldsymbol{\gamma}) + (V(\boldsymbol{\gamma}) - h)G(\boldsymbol{\gamma})$$

on the zero level $H' = 0$.

In Sec. 3.1 of Ch. 3, we showed that under such an orbital isomorphism (resulting from rescaling time), the Jacobi problem is isomorphic to the Clebsch case on the zero level set of the area integral. For $\mathbf{B} = \mathbf{E}$, we have motion on the sphere S^2, and the Hamiltonian (G.4) can be written in the form

$$H = \frac{1}{2}\boldsymbol{M}^2 + V(\boldsymbol{\gamma}), \tag{G.6}$$

which describes the spherical top motion for $(\boldsymbol{M}, \boldsymbol{\gamma}) = 0$. This complete isomorphism (without a rescaling of the time variable along the trajectory) may be understood in various ways, in particular, we can perform direct calculations using Euler angles.

Three-dimensional ellipsoid and sphere (E^3, S^3). Point motion on a three-dimensional ellipsoid E^3 is not directly related to ordinary three-dimensional body dynamics in some potential field. Such an analogy is possible if we consider the motion of a four-dimensional top, i.e., a top in $e(4)$ (in the n-dimensional case, the interconnection is between point motion in E^n and motion of a top in $e(n)$ [99, 460]). We shall not consider here this analogy, since the four-dimensional top has no physical meaning (see [95]).

The isomorphism between point motion on S^3 in a certain force field and spherical top dynamics in a related field is more natural. In this case, the field is not necessarily axially symmetric, and the system has three degrees of freedom.

Material point motion on the three-dimensional sphere given by the equation $\sum_{i=0}^{3} q_i^2 = 1$ looks as follows in the Lagrangian form:

$$\ddot{q}_i = -\frac{\partial V}{\partial q_i} + \lambda q_i, \qquad \lambda = \sum_{i=0}^{3}\left(q_i \frac{\partial V}{\partial q_i} - \dot{q}_i^2\right), \quad i = 0, \dots, 3, \tag{G.7}$$

where V is the potential.

In the redundant canonical variables (p, q) on $\mathbb{R}^4 \times \mathbb{R}^4$, the Hamiltonian has the form

$$H = \frac{1}{2}\left(p^2 - \frac{(p, q)^2}{(q, q)}\right) + V(q), \tag{G.8}$$

where $(\cdot, \cdot)$ denotes the standard inner product on $\mathbb{R}^4$.

Let us define the new variables M_i and λ_μ by the formulas

$$\begin{gathered} M_i = \frac{1}{2}(\pi_i - L_i), \quad \lambda_\mu = q_\mu, \\ L_i = \frac{1}{2}\varepsilon_{ijk}\widetilde{M}_{jk}, \quad \pi_i = \widetilde{M}_{0i}, \quad \boldsymbol{\pi} = (\pi_1, \pi_2, \pi_3), \quad \boldsymbol{L} = (L_1, L_2, L_3), \\ \widetilde{M}_{\mu\nu} = q_\mu p_\nu - q_\nu p_\mu, \quad i, j, k = 1, 2, 3, \quad \mu, \nu = 0, \dots, 3. \end{gathered} \tag{G.9}$$

It is easy to see that the commutation relations between the variables $\boldsymbol{M}$ and λ are analogous to those of the quaternion variables in rigid body dynamics (see (1.65) in Sec. 1.4 of Ch. 1), while the Hamiltonian function (G.8) has the form

$$H = \frac{1}{2}\boldsymbol{M}^2 + V(\lambda) \tag{G.10}$$

and so describes spherical top motion in an arbitrary potential field.

The following relations hold for the vectors $\boldsymbol{\pi}$, $\boldsymbol{L}$, and $\boldsymbol{M}$:

$$\begin{aligned} \boldsymbol{\pi} &= 2(q_0 \boldsymbol{q} \times \boldsymbol{M} + \boldsymbol{q}(\boldsymbol{M}, \boldsymbol{q}) + q_0^2 \boldsymbol{M}), \\ \boldsymbol{L} &= 2(q_0 \boldsymbol{q} \times \boldsymbol{M} + \boldsymbol{q}(\boldsymbol{M}, \boldsymbol{q}) - \boldsymbol{q}^2 \boldsymbol{M}), \\ \boldsymbol{M} &= \frac{1}{2}(\boldsymbol{\pi} - \boldsymbol{L}); \end{aligned} \tag{G.11}$$

where $\boldsymbol{q} = (q_1, q_2, q_3)$. Geometrically, the relations (G.11) define the parameterization of a singular orbit in the coalgebra $e(4)$ composed of the variables $(\boldsymbol{\pi}, \boldsymbol{L}, q_0, \boldsymbol{q})$ and $\boldsymbol{q} = (q_1, q_2, q_3)$. This orbit is homeomorphic to the tangent bundle TS^3 of the three-dimensional sphere TS^3 (see Appendix B).

Remark. The relation $\boldsymbol{M}^2 = \boldsymbol{N}^2$ holds for the components of $\boldsymbol{N} = \frac{1}{2}(\boldsymbol{\pi} + \boldsymbol{L})$, $\boldsymbol{N} = (N_1, N_2, N_3)$, and $\boldsymbol{N}$ and λ also satisfy quaternionic commutation relations on the singular orbit. For the spherical top, they correspond to the projections of the angular momentum on the fixed axes. The simplest recalculation formulas for $\boldsymbol{M}$, $\boldsymbol{N}$, $\boldsymbol{\pi}$, and $\boldsymbol{L}$ are as follows:

$$\boldsymbol{M} = \frac{1}{2}(\boldsymbol{\pi} - \boldsymbol{L}), \quad \boldsymbol{N} = \frac{1}{2}(\boldsymbol{\pi} + \boldsymbol{L}), \quad \boldsymbol{\pi} = \boldsymbol{M} + \boldsymbol{N}, \quad \boldsymbol{L} = \boldsymbol{N} - \boldsymbol{M}. \tag{G.12}$$

The above-presented analogies with point motion allow us to extend to rigid body dynamics a number of integrable problems (and various methods in general) of celestial mechanics in constant curvature spaces (in particular, in S^2 and S^3). In [95], the authors discuss the general formalism of the equations of motion for "curved" celestial mechanics and also present a number of its qualitative distinctive features as compared with the usual one. However, the majority of integrable problems of flat celestial mechanics may be extended to the curved situation. They are considered in the next section in detail.

Let us consider two integrable systems on S^2 and S^3, one of which is connected with a number of variations on the Jacobi and Neumann problems and the other is a newer one, having an additional third-degree integral. To discuss the first problem, we first consider an analog of the elastic oscillator potential on the sphere.

2 Harmonic oscillator on S^2 and S^3. A generalization of the Neumann and Jacobi problems

Hooke centers on sphere. In flat space, the Bertrand theorem holds; according to this theorem, there exist only two central force laws for which all bounded trajectories are closed. One of them is Newton's law of gravitation, and the other is Hooke's law for elastic force; all closed trajectories are always ellipses. A similar assertion holds on the three-dimensional sphere with the Newtonian potential replaced by $V = \mu \cot \theta$, $\mu = \text{const.}$ [76] and the Hooke potential by $V = \mu \tan^2 \theta$, $\mu = \text{const.}$

In redundant coordinates, the Hooke potential has the form $V = \frac{c}{q_0^2}$, i.e., it has the form of one of those singular additions that are ubiquitous in rigid body dynamics (see Sec. 5.1 of Ch. 5).

In the flat scenario, there exists a redundant collection of quadratic integrals which make up the so-called "*Fradkin tensor*" (in physicists' terminology) [241]. We can find the analogous quantities for the sphere. First let us present explicit expressions for the two-dimensional sphere in a more general situation having Hooke centers of arbitrary intensities situated at three mutually perpendicular poles, and then we present the corresponding formulas for the n-dimensional sphere.

1. *Two-dimensional sphere S^2*. Hamilton's equations on the algebra $e(3)$ with Hamiltonian

$$H = \frac{1}{2}\boldsymbol{M}^2 + \frac{1}{2}\sum \frac{c_i}{\gamma_i^2} = \frac{1}{2}\boldsymbol{M}^2 + \frac{1}{2}\sum c_i \frac{\gamma_j^2 + \gamma_k^2}{\gamma_i^2} + \text{const.} \tag{G.13}$$

have the following three integrals on the level set $(\boldsymbol{M}, \boldsymbol{\gamma}) = 0$:

$$\begin{gathered} F_1 = M_1^2 + c_2\frac{\gamma_3^2}{\gamma_2^2} + c_3\frac{\gamma_2^2}{\gamma_3^2}, \quad F_2 = M_2^2 + c_2\frac{\gamma_3^2}{\gamma_1^2} + c_3\frac{\gamma_1^2}{\gamma_3^2}, \\ F_3 = M_3^2 + c_2\frac{\gamma_2^2}{\gamma_1^2} + c_3\frac{\gamma_1^2}{\gamma_2^2}. \end{gathered} \tag{G.14}$$

The Poisson bracket of these integrals on the level set $(\boldsymbol{M}, \boldsymbol{\gamma}) = 0$ leads to one more cubic integral

$$\{F_i, F_j\}\big|_{(\boldsymbol{M},\boldsymbol{\gamma})=0} = (-1)^{(ijk)} 4F_0,$$
$$F_0 = M_1 M_2 M_3 - \gamma_1\gamma_2\gamma_3 \left(c_1 \frac{M_1}{\gamma_1^3} + c_2 \frac{M_2}{\gamma_2^3} + c_3 \frac{M_3}{\gamma_3^3} \right). \tag{G.15}$$

It is interesting to notice the similarity of the integral F_0 to the additional integral in the Gaffet system: in its homogeneous form, it admits a generalization to the whole bracket pencil $\mathcal{L}_x$ (see below).

2. *n-dimensional sphere* S^n. In this case, it is more convenient to use the Lagrangian equations on the sphere:

$$S^n = \{(\boldsymbol{q}, \boldsymbol{q}) = 1\}, \quad \boldsymbol{q} = (q_0, q_1, \ldots, q_n) \in \mathbb{R}^{n+1}$$

with Lagrange multiplier. If we represent the potential in the form $V(\boldsymbol{q}) = \frac{1}{2}\sum_{i=0}^{n} \frac{c_i}{q_i^2}$, then the equations of motion have the form

$$\ddot{q}_i = \frac{c_i}{q_i^3} + \lambda q_i, \quad i = 0, 1, \ldots, n, \quad \lambda = -\dot{\boldsymbol{q}}^2 - 2V(\boldsymbol{q}). \tag{G.16}$$

In the $2n$-dimensional space, there is a family of $n(n+1)$ quadratic integrals (which is a direct generalization of the family (G.14))

$$F_{ij} = (\dot{q}_i q_j - \dot{q}_j q_i)^2 + q_i^2 \frac{c_j}{q_j^2} + q_j^2 \frac{c_i}{q_i^2}, \quad i, j = 0, 1, \ldots, n, \quad i < j. \tag{G.17}$$

They were presented in this form by V. V. Kozlov and Y. N. Fedorov in [354]. Amongst the F_{ij}, $(2n-1)$ are independent, which leads to the closedness of all trajectories. The system is said to be superintegrable.

Generalization of the Neumann problem on S^2. The integrability of Eqs. (G.16) was established much earlier by E. Rosochatius in 1877 [495]. This apparently passed unnoticed by the authors of [354]. In fact, Rosochatius considered the even more general potential

$$V = \frac{1}{2} k(\mathbf{A}\boldsymbol{q}, \boldsymbol{q}) + \frac{1}{2} \sum_{i=0}^{n} \frac{c_i}{q_i^2}, \tag{G.18}$$

where $\mathbf{A} = \operatorname{diag}(a_0, a_1, \ldots, a_n)$. The system (G.18) is a combination of an n-dimensional Neumann system (which is the same as an anisotropic oscillator) and a collection of Hooke potential sources, each placed at points of intersection of the sphere with $n+1$ mutually orthogonal lines.

Rosochatius proved the integrability of the system (G.19) by separation in spherical coordinates. (This is analogous to the n-dimensional Neumann system; see Sec. 1.7 of Ch. 1). A Lax representation and a complete set of involutive integrals were presented by J. Moser in [432].

As illustrative examples we present the integrals for the three- and two-dimensional spheres; the second of these can be considered as a natural generalization of both the usual Neumann system and the Clebsch cases of the Kirchhoff equations (Sec. 3.1 of Ch. 3). In this case, it is necessary to assume that $(\boldsymbol{M}, \boldsymbol{\gamma}) = 0$, i.e., that the additional integral F is not general, but partial (i.e., $\{H, F\}|_{(\boldsymbol{M},\boldsymbol{\gamma})=0} = 0$). On the algebra $e(3)$, the Hamiltonian and the partial integral F can be written as

$$
\begin{gathered}
H = \frac{1}{2}\boldsymbol{M}^2 + \frac{1}{2}k(\mathbf{A}\boldsymbol{\gamma}, \boldsymbol{\gamma}) + \frac{1}{2}\sum_{i=1}^{3}\frac{c_i}{\gamma_i^2}, \\
F = \frac{1}{2}(\boldsymbol{M}, \mathbf{A}\boldsymbol{M}) - \frac{1}{2}k \det \mathbf{A}(\mathbf{A}\boldsymbol{\gamma}, \boldsymbol{\gamma}) + \frac{1}{2}\sum_{i,j,k} a_i\left(\frac{c_j\gamma_k^2}{\gamma_j^2} + \frac{c_k\gamma_j^2}{\gamma_k^2}\right), \\
\mathbf{A} = \operatorname{diag}(a_1, a_2, a_3), \quad k, c_i = \text{const.}
\end{gathered} \tag{G.19}
$$

In a more symmetric form, this family can be represented using the following commuting functions:

$$
\begin{aligned}
L_1 &= k\gamma_1^2 + \frac{1}{a_1 - a_2}\left(M_3^2 + c_1\frac{\gamma_2^2}{\gamma_1^2} + c_2\frac{\gamma_1^2}{\gamma_2^2}\right) + \frac{1}{a_1 - a_3}\left(M_2^2 + c_1\frac{\gamma_3^2}{\gamma_1^2} + c_3\frac{\gamma_1^2}{\gamma_3^2}\right), \\
L_2 &= k\gamma_2^2 + \frac{1}{a_2 - a_1}\left(M_3^2 + c_2\frac{\gamma_1^2}{\gamma_2^2} + c_1\frac{\gamma_2^2}{\gamma_1^2}\right) + \frac{1}{a_2 - a_3}\left(M_1^2 + c_2\frac{\gamma_3^2}{\gamma_2^2} + c_3\frac{\gamma_2^2}{\gamma_3^2}\right), \\
L_3 &= k\gamma_3^2 + \frac{1}{a_3 - a_1}\left(M_2^2 + c_3\frac{\gamma_1^2}{\gamma_3^2} + c_1\frac{\gamma_3^2}{\gamma_1^2}\right) + \frac{1}{a_3 - a_2}\left(M_1^2 + c_3\frac{\gamma_2^2}{\gamma_3^2} + c_2\frac{\gamma_3^2}{\gamma_2^2}\right), \\
&a_i, c_i = \text{const.}
\end{aligned} \tag{G.20}
$$

Note that due to the addition of the Neumann potential ($k \neq 0$) in (G.18) the system is no longer degenerate and the trajectories are no longer closed. Moreover, we can say that a dynamically symmetric (but not spherical) top of Goryachev – Chaplygin or Kovalevskaya type admits the addition of only one integrability-preserving singular term $\frac{c}{\gamma_3^2}$ corresponding to the distinguished symmetry axis for $(\boldsymbol{M}, \boldsymbol{\gamma}) = 0$, while the (spherical) Clebsch case (Neumann problem) admits the addition of three (mutually orthogonal) integrability-preserving singular terms.

Generalization of the Jacobi problem on E^2. The Jacobi problem of geodesics on an ellipsoid may also be modified by the addition of singular terms to the Hamiltonian. Let us present the form of the Hamiltonian and the additional integral for the two-dimensional case, i.e., for the Hamiltonian system (G.4) on $e(3)$ and on the level set $(\boldsymbol{M}, \boldsymbol{\gamma}) = 0$:

$$
\begin{gathered}
H = \frac{1}{2}\frac{(\mathbf{A}\boldsymbol{M}, \boldsymbol{M})}{(\mathbf{A}\boldsymbol{\gamma}, \boldsymbol{\gamma})} + \frac{k}{2}(\mathbf{A}\boldsymbol{\gamma}, \boldsymbol{\gamma}) + \frac{1}{2}\sum_{i=0}^{3}\frac{c_i}{a_i\gamma_i^2}, \\
F = \frac{(\mathbf{A}\boldsymbol{M} \times \boldsymbol{\gamma})^2}{(\mathbf{A}\boldsymbol{\gamma}, \boldsymbol{\gamma})} + (\mathbf{A}\boldsymbol{\gamma}, \boldsymbol{\gamma})\left(k - \sum_{i=0}^{3}\frac{c_i}{a_i^2\gamma_i^2}\right).
\end{gathered} \tag{G.21}
$$

The integral F generalizes the Ioachimsthal integral; it was presented in [343] in terms of the redundant variables on the sphere. On the ellipsoid in $\mathbb{R}^3$ with the equation

$(\boldsymbol{q}, \mathbf{A}^{-1}\boldsymbol{q}) = 1$, the potential (G.21) corresponds to

$$U(\boldsymbol{q}) = \frac{1}{2}k(\boldsymbol{q}, \boldsymbol{q}) + \frac{1}{2}\sum_{i=0}^{3}\frac{c_i}{q_i^2}, \tag{G.22}$$

i.e., to the combination of the standard elastic potential energy (in Euclidean space) and some singular terms. The case $c_i = 0$ and $k \neq 0$ was already known to Jacobi. On the level set $(\boldsymbol{M}, \boldsymbol{\gamma}) = 0$, the system (G.21) is orbitally isomorphic to a spherical top in a potential field quartic in $\boldsymbol{\gamma}$ and with addition of a singular term. The separation of variables on S^2 (and on S^n in general) motivates the problem of finding potentials for which the system is separable. In the case of polynomial potentials, this problem was solved in [58, 616]. As a rule, the corresponding systems describing the spherical top motion are not physically realizable. For even more sophisticated potentials, see [620].

Generalization of the Neumann system to S^3. We first consider the Neumann system with singular terms on S^3 and also the corresponding spherical top. In the quaternion variables $\boldsymbol{M}, \lambda_0, \lambda_1, \lambda_2$, and λ_3 (G.9), the Hamiltonian has the form

$$H = \frac{1}{2}\boldsymbol{M}^2 + \frac{1}{2}k(\mathbf{A}\boldsymbol{\lambda}, \boldsymbol{\lambda}) + \frac{1}{2}\sum_{i=0}^{3}\frac{c_i}{\lambda_i^2}, \qquad \mathbf{A} = \operatorname{diag}(a_0, a_1, a_2, a_3). \tag{G.23}$$

Let us represent the additional integrals in the following symmetric form using a redundant involutive family analogous to (G.20):

$$\begin{aligned}
G_0 &= k\lambda_0^2 + \frac{\pi_1^2 + c_0\lambda_0^{-2}\lambda_1^2 + c_1\lambda_1^{-2}\lambda_0^2}{a_0 - a_1} \\
&\quad + \frac{\pi_2^2 + c_0\lambda_0^{-2}\lambda_2^2 + c_2\lambda_2^{-2}\lambda_0^2}{a_0 - a_2} + \frac{\pi_3^2 + c_0\lambda_0^{-2}\lambda_3^2 + c_3\lambda_3^{-2}\lambda_0^2}{a_0 - a_3}, \\
G_1 &= k\lambda_1^2 + \frac{\pi_1^2 + c_1\lambda_1^{-2}\lambda_0^2 + c_0\lambda_0^{-2}\lambda_1^2}{a_1 - a_0} \\
&\quad + \frac{L_3^2 + c_1\lambda_1^{-2}\lambda_2^2 + c_2\lambda_2^{-2}\lambda_1^2}{a_1 - a_2} + \frac{L_2^2 + c_1\lambda_1^{-2}\lambda_3^2 + c_3\lambda_3^{-2}\lambda_1^2}{a_1 - a_3}, \\
G_2 &= k\lambda_2^2 + \frac{\pi_2^2 + c_2\lambda_2^{-2}\lambda_0^2 + c_0\lambda_0^{-2}\lambda_2^2}{a_2 - a_0} \\
&\quad + \frac{L_3^2 + c_2\lambda_2^{-2}\lambda_1^2 + c_1\lambda_1^{-2}\lambda_2^2}{a_2 - a_1} + \frac{L_1^2 + c_2\lambda_2^{-2}\lambda_3^2 + c_3\lambda_3^{-2}\lambda_2^2}{a_2 - a_3}, \\
G_3 &= k\lambda_3^2 + \frac{\pi_3^2 + c_3\lambda_3^{-2}\lambda_0^2 + c_0\lambda_0^{-2}\lambda_3^2}{a_3 - a_0} \\
&\quad + \frac{L_2^2 + c_3\lambda_3^{-2}\lambda_1^2 + c_1\lambda_1^{-2}\lambda_3^2}{a_3 - a_1} + \frac{L_1^2 + c_3\lambda_3^{-2}\lambda_2^2 + c_2\lambda_2^{-2}\lambda_3^2}{a_3 - a_2},
\end{aligned} \tag{G.24}$$

where π_i and L_i are defined by formulas (G.9). The integrals G_μ satisfy two relations

$$\sum_{\mu=0}^{3} G_\mu = k\lambda^2, \quad \sum_{\mu=0}^{3} a_\mu G_\mu = H,$$

where λ^2 is the quaternion norm and H is the Hamiltonian (G.23).

Rational separable potentials. There is another generalization of the Neumann problem on the sphere to the case of polynomial and rational potentials admitting separation of variables in spherical coordinates. As shown previously, such systems induce particular integrable cases for rigid body dynamics.

On the n-dimensional sphere (in this case, the dimension plays no role) embedded in $\mathbb{R}^{n+1}$ in the standard way $\left(\sum_{\mu=0}^{n} x_\mu = 1\right)$, let us consider systems having a Lagrangian function of "natural form":

$$L = \frac{1}{2}\sum_{\mu=0}^{n} \dot{x}_\mu^2 - V(\mathbf{x}). \tag{G.25}$$

We can show that for this Lagrangian the separation of variables in spherical-conical coordinates (see Sec. 1.7 of Ch. 1) is possible for any potential having the form

$$\Phi_z(\mathbf{x}) = \left(\prod_{\mu=0}^{n}(z - a_\mu)\sum_{\mu=0}^{n}\frac{x_\mu^2}{z - a_\mu}\right)^{-1}. \tag{G.26}$$

The function $\Phi_z(\mathbf{x})$ is a generating function for integrable potentials [514] in the following sense: the coefficients of the series expansion of $\Phi_z(\mathbf{x})$ in powers of $1/z$ define rational integrable potentials presented in [616, 58]; the expansion of (G.26) in powers of z leads to polynomial integrable potentials.

We give here some simple examples of integrable polynomial potentials (among these there is the potential of the Neumann problem) [616, 58]:

$$V_2 = \sum_{\mu=0}^{n} a_\mu x_\mu^2,$$

$$V_4 = \left(\sum_{\mu=0}^{n} a_\mu x_\mu^2\right)^2 - \sum_{\mu=0}^{n} a_\mu^2 x_\mu^2,$$

$$V_6 = \left(\sum_{\mu=0}^{n} a_\mu x_\mu^2\right)^3 - 2\left(\sum_{\mu=0}^{n} a_\mu x_\mu^2\right)\left(\sum_{\mu=0}^{n} a_\mu^2 x_\mu^2\right) + \sum_{\mu=0}^{n} a_\mu^3 x_\mu^2.$$

The simplest rational potentials are as follows:

$$V_{-2} = \left(\sum_{\mu=0}^{n}\frac{x_\mu^2}{a_\mu}\right)^{-1},$$

$$V_{-4} = \left(\sum_{\mu=0}^{n}\frac{x_\mu^2}{a_\mu}\right)^{-2}\left(\left(\sum_{\mu=0}^{n}\frac{1}{a_\mu}\right)\left(\sum_{\mu=0}^{n}\frac{x_\mu^2}{a_\mu}\right) - \sum_{\mu=0}^{n}\frac{x_\mu^2}{a_\mu^2}\right).$$

The potential V_{-2} was presented by Braden (1982) [105] and, independently, in [622].

3 The problem of *n* Hooke centers on a sphere

The most recent known integrable variant of the Hooke potentials $\frac{c_i}{(\boldsymbol{\gamma},\boldsymbol{r}_i)^2}$ differs from (G.18) in that the Hooke centers of attraction $\boldsymbol{r}_i$, $i = 1, \dots, n$, are placed not on the mutually orthogonal axes, but are arbitrarily situated on the same equator. This leads to a system that, for simplicity, is now described for the case of the two-dimensional sphere (i.e., the corresponding system on $e(3)$).

The Hamiltonian and the additional integral (for $(\boldsymbol{M}, \boldsymbol{\gamma}) = 0$) have the form

$$\begin{aligned} H &= \frac{1}{2}\boldsymbol{M}^2 + \frac{1}{2}\sum_{i=1}^{n} \frac{c_i}{(\boldsymbol{r}_i, \boldsymbol{\gamma})^2} + U(\gamma_3), \\ F &= M_3^2 + (1-\gamma_3^2)\sum_{i=1}^{n} \frac{c_i}{(\boldsymbol{r}_i, \boldsymbol{\gamma})^2}. \end{aligned} \tag{G.27}$$

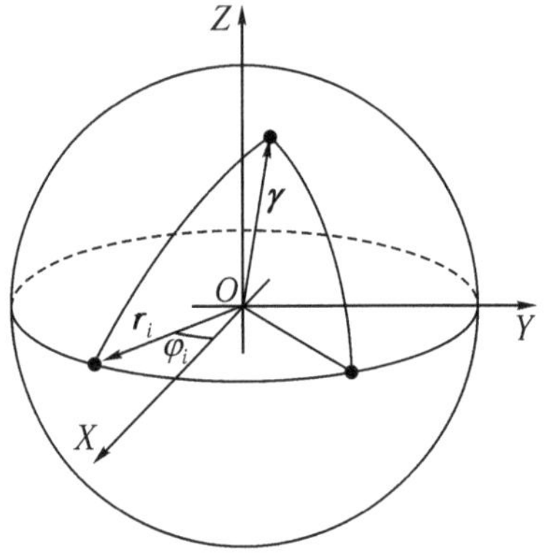

Fig. G.1.

In the expression (G.27) there is an arbitrary function $U(\gamma_3)$, which means the addition of an arbitrary "central" field whose center is situated on the axis perpendicular to the plane of Hooke potentials (see Fig. G.1). In particular, we can place one more Hooke center at the pole. In this case, upon reduction in the variable $\psi \pm \varphi$ (see Sec. 5.1 of Ch. 5 and Sec. 4.1 of Ch. 4), it immediately follows that the spatial problem, i.e., the problem of point motion on the three-dimensional sphere S^3 under the action of n Hooke centers situated on the equator, is also integrable.

Note that the Euclidean analog of the problem considered is trivial: the separation is already possible in the Cartesian coordinates (n linear connected oscillators are obtained). In this case, the location of the Hooke centers is arbitrary. In the curvilinear situation, even on the two-dimensional sphere the problem of motion in the field of three arbitrarily situated Hooke centers is not integrable. Although this is not fully proved, it is easy to perform experiments which demonstrate chaotic behavior. The quadratic integral F in (G.27) is related to the separability of the problem in spherical coordinates (θ, φ). Indeed, we can write the Hamiltonian H as follows:

$$
\begin{aligned}
H &= \frac{1}{2}\left(p_\theta^2 + \frac{p_\varphi^2}{\sin^2\theta}\right) + \frac{1}{2}\sum_{i=1}^{n} \frac{c_i}{\sin^2\theta\cos^2(\varphi-\varphi_i)} + U(\theta) \\
&= \frac{1}{2}p_\theta^2 + \frac{1}{2\sin^2\theta}\left(p_\varphi^2 + \sum_{i=1}^{n}\frac{c_i}{\cos^2(\varphi-\varphi_i)}\right) + U(\theta),
\end{aligned}
$$

where θ and φ are the coordinates of a moving material point and φ_i defines the position of the ith Hooke center on the equator (Fig. G.1). The expression in the parentheses is the additional integral of motion (G.27). We also note that if the Hooke centers are not situated on a great circle of the sphere, then the problem is no longer integrable.

4 Gaffet system

In conclusion, we consider a somewhat exotic system on $S^2\{\boldsymbol{y}^2=1\}$ with the potential $U = \varepsilon(y_1y_2y_3)^{-2/3}$, ε = const, which was found and studied by B. Gaffet [199, 200]. This is a Hamiltonian system on $e(3)$ and it is partially integrable. The Hamiltonian H and the partial integral F (for $(\boldsymbol{M}, \boldsymbol{y}) = 0$) have the form

$$
\begin{aligned}
H &= \frac{1}{2}(M_1^2 + M_2^2 + M_3^2) - \frac{1}{2}\frac{a^2(y_1^2+y_2^2+y_3^2)}{(y_1y_2y_3)^{2/3}}, \quad a = \text{const}, \\
F &= M_1M_2M_3 + a^2\left(\frac{M_1}{y_1} + \frac{M_2}{y_2} + \frac{M_3}{y_3}\right)(y_1y_2y_3)^{1/3}.
\end{aligned} \tag{G.28}
$$

Despite a number of partial results [200, 580], no explicit integration of the system (G.28) has been carried out. The sum $y_1^2 + y_2^2 + y_3^2$, which is a Casimir and can be assumed without loss of generality to be equal to 1, has been left in the formula for H for the sake of homogeneity.

Remark. The **L**-**A** pair for system (G.28) was presented in [580]. It has the form

$$
\begin{gathered}
\frac{d}{dt}\mathbf{L} = [\mathbf{L}, \mathbf{A}], \\
\mathbf{L} = \begin{pmatrix} \lambda & M_3 + ay_3 & M_2 - ay_2 \\ M_3 - ay_3 & \lambda & M_1 + ay_1 \\ M_2 + ay_2 & M_1 - ay_1 & \lambda \end{pmatrix}, \\
\mathbf{A} = \frac{2a}{3}(\boldsymbol{y}, \boldsymbol{y}) \begin{pmatrix} 0 & y_3^{-1} & y_2^{-1} \\ y_3^{-1} & 0 & y_1^{-1} \\ y_2^{-1} & y_1^{-1} & 0 \end{pmatrix},
\end{gathered}
$$

where

$$
\boldsymbol{y} = \frac{\boldsymbol{\gamma}}{(\gamma_1\gamma_2\gamma_3)^{1/3}}.
$$

We also note that the integrability of the system and the form of integrals (G.28) is preserved on the whole bracket pencil (analogously to the Goryachev – Chaplygin case):

$$\{M_i, M_j\} = \varepsilon_{ijk} M_k, \quad \{M_i, \gamma_j\} = \varepsilon_{ijk} \gamma_k, \quad \{\gamma_i, \gamma_j\} = x \varepsilon_{ijk} M_k \tag{G.29}$$

and also on the level set $(\boldsymbol{M}, \boldsymbol{\gamma}) = 0$.

This is a special case of a family of partially integrable systems on the two-dimensional sphere presented by Gaffet in [204]. In general the integral has degree six in momenta.

H On the Motion of a Heavy Rigid Body in an Ideal Fluid with Circulation

1 Introduction and a survey of known results

The general problem concerning the forces and moments acting on an arbitrary rigid body executing plane-parallel motion in an infinite volume of an ideal incompressible fluid was considered by S. A. Chaplygin in [125]. Particular versions of the problem had previously been studied by N. E. Zhukovskii [648, 645], who considered the application of his formula for lifting force in order to describe the motion of a heavy body in a fluid. Zhukovskii's work led to the nonphysical conclusion that rotational and translational motions are independent, and thus the problem required a more complete analysis. It was precisely this analysis that was carried out by S. A. Chaplygin.

The general assumptions of Chaplygin were that the fluid executes vortex-free motion and is at rest at infinity, while the circulation of the fluid around the body is constant. The general conclusions, which can be obtained from the study of these equations, have a remarkable mechanical interpretation. They are of theoretical, as well as practical, interest even from the viewpoint of modern applied hydro- and aerodynamics.

We now return to the main problem of this appendix, that of plane-parallel motion of a rigid body in a fluid in the presence of circulation. The equations of motion for this problem were derived by Chaplygin in [125]. Addition of circulation makes the motion considerably more complicated. Under certain particular assumptions (as compared with [125]), it has been studied in [346]. In [346], the steady motions were found and their stability was studied, the integrable case was presented, and certain classes of particular solutions analogous to the Zhukovskii solutions [648, 645] were given.

In this appendix, we present the Hamiltonian form of the general equations, new integrable cases, and classes of new particular solutions. We also show that in the general case [125] are not integrable and the system exhibits chaotic behavior.

2 General equations of motion. Lagrangian and Hamiltonian descriptions

We define a moving Cartesian coordinate system $O_1x_1x_2$ attached to the body and characterize its position with respect to the fixed coordinate system by the coordinates (x, y) of its center O_1 and the angle of rotation φ (Fig. H.1). Let (ξ, η) denote the Cartesian coordinates of the center of mass C in the moving axes $O_1x_1x_2$, I be the constant circulation around the body, ρ be the fluid density, and v_1 and v_2 be the projections of the linear velocity of the center of mass O_1 on the same axes. Also, we denote by Q_1 and Q_2 the projections of the external forces on the moving axes, and let ω be the

DOI 10.1515/9783110544442-014

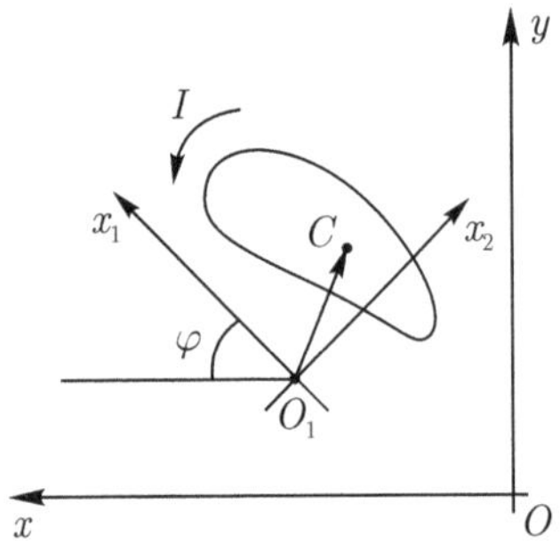

Fig. H.1.

angular velocity of the body and M_1 be the moment of the external forces with respect to the center of mass.

By an explicit calculation of the forces and moments related to the circular flow around the body, S. A. Chaplygin [125] obtained the following equations, which are of a similar form to the Kirchhoff equations [365]:

$$\begin{aligned} a_1\dot{v}_1 &= a_2\omega v_2 - \lambda v_2 - \zeta\omega - Q_1, \quad a_2\dot{v}_2 = -a_1\omega v_1 + \lambda v_1 + \chi\omega - Q_2, \\ b\dot{\omega} &= (a_1 - a_2)v_1 v_2 + \zeta v_1 - \chi v_2 - M_1 - \xi Q_2 + \eta Q_1, \\ \dot{\varphi} &= \omega, \quad \dot{x} = (\boldsymbol{\alpha}, \boldsymbol{v}) = v_1\cos\varphi - v_2\sin\varphi, \\ \dot{y} &= (\boldsymbol{\gamma}, \boldsymbol{v}) = v_1\sin\varphi + v_2\cos\varphi, \end{aligned} \tag{H.1}$$

where $\boldsymbol{\alpha} = (\cos\varphi, -\sin\varphi)$, $\boldsymbol{\gamma} = (\sin\varphi, \cos\varphi)$, and $\boldsymbol{v} = (v_1, v_2)$.

In Eqs. (H.1), a_1 and a_2 are added masses, b is the added moment of inertia, $\lambda = I\rho$, $\zeta = I\rho\mu_1$, and $\chi = I\rho\lambda_1$. The parameters ζ and χ proportional to the circulation I are related to the body asymmetry, and their calculation was presented in [125]. In [346], ζ and χ were both set to be zero. In the general case including the effects of both circulation and body asymmetry, $\zeta \neq 0$ and $\chi \neq 0$, but then these parameters cannot be eliminated by a suitable choice of the coordinate system attached to the body.

REMARK. Equations (H.1) were obtained by S. A. Chaplygin in 1926, and slightly later (and probably independently) by H. Lamb [365] and H.Glauert in 1929. In contrast to H. Lamb, Chaplygin [125] also analyzed the behavior of the solutions of these equations.

Later the same equations (H.1) were considered by G. Couchet [142] and P. Capodanno [121]. In [121], the reduction of the equations of motion to a system of two second-order equations that do not explicitly contain time was carried out.

Note that the equations of motion for the system describing planar rigid body motion in an ideal fluid of uniform vorticity are close in their form to (H.1). An analysis of the motion of a circular cylinder in a fluid of uniform vorticity (for various boundary conditions) was carried out by Prudman and Taylor [365]. In this case, the equations of motion reduce to the simplest linear form. The analysis of the general motion of an arbitrary rigid body is very complicated (even in the plane-parallel case) and has never been carried out.

If the forces are potential, then

$$\begin{gathered} Q_1 = \frac{\partial U}{\partial x}\cos\varphi + \frac{\partial U}{\partial y}\sin\varphi, \quad Q_2 = -\frac{\partial U}{\partial x}\sin\varphi + \frac{\partial U}{\partial y}\cos\varphi, \\ M_1 - \xi Q_2 + \eta Q_1 = \frac{\partial U}{\partial \varphi}, \end{gathered} \tag{H.2}$$

where the function $U = U(x, y, \varphi)$ is the potential. In this case, Eqs. (H.1) have the energy integral

$$H = T + U = \frac{1}{2}(a_1 v_1^2 + a_2 v_2^2 + b\omega^2) + U, \tag{H.3}$$

where the kinetic energy T of the body-fluid system is written in diagonal form, which is attained via shifts and rotations of the moving coordinate system.

We can show that the forces due to the circulation possess a vector potential, and the equations of motion (in cases where the external forces are potential) can be written in the Lagrangian form

$$\begin{aligned} \frac{d}{dt}\left(\frac{\partial L}{\partial v_1}\right) &= \omega\frac{\partial L}{\partial v_2} + \frac{\partial L}{\partial x}\alpha_1 + \frac{\partial L}{\partial y}\gamma_1 \\ \frac{d}{dt}\left(\frac{\partial L}{\partial v_2}\right) &= -\omega\frac{\partial L}{\partial v_1} + \frac{\partial L}{\partial x}\alpha_2 + \frac{\partial L}{\partial y}\gamma_2 \\ \frac{d}{dt}\left(\frac{\partial L}{\partial \omega}\right) &= v_2\frac{\partial L}{\partial v_1} - v_1\frac{\partial L}{\partial v_2} + \frac{\partial L}{\partial \varphi} \\ \dot\varphi = \omega, \qquad & \dot x = (\boldsymbol{\alpha}, \boldsymbol{v}), \qquad \dot y = (\boldsymbol{\gamma}, \boldsymbol{v}), \end{aligned} \tag{H.4}$$

where the Lagrangian function has the form

$$L = T - U - (x\sin\varphi - y\cos\varphi)\left(\frac{\lambda}{2}v_1 + \chi\omega\right) - (x\cos\varphi + y\sin\varphi)\left(\frac{\lambda}{2}v_2 + \zeta\omega\right). \tag{H.5}$$

REMARK. The gauge has been chosen so that the Lagrangian function is symmetric with respect to v_1 and v_2: recall that the terms linear in velocities are defined modulo the addition of the total differential of an arbitrary function of the position variables.

Equations (H.4) are the Poincaré equations on the group of motions of the plane $E(2)$ [77]; by using the Legendre transform, they can be represented in Hamiltonian form (Poincaré – Chetaev equations) with Hamiltonian containing additional terms linear in momenta. It turns out [72] that for systems with vector potential it is more convenient to represent the equations of motion in the slightly modified variables

$$p_1 = \frac{\partial T}{\partial v_1} = a_1 v_1, \qquad p_2 = \frac{\partial T}{\partial v_2} = a_2 v_2, \qquad M = \frac{\partial T}{\partial \omega} = b\omega.$$

Applying the Legendre transform in the new variables, we find the Hamiltonian:

$$H = (\boldsymbol{p}, \boldsymbol{v}) - L = \frac{1}{2}\left(\frac{p_1^2}{a_1} + \frac{p_2^2}{a_2} + \frac{M^2}{b}\right) + U(x, y, \varphi). \tag{H.6}$$

In this case, the equations of motion are close (in their form) to the Poincaré – Chetaev equations:

$$
\begin{aligned}
\dot{p}_1 &= p_2\frac{\partial H}{\partial M} - \frac{\partial H}{\partial x}\alpha_1 - \frac{\partial H}{\partial y}\gamma_1 - \lambda\frac{\partial H}{\partial p_2} - \zeta\frac{\partial H}{\partial M},\\
\dot{p}_2 &= -p_1\frac{\partial H}{\partial M} - \frac{\partial H}{\partial x}\alpha_2 - \frac{\partial H}{\partial y}\gamma_2 + \lambda\frac{\partial H}{\partial p_1} + \chi\frac{\partial H}{\partial M},\\
\dot{M} &= p_1\frac{\partial H}{\partial p_2} - p_2\frac{\partial H}{\partial p_1} - \frac{\partial H}{\partial \varphi} + \zeta\frac{\partial H}{\partial p_1} - \chi\frac{\partial H}{\partial p_2},\\
\dot{\varphi} &= \frac{\partial H}{\partial M}, \qquad \dot{x} = \left(\boldsymbol{\alpha}, \frac{\partial H}{\partial \boldsymbol{p}}\right), \qquad \dot{y} = \left(\boldsymbol{\gamma}, \frac{\partial H}{\partial \boldsymbol{p}}\right),
\end{aligned}
\tag{H.7}
$$

whereas the Poisson bracket in these variables contains "circulation additions" (gyroscopic terms):

$$
\begin{gathered}
\{M, p_1\} = -p_2 + \zeta, \quad \{M, p_2\} = p_1 - \chi, \quad \{p_1, p_2\} = -\lambda,\\
\{M, \varphi\} = -1, \quad \{M, x\} = \{M, y\} = 0,\\
\{p_1, x\} = -\alpha_1, \quad \{p_2, x\} = -\alpha_2, \quad \{p_1, y\} = -\gamma_1, \quad \{p_2, y\} = -\gamma_2
\end{gathered}
\tag{H.8}
$$

The rank of this Poisson structure is equal to 6, and hence the system (H.7) can be reduced to the canonical system with three degrees of freedom.

3 Motion in a gravitational field

In this case, the potential energy of the system can be written in the form $U = \mu(y + \xi\sin\varphi + \eta\cos\varphi)$ and the Hamiltonian (H.7) has the form

$$
H = \frac{1}{2}\left(\frac{p_1^2}{a_1} + \frac{p_2^2}{a_2} + \frac{M^2}{b}\right) + \mu(y + \xi\sin\varphi + \eta\cos\varphi). \tag{H.9}
$$

The system (H.7), (H.9) admits one autonomous integral and one nonautonomous integral that correspond to projections of the linear momentum on the fixed axes Ox and Oy (see Fig. H.1):

$$
\begin{aligned}
p_x &= p_1\cos\varphi - p_2\sin\varphi + \lambda y + \zeta\sin\varphi - \chi\cos\varphi = c_1 = \text{const.},\\
p_y &= p_1\sin\varphi + p_2\cos\varphi - \lambda x - \zeta\cos\varphi - \chi\sin\varphi = -\mu t + c_2, \quad c_2 = \text{const.},
\end{aligned}
\tag{H.10}
$$

which are not commutative: $\{p_x, p_y\} = \lambda$. Fixing the integral $p_x = c_1$, we can perform a reduction by one degree of freedom; for this purpose, let us use the reduction condition to substitute for y in the Hamiltonian (H.9):

$$
\begin{aligned}
H \;=\; & \frac{1}{2}\left(\frac{p_1^2}{a_1} + \frac{p_2^2}{a_2} + \frac{M^2}{b}\right) + \frac{\mu}{\lambda}(-p_1\cos\varphi + p_2\sin\varphi)\\
& + \mu\left(\left(\xi - \frac{\zeta}{\lambda}\right)\sin\varphi + \left(\eta + \frac{\chi}{\lambda}\right)\cos\varphi\right).
\end{aligned}
\tag{H.11}
$$

This reduced Hamiltonian depends only on the variables p_1, p_2, M, and φ, and according to (H.8), the Poisson brackets between these variables compose a closed subalgebra of rank 4. Thus, we have obtained the reduced system with two degrees of freedom, which can be written in canonical form (see below). However, in what follows, we do not use the canonical form of the equations, and, moreover, we even algebraicize the reduced system further using the dependent variables $\gamma_1 = \sin\varphi$ and $\gamma_2 = \cos\varphi$ in which the Hamiltonian and the equations of motion have the simpler form

$$
\begin{gathered}
H = \frac{1}{2}\left(\frac{p_1^2}{a_1} + \frac{p_2^2}{a_2} + \frac{M^2}{b}\right) - \frac{\mu}{\lambda}\left((\boldsymbol{p}-\boldsymbol{s}), \mathbf{J}\boldsymbol{\gamma}\right) + \mu(\boldsymbol{r}, \boldsymbol{\gamma}), \\
\dot{\boldsymbol{p}} = \frac{\partial H}{\partial M}\mathbf{J}(\boldsymbol{p}-\boldsymbol{s}) - \mu\boldsymbol{\gamma} - \lambda\mathbf{J}\frac{\partial H}{\partial \boldsymbol{p}}, \\
\dot{M} = \left(\boldsymbol{p}-\boldsymbol{s}, \mathbf{J}\frac{\partial H}{\partial \boldsymbol{p}}\right) - \mu(\boldsymbol{r}, \mathbf{J}\boldsymbol{\gamma}), \qquad \dot{\boldsymbol{\gamma}} = \frac{\partial H}{\partial M}\mathbf{J}\boldsymbol{\gamma},
\end{gathered}
\tag{H.12}
$$

where $\boldsymbol{p} = (p_1, p_2)$, $\boldsymbol{\gamma} = (\gamma_1, \gamma_2)$, $\mathbf{J} = \begin{Vmatrix} 0 & 1 \\ -1 & 0 \end{Vmatrix}$, $\boldsymbol{s} = (\chi, \zeta)$, and $\boldsymbol{r} = (\xi, \eta)$.

The system has the following two obvious integrals: the energy integral H and the geometric integral $\gamma_1^2 + \gamma_2^2 = 1$; because of the Hamiltonian property, for this system to be integrable, one more additional integral is needed (although it can be integrated by using the Euler – Jacobi method, since Eqs. (H.12) preserve the standard invariant measure).

Canonical variables. By using the algorithm described in [77], it is easy to construct the canonical variables analogous to the Andoyer variables of rigid body dynamics. Using the standard notation, we can write

$$
\begin{gathered}
M = L, \qquad p_1 - \chi = \sqrt{2\lambda(G-L)}\sin l, \quad p_2 - \zeta = \sqrt{2\lambda(G-L)}\cos l, \\
M + \frac{(p_1-\chi)^2 + (p_2-\zeta)^2}{2\lambda} = G, \qquad \varphi = g,
\end{gathered}
$$

where l, L and g, G are two pairs of canonically conjugate variables ($\{l, L\} = \{g, G\} = 1$). These variables are useful for introducing new variables of action-angle type and for applying methods of Hamiltonian perturbation theory.

REMARK. Note that in the case where there is no circulation, $\lambda = \chi = \zeta = 0$, reduction to system (H.12) is impossible, since $\frac{\mu}{\lambda} = \infty$. However, using the integrals (H.10), we can eliminate p_1 and p_2 from the equations (but not from the Hamiltonian) and (analogously to Chaplygin [125]) obtain the following nonautonomous equation of "pendulum" type for $\ddot{\varphi}$ (see also [121, 345]):

$$
\begin{aligned}
\ddot{\varphi} = {} & \frac{\mu^2 t^2}{2b}(a_2^{-1} - a_1^{-1})\sin 2\varphi - \frac{\mu t}{b}(a_2^{-1} - a_1^{-1})(c_1\cos 2\varphi + c_2\sin 2\varphi) \\
& - \frac{\mu}{b}(\xi\cos\varphi - \eta\cos\varphi) + \frac{(a_1^{-1} - a_2^{-1})}{2b}\left((c_1^2 - c_2^2)\sin 2\varphi - 2c_1c_2\cos 2\varphi\right),
\end{aligned}
\tag{H.13}
$$

which was studied in [345, 153, 156] for certain particular cases.

Note that the system (H.12) has an important mechanical significance, it is as simple in its presentation as the Euler – Poisson equations, and analogous problem statements can be formulated for this system, so the first problem concerns integrability of the system (H.12).

4 Integrable cases

The following two integrable cases are known.

1. $\mu = 0$ (S. A. Chaplygin, 1926 [125]), means the absence of a gravitational field. The additional quadratic integral has the form

$$F = \frac{1}{2}\left((p_1 - \chi)^2 + (p_2 - \zeta)^2\right) + \lambda M, \qquad \text{(H.14)}$$

and analogously to the Euler – Poinsot case in the classical problem, the system (H.12) reduces to a system of three equations in the variables M, p_1, and p_2. In [125], Chaplygin showed that explicit integration of this system reduces to a certain rather complicated quadrature expressed in elliptic functions under the condition $\chi = \zeta = 0$.

2. $\zeta = \chi = 0$ and $a_1 = a_2 = 1$ (V. V. Kozlov, 1993. [346]), is the case of dynamical symmetry. This case is analogous to the Lagrange case, although the additional integral is quadratic in momenta:

$$F = \frac{1}{2}\frac{M^2}{b} + \mu(\mathbf{r}, \boldsymbol{\gamma}). \qquad \text{(H.15)}$$

This case was integrated by quadratures in [161].

We present the integrable case found by us recently in [161]:

3. Let $\chi = -\lambda\eta$, $\zeta = \lambda\xi$ (i.e., $\mathbf{s} = -\lambda\mathbf{J}\mathbf{r}$) and let $a_1 = a_2 = 1$. The additional quadratic integral has the form

$$F = \frac{1}{2}M^2 - \lambda b(M + (\mathbf{p}, \mathbf{J}\mathbf{r})) + \mu b(\mathbf{r}, \boldsymbol{\gamma}). \qquad \text{(H.16)}$$

The system has no other (general) integrable cases with additional integral linear and quadratic in the phase variables, other than the three cases presented above.

This simple assertion can be proved directly by using the undetermined coefficient method. The problem of the existence of integrals of higher degree remains open.

Here is a linear invariant relation analogous to the Hess case for the Euler – Poisson equations.

4. Let $\zeta = 0$, $\xi = 0$, and let $\eta = \pm\sqrt{b\left(a_2^{-1} - a_1^{-1}\right)}$. In this case,

$$F = M \pm \sqrt{b\left(a_2^{-1} - a_1^{-1}\right)}\left(p_1 - \chi\frac{a_1}{a_1 - a_2}\right) - \frac{\lambda b}{a_2} \qquad \text{(H.17)}$$

and $\dot{F} = 0$ on the level set of $F = 0$.

5 The Chaplygin case. Bifurcation analysis

Let us consider the system (H.9) for $\mu = 0$ and $\zeta = \chi = 0$. As shown previously, the equations for the variables M, p_1, and p_2 decouple from those for all other variables and have the form

$$\dot{p}_1 = \frac{p_2 M}{b} - \lambda\frac{p_2}{a_2}, \quad \dot{p}_2 = -\frac{p_1 M}{b} + \lambda\frac{p_1}{a_1}, \quad \dot{M} = \left(\frac{1}{a_2} - \frac{1}{a_1}\right) p_1 p_2. \tag{H.18}$$

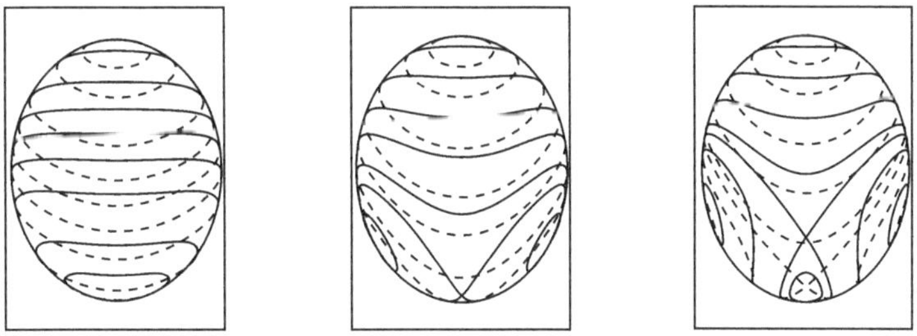

(a) $h = 1/2$ (b) $h = 3$ (c) $h = 7$

Fig. H.2. Trajectories of the system in the Chaplygin case in the space of variables p_1, p_2, and M. The parameter values are as follows: $\lambda = 1$, $b = 2$, $a_1 = 1$, $a_2 = 1/2$, and $\chi = \zeta = 0$.

The common level set of the first integrals

$$H = \frac{1}{2}\left(\frac{p_1^2}{a_1} + \frac{p_2^2}{a_2} + \frac{M^2}{b}\right) = h = \text{const.}, \qquad F = M + \frac{p_1^2 + p_2^2}{2\lambda} = c = \text{const.} \tag{H.19}$$

consists of closed curves, which are the intersections of ellipsoids and elliptic paraboloids; they are analogous to polhodes in the classical Euler – Poisson problem (Fig. H.2). Also, it is easy to describe particular solutions of (H.18) analogous to permanent rotations in the latter problem.

For definiteness, we set $\lambda > 0$ and $a_1 > a_2$.

Solution I. $p_1 = p_2 = 0$, $M = c = \text{const.}$, and $h = \frac{1}{2b}c^2$. In this case, the body executes the uniform rotation ($\dot{\varphi} = \frac{c}{b} = \text{const.}$) about the origin O_1 of the moving coordinate system (see Fig. H.1), which is fixed (and does not coincide with the center of mass). This solution is unstable if $\frac{\lambda b}{a_1} < c < \frac{\lambda b}{a_2}$ and is stable in other cases (the stability on the boundaries requires a separate analysis).

Solutions II and III. There exists a pair of solutions analogous to one another, having the form

$$p_i = 0, \qquad M = \frac{\lambda}{a_j} b = \text{const.}, \qquad p_j^2 = 2\lambda\left(c - \frac{b\lambda}{a_j}\right), \tag{H.20}$$

where $i = 1$ and $j = 2$ in one case and $i = 2$ and $j = 1$ in the other case. Each of these solutions exists under the condition $c > \frac{b\lambda}{a_j}$, and the constants of integrals h and c for them are related by

$$h = c\frac{\lambda}{a_j} - \frac{\lambda^2 b}{2a_j^2}. \tag{H.21}$$

For such a motion, the rigid body rotates uniformly about the origin of the fixed coordinate system, and the body's principal axes pass through this fixed center. For one of these solutions the body is oriented in such a way that its wide side faces the center, while for the other solution the narrow side faces the center. The solution $p_2 = 0$, $h = \frac{\lambda}{a_1}c - \frac{\lambda^2 b}{2a_1^2}$ is always stable, whereas the solution $p_1 = 0$, $h = \frac{\lambda}{a_2}c - \frac{\lambda^2 b}{2a_2^2}$ is always unstable. Thus, motions for which the narrow side of the body faces the fixed point are stable ones.

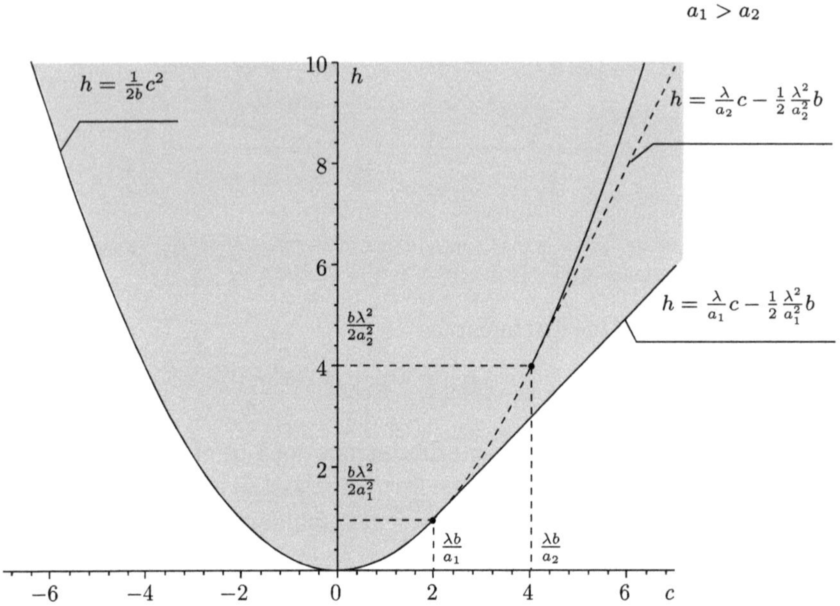

Fig. H.3. Bifurcation diagram of the Chaplygin case for $a_1 > a_2$. Gray denotes the domain of physically admissible values of the integrals c and h. The solid curves correspond to stable "permanent rotations", while the dashed curves correspond to unstable rotations. The values of the parameters are as follows: $\lambda = 1$, $b = 2$, $a_1 = 1$, $a_2 = \frac{1}{2}$.

The bifurcation diagram is presented in Fig. H.3. The lines correspond to the particular solutions II and III; they are tangent to the parabola corresponding to the solution I at those points at which the admissible values of the integral c start. The domain of possible motions is shown in gray.

In the bifurcation diagram there are three different intervals

$$\left(0, \frac{b\lambda^2}{2a_1^2}\right), \quad \left(\frac{b\lambda^2}{2a_1^2}, \frac{b\lambda^2}{2a_2^2}\right), \quad \left(\frac{b\lambda^2}{2a_2^2}, \infty\right)$$

of the energy values h for which the location of the trajectories corresponding to different values of c is qualitatively of the same type (Fig. H.2).

Let us present explicit formulas for the doubly asymptotic motions for the solution I, which is unstable for $\frac{\lambda b}{a_1} < c < \frac{\lambda b}{a_2}$. These solutions are homoclinic (see Fig. H.2 b) and have the form

$$M = c - \frac{2Bd_1^2 d_2^2}{\cosh 2\tau - d_1^2 + d_2^2},$$
$$p_1 = \pm\sqrt{\frac{8\lambda B}{d_1^2}}\frac{d_1^2 d_2^2 \cosh\tau}{\cosh 2\tau - d_1^2 + d_2^2}, \quad p_2 = \sqrt{\frac{8\lambda B}{d_2^2}}\frac{d_1^2 d_2^2 \sinh\tau}{\cosh 2\tau - d_1^2 + d_2^2}, \tag{H.22}$$
$$\varphi = \frac{c}{b}(t - t_0) - 2\arctan\left(\frac{1 + d_1^2 - d_2^2}{2d_1 d_2}\tanh\tau\right), \quad \tau = \frac{\sqrt{b_1 b_2}}{2b}t,$$

where $d_1^2 = \frac{b_1}{B}$, $d_2^2 = \frac{b_2}{B}$, $b_1 = 2(c - \lambda b a_1^{-1})$, $b_2 = 2(\lambda b a_2^{-1} - c)$, and $B = b_1 + b_2$, and the different signs correspond to two different separatrices.

6 Perturbation of the Chaplygin case. Separatrix splitting

Let us consider the "perturbed" Hamiltonian (H.9): $H = H_0 + \mu H_1$, assuming that μ is a small parameter. In this case, H_0 is the Hamiltonian of the integrable Chaplygin problem and $H_1 = \frac{U}{\mu} = y + \xi\sin\varphi + \eta\cos\varphi$ is the perturbing function. One of the dynamical effects that yields an obstruction to the existence of an additional analytic integral for the perturbed system is the splitting of separatrices, which are doubled for the unperturbed system. In the first order of perturbation theory, for small μ, the magnitude of separatrix splitting is determined by the value of the Poincaré – Mel'nikov integral [335]

$$J = \int_{-\infty}^{\infty} \{F_0, H_1\}\, dt \tag{H.23}$$

calculated along the asymptotic solutions of the unperturbed problem. Here, F_0 is the integral of the unperturbed system.

If the integral J has a simple zero (depending on the separatrix parameter), then the separatrix splitting is transversal; in the case of two degrees of freedom, this leads to the nonintegrability of the perturbed problem [335], and, moreover, as F_0 we take the integral (H.19) $F = M^2 + \frac{p_1^2 + p_2^2}{2\lambda}$.

On the homoclinic asymptotic solutions (H.22), the integral (H.23) can be explicitly calculated by using residues:

$$\begin{aligned} J(g_0) &= J_Q + J_R, \\ J_Q &= -2\pi\frac{\sqrt{2}b}{\sqrt{\lambda B}}\frac{e^{\pi S/2}\sinh\left(\frac{\pi}{2}-\alpha\right)S}{\sinh \pi S}\sin g_0, \\ J_R &= -4\pi\frac{b}{B}\frac{e^{\pi S/2}\cosh\left(\frac{\pi}{2}-\alpha\right)S}{\sinh \pi S}(\xi\cos g_0 - \eta\sin g_0), \end{aligned} \tag{H.24}$$

where $S = 2c(b_1b_2)^{-1/2}$, $\cos\alpha = d_1$, $\sin\alpha = d_2$, and g_0 is the separatrix parameter.

The integral (H.24) has a simple zero, except for the singular case $B = 0$. From this we can conclude that the perturbed system (H.9) is nonintegrable, except for the case where $a_1 = a_2$. Note that it is precisely under this condition that the integrals (H.15) and (H.16) are obtained. This condition is a necessary (for $\mu \neq 0$), but not sufficient condition for integrability.

The chaotic behavior of the system (H.9) related to the separatrix splitting and the nonintegrability is illustrated in Fig. H.4, which is the Poincaré section of the phase flow of system (H.12) on the energy level H = const. In this figure we show the split separatrices and also the stochastic layer arising near to them. In the figure we show the projection of the section of the energy level $H = h$ = const on the plane (p_1, p_2), which is defined by the relation $\varphi = \pi$ = const. As above, we fix the following values of the parameters: $\lambda = 1$, $b = 2$, $a_1 = 2a_2 = 1$, and $\chi = \zeta = 0$, and for these parameters we set $h = 2$ for the energy level.

Remark. If we consider perturbations for which ζ and χ are also assumed to be small, then the expression (H.24) for the Poincaré – Mel'nikov integral does not change.

Let us consider the problem concerning the possibility of the body falling in the presence of circulation. Recall that in the case where there is no circulation, the body motion tends asymptotically to that of the body falling with its widest side pointing downwards [334].

Proposition. *If $\lambda \neq 0$, then the body motion remains in a band of finite thickness parallel to the axis x.*

Indeed, according to (H.10), we have

$$\begin{aligned} x &= \frac{\mu}{\lambda}t + \lambda^{-1}\left((p_1-\chi)\sin\varphi + (p_2-\zeta)\cos\varphi - c_2\right), \\ y &= -\lambda^{-1}\left((p_1-\chi)\cos\varphi - (p_2-\zeta)\sin\varphi - c_1\right) \\ & c_1, c_2 = \text{const}. \end{aligned} \tag{H.25}$$

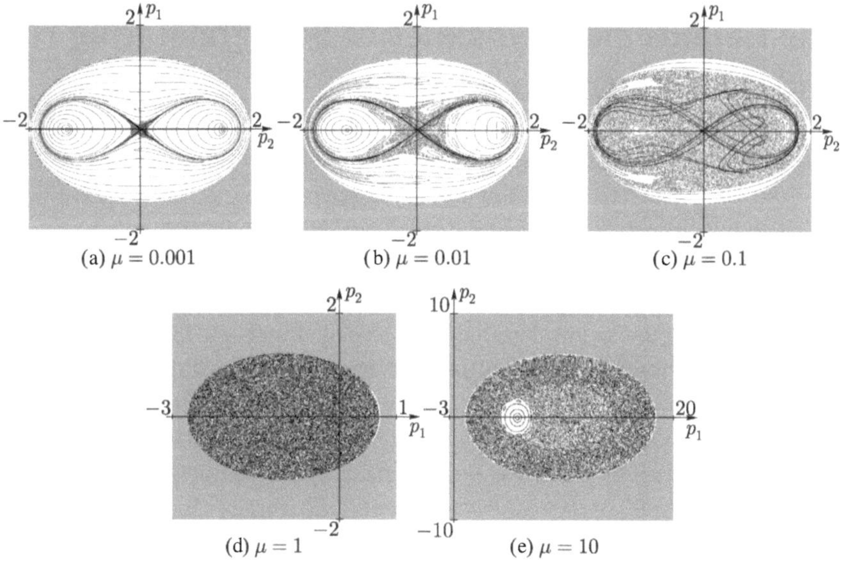

Fig. H.4. Transition to chaos in the Chaplygin system with increasing μ. In the first three pictures, we present the trajectories and one pair of separatrices; in the last two pictures, the separatrices are not presented, since the corresponding fixed point becomes stable for these parameter values. (Gray regions indicate values of the variables in the plane for which the motion is impossible with these parameter values. The scale of the last picture is different from that of the other pictures).

Let us show that p_1 and p_2 are bounded. For this purpose, we rewrite (H.11) as

$$\frac{1}{2}\left(\frac{1}{a_1}\left(p_1 - \frac{\mu a_1}{\lambda}\cos\varphi\right)^2 + \frac{1}{a_2}\left(p_2 + \frac{\mu a_2}{\lambda}\sin\varphi\right)^2 + \frac{\mu^2}{b}\right)$$
$$= h + \frac{\mu^2}{\lambda^2}\left(a_1\cos^2\varphi + a_2\sin^2\varphi\right) - \mu\left(\left(\xi - \frac{\zeta}{\lambda}\right)\sin\varphi + \left(\eta + \frac{\chi}{\lambda}\right)\cos\varphi\right),$$

where h is the constant value of the integral (H.11). Obviously, for $\lambda \neq 0$, the right-hand side of this relation is a bounded function, and hence p_1 and p_2 are also bounded.

According to (H.25), the body moves in the horizonal direction with mean speed $\frac{\mu}{\lambda}$. For $\chi = \zeta = 0$, this result was obtained in [479].

I The Hamiltonian Dynamics of Self-gravitating Fluid and Gas Ellipsoids

This appendix includes our literature review published in 2009 and devoted to the dynamics of self-gravitating fluid and gas ellipsoids. There is an extensive body of research on this branch of mechanics, carried out mostly independently by scientists from different countries and published in various journals, not always readily available; certain results in this area are still controversial.

The first part of the review describes, in a closed, systematic form, basic results on the dynamics of ellipsoidal figures of an ideal, incompressible fluid, starting with the seminal work of Dirichlet and Riemann. As is well known, these studies were preceded by investigations of the static equilibrium of a rotating fluid mass, which were motivated by a scientific interest in the figure of the Earth and can be traced back to Newton's *Mathematical Principles of Natural Philosophy* and the work of Clairaut and Maclaurin. A historical survey of this period (also covering the work of Jacobi, Mayer, and Liouville) can be found in a monograph by Chandrasekhar [123]. Fundamental discoveries by Dirichlet [160] and Riemann [487] (1858 – 6181) were a turning point in developing the theory of figures of equilibrium; Dirichlet and Riemann were the first to investigate the dynamics of fluid ellipsoids. They noted the existence of a finite-dimensional solution to the Euler equations of ideal fluid dynamics according to which the ellipsoid preserves its shape but deforms. This is an exact hydrodynamic solution, so that the question of investigating the dynamics of such fluid ellipsoids can be correctly posed.

We first derive equations of motion of this system, find integrals of motion, and discuss the Lagrangian and Hamiltonian forms of the equations, which makes it possible to naturally carry out a symmetry group reduction of the system. Next, we analyze partial solutions starting from the standard configurations found for the first time by Maclaurin, Jacobi, and Riemann. We present a new class of chaotic motions of ellipsoids in the form of irregular pulsations, with the axes of the ellipsoid remaining fixed in absolute space. For this class of motions we numerically construct separatrices that intersect with each other, and thus prove analytical nonintegrability of the system.

The second part of the review is concerned with the dynamics of a gas cloud with ellipsoidal stratification. The most complete survey of early results on equilibrium and stability of compressible liquid and gas masses can be found in the fundamental essay [275] by Jeans, which was the winner of the Adams Prize for 1917. This book also considers possible applications of these studies to the problems of cosmogony. Note that the most elementary model which is different from the classical incompressible model of figures of equilibrium was given by Roche [490] (see [275], p. 147), who used it to construct his cosmogonic theory generalizing the Laplace hypothesis. The Roche model (as we call it following [275]) contains a gravitating Newtonian center surrounded by weightless atmosphere inside which self-gravitation is neglected. In this part, in a unified form, we present results obtained by Ovsyannikov, Dyson, Linden-Bell,

DOI 10.1515/9783110544442-015

Zel'dovich, Fujimoto, etc. Various forms of the equations of motion of the gas cloud are derived under certain thermodynamic assumptions. Some problems of Lagrangian and Hamiltonian formalism are also discussed. A generalized Dedekind law of reciprocity is formulated. An analog of the Riemann equation for the case of a compressible fluid or gas is obtained. A particular case of the expansion of an ellipsoidal cloud of an ideal monoatomic gas in the absence of gravitation is analyzed in detail. This case has recently been considered by Gaffet, who found new integrals of this system. The involutivity of these integrals is revealed and the Liouville integrability of the system is shown. An analogy between the general system considered by Gaffet and the generalized Euler – Calogero system is revealed. The question of a Lax representation in this case is discussed.

In addition, some partial solutions that are axisymmetric in the presence and absence of gravitation are presented. For the case of zero gravitation, the system is reduced to quadratures under arbitrary initial conditions. For the case of nonzero gravitation, chaotic motions are presented (their existence is due to the nonintegrability of this system). Chaotic motions are also revealed for more general cases of an oscillating triaxial gaseous ellipsoid.

I.1 Dynamics of a Self-gravitating Fluid Ellipsoid

1 Introduction

The studies by L. Dirichlet on the dynamics of a self-gravitating fluid ellipsoid date back to 1856 – 1857. Dirichlet reported these studies in his lectures in 1857 and simultaneously in the *Proceedings of the Göttingen Mathematical Society* as a brief note [159]. Unfortunately, he had no time to describe and publish his results in full (due to his illness and untimely death in 1859). These studies were prepared for publication and posthumously published by Dedekind in 1861 [160]. The main results of Dirichlet's study can be summarized as follows:

1. A new partial solution of the hydrodynamic equations is presented, which describes the motion of a homogeneous, self-gravitating ellipsoid, and equations of motion (of fluid particles) in fixed axes are derived.
2. Seven first integrals of the resulting equations are found; six of them, linear in velocities, correspond to the conservation laws of vorticity and total momentum, and the seventh integral is the total energy of the moving fluid.
3. The motion of an axisymmetric ellipsoid is integrated by quadratures with the inclusion of Newton's and Maclaurin's spheroids as partial solutions (in this case, Dirichlet also analyzes the possibility of existence of the solutions found in the case of no external pressure, i. e., in vacuum).

It is interesting to note that Dirichlet found the integrals corresponding to the conservation of the vorticity vector prior to the publication of the well-known study by Helmholtz (1858) [256]. As can be judged by the form of the integrals obtained, Dirichlet was aware (before Helmholtz) of the conservation of vorticity not only for a partial solution, but also for the general hydrodynamic equations (Dirichlet's note [159] is also evidence of his awareness). This fact was also noted by Klein in his well-known lectures [303].

Dedekind, while preparing Dirichlet's results for publication, discovered the reciprocity law according to which each solution of the Dirichlet equations corresponds to a reciprocal solution in which the variables that describe the rotation of the ellipsoid and the fluid motion inside it are permutated; in particular, he presented a solution (the Dedekind ellipsoid) reciprocal to the Jacobi ellipsoid, with the coordinate axes remaining fixed in space and with the fluid moving inside this invariable region [147].

An enormous contribution to the investigation of the dynamics of the fluid ellipsoid was made by Riemann in his outstanding work [487], which appeared in 1861, almost immediately after the publication of Dirichlet's studies. The main results of this work can be briefly formulated as follows:

1. Equations of motion in moving axes (the principal axes of the ellipsoid) were obtained, so that the order of the system was lowered and a symmetry reduction was performed. Furthermore, Riemann represented the equations of motion of the reduced system in Hamiltonian form with a linear Lie – Poisson bracket (Riemann himself called this procedure *reduction to a more transparent form*).
2. All partial solutions corresponding to the motion of the ellipsoid without changes in its form were presented and conditions for their existence were analyzed (i. e., the possible lengths of the major semiaxes). All these solutions imply that the ellipsoid rotates about an axis fixed in space. They included all solutions known by that time – those obtained by Newton, Maclaurin, Jacobi, and Dedekind (for which the rotational axis coincides with one of the principal axes) and also new solutions (Riemann ellipsoids) for which the rotational axis lies in one of the principal planes of the ellipsoid.
3. Riemann used the energy integral of the system as the Lyapunov function (in modern terminology) to investigate the stability of shape-preserving motions (in the class of motions preserving the ellipsoidal shape); in this way, he found the Lyapunov stability limits for the Maclaurin spheroids and Jacobi ellipsoids.
4. A particular case was found in which a triaxial ellipsoid rotates (unsteadily) about one of the principal axes, and its semiaxes vary with time. This gives rise to a (Hamiltonian) system with two degrees of freedom for which Riemann noticed an analogy with the motion of a material point on a two-dimensional surface of the form $xyz = \text{const}$ in a potential force field (it is this case that we will consider below in detail).

The results of Riemann's study were unique in that they opened up the possibility of further generalizations; it was well in advance of its time.

There is also a study by Brioschi of 1861 [106], which was devoted to lowering the order in the Dirichlet equations by using a decomposition into potential and vortical components. However, no substantial advance in the problem was associated with this work.

In his 1876 lectures on mechanics, Kirchhoff [301] also considered the motion of self-gravitating fluid ellipsoids. He noted that the d'Alembert principle is applicable to the Dirichlet motion (but he did not use it to derive the equation of motion). Kirchhoff presented a quadrature for the axisymmetric case and separately analyzed the case where the ellipsoid preserves the directions of its axes in space (a particular case of the motion considered by Riemann); Kirchhoff (following Riemann) conjectured that this problem cannot be integrated by quadratures either.

The possibility of applying the variational principle to the derivation of the equations of motion of a fluid ellipsoid was independently shown by Padova in 1871 [458] and by Lipschitz in 1874 [387]. In the latter study [387], the problem of the motion of an elliptic cylinder was formulated and integrated by quadratures.

Betti [43] also used the variational principle to derive the equations of motion of a fluid ellipsoid and represented these equations in Lagrangian and Hamiltonian form. However, as Tedone noted in his extensive survey [568], Betti made a mistake in his study when applying the variational principle to the derivation of the equation of motion of a homogeneous ellipsoid with an ellipsoidal fluid-density stratification. In this case, the hydrodynamic equations for the stratified, self-gravitating ellipsoid do not admit a solution with a linear dependence on the initial coordinates, which Betti considered (in view of the complex dependence of the gravitational potential inside the stratified ellipsoid). Nevertheless, all Betti's results remain valid for constant density. Betti also represented the equations of motion in Hamiltonian form (explicitly using Poisson brackets on the $so(3)$ algebra) with a linear Poisson bracket and performed a symmetry reduction.

The above-listed results are the principal achievements of the classical period of investigations into the dynamics of the Dirichlet ellipsoids.

General problems of the dynamics and statics of fluid ellipsoids, including issues of stability, were investigated in classical treatises by Basset [28], Lamb [367], Thomson and Tait [572], Routh [497], in the books by Appell [10], Lyttleton [397], in certain studies by Basset [29, 30, 31], Duhem [172], Hagen [251], Hicks [259], Hill [260], Love [389, 390], etc. We also mention the following related topics, which form independent lines of research in this area.

1. Investigation of figures of equilibrium bifurcating from the ellipsoid, for example, pear-shaped figures, and the analysis of their stability (Lyapunov [395, 553], Poincaré [470, 471], Darwin [146], Jeans [275], and Sretenskii [544]). As is well known, a comprehensive analysis of this problem led Lyapunov to the develop-

ment of a general theory of stability of motion, which goes under his name. Results concerning the stability of figures of equilibrium were also obtained in classical studies by Giesen [221], Bryan [109], and Liouville [386]. Here, we do not touch upon the theory of stability, where many problems still remain open.

2. Figures of equilibrium of a fluid that is inhomogeneous, but stratified in a special manner. The theoretical analysis reached here its summit with a posthumous study by Lyapunov [553] (published by Stekloff), which still remains poorly understood. These investigations are closely related to the theory of the potential of stratified fluids (see Dyson [178], Ferrers [195], Volterra [605]). In [605], as in Lyapunov's studies on this subject (Lyapunov's results are presented from a more modern, functional analysis standpoint in a book by Lichtenstein [381]), integral equations arise; they were later studied in the framework of functional analysis. In the context of this problem, we mention a valuable, but virtually forgotten study by Veronnet [595], a book by Pizzetti [467], and certain modern works [137, 138, 420] and [266] (the last publications do not contain any references to classical results and are highly controversial). A new period of research in astrophysics was marked by studies by Chandrasekhar and his school (see, e. g., [123, 371, 372]).
3. In many problems that trace back to the classical Plateau experiments and the Bohr – Wheeler model of the atom [62], the surface tension forces are considered instead of Newtonian attraction. For more on this, we refer the reader to the books by Appell [10] and Chandrasekhar [123]; however, it should be emphasized that the theory presented there is much less advanced (see also [493, 238, 492], where a Pade approximation for the potential of the tension surface is given). At least, we are not aware of any results concerning the dynamics of liquid (masses of) drops subjected to the action of surface tension forces. Effects of viscosity on the dynamics in the Dirichlet – Riemann problem are discussed, e. g., in [561]. Early computer investigations of self-gravitating nonellipsoidal figures can be found, e. g., in [185].
4. We mention an interesting study by Narlikar and Larmor (1933) [435], where, probably for the first time, the classical results (by Maclaurin, Jacobi, Dirichlet, Poincaré, etc.) are revised in the context of stellar dynamical problems rather than planetary evolution. These researchers assumed that energy dissipation occurs in the process of stellar evolution. Later, this line of research was further developed by Chandrasekhar.

2 The Dirichlet and Riemann Equations

The Dirichlet Equations. We recall here the principal steps in the derivation of the Dirichlet and Riemann equations and represent them in a modern matrix form.

In the case of potential forces applied to the fluid, the equations of the dynamics of a homogeneous, incompressible, ideal fluid of unit density in Lagrangian form are as follows:

$$\left(\frac{\partial \boldsymbol{x}}{\partial \boldsymbol{a}}\right)^T \ddot{\boldsymbol{x}} = -\frac{\partial(U+p)}{\partial \boldsymbol{a}}, \tag{I.1}$$

where $\boldsymbol{a} = (a_1, a_2, a_3)$ are the initial positions of the material points of the medium (the so-called Lagrangian coordinates), $\boldsymbol{x}(\boldsymbol{a}, t)$ are the coordinates of the points of the medium at time t (i. e., $\boldsymbol{x}(\boldsymbol{a}, 0) = a$), $U(\boldsymbol{a}, t)$ is the density of the potential energy of the external forces, $p(\boldsymbol{a}, t)$ is the pressure, and $\frac{\partial \boldsymbol{x}}{\partial \boldsymbol{a}} = \left\|\frac{\partial x_i}{\partial a_j}\right\|$ is the matrix of partial derivatives. These equations must be supplemented with the incompressibility condition, which in this case can be written as

$$\det\left(\frac{\partial \boldsymbol{x}}{\partial \boldsymbol{a}}\right) = 1. \tag{I.2}$$

Thus, we obtain a system of partial differential equations in which four quantities, viz., x_1, x_2, x_3, and p, are unknown as the functions of the variables $\boldsymbol{a}$ and t. To define them, we have to specify, in addition to the initial conditions ($\boldsymbol{x}(\boldsymbol{a}, 0) = \boldsymbol{a}$, $\dot{\boldsymbol{x}}(\boldsymbol{a}, 0) = \boldsymbol{v}_0(\boldsymbol{a})$), boundary conditions, which in our case reduce to the statement that the pressure has the same value, independent of $\boldsymbol{a}$, everywhere on the free surface.

Dirichlet noted that, if the potential of the external forces $U(\boldsymbol{a}, t)$ is a homogeneous quadratic function of the Lagrangian coordinates, i. e.,

$$U(\boldsymbol{a}, t) = U_0(t) + (\boldsymbol{a}, \mathbf{V}(t)\boldsymbol{a}), \tag{I.3}$$

where $U_0(t)$ is independent of $\boldsymbol{a}$ and $\mathbf{V}(t)$ is a symmetric matrix, then the equations of motion (I.1) and (I.2) admit a partial solution

$$\boldsymbol{x}(\boldsymbol{a}, t) = \mathbf{F}(t)\boldsymbol{a}, \quad \det \mathbf{F}(t) = 1. \tag{I.4}$$

Here, $\mathbf{F}(t)$ is a 3×3 matrix.

In this case, the boundary conditions will be satisfied provided that the fluid has initially an ellipsoidal shape, viz.,

$$(\boldsymbol{a}, \mathbf{A}_0^{-2}\boldsymbol{a}) \leqslant 1, \tag{I.5}$$

where $\mathbf{A}_0 = \operatorname{diag}(A_1^0, A_2^0, A_3^0)$ is the matrix of the initial semiaxes and the pressure has the form

$$p(\boldsymbol{a}, t) = p_0(t) + \sigma(t)(1 - (\boldsymbol{a}, \mathbf{A}_0^{-2}\boldsymbol{a})). \tag{I.6}$$

We substitute (I.3), (I.4), and (I.6) into (I.1) and (I.2) to obtain equations for the matrix $\mathbf{F}(t)$ and the function $\sigma(t)$ in the form

$$\begin{aligned} &\mathbf{F}^T\ddot{\mathbf{F}} = -2\mathbf{V} - 2\sigma\mathbf{A}_0^{-2}, \\ &\det \mathbf{F} = 1. \end{aligned} \qquad \textit{(the Dirichlet equations)} \tag{I.7}$$

As Dirichlet showed, the system of ten equations (I.7) for ten unknown functions $F_{ij}(t)$, $\sigma(t)$, $i, j = 1, 2, 3$, is compatible.

Obviously, the transformation (I.4) changes the original ellipsoid (I.5) to an ellipsoid specified by the quadratic form

$$(\boldsymbol{x}, (\mathbf{F}\mathbf{A}_0^2\mathbf{F}^T)^{-1}\boldsymbol{x}) \leqslant 1. \tag{I.8}$$

The Riemann Equations. Before writing an explicit expression for the potential (I.3) and, accordingly, the right-hand side of equations (I.7), let us show how the equations of motion can be written in Riemannian form. To this end, we pass to the moving system of principal axes of the ellipsoid. It is well known that such a transformation is given by the orthogonal matrix

$$\boldsymbol{\zeta} = \mathbf{Q}\mathbf{x}, \quad \mathbf{Q}^T = \mathbf{Q}^{-1}. \tag{I.9}$$

In the new coordinates $\boldsymbol{\zeta}$, the ellipsoid is specified by the relation

$$(\boldsymbol{\zeta}, \mathbf{A}^{-2}\boldsymbol{\zeta}) \leqslant 1, \tag{I.10}$$

where $\mathbf{A} = \operatorname{diag}(A_1, A_2, A_3)$ is the matrix of the principal semiaxes at the given time.

We also note that, since the transform (I.4) is linear, the fluid particles constantly move over ellipsoids for which

$$(\boldsymbol{\zeta}, \mathbf{A}^{-2}\boldsymbol{\zeta}) = (\boldsymbol{a}, \mathbf{A}_0^{-2}\boldsymbol{a}) = n^2 = \text{const.}, \quad 0 \leqslant n^2 < 1. \tag{I.11}$$

(In particular, the fluid particles that were initially at the boundary remain at the boundary at any time). Therefore, the absolute value of the vector $\mathbf{A}^{-1}\boldsymbol{\zeta}$ does not vary, so that the vectors $\mathbf{A}^{-1}\boldsymbol{\zeta}$ and $\mathbf{A}_0^{-1}\boldsymbol{a}$ are also related by the orthogonal transformation

$$\mathbf{A}^{-1}\boldsymbol{\zeta} = \boldsymbol{\Theta}\mathbf{A}_0^{-1}\boldsymbol{a}, \quad \boldsymbol{\Theta}^T = \boldsymbol{\Theta}^{-1}. \tag{I.12}$$

Thus, we obtain the following decomposition of the matrix $\mathbf{F}$:

$$\mathbf{F} = \mathbf{Q}^T\mathbf{A}\boldsymbol{\Theta}\mathbf{A}_0^{-1}. \tag{I.13}$$

REMARK 1. Multiplying the matrix $\mathbf{A}_0$ by a constant yields a decomposition of the form

$$\mathbf{F}\mathbf{A}_0 = \mathbf{Q}^T\mathbf{A}\boldsymbol{\Theta} \tag{I.14}$$

known in linear algebra as a singular decomposition [186].

We introduce the angular velocities corresponding to the orthogonal transformations,

$$\mathbf{w} = \dot{\mathbf{Q}}\mathbf{Q}^T, \quad \boldsymbol{\omega} = \dot{\boldsymbol{\Theta}}\boldsymbol{\Theta}^T, \tag{I.15}$$

which are known to be skew-symmetric matrices [97]. Substituting (I.15) into (I.7) yields Riemann equations, which can be written in the following matrix form:

$$\begin{aligned} &\dot{\mathbf{v}} - \mathbf{w}\mathbf{v} + \mathbf{v}\boldsymbol{\omega} = -2\hat{\mathbf{V}}\mathbf{A} + 2\sigma\mathbf{A}^{-1}, \\ &\mathbf{v} = \dot{\mathbf{A}} - \mathbf{w}\mathbf{A} + \mathbf{A}\boldsymbol{\omega}, \qquad \textit{(Riemann equations)} \\ &A_1A_2A_3 = 1, \end{aligned} \tag{I.16}$$

where $\hat{\mathbf{V}} = \mathbf{A}^{-1}\boldsymbol{\Theta}\mathbf{A}_0\mathbf{V}\mathbf{A}_0\boldsymbol{\Theta}^T\mathbf{A}^{-1}$.

We complement this system with the equations of evolution of the matrices $\mathbf{Q}$ and $\boldsymbol{\Theta}$,

$$\dot{\mathbf{Q}} = \mathbf{wQ}, \quad \dot{\boldsymbol{\Theta}} = \boldsymbol{\omega\Theta}, \tag{I.17}$$

to obtain a complete system of equations of motion describing the dynamics of the fluid ellipsoid.

REMARK 2. For an arbitrary matrix $\mathbf{G} \in GL(3)$ with differing eigenvalues, the decomposition $\mathbf{G} = \mathbf{QA}\boldsymbol{\Theta}$ with $\mathbf{Q}, \boldsymbol{\Theta} \in SO(3)$, $\mathbf{A} = \operatorname{diag}(a_1, a_2, a_3)$, $a_1 > a_2 > a_3$ is not unique. Indeed, the matrices $\mathbf{Q}$ and $\boldsymbol{\Theta}$ admit discrete transforms of the form [186]

$$\begin{gathered}\mathbf{Q}' = \mathbf{QR}_i, \quad \boldsymbol{\Theta}' - \boldsymbol{\Theta}\mathbf{R}_i, \quad i = 0, 1, 2, 3, \\ \mathbf{R}_0 = \mathbf{E}, \quad \mathbf{R}_1 = \operatorname{diag}(1, -1, -1), \quad \mathbf{R}_2 = \operatorname{diag}(-1, 1, -1), \quad \mathbf{R}_3 = \operatorname{diag}(-1, -1, 1).\end{gathered} \tag{I.18}$$

Therefore, the space $\mathbb{R}^2 \otimes SO(3) \otimes SO(3)$ (which is diffeomorphic to the configuration space of the Riemann system) is a four-sheeted covering of the real configuration space $SL(3)$. A similar procedure is used for a quaternion representation of rigid body equations [97].

Gravitational potential. We now determine the right-hand sides of equations (I.7) and (I.17). We use the known representation of the gravitational potential for the interior of the ellipsoid in the system of principal axes [320]

$$U(\boldsymbol{\zeta}) = -\frac{3}{4}mG \int_0^\infty \frac{d\lambda}{\Delta(\lambda)} \left(1 - \sum_i \frac{\zeta_i^2}{A_i^2 + \lambda}\right), \quad \Delta^2(\lambda) = \prod_i (A_i^2 + \lambda), \tag{I.19}$$

where G is the gravitational constant and $m = \frac{4}{3}\pi\rho A_1 A_2 A_3$ is the mass of the ellipsoid.

It is now necessary to represent (I.19) in terms of the elements of the transformation matrix $\mathbf{F}$ and in the Lagrangian coordinates $\boldsymbol{a}$. We use (I.13) to find $\mathbf{A} = \mathbf{QFA}_0\boldsymbol{\Theta}^T$ and obtain

$$\begin{aligned}&\mathbf{A}^2 = \mathbf{AA}^T = \mathbf{QFA}_0^2\mathbf{F}^T\mathbf{Q}^T, \\ &\Delta^2(\lambda) = \det(\mathbf{A}^2 + \lambda\mathbf{E}) = \det(\mathbf{FA}_0^2\mathbf{F}^T + \lambda\mathbf{E}), \\ &\sum_i \frac{\zeta_i^2}{A_i^2 + \lambda} = (\boldsymbol{\zeta}, (\mathbf{A}^2 + \lambda\mathbf{E})^{-1}\boldsymbol{\zeta}) = (\boldsymbol{a}, \mathbf{F}^T(\mathbf{FA}_0^2\mathbf{F}^T + \lambda\mathbf{E})^{-1}\mathbf{F}\boldsymbol{a}).\end{aligned} \tag{I.20}$$

Thus, we find the following representation for the matrix $\mathbf{V}$ in the Dirichlet equations:

$$\mathbf{V} = \varepsilon \int_0^\infty \frac{d\lambda}{\sqrt{\det(\mathbf{FA}_0^2\mathbf{F}^T + \lambda\mathbf{E})}} \mathbf{F}^T(\mathbf{FA}_0^2\mathbf{F}^T + \lambda\mathbf{E})^{-1}\mathbf{F}, \quad \varepsilon = \frac{3}{4}mG. \tag{I.21}$$

It can be shown by direct calculations (see [160]) that $\mathbf{V}$ depends on the elements of the matrix $\mathbf{F}$ only through symmetric combinations of the form $\Phi_{ij} = \sum_k F_{ik}F_{jk}$, which are the dot products of columns of the matrix $\mathbf{F}$.

Using the relation $\frac{\partial \boldsymbol{a}}{\partial \boldsymbol{\zeta}} = \mathbf{A}_0 \boldsymbol{\Theta}^T \mathbf{A}^{-1}$, one can easily show that, in the Riemann equations, $\hat{\mathbf{V}} = \mathrm{diag}(\hat{V}_1, \hat{V}_2, \hat{V}_3)$, where

$$\hat{V}_i = \varepsilon \int_0^\infty \frac{1}{\lambda + A_i^2} \frac{d\lambda}{\Delta(\lambda)} = \frac{1}{A_i} \frac{\partial}{\partial A_i} \varepsilon \int_0^\infty \frac{d\lambda}{\Delta(\lambda)}. \tag{I.22}$$

The Roche Problem. By the Roche problem, according to Jeans's terminology [275] (see also [146, 123]) we mean the problem of the interaction of a deformable body (satellite) and a spherical rigid body which moves along circular Keplerian orbits. In fact, in [490] Roche considered the motion of the liquid mass under the action of a gravitating center [490].

Let a self-gravitating fluid mass move in the field of a spherically symmetric rigid body and let both of these bodies rotate about their common center of mass in circular orbits. We choose a (moving) coordinate system $Ox_1x_2x_3$ with its origin at the center of mass of the ellipsoid and direct the Ox_1 axis toward the common center and the Ox_3 axis normally to the plane of rotation (see Fig. I.1).

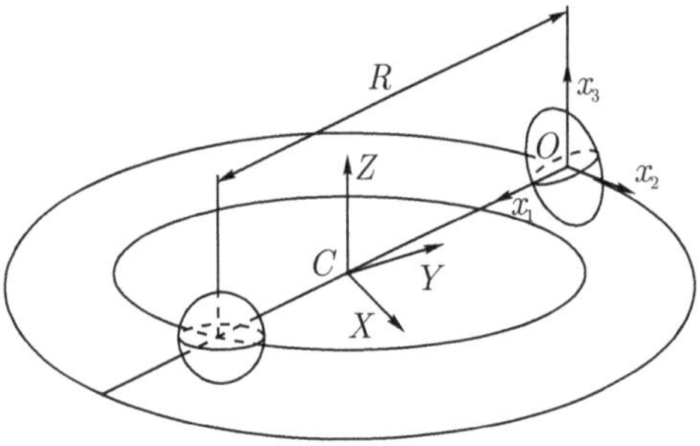

Fig. I.1.

In this case, the equations of motion of an incompressible fluid can be written in the following Lagrangian form:

$$\left(\frac{\partial \boldsymbol{x}}{\partial \boldsymbol{a}}\right)^T (\ddot{\boldsymbol{x}} + 2\omega \boldsymbol{e}_3 \times \dot{\boldsymbol{x}})$$
$$= -\frac{\partial}{\partial \boldsymbol{a}} \left(p + U + U_s - \frac{1}{2}\omega^2(x_1^2 + x_2^2) + \omega^2 \frac{m_s}{m_e + m_s} R x_1\right), \tag{I.23}$$

$$\det \left(\frac{\partial \boldsymbol{x}}{\partial \boldsymbol{a}}\right) = 1, \tag{I.24}$$

where, as before, $\boldsymbol{a}$ are the Lagrangian coordinates of fluid elements, $\boldsymbol{x}(\boldsymbol{a}, t)$ are their positions at the given time, $p(\boldsymbol{a}, t)$ is the pressure, R is the distance between the centers of mass of the bodies, m_e and m_s are the masses of the ellipsoid and the sphere, respectively, ω is the angular velocity of rotation of the systems about their common center of mass, and U is the gravitational potential (I.19). The gravitational potential of a spherical body U_s has the form

$$\begin{aligned} U_s &= -\frac{m_s G}{\sqrt{(x_1 - R)^2 + x_2^2 + x_3^2}} \\ &= -\frac{m_s G}{R}\left(1 + \frac{x_1}{R} + \frac{1}{2}\frac{1}{R^2}\left(2x_1^2 - x_2^2 - x_3^2\right) + \dots\right), \end{aligned}$$

where G is the gravitational constant.

We omit higher-order terms in $\frac{|x|}{R}$ and use the well-known relation for a circular Keplerian orbit $R^3\omega^2 = G(m_e + m_s)$ to obtain finally (after collecting terms) the equation

$$\left(\frac{\partial \boldsymbol{x}}{\partial \boldsymbol{a}}\right)^T (\ddot{\boldsymbol{x}} + 2\omega \boldsymbol{e}_3 \times \dot{\boldsymbol{x}}) = -\frac{\partial}{\partial \boldsymbol{a}}\left(p + U - \frac{1}{2}\omega^2(\mathbf{x}, \mathbf{B}\mathbf{x})\right), \quad \det\left(\frac{\partial \boldsymbol{x}}{\partial \boldsymbol{a}}\right) = 1, \qquad (I.25)$$

where $\mathbf{B} = \operatorname{diag}\left(\frac{3m_s + m_e}{m_e + m_s}, \frac{m_e}{m_e + m_s}, -\frac{m_s}{m_e + m_s}\right)$. In the limiting case of a motionless Newtonian center $\left(\frac{m_e}{m_s} \to 0\right)$ we have $\mathbf{B} = \operatorname{diag}(3, 0, -1)$.

Substituting (I.4) and (I.6) into this equation, we obtain the equations of motion in Roche's problem in the form

$$\begin{aligned} \mathbf{F}^T(\ddot{\mathbf{F}} + 2\boldsymbol{\Omega}\dot{\mathbf{F}}) &= -2\mathbf{V} + 2\sigma\mathbf{A}_0^{-2} + \omega^2\mathbf{F}^T\mathbf{B}\mathbf{F}, \\ \det\mathbf{F} &= 1, \end{aligned} \qquad (I.26)$$

where $\boldsymbol{\Omega} = \| - \omega\varepsilon_{ijk}\|$ is the matrix of the rotational velocity.

REMARK 3. Equations (I.25) are presented in the book by Chandrasekhar, who uses them only to find hydrostatic equilibrium configurations of fluid masses and analyze their stability. Chandrasekhar does not present the dynamical equations (I.26).

Obviously, equations (I.26) can also be written in Riemannian form as

$$\begin{aligned} &\dot{\mathbf{v}} - \mathbf{w}\mathbf{v} - \mathbf{v}\mathbf{w} + 2\bar{\boldsymbol{\Omega}}\mathbf{v} = -2\dot{\mathbf{V}}\mathbf{A} + 2\sigma\mathbf{A}^{-1} + \omega^2\mathbf{A}\bar{\mathbf{B}}\mathbf{A}, \\ &\mathbf{v} = \dot{\mathbf{A}} - \mathbf{w}\mathbf{A} + \mathbf{A}\mathbf{w}, \\ &A_1A_2A_3 = 1, \end{aligned} \qquad (I.27)$$

where $\bar{\boldsymbol{\Omega}} = \mathbf{Q}\boldsymbol{\Omega}\mathbf{Q}^T$, $\bar{\mathbf{B}} = \mathbf{Q}\mathbf{B}\mathbf{Q}^T$ are the matrices reduced to the principal axes of the ellipsoid. As before, equations (I.17) should be added.

Note that equations (I.27) in this case do not form a closed system (in contrast to the Riemann equations), and the system closes only upon adding equations (I.17) for the evolution of the matrix **Q**.

3 First Integrals

The first integrals of the equations, linear in the velocities, can be obtained from the conservation laws for vorticity and angular momentum (the law of areas).

Vorticity. Writing the law of conservation of vorticity for the hydrodynamic equations in Lagrangian form (I.1), we obtain

$$\sum_i \left(\frac{\partial x_i}{\partial a_k} \frac{\partial \dot{x}_i}{\partial a_l} - \frac{\partial x_i}{\partial a_l} \frac{\partial \dot{x}_i}{\partial a_k} \right) = \xi_{kl} = \text{const}, \tag{I.28}$$

where the condition $\xi_{kl} = -\xi_{lk}$ is satisfied. We denote this skew-symmetric matrix as $\Xi = \|\xi_{kl}\|$ and find for the Dirichlet equations (I.7) that

$$\Xi = \mathbf{F}^T \dot{\mathbf{F}} - \dot{\mathbf{F}}^T \mathbf{F} = \text{const}. \tag{I.29}$$

In the Riemannian variables, we obtain

$$\Xi' = \mathbf{A}_0 \Xi \mathbf{A}_0 = \boldsymbol{\Theta}^T (\mathbf{A}^2 \boldsymbol{\omega} + \boldsymbol{\omega} \mathbf{A}^2 - 2\mathbf{A}\mathbf{w}\mathbf{A}) \boldsymbol{\Theta} = \text{const}. \tag{I.30}$$

A straightforward proof of the conservation of vorticity Ξ based on the Dirichlet equations (I.7) is obvious (since the right-hand side is a symmetric matrix).

As already mentioned, the conservation of vorticity in this problem was noted by Dirichlet even before the appearance of the classical study by Helmholtz in which this law was extended to the whole ideal fluid dynamics.

Momentum. The angular momentum relative to the center of the ellipsoid can be represented as

$$M_{ij} = \int (x_i \dot{x}_j - x_j \dot{x}_i) d^3\mathbf{x} = \frac{m}{5} \sum_k (F_{ik} \dot{F}_{jk} - F_{jk} \dot{F}_{ik})(A_k^0)^2. \tag{I.31}$$

In matrix form, omitting the inessential multiplier, we obtain

$$\mathbf{M}' = \mathbf{F}\mathbf{A}_0^2 \dot{\mathbf{F}}^T - \dot{\mathbf{F}}\mathbf{A}_0^2 \mathbf{F}^T = \text{const}, \tag{I.32}$$

where $\mathbf{M}' = \|\frac{5}{m} M_{ij}\|$. Similarly, in the Riemannian variables, we have

$$\mathbf{M}' = \mathbf{Q}^T (\mathbf{A}^2 \mathbf{w} + \mathbf{w} \mathbf{A}^2 - 2\mathbf{A}\boldsymbol{\omega}\mathbf{A}) \mathbf{Q} = \text{const}. \tag{I.33}$$

The use of the Riemann equations (I.16) is convenient in proving the invariability of the momentum $\mathbf{M}$ (the relevant calculations are also straightforward in this case).

Energy. In addition to the linear integrals, the equations of motion admit another, quadratic integral, viz., the total energy of the system. Integrating the kinetic and the potential energy of the fluid particles over the volume of the ellipsoid yields

$$\begin{aligned} \mathcal{E} &= \frac{m}{5}(T_e + U_e), \\ T_e &= \frac{1}{2} \operatorname{Tr}(\dot{\mathbf{F}}\mathbf{A}_0^2 \dot{\mathbf{F}}^T) = \frac{1}{2} \operatorname{Tr}(\dot{\mathbf{A}}^2 - \mathbf{w}^2 \mathbf{A}^2 - \boldsymbol{\omega}^2 \mathbf{A}^2 + 2\mathbf{A}\mathbf{w}\mathbf{A}\boldsymbol{\omega}), \\ U_e &= -2\varepsilon \int_0^\infty \frac{d\lambda}{\sqrt{(\lambda + A_1^2)(\lambda + A_2^2)(\lambda + A_3^2)}}. \end{aligned} \tag{I.34}$$

4 Lagrangian and Hamiltonian formalism

The Hamilton Principle and Lagrangian Formalism. As is well known (see, e. g., [301]), the motion of an ideal fluid satisfies the Hamilton principle; therefore, Dirichlet's solution also satisfies this principle. This makes it possible to represent the equations of motion in Lagrangian and then in Hamiltonian form. The Hamilton principle for the problem considered was used for the first time by Lipschitz [387] and Padova [458].

As the Lagrangian function, it is necessary to choose the difference between the kinetic and potential energies of the fluid in the ellipsoid; omitting the inessential multiplier, we have

$$L = T_e - U_e, \tag{I.35}$$

where T_e and U_e were defined above in (I.34). The elements of the matrix $\mathbf{F}$ appear as generalized coordinates. Writing the Lagrange – Euler equations taking into account the constraint $\det \mathbf{F} = 1$, we obtain

$$\left(\frac{\partial L}{\partial \dot{\mathbf{F}}}\right)^{\cdot} - \frac{\partial L}{\partial \mathbf{F}} = \kappa \frac{\partial \varphi}{\partial \mathbf{F}}, \tag{I.36}$$

where $\varphi = \det \mathbf{F}$, and use the following matrix notation for any function: $\frac{\partial f}{\partial \mathbf{F}} = \left\| \frac{\partial f}{\partial F_{ij}} \right\|$, $\frac{\partial f}{\partial \dot{\mathbf{F}}} = \left\| \frac{\partial f}{\partial \dot{F}_{ij}} \right\|$, κ being the undefined Lagrange multiplier. Differentiating in (I.36) and taking into account the formula $\left(\frac{\partial \varphi}{\partial \mathbf{F}}\right)^T = \varphi \mathbf{F}^{-1}$ yields

$$\ddot{\mathbf{F}}\mathbf{A}_0^2 = 2\varepsilon \frac{\partial}{\partial \mathbf{F}} \int_0^\infty \frac{d\lambda}{\sqrt{\det(\mathbf{F}\mathbf{A}_0^2\mathbf{F}^T + \lambda \mathbf{E})}} + \kappa (\mathbf{F}^{-1})^T \det \mathbf{F}. \tag{I.37}$$

We can easily verify that these equations coincide with the Dirichlet equations (I.7) if we set $\kappa = 2\sigma$.

The matrix of the initial semiaxes $\mathbf{A}_0$ appears in the Lagrangian function and the equations of motion of the system as a set of parameters. Obviously, these parameters can be transferred to the initial conditions; indeed, upon the substitution $\mathbf{G} = \mathbf{F}\mathbf{A}_0$ (suggested by Dedekind [147]), the Lagrangian function and the constraint equation can be written as

$$L = \frac{1}{2} \operatorname{Tr}(\dot{\mathbf{G}}\dot{\mathbf{G}}^T) + 2\varepsilon \int_0^\infty \frac{d\lambda}{\sqrt{\det(\mathbf{G}\mathbf{G}^T + \lambda \mathbf{E})}}, \tag{I.38}$$
$$\varphi = \det \mathbf{G} = \det \mathbf{A}_0 = \text{const}.$$

The initial conditions have obviously the form $\mathbf{G}|_{t=0} = \mathbf{A}_0$, and the equation of motion preserves its form, $\left(\frac{\partial L}{\partial \dot{\mathbf{G}}}\right)^{\cdot} - \frac{\partial L}{\partial \mathbf{G}} = \widetilde{\mathcal{H}} \frac{\partial \varphi}{\partial \mathbf{G}}$.

It can also be shown that the substitution

$$\mathbf{G} \to (\det \mathbf{A}_0)^{1/3} \mathbf{G}, \quad t \to \frac{(\det \mathbf{A}_0)^{1/3}}{2\varepsilon} t$$

reduces the system (I.38) to the case $\varepsilon = 1/2, \varphi = 1$. Thus, the dynamics of the self-gravitating fluid ellipsoid is described by a natural Lagrangian system without parameters on the $SL(3)$ group.

The first integrals — vorticity (I.30), momentum (I.32), and energy (I.34) — can be represented in the form

$$\boldsymbol{\Xi} = \mathbf{G}^T\dot{\mathbf{G}} - \dot{\mathbf{G}}^T\mathbf{G}, \quad \mathbf{M} = \mathbf{G}\dot{\mathbf{G}}^T - \dot{\mathbf{G}}\mathbf{G}^T,$$
$$\mathcal{E} = \frac{1}{2}\operatorname{Tr}(\dot{\mathbf{G}}\dot{\mathbf{G}}^T) - 2\varepsilon\int_0^\infty \frac{d\lambda}{\sqrt{\det(\mathbf{G}\mathbf{G}^T + \lambda\mathbf{E})}}. \tag{I.39}$$

Riemann used the decomposition (I.13) to represent the equations of motion on the configuration space $\mathbb{R}^2 \otimes \mathrm{SO}(3) \otimes \mathrm{SO}(3)$ (the direct product of the Abel group of translations and two copies of rotation groups of three-dimensional space), with the elements of the matrices $\mathbf{w}$ and $\boldsymbol{\omega}$ corresponding to the velocity components with respect to the basis of left-invariant vector fields. The equations of motion take the form of the Poincaré equations on the Lie group [97]; in view of the fact that the Lagrangian function (I.38) is independent of the elements of the matrices $\mathbf{Q}$ and $\boldsymbol{\Theta}$ and taking into account the constraint $\varphi = A_1A_2A_3 = \text{const}$, we obtain the following representation of the Riemann equations:

$$\left(\frac{\partial L}{\partial \dot{A}_i}\right)^{\cdot} = \frac{\partial L}{\partial A_i} + \tilde{\kappa}\frac{\partial \varphi}{\partial A_i},$$
$$\left(\frac{\partial L}{\partial \mathrm{w}_i}\right)^{\cdot} = \sum_{j,k}\varepsilon_{ijk}\frac{\partial L}{\partial \mathrm{w}_j}\mathrm{w}_k, \quad \left(\frac{\partial L}{\partial \omega_i}\right)^{\cdot} = \sum_{j,k}\varepsilon_{ijk}\frac{\partial L}{\partial \omega_j}\omega_k, \tag{I.40}$$

where $\tilde{\kappa}$ is the undetermined Lagrange multiplier (which coincides with σ up to a factor) and ε_{ijk} is the Levi – Civita antisymmetric tensor.

Here and in what follows, the components w_i and ω_i are related to the elements of the skew-symmetric matrices (I.15) according to the standard rule

$$\mathrm{w}_{ij} = \varepsilon_{ijk}\mathrm{w}_k, \quad \omega_{ij} = \varepsilon_{ijk}\omega_k. \tag{I.41}$$

Symmetry Group and the Dedekind Reciprocity Law. The Lagrangian representation of the Dirichlet equations (I.36) offers a very simple way to find the symmetry group of the system. Indeed, it can be shown that the Lagrangian with the constraint (I.38) and, therefore, the equations of motion are invariant under transformations of the form

$$\mathbf{G}' = \mathbf{S}_1\mathbf{G}\mathbf{S}_2, \quad \mathbf{S}_1, \mathbf{S}_2 \in SO(3). \tag{I.42}$$

Thus, the system is invariant under the group $\Gamma = SO(3) \otimes SO(3)$.

Clearly, the Noether integrals corresponding to the transformations (I.42) are the integrals of vorticity and total momentum (I.39). Accordingly, as will be shown below, the Riemann equations describe a system reduced using the above symmetry group.

Furthermore, it can easily be shown using (I.38) that the equations of motion are invariant under a discrete transformation of the transposition of matrices:

$$\mathbf{G}' = \mathbf{G}^T.$$

Therefore, we have

Theorem 1 (The Dedekind reciprocity law). *Any solution,* $\mathbf{G}(t)$*, of the Dirichlet equations can be placed in correspondence with the solution* $\mathbf{G}'(t) = \mathbf{G}^T(t)$ *for which the rotation of the ellipsoid and the rotation of the fluid inside the ellipsoid (i. e.,* $\boldsymbol{\Theta}$ *and* $\mathbf{Q}$*; see* (I.13)*) are interchanged.*

The most well-known example is the Dedekind ellipsoid reciprocal to the Jacobi ellipsoid. In this case, the axes of the triaxial ellipsoid are fixed in space and the fluid inside it moves about the minor axis in closed ellipses [147, 487].

Hamiltonian Formalism and Symmetry Reduction. We represent the Riemann equations in Hamiltonian form. To this end, we first use the constraint equation $\varphi = \text{const.}$ to find a representation of one semiaxis,

$$A_3 = \frac{v_0}{A_1 A_2}, \tag{I.43}$$

where v_0 is the volume of the ellipsoid (up to a factor). We perform the Legendre transformation

$$\begin{gathered} p_i = \frac{\partial L}{\partial \dot{A}_i}, \quad m_k = \frac{\partial L}{\partial \mathrm{w}_k}, \quad \mu_k = \frac{\partial L}{\partial \omega_k}, \quad i = 1, 2, \quad k = 1, 2, 3, \\ H = \sum_i p_i \dot{A}_i + \sum_k (m_k \mathrm{w}_k + \mu_k \omega_k) - L\,|_{\dot{A},\omega,\mathrm{w}\to p,m,\mu}. \end{gathered} \tag{I.44}$$

Using the expressions for the integrals, (I.30) and (I.33), it can be shown that the vectors $\boldsymbol{m} = (m_1, m_2, m_3)$ and $\boldsymbol{\mu} = (\mu_1, \mu_2, \mu_3)$ are related to the momentum and vorticity of the ellipsoid by

$$\boldsymbol{m} = \mathbf{Q}^T \boldsymbol{M}', \qquad \boldsymbol{\mu} = \boldsymbol{\Theta}^T \boldsymbol{\xi}', \tag{I.45}$$

where the vectors $\boldsymbol{M}'$ and $\boldsymbol{\xi}'$ are composed of the components of the antisymmetric matrices $\mathbf{M}'$ and Ξ' according to the standard rule (I.41). In terms of the new variables, the equations of motion become

$$\begin{gathered} \dot{A}_i = \frac{\partial H}{\partial \dot{p}_i}, \quad \dot{p}_i = \frac{\partial H}{\partial A_i}, \quad i = 1, 2, \\ \dot{\boldsymbol{m}} = \boldsymbol{m} \times \frac{\partial H}{\partial \boldsymbol{m}}, \quad \dot{\boldsymbol{\mu}} = \boldsymbol{\mu} \times \frac{\partial H}{\partial \boldsymbol{\mu}}, \end{gathered} \tag{I.46}$$

Here, the Hamiltonian is

$$\begin{aligned} H &= H_A + H_{m\mu} + U_e, \\ H_A &= \frac{1}{2}\frac{A_3^{-2}(p_1^2+p_2^2) + (p_1A_2^{-1} - p_2A_1^{-1})^2}{\sum A_i^{-2}}, \\ H_{m\mu} &= \frac{1}{4}\sum_{\text{cycle}}\left(\frac{m_i+\mu_i}{A_j-A_k}\right)^2 + \left(\frac{m_i-\mu_i}{A_j+A_k}\right)^2, \end{aligned} \tag{I.47}$$

where U_e is specified by (I.34) and it is assumed that A_3 is defined according to (I.43).

In addition, equations (I.46) must necessarily be supplemented with equations governing the evolution of the matrices $\mathbf{Q}$ and $\boldsymbol{\Theta}$; they have the form

$$\dot{Q}_{ij} = \sum_{k,l}\varepsilon_{ikl}Q_{kj}\frac{\partial H}{\partial m_l}, \quad \dot{\Theta}_{ij} = \sum_{k,l}\varepsilon_{ikl}\Theta_{kj}\frac{\partial H}{\partial \mu_l}. \tag{I.48}$$

Equations (I.46) and (I.48) form a Hamiltonian system with eight degrees of freedom and noncanonical Poisson brackets:

$$\{A_i, p_j\} = \delta_{ij}, \quad \{m_i, m_j\} = \varepsilon_{ijk}m_k, \quad \{\mu_i, \mu_j\} = \varepsilon_{ijk}\mu_k, \tag{I.49}$$

$$\{m_r, Q_{jk}\} = \varepsilon_{ikl}Q_{jl}, \quad \{\mu_i, \Theta_{jk}\} = \varepsilon_{ikl}\Theta_{jl}, \tag{I.50}$$

where zero brackets are omitted.

REMARK 4. The elimination of one of the semiaxes (I.43) results in the loss of symmetry of the Hamiltonian (I.47); therefore, the equations for the semiaxes A_i are usually left in Lagrangian form with an undetermined multiplier [487, 123].

It can be seen from the above relations that the system of equations (I.46), which governs the evolution of the variables $A_i, p_i, \boldsymbol{m}$, and $\boldsymbol{\mu}$, decouples; in addition, the Poisson bracket of these variables, (I.49), is also closed. It is not difficult to show that equations (I.46) describe a system reduced using the symmetry group (I.42).

Limitation: the brackets (I.49) obviously have two Casimir functions

$$\Phi_m = (\boldsymbol{m}, \boldsymbol{m}), \quad \Phi_\mu = (\boldsymbol{\mu}, \boldsymbol{\mu}), \tag{I.51}$$

and have rank eight (provided that $\Phi_m \neq 0$, $\Phi_\mu \neq 0$).

Therefore, the reduced system has generally four degrees of freedom.

In particular cases where *one of the integrals* (I.51) *is zero, the reduced system has three degrees of freedom.* These are the so-called irrotational ($\Phi_\mu = 0$) and momentum-free ($\Phi_m = 0$) ellipsoids.

If both of the integrals (I.51) *vanish, the reduced system has two degrees of freedom* and describes oscillations of the ellipsoid without changes in the directions of the axes and without inner flows (this case will be considered below in detail).

REMARK 5. The linear transformation

$$\boldsymbol{L} = \boldsymbol{m} + \boldsymbol{\mu}, \quad \boldsymbol{\pi} = \boldsymbol{m} - \boldsymbol{\mu}$$

reduces the angular part of the Hamiltonian (I.47) to the diagonal form

$$\begin{gathered} H_{m\mu} = \frac{1}{4}(\boldsymbol{L}, \boldsymbol{\Lambda L}) + \frac{1}{4}(\boldsymbol{\pi}, \boldsymbol{\Pi \pi}), \\ \boldsymbol{\Lambda} = \operatorname{diag}\left(\frac{1}{(A_2 - A_3)^2}, \frac{1}{(A_3 - A_1)^2}, \frac{1}{(A_1 - A_2)^2}\right), \\ \boldsymbol{\Pi} = \operatorname{diag}\left(\frac{1}{(A_2 + A_3)^2}, \frac{1}{(A_3 + A_1)^2}, \frac{1}{(A_1 + A_2)^2}\right), \end{gathered} \tag{I.52}$$

in this case, the Poisson brackets reduce to the form

$$\{L_i, L_j\} = \varepsilon_{ijk} L_k, \quad \{L_i, \pi_j\} = \varepsilon_{ijk}\pi_k, \quad \{\pi_i, \pi_j\} = \varepsilon_{ijk} L_k$$

and correspond, as is well known, to an $so(4)$ algebra. The corresponding equations of motion can be represented in the matrix form

$$\dot{\mathbf{X}} = [\mathbf{X}, \boldsymbol{\Omega}], \tag{I.53}$$

where

$$\mathbf{X} = \begin{Vmatrix} 0 & L_3 & -L_2 & \pi_1 \\ -L_3 & 0 & L_1 & \pi_2 \\ L_2 & -L_1 & 0 & \pi_3 \\ -\pi_1 & -\pi_2 & -\pi_3 & 0 \end{Vmatrix}, \quad \boldsymbol{\Omega} = \begin{Vmatrix} 0 & \frac{\partial H}{\partial L_3} & -\frac{\partial H}{\partial L_2} & \frac{\partial H}{\partial \pi_1} \\ -\frac{\partial H}{\partial L_3} & 0 & \frac{\partial H}{\partial L_1} & \frac{\partial H}{\partial \pi_2} \\ \frac{\partial H}{\partial L_2} & -\frac{\partial H}{\partial L_1} & 0 & \frac{\partial H}{\partial \pi_3} \\ -\frac{\partial H}{\partial \pi_1} & -\frac{\partial H}{\partial \pi_2} & -\frac{\partial H}{\partial \pi_3} & 0 \end{Vmatrix}, \tag{I.54}$$

and the equations for A_i and p_i preserve their previous form (I.46).

Equations (I.53) coincide in their form with the equations of motion of a free four-dimensional rigid body. In rigid body dynamics, the momentum and angular velocity are in this case related by the linear equations

$$\mathbf{X} = \frac{1}{2}(\mathbf{J}\boldsymbol{\Omega} + \boldsymbol{\Omega}\mathbf{J}), \tag{I.55}$$

where the constant symmetric matrix $\mathbf{J}$ is the moment of inertia of the body with respect to the body-fixed axes. It can easily be shown that the matrices (I.54) for the Dirichlet–Riemann problem do not satisfy the relation (I.55) at any matrix $\mathbf{J}$, i. e., the analogy with rigid body dynamics is purely formal in this case. Recall that the matrices $\boldsymbol{\Lambda}$ and $\boldsymbol{\Pi}$ for a four-dimensional rigid body have the form [97]

$$\boldsymbol{\Lambda} = \operatorname{diag}\left(\frac{1}{\lambda_2 + \lambda_3}, \frac{1}{\lambda_1 + \lambda_3}, \frac{1}{\lambda_1 + \lambda_2}\right), \quad \boldsymbol{\Pi} = \operatorname{diag}\left(\frac{1}{\lambda_0 + \lambda_1}, \frac{1}{\lambda_0 + \lambda_2}, \frac{1}{\lambda_0 + \lambda_3}\right).$$

This form of equations (I.53) was pointed out by Dyson [177] for the case of the dynamics of a compressible ellipsoid (see below).

The canonical variables in the Riemann equations were introduced for the first time by Betti [43], who used the commutation representations of the $so(4)$ algebra long before the advent of the modern theory of Hamiltonian systems on the Lie algebras. Using commutation, he introduced, in a quite modern way, canonical variables to reduce the integration of the Riemann equations to the integration of the Hamilton – Jacobi equations. The Hamiltonian nature of the Riemann equations is also considered in modern studies [493, 494, 238], which are related to the representation of the equations of motion on an extended Lie algebra for which the real motions lie on special orbits; the value of such a calculation for dynamics is not yet clear to us. A more formal procedure of reduction and Hamiltonization of the Riemann equations relevant to our study is described in [186]. A similar analysis is made in [261] in the context of the Dirichlet motions in ideal magnetohydrodynamics. An alternative approach to the Hamiltonian nature, which also should be discussed, is presented in [44].

5 Particular Cases of motion

Shape-preserving Motions of the Ellipsoid. The simplest motions of the fluid ellipsoid are represented by a family of solutions for which all the three axes of the ellipsoid are time-independent,

$$A_i = \text{const.}, \quad i = 1, 2, 3. \tag{I.56}$$

Clearly, the Maclaurin and Jacobi ellipsoids are examples of such motions. In these cases, the ellipsoid rotates as a rigid body about the principal axis (the symmetry axis for the Maclaurin ellipsoid and the shortest axis for the Jacobi ellipsoid).

The Dedekind ellipsoid provides another example of such motions, the axes being invariable in both their lengths and directions. As noted above, the Dedekind ellipsoid is reciprocal to the Jacobi ellipsoid in the sense of Theorem 1 (while the Maclaurin ellipsoid is self-reciprocal).

For all the above-mentioned solutions (the Maclaurin, Jacobi, and Dedekind ellipsoids), two pairs of components of the vectors $\boldsymbol{m}$ and $\boldsymbol{\mu}$ vanish, and the remaining components are constant (for example, it can be assumed without loss of generality that $m_1 = \mu_1 = m_2 = \mu_2 = 0$, $m_3 = \text{const.}$, $\mu_3 = \text{const.}$).

Riemann [487] has proved the following, more general result:

Theorem 2. *Let* (I.56) *be satisfied and let all the* A_i *be different. Then* $\boldsymbol{m}$ *and* $\boldsymbol{\mu}$ *are time-independent and at least one pair of components of these vectors vanishes* (*i. e.,* $m_i = \mu_i = 0$ *for some* i).

As a consequence, we find that any motion of a shape-preserving fluid ellipsoid whose axes do not coincide is a fixed point of the reduced system (I.46) or, which is the same, of the Riemann equations (I.27). Another proof of this statement is given in [458].

Riemann also found new solutions — the Riemann ellipsoids — for the case where only one pair of components of $\boldsymbol{m}$ and $\boldsymbol{\mu}$ vanishes (i. e., $m_1 = \mu_1 = 0, \mu_2, m_2, \mu_3, m_3 \neq 0$).

V. A. Stekloff [551, 552] analyzed in detail the case of equality of a pair of axes ($A_i = A_j \neq A_k$) and showed that no shape-preserving motions other than Maclaurin ellipsoids (spheroids) exist in this case. In this sense, he generalized the Riemann result to the axisymmetric case (Riemann himself gave no detailed proof for this case). An attempt to revise Riemann's results was made in [419].

Axisymmetric Case (Dirichlet [160]) . It can easily be shown that the equations of motion defined by the Lagrangian function (I.38) admit a (two-dimensional) invariant manifold that consists of matrices of the form

$$\mathbf{G} = \begin{Vmatrix} u & v & 0 \\ -v & u & 0 \\ 0 & 0 & w \end{Vmatrix},$$

where $\det \mathbf{G} = (u^2 + v^2)w = v_0 = \text{const.}$ is the volume of the ellipsoid. This manifold corresponds to an axisymmetric motion of the fluid ellipsoid (see [160]). In this case, the matrix of the principal semiaxes is

$$\mathbf{A} = (\mathbf{G}\mathbf{G}^T)^{1/2} = \operatorname{diag}(\sqrt{u^2 + v^2}, \sqrt{u^2 + v^2}, w).$$

Taking into account the condition $\det \mathbf{G} = v_0$, we make the change of variables

$$u = v_0^{1/3} r \cos\psi, \quad v = v_0^{1/3} r \sin\psi, \quad w = \frac{v_0^{1/3}}{r^2}$$

and find that the Lagrangian function (I.38) is

$$L = v_0^{2/3}\left(\left(1 + \frac{2}{r^6}\right)\dot{r}^2 + r^2\dot{\psi}^2 + U_s\right),$$

where

$$U_s = -\frac{2\varepsilon}{v_0}\int_0^\infty \frac{d\lambda}{(\lambda + r^2)\sqrt{\lambda + 1/r^4}} = -\frac{2\varepsilon}{v_0}r^2 \times \begin{cases} \dfrac{2\arctan\sqrt{r^6 - 1}}{\sqrt{r^6 - 1}}, & r > 1, \\[2ex] \dfrac{\ln\left(\frac{1+\sqrt{1-r^6}}{1-\sqrt{1-r^6}}\right)}{\sqrt{1 - r^6}}, & r < 1. \end{cases}$$

The variable ψ is cyclic; therefore, we have a first integral of the form

$$p_\psi = \frac{1}{v_0^{2/3}}\frac{\partial L}{\partial\dot{\psi}} = 2r^2\dot{\psi},$$

which coincides up to a factor with the single nonzero component of the momentum M'_{12} (I.32). Using the energy integral (I.34), we obtain a quadrature that specifies the evolution of r:

$$\left(1 + \frac{2}{r^6}\right)\dot{r}^2 = h - U_*, \quad U_* = U_s + \frac{c}{r^2},$$

where $h = \frac{\mathcal{E}}{mv_0^{2/3}}$ and $c = \frac{p_\psi}{4}$ are fixed values of the energy and momentum integrals. The minimum of the reduced potential $U_\star$ corresponds to the Maclaurin spheroid.

Riemannian Case [487]. There is an invariant manifold that is more general than that described above. It is specified by the block – diagonal matrix of the general form

$$\mathbf{G} = \begin{Vmatrix} u_1 & v_1 & 0 \\ u_2 & v_2 & 0 \\ 0 & 0 & w_3 \end{Vmatrix}. \tag{I.57}$$

Calculating the integrals (I.30) and (I.32), we obtain

$$M'_{12} = u_1\dot{u}_2 - u_2\dot{u}_1 + v_1\dot{v}_2 - v_2\dot{v}_1, \quad M'_{23} = M'_{13} = 0,$$
$$\xi'_{12} = u_1\dot{v}_1 - v_1\dot{u}_1 + u_2\dot{v}_2 - v_2\dot{u}_2, \quad \xi'_{23} = \xi'_{13} = 0.$$

It is also obvious that $\mathbf{Q}$ and $\boldsymbol{\Theta}$ have in this case a block – diagonal form similar to (I.57); therefore, this case corresponds to that found by Riemann, for which, in equations (I.46), we have to set

$$m_1 = m_2 = 0, \qquad m_3 = \text{const},$$
$$\mu_1 = \mu_2 = 0, \qquad \mu_3 = \text{const}.$$

Thus, we obtain a Hamiltonian system with two degrees of freedom, which describes the evolution of the principal semiaxes A_1 and A_2; its Hamiltonian is

$$H = \frac{1}{2}\frac{A_3^{-2}(p_1^2 + p_2^2) + (p_1 A_2^{-1} - p_2 A_1^{-1})^2}{\sum A_i^{-2}} + U_\star(A_1, A_2), \tag{I.58}$$

where the reduced potential is

$$U_\star = U_e + \frac{c_1^2}{(A_1 - A_2)^2} + \frac{c_2^2}{(A_1 + A_2)^2},$$

and $c_1^2 = \frac{1}{4}(m_3 + \mu_3)^2$, $c_2^2 = \frac{1}{4}(m_3 - \mu_3)^2$ are fixed constants of the integrals.

The particular case of the system (I.58) for $c_1 = c_2 = 0$ (i. e., for invariable directions of the principal axes of the ellipsoid) was also found by Kirchhoff [301], who suggested that the problem does not reduce to quadratures.

At $U_\star = 0$, the Hamiltonian (I.58) describes a geodesic flow on the cubic $A_1A_2A_3 =$ const. This remarkable analogy between two different dynamical systems was also pointed out by Riemann.

Elliptic Cylinder (Lipschitz [387]). This case can be obtained through a limiting process in the Riemannian case by letting one axis of the ellipsoid tend to infinity ($A_3 \to \infty$). However, it is more convenient to start by considering the case of a two-dimensional motion of fluid assuming that the matrix $\mathbf{F}$ has the form

$$\mathbf{F} = \begin{Vmatrix} \bar{\mathbf{F}} & 0 \\ 0 & 1 \end{Vmatrix}, \qquad \det \bar{\mathbf{F}} = 1, \tag{I.59}$$

where $\bar{\mathbf{F}}$ is a 2×2 matrix with a unit determinant.

Obviously, the considerations on which the derivation of the Dirichlet equations [395] was based can be applied to this case without modifications; only the right-hand side of the equations should be properly changed. To this end, it is necessary to use the well-known representation of the potential of the interior points of the elliptic cylinder with a large length l in the system of principal axes

$$U(\boldsymbol{\zeta}) = \bar{\varepsilon}\left(U_0(l) - \frac{\zeta_1^2}{A_1(A_1 + A_2)} - \frac{\zeta_2^2}{A_2(A_1 + A_2)}\right) + O(1/l),$$

where $\bar{\varepsilon} = G\bar{m}$, G is the gravitational constant and $\bar{m} = \pi\rho A_1 A_2$ is the mass per unit length of the cylinder. The constant $U_0(l) \underset{l\to\infty}{\longrightarrow} \infty$ does not appear in the equations of motion and can be omitted.

By analogy with the above considerations, we pass to the Lagrangian representation and make the substitution $\bar{\mathbf{G}} = \bar{\mathbf{F}}\bar{\mathbf{A}}_0$, where $\bar{\mathbf{A}}_0 = \operatorname{diag}(A_1^0, A_2^0)$, to obtain the Lagrangian of the system in the form

$$L = \frac{1}{2}\operatorname{Tr}\left(\dot{\bar{\mathbf{G}}}\dot{\bar{\mathbf{G}}}^T\right) - \bar{U}_e,$$
$$\bar{U}_e = -2\bar{\varepsilon}\ln(A_1 + A_2)^2 = -2\bar{\varepsilon}\ln(\operatorname{Tr}(\bar{\mathbf{G}}\bar{\mathbf{G}}^T) + 2\det\bar{\mathbf{G}}).$$

Explicitly substituting

$$\bar{\mathbf{Q}} = \begin{Vmatrix}\cos\phi & -\sin\phi \\ \sin\phi & \cos\phi\end{Vmatrix}, \quad \bar{\boldsymbol{\Theta}} = \begin{Vmatrix}\cos\psi & -\sin\psi \\ \sin\psi & \cos\psi\end{Vmatrix}, \quad \mathbf{A} = \begin{Vmatrix}A_1 & 0 \\ 0 & A_2\end{Vmatrix}$$

into the singular decomposition of the matrix $\bar{\mathbf{G}} = \bar{\mathbf{Q}}^T\bar{\mathbf{A}}\bar{\boldsymbol{\Theta}}$, we obtain a Lagrangian function in the form

$$L = \frac{1}{2}\left(\dot{A}_1^2 + \dot{A}_2^2 + (A_1\dot{\phi} - A_2\dot{\psi})^2 + (A_2\dot{\phi} - A_1\dot{\psi})^2\right) - \bar{U}_e(A_1, A_2).$$

We can see that the variables ϕ and ψ are cyclic and that there are two linear integrals

$$\frac{\partial L}{\partial\dot{\phi}} = p_\phi, \quad \frac{\partial L}{\partial\dot{\psi}} = p_\psi. \tag{I.60}$$

We parameterize the relation $A_1 A_2 = \bar{v}_0$ using hyperbolic functions:

$$A_1 = \bar{v}_0^{1/2}(\operatorname{ch} u + \operatorname{sh} u), \quad A_2 = \bar{v}_0^{1/2}(\operatorname{ch} u - \operatorname{sh} u).$$

Using the energy integral and the integrals (I.60), we obtain a quadrature for the variable u:

$$\bar{v}_0(\operatorname{ch} 2u)\dot{u}^2 = h - \bar{U}_*,$$
$$\bar{U}_* = 2\bar{\varepsilon}\ln(\operatorname{ch} u) + \frac{\bar{c}_1^2}{\operatorname{ch}^2 u} + \frac{\bar{c}_2^2}{\operatorname{sh}^2 u},$$

where $\bar{c}_1^2 = \frac{1}{16}(p_\phi - p_\psi)^2$, $\bar{c}_2^2 = \frac{1}{16}(p_\phi + p_\psi)^2$, and h are fixed constants of the first integrals.

6 Chaotic oscillations of a Triaxial ellipsoid

Let us consider in more detail the oscillations (pulsations) of a fluid ellipsoid in the Riemannian case (I.57). We now represent the equations of motion of the system (I.58) in a Hamiltonian form most convenient for a numerical analysis of the system. We parameterize the surface $A_1A_2A_3 = v_0$ using cylindrical coordinates

$$A_1 = r\cos\phi, \quad A_2 = r\sin\phi, \quad A_3 = \frac{2v_0}{r^2\sin^2 2\phi}, \\ p_1 = p_r\cos\phi - \frac{p_\phi}{r}\sin\phi, \quad p_2 = p_r\sin\phi - \frac{p_\phi}{r}\cos\phi. \tag{I.61}$$

The Hamiltonian (I.58) can be represented as

$$H = \frac{1}{2}\left(1 + \frac{c_0^2}{r^6\sin^4 2\phi}\right)^{-1}\left(p_r^2 + \frac{p_\phi^2}{r^2} + \frac{c_0^2}{r^6\sin^4 2\phi}\left(p_r\cos 2\phi - \frac{p_\phi}{r}\sin 2\phi\right)^2\right) \\ + U_*(r,\phi), \tag{I.62}$$

where $c_0 = 4v_0$.

Since the original system is defined in the square $A_1 > 0, A_2 > 0, A_3 > 0$, for this case we have $0 < \phi < \pi/2$. In this system, making a transformation of variables

$$\rho = r^2, \quad \psi = 2\phi, \tag{I.63}$$

we obtain a Hamiltonian in the form

$$H = \frac{2(\rho^2(c_0^2\cos^2\psi + \rho^3\sin^4\psi)p_\rho^2 + \sin^2\psi(c_0^2 + \rho^3\sin^2\psi)p_\phi^2 - 2\rho c_0^2\cos\psi\sin\psi p_\psi p_\phi)}{\rho(c_0^2 + \rho^3\sin^4\psi)} \\ + U_*(\rho,\psi). \tag{I.64}$$

Passing to new Cartesian coordinates according to the formulas

$$x = \rho\cos\psi, \quad y = \rho\sin\psi, \tag{I.65}$$

we obtain

$$H = 2\rho\left(p_x^2 + \frac{y^4 p_y^2}{y^4 + \bar{c}_0^2\rho}\right) + U_*(x,y), \tag{I.66}$$

where $\rho = \sqrt{x^2 + y^2}$; obviously, the system (I.66) is defined in the upper semiplane ($y > 0$). In this case, as we can see, the kinetic energy of the system has the simplest form.

REMARK 6. The transformation (I.63) is the Levi – Civita transformation (also called the Bolin transformation) known in celestial mechanics, which is usually written in the complex form

$$x + iy = \rho e^{i\psi} = (A_1 + iA_2)^2.$$

As already noted above, at $U_\star = 0$, the Hamiltonian (I.66) describes a geodesic flow on the cubical surface $A_1A_2A_3 = \text{const}$, embedded in Euclidian space $\mathbb{R}^3$. Almost all trajectories (geodesics) of this system are not compact; therefore, computer simulations for a numerical proof of nonintegrability at $U_\star = 0$ cannot be performed. As shown recently by S. L. Ziglin [654], this system (at $U_\star = 0$, i. e., a geodesic flow) does not admit a meromorphic additional integral.

REMARK 7. Various algebraic surfaces in three-dimensional space and singular lines on them (asymptotic lines, lines of curvature, and geodesics) were actively investigated by mathematicians in the 19th century. A tremendous surge of interest in this subject was spurred by the discovery, due to Jacobi, of the integrability of the problem of geodesics on the ellipsoid and on quadrics in general. This integrability also holds in multidimensional cases. The problem at hand is a classical example of separation of variables. There is an extensive literature devoted to studying this problem from both an analytical standpoint (integration in terms of theta functions) and from a qualitative point of view. However, the mathematicians of the 19th century did not succeed much in finding geodesic flows on higher-order surfaces, and those of the 20th century contributed almost nothing to it. It was probably for this problem that Darboux developed the theory of orthogonal families of surfaces and investigated interesting surfaces of the third (Darboux cyclides) and fourth degree (which were also discovered by Roberts [489] and Wangerin [607]). A distinctive feature of these families of surfaces is that they are Lamé families, form a triorthogonal network in three-dimensional space, and a solution for asymptotic and curvature lines can be written for them in the form of elliptic quadratures. However, geodesics for these surfaces have not been found. As our numerical simulations show, the reason for this fact is the nonintegrability of the geodesic flow. In the context of this problem, we also mention the studies by Schläfli [523] and Cayley [122] devoted to the classification of various cubic surfaces in space, which, following the fundamental work of Jacobi on geodesics in quadrics, laid the foundation of modern algebraic geometry. Recently, V. V. Kozlov [338] has found topological obstructions to the integrability of geodesic flows on noncompact algebraic surfaces (in particular, those of the third and fourth degrees). Unfortunately, his results do not apply to Riemannian surfaces $A_1A_2A_3 = \text{const}$.

Figure I.2 shows phase portraits of the system (I.66). The plane $x = 1$ has been chosen as the plane of the Poincaré section. It can be seen from the diagrams that the phase portrait is almost regular at energies close to the minimum energy (see Figs. I.2a and I.2c). As the energy is increased, the phase portrait becomes chaotic, which can be clearly seen from Figs. 1b and 1d. Figures 1e and 1f also show an intersection of unstable invariant manifolds (separatrices), which can serve as a numerical proof of the nonintegrability of this system.

REMARK 8. In the numerical integration of the equations of motion, it is convenient to use the representation of the potential U_e (I.34) and its derivatives in terms of elliptic

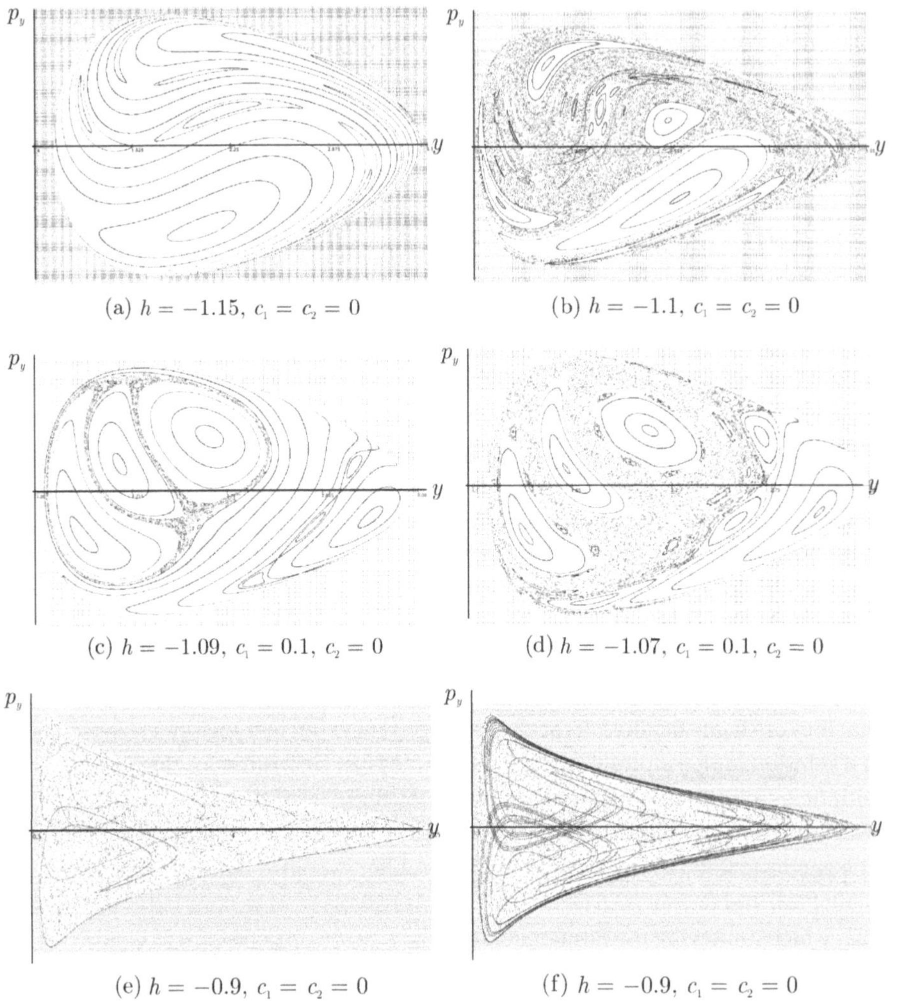

Fig. I.2. The Poincaré map of system (I.66). For all panels, $c_0 = 1, \varepsilon = 0, 6$; the planes $x = 1$ (a – d) and $x = 0.1$ (e ,f) have been chosen as the plane of the Poincaré section.

functions,

$$I(A_1, A_2, A_3) = \int_0^\infty \frac{d\lambda}{\Delta(\lambda)} = \frac{2}{æ} F(\varphi, k),$$

$$æ = \sqrt{A_1^2 - A_3^2}, \quad k = \sqrt{\frac{A_1^2 - A_2^2}{A_1^2 - A_3^2}}, \quad \varphi = \arcsin\left(\sqrt{\frac{A_1^2 - A_3^2}{A_1^2}}\right),$$

where $F(\varphi, k) = \int_0^{\varphi} \frac{d\alpha}{\sqrt{1 - k^2 \sin^2 \alpha}}$ is an elliptic integral of the first kind.

Due to the invariance under permutations of the quantities A_1, A_2, and A_3, they can be ordered so as to make all values of æ, φ, and k real. To find the derivatives (which are obviously not invariant under permutations of A_i), a necessary ordering of A_1, A_2, and A_3 should be made first, after which the derivative of the integral I with respect to the corresponding argument should be calculated.

For the derivatives $F(\varphi, k)$, the following relations hold:

$$\frac{\partial F}{\partial \varphi} = \left(1 - k^2 \sin^2 \varphi\right)^{-1/2},$$

$$\frac{\partial F}{\partial k} = \frac{1}{1 - k^2} \left(\frac{E(\varphi, k) - (1 - k^2) F(\varphi, k)}{k} - \frac{k \sin \varphi \cos \varphi}{\sqrt{1 - k^2 \sin^2 \varphi}} \right),$$

where $E(\varphi, k) = \int_0^{\varphi} \sqrt{1 - k^2 \sin^2 \alpha}\, d\alpha$ is the second-kind elliptic integral.

I.2 Dynamics of a Gas Cloud with Ellipsoidal Stratification

1 Introduction

The investigation of the dynamics of gas ellipsoids traces back to a study by L. V. Ovsyannikov [457] (1956), who analyzed the most general equations describing the motion of an ideal polytropic gas, without taking into account gravitation with a velocity field linear in the coordinates of the gas particles (here and in what follows, by the gas ellipsoid we mean the Dirichlet solution generalized to various models of a compressible fluid). We note that the paper [457] is very brief and purely mathematical: in fact, the author obtains equations of motion, points out several possible cases in which the solution obtained exists, and presents an incomplete set of first integrals for two cases. It is interesting that Ovsyannikov's paper contains no references, so that no relation is revealed between his and Dirichlet's solutions.

Later, D. Lynden-Bell [396] (1962) demonstrated, also without any references, the existence of a solution in the form of a spheroid for a self-gravitating dust cloud (i. e., for a medium not resisting deformations, $p \equiv 0$).

Ya. B. Zel'dovich [643] (1965) obtained the equations of motion of a self-gravitating dust ellipsoid for the general case and studied (on a physical level of rigor) the possibility of collapse and expansion in this problem. Apparently, Ya. B. Zel'dovich also overlooked the relationship of this problem to the Dirichlet – Riemann problem, since the model of a dust cloud can be obtained simply by setting $p = 0$ in the Dirichlet equations.

Independently of Ovsyannikov (at least without any reference to the latter), F. Dyson [177] (1968) obtained the equations of motion of an ideal gas cloud in the case of an isothermal flow (but without the assumption of polytropic behavior of the gas); a Gaussian density distribution with an ellipsoidal stratification was found. Dyson showed a relation between the solution obtained and the Dirichlet problem and wrote the equations of motion of the gas ellipsoid in Riemannian form.

Also independently of Ovsyannikov, M. Fujimoto [198] (1968) described the model of a cooling ellipsoidal gas cloud; in essence, he obtained a generalization of a case considered by Ovsyannikov (if we assume the cooling parameter to be $æ = 0$, we obtain Ovsyannikov's equations). In addition, in Fujimoto's model, the density is constant, which enabled him to take into account the gravitational interaction between the particles of the cloud. Fujimoto also pointed out a relationship of this problem to the Dirichlet problem and, in studying it, used the techniques developed by Chandrasekhar [123] and Rossner [496].

We also mention a study by Anisimov [9] (1970). Following [457] and [177], he considered two integrable cases of the dynamics of a gas ellipsoid without taking account of gravitation, but under the additional condition of the monoatomic structure of the gas (a polytropic index of $\gamma = \frac{5}{3}$). One is the case of motion of an axisymmetric ellipsoid, the other is that of an elliptic cylinder. A nonautonomous Jacobi integral was found (which is due to the uniformity of the potential with uniformity degree -2). This integral is essentially necessary for integration of the cases under study; as we will show below, these systems are nonintegrable in the general case.

Bogoyavlenskii [55] (1976) analyzed the dynamics of a gas ellipsoid on a physical level of rigor taking into account gravitation (i. e., he considered the Fujimoto model without cooling). Explicit Lagrangian and Hamiltonian representations of the system are used.

Gaffet [201, 202, 203] showed that the system that describes irrotational gas ellipsoids without taking into account gravitation, for a monoatomic gas $\left(\gamma = \frac{5}{3}\right)$, satisfies the Painlevé property; in these studies, he found first integrals and carried out integration by quadratures for some particular cases.

There are also studies analyzing the spherically symmetric motion of a gas cloud; one of the most general solutions is described by Lidov [382], who considers the one-dimensional, spherically symmetric, unsteady adiabatic motion of a self-gravitating mass of a perfect gas.

Nemchinov [444] uses a solution that describes the ellipsoidal expansion of a gas cloud to find characteristic features of nonspherical explosions (in particular, he points out an increase in the flow rate in the direction of one of the principal axes as compared to a similar spherical explosion); the effect of the heating of the cloud on the expansion speed is also investigated.

Finally, we mention a series of studies (see [157] and references therein) generalizing the problem of the expansion of an ellipsoidal cloud to vacuum (or the collapse of an ellipsoidal cavity) in the presence of a rarefaction (compression) wave.

2 Equations of motion of a gas cloud with a linear velocity field

We now consider in a similar way the case where the motion of a *compressible fluid* (*gas*) is also defined by a linear transformation of the Lagrangian coordinates

$$\boldsymbol{x}(\boldsymbol{a}, t) = \mathbf{F}(t)\boldsymbol{a}; \tag{I.67}$$

for the compressible medium, the condition $\det \mathbf{F} = 1$ obviously does not hold. Clearly, the velocity field is linear in the coordinates of the fluid particles:

$$\boldsymbol{v}(\boldsymbol{x}, t) = \dot{\boldsymbol{x}} = \dot{\mathbf{F}}(t)\mathbf{F}^{-1}(t)\boldsymbol{x}.$$

In this case, the equations that describe the flow (for potential forces) have the following Lagrangian representation:

$$\left(\frac{\partial \boldsymbol{x}}{\partial \boldsymbol{a}}\right)^T \ddot{\boldsymbol{x}} = -\frac{\partial U}{\partial \boldsymbol{a}} - \frac{1}{\rho}\frac{\partial p}{\partial \boldsymbol{a}}, \tag{I.68}$$

and the continuity equation in Lagrangian form is

$$\dot{\rho} + \rho \operatorname{Tr}\left(\left(\frac{\partial \boldsymbol{x}}{\partial \boldsymbol{a}}\right)^{-1} \frac{\partial \dot{\boldsymbol{x}}}{\partial \boldsymbol{a}}\right) = 0, \tag{I.69}$$

where $\operatorname{Tr}\left(\left(\frac{\partial \boldsymbol{x}}{\partial \boldsymbol{a}}\right)^{-1} \frac{\partial \dot{\boldsymbol{x}}}{\partial \boldsymbol{a}}\right) = \operatorname{div} \boldsymbol{v}(\boldsymbol{x}, t)$.

For the flow of the structure under study (I.67), the continuity equations can easily be integrated. Indeed, if we introduce the notation $\varphi(t) = \det \mathbf{F}(t)$ using the relation $\left(\frac{\partial \varphi}{\partial \mathbf{F}}\right)^T = \varphi \mathbf{F}^{-1}$, we find that $\operatorname{Tr}\left(\left(\frac{\partial \boldsymbol{x}}{\partial \boldsymbol{a}}\right)^{-1} \frac{\partial \dot{\boldsymbol{x}}}{\partial \boldsymbol{a}}\right) = \frac{\dot{\varphi}}{\varphi}$; therefore,

$$\rho(\boldsymbol{a}, t) = \frac{f(\boldsymbol{a})}{\varphi(t)}, \tag{I.70}$$

where the function $f(\boldsymbol{a})$ is time-independent.

In addition to the four functions $\boldsymbol{x}(\boldsymbol{a}, t), p(\boldsymbol{a}, t)$, the medium of interest is described by three additional scalar parameters: density $\rho(\boldsymbol{a}, t)$, specific internal energy $U_{\text{in}}(\boldsymbol{a}, t)$, and temperature $T(a, t)$. Therefore, it is necessary to complement the system of equations (I.68) and (I.69) with three other equations. As is well known [368], these additional equations are of a thermodynamical rather than mechanical nature and depend essentially on our assumptions concerning the properties of the medium and on the character of the flow.

Depending on these assumptions, various gas dynamics models can be obtained. We consider three of the most well-known models, emphasizing the explicit assumptions. Unless otherwise specified, we assume in what follows that the potential of the external forces applied to the system, U, is zero.

The Ovsyannikov Model [457].

1°. *The gas is ideal and can be described by the equation of state*

$$p = \rho R T, \tag{I.71}$$

where R is the universal gas constant.

2°. *The gas is polytropic, and its internal energy depends linearly on the temperature,*

$$U_{\text{in}} = c_V T, \tag{I.72}$$

where c_V = const is the specific heat at constant volume.

3°. *The gas flow is adiabatic (i. e., there is no heat exchange between different parts of the gas volume); therefore, the energy variations are described by the equation*

$$\dot{U}_{\text{in}} = -p \left(\frac{1}{\rho}\right)^{\cdot}. \tag{I.73}$$

REMARK 1. Equation (I.73) is a consequence of the first principle of thermodynamics, $\delta Q = dU_{\text{in}} + p\,dV$, where, in view of assumption 3°, one has to set $\delta Q = 0$, $(V = \frac{1}{\rho})$.

REMARK 2. Recall that, due to the well-known thermodynamic identity

$$\left(\frac{dU_{\text{in}}}{dV}\right)_T = \left(T\left(\frac{\partial p}{\partial T}\right)_V - p\right),$$

the internal energy of the ideal gas (I.71) depends only on the temperature, $U_{\text{in}} = U_{\text{in}}(T)$.

Using equations (I.71) – (I.73) and taking into account (I.70), we find that

$$\frac{\dot{p}}{p} + \gamma \frac{\dot{\varphi}}{\varphi} = 0,$$

where the dimensionless constant $\gamma = 1 + \frac{R}{c_V}$ is the adiabatic index. Thus, for the thermodynamic quantities, we have

$$p(\boldsymbol{a}, t) = \frac{g(\boldsymbol{a})}{\varphi^{\gamma}(t)}, \quad U_{\text{in}}(\boldsymbol{a}, t) = \frac{1}{\gamma - 1} R T(\boldsymbol{a}, t) = \frac{1}{\gamma - 1} \varphi^{1-\gamma}(t) \frac{g(\boldsymbol{a})}{f(\boldsymbol{a})},$$

where $g(\boldsymbol{a})$ is an arbitrary, time-independent quantity.

Thus, for the existence of a solution of the form (I.67) in this case one has to require that

$$\frac{1}{f(\boldsymbol{a})}\nabla_a g(\boldsymbol{a}) = \mathbf{V}\boldsymbol{a}, \tag{I.74}$$

where $\mathbf{V}$ is some constant matrix and $\nabla_a = \left(\frac{\partial}{\partial a_1}, \frac{\partial}{\partial a_2}, \frac{\partial}{\partial a_3}\right)$. Then the equations of motion for $\mathbf{F}(t)$ are

$$\mathbf{F}^T\ddot{\mathbf{F}} + (\det\mathbf{F})^{1-\gamma}\mathbf{V} = 0 \qquad \textit{(the Ovsyannikov equations)}. \tag{I.75}$$

As Ovsyannikov has shown in [457], it is sufficient to consider the following solutions of equation (I.74).

Theorem 3. *Any solution of equation* (I.74) *can be reduced via a linear transformation of the Lagrangian coordinates to one of the following four types (depending on the rank of the matrix* $\mathbf{V}$*):*

I. $\mathbf{V} = \operatorname{diag}(\varepsilon_1, \varepsilon_2, \varepsilon_3)$, $\varepsilon_i = \pm 1 \quad (i = 1, 2, 3)$,
$g(\boldsymbol{a}) = g(s), \quad f(\boldsymbol{a}) = 2g'(s), \quad s = (\boldsymbol{a}, \mathbf{V}\boldsymbol{a})$;

II. $\mathbf{V} = \begin{Vmatrix} \varepsilon_1 & \delta & 0 \\ 0 & \varepsilon_2 & 0 \\ 0 & 0 & 0 \end{Vmatrix}, \quad \varepsilon_i = \pm 1 \quad (i = 1, 2)$,
if $\delta \neq 0$, *then* $g(\boldsymbol{a}) = g(s), \quad f(\boldsymbol{a}) = \frac{a_1 g'(s)}{(\boldsymbol{a}, \mathbf{V}\boldsymbol{a})}, \quad s = a_1 \bar{s}\left(\frac{a_2}{a_1}\right)$,
$\ln \bar{s}(\lambda) = \int \frac{\varepsilon_2 \lambda\, d\lambda}{\varepsilon_1 + \delta\lambda + \varepsilon_2\lambda}$;
if $\delta = 0$, *then* $g(\boldsymbol{a}) = g(s), \quad f(\boldsymbol{a}) = 2g'(s), \quad s = (\boldsymbol{a}, \mathbf{V}\boldsymbol{a})$;

III. $\mathbf{V} = \operatorname{diag}(\varepsilon, 0, 0)$, $\varepsilon = \pm 1, \quad g(\boldsymbol{a}) = g(a_1), \quad f(\boldsymbol{a}) = \frac{\varepsilon}{a_1} g'(a_1)$;

IV. $\mathbf{V} = 0, \quad g(\boldsymbol{a}) = \text{const}, \quad f(\boldsymbol{a})$ *is an arbitrary function.*

Proof. We outline the main steps of the proof. Rewriting the solvability condition (I.74) as $\nabla_a \times \nabla_a g(\boldsymbol{a}) = \nabla_a \times f(\boldsymbol{a})\mathbf{V}\boldsymbol{a} = 0$, we obtain

$$\nabla_a f(\boldsymbol{a}) \times (\mathbf{V}\boldsymbol{a}) = f(\boldsymbol{a})\boldsymbol{\omega}_V, \tag{I.76}$$

where $\boldsymbol{\omega}_V = (V_{23} - V_{32}, V_{31} - V_{13}, V_{12} - V_{21})$. We perform scalar multiplication by $\mathbf{V}\boldsymbol{a}$ and represent the solvability condition in the form of the vector equation

$$\mathbf{V}^T\boldsymbol{\omega}_V = 0. \tag{I.77}$$

In addition, it is clear that equations (I.75) are invariant under nonsingular transformations of the Lagrangian variables, for which

$$\boldsymbol{a} = \mathbf{S}\boldsymbol{a}', \quad \mathbf{V}' = \mathbf{S}^T\mathbf{V}\mathbf{S}, \quad \mathbf{F}' = (\det\mathbf{S})^{\frac{1-\gamma}{1+\gamma}}\mathbf{F}\mathbf{S}, \quad \det\mathbf{S} \neq 0. \tag{I.78}$$

If $\operatorname{rank}\mathbf{V} < 3$, the most general matrix of the corresponding rank can be reduced via the transformations (I.78) to the form given in cases II – IV. If $\operatorname{rank}\mathbf{V} = 3$, then $\boldsymbol{\omega}_V = 0$; hence, $\mathbf{V}$ is symmetric and reduces to the form of case I.

Relations (I.74) and (I.76) make it possible to easily find the corresponding functions $f(\boldsymbol{a})$ and $g(\boldsymbol{a})$ from the known matrix $\mathbf{V}$. □

Case I with the sign-definite matrix $\mathbf{V}$ is the most interesting from a physical point of view. In particular, if we set $\mathbf{V} = \mathrm{diag}(-1, -1, -1)$ (i. e., $s = -(\boldsymbol{a}, \boldsymbol{a})$) and choose a linear function $g(s)$, we obtain

$$g(\boldsymbol{a}) = \frac{1}{2}\rho_0(d_0^2 - (\boldsymbol{a}, \boldsymbol{a})), \quad f(\boldsymbol{a}) = \rho_0, \tag{I.79}$$

where we must set $\rho_0 > 0$ (since the density of the gas is positive). In this case, the gas is distributed with constant density $\rho = \rho_0 / \det(\mathbf{F})$ inside the finite ellipsoidal volume

$$(\boldsymbol{a}, \boldsymbol{a}) = (\boldsymbol{x}, (\mathbf{F}\mathbf{F}^T)^{-1}\boldsymbol{x}) \leqslant d_0^2. \tag{I.80}$$

Therefore, the solution of the form (I.67) in this case remains valid upon adding gravitational forces (the right-hand side of equation (I.75) is supplemented with the matrix (I.21), in which one has to set $\mathbf{A}_0 = \mathbf{E}$). The problem of the motion of a compressible self-gravitating gas cloud was formulated in [55]. However, the analysis presented in [55] seems somewhat naive in the light of the modern knowledge of regular and chaotic motions in dynamical systems.

The Dyson model [177]. *Assumptions* 1° *and* 3° *coincide with those of the previous case, and instead of the polytropic behavior we assume that*

2°. *The gas is isothermal at the initial time, i. e.,* $T(\boldsymbol{a}, t = 0)$ *does not depend on* $\boldsymbol{a}$.

Substituting the pressure from the equation of state (I.71) into (I.73) and using (I.70), we obtain

$$\frac{\dot{U}_{\mathrm{in}}}{RT} = -\frac{\dot{\varphi}}{\varphi}. \tag{I.81}$$

At the same time, as mentioned above (see Remark 9), the internal energy depends only on the temperature, and the right-hand side of (I.81) does not depend on $\boldsymbol{a}$; therefore, the gas remains isothermal at all later times and (I.81) can be represented as

$$\varphi \frac{dU_{\mathrm{in}}}{d\varphi} + RT = 0. \tag{I.82}$$

Integrating this equation and taking (I.72) into account, we obtain a relationship between T and φ:

$$\varphi = \varphi_0 \exp\left(-\int (RT)^{-1}\left(\frac{dU_{\mathrm{in}}}{dT}\right) dT\right). \tag{I.83}$$

Thus, according to (I.70), (I.71) and (I.82), the pressure can ultimately be written as

$$p(\boldsymbol{a}, t) = \frac{RT(\varphi(t))}{\varphi(t)} f(\boldsymbol{a}). \tag{I.84}$$

Let us substitute (I.84) into (I.68) and restrict ourselves to the case where there are no external forces (i. e., $U = 0$). We find that for a solution of the form (I.67) to exist requires that $\ln f(\boldsymbol{a})$ be a uniform quadratic function of the Lagrangian coordinates.

Since the Lagrangian coordinates are defined up to a nonsingular linear substitution (I.78), we can represent $f(\boldsymbol{a})$ in the form

$$f(\boldsymbol{a}) = \frac{m}{(2\pi)^{3/2}} \exp(-\frac{1}{2}(\boldsymbol{a}, \boldsymbol{a})), \tag{I.85}$$

where $m = \int \rho(\boldsymbol{x}) d^3\boldsymbol{x} = \int f(\boldsymbol{a}) d^3\boldsymbol{a}$ is the mass of the gas.

Finally, for the elements of the matrix $\mathbf{F}$, we obtain the equation of motion

$$\mathbf{F}^T \ddot{\mathbf{F}} = RT(\varphi)\mathbf{E} \qquad \textit{(the Dyson equations)}. \tag{I.86}$$

REMARK 3. According to (I.85), we find that, as the gas moves, it has an ellipsoidal density stratification of the form

$$\rho(\boldsymbol{x}, t) = \frac{1}{\varphi(t)} f(n^2), \quad n^2 = (\boldsymbol{x}, (\mathbf{F}\mathbf{F}^T)^{-1}\boldsymbol{x}). \tag{I.87}$$

Thus, everywhere on the ellipsoid n^2 = const, the density ρ has the same value. As is well known, the gravitational potential of such bodies is not a uniform quadratic function of the coordinates; therefore, no solution of the form (I.67) exists in the presence of gravitation.

Model of a cooling gas cloud (Fujimoto [198]). *In this model, assumptions* 1° *and* 2° *coincide with those made in Ovsyannikov's case, i. e., the gas is assumed to be ideal and polytropic, while the third assumption in this case has the form*

3°. *The motion of the gas is not adiabatic, and the variations in the internal energy satisfy the equation*

$$\rho \dot{U}_{\text{in}} + p \operatorname{Tr}\left(\left(\frac{\partial \boldsymbol{x}}{\partial \boldsymbol{a}}\right)^{-1} \frac{\partial \dot{\boldsymbol{x}}}{\partial \boldsymbol{a}}\right) = -æ\rho^n T^m. \tag{I.88}$$

REMARK 4. Equations (I.88) differ from the equations of an adiabatic process (I.73) by the terms $-æ\rho^n T^m$.

Using equations (I.70) – (I.72), we eliminate U_{in} from (I.88) and find

$$\frac{\dot{p}}{p} + \gamma \frac{\dot{\varphi}}{\varphi} = -\frac{æ(\gamma - 1)}{R} \rho^{n-1} T^{m-1}. \tag{I.89}$$

To obtain a solution in the form (I.67), we additionally require that

$$m = 1, \quad \rho(\boldsymbol{a}, t) = \frac{\rho_0}{\varphi(t)},$$

where ρ_0 = const is independent of $\boldsymbol{a}$, i. e., the density is constant inside the cloud. The solution of equation (I.89) in this case has the form $p(\boldsymbol{a}, t) = \sigma(t) g(\boldsymbol{a})$, where $\sigma(t)$ satisfies the equation

$$\frac{\dot{\sigma}}{\sigma} + \gamma \frac{\dot{\varphi}}{\varphi} = -\bar{æ}\varphi^{1-n}, \quad \bar{æ} = \frac{æ(\gamma - 1)}{R} \rho_0^{n-1}. \tag{I.90}$$

The function $g(\boldsymbol{a})$ should obviously satisfy equation (I.74), and it can easily be shown that, according to Theorem 3, we may choose

$$g(\boldsymbol{a}) = 1 - (\boldsymbol{a}, \boldsymbol{a}), \quad f(\boldsymbol{a}) = \text{const.}$$

The condition of the finiteness of the total gas mass implies that $\mathbf{V} = \operatorname{diag}(-1, -1, -1)$ and that the gas occupies initially the region $(\boldsymbol{a}, \boldsymbol{a}) \leqslant 1$ (in the original physical variables, this inequality specifies an ellipsoid of the form $(\boldsymbol{x}, (\mathbf{F}, \mathbf{F})^{-1}|_{t=0}\, \boldsymbol{x}) \leqslant 1$).

Finally, we obtain a system of equations describing the dynamics of the cooling cloud in the form

$$\mathbf{F}^T\ddot{\mathbf{F}} = \frac{2\sigma\varphi}{\rho_0}\mathbf{E}\left(+2\varepsilon\int_0^\infty \mathbf{F}^T(\mathbf{F}\mathbf{F}^T + \lambda\mathbf{E})^{-1}\mathbf{F}\frac{d\lambda}{\sqrt{\det(\mathbf{F}\mathbf{F}^T + \lambda\mathbf{E})}}\right),$$

$$(\ln(\sigma\varphi^\gamma))^\cdot = -\bar{æ}\varphi^{1-n}.$$

The parenthesized term describes the gravitational interaction between the particles of the cloud. In this case, the gravitational interaction in the solution of the form (I.68) can be taken into account due to the uniformity of gas in the cloud ($\rho_0 = \text{const}$).

The numerical results of [198] demonstrate the possibility of gravitational collapse in this system (for $æ > 0$).

Model of a dust cloud (Gravitational collapse).
1°. *The medium (dust) does not resist deformations,*

$$p \equiv 0.$$

2°. *At the initial time, the particles are distributed uniformly (inside the ellipsoid),*

$$\rho(t, \boldsymbol{a})\big|_{t=0} = \rho_0 = \text{const.}$$

For a solution of the form (I.68), the density obviously does not depend on the coordinates at all subsequent times and is determined by

$$\rho(t) = \frac{\rho_0}{\det \mathbf{F}(t)}.$$

Therefore, it is possible to incorporate the gravitational attraction of cloud particles into this model within the framework of the linear solution (I.68), and the equations of motion can be written as

$$\mathbf{F}^T\ddot{\mathbf{F}} = 2\varepsilon\int_0^\infty \mathbf{F}^T(\mathbf{F}\mathbf{F}^T + \lambda\mathbf{E})^{-1}\mathbf{F}\frac{d\lambda}{\sqrt{\det(\mathbf{F}\mathbf{F}^T + \lambda\mathbf{E})}}. \tag{I.91}$$

This model is used in astrophysics to describe the gravitational collapse [643]. In particular, it is applied in [396] to the description of the collapse of an elliptic gas cloud at zero temperature.

3 Lagrangian Formalism, Symmetries, and First Integrals

We now show that the Dyson equations (I.86), the Ovsyannikov equations (I.75) under the condition (I.79), and the equations of a dust cloud (I.91) admit a natural Lagrangian representation. It can be shown by direct calculations that the equations of motion can be written as

$$\left(\frac{\partial L}{\partial \dot{\mathbf{F}}}\right)^{\cdot} - \frac{\partial L}{\partial \mathbf{F}} = 0, \qquad L = \frac{1}{2}\operatorname{Tr}(\dot{\mathbf{F}}\dot{\mathbf{F}}^T) - U_g(\mathbf{F}), \tag{I.92}$$

where

$$U_g(\mathbf{F}) = \begin{cases} U_{\text{in}}(\varphi) - \text{for the Dyson model}, \\ \dfrac{1}{\gamma-1}\varphi^{1-\gamma} - 2\varepsilon \displaystyle\int_0^\infty \frac{d\lambda}{\sqrt{\det(\mathbf{F}\mathbf{F}^T + \lambda\mathbf{E})}} \quad \text{— for the Ovsyannikov model with, gravitation} \\ -2\varepsilon \displaystyle\int_0^\infty \frac{d\lambda}{\sqrt{\det(\mathbf{F}\mathbf{F}^T + \lambda\mathbf{E})}} \text{ — for the dust cloud model,} \end{cases} \tag{I.93}$$

where, as above, $\varphi = \det \mathbf{F}$, $\mathbf{F} \in GL(3)$.

Remark 5. In the Dyson model, the Lagrangian representation (I.92) can be directly obtained from Hamilton's principle for barotropic flows (see [301])

$$\delta \int_{t_1}^{t_2} (T - U)\, dt = \delta \int_{t_1}^{t_2} W\, dt, \tag{I.94}$$

where T and U are the kinetic and the potential energy of the fluid, respectively, and W is the barotropic potential which satisfies the equation

$$\delta W = \int p \frac{\delta \rho}{\rho} d^3\mathbf{x}. \tag{I.95}$$

Based on the above assumption, we obtain for our case, up to a constant,

$$W = \int RT \ln \rho\, d^3\mathbf{x} = U_{\text{in}}. \tag{I.96}$$

These considerations can also be extended to the Ovsyannikov model.

By analogy with the fluid ellipsoid (see Sec. 2.4), we conclude that the system (I.92) is invariant under linear transformations of the form

$$\mathbf{F}' = \mathbf{S}_1\mathbf{F}\mathbf{S}_2, \quad \mathbf{S}_1, \mathbf{S}_2 \in SO(3), \tag{I.97}$$

which form a symmetry group $\Gamma = SO(3) \otimes SO(3)$.

The Dedekind reciprocity law (Teorem 1 in Part 1), which corresponds to a discrete transformation $\mathbf{F}' = \mathbf{F}^T$, *is also valid in the dynamics of gas clouds.*

According to the Noether theorem, integrals of motion linear in velocity — the vorticity and total angular momentum of the system – correspond to the transformations (I.97) and can be represented in the matrix form

$$\boldsymbol{\Xi} = \mathbf{F}^T\dot{\mathbf{F}} - \dot{\mathbf{F}}^T\mathbf{F}, \quad \mathbf{M} = \mathbf{F}\dot{\mathbf{F}}^T - \dot{\mathbf{F}}\mathbf{F}^T. \tag{I.98}$$

In addition, there is also a quadratic integral, the total energy of the system

$$\mathcal{E} = \frac{1}{2}\operatorname{Tr}(\dot{\mathbf{F}}\dot{\mathbf{F}}^T) + U_g(\mathbf{F}). \tag{I.99}$$

4 Symmetry Reduction and Hamiltonian Formalism

It is not difficult to perform a reduction using the linear integrals (I.98). To this end, we make use of the Riemannian decomposition

$$\mathbf{F} = \mathbf{Q}^T\mathbf{A}\boldsymbol{\Theta}, \quad \mathbf{Q}, \boldsymbol{\Theta} \in SO(3), \quad \mathbf{A} = \operatorname{diag}(A_1, A_2, A_3).$$

For the Lagrangian function of the gas cloud (I.92), in view of the equations

$$\dot{\mathbf{Q}} = \mathbf{w}\mathbf{Q}, \quad \dot{\boldsymbol{\Theta}} = \boldsymbol{\omega}\boldsymbol{\Theta}$$

we obtain the expression

$$L = \frac{1}{2}\sum \dot{A}_i^2 + \frac{1}{4}\sum (A_j + A_k)^2(\mathrm{w}_i - \omega_i)^2 + (A_j - A_k)^2(\mathrm{w}_i + \omega_i)^2 - U_g(\mathbf{A}).$$

We denote the three-dimensional vector of the semiaxes as $\boldsymbol{q} = (A_1, A_2, A_3)$ and represent the equations of motion in the form

$$\left(\frac{\partial L}{\partial \dot{\boldsymbol{q}}}\right)^{\cdot} - \frac{\partial L}{\partial \boldsymbol{q}} = 0,$$

$$\left(\frac{\partial L}{\partial \mathbf{w}}\right)^{\cdot} = \frac{\partial L}{\partial \mathbf{w}} \times \mathbf{w}, \quad \left(\frac{\partial L}{\partial \boldsymbol{\omega}}\right)^{\cdot} = \frac{\partial L}{\partial \boldsymbol{\omega}} \times \boldsymbol{\omega}.$$

This is an analog of the Riemann equations (I.16), (I.40) for the case of a gas cloud (the difference is that there is no term containing pressure). These equations can easily be written in a matrix form similar to that of (I.16) [177].

The Lagrangian transformation

$$\boldsymbol{p} = \frac{\partial L}{\partial \dot{\boldsymbol{q}}}, \quad \boldsymbol{m} = \frac{\partial L}{\partial \dot{\mathbf{w}}}, \quad \boldsymbol{\mu} = \frac{\partial L}{\partial \dot{\boldsymbol{\omega}}}$$

yields the Hamiltonian system

$$\begin{gathered} \boldsymbol{q} = \frac{\partial H}{\partial \dot{\boldsymbol{p}}}, \quad \boldsymbol{p} = -\frac{\partial H}{\partial \dot{\boldsymbol{q}}}, \quad \dot{\boldsymbol{m}} = \boldsymbol{m} \times \frac{\partial H}{\partial \boldsymbol{m}}, \quad \dot{\boldsymbol{\mu}} = \boldsymbol{\mu} \times \frac{\partial H}{\partial \boldsymbol{\mu}}, \\ H = \frac{1}{2}\sum p_i^2 + \frac{1}{4}\sum \left(\frac{m_i + \mu_i}{q_j - q_k}\right)^2 + \left(\frac{m_i - \mu_i}{q_j + q_k}\right)^2 + U_g(\boldsymbol{q}). \end{gathered} \tag{I.100}$$

The Poisson structure of the system (I.100) has the form

$$\{q_i, p_j\} = \delta_{ij}, \quad \{m_i, m_j\} = \varepsilon_{ijk} m_k, \quad \{\mu_i, \mu_j\} = \varepsilon_{ijk} \mu_k, \tag{I.101}$$

where the zero brackets are omitted. As above, the bracket (I.101) has two Casimir functions

$$\Phi_m = (\boldsymbol{m}, \boldsymbol{m}), \quad \Phi_\mu = (\boldsymbol{\mu}, \boldsymbol{\mu}),$$

which correspond to the squared total momentum and the vorticity of the system.

In the general case ($\Phi_m \neq 0$, $\Phi_\mu \neq 0$), we have a Hamiltonian system with five degrees of freedom.

In the particular case $\Phi_m = 0$ or $\Phi_\mu = 0$, we have a system with four degrees of freedom.

If $\Phi_m = \Phi_\mu = 0$, we obtain a system with three degrees of freedom similar to the problem of the motion of a unit mass point in $\mathbb{R}^3 = \{\boldsymbol{q}\}$.

5 Particular Cases of Motion

The case $\gamma = \frac{5}{3}$ (monoatomic gas). We consider in more detail the case of the expansion of an ellipsoidal cloud of an ideal monoatomic gas in the absence of gravitation; we will show that in this case the system has additional symmetries. In this case, as is well known, $c_V = \frac{3}{2}R$, and hence $\gamma = \frac{5}{3}$.

Using (I.92), we represent the Lagrangian of the system as

$$L = \frac{1}{2}\operatorname{Tr}(\dot{\mathbf{F}}\dot{\mathbf{F}}^T) - U_g(\mathbf{F}), \quad U_g(\mathbf{F}) = \frac{3}{2}k\frac{1}{(\det \mathbf{F})^{2/3}}, \tag{I.102}$$

where $k = \text{const}$ is a positive constant (introduced for convenience). The integrals — vorticity Ξ, momentum $\mathbf{M}$, and energy $\mathcal{E}$ — were presented above in (I.98) and (I.99).

We denote the eigenvalues of the matrix $\mathbf{F}\mathbf{F}^T$ by A_1^2, A_2^2, A_3^2 and call A_i the principal semiaxes of the gas ellipsoid (A_i coincides with the semiaxes of the gas ellipsoid in Ovsyannikov's model at the pressure and density distribution (I.79); as for Dyson's model with a normal density distribution (I.85), the above name is used also, although it does not reflect what this model really implies. We define an analog of the central moment of inertia of the system by the formula

$$I = \operatorname{Tr}\mathbf{F}\mathbf{F}^T = A_1^2 + A_2^2 + A_3^2. \tag{I.103}$$

As we can see, according to (I.102), the dynamics of the cloud can be described in this case by a natural Lagrangian system with a uniform potential of uniformity degree $\alpha = -2$ (for an arbitrary γ, the uniformity degree is $\alpha = 3(1-\gamma)$). Using the Lagrange–Jacobi formula for uniform systems [271], we obtain

$$\ddot{I} = 4\mathcal{E} = \text{const},$$

where $\mathcal{E}$ is the energy of the system (for an arbitrary γ, we find $\ddot{I} = 4\mathcal{E} - 2(3(1-\gamma)+2)U_g$).

Integrating this relation yields

$$I = 2\mathcal{E}t^2 + at + b, \tag{I.104}$$

where the integration constants a and b can be expressed in terms of the phase variables and time according to the formulas

$$a = 2\,\mathrm{Tr}(\mathbf{F}^T\dot{\mathbf{F}}) - 4\mathcal{E}t, \quad b = 2\mathcal{E}t^2 - 2\,\mathrm{Tr}(\mathbf{F}^T\dot{\mathbf{F}})t + I. \tag{I.105}$$

In fact, a and b are nonautonomous (explicitly time-dependent) integrals of the system considered.

Jacobi was the first to find the integral (I.104) for uniform systems of degree −2 in the problem of the motion of particles in a straight line. For the problem of the motion of a gas cloud, the Jacobi integral was found in [9]. The symmetries corresponding to the integrals (I.105) in the particular case $\Xi = 0$ were pointed out in [202].

Since $\mathcal{E} > 0$ for the system (I.102), we find from (I.104) that

Proposition 1. *As $t \to \pm\infty$, at least one of the semiaxes of the gas cloud, A_i, goes to infinity.*

In addition to the nonautonomous integrals (I.105), the systems in this case admits an autonomous quadratic integral that is independent of the energy integral,

$$J = 2I\mathcal{E} - [\mathrm{Tr}(\mathbf{F}^T\dot{\mathbf{F}})]^2. \tag{I.106}$$

For uniform systems of degree −2, this integral was found in a general form in [6]. For the system (I.102) in the particular case of $\Xi = 0$, it is also given in [202].

For uniform natural systems of degree −2, a special reduction can be performed to lower the number of degrees of freedom by unity. We describe it in relation to the system of interest (I.102).

We rescale time and make a (projective) change of variables

$$dt = I\,d\tau, \quad \mathbf{G} = I^{-1/2}\mathbf{F}. \tag{I.107}$$

It can easily be shown by direct calculations that the evolution of the matrix $\mathbf{G}(t)$ can be described by a Lagrangian system with a constraint φ in the following form:

$$L = \frac{1}{2}\,\mathrm{Tr}\left(\frac{d\mathbf{G}}{d\tau}\frac{d\mathbf{G}^T}{d\tau}\right) - \bar{U}_g(\mathbf{G}), \quad \bar{U}_g(\mathbf{G}) = \frac{3}{2}k\frac{1}{(\det\mathbf{G})^{2/3}}, \tag{I.108}$$
$$\varphi = \mathrm{Tr}(\mathbf{G}\mathbf{G}^T) = 1.$$

A relationship between the "old" time t and the "new" time τ can be found using (I.104). Note that the system (I.108) differs from the Dirichlet system, since the constraint φ is different in this case (in the Dirichlet problem, $\det\mathbf{G} = 1$). It is interesting that the energy integral for the system (I.108) coincides with the integral (I.106):

$$\bar{\mathcal{E}} = \frac{1}{4}J = \frac{1}{2}\,\mathrm{Tr}\left(\frac{d\mathbf{G}}{d\tau}\frac{d\mathbf{G}^T}{d\tau}\right) - \bar{U}_g(\mathbf{G}).$$

The linear integrals in the system (I.108) remain the same:

$$\Xi = \mathbf{G}^T \frac{d\mathbf{G}}{d\tau} - \frac{d\mathbf{G}^T}{d\tau}\mathbf{G}, \quad \mathbf{M} = \mathbf{G}\frac{d\mathbf{G}^T}{d\tau} - \frac{d\mathbf{G}}{d\tau}\mathbf{G}^T,$$

moreover, the system (I.108) is invariant under the same transformations (I.97), which form the group $\Gamma = SO(3) \otimes SO(3)$. Therefore, a symmetry reduction similar to that described above is possible (see Part II, Section 3), with the only difference that, in this case, the following relation between the semiaxes is satisfied:

$$\bar{A}_1^2 + \bar{A}_2^2 + \bar{A}_3^2 = 1. \tag{I.109}$$

Using the Riemann decomposition of the matrix $\mathbf{G} = \mathbf{Q}^T \bar{\mathbf{A}} \boldsymbol{\Theta}$, $\mathbf{Q}, \boldsymbol{\Theta} \in SO(3)$, $\bar{\mathbf{A}} = \operatorname{diag}(\bar{A}_1, \bar{A}_2, \bar{A}_3)$, we obtain a system similar to (I.100), but with the additional constraint (I.109). To take this constraint into account and to represent the equations in the most symmetric form, we define variables $\boldsymbol{q}$ and $\boldsymbol{K}$ according to the formulas

$$q_i = A_i, \quad \boldsymbol{K} = \boldsymbol{q} \times \frac{d\boldsymbol{q}}{d\tau}. \tag{I.110}$$

Then we finally obtain a reduced system in the form

$$\begin{gathered}
\frac{d\boldsymbol{K}}{d\tau} = \boldsymbol{K} \times \frac{\partial \bar{H}}{\partial \boldsymbol{K}} + \boldsymbol{q} \times \frac{\partial \bar{H}}{\partial \boldsymbol{q}}, \quad \frac{d\boldsymbol{q}}{d\tau} = \boldsymbol{q} \times \frac{\partial \bar{H}}{\partial \boldsymbol{K}}, \\
\frac{d\boldsymbol{m}}{d\tau} = \boldsymbol{m} \times \frac{\partial \bar{H}}{\partial \boldsymbol{m}}. \quad \frac{d\boldsymbol{\mu}}{d\tau} = \boldsymbol{\mu} \times \frac{\partial \bar{H}}{\partial \boldsymbol{\mu}}, \\
\bar{H} = \frac{1}{2}(\boldsymbol{K}, \boldsymbol{K}) + \frac{1}{4}\sum \left(\frac{m_i + \mu_i}{q_j - q_k}\right)^2 + \left(\frac{m_i - \mu_i}{q_j + q_k}\right)^2 + U_g(\boldsymbol{q}).
\end{gathered} \tag{I.111}$$

The (nonzero) Poisson brackets corresponding to the system (I.110) are

$$\begin{aligned}
\{K_i, K_j\} &= \varepsilon_{ijk} K_k, \quad \{K_i, q_j\} = \varepsilon_{ijk} q_k, \\
\{m_i, m_j\} &= \varepsilon_{ijk} m_k, \quad \{\mu_i, \mu_j\} = \varepsilon_{ijk} \mu_k.
\end{aligned}$$

As is well known, this Poisson structure corresponds to the algebra $e(3) \oplus so(3) \oplus so(3)$ and has four Casimir functions

$$\begin{aligned}
\Phi_K &= (\boldsymbol{K}, \boldsymbol{q}), \quad \Phi_q = (\boldsymbol{q}, \boldsymbol{q}), \\
\Phi_m &= (\boldsymbol{m}, \boldsymbol{m}), \quad \Phi_\mu = (\boldsymbol{\mu}, \boldsymbol{\mu});
\end{aligned}$$

in view of the definition (I.110) of the variables $\boldsymbol{K}$ and $\boldsymbol{q}$, we have

$$\Phi_K = 0, \quad \Phi_q = 1.$$

Thus, we ultimately conclude that

1. *if* $\Phi_m, \Phi_\mu \neq 0$, *equations* (I.100) *correspond to a Hamiltonian system with four degrees of freedom;*
2. *if* $\Phi_m = 0$ (*or* $\Phi_\mu = 0$), *we obtain a system with three degrees of freedom;*

3. *if* $\Phi_m = \Phi_\mu = 0$*, we obtain a system with two degrees of freedom.*

As already mentioned above, Gaffet [203] found, for the case $\Phi_\mu = 0$, two additional first integrals (of degree 6 in the velocities) independent of the energy integral and put forward the hypothesis of the integrability of the system in this case.

Moreover, it was stated in [203] that for $\Xi = 0$ (or $\mathbf{M} = 0$) the system (I.102) is Liouville integrable; the missing integrals are presented, although their commutativity is not shown. The missing integrals are polynomials of degree 6 in momenta and have the form

$$\begin{aligned}I_6 &= 36k^2\left(Y_0Y_2 - \frac{1}{4}Y_1^2 + 3X_2 + T(X_0 + Y_0^2)\right)\\&\quad +6k(4T^2Y_0 + 3PY_1 + 6TY_2) + 27P^2 + 4T^3,\\L_6 &= \left(\mathbf{A}^2m, \mathbf{V}_0\mathbf{A}^2m \times (\mathbf{V}_0^2\mathbf{A}^2m + \frac{3k}{(q_1q_2q_3)^{2/3}}m)\right),\end{aligned}$$

where $\mathbf{A} = \operatorname{diag}(q_1, q_2, q_3)$ and the quantities X_i, Y_i, P, T can be expressed in terms of the symmetric matrix

$$\mathbf{V}_0 = \begin{pmatrix} \frac{1}{3}\sum\limits_{i=1}^{3}\frac{K_i}{q_i} - \frac{K_1}{q_1} & \frac{m_3}{q_1^2-q_2^2} & \frac{m_3}{q_3^2-q_1^3} \\ \frac{m_3}{q_1^2-q_2^2} & \frac{1}{3}\sum\limits_{i=1}^{3}\frac{K_i}{q_i} - \frac{K_2}{q_2} & \frac{m_1}{q_2^2-q_3^2} \\ \frac{m_2}{q_3^2-q_1^2} & \frac{m_1}{q_2^2-q_3^2} & \frac{1}{3}\sum\limits_{i=1}^{3}\frac{K_i}{q_i} - \frac{K_3}{q_3} \end{pmatrix}$$

as follows:

$$\begin{aligned}X_k &=(q_1q_2q_3)^{2(k-1)/3}\operatorname{Tr}(\mathbf{V}_0^k\mathbf{A}^2), & Y_k &=(q_1q_2q_3)^{2(k+1)/3}\operatorname{Tr}(\mathbf{V}_0^k\mathbf{A}^{-2}),\\ T &= -\frac{1}{2}(q_1q_2q_3)^{4/3}\operatorname{Tr}(\mathbf{V}_0^2), & P &=(q_1q_2q_3)^2\det\mathbf{V}_0.\end{aligned}$$

In the case $\Phi_\mu = 0$, the system (I.100) is an Euler – Calogero – Moser system of type D_3 [460, 616] with potential U_g. The Lax representation of this system without a potential can be found, e. g., in [298]. In the case where $U_g = \frac{3}{2}\frac{k}{(q_1q_2q_2)^{2/3}}$, equations (I.100) can be written as

$$\begin{cases}\dot{\mathbf{L}} = [\mathbf{L}, \mathbf{A}] + \dfrac{k}{(q_1q_2q_2)^{2/3}}\mathbf{D}^{-1},\\ \dot{\mathbf{D}} = [\mathbf{D}, \mathbf{A}] + \mathbf{L},\\ \dot{\mathbf{l}} = [\mathbf{l}, \mathbf{A}],\end{cases} \tag{I.112}$$

where the matrices

$$\mathbf{L} = \begin{pmatrix} p_1 & \frac{m_3}{q_2-q_1} & \frac{-m_2}{q_3-q_1} & \frac{-m_2}{q_3+q_1} & \frac{m_3}{q_2+q_1} & 0 \\ \frac{m_3}{q_2-q_1} & p_2 & \frac{m_1}{q_3-q_2} & \frac{m_1}{q_3+q_2} & 0 & \frac{-m_3}{q_2+q_1} \\ \frac{-m_2}{q_3-q_1} & \frac{m_1}{q_3-q_2} & p_3 & 0 & \frac{-m_1}{q_3+q_2} & \frac{m_2}{q_3+q_1} \\ \frac{-m_2}{q_3+q_1} & \frac{m_1}{q_3+q_2} & 0 & -p_3 & \frac{-m_1}{q_3-q_2} & \frac{m_2}{q_3-q_1} \\ \frac{m_3}{q_2+q_1} & 0 & \frac{-m_1}{q_3+q_2} & \frac{-m_1}{q_3-q_2} & -p_2 & \frac{-m_3}{q_2-q_1} \\ 0 & \frac{-m_3}{q_2+q_1} & \frac{m_2}{q_3+q_1} & \frac{m_2}{q_3-q_1} & \frac{-m_3}{q_2-q_1} & -p_1 \end{pmatrix},$$

$$\mathbf{l} = \begin{pmatrix} 0 & m_3 & -m_2 & m_2 & -m_3 & 0 \\ -m_3 & 0 & m_1 & -m_1 & 0 & m_3 \\ m_2 & -m_1 & 0 & 0 & m_1 & -m_2 \\ -m_2 & m_1 & 0 & 0 & -m_1 & m_2 \\ m_3 & 0 & -m_1 & m_1 & 0 & -m_3 \\ 0 & -m_3 & m_2 & -m_2 & m_3 & 0 \end{pmatrix},$$

$$\mathbf{A} = \begin{pmatrix} 0 & \frac{m_3}{(q_2-q_1)^2} & \frac{-m_2}{(q_3-q_1)^2} & \frac{m_2}{(q_3+q_1)^2} & \frac{-m_3}{(q_2+q_1)^2} & 0 \\ \frac{-m_3}{(q_2-q_1)^2} & 0 & \frac{m_1}{(q_3-q_2)^2} & \frac{-m_1}{(q_3+q_2)^2} & 0 & \frac{m_3}{(q_2+q_1)^2} \\ \frac{m_2}{(q_3-q_1)^2} & \frac{-m_1}{(q_3-q_2)^2} & 0 & 0 & \frac{m_1}{(q_3+q_2)^2} & \frac{m_2}{(q_3+q_1)^2} \\ \frac{-m_2}{(q_3+q_1)^2} & \frac{m_1}{(q_3+q_2)^2} & 0 & 0 & \frac{-m_1}{(q_3-q_2)^2} & \frac{m_2}{(q_3-q_1)^2} \\ \frac{m_3}{(q_2+q_1)^2} & 0 & \frac{-m_1}{(q_3+q_2)^2} & \frac{m_1}{(q_3-q_2)^2} & 0 & \frac{-m_3}{(q_2-q_1)^2} \\ 0 & \frac{-m_3}{(q_2+q_1)^2} & \frac{m_2}{(q_3+q_1)^2} & \frac{-m_2}{(q_3-q_1)^2} & \frac{m_3}{(q_2-q_1)^2} & 0 \end{pmatrix}$$

form an $L - A$ pair for the system without a potential, and the matrix $\mathbf{D}$ has the form $\mathbf{D} = \operatorname{diag}(q_1, q_2, q_3, -q_3, -q_2, -q_1)$.

The case of axial symmetry. As in the case of a fluid ellipsoid, it is easy to show that the system (I.92) admits a three-dimensional invariant manifold formed by matrices of the form

$$F = \begin{pmatrix} u & v & 0 \\ -v & u & 0 \\ 0 & 0 & w \end{pmatrix}. \tag{I.113}$$

We have not been able to write the general equations (I.112) in the form of a normal $L-A$ pair. An $L-A$ pair for the reduced system with a third-degree integral was obtained in [586] using a completely different technique. This makes it appropriate to generalize the construction of this $L - A$ pair to the Gaffet system. In this case, the linear integrals (I.98) simplify to

$$\Xi_{12} = -M_{12} = 2(u\dot{v} - v\dot{u}), \quad \Xi_{13} = \Xi_{23} = M_{13} = M_{23} = 0. \tag{I.114}$$

Let us consider the Ovsyannikov model with gravitation (I.93) and make the change of variables

$$u = \frac{1}{\sqrt{2}} r\cos\psi, \quad v = \frac{1}{\sqrt{2}} r\sin\psi, \quad w = z.$$

Then the Lagrangian function of the system takes the form

$$\begin{aligned} L &= \frac{1}{2}(\dot{r}^2 + r^2\dot{\psi}^2 + \dot{z}^2) - U_g(r, z), \\ U_g &= \frac{k}{\gamma - 1}\frac{1}{(r^2 z)^{\gamma-1}} + U_e(r, z), \end{aligned} \tag{I.115}$$

where the energy of the gravitational field U_e can be expressed in terms of elementary functions:

$$U_e = -2\varepsilon \int_0^\infty \frac{d\lambda}{(\lambda + \frac{r^2}{2})\sqrt{\lambda + z^2}} = -\frac{2\varepsilon}{z} \times \begin{cases} \dfrac{2\arctan\sqrt{\chi^2 - 1}}{\sqrt{\chi^2 - 1}}, & \chi > 1, \\ \dfrac{\ln\left(\frac{1+\sqrt{1-\chi^2}}{1-\sqrt{1-\chi^2}}\right)}{\sqrt{1-\chi^2}}, & \chi < 1, \end{cases}$$

where $\chi = \frac{1}{\sqrt{2}}\frac{r}{z}$ is the ratio between the semiaxes. Since the Lagrangian (I.115) is independent of ψ, there is the cyclic integral

$$\frac{\partial L}{\partial \dot{\psi}} = r^2\dot{\psi} = c = \text{const.},$$

which coincides with the integrals (I.114) up to a factor.

For a fixed value of this integral, we make the Legendre transformation $p_r = \frac{\partial L}{\partial \dot{r}} = \dot{r}$, $p_z = \frac{\partial L}{\partial \dot{z}} = \dot{z}$ and obtain a Hamiltonian system with two degrees of freedom in the canonical form

$$H = \frac{1}{2}(p_r^2 + p_z^2) + U_\star(r, z), \quad U_\star = \frac{c^2}{2r^2} + U_g(r, \psi), \tag{I.116}$$

where $U_\star$ is the reduced potential.

Consider the simplest (integrable) case, the motion of a monoatomic gas ($\gamma = \frac{5}{3}$), ignoring gravitation (i. e., $U_e = 0$; see also the previous section).

It was shown above that in this case the system admits a reduction by one more degree of freedom and hence reduces to a quadrature. Indeed, we make a change of variables and rescale time as follows:

$$r = R\cos\theta, \quad z = R\sin\theta, \quad dt = R^2\,d\tau,$$

where, due to the conditions $r > 0$ and $z > 0$, we have $\theta \in (0, \pi/2)$. We obtain the following equations for R and θ:

$$\frac{d^2}{dt^2}(R^2) = 4H = \text{const},$$
$$\frac{1}{2}\left(\frac{d\theta}{dt}\right)^2 + \frac{1}{2}\frac{c^2}{\cos^2\theta} + \frac{3}{2}\frac{k}{(\cos^2\theta\sin\theta)^{2/3}} = h_1 = \text{const}.$$

The quadrature for θ with $c = 0$ and certain restrictions on the initial conditions was obtained in [9]. As we can see, the evolution of $\theta(t)$ is governed by the reduced potential

$$\bar{U}_\star(\theta) = \frac{1}{2}\frac{c^2}{\cos^2\theta} + \frac{3}{2}\frac{k}{(\cos^2\theta\sin\theta)^{2/3}}.$$

For all values of the parameters c and k, this function has one critical value θ_0 in the interval $(0, \pi/2)$, in which $\bar{U}_\star$ reaches its minimum. This value corresponds to the self-similar expansion of a spheroidal gas cloud. In other cases, the expansion of the cloud is accompanied by fluctuations in the lengths of the semiaxes, with θ varying in the interval (θ_1, θ_2), where θ_i are the roots of the equation $\bar{U}_\star(\theta) = h_1$.

In the general case, $U_e \neq 0$, the trajectories of the system (I.116) are not bounded. However, it is easy to show that, for $k > \frac{1}{96} 2^{2/3}(\sqrt{665} - 21)c^2 \approx 0,43c^2$, the reduced potential has a minimum at the point

$$\theta_0 = \arctan \frac{1}{\sqrt{2}}, \quad R_0 = \frac{\sqrt{3}}{8\varepsilon}(c^2 + 3 \cdot 2^{1/3} k).$$

Therefore, near the minimum of the energy $U_\star(\theta_0, R_0)$, the trajectories of the system are bounded and a Poincaré map can be constructed. Such a map in the plane $\theta = \frac{\pi}{4}$ as the plane of section is shown in Fig. I.3. A chaotic layer that arises from the splitting of resonant tori is clearly seen, which suggests that the system (I.116) is nonintegrable.

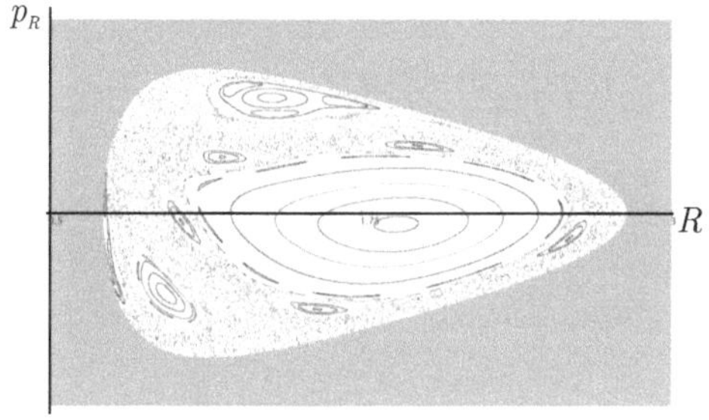

Fig. I.3. The Poincaré map of the system (I.116) for $k = c = \varepsilon = 1$ in the section plane $\theta = \frac{\pi}{4}$

Generalization of the Riemannian case. An invariant manifold of the form (I.57) also exists for gas ellipsoids, i. e.,

$$\mathbf{F} = \begin{Vmatrix} u_1 & v_1 & 0 \\ u_2 & v_2 & 0 \\ 0 & 0 & w_3 \end{Vmatrix}. \tag{I.117}$$

As in the Riemannian case (the case of a fluid ellipsoid), it can be shown that, in the case of gas, the following relations are also satisfied:

$$m_1 = m_2 = \mu_1 = \mu_2 = 0, \quad m_3 = \text{const}, \quad \mu_3 = \text{const}.$$

Thus, we conclude that, according to (I.100), the evolution of the semiaxes $A_i = q_i$, $i = 1, 2, 3$ is governed by the three-degree-of-freedom Hamiltonian system

$$H = \frac{1}{2}\boldsymbol{p}^2 + U_\star(q), \quad U_\star = \frac{c_1^2}{(q_1 - q_2)^2} + \frac{c_2^2}{(q_1 + q_2)^2} + U_g(\boldsymbol{q}), \tag{I.118}$$

where $\boldsymbol{q}, \boldsymbol{p}$ are canonically conjugate variables and $c_1 = \frac{1}{2}(m_3 + \mu_3)$, $c_2 = \frac{1}{2}(m_3 - \mu_3)$ are fixed constants.

It was shown above that, for a monatomic gas ($\gamma = \frac{5}{3}$), with gravitation ignored ($U_e = 0$), the system admits a reduction by one more degree of freedom. This yields a system of the form

$$\frac{d\boldsymbol{K}}{d\tau} = \boldsymbol{K} \times \frac{\partial \bar{H}}{\partial \boldsymbol{K}} + \boldsymbol{q} \times \frac{\partial \bar{H}}{\partial \boldsymbol{q}}, \quad \frac{d\boldsymbol{q}}{d\tau} = \boldsymbol{q} \times \frac{\partial \bar{H}}{\partial \boldsymbol{K}},$$

$$\bar{H} = \frac{1}{2}\boldsymbol{K}^2 + \bar{U}_\star(\boldsymbol{q}), \quad \bar{U}_\star(\boldsymbol{q}) = \frac{3}{2}\frac{k}{(q_1 q_2 q_3)^{2/3}} + \frac{c_1^2}{(q_1 - q_2)^2} + \frac{c_2^2}{(q_1 + q_2)^2}.$$

This system is equivalent to the problem of the motion of a spherical top in an axisymmetric potential [97]. As shown in [201], this system is integrable provided that $c_1^2 = c_2^2$. For $c_1 = c_2 = 0$, the additional integral of degree 3 in the velocities has the form

$$F_3 = K_1 K_2 K_3 - 3k \frac{K_1 q_2 q_3 + K_2 q_3 q_1 + K_3 q_1 q_2}{(q_1 q_2 q_3)^{2/3}}.$$

If $c_1 = c_2 = c \neq 0$, we have an additional sixth-degree integral

$$F_6 = (F_3 + F_a)^2 + 4\frac{f(q_1^2\theta + 3kq_3^2)(q_2^2\theta + 3kq_3^2)}{q_3^4},$$

where

$$F_a = \frac{4c^2 q_1 q_2 q_3^2}{q_1^2 - q_2^2} K_3, \quad f = \frac{4c^2 (q_1 q_2 q_3)^{2/3}}{(q_1^2 - q_2^2)^2} q_3^2, \quad \theta = \frac{(q_1 q_2 q_3)^{2/3}}{q_1^2 q_2^2} K_1 K_2 - 3k + f.$$

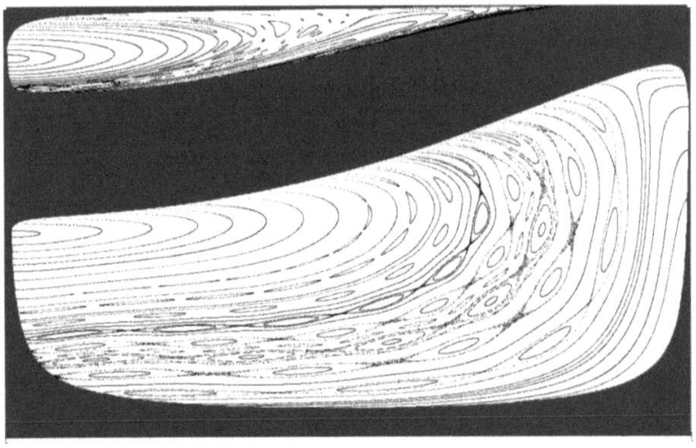

Fig. I.4. Poincaré map of the system (I.117) on the level set of the energy integral $\bar{H} = 30$ for $k = 1/3$, $c_1 = 1$, $c_2 = 0.3$ in the section plane $g = \pi$.

In the more general case, $c_1^2 \neq c_2^2$, the system (I.117) becomes nonintegrable.

Figure I.4 shows the corresponding Poincaré map in the Andoyer variables, which are traditionally used for reductions in problems of rigid body motion with a fixed point [97]. The figure clearly shows a splitting of the resonant tori, which suggests that the problem is nonintegrable.

Bibliography

[1] Adler M., van Moerbeke P. *A new geodesic flow on so(4). Probability, statistical mechanics and number theory.* Advances in mathematics supplementary studies, 1986, v. 9, p. 81–96.

[2] Adler M., van Moerbeke P. *Geodesic flow on so(4) and intersection of quadrics.* Proc. Nat. Acad. Sci. USA, 1984, v. 81, No. 14, p. 4613–4616.

[3] Adler M., van Moerbeke P. *The algebraic integrability of geodesic flow on so(4).* Inventiones mathematicae, 1982, v. 67, No. 2, p. 297–331.

[4] Adler M., van Moerbeke P. *The Kowalewski's and Henon-Heiles motion as Manakov geodesic flow on SO(4) — a two-dimensional family of Lax pairs.* Communications in mathematical physics, 1988, v. 113, No. 4, p. 659–700.

[5] Afonin A. A. *Non-integrability of the problem on sliding of an unbalanced disk over the horizontal plane.* Vestnik Moskovskogo Universiteta. Seriya 1. Matematika, Mekhanika, 1998, No. 4, p. 69–70.

[6] Albouy A., Chenciner A. *Le problème des n corps et les distances mutuelles.* Invent. Math., 1998, v. 131, p. 151–184.

[7] Ampère A. M. *Sur quelques nouvelles propriétés des axes permanents de rotation des corps et des plans directeurs de cas axes.* Mém. Acad. Sci. Paris, 1921, v. 1822, p. 86–152.

[8] Andoyer M. H. *Cours de mécanique céleste.* Paris: Gauthier-Villars, 1923, v. 1.

[9] Anisimov S. I., Lysikov Yu. I. *Expansion of a gas cloud in vacuum.* Journal of Applied Mathematics and Mechanics, 1970, v. 34, No. 5, p. 882–885.

[10] Appell P. *Traité de mécanique rationnelle.* Paris: Gauthier–Villars, ed. 4th, v. 1, 1919, 619 p.; v. 2, 1924, 575 p.

[11] Appell P. *Traité de Mécanique rationnelle*, tome IV: *Figures d'équilibre d'une masse liquide homogène en rotation.* Paris: Gauthier-Villars, 2 ed., 1932 (IV-1), 1937 (IV-2).

[12] Appelrot G. G. *Not entirely symmetrical heavy gyroscopes.* In Rigid Body Motion About a Fixed Point. Collection of papers in memory of S. V. Kovalevskaya. M.-L.: Acad. Sci. USSR, 1940, p. 61–155 (Russian).

[13] Appelrot, G. G., Concerning Section 1 of the Memoir of S. V. Kovalevskaya «Sur le problème de la rotation d'un corps solide autour d'un point fixe», and the Appendix to This Paper, *Mat. Sb.*, 1892, vol. 16, no. 3, pp. 483–507 (Russian).

[14] Appelroth H. H. *Sur les cas particuliers les plus simples du mouvement d'un gyroscope pesant asymmétrique de M-me Kowalewsky,* Mat. Sb., 1910, v. 27, No. 3, p. 262–334 (Russian); No. 4, p. 477–559 (Russian).

[15] Aref H., Jones S. W. *Chaotic motion of a solid through ideal fluid.* Physics of Fluids A, 1993, v. 5, No. 12, p. 3026–3028.

[16] Arkhangelskiy Yu. A. *Analytical dynamics of a rigid body.* Moscow: Nauka, 1977 (Russian).

[17] Arkhangelskiy Yu. A. *Dynamics of a fast rotating rigid body.* Moscow: Nauka, 1985 (Russian).

[18] Arnold V. I. *Mathematical methods of classical mechanics.* New York: Springer-Verlag, 1989 (Russian).

[19] Arnold V. I., Kozlov V. V., Neishtadt A. I. *Mathematical aspects of classical and celestial mechanics.* Berlin, Heidelberg, New York: Springer-Verlag, 2006, v. 3 (Russian).

[20] Arnold V. I. *Proof of a theorem of A. N. Kolmogorov on the invariance of quasi-periodic motions under small perturbations of the Hamiltonian.* Russian Mathematical Surveys, 1963, v. 18, No. 5, p. 9–36.

[21] Arnold V. I., Givental A. B. *Symplectic geometry.* In Dynamical Systems IV. Encyclopaedia of Mathematical Sciences, v. 4, p. 1–136, Berlin, Heidelberg, New York: Springer-Verlag, 1990 (Russian).

DOI 10.1515/9783110544442-016

[22] Arnold V. I. *Small denominators. I: On the mappings of the circumference onto it self.* Transactions of the American Mathematical Society, 1965, v. 46, No. 2, p. 213–284.
[23] Arnold V. I. *The Hamiltonian nature of the Euler equation in the dynamics of rigid body and of an ideal fluid.* Russian Mathematical Surveys, 1969, v. 24, No. 3, p 225–226.
[24] Arribas M., Elipe A. *Attitude dynamics of a rigid body on a Keplerian orbit: a simplificaton.* Celest. Mech. & Dyn. Ast., 1993, v. 55, p. 243–247.
[25] Audin M. *The spinning tops. A course on integrable systems.* Cambridge Univ. Press, 1997.
[26] Barkin Y. V., Borisov A. V. *Non-integrability of the Kirchhoff equations and related problems in rigid body dynamics.* VINITI, 1989, No. 5037–B89.
[27] Barkin I. V. *Periodic and conditionally periodic solutions in the problem of motion of a heavy solid about a fixed point.* Journal of Applied Mathematics and Mechanics, 1981, v. 45, No. 3, p. 391–397.
[28] Basset A. *A Treatise on hydrodynamics: With numerous examples.* Vol. II, Ch. 15., Cambridge: Deighton, Bell and Co., 1888.
[29] Basset A. *On the motion of a liquid ellipsoid under the influence of its own attraction.* Proc. London Math. Soc., 1885, v. 1, No.. 1, p. 255–262.
[30] Basset A. *On the stability of a liquid ellipsoid which is rotating about a principal axis under the influence of its own attraction.* Proc. London Math. Soc., 1887, v. 1, No. 1, p. 46–56.
[31] Basset A. *On the steady motion of an annular mass of rotating liquid.* Amer. J. Math., 1889, v. 11, No. 2, p. 172–181.
[32] Bechlivanidis C., van Moerbeke P. *The Goryachev – Chaplygin top and the Toda lattice.* Comm. Math. Phys., 1987, v. 110, p. 317–324.
[33] Beletsky V. V. *The motion of an artificial satellite about its mass center.* Moscow: Nauka, 1965.
[34] Beletsky V. V. *Regulare und chaotische Bewegung starrer Körper.* Stuttgart: Teubner, 1995.
[35] Belmonte A., Eisenberg H., Moses E. *From flutter to tumble: inertial drag and Froude similarity in falling paper.* Phys. Rev. Lett., 1998, v. 81, p. 345–348.
[36] Belokolos E. P., Bobenko A. I., Enol'skii V. Z., Its A. R., Matveev V. B. *Algebro-geometric approach to nonlinear integrable equations.* Springer, 1994.
[37] Benettin G., Fassò F. *Fast rotations of the rigid body. A study by Hamiltonian perturbation theory,* Part I. Nonlinearity, 1996, v. 9, p. 137–186.
[38] Benettin G., Fassò F., Guzzo M. *Long term stability of proper rotations and local chaotic motions in the perturbed Euler rigid body.* Regul. Chaotic Dyn., 2006, v. 11, No. 1, p. 1–17.
[39] Benvenuti P., Balli R. *Risolubilita per quadrature del problema del moto di un solido soggeto a forze di potenza nulla.* Atti. Acc. Naz., Linsei Rend., Sci. fis. math. e natur., 1974, v. 56, p. 1–12.
[40] Berry M. V. *The levitron: an adiabatic trap for spins.* Proc. Royal Soc. London, Series A, 1996, v. 452, p. 1207–2220.
[41] Berry M. V., Mount K. E. *Semiclassical approximations in wave mechanics.* Rep. Prog. Phys., 1972, vol. 35, pp. 315.
[42] Bertolli M. L., Bolotin S. V. *Doubly asymptotic trajectories of Lagrangian systems in homogeneous force fields.* Ann. di Matem. pura ed. applicata, 1998 (IV), v. CLXXIV, p. 253–275.
[43] Betti E. *Sopra i moti che conservano la figura ellissoidale a una massa fluida eterogenea.* Annali di Matematica Pura ed Applicata, Serie II, 1881, v. X, p. 173–187.
[44] Biello J.A., Lebovitz N.R., Morison P.J. *Hamiltonian Reduction of Incompressible Fluid Ellipsoids, Preprint.* http://people.cs.uchicago.edu/ lebovitz/hamred.pdf.
[45] Birkhoff G. *Dynamical systems.* Amer. Math. Soc. Colloq. Publ., 1927, v. 9.
[46] Birkhoff G. *Hydrodynamics. A study in logic, fact and similitude.* Princeton Univ. Press, 1960.

[47] Bizyaev I. A. Nonintegrability and Obstructions to the Hamiltonianization of a Nonholonomic Chaplygin Top // Doklady Mathematics, 2014, v. 90, No.. 2, p. 631–634.
[48] Bizyaev I. A., Borisov A. V., Mamaev I. S. *The Hess – Appelrot system and its nonholonomic analogs.* Proceedings of the Steklov Institute of Mathematics, 2016, v. 294, No. 1, p. 252–275.
[49] Bizyaev I. A., Borisov A. V., Mamaev I. S. *The Hess – Appelrot case and quantization of the rotation number.* Regular and Chaotic Dynamics, 2017, v. 22, No. 2, p. 180–196.
[50] Bobenko A. I. *Euler equations in the algebras e(3) and so(4). Isomorphisms of integrable cases.* Functional Analysis and Its Applications, 1986, v. 20, No. 1, p. 53-56.
[51] Bobenko A. I., Kuznetsov V. B. *Lax representation and new formulae for the Gorjachev – Chaplygin top.* J. Phys. A, 1988, v. 21, p. 1999–2006.
[52] Bobenko A. I., Reyman A. G., Semenov-Tian-Shansky M. A. *The Kowalewski top 99 years later: a Lax pair, generalizations and explicit solutions.* Comm. Math. Phys., 1989, v. 122, No. 2, p. 321–354.
[53] Bobylev D. K. *On a particular solution of the differential equations of a heavy rigid body rotation around of a fixed point.* The collection of papers of the amateur mathematicians society, 1892, v. 16. No. 3, p. 544–581.
[54] Bogaevskii V. N., Ostrer L. A. *On the fast rotation of a heavy rigid body about a fixed point.* Journal of Applied Mathematics and Mechanics, 1980, v. 44, No. 6, p. 713–720.
[55] Bogoyavlenskii O. I. *Dynamics of a gravitating gaseous ellipsoid.* Journal of Applied Mathematics and Mechanics, 1976, v. 40, No. 2, p. 246–256.
[56] Bogoyavlenskii O. I. *Integrals of the fourth degree for Euler's equations on six-dimensional Lie algebras.* Soviet Math. Dokl., 1983, v. 28, p 545–549.
[57] Bogoyavlenskii O. I. *Integrable problems of the dynamics of coupled rigid bodies.* Russian Acad. Sci., Izv. Math., 1993, v. 41, No. 3, p. 395–416.
[58] Bogoyavlenskii O. I. *Integrable cases of the dynamics of a rigid body, and integrable systems on the spheres S^n.* Mathematics of the USSR-Izvestiya, 1986, v. 27, No. 2, p. 203–218.
[59] Bogoyavlenskii O. I. *Integrable Euler equations on Lie algebras arising in problems of mathematical physics.* Mathematics of the USSR-Izvestiya, 1985, v. 25, No. 2, p. 207–257.
[60] Bogoyavlenskii O. I. *Integrable Euler equations on six-dimensional Lie algebras.* Sov. Math. Dokl., 1983, v. 27, p. 1–5.
[61] Bogoyavlenskii O. I. *Overturning solitons. Nonlinear integrable equations.* Moscow: Nauka, 1991.
[62] Bohr N., Wheeler J. *The Mechanism of Nuclear Fission.* Phys. Rev., 1939, v. 56, p. 426–450.
[63] Bolotin S. V. *Biasymptotic trajectories and integrability conditions for Hamiltonian systems.* Moskovskii Universitet, Vestnik, Matematika, Mekhanika, 1990, No. 1, p. 55–63 (Russian).
[64] Bolotin S. V. *Variational methods for constructing chaotic motions in the dynamics of a rigid body.* J. Appl. Math. Mech., 1992, v. 56, No. 2, p. 198–205.
[65] Bolsinov A. V. *Compatible Poisson brackets on Lie algebras and completeness of families of functions in involution.* Izvest. Akad. Nauk SSSR, Ser. Matem., 1991, v. 55, No. 1, p. 68–92 (Russian).
[66] Bolsinov A. V., Fomenko A. T. *Integrable Hamiltonian Systems: Geometry, Topology, Classification.* London, New York: CRC Press, 2004.
[67] Bolsinov A. V., Borisov A. V. *Compatible Poisson brackets on Lie algebras.* Mathematical Notes, 2002, v. 72, No. 1, p. 10–30.
[68] Bolsinov A. V., Borisov A. V., Mamaev I. S. *Rolling of a ball without spinning on a plane: the absence of an invariant measure in a system with a complete set of integrals.* Regular and Chaotic Dynamics, 2012, v. 17, No. 6, p. 571–579.

[69] Bolsinov A. V., Dullin H. R., Witek A. *Topology of energy surfaces and existence of transversal Poincare section.* J. Phys. A, 1996, v. 29, No. 16, p. 4977–4985.
[70] Bolsinov A. V., Fedorov Yu. N. *Steklov – Lyapunov type systems.* Preprint, 2003.
[71] Borisov A. V., Kilin A. A., Mamaev I. S. *Absolute and Relative Choreographies in Rigid Body Dynamics.* Regular and Chaotic Dynamics, 2008, v. 13, No. 3, p. 204–220.
[72] Borisov A. V., Mamaev I. S. *Systems on a Sphere with a Redundant Set of Integrals.* in *Classic Dynamics in Non-Euclidean Spaces*, A. V. Borisov, I. S. Mamaev (Eds.), Izhevsk: R&C Dynamics, Institute of Computer Science, 2004, p. 167–182 (Russian).
[73] Borisov A. V., Mamaev I. S. *Dirac Brackets in Geometry and Mechanics,* in P. A. M. Dirac, *Lectures on Theoretical Physics,* Izhevsk: R&C Dynamics, Institute of Computer Science, 2000, p. 191–230 (Russian).
[74] Borisov A. V., Mamaev I. S. *Adiabatic Chaos in Rigid Body Dynamics.* Regular and Chaotic Dynamics, 1997, v. 2, No. 2, p. 65–78.
[75] Borisov A. V., Mamaev I. S., Sokolov V. V. *A New Integrable Case on so(4).* Doklady Physics, 2001, v. 46, No. 12, p. 888–889.
[76] Borisov A. V., Mamaev I. S. *Classical dynamics in non-Eucledian spaces.* Moscow–Izhevsk: Institute of Computer Science, 2004.
[77] Borisov A. V., Mamaev I. S. *Dynamics of the Rigid Body.* Regular and Chaotic Dynamics, 2001.
[78] Borisov A. V., Mamaev I. S., Ramodanov S. M. *Dynamics of a circular cylinder interacting with point vortices.* DCDS-B, 2005, v. 5, No. 1, p. 35–50.
[79] Borisov A. V., Mamaev I. S. *Euler – Poisson equations and integrable cases.* Regular and Chaotic Dynamics, 2001, v. 6, No. 3, p. 253–274.
[80] Borisov A. V., Mamaev I. S. *Generalization of the Goryachev – Chaplygin case.* Regular and Chaotic Dynamics, 2002, v. 7, No. 1, p. 21–30.
[81] Borisov A. V., Mamaev I. S. *Integrability of the problem of the motion of a cylinder and a vortex in an ideal fluid.* Mathematical Notes, 2004, v. 75, No. 1, p. 19–22.
[82] Borisov A. V., Mamaev I. S., Kholmskaya A. G. *Kovalevskaya top and generalizations of integrable systems.* Regular and Chaotic Dynamics, 2001, v. 6, No. 1, p. 1–16.
[83] Borisov A. V., Tsygvintsev A. V. *Kovalevskaya's method in rigid body dynamics.* Journal of Applied Mathematics and Mechanics, 1997, v. 61, No. 1, p. 27–32.
[84] Borisov A. V., Tsygvintsev A. V. *Kowalewski exponents and integrable systems of classic dynamics. I, II.* Regular and Chaotic Dynamics, 1996, v. 1, No. 1, p. 15–37.
[85] Borisov A. V., Mamaev I. S., Kholmskaya A. G. *Kovalevskaya top and generalizations of integrable systems.* Regular and Chaotic Dynamics, 2000, v. 6, No.1, p. 1–16.
[86] Borisov A. V., Dudoladov S. L. *Kovalevskaya exponents and Poisson structures.* Regular and Chaotic Dynamics, 1999, v. 4, No. 3, p. 13–20.
[87] Borisov A. V., Mamaev I. S. *Modern methods of the theory of integrable systems.* Moscow–Izhevsk: Institute of Computer Science, 2003.
[88] Borisov A. V., Emelyanov K. V. *Nonintegrability and stochacticity of rigid body dynamics.* Izhevsk: Isd-vo Udm. univ., 1995.
[89] Borisov A. V., Kir'yanov A. I. *Non-integrability of the Kirchhoff equations.* Mathematical Methods in Mechanics, Moscow: Izd. Mosk. Univer., 1990, p. 13–18.
[90] Borisov A. V., Mamaev I. S. *Nonholonomic dynamical systems.* Moscow – Izhevsk: Institute of Computer Science, 2002.
[91] Borisov A. V. *Necessary and sufficient conditions of Kirchhoff equation integrability.* Regular and Chaotic Dynamics, 1996, v. 1, No. 2, p. 61–76
[92] Borisov A. V., Mamaev I. S. *Non-linear Poisson brackets and isomorphisms in dynamics.* Regular and Chaotic Dynamics, 1997, v. 2, No. 3–4, p. 72–89.

[93] Borisov A. V. *On the Liouville problem.* In Numerical Modelling in the Problems of Mechanics. Moscow: Mosk. Gos. Univ., 1991, p. 110–118 (Russian).
[94] Borisov A. V., Fedorov Y. N. *On two modified integrable problems of dynamics.* Vestnik Moskov. Univ. Ser. 1. Mat. Mekh., 1995, No. 6, p. 102–105 (Russian).
[95] Borisov A. V., Mamaev I. S. *Poisson structures and Lie algebras in Hamiltonian mechanics.* Izhevsk: Izd. UdSU, 1999.
[96] Borisov A. V., Simakov N. N. *Period doubling bifurcation in rigid body dynamics.* Regular and Chaotic Dynamics, 1997, v. 2, No. 1, p. 64–74.
[97] Borisov A. V., Mamaev I. S. *Rigid body dynamics. Hamiltonian methods, integrability, chaos.* Moscow–Izhevsk: Institute of Computer Science, 2005, 576 p. (in Russian).
[98] Borisov A. V., Mamaev I. S. *Rolling of a rigid body on plane and sphere. hierarchy of dynamics.* Regular and Chaotic Dynamics, 2002, v. 7, No. 2, p. 177–200.
[99] Borisov A. V., Mamaev I. S. *Some comments to the paper of perelomov A. M. "A note on geodesics on ellipsoid".* Regular and Chaotic Dynamics, 2000, v. 5, No. 1, p. 92–94.
[100] Borisov A. V., Mamaev I. S. *Symmetries and reduction in nonholonomic mechanics.* Regular and Chaotic Dynamics, 2015, v. 20, No.. 5, p. 553–604.
[101] Borisov A. V., Mamaev I. S. *Strange attractors in rattleback dynamics.* Physics-Uspekhi, 2003, v. 46, No. 4, p. 393–403.
[102] Borisov A. V., Mamaev I. S. *The Hess case in the dynamics of a rigid body.* Journal of Applied Mathematics and Mechanics, 2003, v. 67, No.. 2, p. 227–235.
[103] Born M. *Vorlesungen über Atommechanik.* Berlin: Springer, 1925.
[104] Blaschke W. *Nicht-Euklidische geometrie und mechanik,* I, II, III. Hamburger Mathematische Einrelschriften, 1942, Bd. 34, S. 45–47.
[105] Braden H. W. *A completely integrable mechanical systems.* Lett. Math. Phys., 1982, v. 6, p. 449–452.
[106] Brioschi F. *Développements relatifs au Sec. 3 des Recherches de* Dirichlet *sur un problème d'Hydrodynamique, vol. 58, pag. 181 et suivantes de ce Journal.* J. reine angew. Math. (Crelle's Journal), 1861, v. 59, p. 63–73.
[107] Broucke R. A. *On the use of Poincaré surface of section in rigid-body motion.* J. of Astronautical Sc., 1993, v. 41, No. 4, p. 593–601.
[108] Brun F. *Rotation kring fix punkt.* Öfversigt at Kongl. Svenska Vetenskaps Akad. Förhadl. Stokholm, 1893, v. 7, p. 455–468.
[109] Bryan G. H. *On the stability of a rotating spheroid of perfect liquid.* Proc. Roy. Soc. London, 1889–1890, v. 47, p. 367–376.
[110] Bryum A. Z. *Investigation of the regular precession of a heavy rigid body with a fixed point by Lyapunov's first method.* Mechanics of Solids, Naukova Dumka, Kiev, 1987, v. 19, p. 68–72 (Russian).
[111] Buchstaber V. M., Karpov O. V., Tertychniy S. I. *Rotation number quantization effect.* Theoretical and Mathematical Physics, 2010, v. 162, No. 2, p. 211–221.
[112] Burgatti P. *Demostrazione della non esistenza d' integrali algebrici (oltre i noti) nel Problema del moto d'un corpo pesante in torno a un punto fisso.* Rend. Circ. mat. Palermo, 1910, v. 29, p. 369–377.
[113] Burov A. A. *The "restricted" formulation of the problem of the motion of a heavy rigid body.* Journal of Applied Mathematics and Mechanics, 2004, v. 68, No. 6, p. 857–861.
[114] Burov A. A., Karapetian A. V. *On the non-existence of the additional integral in the problem of the motion of a heavy rigid ellipsoid along a smooth plane.* Journal of Applied Mathematics and Mechanics, 1985, v. 49, No. 3, p. 387–389.
[115] Burov A. A., Motte I., Stepanov S. Ya. *On motion of rigid bodies on a spherical surface.* Regular and Chaotic Dynamics, No. 3–4, p. 61–66.

[116] Burov A. A., Subkhankulov G. I. *On the existence of additional integrals of the equations of motion of a magnetizable solid in an ideal fluid, in the presence of a magnetic field.* Journal of Applied Mathematics and Mechanics, 1984, v. 48, No. 5, p. 538–541.
[117] Burov A. A. *On the Nonintegrability of the equations of motion of a gyrostat in a Cardan suspension,* in Research Problems of Stability and Stabilisation of Motion, Moscow: Computing Centre of the USSR Acad. Sci., 1986, p. 3–10 (Russian).
[118] Burov A. A. *On a partial integral in the problem of the motion of a heavy rigid body suspended on a rod,* in Research Problems of Stability and Stabilisation of Motion, Moscow: Computing Centre of the USSR Acad. Sci., 1986, p. 93–95 (Russian).
[119] Burov A. A. *On partial integrals of the equations of motion of a rigid body on a frictionless horizontal plane,* in Research Problems of Stability and Stabilisation of Motion, Moscow: Computing Centre of the USSR Acad. Sci., 1985, p. 118–121 (Russian).
[120] Burov A. A., Rubanovskii V. N. *On a general solution of equations of Kirchhoff– Clebsch type,* in Research Problems of Stability and Stabilisation of Motion, Moscow: Computing Centre of the USSR Acad. Sci., 1987, p. 83–86 (Russian).
[121] Capodanno P. *Sur la réduction des equations du mouvement d'un profil pesant dans un fluide.* Rev. Roumaine math. pures appl., 1987, v. 32, p. 763–770.
[122] Cayley A. *A memoir on cubic surfaces.* Philos. Trans. Roy. Soc. London, 1869, v. 159, p. 231–326.
[123] Chandrasekhar S. *Ellipsoidal figures of equilibrium.* New Haven, London: Yale University Press, 1969.
[124] Chaplygin S. A. *Geometric interpretation of the motion of a body with helical symmetry in a fluid,* in Collected Works: Vol. 3, Moscow: Gostekhizdat, 1950, p. 288–291 (Russian).
[125] Chaplygin S. A. *On the Effect of a Plane-Parallel Air Flow on a Cylindrical Wing Moving in It,* in Collected Works: Vol. 3, Moscow: Akad. Nauk SSSR, 1933, p. 3–64 (Russian).
[126] Chaplygin S. A. *Characteristic function in rigid body dynamics,* in Collected Works: Vol. 3, Moscow: Gostekhizdat, 1950, p. 260–282 (Russian).
[127] Chaplygin S. A. *A new particular solution of the problem of the motion of a rigid body around a fixed point,* Trudy Otdel. Fiz. Nauk Obsh. Liub. Est., 1904, v. 12, No. 1, p. 1–4 (Russian).
[128] Chaplygin S. A. *A new particular solution of the problem of the motion of a rigid body in a liquid.* Trudy Otdel. Fiz. Nauk Obsh. Liub. Est., 1903, v. 11, No. 2, p. 7–10 (Russian).
[129] Chaplygin S. A. *On the motion of heavy bodies in an incompressible fluid,* in Complete Works: Vol. 1, Leningrad: Izd. Akad. Nauk SSSR, 1933, p. 133–150.
[130] Chaplygin S. A. *On a ball's rolling on a horizontal plane,* Regul. Chaotic Dyn., 2002, v. 7, No. 2, p. 131–148; see also: Math. Sb., 1903, v. 24, No. 1, p. 139–168.
[131] Chaplygin S. A. *Some cases of motion of a rigid body in a fluid: 1, 2,* in Collected Works: Vol. 1, Moscow: Gostekhizdat, 1948, p. 136–311 (Russian).
[132] Chaplygin S. A. *Concerning Hess' loxodromic pendulum,* in Collected Works: Vol. 1, Moscow: Gostekhizdat, 1948, p. 133–135 (Russian).
[133] Chenciner A., Montgomery R. *A remarkable periodic solution of the three body problem in the case of equal masses.* Annals of Math., 2000, v. 152, p. 881–901.
[134] Chetaev N. G. *On Poincare's equations.* Appl. Math. and Mech., 1941, v. 5, No. 2, p. 253–262 (Russian).
[135] Chetaev N. G. *On the stability of a rotation of a rigid body with one fixed point in the Lagrange case.* Appl. Math. and Mech., 1954, v. 18, No. 1, p. 123–124 (Russian).
[136] Chetaev N. G. *Motion stability.* Moscow: Nauka, 1990.
[137] Cisneros J. U., Martinez F. J., Montalvo J. D. *On the Stability of a Self-Gravitating Inhomogeneous Fluid in the form of Two Confocal Spheroids Rotating with Different Angular Velocities.* Rev. Mexicana Astron. Astrof., 2000, v. 36, p. 185–210.

[138] Cisneros J. U., Martinez F. J., Montalvo J. D. *On the stability of a self-gravitating inhomogeneous fluid in the form of two confocal ellipsoids carrying dedekind-type internal currents.* Rev. Mexicana Astron. Astrof., 2004, v. 40, p. 167–180.

[139] Clebsch A. *Über die Bewegung eines Körpers in einer Flüssigkeit.* Math. Annalen, 1871, Bd. 3, S. 238–262.

[140] Coddington E. A., Levinson N. *Theory of ordinary differential equations.* New York–Toronto–London: McGraw-Hill, 1955.

[141] Correia A. C. M., Laskar J., Néron de Surgy O. *Long-term evolution of the spin of Venus,* I. *Theory.* Icarus, 2003, v. 163, p. 1–3.

[142] Couchet G. *Sur un cas d'integration par quadratures du mouvement d'un profil au sein d'un fluide parfait incompressible en mouvement a irrotational.* C. R. Acad. Sci. Paris, 1964, v. 258, p. 1722–1724.

[143] Cushman R. *Examples of nonintegrable analytic Hamiltonian vector fields with no small divisors.* Trans. Amer. Math. Soc., 1978, v. 238, p. 45–55.

[144] Darboux G. *Leçons sur la théorie des surfaces.* Paris, 1887, t. 1, chap. 2.

[145] Darboux G. *Sur la théorie de Poinsot et sur des mouvements correspondants á la méme polhodie.* C. r. Acad. Sci. 1885, v. 101, p. 1555–1561.

[146] Darwin G. H. *On the figure and stability of a liquid satellite.* Phil. Trans. Roy. Soc. London, 1906, v. 206, p. 161–248.

[147] Dedekind R. *Zusatz zu der vorstehenden Abhandlung.* J. reine angew. Math. (Crelle's Journal), 1861, v. 58, p. 217–228.

[148] Delone N. B. *On the geometric interpretation of integrals of rigid-body motion about a fixed point given by SV Kovalevskaya.* Mat. Sb. Kruzhka Lyubit. Mat. Nauk, 1892, v. 16, No. 2, p. 346–351 (Russian).

[149] Demin V. G., Kiselev F. I. *A new class of periodic motions of a solid body with one fixed point in a Newtonian force field.* Dokl. Akad. Nauk SSSR, 1974, v. 214, No. 5, p. 270–272.

[150] Deprit A. *Free rotation of a rigid body studied in the phase plane.* Amer. J. Phys., 1967, v. 35, No. 5, p. 424–428.

[151] Deprit A. *Le probléme des deux centers fixes.* Bull. Nath. Belg., 1962, v. 142, No. 1, p. 12–45.

[152] Deprit A., Elipe A. *Complete reduction of the Euler – Poinsot priblem.* J. of Astron. Sciences, 1993, v. 41, p. 603–628.

[153] Deryabin M. V. *On asymptotics of Chaplygin equation.* Regular and Chaotic Dynamics, 1998, v. 3, No. 1, p. 93–97.

[154] Deryabin M. V. *On stability of uniformly-accelerated motions of an axially-symmetric rigid body in an ideal fluid.* Z. Angew. Math. Mech., 2003, v. 83, No. 3, p. 197–203.

[155] Deryabin M. V. *On the stability of uniformly accelerated rotations of a heavy rigid body in an ideal fluid.* Mechanics of solids, 2002, v. 37, No. 5, p. 23–26.

[156] Deryabin M. V., Kozlov V. V. *On the surfacing effect for a heavy rigid body in a fluid.* Mechanics of solids, 2002, v. 37, No. 1, p. 54–59.

[157] Deryabin S. L. *One-dimension escape of self-gravitating ideal gas into vacuum.* Computational technolgies, 2003, v. 8, No. 4, p. 32–44.

[158] Devaney R. L. *Transversal homoclinic orbits in an integrable system.* Amer. J. Math., 1978, v. 100, No. 3, p. 631–642.

[159] Dirichlet G. L. *Untersuchungen über ein Problem der Hydrodynamik.* Nachrichten von der Gesellschaft der Wissenschaften zu Göttingen (Mathematisch-Physikalische Klasse), Jg. 1857, No. 14, p. 203–207.

[160] Dirichlet G. L. *Untersuchungen über ein Problem der Hydrodynamik (Aus dessen Nachlass hergestellt von Herrn R. Dedekind zu Zürich).* J. reine angew. Math. (Crelle's Journal), 1861, v. 58, p. 181–216.

[161] Dobrokhotov O. S. *Integrable cases for the problem of falling heavy cylindrical rigid body in liquid.* Vestnik Moskovskogo Universiteta. Seriya 1. Matematika, Mekhanika, 1999, No. 1, p. 63–65 (Russian).

[162] Dokshevich A. I. *Solutions in a finite form of the Euler – Poisson equations.* Kiev: Nauk. dumka, 1992 (Russian).

[163] Domogarov A. S. On Free Motion of a Gyroscope: *Dissertation,* St. Petersburg, 1893 (Russian).

[164] Dovbysh S. A. *Some new dynamical effects in the perturbed euler-poinsot problem, due to splitting of separatrices.* Journal of Applied Mathematics and Mechanics, 1989, v. 53, No. 2, p. 165–173.

[165] Dovbysh S. A. *The separatrix of an unstable position of equilibrium of a Hess-Appelrot gyroscope.* Journal of Applied Mathematics and Mechanics, 1992, v. 56, No. 4, p. 534–545.

[166] Dovbysh S. A. *Oscillation properties of plane motions in symmetric rigid body dynamics.* Mekhanika Tverdogo Tela, 1990, v. 25, p. 11–19 (Russian).

[167] Dovbysh S. A. *Intersection of asymptotic surfaces of the perturbed Euler – Poinsot problem.* Journal of Applied Mathematics and Mechanics, 1987, v. 51, No. 3, p. 283–288.

[168] Dovbysh S. A. *Splitting of separatrices of unstable uniform rotations and non-integrability of a perturbed Lagrange problem.* Vestn. MGU, Ser. 1, Mat., Mekh., 1990, v. 3, p. 70–77 (Russian).

[169] Dovbysh S. A. *Numerical investigation of two mechanics problems: transverse intersection of separatrices, Kolmogorov stability,* in Numerical Analysis, Mathematical Simulation and Their Application in Mechanics, Moscow: MGU, 1988, (Russian).

[170] Dubrovin B. A., Fomenko A. T., Novikov S. P. *Modern geometry - methods and applications.* Part I. The Geometry of Surfaces, Transformation Groups, and Fields, New York: Springer-Verlag, 1984; Part II. The Geometry and Topology of Manifolds, New York: Springer-Verlag, 1985,

[171] Dubrovin B. A., Krichever I. M., Novikov S. P. *Integrable systems. I.* Sovremennye problemy matematiki (VINITI), Dynamical systems, 1985, v. 4, p. 179–288 (Russian).

[172] Duhem M. P. *Sur la stabilite de l'équilibre relatif d'une masse fluide animée d'un mouvement de rotation.* J. de Math. Pures et Appl., 1905, v. 7, No. 5, p. 331–350.

[173] Dullin H. R., Easton R. W. *Stability of levitrons.* Physica D, 1999, v. 126, p. 1–17.

[174] Dullin H. R., Matveev V. S. *A new integrable system on the sphere.* Math. Research Letters, 2004, v. 11, No. 6, p. 715–722.

[175] Dullin H. R., Richter P. H., Juhnke M. *Action integrals and energy surfaces of the Kovalevskaya top.* Int. J. of Bif. and Chaos, 1994, v. 4, No. 6, p. 1535–1562.

[176] Dullin H. R., Richter P. H., Veselov A. P. *Action variables of the Kovalevskaya top.* Regular and Chaotic Dynamics, 1998, v. 3, No. 3, p. 18–26.

[177] Dyson F. J. *Dynamics of a spinning gas cloud.* J. Math. Mech., 1968, v. 18, No.. 1, p. 91–101.

[178] Dyson F. W. *The Potentials of ellipsoids of variable densities.* Quart. J. Pure Appl. Math., 1891, v. 25, p. 259–288.

[179] Elipe A. *Rotations of perturbed triaxial rigid bodies.* Int. J. of Theor. Phys. Group Theory and Nonlinear Optics, 2002, v. 8, No. 2, p. 163–196.

[180] Elipe A., Arribas M., Riaguas A. *Complete analysis of bifurcations in the axial gyrostat problem.* J. Phys. A., 1997, v. 30, p. 587–601.

[181] Elipe A., Lanchares V. *Phase flow of an axially symmetrical gyrostat with one constant rotor.* J. Math. Phys., 1997, v. 38, No. 7, p. 3533–3544.

[182] Elipe A., Lanchares V. *Two equivalent problems: gyrostats in free motion and parametric quadratic Hamiltonians.* Mech. Res. Comm., 1997, v. 24, No. 6, p. 583–590.

[183] Engelbrecht J. R., Mirollo R. *Structure of long-term average frequencies for kuramoto oscillator systems.* Physical review letters, 2012, v. 109, No.. 3, p. 034103.

[184] Enos M. J. *On an optimal control problem on $SO(3) \times SO(3)$ and the falling cat.* Fields Institute Communications, 1993, v. 1, p. 75–111.

[185] Eriguchi Y., Muller E. *A General computational method for obtaining equilibria of self-gravitating and rotating gases.* Astron. Astrophys., 1985, v. 146, p. 260–268.

[186] Fasso F., Lewis D. *Stability properties of the Riemann ellipsoids.* Arch. Rational Mech. Anal., 2001, v. 158, p. 259–292.

[187] Fassò F. *The Euler – Poinsot top: a non-commutatively integrable systems without global action-angle coordinates.* J. Appl. Math. Phys. (ZAMP), 1996, v. 47, p. 953–976.

[188] Faye I. *On the meromorphic non-integrability of Euler's equations on $SO(4)$.* Regular and Chaotic Dynamics, 2000, v. 5, No. 4, p. 477–483.

[189] Fedorov Yu. N. *Discrete versions of some algebraic integrable systems related to generalized Jacobians*, side III — *Symmetrics and integrability of difference equations.* Sabaudia, 1998, p. 147–160, CRH Proc. Lect. Notes, v. 25, Amer. Math. Soc., Providence, RI2000.

[190] Fedorov Y. N. *Lax representations with a spectral parameter defined on coverings of hyperelliptic curves.* Mathematical Notes, 1993, v. 54, No. 1, p. 728–738.

[191] Fedorov Yu. N. *Integrable systems, Poisson pencils, and hyperelliptic Lax pairs.* Regular and Chaotic Dynamics, 2000, v. 5, No. 2, p. 171–180.

[192] Fedorov Yu. N., Kozlov V. V. *A Memoir on integrable systems.* Springer, Monographs in Mathematics, 2005 (to appear).

[193] Feigenbaum M. J., Greene J. M., MacKay R. S., Vivaldi V. *Universal behaviour in families of area-preserving maps.* Physica 3D, 1981, p. 468–486.

[194] Feingold M., Peres A. *Regular and chaotic motion of coupled rotators.* Physica D, 1983, v. 9, p. 433–438.

[195] Ferrers N. M. *On the potentials, ellipsoids, ellipsoidal shells, elliptic laminae, and elliptic rings, of variable densities.* Quart. J. Pure Appl. Math., 1875, v. 14, p. 1–23.

[196] Frahm W. *Über gewisse differentialgleichungen.* Math. Annalen, 1875, v. 8, p. 35–44.

[197] Francoise J. P. *Monodromy and the kovalevskaya top.* Astérisque, 1987, 150–151, p. 87–108.

[198] Fujimoto F. *Gravitational collapse of rotating gaseous ellipsoids.* Astrophys. J., 1968, v. 152, No.. 2, p. 523–536.

[199] Gaffet B. J. *A completely integrable Hamiltonian motion on the surface of a sphere.* J. Phys. A, 1998, v. 31, p. 1581–1596.

[200] Gaffet B. J. *An Integrable Hamiltonian motion on a sphere: II. The separation of variables.* J. Phys. A, 1998, v. 31, p. 8341–8354.

[201] Gaffet B. *Expanding gas clouds of ellipsoidal shape: new exact solutions.* J. Fluid Mech., 1996, v. 325, p. 113–144.

[202] Gaffet B. *Sprinning Gas without Vorticity: the Two Missing Integrals.* J. Phys. A: Math. Gen., 2001, v. 34, p. 2087–2095.

[203] Gaffet B. *Sprinning gas clouds: Liouville integrability.* J. Phys. A: Math. Gen., 2001, v. 34, p. 2097–2109.

[204] Gaffet B. *Spinning gas clouds — without vorticity.* J. Phys. A, 2000, v. 33, p. 3929–3946.

[205] Galgani L., Giorgilli A., Strelcyn J.-M. *Chaotic motions and transition to stochasticity in the classical problem of the heavy rigid body with a fixed point.* Nuovo Cimento, 1981, v. 61B, No. 1, p. 1–20.

[206] Galper A. R., Miloh T. *Dynamic equations of motion for a rigid or deformable body in an arbitrary non-uniform potential flow field.* J. Fluid Mech., 1995, v. 295, p. 91–120.

[207] Galper A. R., Miloh T. *Hydrodynamics and stability of a deformable body moving in the proximity of interfaces.* Phys. of fluids, 1999, v. 11, No. 4, p. 795–806.

[208] Gashenenko I. N. *Angular velocity of the Kovalevskaya top.* Regular and Chaotic Dynamics, 2000, v. 5, No. 1, p. 104–113.

[209] Gashenenko I. N., Richter P. *Enveloping surfaces and admissible velocities of heavy rigid bodies.* Int. J. of Bifurcation and Chaos, 2004, v. 14, No. 8, p. 2525–2553.
[210] Gashenenko I. N. *Bifurcations of the levels of the first integrals of the problem of the motion of a heavy gyrostat.* Rigid Body Mechanics, 2004, No. 34, p. 27–37 (Russian).
[211] Gashenenko I. N. *Invariant sets in the space of angular velocities of a heavy gyrostat.* Rigid Body Mechanics, 2000, No. 30, p. 79–97 (Russian).
[212] Gashenenko I. N. *Integral manifolds in the problem of the motion of a heavy rigid body.* Rigid Body Mechanics, 2003, No. 33, p. 20–32 (Russian).
[213] Gashenenko I. N. *Enveloping surfaces in the problem of the motion of a heavy gyrostat.* Rigid Body Mechanics, 2002, No. 32, p. 39–49 (Russian).
[214] Gashenenko I. N., Kucher E. Yu. *Analysis of isoenergy surfaces for exact solutions of the problem of the motion of a rigid body.* Rigid Body Mechanics, 2001, No. 31, p. 18–30 (Russian).
[215] Gashenenko I. N., Kucher E. Yu. *Characteristic exponents of periodic solutions of Euler-Poisson equations.* Rigid Body Mechanics, 2002, No. 32, p. 50–59 (Russian).
[216] Gashenenko I. N. *One case of integrability of the equations of motion of gyrostat,* Mekh. Tverd. Tela, 1992, No. 24, p. 1–4 (Russian).
[217] Gashenenko I. N. *On a fixed hodograph of angular velocity in goryachev – chaplygin's solution,* Mekh. Tverd. Tela, 1989, No. 21, p. 9–18 (Russian).
[218] Gavrilov L. N. *On the geometry of Gorjatchev – Tchaplygin top.* C. R. Acad. Bulg. Sci., 1987, v. 40, p. 33–36.
[219] Gavrilov L. N. *Remarks on the equations of heavy gyrostat.* C. R. Acad. Bulg. Sci., 1989, vol. 42, No. 5, p. 17–20.
[220] Gavrilov L. N., Ouazzani-Jamil M., Caboz R. *Bifurcation diagrams and Fomenko's surgery on Liouville tori of the Kolossoff potential* $U = \rho + \frac{1}{\rho} - k\cos\varphi$. Ann. Scient. Ec. Norm. Sup., 4 serie, 1993, v. 26, p. 545–564.
[221] Giesen, A. *Über die Stabilitat des Gleichgewichtes einer nur der Gravitation unterworfenen Fliissigkeit.* Jahres-Bericht fiber die hohere Schule in Opladen, 1872–1873.
[222] Glutsyuk A. A., Kleptsyn V. A., Filimonov D. A. *On the adjacency quantization in an equation modeling the Josephson effect.* Functional Analysis and Its Applications, 2014, v. 48, No. 4, p. 272–285.
[223] Golo V. L. *Nonlinear regimes in spin dynamics of superfluide* ^{3}He. Lett. Math. Phys., 1981, v. 5, p. 155–159.
[224] Golubev V. V. *Lectures on integration of the equations of motion of a rigid body about a fixed point.* M.: State Publishing House of Theoretical Technical Literature, 1953 (Russian).
[225] Goriely A., Nizette M. *Kovalevskaya rods and Kovalevskaya waves.* Regular and Chaotic Dynamics, 2000, v. 5, No. 1, p. 95–106.
[226] Gorr G. V., Iljukhin A. A. *Cases of a constancy of the module of angular momentum of gyrostat.* Mekh. Tverd. Tela, 1974, v. 6, p. 9–15 (Russian).
[227] Gorr G. V., Iliukhin A. A., Kovalev A. M. *Nonlinear analysis of the behavior of mechanical systems.* Kiev: Naukova Dumka, 1984 (Russian).
[228] Gorr G. V., Kudryashova L. V., Stepanova L. A. *Classical problems of rigid body dynamics.* Kiev: Naukova Dumka, 1978 (Russian).
[229] Gorr G. V., Levitskaya G. D. *A periodic motion of a Goryachev-Chaplygin gyroscope.* Mekhanika Tverdogo Tela, 1971, No. 3, p. 101–106 (Russian).
[230] Gorr G. V., Savchenko A. Ya. On One Periodic Motion in the Kovalevskaya Solution. Mekhanika Tverdogo Tela, 1971, No. 3, p. 64–69 (Russian).
[231] Gorr G. V. *A motion of a rigid body in the Goryachev – Chaplygin case.* Prikl Mat Mekh, 1970, v. 34, No. 6, p. 1139–1143.

[232] Goryachev D. N. *On the motion of a heavy rigid body in a fluid,* Izv. Imper. Ob-va (Mosk. Imper. Univ.), 1893, v. 78, No. 2, p. 59–61 (Russian).

[233] Goryachev D. N. *Certain general integrals in the problem on the motion of a rigid body,* Warsaw, 1910 (Russian).

[234] Goryachev D. N. *A new particular solution of the problem of motion of a heavy rigid body about a fixed point,* Tr. Otdel. Fiz. Nauk Ob-va Lubit. Estestvozn., 1899, v. 10, No. 1, p. 23–24 (Russian).

[235] Goryachev D. N. *New cases of motion of a rigid body around a fixed point,* Warshav. Univ. Izv., 1915, v. 3, p. 3–14 (Russian).

[236] Goryachev D. N. *New cases of integrability of Euler's dynamical equations.* Warshav. Univ. Izv., 1916, v. 3, p. 3–15 (Russian).

[237] Goriatchev D. N. *Sur le mouvement d'un solide pesant autour d'un point fixe dans le cas* $A = B = 4C$, Mat. Sb., 1900, v. 21, No. 3, p. 431–438 (Russian).

[238] Graber J. L., Rosensteel G. *Circulation of a triaxial, charged ellipsoidal droplet.* Phys. Rev. C, 2002, v. 66, p. 034309.

[239] Grammel R. *Der Kreisel. Seine Theorie und seine Anwedungen.* Berlin: Springer-Verlag, 1950, Bd. 1, 2.

[240] Granovskii Y. I., Zhedanov A. S. *Solutions of domain type in anisotropic magnetic chains.* Theoretical and Mathematical Physics, 1987, v. 71, No. 1, p. 438–446.

[241] Granovskii Y. I., Zhedanov A. S., Lutsenko I. M. *Quadratic algebras and dynamics in curved spaces. I. Oscillator; II. The Kepler problem.* Theoretical and Mathematical Physics, 1992, v. 91, No. 2, p. 474–480; No. 3, p. 604–612.

[242] Greenhill A. G. *On the motion of a top and allied problems in dynamics.* Quart. J., 1877, v. 11, p. 176–194.

[243] Grioli G. *Esistenza e determinazione delle prezessioni regolari dinamicamente possibili per un solido pesante asimmetrico.* Ann. mat. pura e appl., 1947, v. 26, No. 1, p. 271–281.

[244] Grioli G. *Moto attorno al baricento un giroscopio soggeto a forze potenza nulla.* Rend. di Math. e dilla sue appl., Univ. Roma, Ist. Naz. Alta. Math., 1947, No. 5–6.

[245] Grioli G. *Qualche teorema di cinematica dei moti rigidi.* Rend. Acad. naz. Lencei, Ser. 8, 1963, v. 34, p. 636–641.

[246] Grioli G. *Sul moto di un corpo rigido asimmetrico soggeto a forze potenza nulla.* Rend. del Sem. Math. Univ. de Padova, 1957, v. 27, p. 90–102.

[247] Guckenheimer J., Holmes P. *Nonlinear oscillations, dynamical systems, and bifurcations of vector fields.* Berlin: Springer, 1983.

[248] Gulyaev M. P. *On pendulum-like motions of a heavy rigid body having one fixed point.* Tr. Mekh.-Mat. Fak. Kazakhsk. Univ., 1960, v. 1, No. 2, p. 8–17 (Russian).

[249] Gulyaev M. P. *On a new particular solution of the equations of motion of a heavy rigid body having one fixed point.* Vestn. Mosk. Univ. Ser. Fiz. Mat. i Estestv. Nauk, 1955, No. 3, p. 15–21 (Russian).

[250] Hadeler K. P., Selivanova E. N. *On the case of Kovalevskaya and new examples of integrable conservative systems on* S^2. Regular and Chaotic Dynamics, 1999, v. 4, No. 3, p. 45–52.

[251] Hagen J. *Ueber die Stabilitat des Gleichgewichtes einer auf einem dreiaxigen Ellipsoid mit kleinen Excentricitaten ausgebreiteten Flussigkeit.* Zeitschrift für Mathematik und Physik, 1877, v. 22, p. 65–86.

[252] Haine L. *Geodesic flow on* $SO(4)$ *and abelian surfaces.* Math. Ann., 1983, v. 4, p. 435–472.

[253] Haine L., Horozov E. A. *Lax pair for Kowalewski's top.* Physica D, 1987, v. 29, p. 173–180.

[254] Halphen G. *Sur le mouvement d'un solide dans un liquide.* J. math. pure et appl., 1988, v. 4, p. 28–37.

[255] Haßmann H. *Quasi-perodic motions of a rigid body.* Proefschrift, Groningen, 1995.

[256] Helmholtz H. *Über Integrale der hydrodynamischen Gleichungen, welche den Wirbelbewegungen entsprechen.* J. reine angew. Math., 1858, v. 55, p. 25–55.
[257] Hertz H. *Die Prinzipien der Mechanik in neuen Zusammenhange dargestellt.* Ges. Werke, Bd. 3, Leipzig, Barth., 1910.
[258] Hess W. *Über die Eulerschen Bewegungsgleichungen und über eine neue particulare Lösung des Problems der Bewegung eines starren Körpers un einen festen Punkt.* Math. Ann., 1890, v. 37, No. 2, p. 178–180.
[259] Hicks W. M. *On the Motion of a Mass of Liquid under its Own Attraction, When the Initial Form is an Ellipsoid.* Proc. Camb. Phil. Soc., 1883, v. 4, No. Pt. VI, p. 1–4.
[260] Hill M. J. M. *Note on the motion of a fluid ellipsoid under its own Attraction.* Proc. London Math. Soc., 1891, v. 1, No. 1, p. 88–95.
[261] Holm D. D. *Magnetic tornadoes: three-dimensional affine motions in ideal magnetohydrodynamics.* Phys. D, 1983, v. 8, p. 170–182.
[262] Holms P. J., Marsden J. E. *Horseshoes and Arnold diffusion for Hamiltonian systems on Lie groups.* Indiana Univ. Math. J., 1983, v. 32, No. 2, p. 273–309.
[263] Husson E. *Recherche des intégrales algebriques dans le mouvement d'un solide pesant autour d'un point fixe.* Ann. Fac. Sci. Univ. Toulouse, Ser. 2, 1906, v. 8, p. 73–152.
[264] Iacob A. *Invariant manifolds in the motion of a rigid body about a fixed point.* Rev. Roum. Math. Pures et Appl., 1971, v. 16, No. 10, p. 1497–1521.
[265] Ilyashenko Y. S., Ryzhov D. A., Filimonov D. A. *Phase-lock effect for equations modeling resistively shunted Josephson junctions and for their perturbations.* Functional Analysis and Its Applications, 2011, v. 45, No. 3, p. 192–203.
[266] Ipser J. R., Managan R. A. *On the existence and structure of inhomogeneous analogs of the Dedekind and Jacobi ellipsoids.* Astrophys. J., 1981, v. 250, p. 362–372.
[267] Irtegov V., Titorenko T. *Application of computer algebra methods in investigation of quite integrable systems*, SIAM, 2005, v. 1 (to appear).
[268] Ivin E A. *The problem in integrability in the problem of the inertial motion of a coupled array of two rigid bodies.* Vestnik MGU. Ser. 1. Mat., Mech., 1986, v. 2, p. 63–66 (Russian).
[269] Ivin E. A. *Separation of variables in the problem concerning the motion of a gyrostat.* Vestnik MGU. Ser. 1. Mat., Mech., 1985, No. 3, p. 69–72 (Russian).
[270] Ivin E. A. *Direct and conjugate representation of the motion of two coupled rigid bodies: three new integrable cases,* in Geometry, Differential Equations and Mechanics, V. V. Kozlov, A. T. Fomenko (Eds.), Moscow: MGU, 1986, p. 72–77 (Russian).
[271] Jacobi C. G. J. *Problema trium corporum mutuis attractionibus cubis distantiarum inverse proportionalibus recta linea se moventium.* Gesammelte Werke, v. 4, Berlin: Reimer, 1886. p. 531–539.
[272] Jacobi C. G. J. *Second mémoire sur la rotation d'un corps non soumis á des forces accélératrices.* Berlin: Gesammelte Werke, 1882, Bd. 2, p. 427–467.
[273] Jacobi C. G. J. *Sur la rotation d'un corps.* Berlin: Gesammelte Werke, 1882, Bd. 2, p. 289–352.
[274] Jacobi C. G. J. *Vorlesungen über dynamik.* Berlin: G. Reimer, 1884.
[275] Jeans J. H. *Problems of cosmogony and stellar dynamics.* Cambridge University Press, 1919.
[276] Kalnis E. G., Miller W. J. R., Winternitz P. *The group* $so(4)$*, separation of variables and the hydrogen atom.* SIAM, J. Appl. Math., 1976, v. 30, No. 4, p. 630–664.
[277] Kane T. R., Scher M. P. *A dynamical explanation of the falling cat phenomenon.* Intenational J. of Struct. and Solids, 1969, v. 5, p. 663–670.
[278] Kanso E., Marsden J. E., Rowley C. W., Helli-Huber J. *Locomotion of articulated bodies in a perfect fluid.* Preprint, 2004.
[279] Kaper T. J., Kovacic G. *A geometric criterion for adiabatic chaos.* J. Math. Phys., 1994, v. 35, No. 3, p. 1202–1218.

[280] Karapetyan A. V. *Stability of stationary motions.* M.: Editorial URRS, 1998 (Russian).
[281] Katok S. B. *Bifurcation sets and integral manifolds in the problem of motion of a heavy rigid body.* Usp. Mat. Nauk, 1972, v. 27, No. 2, p. 126–133 (Russian).
[282] Kharlamov M. P., Zotev D. V. *Non-degenerate energy surfaces of rigid body in two constant fields.* Regular and Chaotic Dynamics, 2005, v. 10, No. 1, p. 15–19.
[283] Kharlamov M. P. *On one class of motions of a gyrostat.* Mekh. Tverd. Tela., 1983, No. 15, p. 47–56 (Russian).
[284] Kharlamov M. P. *One class of solutions with two invariant relations for the problem of motion of the Kowalevski top in double constant field.* arXiv preprint arXiv:0803.1028., 2008.
[285] Kharlamov M. P. *Symmetry in systems with gyroscopic forces.* arXiv:1402.0663v1.
[286] Kharlamov M. P. *Topological analysis of integrable problems of rigid body dynamics.* Leningrad: Leningrad State University Publishing House, 1988 (Russian).
[287] Kharlamov M. P., Savushkin E. Y. *Separation of variables and integral manifolds in one problem of motion of generalized Kowalevski top.* Ukrainian Mathematical Bulletin, 2004, p. 548–565 (Russian).
[288] Kharlamov M. P., Savushkin A. Y. *An explicit integration of a problem of motion of a generalized Kovalevskaya top.* arXiv:1401.4238v1.
[289] Kharlamov M. P., Savushkin A. Y., Shvedov E. G. *Bifurcation set in one problem of motion of the generalized Kowalevski top.* Mekh. tverd. tela, 2003, No. 33, p. 10–19 (Russian).
[290] Kharlamov M. P., Sergeev E. K. *Construction of a complete solution for one problem of rigid-body dynamics.* Mekhanika Tverdogo Tela, 1982, v. 14, p. 33–38 (Russian).
[291] Kharlamov P. V. *Motion of a body bounded by a multiply connected surface in a fluid.* Zh. Prikl. Mekh. Tekh. Fiz., 1963, No. 4, p. 17–29 (Russian).
[292] Kharlamov M. P. *Reduction in Mechanical Systems with Symmetry,* Mekh. Tverd. Tela, 1976, No. 8, p. 4–18 (Russian).
[293] Kharlamov P. V. *Lectures on the dynamics of a rigid body.* Novosibirsk: NGU, 1965 (Russian).
[294] Kharlamova E. I. *On the Motion of a Rigid Body about a Fixed Point in the Central Newtonian Force Field,* Izv. Sibirsk. Otdel. Akad. Nauk SSSR, 1959, No. 6, p. 7–17 (Russian).
[295] Kharlamova E. I. and Stepanova L. A. *On the isomorphism of some classical problems of the dynamics of a rigid body and on attempts at constructing new solutions by a change of variables.* Mekh. Tverd. Tela, 1988, No. 20, p. 1–13 (Russian).
[296] Kharlamova L. N. *One Solution of the equations of motion of a body under the action of potential and gyroscopic forces.* Mekh. Tverd. Tela, 1990, No. 22, p. 38–42 (Russian).
[297] Kholmskaya A. G. *Motion of a disk within a sphere.* Regular and Chaotic Dynamics, 1998, v. 3, No. 2, p. 74–81.
[298] Khvedelidze A., Mladenov D. *Euler-Calogero-Moser system from SU(2) Yang-Mills theory.* arXiv:hep-th/9906033v3.
[299] Killing W. *Die Mechanik in den Nicht-Euklidischen Raumformen,* J. Reine Angew. Math, 1885, v. 98, p. 1–48.
[300] Kinoshita H. *First-order perturbations of the two finite body problem.* Publ. Astron. Soc. Japan, 1972, v. 24, p. 423–453.
[301] Kirchhoff G., *Vorlesungen über mathematische Physik. Mechanik.* Leipzig: Teubner, 1876.
[302] Kirchhoff G. R. *Vorlesungen über mathematische Physik. Mechanik.* Leipzig: B.J. Teubner, 1874.
[303] Klein F., Ackerman M. *Lectures on the development of mathematics in the XIX Century.* Brookline: Mathematical Sciences Press, v. 1, 1979.
[304] Klein F. *The mathematical theory of the top.* Chelsea, 1896.
[305] Klein F., Sommerfeld A. *Über die Theorie des Kreisels.* New York: e. a. Johnson reprint corp., 1965. 966 p.

[306] Kobb G. *Sur le probléme de la rotation d'un corps autour d'un point fixe*. Bull. Soc. math. France, 1895, v. 23, p. 210–215.
[307] Komarov I. V. *A generalization of the Kovalevskaya top*. Phys. Lett., 1987, v. 123, No. 1, p. 14–15.
[308] Komarov I. V. *Remarks on Kowalevki's top*. J. Phys. A, 2001, v. 34, p. 2111–2120.
[309] Komarov I. V. *Kowalewski basis for the hydrogen atom*. Theoretical and Mathematical Physics, 1981, v. 47, No. 1, p. 320–324.
[310] Komarov I. V., Kuznetsov V. B. *Semiclassical quantization of Kowalewski top*. Theoretical and Mathematical Physics, 1987, v. 73, No. 3, p. 1255–1263.
[311] Komarov I. V., Kuznetsov V. B. *Kowalewski's top on the Lie algebras $O(4)$, $e(3)$ and $O(3,1)$*. J. Phys. A, 1990, v. 23, p. 841–846.
[312] Komarov I. V., Sokolov V. V., Tsiganov A. V. *Poisson maps and integrable deformations of Kowalevski top*. J. Phys. A., 2003, v. 36, p. 8035–8048.
[313] Komarov I. V., Tsiganov A. V. *On a trajectory isomorphism of the Kowalevski gyrostat and the Clebsch problem*. J. Phys. A, 2005, v. 38, p. 2917–2927.
[314] Komarov I. V., Tsiganov A. V. *On integration of the Kowalevski gyrostat and the Clebsch problems*. Regular and Chaotic Dynamics, 2004, v. 9, No. 2, p. 169–187.
[315] Komarov I. V., Kuz'netsov V. B. *Generalization of the Goryachev-Chaplygin gyrostat in quantum mechanics*. Journal of Soviet Mathematics, 1989, v. 47, No. 2, p. 2459–2466.
[316] Kolosov G. V. *Trajectory traced out by the end of the principal angular momentum in the S. V. Kovalevskaya problem of rotation of a heavy rigid body*. Uchen. Zap. Yuryev Univ., 1904, No. 1, p. 1–3 (Russian).
[317] Kolosov G. V. *A Note on the motion of a rigid body in an incompressible fluid in the cases of V. A. Steklov and A. M. Lyapunov*, Izv. Ross. Akad. Nauk, 1919, v. 13, p. 711–713 (Russian).
[318] Kolosov G. V. On Certain Modifications of Hamilton's Principle and Its Application to the Solution of Problems of Mechanics of Solid Bodies: *Master's Thesis*, St. Petersburg, 1903 (Russian).
[319] Kolosov G. V. *On some property of the Kovalevskaya problem on the rotation of a heavy rigid body about a fixed point*, Moscow, 1901 (Russian).
[320] Kondratyev B. P. *The potential theory and equilibrium figures*. Moscow-Izhevsk: Regular and Chaotic Dynamic, 2003 (Russian).
[321] Koshlyakov V. N. *Problems of dynamics of a rigid body and applied theory of gyroscopes*. M.: Nauka, 1985 (Russian).
[322] Kotelnikov A. P. and Fok V. A. *Some applications of Lobachevskii's ideas in mechanics and physics*. Moscow – Leningrad: GITTL, 1950 (Russian).
[323] Kötter F. *Bemerkungen zu F. Kleins und A. Sommerfelds Buch über die Theorie des Kreisels*. Berlin, 1899.
[324] Kötter F. *Die von Steklow und Liapunow entdeckten integralen Fälle der Bewegung eines starren Körpers in einer Flüssigkeit*. Berlin: Sitzungsber. König. Preuss. Akad. Wiss., 1900, Bd. 6, p. 79–87.
[325] Kötter F. *Sur le cas traité par M-me Kowalevski de rotation d'un corps solide autour d'un point fixe*. Acta Math., 1893, v. 17, No.1–2.
[326] Kötter F. *Über die Bewegung eines festen Körpers in einer Flüssigkeit, I, II*. J. Reine Angew. Math., 1892, Bd. 109, p. 51–81, p. 89–111.
[327] Kovalev A. M. *The moving angular velocity hodograph in Hess' solution of the problem of motion of a body with a fixed point*. Journal of Applied Mathematics and Mechanics, 1968, v. 32, No. 6, p. 1129–1137.
[328] Kovalev A. M. *Stability of steady rotations of a heavy gyrostat about its principal axis*. Journal of Applied Mathematics and Mechanics, 1980, v. 44, No. 6, p. 709–712.

[329] Kowalewsky S. *Sur le probléme de la rotation d'un corps solide autor d'un point fixe.* Acta. math., 1889, v. 12, No. 2, p. 177–232.
[330] Kowalewsky S. *Mémoires sur un cas particulies du probléme de la rotation d'un corps pesant autour d'un point fixe, cú l'integration s'effectue á l'aide de fonctions ultraelliptiques du temps.* — Mémoires présentés par divers savants á l'Académie des sciences de l'Institut national de France, Paris, 1890, v. 31, p. 1–62.
[331] Kowalewski N. *Eine neue particulare Lösung der Differentialgleichungen der Bewegung eines schweren starren Körpers um einen festen Punkt.* Math. Ann., 1908, Bd. 65, No. 4, p. 528–537.
[332] Kozlov V. V. *On the integration theory of equations of nonholonomic mechanics.* Adv. in Mech., 1985, v. 8, No. 3, p. 85–107 (Russian). See also: Regul. Chaotic Dyn., 2002, v. 7, No. 2, p. 191–176.
[333] Kozlov V. V. *Methods of qualitative analysis in the dynamics of a rigid body.* Moscow: Moscow State University, 1980 (Russian).
[334] Kozlov V. V. *Nonintegrability of Kirchhoff's equations.* Commun. Int. Congr. Math. Warszawa, sec. 13, Math. Phys. and Mech., 1982, p. 41.
[335] Kozlov V. V. *Symmetries, topology, and resonanses in Hamiltonian mechanics.* Berlin: Springer-Verlag, 1995.
[336] Kozlov V. V. *Splitting of the separatrices in the perturbed Euler – Poinsot problem.* Moscow Univ. Mech. Bull., 1976, v. 31, No. 5–6, p. 55–59.
[337] Kozlov V. V. *Two integrable problems of classical dynamics.* Vestnik Moskov. Univ. Ser. 1. Mat. Mekh., 1981, v. 36, No. 3–4, p. 68–71.
[338] Kozlov V. V. *Topology of real algebraic curves and integrability of geodesic flows on algebraic surfaces.* Functional Analysis and Its Applications, 2008, v. 42, No. 3, p. 245–245.
[339] Kozlov V. V. *Dynamics of variable systems, and Lie groups.* Journal of Applied Mathematics and Mechanics, 2004, v. 68, No. 6, p. 803–808.
[340] Kozlov V. V. *Integrable cases of the problem of motion of a point over a three-dimensional sphere in a force field with fourth-degree potential.* Vestnik Moskov. Univ. Ser. 1. Mat. Mekh., 1985, v. 40, No. 3, p. 48–51.
[341] Kozlov V. V. *To a Problem on rotation of a rigid body in magnetic field.* Izv. AN SSSR. Mekh. tv. tela., 1985, No. 6, p. 28–33 (Russian).
[342] Kozlov V. V. *On the problem of fall of a rigid body in a resisting medium.* Vestnik Moskov. Univ. Ser. 1. Mat. Mekh., 1990, v. 45, No. 1, p. 30–36.
[343] Kozlov V. V. *Some integrable generalizations of the Jacobi problem on geodesics on an ellipsoid.* Journal of Applied Mathematics and Mechanics, 1995, v. 59, No. 1, p. 1–7.
[344] Kozlov V. V. *Symmetry groups of dynamical systems.* Journal of Applied Mathematics and Mechanics, 1988, v. 52, No. 4, p. 413–420.
[345] Kozlov V. V. *On falling of a heavy rigid body in an ideal fluid.* Mechanics of Solids, 1989, No. 5, p. 10–17.
[346] Kozlov V. V. *On a heavy cylindrical body falling in a fluid.* Mechanics of Solids, 1993, No. 4, p. 113–117.
[347] Kozlov V. V. *Polynomial integrals of dynamical systems with one-and-a-half degrees of freedom.* Mathematical Notes, 1989, v. 45, No. 4, p. 296–300.
[348] Kozlov V. V. *The stability of equilibrium positions in a nonstationary force field.* Journal of Applied Mathematics and Mechanics, 1991, v. 55, No. 1, p. 8–13.
[349] Kozlov V. V. *The principle of least action and periodic solutions in problems of classical mechanics.* Journal of Applied Mathematics and Mechanics, 1976, v. 40, No. 3, p. 363–370.
[350] Kozlov V. V., Onoshchenko D. A. *Nonintegrability of Kirchhoff's equations.* Soviet Mathematics. Doklady, 1982, v. 26, No. 6, p. 495–498.

[351] Kozlov V. V., Ramodanov S. M. *On the motion of a body with a rigid hull and changing geometry of masses in an ideal fluid.* Doklady Mathematics, 2002, v. 47, No. 2, p. 132–135.
[352] Kozlov V. V., Ramodanov S. M. *On the motion of a variable body in an ideal fluid.* Journal of Applied Mathematics and Mechanics, 2001, v. 65, No. 4, p. 579–587.
[353] Kozlov V. V., Treschev D. V. *Nonintegrability of the general problem of the rotation of a dynamically symmetric heavy rigid body with a stationary point. I. II.* Vestnik Moskov. Univ. Ser. 1. Mat. Mekh., 1985, No. 6, p. 73–81.
[354] Kozlov V. V., Fedorov Y. N. *Integrable systems on the sphere with elastic interaction potentials.* Mathematical Notes, 1994, v. 56, No. 3, p. 927–930.
[355] Kozlov V. V., Onischenko D. A. *Motion of a body with undeformable shell and variable mass geometry in an unbounded perfect fluid.* Journal of Dynamics and Differential Equations, 2003, v. 15, No. 213, p. 553–570.
[356] Kozlov V. V. *Integral Invariants after Poincaré and Cartan,* in H. Cartan, Integral Invariants, Moscow: URSS, 1998, p. 218–260 (Russian).
[357] Kucher E. Yu. Characteristic exponents of periodic Steklov and Chaplygin solutions Mekh. Tverd. Tela, 2003, v. 33, p. 33–39 (Russian).
[358] Kuzmin P. A. *Note on the Steklov case of motion of a heavy rigid body about a fixed point.* Prikl. Mat. Mekh., 1952, v. 16, No. 3, p. 243–245 (Russian).
[359] Kuzmina R. P. *Bifurcation set in a problem concerning the motion of a heavy solid with a stationary point.* Mechanics of Solids, 1982, v. 17, p. 3–10.
[360] Kuznetsov V. B. *Kowalevski top revisited.* CRM Proc. Lecture Notes, 2002, v. 32, AMS, Providence, RI, p. 181–196.
[361] Kuznetsov V. B. *Simultaneous separation for the Kowalevski and Goryachev – Chaplygin gyrostats.* J. Phys. A, 2002, v. 35, p. 6419–6430.
[362] Kuznetsov V. B., Jorgensen M. F., Christiansen P. L. *New boundary conditions for integrable lattices.* J. Phys., A., 1995, v. 28, p. 4639–4654.
[363] Kuznetsov V. M., Lugostov B. A., Sher Y. N. *On the motive mechanism of snakes and fish.* Arch. Rath. Mech. Analysis, 1967, v. 25, No. 5, p. 367–387.
[364] Lagrange J. L. Mécanique analytique. Oeuvres de Lagrange, v. 12. Paris: Gauthier–Villars, 1889.
[365] Lamb H. *Hydrodynamics,* ed. 6-th. N. Y., Dover publ., 1945.
[366] Lamb H. *Higher mechanics.* Cambridge Univ. Press, 1929.
[367] Lamb H. *Hydrodynamics.* New York: Dover Publications, 1932.
[368] Landau L. D., Lifshitz E. M. *Theoretical physics.* M.: Nauka, v. 6, 1959 (Russian).
[369] Lapporte O. *Note on a Kowalewski's top in quantum mechanics.* Phys. Rev., 1933, v. 43, p. 584–551.
[370] Laskar J., Robutel P. *The chaotic obliquity of the planets.* Nature, 1993, v. 361, No. 18, p. 608–612.
[371] Lebovitz N. R. *On the principle of the exchange of stabilities. I. The Roche Ellipsoids.* Astrophys. J., 1963, v. 138, p. 1214–1217.
[372] Lebovitz N. R. *The mathematical development of the classical ellipsoids.* Intern. Jour. Engineering Science, 1998, v. 36, No. 12, p. 1407–1420.
[373] Leimanis E. *The general problem of the motion of coupled rigid bodies about a fixed point.* Berlin: Springer-Verlag, 1965.
[374] Lerman L. M. *More about the structure of integrable waves for the Landau – Lifshits equations.* Select Math. from Sov., 1993, v. 12, No. 4, p. 333–351.
[375] Lerman L. M. *Nonintegrability and stationary solitons of complex profile for Landau – Lifshitz equation.* JETP Lett., 1990, v. 51, No. 6, p. 383–386.

[376] Lesfari A. *Abelian surfaces and Kowalevski's top*. Ann. Scient. Ec. Nor. Sup., 1988, v. 21, 4 ser., p. 193–223.
[377] Levi-Civita T., Amaldi U. *Lezioni di meccanica razionale*, v. 2. Bologna, 1927.
[378] Lewis D., Ratiu T., Simo J. C., Marsden J. E. *The heavy top: a geometric treatment*. Nonlinearty, 1992, v. 15, p. 1–48.
[379] Liapounoff A. M. *Sur certaines séries de figures d'équilibre d'un liquide hétérogène en rotation: In 2 vols.*, Leningrad: Akad. Nauk SSSR, 1925, 1927.
[380] Liao J. C., Beal D. N., Lauder G. V., Trantafyllou M. S. *Fish exploiting vortices decrease muscle activity*. Science, 2003, v. 32, p. 1566–1569.
[381] Lichtenstein L. *Gleichgewichtsfiguren rotietender Flüssigkeiten*. Berlin: Springer-Verlag, 1933.
[382] Lidov M. L. *Exact solutions of equations of one-dimensional unsteady motions of gas with allowance for Newtonian attraction forces*. Dokl. Akad. Nauk SSSR, 1954, v. 97, No. 3.
[383] Liouville J. *Développements sur un chapitre de la Méchanique de Poisson*. J. Math. Pures et Appl., 1858, v. 3, p. 1–25.
[384] Liouville J. *Formules générales relatives à la question de la Stabilité de l'équilibre d'une masse liquide homogène douée d'un mouvement de rotation autour d'un axe*. J. Math. Pures Appl., 1855, v. 20, p. 164–184.
[385] Liouville R. *Sur le mouvement d'un solide dans un liquide indéfini*. Comp. Rend. Ac. Sc., ser. 2, 1896, p. 874–876.
[386] Liouville R. *Sur la rotation des solides*. Comp. Rend. Acad. Sc., 1895, v. CXX, p. 903–906.
[387] Lipschitz R. *Reduction der Bewegung eines flüssigen homogenen Ellipsoids auf das Variationsproblem eines einfachen Integrals, und Bestimmung der Bewegung für den Grenzfall eines unendlichen elliptischen Cylinders*. J. reine angew. Math. (Crelle's Journal), 1874, v. 78, p. 245–272.
[388] Lloyd N. G. *The number of periodic solutions of the equation* $\dot{z} = z^N + p_1(t)z^{N-1} + \dots + p_N(t)$. Proceedings of the London Mathematical Society, 1973, v. 3, No.. 4, p. 667–700.
[389] Love A. E. H. *On the motion of a liquid elliptic cylinder under its own attraction*. Quart. J. of Pure and Appl. Math., 1889, v. 23, p. 153–158.
[390] Love A. E. H. *The oscillations of a mass of gravitating liquid in the form of an elliptic cylinder which rotates as if rigid about its axis*. Quart. J. of Pure and Appl. Math., n.d., p. 158–165.
[391] Lubowiecki P., Zoladek H. *The Hess-Appelrot system. I. Invariant torus and its normal hyperbolicity*. Journal of Geometric Mechanics, 2012, v. 4, No.. 4, p. 443–467.
[392] Lyapunov A. M. *New case of integrability of equations of motion of a solid body in fluid*. Comm. Soc. Math. Kharkow, 1893, v. 4, No. 1–2, p. 81–85.
[393] Lyapunov A. M. *On constant screw motions of a rigid body in a liquid*. Soobshch. Kharkov. Mat. Obshch., Ser. 2, 1888, v. 1, No. 1–2, p. 7–60 (Russian).
[394] Lyapunov A. M. *On one property of the differential equations of the problem of the motion of a heavy rigid body having a fixed point*. Coll. Works (AS USSR, Moscow), 1954, v. 1, p. 402–417 (Russian).
[395] Lyapunov A. M. *Collected works*. Moscow: Akad. Nauk SSSR, v. 5, 1965.
[396] Lynden-Bell D. *On the gravitational collapse of a cold rotating gaz cloud*. Proc. Camb. Phys. Soc., 1962, v. 58, p. 709–711.
[397] Lyttleton R. A. *The stability of rotating liquid masses*. Cambridge: Cambridge Univ. Press, 1953.
[398] Maciejewski A. J., Przybylska M. *Differential Galois approach to non-integrability of the heavy top problem*. Ann. Fac. Sci. Toulouse Math., 2005, v. 14, No. 1, p. 123–160.
[399] MacKay R. S. *Renormalization in area-preserving maps*. World Scientific, 1993.
[400] Macmillan W. D. *Dynamics of rigid bodies*. N. Y., London, 1936.

[401] Magnus K. *Kreisel. Theorie und Anwendungen.* Springer-Verlag, 1971.

[402] Magyari E., Thomas H., Weber R., Kaufman C., Müller G. *Integrable and nonintegrable classical spin clusters. Integrability criteria and analytic structure of invariants.* Z. Phys. B, Condensed Matter, 1987, v. 65, p. 363–374.

[403] Maki K., Ebisawa H. *Exact magnetic ringing solutions in superfluid* ^{3}He–B. Phys. Rev., 1976, v. 13B, No. 7, p. 2924–2930.

[404] Manakov S. V. *Note on the integration of Euler's equations of the dynamics of an n-dimensional rigid body.* Functional Analysis and Its Applications, 1976, v. 10, No. 4, p. 328–329.

[405] Marey M. *Des mouvements que certains animaux exécutent pour retomber sur leurs pieds, lorsqu'ils sont précipités d'un élevé.* Comp. Rend. Acad. Sci. Paris, 1894, v. 119, p. 714–717.

[406] Marikhin V. G., Sokolov V. V. *Separation of variables on a non-hyperelliptic curve.* Regular and Chaotic Dynamics, 2005, v. 10, No. 1, p. 59–70.

[407] Marinbakh M. A. *Liapunov periodic motions of a heavy rigid body with one fixed point.* Vestnik Moskov. Univ. Ser. 1. Mat. Mekh., 1979, v. 1, p. 75–79 (Russian).

[408] Marinbakh M. A. *The general case of liapunov's periodic motions of a heavy solid body with a single fixed point.* Journal of Applied Mathematics and Mechanics, 1981, v. 45, No. 5, p. 598–603.

[409] Markeev A. P. *On the planar and close to planar rotations of a heavy rigid body around a fixed point.* Mekh. Tverd. Tela, 1988, No. 4, p. 29–36 (Russian).

[410] Markeev A. P. *Theoretical mechanics (edition fourth, revised).* M.–Izhevsk: RCD, 2007 (Russian).

[411] Markeyev A. P. *The dynamics of a body in contact with a solid surface.* Moscow: Nauka, 1992.

[412] Markeev A. P. *On the Steklov case in rigid body dynamics.* Regular and Chaotic Dynamics, 2005, v. 10, No. 1, p. 81–93.

[413] Markeyev A. P. *An algorithm for normalizing Hamiltonian systems in the problem of the orbital stability of periodic motions.* Journal of applied mathematics and mechanics, 2002, v. 66, No. 6, p. 889–896.

[414] Markeyev A. P. *The pendulum-like motions of a rigid body in the Goryachev-Chaplygin case.* Journal of applied mathematics and mechanics, 2004, v. 68, No. 2, p. 249-258.

[415] Markeyev A. P. *The stability of the plane motions of a rigid body in the Kovalevskaya case.* Journal of Applied Mathematics and Mechanics, 2001, v. 65, No. 1, p. 47–54.

[416] Markeev A. P. *On the stability of the regular precession of an asymmetric gyroscope.* Doklady Physics, 2002, v. 47, No. 11, p. 833–837.

[417] Markeev A. P., Moshchuk N. K. *Qualitative analysis of motion of a heavy solid body on a smooth horizontal plane.* Journal of Applied Mathematics and Mechanics, 1983, v. 47, No. 1, p. 22–26.

[418] Marsden J., Ratiu T. S. *Introduction to mechanics and symmetry. A basic exposition of classical mechanical systems.* Springer-Verlag, 1994.

[419] Marshalek E. R. *An overlooked figure of equilibrium of a rotating ellipsoidal self-gravitating fluid and the Riemann theorem.* Phys. Fluids, 1996, v. 8, No.. 12, p. 3414–3422.

[420] Martinez F. J., Cisneros J., Montalvo D. *On equilibrium figures for ideal fluids in the form of confocal ellipsoids rotating with common angular velocity.* Rev. Mexicana Astron. Astrof., 1990, v. 20, p. 15–22. See also: Erratum, *ibid.*, p. 153–154.

[421] Melchior P. *Physique et dynamique planétaires*, v. 4. Vander-éditeur, 1973.

[422] Mertsalov N. I. *Problem of the Motion of a Rigid Body Having a Fixed Point for $A = B = 4C$ and for Area Integral Not Equal To Zero*, Izv. Akad. Nauk SSSR, Otdel. Tekh. Nauk, 1946, No. 5, p. 697–701 (Russian).

[423] Mettler E. *Periodische und asymptotische Bewegungen des unsymmetrischen schweren Kreisels.* Mathematische Zeitschrift, 1937, v. 43, No.. 1, p. 59–100.

[424] Miloh T., Galper A. *Self-propulsion of general deformable shapes in a perfect fluid.* Proc. R. Soc. Lond. A., 1993, v. 442, p. 273–299.
[425] Minkowski H. *Über die Bewegung eines festen Körpers in einer Flüssigkeit.* Berlin: Sitzungsber. König. Preuss. Akad. Wiss., 1888, v. 30, p. 1095–1110.
[426] Mishchenko A. S., Fomenko A. T. *Integration of Eulerian equations on Lie semisimple algebras.* Doklady Akademii Nauk SSSR, 1976, v. 231, No. 3, p. 536–538.
[427] Mlodzieiowski B. C. *Über permanente Axen bei der Bewegung eines schweren festen Körpers um einen Punkt.* Tr. Otdel. Fiz. Nauk Ob-va Lubit. Estestvozn., 1894, v. 7, No. 1, p. 46–48 (Russian).
[428] Mlodzieiowski B. C. *Sur un cas du mouvement d'un corps pesant autour d'un point fixe.* Mat. Sb., 1896, v. 18, No. 1, p. 76–85 (Russian).
[429] Moiseev N. N., Rumyantsev V. V. *Dynamics of a body with cavities containing a fluid.* Moscow: Nauka, 1965 (Russian).
[430] Montgomery R. *Gauge theory of the falling cat.* Field Inst. Commun., 1993, v. 1, p. 193–217.
[431] Morosi C., Tondo G. *Separation of variables in multi-Hamiltonian systems: an application to the Lagrange top.* Theoretical and mathematical physics, 2003, v. 137, No. 2, p. 1550–1560.
[432] Moser J. *Integrable Hamiltonian system and spectral theory.* Piza, Lezioni Fermiane, 1981.
[433] Moser J., Veselov A. P. *Disctete version of some classical integrable systems and factorisation of matrix polinomials.* Comm. Math. Phys., 1991, v. 139, p. 217–243.
[434] Moshchuk N. K. *On the Analogy between Two Classical Problems in Mechanics.* Izv. Akad. Nauk SSSR. Mekh. Tverd. Tela, 1987, No. 1, p. 53–56 (Russian).
[435] Narlikar V. V., Larmor J. *The Kelvin – Poincare problem of stellar evolution.* Proc. Roy. Soc. London Ser. A, 1934, v. 144, No. 851, p. 28–46.
[436] Neistadt A. I. *Probability phenomena due to separatrix crossing.* Chaos, 1991, v. 1, No. 1, p.42–48.
[437] Neistadt A. I., Treshev D. V., Sidorenko V. V. *Stable periodic motions in the problem on passage through a separatrix.* Chaos, 1997, v. 7, No. 3, p. 2–11.
[438] Neishtadt A. I. *On the change in the adiabatic invariant on crossing a separatrix in systems with two degrees of freedom.* Journal of Applied Mathematics and Mechanics, 1987, v. 51, No. 5, p. 586–592.
[439] Neishtadt A. I. *Passage through a separatrix in a resonance problem with a slowly-varying parameter.* Journal of Applied Mathematics and Mechanics, 1975, v. 39, No. 4, p. 594–605.
[440] Neishtadt A. I. *Jumps in the adiabatic invariant on crossing the separatrix and the origin of the 3: 1 Kirkwood gap.* Soviet Physics Doklady, 1987, v. 32, p. 571–573.
[441] Neishtadt A. I. *Variation of adiabatic invariant accompanying transition through a separatrix.* Sov. J. Plasma Phys., 1986, v. 12, No. 8, p. 992–999 (Russian).
[442] Neishtadt A. I. *Evolution of rotation of a solid acted upon by the sum of a constant and a dissipative perturbing moments.* Mechanics of Solids., 1980, v. 15, p. 21–27.
[443] Nekhoroshev N. N. *Action-angle variables, and their generalizations.* Trudy Moskovskogo Matematicheskogo Obshchestva, 1972, v. 26, p. 181–198 (Russian).
[444] Nemchinov I. V. *Expansion of a triaxial gas ellipsoid in a regular mode.* Applied Mathematics and Mechanics, 1965 v. 29, No. 1, p. 926–929.
[445] Nekrassov P. A. *Étude analytique d'un cas du mouvement d'un corps pesant autour d'un point fixe.* Mat. Sb., 1896, v. 18, No. 2, p. 161–274 (Russian).
[446] Neumann C. *De problemate quodam mechanico, quod ad primam integralium ultraellipticorum slassem revocatur.* Rein. und Agew. Math., 1859, v. 56, p. 46–63.
[447] Novikov S. P. *The Hamiltonian formalism and a many-valued analogue of Morse theory.* Russian mathematical surveys, 1982, v. 37, No. 5, p. 1–56.
[448] A) Novikov S. P., Shmel'Tser I. *Periodic solutions of Kirchhoff's equations for the free motion*

of a rigid body in a fluid and the extended theory of Lyusternik-Shnirel'man -morse (LSM). I. Functional Analysis and Its Applications, 1981, v. 15, No. 3, p. 197–207.
B) Novikov S. P. *Variational methods and periodic solutions of Kirchhoff-type equations. II.* Functional Analysis and Its Applications, 1981, v. 15, No. 4, p. 263–274.

[449] Olshanetsky M. A., Perelomov A. M., Reyman A. G., Semenov-Tian-Shansky M. A. *Integrable systems. II.* Itogi Nauki and Techniki, VINITI AN SSSR, Contemporary problems in mathematics, 1987, v. 16, p. 86–226 (Russian).

[450] Olver P. *Applications of Lie groups to differential equations.* Springer, 1986.

[451] Orekhov V. I. *Topological analysis of natural systems with quadratic integrals.* Journal of Applied Mathematics and Mechanics, 1985, v. 49, No. 1, p. 10–16.

[452] Oreshkina L. N. *A Case of Integrability of M. P. Kharlamov's equations.* Mekh. Tverd. Tela, 1989, No. 21, p. 1–9 (Russian).

[453] Oreshkina L. N. *Necessary and sufficient conditions for the existence of a fourth quadratic integral in some problems of the dynamics of a rigid body.* Mekh. Tverd. Tela, 1988, No. 20, p. 18–29 (Russian).

[454] Oreshkina L. N. *Combination of two problems of dynamics of rigid bodies.* Izv. AN SSSR. Mekh. Tverd. Tela, 1986, No. 5, p. 36–43 (Russian).

[455] Oshemkov A. A. *Fomenko invariants for the main integrable cases of the rigid body motion equations.* Advances in Sov. Math., 1991, v. 6, p. 67–146.

[456] Oshemkov A. A. *The topology of surfaces of constant energy and bifurcation diagrams for integrable cases of the dynamics of a rigid body on* $so(4)$. Russian Mathematical Surveys, 1987, v. 42, No. 6, p. 241–242.

[457] Ovsyannikov L. V. *A new solution of the equations of hydrodynamics.* Dold. Akad. Nauk SSSR, 1956, v. 111, p. 47–49 (Russian).

[458] Padova E. *Sul moto di un ellissoide fluido ed omogeneo.* Annali della Scuola Normale Superiore di Pisa, 1871, v. 1, p. 1–87.

[459] Parker T. S., Chua L. O. *Practical numerical algorithms for chaotic systems.* New York: Springer-Verlag, 1989.

[460] Perelomov A. M. *Integrable systems of classical mechanics and Lie algebras.* Springer Verlag, 1990.

[461] Perelomov A. M. *Lax representation for the systems of Kovalevsky type.* Comm. Math. Phys., 1981, p. 239–241.

[462] Perelomov A. M. *Integrable systems of classical mechanics and Lie algebras.* Basel: Birkhauser, 1990.

[463] Perelomov A. M. *A note on geodesics on ellipsoid.* Regular and Chaotic Dynamics, 2000, v. 5, No. 1, p. 89–94.

[464] Perelomov A. M. *Motion of four-dimesional rigid body around a fixed point: an elementary approach* I. math-ph/0502053, 2005.

[465] Perelomov A. M., Ragnisco O., Wojciechowski S. *Integrability of two interacting N-dimensional rigid bodies.* Comm. Math. Phys., 1986, v. 102, p. 573–583.

[466] Reyman A. G. and Semenov-Tian-Shansky M. A. *Integrable systems.* Izhevsk: R&C Dynamics, Institute of Computer Science, 2003 (in Russian).

[467] Pizzetti P. *Principii della teoria meccanica della figura dei pianeti.* Pisa, 1912.

[468] Pliss V. A. *Nonlocal problems of the theory of oscillations.* London: Academic Press, 1966.

[469] Pogosian T. I., Kharlamov M. P. *Bifurcation set and integral manifolds of the problem concerning the motion of a rigid body in a linear force field.* Journal of Applied Mathematics and Mechanics, 1979, v. 43, No. 3, p. 452–462.

[470] Poincaré H. *Sur l'équilibre d'une masse fluide animée d'un mouvement de rotation.* Acta Math., 1885, v. 7, p. 259–380.

[471] Poincaré H., *Figures d'équilibre d'une masse fluide (Leçons professées à la Sorbonne en 1900)*. Paris: Gauthier-Villars, 1902.
[472] Poincaré H. *Sur la precession des corps deformables*. Bull. Astr., 1910, v. 27, p. 321–356.
[473] Poincaré H. *Sur le forme nouvelle des equations de la mecanique*. C. R. Acad. Sci. Paris, 1901, v. 132, p. 369–371.
[474] Poincaré H. *Le méthodes nouvelles de la mécanique céleste*. Paris, Gauthier-Villars, 1892, v. 1.
[475] Poinsot L. *Théorie nouvelle de la rotation des corps*. J. math. pures et appl., 1851, v. 16, p. 9–130, p. 289–336.
[476] Radau R. *Rotation des corps solides*. Annal. scientif. de l'Ecole Normale superieure, v. 6, p. 233–250.
[477] Ramani A., Grammaticos B., Dorizzi B. *On the quantization of the Kowalevskaya top*. Phys. Lett., 1984, v. 101A, No. 2, p. 69–71.
[478] Ramodanov S. M. *Asymptotic solutions of the Chaplygin equations*. Vestnik Moskov. Univ. Ser. 1. Mat. Mekh., 1995, No.. 3, p. 93–97 (Russian).
[479] Ramodanov S. M. *The influence of circulation on the behavior of a rigid body falling in a fluid*. Mechanics of Solids, 1996, v. 31, No. 5, p. 14–18.
[480] Ratiu T., van Moerbeke P. *The Lagrange rigid body motion*. Ann. Inst. Fourier, 1982, v. 32, No. 1, p. 211–234.
[481] Reichl L. E. *The transition to chaos. Conservative classical systems and quantum manifestations*. Springer, 2004.
[482] Reyman A. G. *Integrable Hamiltonian systems associated with graded Lie algebras*. J. Sov. Math., 1982, v. 19, No. 5, p. 1507–1545.
[483] Reyman A. G., Semenov-Tian-Shansky M. A. *Lax representation with a spectral parameter for the kowalevski top and its generalizations*. Functional Analysis and Its Applications, 1988, v. 22, No. 2, p. 158–160.
[484] Reyman A. G., Semenov-Tian-Shansky M. A. *A new integrable case of the motion of the 4-dimensional rigid body*. Comm. Math. Phys., 1986, v. 105, p. 461–472.
[485] Reyman A. G., Semenov-Tian-Shansky M. A. *Lax representation with a spectral parameter for the Kowalewski top and its generalizations*. Lett. Math. Phys., 1987, v. 14, p. 55–61.
[486] Reyman A. G., Semenov-Tian-Shansky M. A. *Reduction of Hamiltonian systems, Lie algebras and Lax equations*, I, II. Invent. Math., 1979, v. 54, p. 81–100; 1981, v. 63, p. 423–432.
[487] Riemann B. *Ein Beitrag zu den Untersuchungen über die Bewegung einer flüssigen gleichartigen Ellipsoïdes*. Abh. d. Königl. Gesell. der Wiss. zu Göttingen, 1861.
[488] Rimmer R. *Generic bifurcations from fixed points of involutory area preserving maps*. Diff. Equations, 29, 1978, p. 329, and P. Math. Res. Paper 79-7, La Trobe U., Melbourne, 1979.
[489] Roberts W. *Application des coordonnées elliptiques à la recherche des surfaces orthogonales*. J. reine angew. Math. (Crelle's Journal), 1863, v. 62, p. 50–60.
[490] Roche E. *Mémoire sur la figure d'une masse fluide, soumise a l'attraction d'un point éloigné*. Acad. des Sci. de Montpellier, 1849–1850, v. 1, p. 243–262, p. 333–348; 1852, v. 2, pag. 21.
[491] Roche E. *Essai sur la Constitution et l'Origine du Systeme Solaire*. Aca. de Montpellier Section des Sciences, 1873, v. 8, p. 235.
[492] Rosenkilde C. E. *Stability of axisymmetric figures of equilibrium of a rotating charged liquid drop*. J. Math. Phys., 1967, v. 8, No. 1, p. 98–118.
[493] Rosensteel G., Tran H. Q. *Hamiltonian dynamics of self-gravitating ellipsoids*. The Astrophysical Journal, 1991, v. 366, p. 30–37.
[494] Rosensteel G. *Gauge theory of Riemann ellipsoids*. J. Phys. A: Math. Gen., 2001, v. 34, p. L1–L10.

[495] Rosochatius E. *Über die Bewegung eines Punktes*. Inaugural Berlin: Dissertation, Univ. Göttingen, 1877.
[496] Rossner L. F. *The Finite-amplitude oscillations of the maclaurin spheroids*. Astrophys. J., 1967, v. 149, p. 145–168.
[497] Routh E. J. *A Treatise on analytical statics*. Cambridge: Cambridge University Press, 1922, v. 2.
[498] Routh E. J. *Dynamics of a system of rigid bodies*, v. 2. New York: Dover Publ., 1960.
[499] Rubanovskii V. N. *New cases of the integrability of the equations of motion of a heavy solid body in fluid*. Vestnik Moskov. Univ., Ser. 1, Mat. Mekh., 1968, v. 23, No. 2, p. 99–106 (Russian).
[500] Rubanovskii V. N. *The bifurcation and stability of steady motions*. Theor. Appl. Mech., 1974, v. 5, No. 1, p. 67–79 (Russian).
[501] Rubanovskii V. N. *Quadratic integrals of the equations of motion of a rigid body in a liquid*. Journal of Applied Mathematics and Mechanics, 1988, v. 52, No. 3, p. 312–322.
[502] Rubanovsky V. N., Samsonov V. A. *Stability of stationary motions in examples and problems*. M.: Nauka, 1988 (Russian).
[503] Rubanovskii V. N. *Application of the small parameter method to the equations of motion of a body in a fluid*. Vestn. Mosk. Univ. Ser. 1. Mat. Mekh., 1967, No. 3, p. 80–87 (Russian).
[504] Rueb Ad. *Specimen inagurale de motu gyratorio corporis rigidi nulle vi acceleratrice sollicitati*. Auctore Adolpho Stephano Rueb, Roterodamonsi (Utrecht), 1834, v. 4, p. 74.
[505] Rueb Ad. *Sur la rotation d'un corps*. Com. rend. Acad. des Sciences, 1849, J. de Crelle, t. XXXIX.
[506] Rumer Yu. B., Ryvkin M. S. *Thermodynamics, statistical physics, and kinetics*. Moscow: Mir, 1980.
[507] Rumyantsev V. V. *On the stability of rotation of a heavy rigid body with a fixed point in the case of SV Kovalevskaya*. Prikl. Math. Mekh., 1954, v. 18, p. 457–458 (Russian).
[508] Rumyantsev V. V. *The stability of permanent rotations of a heavy solid*. Prikl. Math. Mekh., 1956, v. XX, No. 1, p. 51–66 (Russian).
[509] Rumyantsev V. V. *The dynamics of a rigid body suspended by a string*. Mechanics of Solids, 1983, v. 18, p. 5–15.
[510] Rumyantsev V. V. *On Poincare-Chetaev equations*. Doklady Physics, 1994, v. 39, p. 647–649.
[511] Rumyantsev V. V. *On the stability of stationary motions of satellites*. M.: VTs Akad, Nauk SSSR, 1967.
[512] Sadetov S. T. *The fourth algebraic integral of kirchhoff's equations*. Journal of Applied Mathematics and Mechanics, 2000, v. 64, No. 2, p. 229–242.
[513] Sadetov S. T. *Integrability conditions of Kirchhoff equations*. Vestnik Moskov. Univ. Ser. 1. Mat. Mekh., 1990, No. 3, p. 56–62 (Russian).
[514] Sadètov S. T., verbal communication, 2002.
[515] Sadov Yu. A. *The action-angle variables in the Euler-Poinsot problem*. Preprint No. 22 of the Keldysh Institute of Applied Mathemat ics of the Academy of Sciences of the USSR, Moscow, 1970 (in Russian).
[516] Saffman P. O. *The self-propulsion of a deformable body in a perfect fluid*. J. Fluid Mech., 1967, v. 28, p. 385–389.
[517] Salnikova T. V. *Branching of solutions of the perturbed Lagrange problem and nonexistence of single-valued integrals*. Vestnik Moskov. Univ. Ser. 1. Mat. Mekh., 1985, v. 40, No. 2, p. 85–87 (Russian).
[518] Salnikova T. V. *Non integrability of perturbed lagrange problem*. Vestnik Moskov. Univ. Ser. 1. Mat. Mekh., 1984, v. 39, No. 4, p. 62–66 (Russian).
[519] Salnikova T. V. *Integrability of the Kirchhoff equations in the symmetrical case*. Vestnik Moskov. Univ. Ser. 1. Mat. Mekh., 1985, v. 40, No. 4, p. 68–71 (Russian).

[520] Salnikova T. V. *On integrability of the motion of an axisymmetric rigid body*. J. Phys. A., 2001, v. 34, No. 11, p. 2179–2184.

[521] Samsonov V. A. *On the rotation of a body in a magnetic field*. Transaction of the USSR Academy of Sciences, MTT (Mechanics of Solids), 1984, No. 4, p. 32–34 (Russian).

[522] Schottky F. *Über das analytische problem der rotation eines starren Körpers in Raume von vier Dimensionen*. Berlin: Sitzungsber. König. Preuss. Akad. Wiss., 1891, v. XIII, p. 227–232.

[523] Schläfli L. *On the distribution of surfaces of the third order into species, in reference to the absence or presence of singular points, and the reality of their lines*. Philos. Trans. Roy. Soc. London, 1863, v. 153, p. 193–241.

[524] Selivanova E. N. *New examples of integrable conservative systems on S^2 and the case of Goryachev – Chaplygin*. Commun. Math. Phys., 1999, v. 207, p. 641–663.

[525] Selivanova E. N. *New families of conservative systems on S^2 possessing an integral of fourth degree in momenta*. Ann. of Glob. An. and Geom., 1999, v. 17, p. 201–219.

[526] Sergeev V. S. *On periodic solutions to the equations of motion of a heavy rigid body with a fixed point*. Vestnik Moskov. Univ. Ser. 1. Mat. Meh., 1969, No. 1, p. 40–51 (Russian).

[527] Sergeev V. S. *On the stability of permanent rotation of a heavy solid body about a fixed point*. Journal of Applied Mathematics and Mechanics, 1976, v. 40, No. 3, p. 370–378.

[528] Serret J. A. *Mémoiré sur l'emploi de la méthode de la variation des arbitraires dans la théorie des mouvement des rotation*. Mémoires de l'Académies des Sciences de Paris, 1866, v. 35, p. 585–616.

[529] Shashikanth B. N. *Poisson brackets for the dynamically interacting systems of a 2D rigid cylinder and N point vortices: the case of arbitrary smooth cylinder shapes*. Regular and Chaotic Dynamics, 2005, v. 10, No. 1, p. 1–14.

[530] Shashikanth B. N., Marsden J. E., Burdick J. W., Kelly S. D. *The Hamiltonian structure of a 2D rigid circular cylinder interacting dynamically with N point vortices*. Phys. of Fluids, 2002, v. 14, p. 1214–1227.

[531] Simó C. *Dynamical properties of the figure eight solution of the three-body problem*. Celestial mechanics, v. 292, p. 209–228.

[532] Simó C. *Invariant curves of perturbations of non twist integrable area preserving maps*. Regular and Chaotic Dynamics, 1998, No. 3, p. 180–195.

[533] Simó C., Stuchi T. *Central stable/unstable manifolds and the destruction of KAM tori in the planar Hill problem*. Physica D, 2000, v. 140, p. 1–32.

[534] Sklyanin E. K., Takebe T. *Separation of variables in the elliptic Gaudin model*. Commun. Math. Phys., 1999, v. 204, p. 17–38.

[535] Sokolov V. V. *A generalized Kowalewski Hamiltonian and new integrable cases on $e(3)$ and $so(4)$*. In "Kowalevski property", ed. V. B. Kuznetsov, CRM Proceedings and Lect. Notes, AMS, 2002, p. 307–315.

[536] Sokolov V. V., Wolf T. *Integrable quadratic classical Hamiltonians on $so(4)$ and $so(3, 1)$*. Journal of Physics A: Mathematical and General, 2006, v. 39, No. 8, p. 1915–1926.

[537] Sokolov V. V. *A new integrable case for the Kirchhoff equation*. Theoretical and Mathematical Physics, 2001, v. 129, No. 1, p. 1335–1340.

[538] Sokolov V. V. *On a class of quadratic Hamiltonians on $so(4)$*. Dokl. Math., 2004, v. 69, p. 108–111.

[539] Sokolov V. V., Tsiganov A. V. *Commutative Poisson subalgebras for sklyanin brackets and deformations of some known integrable models*. Theoretical and mathematical physics, 2002, v. 133, No. 3, p. 1730–1743.

[540] Sokolov V. V., Tsiganov A. V. *Lax pairs for the deformed Kowalevski and Goryachev – Chaplygin tops*. Theoretical and mathematical physics, 2002, v. 131, No. 1, p. 543–549.

[541] Springborn B. A. *The toy top, an integrable system of rigid body dynamics.* J. of Nonlinear Math. Phys., 2000, v. 7, No. 3, p. 386–410.

[542] Sretenskii L. N. *On certain cases of motion of a heavy rigid body with gyroscope.* Vestnik Moskov. Univ. Ser. 1. Mat. Mekh., 1963, v. 3, p. 60–71.

[543] Sretenskii L. N. *Some cases of integrability of the equations of motion of a gyrostat.* Dokl. Acad. Nauk SSSR, 1663, v. 149, No. 2, p. 292–294 (Russian).

[544] Sretenskii L. N. *Theory of equilibrium figures of liquid rotating mass.* Uspekhi Matematicheskikh Nauk, 1938, No. 5, p. 187–230 (Russian).

[545] Sretenskii L. N. *The movement of a Gorjachev-Chaplygin gyroscope.* Izv Akad. Nauk OTN, 1953, No. 1, p. 109–119 (Russian).

[546] Srivastava N., Kaufman C., Müller G., Magyari E., Weber R., Thomas H. *Classical spin clasters: integrability and dynamical properties.* J. Appl. Phys., 1987, v. 61, No. 8, p. 4438–4440.

[547] Srivastava N., Kaufman C., Müller G., Weber R., Thomas H. *Integrable and nonintegrable classical spin clasters.* Z. Phys. B, Condensed Matter, 1988, v. 70, p. 251–268.

[548] Staude O. *Über permanente Rotationaxen bei der Bewegung eines schweren Körpers um einen festen Punkt.* J. reine und andew. Math, 1894, v. 113, No. 4, p. 318–334.

[549] Stekloff V. A. *Remarque sur un probleme de Clebsch sur le mouvement d'un corps solide dans un liqiude indefini en sur le probleme de M. de Brun.* Comp. Rend. Ac. Sc. Paris, 1902, v. 135, p. 526–528.

[550] Stekloff V. A. *Sur le mouvement d'un corps solide ayant une cavite de forme ellipsoidale remple par un liquide incompressible en sur les variations des latitudes.* Ann. de la fac. des Sci. de Toulouse, 1909, v. 1, p. 145–256.

[551] Stekloff W. *Problème du mouvement d'une masse fluide incompressible de la forme ellipsoïdale les parties s'attirent suivant la loi de Newton.* Annales scientifiques de l'É.N.S. 3e série, 1908, v. 25, p. 469–528.

[552] Stekloff W. *Problème du mouvement d'une masse fluide incompressible de la forme ellipsoïdale les parties s'attirent suivant la loi de Newton (Suite).* Annales scientifiques de l'É.N.S. 3e série, 1909, v. 26, p. 275–336.

[553] Steklov V. A. *Posthumous works of Lyapunov on the figures of equilibrium of an inhomogeneous rotating fluid,* in A. M. Lyapunov, Collected Works: Vol. 5, Moscow: Akad. Nauk SSSR, 1965, p. 379–384 (Russian).

[554] Steklov V. A. *Supplements to the Work "On the motion of a rigid body in a fluid".* Kharkov: Kharkov Univ., 1895 (Russian).

[555] Steklov V. A. *New Particular Solution of Differential Equations of Motion of Heavy Rigid Body about Fixed Point.* Tr. Otdel. Fiz. Nauk Ob-va Lubit. Estestvozn., 1899, vol. 10, no. 1, pp. 1–23 (Russian).

[556] Steklov V. A. *On the motion of a rigid body in a fluid.* Kharkov: Kharkov. Univ., 1893 (Russian).

[557] Steklov V. A. *On the motion of a rigid body in a fluid: 1.* Soobshch. Kharkov. Mat. Obshch., Ser. 2, 1891, v. 2, No. 5–6, p. 209–235 (Russian).

[558] Steklov V. A. *One case of motion of heavy rigid body in a fluid.* Tr. Otdel. Fiz. Nauk Obsch. Lyubit. Estestvozn., 1895, v. 7, No. 2, p. 10–21 (Russian).

[559] Steklov V. A. *A certain case of motion of heavy rigid body having fixed point.* Tr. Otdel. Fiz. Nauk Ob-va Lubit. Estestvozn., 1896, v. 8, No. 1, p. 19–21 (Russian).

[560] Stöckmann H. J. *Quantum chaos. An introduction.* Cambridge: Cambridge University Press, 2000.

[561] Sudakov S. N. *About oscillations of rotating gravitating liquid ellipsoids with variable viscosity.* Mekh. Tverd. Tela, 2002, No. 32, p. 217–226 (Russian).

[562] Suslov G. K. *Theoretical mechanics.* Moscow: Gostekhizdat, 1946 (Russian).

[563] Taimanov I. A. *Non-self-itersecting closed extremals of multivalued or not everywhere positive functionals.* Izv. Akad. Nauk SSSR Ser. Mat., 1991, v. 55, No. 2, p. 367–383.
[564] Tanabe Y., Kaneko K. *Behavior of a falling paper.* Phys. Rev. Lett., 1994, v. 73, No. 10, p. 1372–1377.
[565] Tannenberg W. *Sur le mouvement d'un corps solide pesant autour d'un point fixe: cas particulier signalé par M-me Kowalewsky.* Bordeaux: Gounouilhou, 1898.
[566] Tatarinov Ya. V. *Portraits of the Classical Integrals of the Problem of Rotation of a Rigid Body about a Fixed Point.* Vestn. Moskov. Univ., Ser. 1. Mat. Mekh., 1974, v. 6, p. 99–105 (Russian).
[567] Tatarinov Ya. V. *Frequency nondegeneracy of the Lagrange gyroscope and a balanced gyroscope with a gimbal suspension.* Dokl. Acad. Nauk SSSR, 1987, No. 4, p. 30–36 (Russian).
[568] Tedone O. *Il moto di un ellissoide fluido secondo l'ipotesi di Dirichlet.* Annali della Scuola Normale Superiore di Pisa, 1895, v. 7, p. I–IV+1–100.
[569] Tennysonn J. L., Carry J. R., Escande D. F. *Change of the adiabatic invariant due to separatrix crossing.* Phys. Rev. Lett., 1986, v. 56, No. 20, p. 2117–2120.
[570] S. A. Chaplygin, N. I. Mertsalov (Eds.). *The rigid body motion around a fixed point: collection devoted memory S. V. Kowalewskaya.* Moscow: Akad. Nauk SSSR, 1940 (Russian).
[571] Thomson W., Tait P. *Treatise on natural philosophy.* Oxford: Clarendon Press, 1867, v. 1.
[572] Thomson W., Tait P. G. *Treatise on natural philosophy.* Cambridge University Press, 1912, v. 2, (first edition 1883).
[573] Timofeev A. V. *On the question of constancy of the adiabatic invariant under change of the character of movement.* Journal of Experimental and Theoretical Physics, 1978, v. 75, No. 4, p. 1303–1308.
[574] Tisserand F. *Traite de mecanique celeste. Theorie de la figure des corps celestes et de leur mouvement de rotation.* Paris: Gauthier-Villars, 1891.
[575] Tkhai V. N. *The stability of regular Grioli precessions.* Journal of Applied Mathematics and Mechanics, 2000, v. 64, No. 5, p. 811–819.
[576] Treshchev D. V. *Introduction to the theory of perturbations of Hamiltonian systems.* Moscow: PHASIS, 1998 (Russian).
[577] Trofimov V. V., Fomenko A. T. *Algebra and geometry of integrable Hamiltonian differential equations.* Moscow: Faktorial, 1995 (Russian).
[578] Troilo R. *Moti del solido pesanto in cui e constante il modulo del momento delle quantita di moto.* Rend. Semin. mat. univ. e politech. Torino, 1974–75 (1976), v. 33, p. 337–347.
[579] Tsiganov A. V. *A new integrable system in S^2 with second integral quartic in the momenta.* J. Phys. A, 2005, v. 38, p. 3547–3553.
[580] Tsiganov A. V. *Lax representation for an integrable motion on the sphere with a cubic second invariant.* Regular and Chaotic Dynamics, 1999, v. 4, No. 3, p. 21–29.
[581] Tsiganov A. V. *On a family of integrable systems on S^2 with a cubic integral of motion.* J. Phys. A, Math. Gen., 2005, v. 38, p. 921–927.
[582] Tsiganov A. V. *On integrable deformation of the Poincaré system.* Regular and Chaotic Dynamics, 2002, v. 7, No. 3, p. 331–336.
[583] Tsiganov A. V. *On the Kowalevski – Goryachev – Chaplygin gyrostat.* J. Phys. A, 2002, v. 35, p. 309–318.
[584] Tsiganov A. V. *On the Steklov – Lyapunov case in the rigid motion.* Regular and Chaotic Dynamics, 2004, v. 9, No. 2, p. 77–90.
[585] Tsiganov A. V., Goremykin O. V. *Integrable systems on $so(4)$ related to XXX spin chains with boundaries.* J. Phys. A, Math. Gen., 2004, v. 37, p. 4843–4849.
[586] Tsiganov A. V. *An integrable system related to the spherical top and the Toda chain.* Theoretical and Mathematical Physics, 2000, v. 124, No. 2, p. 1121–1131.

[587] Tsiganov A. V. *Integrable systems in the method of separation of variables.* Izhevsk: R&C Dynamics, Institute of Computer Science, 2005, 384 p. (in Russian).

[588] Tsiganov A. V. *Separation of variables in the Kovalevskaya – Goryachev – Chaplygin gyrostat.* Theoretical and mathematical physics, 2003, v. 135, No. 2, p. 651–658.

[589] Uhlenbeck K. *Minimal 2-spheres and tori in S^k.* Preprint, 1975.

[590] Uzbek Ye. K. *A new solution of Kirchhoff's equations of the problem of the motion of a gyrostat acted upon by potential and gyroscopic forces.* Journal of Applied Mathematics and Mechanics, 2004, v. 68, No. 6, p. 863–869.

[591] Vanhaecke P. *Linearising two-dimensional integrable systems and the construction of action-angle variables.* Math. Z., 1992, v. 211, No. 1, p. 265–313.

[592] van der Woude W. *Über die Staudischen Kreiselbewegungen.* Math. Z., 1923, v. 16, p. 170–172.

[593] Varkhalev I. P., Gorr G. V. *A new class of asymptotically uniform motions of a heavy solid body with a fixed point.* Journal of Applied Mathematics and Mechanics, 1982, v. 46, No. 3, p. 308–310.

[594] Varkhalev Yu. P., Gorr G. V. *The first Lyapunov method in the study of asymptotic motions in the dynamics of a rigid body.* Mekh. Tverd. Tela, 1992, No. 24, p. 25–41 (Russian).

[595] Veronnet A. *Rotation de l'ellipsoide hétérogène et figure exacte.* J. de Math. Pures et Appl., 1912, v. 8, No. 6, p. 331–463.

[596] Veselov A. P. *Integration of the stationary problem for a classical spin chain.* Theoretical and Mathematical Physics, 1987, v. 71, No. 1, p. 446–450.

[597] Veselov A. P. *Integrable discrete-time systems and difference operators.* Functional Analysis and its Applications, 1988, v. 22, No. 2, p. 83–93.

[598] Veselov A. P. *On the conditions of integrability of Euler equations on $so(4)$.* Dokl. Akad. Nauk SSSR, 1983, v. 270, No. 6, p. 1298–1300 (Russian).

[599] Veselov A. P. *Parametric resonance and geodesics on an ellipsoid.* Functional Analysis and Its Applications, 1992, v. 26, No. 3, p. 211–213.

[600] Veselov A. P. *The Landau-Lifshits equation and integrable systems of classical mechanics.* Soviet Physics Doklady, 1983, v. 28, p. 458–459.

[601] Veselov A. P., Novikov S. P. *Poisson brackets and complex tori.* Proc. Steklov Inst. Math., 1985, v. 3, p. 53–65.

[602] Veselova L. E. *Problems in the dynamics of a rigid body.* Vestnik Moskovskogo Universiteta, Seriya 1, Matematika, Mekhanika, 1986, No. 5, p. 90–91 (Russian).

[603] Veselova L. E. *The dynamics of a body with an ellipsoid cavity full of a fluid.* Vestnik Moskovskogo Universiteta, Seriya 1, Matematika, Mekhanika, No. 2, p. 64–67 (Russian).

[604] Volterra V. *Sur la theorie des variations des latitudes.* Acta Math., 1899, v. 22, p. 201–358.

[605] Volterra V. *Sur la Stratification d'une Masse Fluide en Équilibre.* Acta Math., 1903, v. 27, No. 1, p. 105–124.

[606] Voronets P. V. *Geometric investigation of Euler's case of rotation of a rigid body about a fixed point.* Kiev. Univ. Izv., 1898, v. 38, No. 4, p. 1–52, No. 5, p. 53–113.

[607] Wangerin A. *Uber ein dreifach orthogonales Flachensystem, gebildet aus gewissen Flächen vierter Ordnung.* J. reine angew. Math. (Crelle's Journal), 1876, v. 82, p. 145–157.

[608] Wangerin A. *Über die Bewegung miteinander verbundener Körper.* Univ.-Schrift Halle, 1889.

[609] Weber H. *Anwendung der Thetafunktionen zweiter Veründlicher auf Theorie der Bewegung eines festen Körpers in einer Flüssigkeit.* Math. Ann., 1879, v. 14, p. 143–206.

[610] Webster A. G. *The Dynamics of particles and of rigid, elastic and fluid bodies.* Leipzig: B. G. Teubner, 1904.

[611] Whittaker E. T. *A treatise on the analytical dynamics,* ed. 3rd. Cambridge Univ. Press, 1927.

[612] Wiggins S. *Adiabatic chaos.* Phys. Lett. A, 1988, v. 128, No. 67, p. 339–342.

[613] Wiggins S. *Chaotic transport in dynamical systems.* New York: Springer, 1992.
[614] Wintner A. *The analytical foundations of celestial mechanics.* Princeton Univ. Press, 1941.
[615] Wittenburg J. *Dynamics of systems of rigid bodies.* Stuttgart: B. G. Teubner, 1977.
[616] Wojciechowski S. *Integrable one-particle potentials related to the Neumann system and the Jacobi problem of geodesic motion on an ellipsoid.* Phys. Lett. A, 1985, v. 107, No. 3, p. 106–111.
[617] Wojciechowski S. *An integrable marriage of the Euler equations with the Calogero-Moser system.* Phys. Lett. A, 1985, v. 111, No. 3, p. 101–103.
[618] Yehia H. M. *Qualitative investigations of plane and similar motions of a solid about a fixed point.* Journal of Applied Mathematics and Mechanics, 1981, v. 45, No. 4, p. 454-458.
[619] Yehia H. M. *New integrable cases of the gyrostat motion.* Vestnik Moskovskogo Universiteta Seriya 1 Matematika Mekhanika, 1987, No. 4, p. 88–90 (Russian).
[620] Yehia H. M. *An integrable motion of a particle on a smooth ellipsoid.* Regular and Chaotic Dynamics, 2003, v. 8, No. 4, p. 463–468.
[621] Yehia H. M. *Equivalent mechanical systems with cyclic coordinates and new integrable problems.* ZAMP, 2001, v. 52, p. 289–316.
[622] Yehia H. M. *Equivalent problems in rigid body dynamics.* I. Cel. Mech., 1988, v. 41, p. 275–288.
[623] Yehia H. M. *Equivalent problems in rigid body dynamics.* II. Cel. Mech., 1988, v. 41, p. 289–295.
[624] Yehia H. M. *Geometric transformations and new integrable problems of rigid body dynamics.* J. Phys. A, 2000, v. 33, p. 4393–4399.
[625] Yehia H. M. *Kovalevskaya's integrable case: generalizations and related new results.* Regular and Chaotic Dynamics, 2003, v. 8, No. 3, p. 337–348.
[626] Yehia H. M. *New generalization of all the known integrable problems in rigid body dynamics.* J. Phys. A., 1999, v. 32, p. 7565–7580.
[627] Yehia H. M. *New generalizations of the integrable problems in rigid body dynamics.* J. Phys. A: Math. Gen, 1997, v. 30, p. 7269–7275.
[628] Yehia H. M. *New integrable cases in the dynamics of rigid bodies.* V. Mech. Research. Com., 1989, v. 16, No. 6, p. 349–352.
[629] Yehia H. M. *New integrable problems in the dynamics of rigid bodies.* Mech. Res. Commun., 1986, v. 13, p. 169–172.
[630] Yehia H. M. *New integrable problems in the dynamics of rigid bodies.* IV. Mech. Res. Commun., 1987, v. 14, p. 177–180.
[631] Yehia H. M. *New integrable problems in the dynamics of rigid bodies with the Kovalevskaya configuration. I – The case of axisymmetric forces.* Mech. Res. Com., 1996, v. 23, No. 5, p. 423–427.
[632] Yehia H. M. *New integrable problems of motion a rigid body with a particle oscillating or bouncing in it.* Mech. Res. Commun., 1977, v. 24, No. 3, p. 243–246.
[633] Yehia H. M. *On a generalization of certain results of Goriatchev, Chaplygin and Sretenskii on the dynamics of rigid bodies.* J. Phys. A, 1996, v. 29, p. 8159–8161.
[634] Yehia H. M. *On certain integrable motions of a rigid body acted upon by gravity and magnetic fields.* Int. J. of Non-lin. Mech., 2001, v. 36, p. 1173–1175.
[635] Yehia H. M. *On certain two-dimensional conservative mechanical systems with a cubic second integral.* J. Phys. A, Mat. Gen., 2002, v. 35, p. 9469–9487.
[636] Yehia H. M. *On the integrability of certain problems in particle and rigid body dynamics.* J. of theor. and applied mech., 1996, v. 5, No. 1, p. 55–71.
[637] Yehia H. M. *On the stability of plane motion of heavy rigid body about a fixed point.* ZAMM, 1987, v. 67, p. 641–648.
[638] Yehia H. M. *Particular integrable cases in rigid body dynamics.* ZAMM, 1988, v. 68, p. 33–37.

[639] Yehia H. M. *Relation between the first and second moments of distributions.* J. Phys. A, 2002, v. 35, p. 6505–6508.

[640] Yehia H. M. *Two-dimensional conservative mechanical systems with quartic second integral.* Regular and Chaotic Dynamics, 2006, v. 11, No. 1, p. 103–122.

[641] Yehia H. M., Bedweihy N. *Certain generalizations of Kovalevskaya's case.* Mansoura Science Bulletin, 1987, v. 14, No. 2, p. 373–386.

[642] Yehia H. M., Elmandouh A. A. *New conditional integrable cases of motion of a rigid body with Kovalevskaya's configuration.* Journal of Physics A: Mathematical and Theoretical, 2011, v. 44, No. 1, 012001 (8 pp).

[643] Zel'dovich Ya. B. *Newtonian and Einsteinian motion of homogeneous matter.* Soviet Astronomy, 1964, v. 8, No. 5, p. 700.

[644] Zhukovsky N. E. *On the motion of the pseudospherical figure across the surface of the pseudosphere,* in Complete Collection of Works, 1937, v. 1, p. 490–535.

[645] Zhukovsky N. E. *On birds' hovering.* Moscow-Leningrad: Glav. red. aviats. lit., 1973, v. 5, p. 7–35 (Russian).

[646] Zhukovskii N. E. *The Hess' loxodromic pendulum.* Tr. Otdel. Fiz. Nauk Ob-va Lubit. Estestvozn., 1893, v. 5, No. 2, p. 37–45 (Russian).

[647] Zhukovskii N. E. *Motion of a rigid body containing a cavity filled with a homogeneous continuous liquid,* in Collected Works: Vol. 2, Moscow: Gostekhteorizdat, 1949, p. 31–152 (Russian).

[648] Zhukovskii N. E. *On the falling in the air of light oblong bodies rotating about their longitudinal axis: Paper 1,* in Collected Works: Vol. 5, Moscow–Leningrad: Gostekhizdat, 1937, p. 72–80.

[649] Zhukovskii N. E. *Geometric interpretation of the case considered by Kovalevskaya of the motion of a heavy rigid body about a fixed point,* in Collected Works: Vol. 1, Moscow: Gostekhizdat, 1948, p. 294–339 (Russian).

[650] Ziglin S. L. *Branching of solutions and nonexistence of first integrals in Hamiltonian mechanics,* I, II. Functional Analysis and Its Applications, 1982, v. 16, No. 3, p. 181–189; 1983, v. 17, No. 1, p. 6–17.

[651] Ziglin S. L. *Nonintegrability of variational equations in quadratures in some dynamic problems.* Doklady Physics, 2002, v. 47, No. 10, p. 750–751.

[652] Ziglin S. L. *The absence of an additional real-analytic first integral in some problems of dynamics.* Functional Analysis and Its Applications, 1997, v. 31, No. 1, p. 3–9.

[653] Ziglin S. L. *Dichotomy of the separatrices, bifurcation of solutions and nonexistence of an integral in the dynamics of a rigid body.* Transactions of the Moscow Mathematical Society, 1980, v. 41, p. 287–303 (Russian).

[654] Ziglin S. L. unpublished, personal communication.

[655] Zotev D. V., Fomenko-Zeischang. *Invariant in the Bogoyavlenskyi integrable case.* Regular and Chaotic Dynamics, 2000, v. 5, No. 4, p. 437–458.

[656] Zwanziger D. *Exactly soluble nonrelativistic model of particles with both electric and magnetic charges.* Phys. Rev., 1968, v. 176, p. 1480–1488.

Index of Names

DOI 10.1515/9783110544442-017

Index

DOI 10.1515/9783110544442-018

De Gruyter Studies in Mathematics

Volume 51
Peter Galenko, Vladimir Ankudinov, Ilya Starodumov
Phase-Field Crystals: Fast Interface Dynamics, 2019
ISBN 978-3-11-058597-1, e-ISBN (PDF) 978-3-11-058809-5,
e-ISBN (EPUB) 978-3-11-058653-4

Volume 50
Sergey Lychev, Konstantin Koifman
Geometry of Incompatible Deformations: Differential Geometry in Continuum Mechanics, 2018
ISBN 978-3-11-056201-9, e-ISBN (PDF) 978-3-11-056321-4,
e-ISBN (EPUB) 978-3-11-056227-9

Volume 49
Vladimir K. Dobrev
Invariant Differential Operators: Volume 3: Supersymmetry, 2018
ISBN 978-3-11-052663-9, e-ISBN (PDF) 978-3-11-052749-0,
e-ISBN (EPUB) 978-3-11-052669-1

Volume 48
Jared Maruskin
Dynamical Systems and Geometric Mechanics: An Introduction, 2018
ISBN 978-3-11-059729-5, e-ISBN (PDF) 978-3-11-059780-6,
e-ISBN (EPUB) 978-3-11-059803-2

Volume 47
Eugene Stefanovich
Elementary Particle Theory: Volume 3: Relativistic Quantum Dynamics, 2018
ISBN 978-3-11-049090-9, e-ISBN (PDF) 978-3-11-049322-1,
e-ISBN (EPUB) 978-3-11-049139-5

Bei Fragen zur Produktsicherheit wenden Sie sich bitte an:
If you have any questions regarding product safety,
please contact:

Walter de Gruyter GmbH
Genthiner Straße 13
10785 Berlin
productsafety@degruyterbrill.com